Methods in Enzymology

Volume 400

PHASE II CONJUGATION ENZYMES AND TRANSPORT SYSTEMS

METHODS IN ENZYMOLOGY

EDITORS-IN-CHIEF

John N. Abelson   Melvin I. Simon

DIVISION OF BIOLOGY
CALIFORNIA INSTITUTE OF TECHNOLOGY
PASADENA, CALIFORNIA

FOUNDING EDITORS

Sidney P. Colowick and Nathan O. Kaplan

Methods in Enzymology

Volume 400

Phase II Conjugation Enzymes and Transport Systems

EDITED BY

Helmut Sies

INSTITUTE OF BIOCHEMISTRY AND MOLECULAR BIOLOGY I
HEINRICH-HEIN UNIVERSITY
DÜSSELDORF, GERMANY

Lester Packer

MOLECULAR PHARMACOLOGY AND TOXICOLOGY
SCHOOL OF PHARMACY
UNIVERSITY OF SOUTHERN CALIFORNIA
LOS ANGELES, CALIFORNIA

AMSTERDAM • BOSTON • HEIDELBERG • LONDON
NEW YORK • OXFORD • PARIS • SAN DIEGO
SAN FRANCISCO • SINGAPORE • SYDNEY • TOKYO
Academic Press is an imprint of Elsevier

ELSEVIER

Elsevier Academic Press
525 B Street, Suite 1900, San Diego, California 92101-4495, USA
84 Theobald's Road, London WC1X 8RR, UK

This book is printed on acid-free paper. ∞

For all information on all Elsevier Academic Press publications
visit our Web site at www.books.elsevier.com

ISBN-13: 978-0-12-182805-9
ISBN-10: 0-12-182805-0

PRINTED IN THE UNITED STATES OF AMERICA
05 06 07 08 09 9 8 7 6 5 4 3 2 1

Table of Contents

CONTRIBUTORS TO VOLUME 400   ix

PREFACE   xv

VOLUMES IN SERIES   xvii

1. UDP-Glucuronosyltransferases: Gene Structures of *UGT1* and *UGT2* Families — IDA S. OWENS, NIKHIL K. BASU, AND RAJAT BANERJEE   1

2. Identification and Characterization of Functional Hepatocyte Nuclear Factor 1-Binding Sites in UDP-Glucuronosyltransferase Genes — DIONE A. GARDNER-STEPHEN, PHILIP A. GREGORY, AND PETER I. MACKENZIE   22

3. Substrate Specificity of Human Hepatic UDP-Glucuronosyltransferaes — BRIAN BURCHELL, DAVID J. LOCKLEY, ADAM STAINES, YOSHIHIRO UESAWA, AND MICHAEL W. H. COUGHTRIE   46

4. UDP-Glucuronosyltransferase 1A6: Structural, Functional, and Regulatory Aspects — KARL WALTER BOCK AND CHRISTOPH KÖHLE   57

5. The Role of Ah Receptor in Induction of Human UDP-Glucuronosyltransferase 1A1 — MEI-FEI YUEH, JESSICA A. BONZO, AND ROBERT H. TUKEY   75

6. Regulation of the Human *UGT1A1* Gene by Nuclear Receptors Constitutive Active/Androstane Receptor, Pregnane X Receptor, and Glucocorticoid Receptor — JUNKO SUGATANI, TATSUYA SUEYOSHI, MASAHIKO NEGISHI, AND MASAO MIWA   92

7. Isoform-Selective Probe Substrates for *In Vitro* Studies of Human UDP-Glucuronosyltransferases — MICHAEL H. COURT   104

8. Structure of UDP-Glucuronosyltransferases in Membranes — ANNA RADOMINSKA-PANDYA, MOHAMED OUZZINE, SYLVIE FOURNEL-GIGLEUX, AND JACQUES MAGDALOU   116

9. Human *SULT1A* Genes: Cloning and Activity Assays of the *SULT1A* Promoters — NADINE HEMPEL, MASAHIKO NEGISHI, AND MICHAEL E. McMANUS   147

10. Vitamin D Receptor Regulation of the Steroid/ Bile Acid Sulfotransferase SULT2A1 — BANDANA CHATTERJEE, IBTISSAM ECHCHGADDA, AND CHUNG SEOG SONG — 165

11. Screening and Characterizing Human NAT2 Variants — MIHAELA R. SAVULESCU, ADEEL MUSHTAQ, AND P. DAVID JOSEPHY — 192

12. Inactivation of Human Arylamine N-Acetyltransferase 1 by Hydrogen Peroxide and Peroxynitrite — JEAN-MARIE DUPRET, JULIEN DAIROU, NOUREDDINE ATMANE, AND FERNANDO RODRIGUES-LIMA — 215

13. Sulfotransferases and Acetyltransferases in Mutagenicity Testing: Technical Aspects — HANSRUEDI GLATT AND WALTER MEINL — 230

14. A Comparative Molecular Field Analysis-Based Approach to Prediction of Sulfotransferase Catalytic Specificity — VYAS SHARMA AND MICHAEL W. DUFFEL — 249

15. Glucuronidase Deconjugation in Inflammation — KAYOKO SHIMOI AND TSUTOMU NAKAYAMA — 263

16. Three-Dimensional Structures of Sulfatases — DEBASHIS GHOSH — 273

17. Estrogen Sulfatase — MASAO IWAMORI — 293

18. Analysis for Localization of Steroid Sulfatase in Human Tissues — TAKASHI SUZUKI, YASUHIRO MIKI, TSUYOSHI FUKUDA, TAISUKE NAKATA, TAKUYA MORIYA, AND HIRONOBU SASANO — 303

19. Metabolism of Phytoestrogen Conjugates — TRACY L. D'ALESSANDRO, BRENDA J. BOERSMA-MALAND, T. GREG PETERSON, JEFF SFAKIANOS, JEEVAN K. PRASAIN, RAKESH P. PATEL, VICTOR M. DARLEY-USMAR, NIGEL P. BOTTING, AND STEPHEN BARNES — 316

20. Sulfation and Glucuronidation of Phenols: Implications in Coenzyme Q Metabolism — NANDITA SHANGARI, TOM S. CHAN, AND PETER J. O'BRIEN — 342

21. Synthesis of Bile Acid Coenzyme A Thioesters in the Amino Acid Conjugation of Bile Acids — ERIN M. SHONSEY, JAMES WHEELER, MICHELLE JOHNSON, DONGING HE, CHARLES N. FALANY, JOSIE FALANY, AND STEPHEN BARNES — 360

22. Bile Acid Coenzyme A: Amino Acid *N*-Acyltransferase in the Amino Acid Conjugation of Bile Acids   ERIN M. SHONSEY, MINDAN SFAKIANOS, MICHELLE JOHNSON, DONGING HE, CHARLES N. FALANY, JOSIE FALANY, DAVID J. MERKLER, AND STEPHEN BARNES   374

23. Multidrug Resistance Protein 1-Mediated Export of Glutathione and Glutathione Disulphide from Brain Astrocytes   JOHANNES HIRRLINGER AND RALF DRINGEN   395

24. The Genetics of ATP-Binding Cassette Transporters   MICHAEL DEAN   409

25. Functional Analysis of Detergent-solubilized and Membrane-Reconstituted ATP-Binding Cassette Transporters   BERT POOLMAN, MARK K. DOEVEN, ERIC R. GEERTSMA, ESTHER BIEMANS-OLDEHINKEL, WIL N. KONINGS, AND DOUGLAS C. REES   429

26. Yeast ATP-Binding Cassette Transporters: Cellular Cleaning Pumps   ROBERT ERNST, ROBIN KLEMM, LUTZ SCHMITT, AND KARL KUCHLER   460

27. High-Speed Screening of Human ATP-Binding Cassette Transporter Function and Genetic Polymorphisms: New Strategies in Pharmacogenomics   TOSHIHISA ISHIKAWA, AKI SAKURAI, YOICHI KANAMORI, MAKOTO NAGAKURA, HIROYUKI HIRANO, YUTAKA TAKARADA, KAZUNARI YAMADA, KAZUHISA FUKUSHIMA, AND MASATO KITAJIMA   485

28. Coordinate Transcriptional Regulation of Transport and Metabolism   JYRKI J. ELORANTA, PETER J. MEIER, AND GERD A. KULLAK-UBLICK   511

29. Uptake and Efflux Transporters for Conjugates in Human Hepatocytes   DIETRICH KEPPLER   531

30. Biliary Transport Systems: Short-Term Regulation   RALF KUBITZ, ANGELIKA HELMER, AND DIETER HÄUSSINGER   542

31. Inborn Errors of Biliary Canalicular Transport Systems   RALF KUBITZ, VERENA KEITEL, AND DIETER HÄUSSINGER   558

32. Epoxide Hydrolases: Structure, Function,   MICHAEL ARAND,
    Mechanism, and Assay   ANNETTE CRONIN,
                                                    MAGDALENA ADAMSKA, AND
                                                    FRANZ OESCH   569

33. Pregnane X Receptor-Mediated Transcription   THOMAS K. H. CHANG AND
                                                    DAVID J. WAXMAN   588

34. Animal Models of Xenobiotic Receptors in   HAIBIAO GONG,
    Drug Metabolism and Diseases   MICHAEL W. SINZ,
                                                    YAN FENG, TAOSHENG CHEN,
                                                    RAMAN VENKATARAMANAN,
                                                    AND WEN XIE   598

35. Cancer and Molecular Biomarkers of Phase 2   KIM DALHOFF,
                                                    KASPAR BUUS JENSEN, AND
                                                    HENRIK ENGHUSEN POULSEN   618

AUTHOR INDEX   629

SUBJECT INDEX   683

Contributors to Volume 400

Article numbers are in parantheses and following the name of contributors.
Affiliations listed are current.

MAGDALENA ADAMSKA (32), *Institut fuer Pharmakologie und Toxikologie, Universitaet Wuerzburg, Wuerzburg, Germany*

MICHAEL ARAND (32), *Institut fuer Pharmakologie und Toxikologie, Universitaet Wuerzburg, Wuerzburg, Germany*

NOUREDDINE ATMANE (12), *CNRS-Unite Mixte de Recherche 7000, Faculte de Medecine, Pitie-Salpetriere, Paris, France*

RAJAT BANERJEE (1), *Section on Genetic Disorders of Drug Metabolism, Heritable Disorders Branch, National Institute of Child Health and Human Development, National Institutes of Health, Bethesda, Maryland*

STEPHEN BARNES (19, 21, 22), *Department of Pharmacology and Toxicology, University of Alabama at Birmingham, Birmingham, Alabama*

NIKHIL K. BASU (1), *Section on Genetic Disorders of Drug Metabolism, Heritable Disorders Branch, National Institute of Child Health and Human Development, National Institutes of Health, Bethesda, Maryland*

ESTHER BIEMANS-OLDEHINKEL (25), *Department of Biochemistry, Groningen Biomolecular Sciences and Biotechnology Institute and Materials Science Centre^plus, University of Groningen, Groningen, The Netherlands*

KARL WALTER BOCK (4), *Institut of Pharmacology and Toxicology, Department of Toxicology, University of Tübingen, Wilhelmstrasse 56, Tübingen, Germany*

BRENDA J. BOERSMA-MALAND (19), *Laboratory of Human Carcinogenesis, National Cancer Institute, Bethesda, Maryland*

JESSICA A. BONZO (5), *Department of Pharmacology, Chemistry & Biochemistry, Laboratory of Environmental Toxicology, University of California, San Diego, La Jolla, California*

NIGEL P. BOTTING (19), *Department of Chemistry, St. Andrews University, St. Andrews Fife, Scotland*

BRIAN BURCHELL (3), *Department of Molecular and Cellular Pathology, Ninewells Hospital and Medical School, University of Dundee, Dundee, Scotland, United Kingdom*

KASPAR BUUS JENSEN (35), *Department of Clinical Pharmacology, H:S Rigshospitalet, Copenhagen, Denmark*

TOM S. CHAN (20), *Department of Pharmaceutical Sciences, University of Toronto, Toronto, Ontario, Canada*

THOMAS K. H. CHANG (33), *Faculty of Pharmaceutical Sciences, The University of British Columbia, Vancouver, Canada*

BANDANA CHATTERJEE (10), *Department of Molecular Medicine, Institute of Biotechnology, University of Texas Health Science Center at San Antonio and South Texas Veterans Health Care System, San Antonio, Texas*

TAOSHENG CHEN (34), *Lead Discovery and Profiling, Bristol-Myers Squibb Company, Wallingford, Connecticut*

MICHAEL W. H. COUGHTRIE (3), *Division of Pathology and Neuroscience, Ninewells Medical School, University of Dundee, Scotland, United Kingdom*

MICHAEL H. COURT (7), *Department of Pharmacology and Experimental Therapeutics, Tufts University School of Medicine, Boston, Massachusetts*

ANNETTE CRONIN (32), *Institut fuer Pharmakologie und Toxikologie, Universitaet Wuerzburg, Wuerzburg, Germany*

JULIEN DAIROU (12), *CNRS-Unite Mixte de Recherche 7000, Faculte de Medecine, Pitie-Salpetriere, Paris, France*

TRACY L. D'ALESSANDRO (19), *Department of Pharmacology and Toxicology, University of Alabama at Birmingham, Birmingham, Alabama*

KIM DALHOFF (35), *Department of Clinical Pharmacology, H:S Rigshospitalet, Copenhagen, Denmark*

VICTOR M. DARLEY-USMAR (19), *Department of Pharmacology and Toxicology, University of Alabama at Birmingham, Birmingham, Alabama*

MICHAEL DEAN (24), *Human Genetics Section, Laboratory of Genomic Diversity, National Cancer Institute, Frederick, Maryland*

MARK K. DOEVEN (25), *Department of Biochemistry, Groningen Biomolecular Sciences and Biotechnology Institute and Materials Science Centre^plus, University of Groningen, Groningen, The Netherlands*

RALF DRINGEN (23), *University of Bremen, Faculty 2 (Biology/Chemistry), Bremen, Germany*

MICHAEL W. DUFFEL (14), *Division of Medicinal and Natural Products Chemistry, College of Pharmacy, University of Iowa, Iowa City, Iowa*

JEAN-MARIE DUPRET (12), *CNRS-Unite Mixte de Recherche 7000, Faculte de Medecine, Pitie-Salpetriere, Paris, France*

IBTISSAM ECHCHGADDA (10), *Department of Molecular Medicine, Institute of Biotechnology, University of Texas Health Science Center at San Antonio, San Antonio, Texas*

JYRKI J. ELORANTA (28), *Division of Gastroenterology and Hepatology, University Hospital, Zürich, Switzerland*

HENRIK ENGHUSEN POULSEN (35), *Department of Clinical Pharmacology, H:S Rigshospitalet, Copenhagen, Denmark*

ROBERT ERNST (26), *Institute of Biochemistry, Membrane Transport Group, Heinrich-Heine University of Düsseldorf, Düsseldorf, Germany*

CHARLES N. FALANY (21, 22), *Department of Pharmacology and Toxicology, University of Alabama, Birmingham, Alabama*

JOSIE FALANY (21, 22), *Department of Pharmacology and Toxicology, University of Alabama, Birmingham, Alabama*

YAN FENG (34), *Department of Pharmaceutical Sciences, University of Pittsburgh, School of Pharmacy, Pittsburgh, Pennsylvania*

SYLVIE FOURNEL-GIGLEUX (8), *Unité Mixte de Recherche, 7561 CNRS-Université Henri Poincaré Nancy 1. School of Medicine, Vandoeuvre-lès-Nancy, France*

TSUYOSHI FUKUDA (18), *Pharmaceuticals Company, Kyowa Hakko Kogyo Company Limited, Chiyoda-ku, Tokyo, Japan*

KAZUHISA FUKUSHIMA (27), *Yokogawa Electric Corporation, Musashino, Tokyo, Japan*

DIONE A. GARDNER-STEPHEN (2), *Department of Clinical Pharmacology, Flinders University School of Medicine, Flinders-Medical Center, Adelaide, Australia*

ERIC R. GEERTSMA (25), *Department of Biochemistry, Groningen Biomolecular Sciences and Biotechnology Institute* and Materials Science Centre^{plus}, University of Groningen, Groningen, The Netherlands

DEBASHIS GHOSH (16), Department of Structural Biology, Hauptman-Woodward Medical Research Institute, Buffalo, New York

HANSRUEDI GLATT (13), Department of Toxicology, German Institute of Human Nutrition, Potsdam-Rehbrücke, Germany

HAIBIAO GONG (34), Center for Pharmacogenetics, University of Pittsburgh, Pittsburgh, Pennsylvania

PHILIP A. GREGORY (2), Department of Clinical Pharmacology, Flinders University School of Medicine, Flinders Medical Center, Adelaide, Australia

DIETER HÄUSSINGER (30, 31), Clinic for Gastroenterology, Heptalogy and Infectiology, Heinrich-Heine University Düsseldorf, Düsseldorf, Germany

DONGNING HE (21, 22), Department of Pharmacology and Toxicology, University of Alabama, Birmingham, Alabama

ANGELIKA HELMER (30), Clinic for Gastroenterology, Heptalogy and Infectiology, Heinrich-Heine University Düsseldorf, Düsseldorf, Germany

NADINE HEMPEL (9), School of Biomedical Science, Faculty of Biological and Chemical Sciences, University of Queensland, Brisbane, Queensland, Australia

HIROYUKI HIRANO (27), GS PlatZ Company Limited, Chuo-ku, Tokyo, Japan

JOHANNES HIRRLINGER (23), Max-Planck Institute for Experimental Medicine, Department of Neurogenetics, Göttingen, Germany

TOSHIHISA ISHIKAWA (27), Department of Biomolecular Engineering, Tokyo Institute of Technology, Graduate School of Bioscience and Biotechnology, Midori-ku, Yokohama, Japan

MASAO IWAMORI (17), Laboratory of Biochemistry, Department of Life Sciences, Faculty of Science and Technology, Kinki University, Higashiosaka, Osaka, Japan

MICHELLE JOHNSON (21, 22), Department of Pathology, University of Alabama, Birmingham, Alabama

P. DAVID JOSEPHY (11), Department of Chemistry and Biochemistry, University of Guelph, Guelph, Ontario, Canada

YOICHI KANAMORI (27), Bio Research Laboratories, Nosan Corporation, Tsukuba, Japan

VERENA KEITEL (31), Klinik für Gastroenterologie, Hepatologie und Infektiologie MNR-Klinik Gebäude, Düsseldorf, Germany

DIETRICH KEPPLER (29), Deutsches Krebsforschungszentrum Abteilung, Tumorbiochemie Im Neuenheimer Feld 280, Heidelberg, Germany

MASATO KITAJIMA (27), Life Science Systems Department, PLM Solutions Division, Fujitsu Kyushu System Engineering Company Limited, Sawaraku, Fukushima, Japan

ROBIN KLEMM (26), Max Plank Institute of Molecular Cell Biology and Genetics, Laboratory Kai Simons, Dresden, Germany

CHRISTOPH KÖHLE (4), Institut für Pharmakologie und Toxikologie, Abteilung Toxikologie, Universität Tübingen, Wilhelmstrasse 56, Tübingen, Germany

WIL N. KONINGS (25), Department of Microbiology, Groningen Biomolecular Sciences and Biotechnology Institute, University of Groningen, Haren NL 9751 NN, The Netherlands

RALF KUBITZ (30, 31), Clinic for Gastroenterology, Heptalogy and Infectiology, Heinrich-Heine University Düsseldorf, Düsseldorf, Germany

KARL KUCHLER (26), *Department of Medical Biochemistry, Division of Molecular Genetics, Max F. Perutz Laboratories, Medical University of Vienna, Campus Vienna Biocenter, Vienna, Austria*

GERD A. KULLAK-UBLICK (28), *Division of Gastroenterology and Hepatology, University Hospital, Zürich, Switzerland*

DAVID J. LOCKLEY (3), *Division of Pathology and Neuroscience, Ninewells Medical School, University of Dundee, Scotland, United Kingdom*

PETER I. MACKENZIE (2), *Department of Clinical Pharmacology, Flinders University School of Medicine, Flinders Medical Centre, Adelaide, Australia*

JACQUES MAGDALOU (8), *Unité Mixte de Recherche 7561 CNRS-Université Henri Poincaré Nancy 1. School of Medicine, Vandoeuvre-lès-Nancy, France*

MICHAEL E. MCMANUS (9), *School of Biomedical Science, Faculty of Biological and Chemical Sciences, University of Queensland, Brisbane, Queensland, Australia*

PETER J. MEIER (28), *Division of Clinical Pharmacology and Toxicology, Department of Internal Medicine, University Hospital, Zürich, Switzerland*

WALTER MEINL (13), *Department of Toxicology, German Institute of Human Nutrition, Potsdam-Rehbrücke, Germany*

DAVID J. MERKLER (22), *Department of Chemistry, University of South Florida, Tampa, Florida*

YASUHIRO MIKI (18), *Department of Pathology, Tohoku University School of Medicine, Sendai, Japan*

MASAO MIWA (6), *Department of Pharmaco-Biochemistry and 21 COE, School of Pharmaceutical Sciences, University of Shizuoka, Shizuoka, Japan*

TAKUYA MORIYA (18), *Department of Pathology, Tohoku University School of Medicine, Sendai, Japan*

ADEEL MUSHTAQ (11), *Department of Biochemistry, University of Washington and Howard Hughes Medical Institute, Seattle, Washington*

MAKOTO NAGAKURA (27), *BioTec Company Limited, Bunkyo-ku, Tokyo, Japan*

TAISUKE NAKATA (18), *Pharmaceuticals Company, Kyowa Hakko Kogyo Company Limited, Chiyoda-ku, Tokyo, Japan*

TSUTOMU NAKAYAMA (15), *Graduate School of Nutritional and Environmental Sciences, University of Shizuoka, Shizuoka, Japan*

MASAHIKO NEGISHI (6, 9), *Pharmacogenetics Section, Laboratory of Reproductive and Developmental Toxicology, National Institute of Environmental Health Sciences, Research Triangle Park, North Carolina*

PETER J. O'BRIEN (20), *Department of Pharmaceutical Sciences, University of Toronto, Toronto, Ontario, Canada*

FRANZ OESCH (32), *Institute of Toxicology, University of Mainz, Mainz, Germany*

MOHAMED OUZZINE (8), *Unité Mixte de Recherche, 7561 CNRS-Université Henri Poincaré Nancy 1. School of Medicine, Vandoeuvre-lès-Nancy, France*

IDA S. OWENS (1), *Section on Genetic Disorders of Drug Metabolism, Heritable Disorders Branch, National Institute of Child Health and Human Development, National Institutes of Health, Bethesda, Maryland*

RAKESH P. PATEL (19), *Department of Pharmacology and Toxicology, University of Alabama at Birmingham, Birmingham, Alabama*

T. GREG PETERSON (19), *Bradley, Arant, Rose, & White, LLP, Birmingham, Alabama*

BERT POOLMAN (25), *Department of Biochemistry, Groningen Biomolecular Sciences and Biotechnology Institute* and Materials Science Centre[plus], University of Groningen, Groningen, The Netherlands

JEEVAN K. PRASAIN (19), Laboratory of Human Carcinogenesis, National Cancer Institute, Bethesda, Maryland

ANNA RADOMINSKA-PANDYA (8), Department of Biochemistry and Molecular Biology, University of Arkansas for Medical Sciences, Little Rock, Arkansas

DOUGLAS C. REES (25), HHMI and Division of Chemistry, California Institute of Technology, Pasadena, California

FERNANDO RODRIGUES-LIMA (12), CNRS-Unite Mixte de Recherche 7000, Faculte de Medecine, Pitie-Salpetriere, Paris, France

AKI SAKURAI (27), Department of Biomolecular Engineering, Tokyo Institute of Technology, Graduate School of Bioscience and Biotechnology, Yokohama, Japan

HIRONOBU SASANO (18), Department of Pathology, Tohoku University School of Medicine, Sendai, Japan

MIHAELA R. SAVULESCU (11), Department of Chemistry and Biochemistry, University of Guelph, Guelph, Ontario, Canada

LUTZ SCHMITT (26), Institute of Biochemistry, Membrane Transport Group, Heinrich-Heine University of Düsseldorf, Düsseldorf, Germany

JEFF SFAKIANOS (19), Department of Cell Biology, Yale University School of Medicine, New Haven, Connecticut

MINDAN SFAKIANOS (22), Department of Molecular Biophysics and Biochemistry, Yale University School of Medicine, New Haven, Connecticut

NANDITA SHANGARI (20), Department of Pharmaceutical Sciences, University of Toronto, Toronto, Ontario, Canada

VYAS SHARMA (14), Division of Medicinal and Natural Products Chemistry, College of Pharmacy, University of Iowa, Iowa City, Iowa

KAYOKO SHIMOI (15), Institute for Environmental Sciences, Graduate School of Nutritional and Environmental Sciences, University of Shizuoka, Shizuoka, Japan

ERIN M. SHONSEY (21, 22), Department of Pharmacology and Toxicology, University of Alabama, Birmingham, Alabama

MICHAEL W. SINZ (34), Metabolism and Pharmacokinetics, Bristol-Myers Squibb Company, Wallingford, Connecticut

CHUNG SEOG SONG (10), Department of Molecular Medicine, Institute of Biotechnology, University of Texas Health Science Center at San Antonio, San Antonio, Texas

ADAM STAINES (3), Department of Molecular and Cellular Pathology, Ninewells Hospital and Medical School, University of Dundee, Dundee, Scotland, United Kingdom

TATSUYA SUEYOSHI (6), Pharmacogenetics Section, Laboratory of Reproductive and Developmental Toxicology, National Institute of Environmental Health Sciences, Research Triangle Park, North Carolina

JUNKO SUGATANI (6), Department of Pharmaco-Biochemistry and 21 COE, School of Pharmaceutical Sciences, University of Shizuoka, Shizuoka, Japan

TAKASHI SUZUKI (18), Department of Pathology, Tohoku University School of Medicine, Sendai, Japan

YUTAKA TAKARADA (27), Biotechnology Frontier Project, Toyobo Company Limited, Tsuruga, Japan

ROBERT H. TUKEY (5), Department of Pharmacology, Chemistry & Biochemistry, Laboratory of Environmental Toxicology, University of California, San Diego, La Jolla, California

YOSHIHIRO UESAWA (3), *Department of Molecular and Cellular Pathology, Ninewells Hospital and Medical School, University of Dundee, Dundee, Scotland, United Kingdom*

RAMAN VENKATARAMANAN (34), *Department of Pharmaceutical Sciences, University of Pittsburgh School of Pharmacy, Pittsburgh, Pennsylvania*

DAVID J. WAXMAN (33), *Department of Biology, Division of Cell and Molecular Biology, Boston University, Boston, Massachusetts*

JAMES WHEELER (21), *Alcon Laboratories, Fort Worth, Texas*

WEN XIE (34), *Center for Pharmacogenetics, University of Pittsburgh School of Pharmacy, Pittsburgh, Pennsylvania*

KAZUNARI YAMADA (27), *GENESHOT Project, R&D Center, NGK Insulators Limited, Mizuho-ku, Nagoya, Japan*

MEI-FEI YUEH (5), *Department of Pharmacology, Chemistry & Biochemistry, Laboratory of Environmental Toxicology, University of California, San Diego, La Jolla, California*

Preface

This volume on conjugation enzymes and transporters serves to bring together current methods and concepts in an interesting, important, and rapidly developing field of cell and systems biology. It focuses on the so-called Phase II enzymes of drug metabolism (xenobiotics), which have important ramifications for endogenous metabolism and nutrition. Also included are aspects on Phase III, transport systems. This volume of *Methods in Enzymology* presents current knowledge and methodology on glucuronidation, sulfation, acetylation, and transport systems in this field of research. Together with the volumes on **Quinones and Quinone Enzymes (volumes 378 and 382)** and **Glutathione Transferases and gamma-Glutamyl Transpeptidases (volume 401)**, the state of knowledge on proteomics and metabolomics of many pathways of (waste) product elimination, enzyme protein induction and gene regulation and feedback control is provided. We trust that this volume will help stimulate future investigations and speed the advance of knowledge in systems biology.

The editors thank the members of the Advisory Committee: Karl W. Bock, Tübingen, Enrique Cadenas, Los Angeles, Toshihisa Ishikawa, Yokohama, Masahiko Negishi, Research Triangle Park, and Gary Williamson, Lausanne, for their valuable suggestions and wisdom in selecting contributions for this volume. We also thank Marlies Scholtes and Cindy Minor for their valuable help.

HELMUT SIES
LESTER PACKER

METHODS IN ENZYMOLOGY

VOLUME I. Preparation and Assay of Enzymes
Edited by SIDNEY P. COLOWICK AND NATHAN O. KAPLAN

VOLUME II. Preparation and Assay of Enzymes
Edited by SIDNEY P. COLOWICK AND NATHAN O. KAPLAN

VOLUME III. Preparation and Assay of Substrates
Edited by SIDNEY P. COLOWICK AND NATHAN O. KAPLAN

VOLUME IV. Special Techniques for the Enzymologist
Edited by SIDNEY P. COLOWICK AND NATHAN O. KAPLAN

VOLUME V. Preparation and Assay of Enzymes
Edited by SIDNEY P. COLOWICK AND NATHAN O. KAPLAN

VOLUME VI. Preparation and Assay of Enzymes *(Continued)*
Preparation and Assay of Substrates
Special Techniques
Edited by SIDNEY P. COLOWICK AND NATHAN O. KAPLAN

VOLUME VII. Cumulative Subject Index
Edited by SIDNEY P. COLOWICK AND NATHAN O. KAPLAN

VOLUME VIII. Complex Carbohydrates
Edited by ELIZABETH F. NEUFELD AND VICTOR GINSBURG

VOLUME IX. Carbohydrate Metabolism
Edited by WILLIS A. WOOD

VOLUME X. Oxidation and Phosphorylation
Edited by RONALD W. ESTABROOK AND MAYNARD E. PULLMAN

VOLUME XI. Enzyme Structure
Edited by C. H. W. HIRS

VOLUME XII. Nucleic Acids (Parts A and B)
Edited by LAWRENCE GROSSMAN AND KIVIE MOLDAVE

VOLUME XIII. Citric Acid Cycle
Edited by J. M. LOWENSTEIN

VOLUME XIV. Lipids
Edited by J. M. LOWENSTEIN

VOLUME XV. Steroids and Terpenoids
Edited by RAYMOND B. CLAYTON

VOLUME XVI. Fast Reactions
Edited by KENNETH KUSTIN

VOLUME XVII. Metabolism of Amino Acids and Amines
(Parts A and B)
Edited by HERBERT TABOR AND CELIA WHITE TABOR

VOLUME XVIII. Vitamins and Coenzymes (Parts A, B, and C)
Edited by DONALD B. MCCORMICK AND LEMUEL D. WRIGHT

VOLUME XIX. Proteolytic Enzymes
Edited by GERTRUDE E. PERLMANN AND LASZLO LORAND

VOLUME XX. Nucleic Acids and Protein Synthesis (Part C)
Edited by KIVIE MOLDAVE AND LAWRENCE GROSSMAN

VOLUME XXI. Nucleic Acids (Part D)
Edited by LAWRENCE GROSSMAN AND KIVIE MOLDAVE

VOLUME XXII. Enzyme Purification and Related Techniques
Edited by WILLIAM B. JAKOBY

VOLUME XXIII. Photosynthesis (Part A)
Edited by ANTHONY SAN PIETRO

VOLUME XXIV. Photosynthesis and Nitrogen Fixation (Part B)
Edited by ANTHONY SAN PIETRO

VOLUME XXV. Enzyme Structure (Part B)
Edited by C. H. W. HIRS AND SERGE N. TIMASHEFF

VOLUME XXVI. Enzyme Structure (Part C)
Edited by C. H. W. HIRS AND SERGE N. TIMASHEFF

VOLUME XXVII. Enzyme Structure (Part D)
Edited by C. H. W. HIRS AND SERGE N. TIMASHEFF

VOLUME XXVIII. Complex Carbohydrates (Part B)
Edited by VICTOR GINSBURG

VOLUME XXIX. Nucleic Acids and Protein Synthesis (Part E)
Edited by LAWRENCE GROSSMAN AND KIVIE MOLDAVE

VOLUME XXX. Nucleic Acids and Protein Synthesis (Part F)
Edited by KIVIE MOLDAVE AND LAWRENCE GROSSMAN

VOLUME XXXI. Biomembranes (Part A)
Edited by SIDNEY FLEISCHER AND LESTER PACKER

VOLUME XXXII. Biomembranes (Part B)
Edited by SIDNEY FLEISCHER AND LESTER PACKER

VOLUME XXXIII. Cumulative Subject Index Volumes I-XXX
Edited by MARTHA G. DENNIS AND EDWARD A. DENNIS

VOLUME XXXIV. Affinity Techniques (Enzyme Purification: Part B)
Edited by WILLIAM B. JAKOBY AND MEIR WILCHEK

VOLUME XXXV. Lipids (Part B)
Edited by JOHN M. LOWENSTEIN

VOLUME XXXVI. Hormone Action (Part A: Steroid Hormones)
Edited by BERT W. O'MALLEY AND JOEL G. HARDMAN

VOLUME XXXVII. Hormone Action (Part B: Peptide Hormones)
Edited by BERT W. O'MALLEY AND JOEL G. HARDMAN

VOLUME XXXVIII. Hormone Action (Part C: Cyclic Nucleotides)
Edited by JOEL G. HARDMAN AND BERT W. O'MALLEY

VOLUME XXXIX. Hormone Action (Part D: Isolated Cells, Tissues, and Organ Systems)
Edited by JOEL G. HARDMAN AND BERT W. O'MALLEY

VOLUME XL. Hormone Action (Part E: Nuclear Structure and Function)
Edited by BERT W. O'MALLEY AND JOEL G. HARDMAN

VOLUME XLI. Carbohydrate Metabolism (Part B)
Edited by W. A. WOOD

VOLUME XLII. Carbohydrate Metabolism (Part C)
Edited by W. A. WOOD

VOLUME XLIII. Antibiotics
Edited by JOHN H. HASH

VOLUME XLIV. Immobilized Enzymes
Edited by KLAUS MOSBACH

VOLUME XLV. Proteolytic Enzymes (Part B)
Edited by LASZLO LORAND

VOLUME XLVI. Affinity Labeling
Edited by WILLIAM B. JAKOBY AND MEIR WILCHEK

VOLUME XLVII. Enzyme Structure (Part E)
Edited by C. H. W. HIRS AND SERGE N. TIMASHEFF

VOLUME XLVIII. Enzyme Structure (Part F)
Edited by C. H. W. HIRS AND SERGE N. TIMASHEFF

VOLUME XLIX. Enzyme Structure (Part G)
Edited by C. H. W. HIRS AND SERGE N. TIMASHEFF

VOLUME L. Complex Carbohydrates (Part C)
Edited by VICTOR GINSBURG

VOLUME LI. Purine and Pyrimidine Nucleotide Metabolism
Edited by PATRICIA A. HOFFEE AND MARY ELLEN JONES

VOLUME LII. Biomembranes (Part C: Biological Oxidations)
Edited by SIDNEY FLEISCHER AND LESTER PACKER

VOLUME LIII. Biomembranes (Part D: Biological Oxidations)
Edited by SIDNEY FLEISCHER AND LESTER PACKER

VOLUME LIV. Biomembranes (Part E: Biological Oxidations)
Edited by SIDNEY FLEISCHER AND LESTER PACKER

VOLUME LV. Biomembranes (Part F: Bioenergetics)
Edited by SIDNEY FLEISCHER AND LESTER PACKER

VOLUME LVI. Biomembranes (Part G: Bioenergetics)
Edited by SIDNEY FLEISCHER AND LESTER PACKER

VOLUME LVII. Bioluminescence and Chemiluminescence
Edited by MARLENE A. DeLUCA

VOLUME LVIII. Cell Culture
Edited by WILLIAM B. JAKOBY AND IRA PASTAN

VOLUME LIX. Nucleic Acids and Protein Synthesis (Part G)
Edited by KIVIE MOLDAVE AND LAWRENCE GROSSMAN

VOLUME LX. Nucleic Acids and Protein Synthesis (Part H)
Edited by KIVIE MOLDAVE AND LAWRENCE GROSSMAN

VOLUME 61. Enzyme Structure (Part H)
Edited by C. H. W. HIRS AND SERGE N. TIMASHEFF

VOLUME 62. Vitamins and Coenzymes (Part D)
Edited by DONALD B. McCORMICK AND LEMUEL D. WRIGHT

VOLUME 63. Enzyme Kinetics and Mechanism (Part A: Initial Rate and
Inhibitor Methods)
Edited by DANIEL L. PURICH

VOLUME 64. Enzyme Kinetics and Mechanism
(Part B: Isotopic Probes and Complex Enzyme Systems)
Edited by DANIEL L. PURICH

VOLUME 65. Nucleic Acids (Part I)
Edited by LAWRENCE GROSSMAN AND KIVIE MOLDAVE

VOLUME 66. Vitamins and Coenzymes (Part E)
Edited by DONALD B. McCORMICK AND LEMUEL D. WRIGHT

VOLUME 67. Vitamins and Coenzymes (Part F)
Edited by DONALD B. McCORMICK AND LEMUEL D. WRIGHT

VOLUME 68. Recombinant DNA
Edited by RAY WU

VOLUME 69. Photosynthesis and Nitrogen Fixation (Part C)
Edited by ANTHONY SAN PIETRO

VOLUME 70. Immunochemical Techniques (Part A)
Edited by HELEN VAN VUNAKIS AND JOHN J. LANGONE

VOLUME 71. Lipids (Part C)
Edited by JOHN M. LOWENSTEIN

VOLUME 72. Lipids (Part D)
Edited by JOHN M. LOWENSTEIN

VOLUME 73. Immunochemical Techniques (Part B)
Edited by JOHN J. LANGONE AND HELEN VAN VUNAKIS

VOLUME 74. Immunochemical Techniques (Part C)
Edited by JOHN J. LANGONE AND HELEN VAN VUNAKIS

VOLUME 75. Cumulative Subject Index Volumes XXXI, XXXII, XXXIV–LX
Edited by EDWARD A. DENNIS AND MARTHA G. DENNIS

VOLUME 76. Hemoglobins
Edited by ERALDO ANTONINI, LUIGI ROSSI-BERNARDI, AND EMILIA CHIANCONE

VOLUME 77. Detoxication and Drug Metabolism
Edited by WILLIAM B. JAKOBY

VOLUME 78. Interferons (Part A)
Edited by SIDNEY PESTKA

VOLUME 79. Interferons (Part B)
Edited by SIDNEY PESTKA

VOLUME 80. Proteolytic Enzymes (Part C)
Edited by LASZLO LORAND

VOLUME 81. Biomembranes (Part H: Visual Pigments and Purple Membranes, I)
Edited by LESTER PACKER

VOLUME 82. Structural and Contractile Proteins (Part A: Extracellular Matrix)
Edited by LEON W. CUNNINGHAM AND DIXIE W. FREDERIKSEN

VOLUME 83. Complex Carbohydrates (Part D)
Edited by VICTOR GINSBURG

VOLUME 84. Immunochemical Techniques (Part D: Selected Immunoassays)
Edited by JOHN J. LANGONE AND HELEN VAN VUNAKIS

VOLUME 85. Structural and Contractile Proteins (Part B: The Contractile Apparatus and the Cytoskeleton)
Edited by DIXIE W. FREDERIKSEN AND LEON W. CUNNINGHAM

VOLUME 86. Prostaglandins and Arachidonate Metabolites
Edited by WILLIAM E. M. LANDS AND WILLIAM L. SMITH

VOLUME 87. Enzyme Kinetics and Mechanism (Part C: Intermediates, Stereo-chemistry, and Rate Studies)
Edited by DANIEL L. PURICH

VOLUME 88. Biomembranes (Part I: Visual Pigments and Purple Membranes, II)
Edited by LESTER PACKER

VOLUME 89. Carbohydrate Metabolism (Part D)
Edited by WILLIS A. WOOD

VOLUME 90. Carbohydrate Metabolism (Part E)
Edited by WILLIS A. WOOD

VOLUME 91. Enzyme Structure (Part I)
Edited by C. H. W. HIRS AND SERGE N. TIMASHEFF

VOLUME 92. Immunochemical Techniques (Part E: Monoclonal Antibodies and General Immunoassay Methods)
Edited by JOHN J. LANGONE AND HELEN VAN VUNAKIS

VOLUME 93. Immunochemical Techniques (Part F: Conventional Antibodies, Fc Receptors, and Cytotoxicity)
Edited by JOHN J. LANGONE AND HELEN VAN VUNAKIS

VOLUME 94. Polyamines
Edited by HERBERT TABOR AND CELIA WHITE TABOR

VOLUME 95. Cumulative Subject Index Volumes 61–74, 76–80
Edited by EDWARD A. DENNIS AND MARTHA G. DENNIS

VOLUME 96. Biomembranes [Part J: Membrane Biogenesis: Assembly and Targeting (General Methods; Eukaryotes)]
Edited by SIDNEY FLEISCHER AND BECCA FLEISCHER

VOLUME 97. Biomembranes [Part K: Membrane Biogenesis: Assembly and Targeting (Prokaryotes, Mitochondria, and Chloroplasts)]
Edited by SIDNEY FLEISCHER AND BECCA FLEISCHER

VOLUME 98. Biomembranes (Part L: Membrane Biogenesis: Processing and Recycling)
Edited by SIDNEY FLEISCHER AND BECCA FLEISCHER

VOLUME 99. Hormone Action (Part F: Protein Kinases)
Edited by JACKIE D. CORBIN AND JOEL G. HARDMAN

VOLUME 100. Recombinant DNA (Part B)
Edited by RAY WU, LAWRENCE GROSSMAN, AND KIVIE MOLDAVE

VOLUME 101. Recombinant DNA (Part C)
Edited by RAY WU, LAWRENCE GROSSMAN, AND KIVIE MOLDAVE

VOLUME 102. Hormone Action (Part G: Calmodulin and Calcium-Binding Proteins)
Edited by ANTHONY R. MEANS AND BERT W. O'MALLEY

VOLUME 103. Hormone Action (Part H: Neuroendocrine Peptides)
Edited by P. MICHAEL CONN

VOLUME 104. Enzyme Purification and Related Techniques (Part C)
Edited by WILLIAM B. JAKOBY

VOLUME 105. Oxygen Radicals in Biological Systems
Edited by LESTER PACKER

VOLUME 106. Posttranslational Modifications (Part A)
Edited by FINN WOLD AND KIVIE MOLDAVE

VOLUME 107. Posttranslational Modifications (Part B)
Edited by FINN WOLD AND KIVIE MOLDAVE

VOLUME 108. Immunochemical Techniques (Part G: Separation and Characterization of Lymphoid Cells)
Edited by GIOVANNI DI SABATO, JOHN J. LANGONE, AND HELEN VAN VUNAKIS

VOLUME 109. Hormone Action (Part I: Peptide Hormones)
Edited by LUTZ BIRNBAUMER AND BERT W. O'MALLEY

VOLUME 110. Steroids and Isoprenoids (Part A)
Edited by JOHN H. LAW AND HANS C. RILLING

VOLUME 111. Steroids and Isoprenoids (Part B)
Edited by JOHN H. LAW AND HANS C. RILLING

VOLUME 112. Drug and Enzyme Targeting (Part A)
Edited by KENNETH J. WIDDER AND RALPH GREEN

VOLUME 113. Glutamate, Glutamine, Glutathione, and Related Compounds
Edited by ALTON MEISTER

VOLUME 114. Diffraction Methods for Biological Macromolecules (Part A)
Edited by HAROLD W. WYCKOFF, C. H. W. HIRS, AND SERGE N. TIMASHEFF

VOLUME 115. Diffraction Methods for Biological Macromolecules (Part B)
Edited by HAROLD W. WYCKOFF, C. H. W. HIRS, AND SERGE N. TIMASHEFF

VOLUME 116. Immunochemical Techniques (Part H: Effectors and Mediators of Lymphoid Cell Functions)
Edited by GIOVANNI DI SABATO, JOHN J. LANGONE, AND HELEN VAN VUNAKIS

VOLUME 117. Enzyme Structure (Part J)
Edited by C. H. W. HIRS AND SERGE N. TIMASHEFF

VOLUME 118. Plant Molecular Biology
Edited by ARTHUR WEISSBACH AND HERBERT WEISSBACH

VOLUME 119. Interferons (Part C)
Edited by SIDNEY PESTKA

VOLUME 120. Cumulative Subject Index Volumes 81–94, 96–101

VOLUME 121. Immunochemical Techniques (Part I: Hybridoma Technology and Monoclonal Antibodies)
Edited by JOHN J. LANGONE AND HELEN VAN VUNAKIS

VOLUME 122. Vitamins and Coenzymes (Part G)
Edited by FRANK CHYTIL AND DONALD B. MCCORMICK

VOLUME 123. Vitamins and Coenzymes (Part H)
Edited by FRANK CHYTIL AND DONALD B. MCCORMICK

VOLUME 124. Hormone Action (Part J: Neuroendocrine Peptides)
Edited by P. MICHAEL CONN

VOLUME 125. Biomembranes (Part M: Transport in Bacteria, Mitochondria, and Chloroplasts: General Approaches and Transport Systems)
Edited by SIDNEY FLEISCHER AND BECCA FLEISCHER

VOLUME 126. Biomembranes (Part N: Transport in Bacteria, Mitochondria, and Chloroplasts: Protonmotive Force)
Edited by SIDNEY FLEISCHER AND BECCA FLEISCHER

VOLUME 127. Biomembranes (Part O: Protons and Water: Structure and Translocation)
Edited by LESTER PACKER

VOLUME 128. Plasma Lipoproteins (Part A: Preparation, Structure, and Molecular Biology)
Edited by JERE P. SEGREST AND JOHN J. ALBERS

VOLUME 129. Plasma Lipoproteins (Part B: Characterization, Cell Biology, and Metabolism)
Edited by JOHN J. ALBERS AND JERE P. SEGREST

VOLUME 130. Enzyme Structure (Part K)
Edited by C. H. W. HIRS AND SERGE N. TIMASHEFF

VOLUME 131. Enzyme Structure (Part L)
Edited by C. H. W. HIRS AND SERGE N. TIMASHEFF

VOLUME 132. Immunochemical Techniques (Part J: Phagocytosis and Cell-Mediated Cytotoxicity)
Edited by GIOVANNI DI SABATO AND JOHANNES EVERSE

VOLUME 133. Bioluminescence and Chemiluminescence (Part B)
Edited by MARLENE DELUCA AND WILLIAM D. MCELROY

VOLUME 134. Structural and Contractile Proteins (Part C: The Contractile Apparatus and the Cytoskeleton)
Edited by RICHARD B. VALLEE

VOLUME 135. Immobilized Enzymes and Cells (Part B)
Edited by KLAUS MOSBACH

VOLUME 136. Immobilized Enzymes and Cells (Part C)
Edited by KLAUS MOSBACH

VOLUME 137. Immobilized Enzymes and Cells (Part D)
Edited by KLAUS MOSBACH

VOLUME 138. Complex Carbohydrates (Part E)
Edited by VICTOR GINSBURG

VOLUME 139. Cellular Regulators (Part A: Calcium- and Calmodulin-Binding Proteins)
Edited by ANTHONY R. MEANS AND P. MICHAEL CONN

VOLUME 140. Cumulative Subject Index Volumes 102–119, 121–134

VOLUME 141. Cellular Regulators (Part B: Calcium and Lipids)
Edited by P. MICHAEL CONN AND ANTHONY R. MEANS

VOLUME 142. Metabolism of Aromatic Amino Acids and Amines
Edited by SEYMOUR KAUFMAN

VOLUME 143. Sulfur and Sulfur Amino Acids
Edited by WILLIAM B. JAKOBY AND OWEN GRIFFITH

VOLUME 144. Structural and Contractile Proteins (Part D: Extracellular Matrix)
Edited by LEON W. CUNNINGHAM

VOLUME 145. Structural and Contractile Proteins (Part E: Extracellular Matrix)
Edited by LEON W. CUNNINGHAM

VOLUME 146. Peptide Growth Factors (Part A)
Edited by DAVID BARNES AND DAVID A. SIRBASKU

VOLUME 147. Peptide Growth Factors (Part B)
Edited by DAVID BARNES AND DAVID A. SIRBASKU

VOLUME 148. Plant Cell Membranes
Edited by LESTER PACKER AND ROLAND DOUCE

VOLUME 149. Drug and Enzyme Targeting (Part B)
Edited by RALPH GREEN AND KENNETH J. WIDDER

VOLUME 150. Immunochemical Techniques (Part K: *In Vitro* Models of B and T Cell Functions and Lymphoid Cell Receptors)
Edited by GIOVANNI DI SABATO

VOLUME 151. Molecular Genetics of Mammalian Cells
Edited by MICHAEL M. GOTTESMAN

VOLUME 152. Guide to Molecular Cloning Techniques
Edited by SHELBY L. BERGER AND ALAN R. KIMMEL

VOLUME 153. Recombinant DNA (Part D)
Edited by RAY WU AND LAWRENCE GROSSMAN

VOLUME 154. Recombinant DNA (Part E)
Edited by RAY WU AND LAWRENCE GROSSMAN

VOLUME 155. Recombinant DNA (Part F)
Edited by RAY WU

VOLUME 156. Biomembranes (Part P: ATP-Driven Pumps and Related Transport: The Na, K-Pump)
Edited by SIDNEY FLEISCHER AND BECCA FLEISCHER

VOLUME 157. Biomembranes (Part Q: ATP-Driven Pumps and Related Transport: Calcium, Proton, and Potassium Pumps)
Edited by SIDNEY FLEISCHER AND BECCA FLEISCHER

VOLUME 158. Metalloproteins (Part A)
Edited by JAMES F. RIORDAN AND BERT L. VALLEE

VOLUME 159. Initiation and Termination of Cyclic Nucleotide Action
Edited by JACKIE D. CORBIN AND ROGER A. JOHNSON

VOLUME 160. Biomass (Part A: Cellulose and Hemicellulose)
Edited by WILLIS A. WOOD AND SCOTT T. KELLOGG

VOLUME 161. Biomass (Part B: Lignin, Pectin, and Chitin)
Edited by WILLIS A. WOOD AND SCOTT T. KELLOGG

VOLUME 162. Immunochemical Techniques (Part L: Chemotaxis
and Inflammation)
Edited by GIOVANNI DI SABATO

VOLUME 163. Immunochemical Techniques (Part M: Chemotaxis
and Inflammation)
Edited by GIOVANNI DI SABATO

VOLUME 164. Ribosomes
Edited by HARRY F. NOLLER, JR., AND KIVIE MOLDAVE

VOLUME 165. Microbial Toxins: Tools for Enzymology
Edited by SIDNEY HARSHMAN

VOLUME 166. Branched-Chain Amino Acids
Edited by ROBERT HARRIS AND JOHN R. SOKATCH

VOLUME 167. Cyanobacteria
Edited by LESTER PACKER AND ALEXANDER N. GLAZER

VOLUME 168. Hormone Action (Part K: Neuroendocrine Peptides)
Edited by P. MICHAEL CONN

VOLUME 169. Platelets: Receptors, Adhesion,
Secretion (Part A)
Edited by JACEK HAWIGER

VOLUME 170. Nucleosomes
Edited by PAUL M. WASSARMAN AND ROGER D. KORNBERG

VOLUME 171. Biomembranes (Part R: Transport Theory: Cells and Model
Membranes)
Edited by SIDNEY FLEISCHER AND BECCA FLEISCHER

VOLUME 172. Biomembranes (Part S: Transport: Membrane Isolation and
Characterization)
Edited by SIDNEY FLEISCHER AND BECCA FLEISCHER

VOLUME 173. Biomembranes [Part T: Cellular and Subcellular Transport:
Eukaryotic (Nonepithelial) Cells]
Edited by SIDNEY FLEISCHER AND BECCA FLEISCHER

VOLUME 174. Biomembranes [Part U: Cellular and Subcellular Transport:
Eukaryotic (Nonepithelial) Cells]
Edited by SIDNEY FLEISCHER AND BECCA FLEISCHER

VOLUME 175. Cumulative Subject Index Volumes 135–139, 141–167

VOLUME 176. Nuclear Magnetic Resonance (Part A: Spectral Techniques and Dynamics)
Edited by NORMAN J. OPPENHEIMER AND THOMAS L. JAMES

VOLUME 177. Nuclear Magnetic Resonance (Part B: Structure and Mechanism)
Edited by NORMAN J. OPPENHEIMER AND THOMAS L. JAMES

VOLUME 178. Antibodies, Antigens, and Molecular Mimicry
Edited by JOHN J. LANGONE

VOLUME 179. Complex Carbohydrates (Part F)
Edited by VICTOR GINSBURG

VOLUME 180. RNA Processing (Part A: General Methods)
Edited by JAMES E. DAHLBERG AND JOHN N. ABELSON

VOLUME 181. RNA Processing (Part B: Specific Methods)
Edited by JAMES E. DAHLBERG AND JOHN N. ABELSON

VOLUME 182. Guide to Protein Purification
Edited by MURRAY P. DEUTSCHER

VOLUME 183. Molecular Evolution: Computer Analysis of Protein and Nucleic Acid Sequences
Edited by RUSSELL F. DOOLITTLE

VOLUME 184. Avidin-Biotin Technology
Edited by MEIR WILCHEK AND EDWARD A. BAYER

VOLUME 185. Gene Expression Technology
Edited by DAVID V. GOEDDEL

VOLUME 186. Oxygen Radicals in Biological Systems (Part B: Oxygen Radicals and Antioxidants)
Edited by LESTER PACKER AND ALEXANDER N. GLAZER

VOLUME 187. Arachidonate Related Lipid Mediators
Edited by ROBERT C. MURPHY AND FRANK A. FITZPATRICK

VOLUME 188. Hydrocarbons and Methylotrophy
Edited by MARY E. LIDSTROM

VOLUME 189. Retinoids (Part A: Molecular and Metabolic Aspects)
Edited by LESTER PACKER

VOLUME 190. Retinoids (Part B: Cell Differentiation and Clinical Applications)
Edited by LESTER PACKER

VOLUME 191. Biomembranes (Part V: Cellular and Subcellular Transport: Epithelial Cells)
Edited by SIDNEY FLEISCHER AND BECCA FLEISCHER

VOLUME 192. Biomembranes (Part W: Cellular and Subcellular Transport: Epithelial Cells)
Edited by SIDNEY FLEISCHER AND BECCA FLEISCHER

VOLUME 193. Mass Spectrometry
Edited by JAMES A. MCCLOSKEY

VOLUME 194. Guide to Yeast Genetics and Molecular Biology
Edited by CHRISTINE GUTHRIE AND GERALD R. FINK

VOLUME 195. Adenylyl Cyclase, G Proteins, and Guanylyl Cyclase
Edited by ROGER A. JOHNSON AND JACKIE D. CORBIN

VOLUME 196. Molecular Motors and the Cytoskeleton
Edited by RICHARD B. VALLEE

VOLUME 197. Phospholipases
Edited by EDWARD A. DENNIS

VOLUME 198. Peptide Growth Factors (Part C)
Edited by DAVID BARNES, J. P. MATHER, AND GORDON H. SATO

VOLUME 199. Cumulative Subject Index Volumes 168–174, 176–194

VOLUME 200. Protein Phosphorylation (Part A: Protein Kinases: Assays,
Purification, Antibodies, Functional Analysis, Cloning, and Expression)
Edited by TONY HUNTER AND BARTHOLOMEW M. SEFTON

VOLUME 201. Protein Phosphorylation (Part B: Analysis of Protein
Phosphorylation, Protein Kinase Inhibitors, and Protein Phosphatases)
Edited by TONY HUNTER AND BARTHOLOMEW M. SEFTON

VOLUME 202. Molecular Design and Modeling: Concepts and Applications
(Part A: Proteins, Peptides, and Enzymes)
Edited by JOHN J. LANGONE

VOLUME 203. Molecular Design and Modeling:
Concepts and Applications (Part B: Antibodies and Antigens, Nucleic Acids,
Polysaccharides,
and Drugs)
Edited by JOHN J. LANGONE

VOLUME 204. Bacterial Genetic Systems
Edited by JEFFREY H. MILLER

VOLUME 205. Metallobiochemistry (Part B: Metallothionein and
Related Molecules)
Edited by JAMES F. RIORDAN AND BERT L. VALLEE

VOLUME 206. Cytochrome P450
Edited by MICHAEL R. WATERMAN AND ERIC F. JOHNSON

VOLUME 207. Ion Channels
Edited by BERNARDO RUDY AND LINDA E. IVERSON

VOLUME 208. Protein–DNA Interactions
Edited by ROBERT T. SAUER

VOLUME 209. Phospholipid Biosynthesis
Edited by EDWARD A. DENNIS AND DENNIS E. VANCE

VOLUME 210. Numerical Computer Methods
Edited by LUDWIG BRAND AND MICHAEL L. JOHNSON

VOLUME 211. DNA Structures (Part A: Synthesis and Physical Analysis of DNA)
Edited by DAVID M. J. LILLEY AND JAMES E. DAHLBERG

VOLUME 212. DNA Structures (Part B: Chemical and Electrophoretic Analysis of DNA)
Edited by DAVID M. J. LILLEY AND JAMES E. DAHLBERG

VOLUME 213. Carotenoids (Part A: Chemistry, Separation, Quantitation, and Antioxidation)
Edited by LESTER PACKER

VOLUME 214. Carotenoids (Part B: Metabolism, Genetics, and Biosynthesis)
Edited by LESTER PACKER

VOLUME 215. Platelets: Receptors, Adhesion, Secretion (Part B)
Edited by JACEK J. HAWIGER

VOLUME 216. Recombinant DNA (Part G)
Edited by RAY WU

VOLUME 217. Recombinant DNA (Part H)
Edited by RAY WU

VOLUME 218. Recombinant DNA (Part I)
Edited by RAY WU

VOLUME 219. Reconstitution of Intracellular Transport
Edited by JAMES E. ROTHMAN

VOLUME 220. Membrane Fusion Techniques (Part A)
Edited by NEJAT DÜZGÜNEŞ

VOLUME 221. Membrane Fusion Techniques (Part B)
Edited by NEJAT DÜZGÜNEŞ

VOLUME 222. Proteolytic Enzymes in Coagulation, Fibrinolysis, and Complement Activation (Part A: Mammalian Blood Coagulation Factors and Inhibitors)
Edited by LASZLO LORAND AND KENNETH G. MANN

VOLUME 223. Proteolytic Enzymes in Coagulation, Fibrinolysis, and Complement Activation (Part B: Complement Activation, Fibrinolysis, and Nonmammalian Blood Coagulation Factors)
Edited by LASZLO LORAND AND KENNETH G. MANN

VOLUME 224. Molecular Evolution: Producing the Biochemical Data
Edited by ELIZABETH ANNE ZIMMER, THOMAS J. WHITE, REBECCA L. CANN, AND ALLAN C. WILSON

VOLUME 225. Guide to Techniques in Mouse Development
Edited by PAUL M. WASSARMAN AND MELVIN L. DEPAMPHILIS

VOLUME 226. Metallobiochemistry (Part C: Spectroscopic and
Physical Methods for Probing Metal Ion Environments in Metalloenzymes
and Metalloproteins)
Edited by JAMES F. RIORDAN AND BERT L. VALLEE

VOLUME 227. Metallobiochemistry (Part D: Physical and Spectroscopic
Methods for Probing Metal Ion Environments in Metalloproteins)
Edited by JAMES F. RIORDAN AND BERT L. VALLEE

VOLUME 228. Aqueous Two-Phase Systems
Edited by HARRY WALTER AND GÖTE JOHANSSON

VOLUME 229. Cumulative Subject Index Volumes 195–198, 200–227

VOLUME 230. Guide to Techniques in Glycobiology
Edited by WILLIAM J. LENNARZ AND GERALD W. HART

VOLUME 231. Hemoglobins (Part B: Biochemical and Analytical Methods)
Edited by JOHANNES EVERSE, KIM D. VANDEGRIFF, AND ROBERT M. WINSLOW

VOLUME 232. Hemoglobins (Part C: Biophysical Methods)
Edited by JOHANNES EVERSE, KIM D. VANDEGRIFF, AND ROBERT M. WINSLOW

VOLUME 233. Oxygen Radicals in Biological Systems (Part C)
Edited by LESTER PACKER

VOLUME 234. Oxygen Radicals in Biological Systems (Part D)
Edited by LESTER PACKER

VOLUME 235. Bacterial Pathogenesis (Part A: Identification and Regulation of
Virulence Factors)
Edited by VIRGINIA L. CLARK AND PATRIK M. BAVOIL

VOLUME 236. Bacterial Pathogenesis (Part B: Integration of Pathogenic
Bacteria with Host Cells)
Edited by VIRGINIA L. CLARK AND PATRIK M. BAVOIL

VOLUME 237. Heterotrimeric G Proteins
Edited by RAVI IYENGAR

VOLUME 238. Heterotrimeric G-Protein Effectors
Edited by RAVI IYENGAR

VOLUME 239. Nuclear Magnetic Resonance (Part C)
Edited by THOMAS L. JAMES AND NORMAN J. OPPENHEIMER

VOLUME 240. Numerical Computer Methods (Part B)
Edited by MICHAEL L. JOHNSON AND LUDWIG BRAND

VOLUME 241. Retroviral Proteases
Edited by LAWRENCE C. KUO AND JULES A. SHAFER

VOLUME 242. Neoglycoconjugates (Part A)
Edited by Y. C. LEE AND REIKO T. LEE

VOLUME 243. Inorganic Microbial Sulfur Metabolism
Edited by HARRY D. PECK, JR., AND JEAN LEGALL

VOLUME 244. Proteolytic Enzymes: Serine and Cysteine Peptidases
Edited by ALAN J. BARRETT

VOLUME 245. Extracellular Matrix Components
Edited by E. RUOSLAHTI AND E. ENGVALL

VOLUME 246. Biochemical Spectroscopy
Edited by KENNETH SAUER

VOLUME 247. Neoglycoconjugates (Part B: Biomedical Applications)
Edited by Y. C. LEE AND REIKO T. LEE

VOLUME 248. Proteolytic Enzymes: Aspartic and Metallo Peptidases
Edited by ALAN J. BARRETT

VOLUME 249. Enzyme Kinetics and Mechanism (Part D: Developments in Enzyme Dynamics)
Edited by DANIEL L. PURICH

VOLUME 250. Lipid Modifications of Proteins
Edited by PATRICK J. CASEY AND JANICE E. BUSS

VOLUME 251. Biothiols (Part A: Monothiols and Dithiols, Protein Thiols, and Thiyl Radicals)
Edited by LESTER PACKER

VOLUME 252. Biothiols (Part B: Glutathione and Thioredoxin; Thiols in Signal Transduction and Gene Regulation)
Edited by LESTER PACKER

VOLUME 253. Adhesion of Microbial Pathogens
Edited by RON J. DOYLE AND ITZHAK OFEK

VOLUME 254. Oncogene Techniques
Edited by PETER K. VOGT AND INDER M. VERMA

VOLUME 255. Small GTPases and Their Regulators (Part A: Ras Family)
Edited by W. E. BALCH, CHANNING J. DER, AND ALAN HALL

VOLUME 256. Small GTPases and Their Regulators (Part B: Rho Family)
Edited by W. E. BALCH, CHANNING J. DER, AND ALAN HALL

VOLUME 257. Small GTPases and Their Regulators (Part C: Proteins Involved in Transport)
Edited by W. E. BALCH, CHANNING J. DER, AND ALAN HALL

VOLUME 258. Redox-Active Amino Acids in Biology
Edited by JUDITH P. KLINMAN

VOLUME 259. Energetics of Biological Macromolecules
Edited by MICHAEL L. JOHNSON AND GARY K. ACKERS

VOLUME 260. Mitochondrial Biogenesis and Genetics (Part A)
Edited by GIUSEPPE M. ATTARDI AND ANNE CHOMYN

VOLUME 261. Nuclear Magnetic Resonance and Nucleic Acids
Edited by THOMAS L. JAMES

VOLUME 262. DNA Replication
Edited by JUDITH L. CAMPBELL

VOLUME 263. Plasma Lipoproteins (Part C: Quantitation)
Edited by WILLIAM A. BRADLEY, SANDRA H. GIANTURCO, AND JERE P. SEGREST

VOLUME 264. Mitochondrial Biogenesis and Genetics (Part B)
Edited by GIUSEPPE M. ATTARDI AND ANNE CHOMYN

VOLUME 265. Cumulative Subject Index Volumes 228, 230–262

VOLUME 266. Computer Methods for Macromolecular Sequence Analysis
Edited by RUSSELL F. DOOLITTLE

VOLUME 267. Combinatorial Chemistry
Edited by JOHN N. ABELSON

VOLUME 268. Nitric Oxide (Part A: Sources and Detection of NO; NO
Synthase)
Edited by LESTER PACKER

VOLUME 269. Nitric Oxide (Part B: Physiological and
Pathological Processes)
Edited by LESTER PACKER

VOLUME 270. High Resolution Separation and Analysis of Biological
Macromolecules (Part A: Fundamentals)
Edited by BARRY L. KARGER AND WILLIAM S. HANCOCK

VOLUME 271. High Resolution Separation and Analysis of Biological
Macromolecules (Part B: Applications)
Edited by BARRY L. KARGER AND WILLIAM S. HANCOCK

VOLUME 272. Cytochrome P450 (Part B)
Edited by ERIC F. JOHNSON AND MICHAEL R. WATERMAN

VOLUME 273. RNA Polymerase and Associated Factors (Part A)
Edited by SANKAR ADHYA

VOLUME 274. RNA Polymerase and Associated Factors (Part B)
Edited by SANKAR ADHYA

VOLUME 275. Viral Polymerases and Related Proteins
Edited by LAWRENCE C. KUO, DAVID B. OLSEN, AND STEVEN S. CARROLL

VOLUME 276. Macromolecular Crystallography (Part A)
Edited by CHARLES W. CARTER, JR., AND ROBERT M. SWEET

VOLUME 277. Macromolecular Crystallography (Part B)
Edited by CHARLES W. CARTER, JR., AND ROBERT M. SWEET

VOLUME 278. Fluorescence Spectroscopy
Edited by LUDWIG BRAND AND MICHAEL L. JOHNSON

VOLUME 279. Vitamins and Coenzymes (Part I)
Edited by DONALD B. MCCORMICK, JOHN W. SUTTIE, AND CONRAD WAGNER

VOLUME 280. Vitamins and Coenzymes (Part J)
Edited by DONALD B. MCCORMICK, JOHN W. SUTTIE, AND CONRAD WAGNER

VOLUME 281. Vitamins and Coenzymes (Part K)
Edited by DONALD B. MCCORMICK, JOHN W. SUTTIE, AND CONRAD WAGNER

VOLUME 282. Vitamins and Coenzymes (Part L)
Edited by DONALD B. MCCORMICK, JOHN W. SUTTIE, AND CONRAD WAGNER

VOLUME 283. Cell Cycle Control
Edited by WILLIAM G. DUNPHY

VOLUME 284. Lipases (Part A: Biotechnology)
Edited by BYRON RUBIN AND EDWARD A. DENNIS

VOLUME 285. Cumulative Subject Index Volumes 263, 264, 266–284, 286–289

VOLUME 286. Lipases (Part B: Enzyme Characterization and Utilization)
Edited by BYRON RUBIN AND EDWARD A. DENNIS

VOLUME 287. Chemokines
Edited by RICHARD HORUK

VOLUME 288. Chemokine Receptors
Edited by RICHARD HORUK

VOLUME 289. Solid Phase Peptide Synthesis
Edited by GREGG B. FIELDS

VOLUME 290. Molecular Chaperones
Edited by GEORGE H. LORIMER AND THOMAS BALDWIN

VOLUME 291. Caged Compounds
Edited by GERARD MARRIOTT

VOLUME 292. ABC Transporters: Biochemical, Cellular, and
Molecular Aspects
Edited by SURESH V. AMBUDKAR AND MICHAEL M. GOTTESMAN

VOLUME 293. Ion Channels (Part B)
Edited by P. MICHAEL CONN

VOLUME 294. Ion Channels (Part C)
Edited by P. MICHAEL CONN

VOLUME 295. Energetics of Biological Macromolecules (Part B)
Edited by GARY K. ACKERS AND MICHAEL L. JOHNSON

VOLUME 296. Neurotransmitter Transporters
Edited by SUSAN G. AMARA

VOLUME 297. Photosynthesis: Molecular Biology of Energy Capture
Edited by LEE MCINTOSH

VOLUME 298. Molecular Motors and the Cytoskeleton (Part B)
Edited by RICHARD B. VALLEE

VOLUME 299. Oxidants and Antioxidants (Part A)
Edited by LESTER PACKER

VOLUME 300. Oxidants and Antioxidants (Part B)
Edited by LESTER PACKER

VOLUME 301. Nitric Oxide: Biological and Antioxidant Activities (Part C)
Edited by LESTER PACKER

VOLUME 302. Green Fluorescent Protein
Edited by P. MICHAEL CONN

VOLUME 303. cDNA Preparation and Display
Edited by SHERMAN M. WEISSMAN

VOLUME 304. Chromatin
Edited by PAUL M. WASSARMAN AND ALAN P. WOLFFE

VOLUME 305. Bioluminescence and Chemiluminescence (Part C)
Edited by THOMAS O. BALDWIN AND MIRIAM M. ZIEGLER

VOLUME 306. Expression of Recombinant Genes in
Eukaryotic Systems
Edited by JOSEPH C. GLORIOSO AND MARTIN C. SCHMIDT

VOLUME 307. Confocal Microscopy
Edited by P. MICHAEL CONN

VOLUME 308. Enzyme Kinetics and Mechanism (Part E: Energetics of
Enzyme Catalysis)
Edited by DANIEL L. PURICH AND VERN L. SCHRAMM

VOLUME 309. Amyloid, Prions, and Other Protein Aggregates
Edited by RONALD WETZEL

VOLUME 310. Biofilms
Edited by RON J. DOYLE

VOLUME 311. Sphingolipid Metabolism and Cell Signaling (Part A)
Edited by ALFRED H. MERRILL, JR., AND YUSUF A. HANNUN

VOLUME 312. Sphingolipid Metabolism and Cell Signaling (Part B)
Edited by ALFRED H. MERRILL, JR., AND YUSUF A. HANNUN

VOLUME 313. Antisense Technology (Part A: General Methods, Methods of
Delivery, and RNA Studies)
Edited by M. IAN PHILLIPS

VOLUME 314. Antisense Technology (Part B: Applications)
Edited by M. IAN PHILLIPS

VOLUME 315. Vertebrate Phototransduction and the Visual Cycle (Part A)
Edited by KRZYSZTOF PALCZEWSKI

VOLUME 316. Vertebrate Phototransduction and the Visual Cycle (Part B)
Edited by KRZYSZTOF PALCZEWSKI

VOLUME 317. RNA–Ligand Interactions (Part A: Structural Biology Methods)
Edited by DANIEL W. CELANDER AND JOHN N. ABELSON

VOLUME 318. RNA–Ligand Interactions (Part B: Molecular Biology Methods)
Edited by DANIEL W. CELANDER AND JOHN N. ABELSON

VOLUME 319. Singlet Oxygen, UV-A, and Ozone
Edited by LESTER PACKER AND HELMUT SIES

VOLUME 320. Cumulative Subject Index Volumes 290–319

VOLUME 321. Numerical Computer Methods (Part C)
Edited by MICHAEL L. JOHNSON AND LUDWIG BRAND

VOLUME 322. Apoptosis
Edited by JOHN C. REED

VOLUME 323. Energetics of Biological Macromolecules (Part C)
Edited by MICHAEL L. JOHNSON AND GARY K. ACKERS

VOLUME 324. Branched-Chain Amino Acids (Part B)
Edited by ROBERT A. HARRIS AND JOHN R. SOKATCH

VOLUME 325. Regulators and Effectors of Small GTPases (Part D: Rho Family)
Edited by W. E. BALCH, CHANNING J. DER, AND ALAN HALL

VOLUME 326. Applications of Chimeric Genes and Hybrid Proteins (Part A:
Gene Expression and Protein Purification)
Edited by JEREMY THORNER, SCOTT D. EMR, AND JOHN N. ABELSON

VOLUME 327. Applications of Chimeric Genes and Hybrid Proteins (Part B:
Cell Biology and Physiology)
Edited by JEREMY THORNER, SCOTT D. EMR, AND JOHN N. ABELSON

VOLUME 328. Applications of Chimeric Genes and Hybrid Proteins (Part C:
Protein–Protein Interactions and Genomics)
Edited by JEREMY THORNER, SCOTT D. EMR, AND JOHN N. ABELSON

VOLUME 329. Regulators and Effectors of Small GTPases (Part E: GTPases
Involved in Vesicular Traffic)
Edited by W. E. BALCH, CHANNING J. DER, AND ALAN HALL

VOLUME 330. Hyperthermophilic Enzymes (Part A)
Edited by MICHAEL W. W. ADAMS AND ROBERT M. KELLY

VOLUME 331. Hyperthermophilic Enzymes (Part B)
Edited by MICHAEL W. W. ADAMS AND ROBERT M. KELLY

VOLUME 332. Regulators and Effectors of Small GTPases (Part F: Ras Family I)
Edited by W. E. BALCH, CHANNING J. DER, AND ALAN HALL

VOLUME 333. Regulators and Effectors of Small GTPases (Part G: Ras Family II)
Edited by W. E. BALCH, CHANNING J. DER, AND ALAN HALL

VOLUME 334. Hyperthermophilic Enzymes (Part C)
Edited by MICHAEL W. W. ADAMS AND ROBERT M. KELLY

VOLUME 335. Flavonoids and Other Polyphenols
Edited by LESTER PACKER

VOLUME 336. Microbial Growth in Biofilms (Part A: Developmental and Molecular Biological Aspects)
Edited by RON J. DOYLE

VOLUME 337. Microbial Growth in Biofilms (Part B: Special Environments and Physicochemical Aspects)
Edited by RON J. DOYLE

VOLUME 338. Nuclear Magnetic Resonance of Biological Macromolecules (Part A)
Edited by THOMAS L. JAMES, VOLKER DÖTSCH, AND ULI SCHMITZ

VOLUME 339. Nuclear Magnetic Resonance of Biological Macromolecules (Part B)
Edited by THOMAS L. JAMES, VOLKER DÖTSCH, AND ULI SCHMITZ

VOLUME 340. Drug–Nucleic Acid Interactions
Edited by JONATHAN B. CHAIRES AND MICHAEL J. WARING

VOLUME 341. Ribonucleases (Part A)
Edited by ALLEN W. NICHOLSON

VOLUME 342. Ribonucleases (Part B)
Edited by ALLEN W. NICHOLSON

VOLUME 343. G Protein Pathways (Part A: Receptors)
Edited by RAVI IYENGAR AND JOHN D. HILDEBRANDT

VOLUME 344. G Protein Pathways (Part B: G Proteins and Their Regulators)
Edited by RAVI IYENGAR AND JOHN D. HILDEBRANDT

VOLUME 345. G Protein Pathways (Part C: Effector Mechanisms)
Edited by RAVI IYENGAR AND JOHN D. HILDEBRANDT

VOLUME 346. Gene Therapy Methods
Edited by M. IAN PHILLIPS

VOLUME 347. Protein Sensors and Reactive Oxygen Species (Part A: Selenoproteins and Thioredoxin)
Edited by HELMUT SIES AND LESTER PACKER

VOLUME 348. Protein Sensors and Reactive Oxygen Species (Part B: Thiol Enzymes and Proteins)
Edited by HELMUT SIES AND LESTER PACKER

VOLUME 349. Superoxide Dismutase
Edited by LESTER PACKER

VOLUME 350. Guide to Yeast Genetics and Molecular and Cell Biology (Part B)
Edited by CHRISTINE GUTHRIE AND GERALD R. FINK

VOLUME 351. Guide to Yeast Genetics and Molecular and Cell Biology (Part C)
Edited by CHRISTINE GUTHRIE AND GERALD R. FINK

VOLUME 352. Redox Cell Biology and Genetics (Part A)
Edited by CHANDAN K. SEN AND LESTER PACKER

VOLUME 353. Redox Cell Biology and Genetics (Part B)
Edited by CHANDAN K. SEN AND LESTER PACKER

VOLUME 354. Enzyme Kinetics and Mechanisms (Part F: Detection and Characterization of Enzyme Reaction Intermediates)
Edited by DANIEL L. PURICH

VOLUME 355. Cumulative Subject Index Volumes 321–354

VOLUME 356. Laser Capture Microscopy and Microdissection
Edited by P. MICHAEL CONN

VOLUME 357. Cytochrome P450, Part C
Edited by ERIC F. JOHNSON AND MICHAEL R. WATERMAN

VOLUME 358. Bacterial Pathogenesis (Part C: Identification, Regulation, and Function of Virulence Factors)
Edited by VIRGINIA L. CLARK AND PATRIK M. BAVOIL

VOLUME 359. Nitric Oxide (Part D)
Edited by ENRIQUE CADENAS AND LESTER PACKER

VOLUME 360. Biophotonics (Part A)
Edited by GERARD MARRIOTT AND IAN PARKER

VOLUME 361. Biophotonics (Part B)
Edited by GERARD MARRIOTT AND IAN PARKER

VOLUME 362. Recognition of Carbohydrates in Biological Systems (Part A)
Edited by YUAN C. LEE AND REIKO T. LEE

VOLUME 363. Recognition of Carbohydrates in Biological Systems (Part B)
Edited by YUAN C. LEE AND REIKO T. LEE

VOLUME 364. Nuclear Receptors
Edited by DAVID W. RUSSELL AND DAVID J. MANGELSDORF

VOLUME 365. Differentiation of Embryonic Stem Cells
Edited by PAUL M. WASSAUMAN AND GORDON M. KELLER

VOLUME 366. Protein Phosphatases
Edited by SUSANNE KLUMPP AND JOSEF KRIEGLSTEIN

VOLUME 367. Liposomes (Part A)
Edited by NEJAT DÜZGÜNEŞ

VOLUME 368. Macromolecular Crystallography (Part C)
Edited by CHARLES W. CARTER, JR., AND ROBERT M. SWEET

VOLUME 369. Combinational Chemistry (Part B)
Edited by GUILLERMO A. MORALES AND BARRY A. BUNIN

VOLUME 370. RNA Polymerases and Associated Factors (Part C)
Edited by SANKAR L. ADHYA AND SUSAN GARGES

VOLUME 371. RNA Polymerases and Associated Factors (Part D)
Edited by SANKAR L. ADHYA AND SUSAN GARGES

VOLUME 372. Liposomes (Part B)
Edited by NEJAT DÜZGÜNEŞ

VOLUME 373. Liposomes (Part C)
Edited by NEJAT DÜZGÜNEŞ

VOLUME 374. Macromolecular Crystallography (Part D)
Edited by CHARLES W. CARTER, JR., AND ROBERT W. SWEET

VOLUME 375. Chromatin and Chromatin Remodeling Enzymes (Part A)
Edited by C. DAVID ALLIS AND CARL WU

VOLUME 376. Chromatin and Chromatin Remodeling Enzymes (Part B)
Edited by C. DAVID ALLIS AND CARL WU

VOLUME 377. Chromatin and Chromatin Remodeling Enzymes (Part C)
Edited by C. DAVID ALLIS AND CARL WU

VOLUME 378. Quinones and Quinone Enzymes (Part A)
Edited by HELMUT SIES AND LESTER PACKER

VOLUME 379. Energetics of Biological Macromolecules (Part D)
Edited by JO M. HOLT, MICHAEL L. JOHNSON, AND GARY K. ACKERS

VOLUME 380. Energetics of Biological Macromolecules (Part E)
Edited by JO M. HOLT, MICHAEL L. JOHNSON, AND GARY K. ACKERS

VOLUME 381. Oxygen Sensing
Edited by CHANDAN K. SEN AND GREGG L. SEMENZA

VOLUME 382. Quinones and Quinone Enzymes (Part B)
Edited by HELMUT SIES AND LESTER PACKER

VOLUME 383. Numerical Computer Methods (Part D)
Edited by LUDWIG BRAND AND MICHAEL L. JOHNSON

VOLUME 384. Numerical Computer Methods (Part E)
Edited by LUDWIG BRAND AND MICHAEL L. JOHNSON

VOLUME 385. Imaging in Biological Research (Part A)
Edited by P. MICHAEL CONN

VOLUME 386. Imaging in Biological Research (Part B)
Edited by P. MICHAEL CONN

VOLUME 387. Liposomes (Part D)
Edited by NEJAT DÜZGÜNEŞ

VOLUME 388. Protein Engineering
Edited by DAN E. ROBERTSON AND JOSEPH P. NOEL

VOLUME 389. Regulators of G-Protein Signaling (Part A)
Edited by DAVID P. SIDEROVSKI

VOLUME 390. Regulators of G-protein Sgnalling (Part B)
Edited by DAVID P. SIDEROVSKI

VOLUME 391. Liposomes (Part E)
Edited by NEJAT DÜZGÜNEŞ

VOLUME 392. RNA Interference
Edited by ENGELKE ROSSI

VOLUME 393. Circadian Rhythms
Edited by MICHAEL W. YOUNG

VOLUME 394. Nuclear Magnetic Resonance of Biological Macromolecules
(Part C)
Edited by THOMAS L. JAMES

VOLUME 395. Producing the Biochemical Data (Part B)
Edited by ELIZABETH A. ZIMMER AND ERIC H. ROALSON

VOLUME 396. Nitric Oxide (Part E)
Edited by LESTER PACKER AND ENRIQUE CADENAS

VOLUME 397. Environmental Microbiology
Edited by JARED R. LEADBETTER

VOLUME 398. Ubiquitin and Protein Degradation (Part A)
Edited by RAYMOND J. DESHAIES

VOLUME 399. Ubiquitin and Protein Degradation (Part B)
Edited by RAYMOND J. DESHAIES

VOLUME 400. Phase II Conjugation Enzymes and Transport Systems
Edited by HELMUT SIES AND LESTER PACKER

VOLUME 401. Glutathione Transferases and Gamma Glutamyl Transpeptidases
Edited by HELMUT SIES AND LESTER PACKER

VOLUME 402. Biological Mass Spectrometry (in preparation)
Edited by A. L. BURLINGAME

VOLUME 403. GTPases Regulating Membrane Targeting and Fusion
(in preparation)
Edited by WILLIAM E. BALCH, CHANNING J. DER, AND ALAN HALL

VOLUME 404. GTPases Regulating Membrane Dynamics (in preparation)
Edited by WILLIAM E. BALCH, CHANNING J. DER, AND ALAN HALL

VOLUME 405. Mass Spectrometry: Modified Proteins and Glycoconjugates
(in preparation)
Edited by BURLINGAME

VOLUME 406. Regulators and Effectors of Small GTPases: Rho Family
(in preparation)
Edited by WILLIAM E. BALCH, CHANNING J. DER, AND ALAN HALL

[1] UDP-Glucuronosyltransferases: Gene Structures of *UGT1* and *UGT2* Families

By Ida S. Owens, Nikhil K. Basu, and Rajat Banerjee

Abstract

In human, rat, and mice, a *UGT1* complex locus provides for developmental-, inducer-, and cell-specific synthesis of a family of chemical-detoxifying isozymes, UDP-glucuronosyltransferases, which prevent toxicities, mutagenesis, and/or carcinogenesis. Between 10 and 14 first exons with individual promoter elements are tandemly arrayed upstream of 4 shared exons so as to synthesize independently as many overlapping primary transcripts. RNA splice sites allow a lead exon to join the common exons to generate mRNAs with unique 5′ ends, but common 3′ ends. Intra- and interspecies comparisons of amino acid sequences encoded by first exons show an evolutionary continuum; also, recognizable bilirubin- and phenol-specific catalytic units are differentially regulated by model compounds, phenobarbital, and/or aromatic hydrocarbons. Whereas *UGT1* loci allow minimal changes to achieve new isozymes, a single deleterious mutation in a common exon negatively impacts the arrangement by inactivating the entire family of isozymes compared to an event at independent loci as seen in the *UGT2* family. In humans, lethal hyperbilirubinemic Crigler-Najjar type 1 and milder diseases/syndromes are due to deleterious to mildly deleterious mutations in the bilirubin-specific *UGT1A1* or a common exon. In addition, the number of TA repeats (N_{5-8}) in the *UGT1A1* proximal TATA box affects transcriptional rate and, thus, activity. Evidence also shows that polymorphisms in nonbilirubin-specific first exons also impact chemical detoxifications and other diseases.

Introduction

UDP-glucuronosyltransferases (UGT) are a superfamily of enzymes that catalyze the attachment of glucuronic acid to a vast number of lipophilic endobiotics and xenobiotics, thereby converting them to water-soluble glucuronides to hasten their excretion from the cellular milieu and out of the system (Dutton, 1980). Glucuronidation of chemicals prevents tissue and blood accumulation of toxic metabolites such as bilirubin, many dietary-based flavonoid- and anthraquinone-related chemicals, environmental agents typified by benzo[a]pyrene, and many phenolic medications such as acetaminophen. Although bilirubin is the critical endogenous toxin,

METHODS IN ENZYMOLOGY, VOL. 400   0076-6879/05 $35.00
                        DOI: 10.1016/S0076-6879(05)00001-7

many exogenous chemicals are mutagens, carcinogens, or can undergo further metabolism by cytochrome P450 systems in the body to become carcinogens (Levin et al., 1976). Thus, the action of UGT isozymes prevents mutagenesis, carcinogenesis, and other forms of chemical toxicity.

Due to their intrinsic association with the endoplasmic reticulum, UGTs have not been stably purified away from membranes for structural analyses to determine their catalytic mechanism(s). Cloning various UGTcDNAs has allowed DNA-based expression of individual isozymes to use for analyzing substrate selections. Moreover, this technology has led to the identification of the two primary families of UGT genes, UGT1 and UGT2 (Emi et al., 1995; Mackenzie and Rodbourn, 1990; Ritter et al., 1992a) and their flanking regulatory regions. This advancement has given rise to ongoing studies addressing the genetic basis of UGT regulation that includes cell-specific expression that responds to environmental/foreign (Wells et al., 2004) and developmental stimuli. Discovery and characterization of the human UGT1 gene complex (Ritter et al., 1992a) identified a novel multigenic locus that encodes an entire family of UGTs. Moreover, it harbors the bilirubin-specific UGT1A1 gene, which has a range of defects responsible for mild to lethal hyperbilirubinemias, including Gilbert's syndrome and Crigler-Najjar (CN) diseases (see reviews by Owens and Ritter, 1992; Mackenzie et al., 1997). In addition, many polymorphisms uncovered at the UGT1 locus (Guillemette, 2003) and fewer at UGT2 loci (Lin et al., 2005) are shown to impact human health. Regarding substrate specificities, individual UGT1 isozymes are shown to convert a vast number of endo- and xenobiotics (Basu et al., 2004).

Although an array of molecular biology techniques has greatly clarified our understanding of UGTs tissue distribution (Strassburg et al., 2000) and cellular localization (Basu et al., 2004), many biochemical and structural features of UGT remain unclear. Whether UGTs undergo hetero- and/or homodimerization is also under investigation (Ghosh et al., 2001; Meech and Mackenzie, 1997). Studies show that the proteins undergo dimerization, which is controlled by a specific domain(s). Ongoing analyses of transcriptional regulation of UGT genes are also under extensive investigation (Münzel et al., 2003; Wells et al., 2004). Results show that a variety of transcriptional factors are involved in specific mRNA synthesis. Moreover, it has been demonstrated that UGTs require phosphorylation (Basu et al., 2003, 2005), which involves signaling to control conjugating activity. When genetic regulation of UGTs is considered in the context of phosphorylation, it appears that there are short- and long-term controls on UGT activity. Furthermore, if one considers the enormous substrate selection (Basu et al., 2004) of the critical bilirubin-metabolizing UGT1A1 and the potential for competition, it is not clear how it maintains normal serum bilirubin levels. As molecular biology studies are gene based, the aim of this chapter is to present and compare representative structures of the UGT1 and *UGT2* gene families in human, rat, and mouse for which complete genome sequence data are presently available. Further, we considered the impact of the *UGT1* gene structure on detoxification in humans.

Description of the Human UGT1 Complex Locus and Comparison to that in Rat and Mouse

Because lethal hyperbilirubinemic CN-1 disease is caused by defects at the bilirubinemic-specific gene, we cloned the human isozyme and completed characterization of the novel single copy *UGT1* complex locus (Gong *et al.*, 2001; Ritter *et al.*, 1991, 1992a) located on chromosome 2q37. Because major and minor bilirubin transferase cDNAs (Ritter *et al.*, 1991) and a phenol transferase cDNA (Harding *et al.*, 1988) were isolated and shown to contain unique 5′ ends, but shared identical 3′ ends, it was predicted a complex locus was necessary to account for the cDNAs. Based on Southern blot hybridization to unique ends of bilirubin cDNAs, a novel *UGT1* gene complex in three overlapping cosmid clones was identified and characterized (Ritter *et al.*, 1992a). A 95-kb locus (Fig. 1, top) was shown to encode bilirubin (UGT1A1 and UGT1A4) and phenol (UGT1A6) transferase isozymes, as well as three bilirubin-like (*UGT1A3, UGT1A5*, and *UGT1A2P*) genes not identified previously. In the 5′ region of the locus, six unique first exons with individual proximal promoter elements were arranged in a tandem array upstream of four common exons. A first exon encodes the entire 287±2 amino acids in the amino-terminal region of an isozyme, and the common exons code for the identical 245 amino acid carboxyl terminus shared by the six isozymes. Each first exon possibly determines substrate specificity, whereas the common exons most likely determine interaction with the common substrate, UDP-glucuronic acid. Based on exon arrangement and distribution of consensus RNA splice sites, it was predicted that six independently regulated and overlapping primary transcripts are synthesized and that each lead exon is joined to the four common exons to generate six mRNAs with unique 5′ ends, but common 3′ ends. Each exon 1 defines a separate gene linked to the common exons; hence, the complex contains six genes arrayed 5′ to 3′ and originally designated *UGT1F* to *UGT1A*. More recently the genes were renamed *UGT1A6* through *UGT1A1*, in turn, upon updating the nomenclature of UDP-glucuronosyltransferases with distinctions for certain species (Mackenzie *et al.*, 1997).

Subsequently, we extended the *UGT1* complex to 200 kb (Gong *et al.*, 2001). *UT1A7, 1A8, 1A9,* and *1A10* are among seven additional first exons uncovered; *UGT1A11P, 1A12P*, and *1A13P* are pseudo, similar to *1A2P*. In order to span the entire *UGT1* complex without interruption, we identified BAC and PAC clones containing the locus. C20-, 71-, and D15-BAC clones, along with D8- and L20-PAC, were digested with *Xho*I, *Hpa*I,

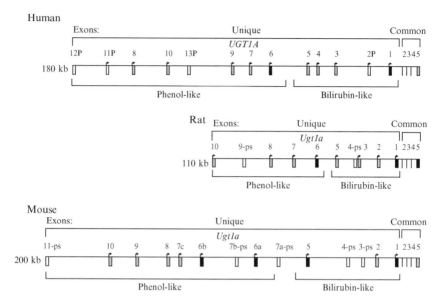

FIG. 1. Comparison of the *UGT1* locus in human, rat, and mouse. Relative positions of unique first exons at the *UGT1* complex locus in human (Ritter *et al.*, 1992), rat (Emi *et al.*, 1995), and mouse are given. The relative positions are based on the availability of complete sequence data for genomes (Gong *et al.*, 2001; accession #AF297093; rodent GenBank databases). Changes in exon positions at the rat *Ugt1* locus are based on more precise annotations due to new genome sequence data as indicated in Table I, and arrangement of the mouse *Ugt1* locus is based on the annotation of complete sequence data in the databases with nomenclature guided by the UGT and the HUGO nomenclature committees. Rodent sequence data were also compared to that in an earlier report (Zhang *et al.*, 2004). A gene for each family A member of a *UGT1* complex locus is defined by one in a series of unique first exons in the 5′ region and is designated *UGT1A1, UGT1A2P, UGT1A3*, and so on; each unique exon shares a set of common exons in the 3′ region designated 2, 3, 4, and 5 that generates a unique mRNA with a common 3′ end. All clear/unfilled exons are pseudo. The location of a promoter element (broken arrow) upstream of each unique exon with strategically distributed donor and acceptor RNA splice sites allows synthesis of overlapping primary transcripts that undergo processing by joining the lead exon in each to common exons forming unique mature messenger RNAs. [Individual promoter elements are predicted to exist for mouse *Ugt1* genes based on tissue-specific expression of unique mRNAs, except for *Ugt1a8* (Zhang *et al.*, 2004).] The position of first exons is drawn to scale. Human, rat, and mouse loci are located on chromosomes 2, 9, and 1, respectively.

and *Spe*I; the fragments were separated using pulse-field and/or regular agarose gel electrophoresis. Assembly of the restriction enzyme map for C20 was facilitated by analysis of shorter PAC/BAC clones spanning the same region. In addition, cos 46, 48, and 49 clones, which contained the first exons of *UGT1A7, 1A8, 1A10, 1A11P*, and *1A12P* were mapped and subcloned to facilitate sequencing the new exons. From C20 BAC, 19- and

24-kb *Xho*I fragments containing *UGT1A13P* and *1A9* were subcloned into pZeroBackground vector (InVitrogen, Carlsbad, CA) to generate a contiguous *UGT1* locus. The 19-kb *Xho*I insert was digested with *Hpa*I to generate 4.8- and 14.2-kb fragments; the 14.2-kb fragment containing exon 1 for *UGT1A13P* was subcloned into the *Eco*RV/*Xho*I-digested pZero-Background vector. Primers used for sequencing first exons in *UGT1A7* through *1A13P* are described (Gong *et al.*, 2001). Having identified, arranged, and sequenced all exons at the complex (Fig. 1, top), the entire locus was sequenced (accession #AF297093), and the relative identity of first exons was reported (Gong *et al.*, 2001).

To assess whether more distantly related exons exist upstream of the *UGT1A12P* gene, Southern blot analysis with a mixture of probes specific for *UGT1A1, UGT1A4, UGT1A6*, UGT1A9, and *UGT2B7* was carried out with *Xho*I-digested 71 and D15 clones. The two clones extended 100 kb upstream of *1A12P*. Sequences used were 5′-tatggttttgttggtggaatcaactgcctt-ca-3′ for UGT1A1, 5′-atggtcttcattgggggcatcaactgtgccaac-3′ for UGT1A4, 5′-atggtcttcattggaggtatcaactgtaagaag-3′ for UGT1A6, 5′-atatattctctattaat-gggttcatacaatgac-3′ for UGT1A9, and 5′-gttgattttgttggaggactccactgcaaa-cct-3′ for UGT2B7. Except for UGT1A9, the oligonucleotides represent a highly conserved amino acid sequence from position 261 to 282 found in all UDP-glucuronosyltransferases (Ciotti *et al.*, 1995). Although the conserved probe mixture detected all of the known exons, it did not detect any upstream of *UGT1A12P*.

Alignments and nucleotide sequence comparisons (Gong *et al.*, 2001) showed that the first exons for *UGT1A2* through *1A5* genes are 86 to 92% identical to each other and overall 60% identical to the *UGT1A1* first exon, the bilirubin-specific gene. First exons for *UGT1A7* through *1A10* are between 71 and 92% identical and overall 55% identical to the first exon for *UGT1A6*, which is phenol like. Hence, *UGT1A1* through *1A5* genes are more bilirubin like, and *UGT1A6* through *1A10* are more phenol like. In this report, amino acid sequence data specified by first exons were used for comparisons of human to that of rat and mouse.

Similarly, rat (Emi *et al.*, 1995) and mouse (GenBank) have a *Ugt1* complex locus composed of unique first exons linked to 4 common exons that specify proteins each with a unique amino terminus, but a common carboxyl terminus. (We have annotated more precisely the *Ugt1* loci for rat and mouse from the genome databases in order to arrange exons and introns.) In rat, 9 first exons were initially uncovered that formed a cluster of 5 bilirubin- and 4 phenol-like exons based on amino acid sequence similarities (Fig. 1, center) (Emi *et al.*, 1995). Sequence data for the entire rat and mouse genomes appearing in the database provide additional information about rodents *Ugt1* loci (Fig. 1, center and bottom). At the rat nearly 110-kb locus, a first exon for *Ugt1a10* at the most 5′ position and an unidentified exon fragment between first exons 2 and 3 were uncovered yielding a total of 10 first exons, which increases the number of potentially viable exons to 8 (see Emi *et al.*, 1995). Another report also details similar findings (Zhang *et al.*, 2004). Like the human *UGT1* locus, each first exon has a proximal promoter element (Emi *et al.*, 1995), and unlike the human, a xenobiotic response element (XRE) appears to have been inserted upstream of *Ugt1a6* (previously *UGT1A6*) (Emi *et al.*, 1996). At the mouse *Ugt1* locus, 14 first exons, predictably with individual promoter sequences (Zhang *et al.*, 2004), and 4 common exons were uncovered (Fig. 1, bottom); 9 viable and 5 pseudo first exons are distributed over 200 kb. Although an individual promoter element upstream of each first exon was not physically established, expression of eight different mRNAs is strong evidence that each has such a sequence. Polymerase chain reaction-based amplification of mRNAs isolated from mouse liver, esophagus, stomach, intestine, colon, kidney, lung, spleen, cerebrum, and cerebellum shows that each gene, except *Ugt1a8*, is expressed (Zhang *et al.*, 2004).

To compare sequence similarities of amino acids encoded by first exons at rat and mouse *Ugt1* complexes, computer analyses were carried out. Before making comparisons, rodent genes were defined in Table I similar to those shown for human with cross-references to original designations. Similar to earlier characterizations (Emi *et al.*, 1995), comparison of rat exons to each other (Table IIA) shows that *Ugt1a2, 1a3* and *1a5* form a bilirubin-like cluster that is 68 to 78% similar and that *Ugt1a7, 1a8*, and *1a10* form a phenol-like cluster that is 69 to 79% similar. Bilirubin-specific *Ugt1a1* and phenol-specific *Ugt1a6* are unique. The limited number of previously identified mouse *Ugt1* genes is also cross-referenced for clarity (Table IB). [Distinctions in the nomenclature for rat and mouse genes have been reported elsewhere (Mackenzie *et al.*, 1997)]. For mouse, a different type of clustering is evident as shown in Table IIB. Four separate clusters are as follows: 71% similarity for *Ugt1a2* and *Ugt1a7c*; 94% for *Ugt1a6a* and *Ugt1a6b*; 76% for *Ugt1a8* and *1a7b*; 75/74% for *Ugt1a9/1a10* to *1a7c* and *1a8*; and, finally, 86% for *Ugt1a10* and *1a9*. We did not assign pseudo exons to clusters. Based on level of similarities, *Ugt1a6a* and *1a6b* represent the most recent gene duplication, followed by *Ugt1a9* and *1a10* and, finally, by *Ugt1a7c* and *1a8*. Bilirubin-specific *Ugt1a1* is again unique. A comparison of rat versus mouse sequences (Table IIC; Fig. 2) shows that bilirubin-specific *Ugt1a1* genes are 86% similar to each other and less than 50% similar to the other mouse genes. Comparisons of other rat forms to mouse are as follows: *Ugt1a2, 1a3*, and *1a5* are 59 to 85% similar to *1a2* and *1a5; Ugt1a6* is 88% similar to *1a6a* and *1a6b; Ugt1a7* is 84% similar to *1a7c; Ugt1a8* is 88% similar to *1a8;* and, finally, *Ugt1a10* is 75% similar to *1a9* and *1a10*. Notably, rat homolog *Ugt1a3* is the least similar to the mouse isozymes (Table IIC). Moreover, six rat isozymes are between 84 and 88% similar to

TABLE I
DEFINITION OF RODENT *Ugt1* GENES[a]

A. Rat

New[b]	1a1	1a2		1a3	1a4-ps	1a5	1a6	1a7	1a8	1a9-ps	1a10
Zhang et al.[c]	1A1	1A2	1A3p	1A4	1A5p	1A6	1A7	1A8	1A9	1A10p	1A11
Old[d]	1B1E	1B2F	—	1B3G	1B4H	1B5D	A1A	A2B	A3C	A4D	—

B. Mouse

New[b]	1a1	1a2	1a3-ps	1a4-ps	1a5	1a6a	1a7-ps	1a7a-ps	1a6b	1a7c	1a8	1a9	1a10	1a11-Ps
Zhang et al.[c]	1A1	1A2	1A3p	1A4p	1A5	1A7	1A6p	1A8p	1A9	1A10	1A11	1A12	1A13	1A14p
Old	mUGT Br1[e] / Ugt1a1[f]				UGT br2-like[g]	UGT 1-06[h] / UGT 1.6[i] / UGP 1a1[f]						mUGT Br/p[j]		

[a] Because complete sequence data for rodent genomes uncovered additional exons at the rat *Ugt1* locus and made information available to organize the mouse *Ugt1*, it was necessary for clarity to redefine first exons with reference to original trivial names as shown. For rat, an additional viable exon *Ugt1a10* has been uncovered (Zhang et al., 2004). Sequence data support a total of 14 first exons for mouse *Ugt1* as shown. The relationships of new and old designations are based on sequence identities of data in databases following computer blasts. Also, nomenclature reported by Zhang et al. (2004) is shown.

[b] UGT and the HUGO Nomenclature Committee, 2004.

[c] Zhang et al., 2004.

[d] Emi et al., 1995.

[e] Kong et al., 1993; Acc. #S64760.

[f] Koiwai et al., 1995; Acc. #D87866.

[g] Chu et al., 1997.

[h] Acc. #U16818.

[i] Lamb et al., 1994; Acc. #U09930.

[j] Kong et al., 1993; Acc. #L27122.

TABLE II
COMPARISONS OF FIRST EXONS

A. Similarity of rat first exons at the *Ugt1* locus

	1a1	1a2	1a3	1a5	1a6	1a7	1a8	1a10
1a1	–	44	45	46	38	38	39	37
1a2		–	68	78	38	37	39	35
1a3			–	68	41	34	37	36
1a5				–	40	37	38	35
1a6					–	36	38	34
1a7						–	79	69
1a8							–	74

B. Similarity of mouse first exons at the *Ugt1* locus

	1a1	1a2	1a5	1a6a	1a6b	1a7c	1a8	1a9	1a10
1a1	–	42	42	36	37	34	37	38	37
1a2		–	71	74	35	35	36	38	36
1a5			–	37	38	34	35	36	34
1a6a				–	94	34	38	39	36
1a6b					–	36	39	39	37
1a7c						–	76	75	73
1a8							–	74	74
1a9								–	86

C. Similarity of first exons at rodents *Ugt1* loci mouse/rat

	Mouse/rat							
	1a1	1a2	1a3	1a5	1a6	1a7	1a8	1a10
1a1	86	44	46	46	39	36	38	36
1a2	44	85	64	72	37	36	37	34
1a5	44	71	59	77	39	37	35	32
1a6a	37	36	38	37	88	36	38	34
1a6b	38	37	38	38	88	37	39	36
1a7c	36	36	34	36	36	84	78	66
1a8	38	38	36	38	39	74	88	69
1a9	40	39	37	36	36	73	75	75
1a10	38	36	36	35	36	73	76	75

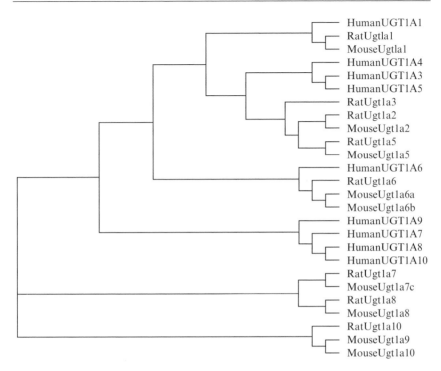

FIG. 2. Phylogenetic relationship among human, rat, and mouse genes encoded at *UGT1* loci.

one or two mouse isozymes, including the bilirubin-specific *Ugt1a1*, and two are 77 or 78% similar. Analysis of *UGT1* loci by the dendogram (Fig. 2) demonstrates that rat and mouse are more similar to each other than either is to human; the bilirubin-specific *UGT1A1* and phenolic-specific *UGT1A6* show the highest degree of conservation between rodents and human (Table III); clustering between rodents and human forms shows two distinct groups (Table III), which is also evident in Fig. 2. However, rat appears to be more related to human than to mouse. We show the correspondence of rodent and mouse genes in Table III (bottom).

Prototypical Inducers of UGTs

To understand the control of UGT function, cloning studies have also made possible genetic analysis of inducer-specific expression of UGTs. Traditionally, induction studies have characterized the types of UGTs, including their predicted substrates of both endogenous and exogenous sources. Because UGTs encoded at *UGT1* loci are known to be inducible by prototypical aryl hydrocarbons (Ah) via the xenobiotic responsive

TABLE III

A. Similarity of human and rodent first exons at *UGT1* loci

Rat/human

	1A1	1A3	1A4	1A5	1A6	1A7	1A8	1A9	1A10
1a1	68	43	44	43	38	38	38	40	41
1a2	46	64	63	62	38	39	39	40	40
1a3	47	63	61	61	40	35	36	38	36
1a5	47	64	62	60	39	37	37	38	38
1a6	37	36	38	36	72	38	38	41	37
1a7	35	35	34	35	39	67	68	68	69
1a8	39	36	36	36	40	68	70	70	69
1a10	38	33	33	35	35	62	64	65	63

Mouse/human

	1A1	1A3	1A4	1A5	1A6	1A7	1A8	1A9	1A10
1a1	66	42	43	42	37	36	39	41	38
1a2	44	62	60	59	39	38	37	37	37
1a5	44	60	62	59	40	36	36	38	37
1a6a	36	34	35	35	70	40	38	41	38
1a6b	37	34	36	36	69	40	39	41	40
1a7c	34	34	34	35	37	65	67	67	67
1a8	37	37	37	37	40	67	67	66	66
1a9	37	37	37	38	39	67	68	68	67
1a10	37	35	35	36	39	68	68	68	67

B. Comparisons of first exons at the *UGT1* locus of human, rat, and mouse

Human	1A1	1A3	1A4	1A5		1A6	1A7	1A8	1A9	1A10
Rat	1a1	1a2	1a2	1a2	1a5	1a6	1a8	1a8	1a8	1a8
		1a5	1a5	1a3			1a7	1a7	1a7	1a7
		1a3	1a3	1a5			1a10	1a10	1a10	1a10
Mouse	1a1	1a2	1a5	1a2	1a5	1a6a	1a10	1a10	1a10	1a10
		1a5	1a2	1a5	1a2	1a6b	1a9	1a9	1a9	1a9
							1a8	1a8	1a8	1a7c
							1a7c	1a7c	1a7c	1a8

element (XRE) (Fujisawa-Sehara *et al.*, 1988) and/or by phenobarbital (PB), predictably, via the PB responsive enhancer module (PBREM) (Sueyoshi and Negishi, 2001) located in the upstream region of genes, we carried out computer searches for the appropriate consensus sequences. 3-Methylcholanthrene (3-MC) and 2,3,7,8-tetrachlorodibenzo-*p*-dioxin (TCDD) are typical inducers that bind the Ah receptor for translocation to the nucleus to interact with the *cis*-acting XRE to recruit transcriptional factors to increase specific mRNA synthesis. Results from computer searches in Fig. 3 show the locations of the conserved sequence, gcgtg, in the upstream regions of human, rat, and mouse genes. Each of the human genes, except *UGT1A8*, has a conserved copy of the pentamer (Fig. 3) within −2 kb of the identifiable translational start site. The relative effectiveness of Ah receptor-mediated induction of human *UGT1A1* via targeting gcgtg versus the recently identified cacgca sequence (Yueh *et al.*, 2003) in the upstream region is not known. Studies have shown Ah inducibility of human *UGT1A6* (Münzel *et al.*, 2003) and rat *Ugt1a6* (Emi *et al.*, 1995) and *1a7* genes. In the case of rat, the conserved pentamer is present upstream of each viable gene, although its location is quite distal for *Ugt1a7*, *1a10*, and *1a8* genes at −6, −12, and −13.8 kb upstream, respectively (Fig. 3). While rat *Ugt1a6* (previously designated *A1*) and *Ugt1a7* (previously designated *A2*) have been shown to respond to 3MC treatment (Emi *et al.*, 1996), it remains to be seen whether other genes respond to Ah inducers. Each viable mouse gene also has a perfectly conserved XRE sequence (Fig. 3); the sequence for *Ugt1a6b* and *1a7c* is −6 and −10 kb upstream, respectively. Whether these genes account for UGT responses to 3MC treatment of mice is not known (Malik and Owens, 1981; Owens, 1977); induction profiles of UGT activity showed both Ah receptor-dependent and -nondependent responses.

In addition, we searched for PBREM (Sugatani *et al.*, 2001) elements at human, rat, and mouse *UGT1* complex loci. Certain orphan receptors and nuclear factors are also shown to be required for PB responses (Sueyoshi and Negishi, 2001). Although computer analysis using the 290-bp enhancer sequence PBREM as a probe detected itself upstream of *UGT1A1*, no other copy or variant was detected at other human *UGT1* genes. We also probed the rat *Ugt1* locus for PB-responsive elements with the human PBREM that contains the mouse 52-bp PBREM core unit identified previously at the mouse *cyp2b10* gene (Honkakoski *et al.*, 1998). Results are shown in Fig. 4A. Additionally, we probed the rat *Ugt1* locus with the phenobarbital response unit (PBRU) found in the upstream region of the rat *CYP2B2* gene (Beaudet *et al.*, 2005). To detect mouse PB-responsive *Ugt1* genes, the complex was probed using the 52-bp PBREM core unit at the mouse *cyp2b10* gene. We again detected short overlapping fragments from the same region of the 52-bp core PBREM at the mouse locus (Fig. 4B). Elements are distributed so as to appear upstream of particular genes—*1a1*,

Human

UGT1A1	(-594)	CTCATGGC**GCGTG**CTCGTGTG	(-574)
UGT1A1	(-6542)	ACAGTTTT**GCGTG**GCCCTTTG	(-6522)
UGT1A2	(-176)	GAGGCACA**GCGTG**GGGTGGAC	(-156)
UGT1A2	(-546)	ATTCATGA**GCGTG**AATGTGGA	(-526)
UGT1A3	(-40)	CAGGCACA**GCGTG**GGGTGGAC	(-20)
UGT1A3	(-1393)	GCCATCCT**GCGTG**TGCTGCCC	(-1373)
UGT1A3	(-1604)	ACTACCAG**GCGTG**TTCCACCC	(-1584)
UGT1A4	(-48)	CAGGCACA**GCGTG**GGGTGGAC	(-28)
UGT1A4	(-214)	CAAGATAG**GCGTG**ATTGGTCT	(-194)
UGT1A5	(-1105)	TCTCATGG**GCGTG**AGACCATT	(-1085)
UGT1A5	(-4382)	ACATGGTA**GCGTG**TGCCTGTA	(-4362)
UGT1A6	(-1511)	AGGAACTC**GCGTG**CCAGCCAG	(-1491)
UGT1A7	(-545)	GCATGGCA**GCGTG**CGCCTGTA	(-525)
UGT1A9	(-687)	TTCCCACT**GCGTG**CGATGTAT	(-667)
UGT1A10	(-263)	CACTTGCA**GCGTG**CTCTCCCT	(-243)

Rat

Ugt1a1	(-1767)	ACAGGACA**GCGTG**GTGGGATC	(-1747)
Ugt1a1	(-7540)	GCGCTCAT**GCGTG**TGTAAGGA	(-7520)
Ugt1a2	(-2876)	GCGCACGC**GCGTG**CACATGTT	(-2856)
Ugt1a3	(-158)	CAGGCCTG**GCGTG**GGCTTCTT	(-138)
Ugt1a5	(-2985)	GCAACTCA**GCGTG**GAGATTGC	(-2965)
Ugt1a6	(-1317)	CCCCTGGA**GCGTG**GCTTACTG	(-1297)
Ugt1a7	(-6239)	GCGTGCAT**GCGTG**TGTCTGTG	(-6219)
Ugt1a8	(-13850)	GGCTTTGG**GCGTG**CAGGCAAG	(-13830)
Ugt1a10	(-12021)	GCAGATGA**GCGTG**GGTTGAGTA	(-12001)

Mouse

Ugt1a1	(-2987)	ACGTGTGC**GCGTG**CACGCTCA	(-2967)
Ugt1a2	(-502)	TCCTCAGA**GCGTG**GCTTCTCT	(-482)
Ugt1a5	(-89)	ACCCAGTG**GCGTG**ACAAGGTT	(-69)
Ugt1a5	(-3048)	TGGTGTGT**GCGTG**TAAACAGA	(-3028)
Ugt1a6a	(-1494)	AAGTGTCT**GCGTG**TGTTTGGG	(-1474)
Ugt1a6b	(-6381)	GTGTGTAT**GCGTG**TATAAGCA	(-6361)
Ugt1a7c	(-10369)	TCTATTCT**GCGTG**GTCAAATC	(-10349)
Ugt1a8	(-2935)	TCTATTCT**GCGTG**GTCAAATC	(-2915)
Ugt1a9	(-2594)	GTTTCTGG**GCGTG**TCTGTGGT	(-2574)
Ugt1a10	(-2358)	CTGTGTAA**GCGTG**GAGAAAAT	(-2338)

FIG. 3. Location of the prototypical aryl hydrocarbon/xenobiotic responsive element at the *UGT1* locus of human, rat, and mouse. The xenobiotic responsive element (XRE), gcgtg, is a conserved consensus sequence present in human, rat, and mouse *UGT1* genes as shown. 3-Methylcholanthrene and 2,3,7,8-tetrachlorodibenzo-*p*-dioxin are examples of ligands that bind the aryl hydrocarbon receptor, which interacts with the XRE element upstream of responsive genes to elicit an increase in pre-mRNA synthesis. Computer searches identified the sequences shown.

1a2, *1a5*, *1a9*, and *1a10*—and also downstream of genes—*1a2*, *1a5*, and *1a10* (Fig. 4B). The limited sequence identity of PBRUs between cytochrome P450 and *Ugt1* genes suggests that the units differ significantly between the two gene families and that other elements remain to be identified.

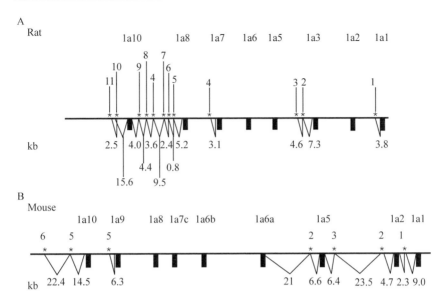

FIG. 4. Identification of components of the phenobarbital responsive enhancer module at the *Ugt1* locus of rat and mouse. First exons are named to preserve designations already in use for rodent forms. Isozymes with high sequence similarity were given subscripts in order to facilitate and maintain their earlier designations. Sequences from the human phenobarbital responsive enhancer module (PBREM) (Sugatani *et al.*, 2001), which is present in the rat phenobarbital responsive unit (162 bp) (PBRU) (Beaudet *et al.*, 2005), and mouse PBREM (Honkakoski and Negishi, 1998) were detected by a computer search. Mouse and human PBREM units are nearly identical. The human probe (Sugatani *et al.*, 2001) detected only itself upstream of *UGT1A1*. The rat PBRU probe detected conserved sequences at rat *Ugt1*, and the mouse probe detected conserved sequences at mouse *Ugt1*. Distances (kb) upstream or downstream of a gene or between elements (*) are placed below exons. Conserved sequences detected at the rat locus were detected with the human 290-bp PBREM (Sugatani *et al.*, 2001) and are designated numerically as follows: 1, tggcacttggtaaacacgaaataaacagt; 2, acatcaaaggaa; 3, agaacaaacttc; 4, acataacctgaaa; 5, tgtgaacaaagc; 6, acgcaatgaaca; 7, aatgattaacca; 8, agttaggggaacagca; 9, aaacttctgagt; 10, tttatataacct; and 11, agtttatataacc. In the case of mouse, first exons, which represent genes, are designated serially beginning with *1a1*. Distances (kb) upstream or downstream of a gene or between elements (*) are placed below exons. Conserved sequences detected within the 52-bp core PBREM found upstream of mouse *cyp2b10* gene are designated numerically as follows: 1, accttggcaca; 2, tggcacagtgcc; 3, gaccttggcaca; 4, cttggcacagtg; 5, aaagaacattct; and 6, ccttggcacag.

Impact of Mutations in Relationship to Structure of the UGT1 Locus

Because each gene shares four common exons, it was predicted (Owens and Ritter, 1992) that a deleterious defect in any common exon would inactivate the entire locus, whereas such a defect in a unique exon would affect a single UGT protein (Figs. 1 and 5). The first defect, a 13-bp deletion in

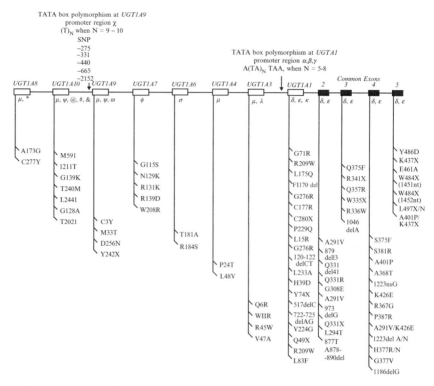

FIG. 5. Locations of mutations and polymorphisms at the human *UGT1* locus. Because any alteration, especially deleterious ones in a common exon and the first exon of *UGT1A1* at the *UGT1* locus depicted, is visible phenotypically and has the greatest impact on bilirubin toxicity compared to other substrates, overwhelming data have accumulated for that region of the locus; most mutations presented therein lead to Crigler-Najjar (CN) type 1 disease; a few cause CN-2 disease or Gilbert's syndrome (Guillemette, 2003; Mackenzie *et al.*, 1997). Alterations in first exons, excluding that for *UGT1A1*, are designated polymorphic. Despite the overwhelming focus on accumulating data on the bilirubin-specific gene, recent focus on polymorphisms at other first exons shows *UGT1A7* and *1A9* presently have the greatest number. Polymorphisms at TATA boxes show (a) that the TA repeat in A(TA)₆TAA of *UGT1A1* varies between 5 and 8 in the population with decreasing activity with increasing TAs and (b) that the consecutive 9 Ts that form the regulatory unit of *UGT1A9* have a version with 10 Ts. Whether the unit with 10 Ts has altered transcriptional activity is not clear (Girard *et al.*, 2004; Yamanaka *et al.*, 2004). Reference citations are as follows: α, Bosma *et al.*, 1995; β, Beutler *et al.*, 1998; γ, Hall *et al.*, 1999; δ, Mackenzie *et al.*, 1997; ε, Guillemette, 2003; κ, Akaba *et al.*, 1999; λ, Iwai *et al.*, 2004; μ, Ehmer *et al.*, 2004; σ, Ciotti *et al.*, 1997; ϕ, Villeneuve *et al.*, 2003; χ, Yamanaka *et al.*, 2004; ψ, Jinno *et al.*, 2003a; ω, Saeki *et al.*, 2003; @, Jinno *et al.*, 2003b; #, Martineau *et al.*, 2004; &, Elahi *et al.*, 2003; and *, Huang *et al.*, 2002. Because this compilation of data was to determine which sites at the locus have undergone changes, we have not attempted to carry out associations with demographics, ethnicities, or diseases (except for hyperbilirubinemias).

exon 2 (Ritter *et al.*, 1992b) (Fig. 5), was uncovered in a CN-I patient that completely inactivated the entire locus. The utilization of exons at the locus explained the heterogeneity observed for bilirubin-conjugating activity among CN-I patients (van Es *et al.*, 1990) in which some had only defective bilirubin glucuronidation and others were defective for both bilirubin and 4-nitrophenol activities. Finally, it was uncovered that deleterious defects exist in every exon of the bilirubin *UGT1A1* gene (Guillemette, 2003; Mackenzie *et al.*, 1997; Owens and Ritter, 1992). Because hyperbilirubine-mic/jaundice phenotypes are generally evident and have potentially detri-mental effects on the central nervous system, mutations that affect UGT1A1 have been evaluated extensively (Fig. 5). Thus, all deleterious alterations in the common exons and in the first exon of *UGT1A1* are associated with CN-1 disease. Less deleterious mutations distributed throughout the bilirubin *UGT1A1* gene account for CN-2 disease, which is associated with approximately 1 to 10% normal bilirubin-conjugating activity assayed with hepatic microsomes (see review by Owens and Ritter, 1992). G71R is found in the Japanese population in association with Gilbert's syndrome (Akaba *et al.*, 1999; Aono *et al.*, 1995).

Moreover, it was uncovered that the unusual TATA box [ATATATA-TATATATAA, $A(TA)_6TAA$] found upstream of the bilirubin-specific *UGT1A1* gene (Ritter *et al.*, 1992b) is a target for TA insertions (Bosma *et al.*, 1995) (Fig. 5). An insertion reduces transcriptional activity, resulting in mild hyperbilirubinemia described as Gilbert's syndrome (see review by Owens and Ritter, 1992), which is characterized as having between 10 and 33% normal bilirubin glucuronidation. Given the structure of the human *UGT1* gene, it is possible to account for the range of defective bilirubin UGT activity seen in populations (Owens and Ritter, 1992; Ritter *et al.*, 1992b). It is interesting that the TA_5 promoter (Beutler *et al.*, 1998) has greater activity than "normal" TA_6, which prompts questions as follows: (a) why is normal/wild type less active and (b) is this site that harbors dinucleotide repeats subject to further mutations as seen for many other such genes (Sutherland and Richards, 1995; Tautz and Schlotterer, 1994)? Consistent with gene alterations surrounding consecutive nucleotide repeats, the normal TATA box for *UGT1A9* shows that 9 consecutive Ts have increased to 10 in polymorphic alleles (Yamanaka *et al.*, 2004), which may affect transcription-al activity. Multiple single nucleotide polymorphisms (SNP) have also been identified in the upstream region of *UGT1A9* (Girard *et al.*, 2004).

Ongoing analysis of the *UGT1* locus shows that defects and/or poly-morphisms exist in each exon (Guillemette, 2003). Among the more than 57 changes described, more than 37 affect a common exon, which impacts the entire locus. Hence, most of the mutations/polymorphisms uncovered are predicted to have an effect on all proteins encoded at the *UGT1* locus.

Among the nonbilirubin-specific first exons, it appears that *UGT1A7* and *1A9* have undergone the greatest number of SNPs. Although *UGT1A5* has shown some 5 SNP in alleles (Guillemette, 2003), that gene was not considered here because its impact on detoxification has not yet been established. It is possible, however, that the *UGT1A5* changes represent an evolutionary process toward or away from viability. To date, it can be said that this locus has undergone many changes at sites that have an amplifying effect on loss of UGT activities, implying a negative impact of gene structure.

Whereas we compared the TATA box upstream of the human bilirubin-specific *UGT1A1* with that predicted for the rat and mouse gene, we find evidence for very different sequences. Relative to the protein translation start site, rat has potential TATA boxes at −60 and −238 bp of TTTATT and TATATT, respectively; and mouse has potential boxes at −190, −246, and −342 bp of TTATTT, TATATT, and AAAATATATAT-GAAA, respectively. For the two rodents, sequences located closest to the protein translation start site are highly conserved. Hence, it is unlikely that rat and mouse encountered evolutionary pressure to expand TA inserts at this regulatory site of *Ugt1a1* genes. Despite any potential benefit of the unusual TATA box, the human sequence, A(TA)$_6$TAA, could be at risk for continued mutations considered to be due to slipped-strand mispairing during DNA replication (Tautz and Schlotterer, 1994). Based on our determination of sequence identities of mouse UGTcDNAs in databases and those of the first exons at the mouse *Ugt1* locus, it is likely that the mouse *Ugt1a1* gene designated TATA-box-less (Bernard *et al.*, 1999) is mouse *Ugt1a5* gene.

Human and Rat UGT2 Loci

In addition to the *UGT1* complex locus, each species has members of the *UGT2* gene with subfamilies *UGT2B* and *UGT2A*. All members of the subfamilies are independently encoded at loci on human chromosome 4q13, as shown in Fig. 6A. After annotating the human genome data more precisely, we show the relative positions of seven *UGT2B*, five pseudo *UGT2B*, and one *UGT2A* gene. *UGT2B29P, 2B17, 2B15, 2B10, 2B27P, 2B26P, 2B7, 2B11, 2B28, 2B25P, 2B24P, 2B4*, and *2A1* are distributed over a 1500-kb region. Two sets of pseudo genes appear closely situated to each other. Although intergenic distances vary greatly, fewer variations exist for relative intronic regions within the 24-kb *UGT2B15* and 16.5-kb *UGT2B7* genes (Fig. 6A). Also, it is shown that transcription occurs on both strands for the 2B family due to differential location of the sense strands for the

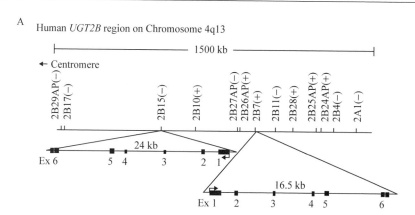

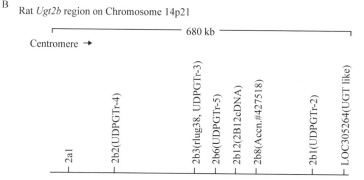

FIG. 6. (A) Relative positions of human *UGT2B* and *2A* genes on chromosome 4q13. A 1500-kb region of chromosome 4q13 encodes members of the UGT2B and 2A subfamilies as shown. Expanded details of the *2B15* and *2B7* genes show the relative position of exons and introns whereby transcription proceeds in the opposite direction for the two genes, indicating that sense coding is on opposite strands. Pluses and minuses above each gene indicate the direction of pre-mRNA synthesis. (B) Relative position of rat *Ugt2b* genes on chromosome 14p21.

genes. The cDNA for the *UGT2A1* gene has not been reported. To specify the arrangement of rat *Ugt2b* genes, cDNA sequences were used to blast genes in the rat database and gave perfect matches as follows: *Ugt2b2, 2b3, 2b6, 2b12, 2b8,* and *2b1.* LOC305264 did not show a perfect match, but was established as UGT like with respect to signature sequences. Designation of rat *Ugt*2a1 is based on sequence similarity to human *UGT2A1.*

Although sequences for mouse *Ugt2b* genes are in the database, the presence of mRNA data for a single isozyme, *Ugt2b5,* makes characterization of *2b* genes very difficult to carry out. Hence, we do not present a general map of mouse *Ugt2b* genes.

Conclusions

While the combination of a promoter element upstream of each of a series of unique first exons with common exons located downstream at human, rat, and mouse *UGT1* loci allows independent synthesis of unique primary RNA transcripts, it likely emerged following a selective evolutionary pressure(s). The resulting structure is inherently flawed because of its susceptibility to complete inactivation following a single strategic mutation.

The fact that 5′ regions of each first exon have the greatest differences among each other and their 3′ regions have high sequence identity that effectively extends the common end suggests minimal changes gave rise to new *UGT1* isozymes. The need for minimal changes could provide a basis for such a multigenic structure. Retention of these gene clusters with prototypical types of regulatory chemicals for human, rat, and mouse *UGT1* loci suggests continued pressure for detoxification of similar substrates over the millennia, typified by the common bilirubin metabolite. The structure of this locus with the distribution of defects can account for all hyperbilirubinemias described in patients; the impact of polymorphisms on chemical detoxifications, including medications, remains to be established. UGT synthesis overlaid with required phosphorylation (Basu *et al.*, 2003, 2005), which involves signaling, suggests that the detoxification system undergoes continual adaptations and/or is involved in other pivotal cellular events.

Annotated sequences at human and rat *UGT2* genes confirm that each locus encodes a single gene as indicated by many cloned 2B cDNAs.

Acknowledgment

We greatly appreciate the assistance of Jack Chen from our IT division for help with aligning and annotating nucleotide and amino acid sequence data from databases.

References

Akaba, K., Kimura, T., Sasaki, A., Tanabe, S., Wakabayashi, T., Hiroi, M., Yasumura, S., Maki, K., Aikawa, S., and Hayasaka, K. (1999). Neonatal hyperbilirubinemia and a common mutation of the bilirubin uridine diphosphate-glucuronosyltransferase gene in Japanese. *J. Hum. Gene.* **44,** 22–25.

Aono, S., Adachi, Y., Uyama, E., Yamada, Y., Keino, H., Nanno, T., Koiwai, O., and Sato, H. (1995). Analysis of genes for bilirubin UDP-glucuronosyltransferase in Gilbert's syndrome. *Lancet* **345,** 958–959.

Basu, N. K., Ciotti, M., Hwang, M. S., Kole, L., Mitra, P. S., Cho, J. W., and Owens, I. S. (2004). Differential and special properties of the major human *UGT1*-encoded gastrointestinal UDP-glucuronosyltransferases enhance potential to control chemical uptake. *J. Biol. Chem.* **279,** 1429–1441.

Basu, N. K., Kole, L., and Owens, I. S. (2003). Evidence for phosphorylation requirement for human bilirubin UDP-glucuronosyltransferase (UGT1A1) activity. *Biochem. Biophys. Res. Commun.* **303**, 98–104.

Basu, N. K., Kovarova, M., Garza, A., Kubota, S., Saha, T., Mitra, P. S., Banerjee, R., Rivera, J., and Lwens, I. S. (2005). Phosphorylation of a UDP-glucuronosyltransferase regulates substrate specificity. *Proc. Natl. Acad. Sci. USA* **102**, 6285–6290.

Beaudet, M.-J., Desrochers, M., Lachaud, A. A., and Anderson, A. (2005). The *CYP2B2* phenobarbital response unit contains binding sites for hepatocyte nuclear factor 4, PBX-PREP1, the thyroid hormone receptor β, and the liver X receptor. *Biochem. J.* **388**, 407–418.

Bernard, P., Goudonnet, H., Artur, Y., Desvergne, B., and Wahli, W. (1999). Activation of the mouse TATA-less and human TATA-containing UDP-glucuronosyltransferase *1A1* promoters by hepatocyte nuclear factor 1. *Mol. Pharmacol.* **56**, 526–536.

Beutler, E., Gelbart, T., and Demina, A. (1998). Racial variability in the UDP-glucuronosyltransferase 1 (UGT1A1) promoter: A balanced polymorphism for regulation of bilirubin metabolism? *Proc. Natl. Acad. Sci. USA* **95**, 8170–8174.

Bosma, P. J., Chowdhury, J. R., Bakker, C., Gantla, S., de Boer, A., Oostra, B. A., Lindhout, D., Tytgat, G. N., Jansen, P. L., and Oude Elferink, R. P. (1995). The genetic basis of the reduced expression of bilirubin UDP-glucuronosyltransferase I in Gilbert's syndrome. *N. Engl. J. Med.* **333**, 1171–1175.

Chu, Q., Lei, W., Gudehithlu, K., Matwyshyn, G.A, Bocchetta, M., and Kong, A. N. (1997). cDNA cloning of the mouse bilirubin/phenol family of UDP-glucuronosyltransferase (mUGTbr2-like). *Pharm. Res.* **14**, 662–666.

Ciotti, M., Marrone, A., Potter, C., and Owens, I. S. (1997). Genetic polymorphism in the human UGT1A6 (planar phenol) UDP-glucuronosyltransferase: Pharmacological implications. *Pharmacogenetics* **7**, 485–495.

Ciotti, M., Yeatman, M. T., Sokol, M. J., and Owens, I. S. (1995). Altered coding for a strictly conserved di-glycine in the major bilirubin UDP-glucuronosyltransferase of a Crigler-Najjar type I patient. *J. Biol. Chem.* **270**, 3284–3291.

G. J. Dutton (ed.) (1980). "Glucuronidation of Drugs and Other Compounds." CRC Press, Boca Raton, FL.

Ehmer, U., Vogel, A., Schütte, J. K., Krone, B., Manns, M. P., and Strassburg, C. P. (2004). Variation of hepatic glucuronidation: Novel funtional polymorphisms of the UDP-glucuronosyltransferase UGT1A4. *Hepatology* **39**, 970–977.

Elahi, A., Bendaly, J., Zheng, Z., Muscat, J. E., Richie, J. P., Jr., Schantz, S. P., and Lazarus, P. (2003). Detection of UGT1A10 polymorphisms and their association with orolaryngeal carcinoma risk. *Cancer* **98**, 872–880.

Emi, Y., Ikushiro, S., and Iyanagi, T. (1995). Drug-responsive and tissue specific alternative expression of multiple first exons in rat UDP-glucuronosyltransferase family (*UGT1*) gene complex. *J. Biochem. (Tokyo)* **117**, 392–399.

Emi, Y., Ikushiro, S., and Iyanagi, T. (1996). Xenobiotic response element-mediated transcriptional activation in the UDP-glucuronosyltransferase family I gene complex. *J. Biol. Chem.* **271**, 3952–3958.

Fujisawa-Sehara, A., Yamane, M., and Fujii-Kuriyama, Y. (1988). A DNA-binding factor specific for xenobiotic-responsive elements of P-450c gene exists as a cryptic form in cytoplasm: Its possible translocation to nucleus. *Proc. Natl. Acad. Sci. USA* **85**, 5859–5863.

Girard, H., Court, M. H., Bernard, O., Fortier, L.-C., Villeneuve, L., Hao, Q., Greenbelt, D. J., von Moltke, L. L., Perussed, L., and Guillemette, C. (2004). Identification of common polymorphisms in the promoter of the *UGT1A9* gene: Evidence that UGT1A9 protein and activity levels are strongly genetically controlled in the liver. *Pharmacogenetics* **14**, 501–515.

Gong, Q.-H., Cho, J. W., Huang, T., Potter, C., Gholami, N., Basu, N. K., Kubota, S., Carvalho, S., Pennington, M. W., Owens, I. S., and Popescu, N. (2001). Thirteen UDPglucuronosyltransferase genes are encoded at the human *UGT1* gene complex locus. *Pharmacogenetics* **11**, 357–368.

Ghosh, S. S., Sappal, B. S., Kalpana, G. V., Lee, S. W., Chowdhury, J. R., and Chowdhury, N. R. (2001). Homodimerization of human bilirubin-uridine-diphosphoglucuronate glucuronosyltransferase-1 (UGT1A1) and its functional implications. *J. Biol. Chem.* **276**, 42108–42115.

Guillemette, C. (2003). Pharmacogenomics of human UDP-glucuronosyltransferase enzymes. *Pharmacogenom. J.* **3**, 136–158.

Hall, D., Ybazeta, G., Destro-Bisol, G., Petzl-Erler, M. L., and Di Rienzo, A. (1999). Variability at the uridine-diphosphate glucuronosyltransferase 1A1 promoter in human populations and primates. *Pharmacogenetics* **9**, 591–599.

Harding, D., Fournel-Gigleux, S., Jackson, M. R., and Burchell, B. (1988). Cloning and substrate specificity of a human phenol UDP-glucuronosyltransferase expressed in COS-1 cells. *Proc. Natl. Acad. Sci. USA* **85**, 8381–8385.

Honkakoski, P., Moore, R., Washburn, K. A., and Negishi, M. (1998). Activation by diverse xenochemicals of the 51-base pair phenobarbital-responsive enhancer module in the *CYP2B10* gene. *Mol. Pharmacol.* **53**, 597–601.

Huang, Y. H., Galijatovic, A., Nguyen, N., Geske, D., Beaton, D., Green, J., Green, M., Peters, W. H., and Tukey, R. H. (2002). Identification and functional characterization of UDP-glucuronosyltransferases UGT1A8*1, UGT1A8*2 and UGT1A8*3. *Pharmacogenetics* **12**, 287–297.

Iwai, M., Maruo, Y., Ito, M., Yamamoto, K., Sato, H., and Takeuchi, Y. (2004). Six novel UDP-glucuronosyltransferase (UGT1A3) polymorphisms with varying activity. *J. Hum. Genet.* **49**, 123–128.

Jinno, H., Saeki, M., Saito, Y., Tanaka-Kagawa, T., Hanioka, N., Sai, K., Kaniwa, N., Ando, M., Shirao, K., Minami, H., Ohtsu, A., Yoshida, T., Saijo, N., Ozawa, S., and Sawada, J.-I. (2003a). Functional characterization of human UDP-glucuronosyltransferase 1A9 variant, D256N found in Japanese cancer patients. *J. Pharmacol. Exp. Ther.* **306**, 688–693.

Jinno, H., Saeki, M., Tanaka-Kagawa, T., Hanioka, N., Saito, Y., Ozawa, S., Ando, M., Shirao, K., Minami, H., Ohtsu, A., Yoshida, T., Saijo, N., and Sawada, J.-I. (2003b). Functional characterization of wild-type and variant (T202I) and M59I) Human UDP-glucuronosyltransferase 1A10. *Drug Metab. Disp.* **31**, 528–532.

Koiwai, O., Hasada, K., Yasui, Y., Sakai, Y., Sato, H., and Watanabe, T. (1995). Isolation of cDNAs for mouse phenol and bilirubin UDP-glucuronosyltransferases and mapping of the mouse gene for phenol UDP-glucuronosyltransferase (Ugt1a1) to chromosome 1 by restriction fragment length variations. *Biochem. Genet.* **33**, 111–122.

Kong, A.-N. T., Ma, M., Tao, D., and Yang, L. (1993). Molecular cloning of two cDNAs encoding the mouse bilirubin/phenol family of UDP-glucuronosyltransferases (mUGTBr/p). *Pharm. Res.* **10**, 461–465.

Lamb, J. G., Straub, P., and Tukey, R. H. (1994). Cloning and characterization of cDNAs encoding Ugt.6 and rabbit UGT1.6: Differential induction by 2,3,7,8-tetrachlorodibenzo-p-dioxin. *Biochemistry* **33**, 10513–10520.

Levin, W., Wood, A. W., Yagi, H., Dansette, P. M., Jerina, D. M., and Conney, A. H. (1976). Carcinogenicity of benzo(a)pyrene 4,5-, 7,8-, and 9,10-oxides on mouse skin. *Proc. Natl. Acad. Sci. USA* **73**, 243–247.

Lin, G.-F., Guo, W.-C., Chen, J.-G., Qin, Y.-Q., Golka, K., Xiang, C.-Q., Ma, Q.-W., Lu, D.-R., and Shen, J.-H. (2005). An association of UDP-glucuronosyltransferase 2B7 $C_{802}T$ (His

268Tyr) polymorphism with bladder cancer in benzidine-exposed workers in China. *Toxicol. Sci.* **85**, 502–506.

Mackenzie, P. I., Owens, I. S., Burchell, B., Bock, K. W., Bairoch, A., Belanger, A., Fournel-Gigleux, S., Green, M., Hum, D. W., Iyanagi, T., Lancet, D., Louisot, P., Magdalou, J., Chowdhury, J. R., Ritter, J. K., Schachter, H., Tephly, T. R., Tipton, K. F., and Nebert, D. W. (1997). The UDP-glycosyltransferase gene superfamily: Recommended nomenclature update based on evolutionary divergence. *Pharmacogenetics* **7**, 255–269.

Mackenzie, P. I., and Rodbourn, L. (1990). Organization of the rat UDP-glucuronosyltransferase, UDPGTr-2, gene and characterization of its promoter. *J. Biol. Chem.* **265**, 11328–11332.

Malik, N., and Owens, I. S. (1981). Genetic regulation of bilirubin UDP-glucuronosyltransferase induction by polycyclic aromatic compounds and phenobarbital in mice. *J. Biol. Chem.* **256**, 9599–9604.

Martineau, I., Tchernof, A., and Belanger, A. (2004). Amino acid residue ILE211 is essential for the enzymatic activity of human UDP-glucuronosyltransferase 1A10 (UGT1A10). *Drug Metab. Disp.* **32**, 455–459.

Meech, R., and Mackenzie, P. I. (1997). UDP-glucuronosyltransferase, the role of the amino terminus in dimerization. *J. Biol. Chem.* **272**, 26913–26917.

Münzel, P. A., Schmohl, S., Buckler, F., Jaehrling, J., Raschko, F. T., Köhle, C., and Bock, K. W. (2003). Contribution of the Ah receptor to the phenolic antioxidant-mediated expression of human and rat UDP-glucuronosyltransferase UGT1A6 in Caco-2 and rat hepatoma 5L cells. *Biochem. Pharmacol.* **66**, 841–847.

Owens, I. S. (1977). Genetic regulation of UDP-glucuronosyltransferase induction by polycyclic aromatic compounds in mice. *J. Biol. Chem.* **252**, 2827–2833.

Owens, I. S., and Ritter, J. K. (1992). The novel bilirubin/phenol UDP-glucuronosyltransferase *UGT1* gene locus: Implications for multiple nonhemolytic familial hyperbilirubinemia phenotypes. *Pharmacogenetics* **2**, 93–108.

Ritter, J. K., Chen, F., Sheen, Y. Y., Tran, H. M., Kimura, S., Yeatman, M. T., and Owens, I. S. (1992a). A novel complex locus *UGT1* encodes human bilirubin, phenol, and other UDP-glucuronosyltransferase isozymes with identical carboxyl termini. *J. Biol. Chem.* **267**, 3257–3261.

Ritter, J. K., Crawford, J. M., and Owens, I. S. (1991). Cloning of two human liver bilirubin UDP-glucuronosyltransferase cDNAs with expression in COS-1 cells. *J. Biol. Chem.* **266**, 1043–1047.

Ritter, J. K., Yeatman, M. T., Ferreira, P., and Owens, I. S. (1992b). Identification of a genetic alteration in the code for bilirubin UDP-glucuronosyltransferase in the *UGT1* gene complex of a Crigler-Najjar type I patient. *J. Clin. Invest.* **90**, 150–155.

Saeki, M., Saito, Y., Jinno, H., Sai, k., Komamura, K., Ueno, K., Kamakura, S., Kitakaze, M., Shirao, K., Minami, H., Ohtsu, A., Yoshida, T., Saijo, N., Ozawa, S., and Sawada, J. (2003). Three novel single nucleotide polymorphisms in UGT1A9. *Drug Metab. Pharmacokin.* **18**, 146–149.

Strassburg, C. P., Kneip, S., Topp, J., Obermayer-Straub, P., Barut, A., Tukey, R. H., and Manns, M. P. (2000). Polymorphic gene regulation and interindividual variation of UDP-glucuronosyltransferase activity in human small intestine. *J. Biol. Chem.* **275**, 36164–36171.

Sueyoshi, T., and Negishi, M. (2001). Phenobarbital response elements of cytochrome P450 genes and nuclear factors. *Annu. Rev. Pharmacol. Toxicol.* **41**, 123–143.

Sugatani, J, Kojima, H., Ueda, A., Kakizaki, S., Yoshinari, K., Gong, Q.-H., Owens, I. S., Negishi, M., and Sueyoshi, T. (2001). The phenobarbital response enhancer module in the human bilirubin UDP-glucuronosyltransferase *UGT1A1* gene and regulation by the nuclear receptor CAR. *Hepatology* **33,** 1232–1238.

Sutherland, G. R., and Richards, R. I. (1995). Simple tandem DNA repeats and human genetic disease. *Proc. Natl. Acad. Sci. USA* **92,** 3636–3641.

Tautz, D., and Schlotterer, C. (1994). Simple sequences. *Curr. Opin. Genet. Dev.* **4,** 832–837.

van Es, H. H., Goldhoorn, B. G., Paul-Abrahamse, M., Oude Elferink, R. P., and Jansen, P. L. (1990). Immunochemical analysis of uridine diphosphate-glucuronoyltransferase in four patients with the Crigler Najjar syndrome type I. *J. Clin. Invest.* **85,** 1199–1205.

Villeneuve, L., Girard, H., Fortier, L. C., Gagné, J. F., and Guillemette, C. (2003). Novel functional polymorphisms in the UGT1A7 and UGT1A9 glucuronidating enzymes in Caucasians and African-American subjects and their impact on the metabolism of 7 ethyl-10-hydroxycamptothecin and flavopiridol anticancer drugs. *J. Pharmacol. Exp. Ther.* **307,** 117–128.

Wells, P. G., Mackenzie, P. I., Chowdhury, J. R., Guillemette, C., Gregory, P. A., Ishii, Y., Hansen, A. J., Kessler, F. K., Kim, P. M., Chowdhury, N. R., and Ritter, J. K. (2004). Glucuronidation and the UDP-glucuronosyltransferases in health and disease. *Drug Metab. Disp.* **32,** 281–290.

Yamanaka, H., Nakajima, M., Katoh, M., Hara, Y., Tachibana, O., Yamashita, J., McLeod, H. L., and Yokoi, T. (2004). A novel polymorphism in the promoter region of human *UGT1A9* gene (*UGT1A9*22*) and its effects on the transcriptional activity. *Pharmacogenetics* **14,** 329–332.

Yueh, M.-F., Huang, Y.-H., Hiller, A., Chen, S., Nguyen, N., and Tukey, R. H. (2003). Involvement of the xenobiotic response element (XRE) in Ah receptor-mediated induction of human UDP-glucuronosyltransferase 1A1. *J. Biol. Chem.* **278,** 15001–15006.

Zhang, T., Haws, P., and Wu, Q. (2004). Multiple variable first exons: A mechanism for cell and tissue-specific regulation. *Genome Res.* **14,** 79–89.

[2] Identification and Characterization of Functional Hepatocyte Nuclear Factor 1-Binding Sites in UDP-Glucuronosyltransferase Genes

By DIONE A. GARDNER-STEPHEN, PHILIP A. GREGORY, and PETER I. MACKENZIE

Abstract

The hepatocyte nuclear factor 1 (HNF1) transcription factor family is composed of two closely related homeodomain proteins with similar but distinct expression profiles. Homodimers and heterodimers of these transcription factors, HNF1α and HNF1β, increase transcription from target genes through direct physical interaction with one or more elements of sufficient similarity to a 13 nucleotide-inverted dyad consensus-binding sequence. Potential HNF1-binding sites have been found in the proximal

METHODS IN ENZYMOLOGY, VOL. 400
0076-6879/05 $35.00
DOI: 10.1016/S0076-6879(05)00002-9

upstream regulatory regions of most known human UDP-glucuronosyltransferase (UGT) genes. As the liver and gastrointestinal tract are both
important sites of glucuronidation and express significant levels of one or
both HNF1 proteins, it is thought that these homeoproteins may play a role
in transcriptional regulation of UGTs. This chapter explores the current
evidence that HNF1 transcription factors are explicitly involved in the
transcription of mammalian UGT genes. Most data supporting this hypothesis come from *in vitro* reporter assays, site-directed mutagenesis, and
electrophoretic mobility-shift assays, for which methods are detailed. However, as *in vitro* functionality of transcription factors does not necessarily
imply significance *in vivo*, some of the limitations of these techniques
are also examined. In addition, available *in vivo* data are discussed, with
particular attention given to contributions made by HNF1α knockout
mouse models and microarray studies of human tissue. Finally, possible
scenarios in which HNF1-mediated regulation of UGT expression may be
clinically relevant are suggested.

Introduction

Hepatocyte nuclear factor 1 (HNF1) homeoproteins are recognized as
key regulators of an increasingly large assortment of diverse vertebrate
genes. In the main, HNF1 responsive genes recognized so far are those
expressed predominantly in the liver, but roles for HNF1 as a transcriptional regulator in other tissues such as the pancreas, kidney, and intestine
are also becoming evident (Boudreau *et al.*, 2001; Gregori *et al.*, 2002;
Mouchel *et al.*, 2004; Pontoglio *et al.*, 1996). Mammalian UDP-glucuronosyltransferase (UGT) genes are among the many recently identified HNF1-
responsive genes, with mounting evidence that most members of the
human *UGT1A* and *UGT2B* families possess potential HNF1 sites directly
upstream of their first exon. However, not all possible binding sites for a
transcription factor are necessarily functional, and the challenging task
ahead is to characterize the nature of the identified sites. Whether a
particular HNF1 site can facilitate gene transcription will depend on its
context in each individual gene and the network of factors available in
various cell types (Gregori *et al.*, 2002; Locker *et al.*, 2002; Mouchel *et al.*,
2004; Tronche *et al.*, 1997).

HNF1 Transcription Factors

The two known members of the HNF1 transcription factor family,
HNF1α and HNF1β, share substantial homology, with their amino-terminal
dimerization and internal DNA-binding domains having about 75 and 93%

identity, respectively, in both rat and mouse (Mendel *et al.*, 1991; Rey-Campos *et al.*, 1991; Tronche and Yaniv, 1992). As a result, the proteins, which bind DNA as dimers, can heterodimerize readily. Homodimers and heterodimers all recognize the same inverted dyad DNA element with the consensus sequence GTTAATNATTAAC (Locker *et al.*, 2002; Tronche and Yaniv, 1992; Tronche *et al.*, 1997). HNF1 dimers are stabilized by the formation of tetramers containing two copies of the dimerization cofactor of HNF1 (DCoH), also known as pterin-4α-carbinolamine dehydratase. As well as augmenting HNF1 activity through stabilization of the dimers, DCoH serves to modify the binding specificity of HNF1. HNF1 activity is especially dependent on DCoH when the homeoprotein concentration is low (Mendel *et al.*, 1991; Rhee *et al.*, 1997). Like HNF1, the expression of DCoH is tissue restricted, and there is considerable overlap in the localization of the two tetramer constituents (Mendel *et al.*, 1991; Pogge von Strandmann *et al.*, 1998).

In contrast to the highly homologous dimerization and DNA-binding domains, the carboxyl-terminal activation domains of the HNF1α and HNF1β are more divergent, with only 47% identity between the two proteins. The activation domain of HNF1α is also considerably longer than that of HNF1β (Mendel *et al.*, 1991; Rey-Campos *et al.*, 1991; Tronche and Yaniv, 1992). Thus they are not equal in their ability to transactivate all genes, although there is significant overlap in activity (Bernard *et al.*, 1999; De Simone *et al.*, 1991; Liu and Gonzalez, 1995; Mendel *et al.*, 1991; Pontoglio *et al.*, 1996; Rey-Campos *et al.*, 1991; Song *et al.*, 1998). Additionally, *HNF1*α and *HNF1*β have similar but distinct temporal and spatial expression profiles, with HNF1β expression preceding that of HNF1α (De Simone *et al.*, 1991; Rey-Campos *et al.*, 1991; Ryffel, 2001). Whereas both isoforms are found at comparable levels in the adult kidney, HNF1α is the predominant form in the liver and HNF1β is expressed exclusively in the lung. Tissues other than the kidney where both HNF1 variants are found include the intestine, stomach, and pancreas (De Simone *et al.*, 1991; Mendel *et al.*, 1991; Pontoglio *et al.*, 1996; Rey-Campos *et al.*, 1991). HNF1α is known to be well conserved between species, with high similarities in nucleotide and amino acid sequence, as well as in genomic structure (Bach *et al.*, 1991, 1992).

HNF1α Serves a Dual Purpose in Gene Transcription

Genomic DNA is subject to a hierarchy of control measures that coordinate gene expression as appropriate (reviewed in Schrem *et al.*, 2002). There is increasing evidence that HNF1α can influence transcription at several of these levels, including chromatin remodeling and recruitment of the transcription machinery.

A compacted chromatin structure is an effective inhibitor of gene transcription (Schrem *et al.*, 2002). In the *in vivo* setting, HNF1α increases the accessibility of promoter elements to other transcription factors and nuclear receptors through modification of the chromatin environment. One study has shown that developmental demethylation of certain genes appears to be under the influence of HNF1α (Pontoglio *et al.*, 1997). In addition, HNF1α is thought to induce repositioning or modification of nucleosomes through recruitment and activation of histone acetyltransferases (HATs) such as p300/CBP [cAMP response element-binding protein (CREB)-binding protein)] and P/CAF (CBP-associated factor) (Parrizas *et al.*, 2001; Pontoglio *et al.*, 1997; Rollini *et al.*, 1999; Soutoglou *et al.*, 2000, 2001). This HNF1α-mediated hyperacetylation of histones in target genes is cell type specific. For example, while HNF1α interacts with the mouse *glut2* promoter chromatin template in both liver and pancreatic islet cells, only the latter are dependent on HNF1α for hyperacetylation and transcriptional activity of the *glut2* gene (Parrizas *et al.*, 2001). HNF1β has also been shown to be able to interact with both p300/CBP and P/CAF HAT factors (Barbacci *et al.*, 2004; Hiesberger *et al.*, 2005). The importance of these aspects of HNF1-mediated gene regulation is not addressed by the *in vitro* techniques described herein and very little has been published to date concerning this issue specifically in the context of *UGT* genes.

Apart from altering the chromatin higher-order structure, HNF1α and HNF1β can, in concert with appropriate combinations of other transcription factors and coactivators, increase the rate of transcription from promoters containing HNF1 sites. This is thought to be mediated by interaction, either directly or indirectly, with components of the general transcription machinery, providing recruitment and positioning services for the preinitiation complex (Schrem *et al.*, 2002; Vorachek *et al.*, 2000). This may be particularly important in promoters that lack a TATA box but still have a well-defined transcription start site, such as the mouse *UGT1A1* gene (Bernard *et al.*, 1999). As this function can be observed in episomal DNA, in which nucleosomal organization is considered relatively random compared to the highly organized nature of chromatin (Akiyama and Gonzalez, 2003; Archer *et al.*, 1992; Liu and Gonzalez, 1995; Soutoglou *et al.*, 2000), it is often considered in isolation from the ability to direct histone acetylation. However, the two functions involve many of the same proteins and, in many cases, are likely to be profoundly linked *in vivo* (Soutoglou *et al.*, 2000).

HNF1α bound to sites in proximal promoters may direct the assembly of the preinitiation complex by either interacting directly with components of the general transcription machinery such as TFIIB (Ktistaki and Talianidis, 1997) or through coactivator proteins that provide a bridge between the two. In addition to their intrinsic HAT activity mentioned earlier, p300/CBP and P/CAF have well-researched roles as coactivators, linking HNF1 dimers to the transcription apparatus (Dallas *et al.*, 1997; Dohda *et al.*, 2004; Schrem *et al.*, 2002; Soutoglou *et al.*, 2000). Other proteins that are considered transcription factors in their own right, such as Octamer transcription factor-1 (Oct-1), may also be able to act as coactivators for HNF1α in some circumstances. Oct-1 upregulates transcription from the *UGT2B7* promoter independently of direct DNA contact, but in a manner that requires HNF1α to be bound to its element (Ishii *et al.*, 2000). Oct-1 is known be able to influence formation of a functional preinitiation complex through physical interaction with TFIIB (Nakshatri *et al.*, 1995). *In vitro* transcription and cotransfection assays are well suited to investigating interactions of this nature and, despite the limitations imposed by lack of genomic context, can also provide information about the cell type-dependent nature of transcription initiation in many instances (Gregori *et al.*, 1998; Gregory *et al.*, 2000; Hu and Perlmutter, 1999; Liu and Gonzalez, 1995).

HNF1 Regulation of Mammalian UDP-Glucuronosyltransferases

The expression pattern of UGT1A and UGT2B enzymes overlaps considerably with that of HNF1 homeoproteins. The liver and intestine are both important sites of glucuronidation (Tukey and Strassburg, 2001) and both contain significant levels of HNF1. The kidneys and the stomach are further examples of tissues that contain at least one of the HNF1 isoforms and also express a subset of UGTs. It is therefore feasible that HNF1 proteins could play an important role in regulating the expression of some, if not all, of the *UGT* genes.

Although functional HNF1 sites can also be located within introns, exons, and distal enhancer regions of various genes (Gregori *et al.*, 2002; Mouchel *et al.*, 2004; Schrem *et al.*, 2002), the only HNF1 responsive elements verified to date in *UGT* genes are within proximal promoters. Evidence that HNF1 factors influence UGT expression in humans and other vertebrates comes from a variety of methods, but the vast majority of data is from cotransfection experiments and electrophoretic mobility-shift assays (EMSA). Since 1997, when our laboratory demonstrated that the rat *UGT2B1* promoter is influenced by HNF1α *in vitro*, promoter studies have shown that the rat *UGT1A6* and *UGT1A7* promoters, the mouse *UGT1A1* regulatory region, and the human *UGT2B7, UGT2B17, UGT1A1, UGT1A8, UGT1A9,* and *UGT1A10* promoters all contain functional HNF1 sites (Auyeung *et al.*, 2003; Bernard *et al.*, 1999; Gregory and Mackenzie, 2002; Gregory *et al.*, 2000, 2004; Hansen *et al.*, 1997; Ishii *et al.*, 2000; Metz *et al.*, 2000). When promoters from *UGT* genes of the same subfamily clusters are aligned, it is evident that HNF1 sites are well conserved within clusters of closely related genes (Fig. 1) (Gregory and Mackenzie, 2002). Additional support for the notion that HNF1α and/or HNF1β is an important regulator of *UGT* promoters comes from transgenic mouse models, *in silico* studies, microarray analysis, and data collected from human subjects, as discussed later.

Two groups have independently developed HNF1α knockout mouse models that have contributed to our knowledge of the relationship between HNF1α and UGT expression (Lee *et al.*, 1998; Pontoglio *et al.*, 1996). HNF$^{-/-}$ mice from the line developed by Pontoglio and colleagues suffer a mild hyperbilirubinemia, which is attributed to insufficient levels of UGT1A1 expression (Bernard *et al.*, 1999; Pontoglio *et al.*, 1996). Homozygous-null mutants from the HNF1α knockout model of Gonzalez and co-workers exhibit a changed liver gene expression pattern, as shown by microarray analysis. This study indicates that *UGT* was one of the many gene families that were downregulated in the livers of adult HNF$^{-/-}$ mice relative to wild-type littermates (Shih *et al.*, 2001). Further to these studies, as mouse models of human UGT expression are developed, crossbreeding with these HNF$^{-/-}$ transgenic models will likely

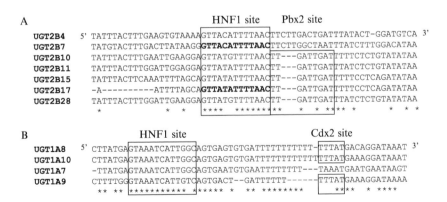

FIG. 1. Alignment of the human *UGT1A7-10* and *UGT2B* proximal promoter regions containing putative HNF1 sites. (A) Putative HNF1 sites of the *UGT2B* genes are boxed, with the HNF1 sites confirmed experimentally indicated in bold. Also indicated are locations of the published Pbx2 sites (boxed) (Gregory and Mackenzie, 2002). (B) The HNF1 and Cdx2 sites of the *UGT1A7-10* gene cluster as published (Gregory *et al.*, 2004) are indicated by their respective boxes.

provide useful information about the role of HNF1α in the expression of human genes.

As mentioned previously, *in silico* databases are another resource that can provide evidence that HNF1 transcription factors are involved in human UGT expression. An early publication highlighted three potential HNF1 sites on the human *UGT1A* locus, using genomic data available at the time. The first is in the regulatory region of *UGT1A4* and the remaining two are intronic sites, both falling between the unique exon of *UGT1A1* and the first common exon, known as exon 2 (Tronche *et al.*, 1997). Experimental verification of these sites is still to be published, although our laboratory has confirmed the functional nature of the *UGT1A4* promoter site (manuscript in preparation).

Recent support for the functional nature of HNF1 sites in the human *UGT1A1, UGT2B11,* and *UGT2B15* promoters comes from a microarray study that was designed to detect HNF1α occupation of 13,000 candidate promoter regions in primary hepatocytes and pancreatic islet cells. HNF1α was found to occupy *UGT1A1, UGT2B11,* and *UGT2B15* in liver extracts but not in pancreatic tissue (Odom *et al.*, 2004). Although binding does not necessarily correlate with an absolute requirement for HNF1α in transcription, even *in vivo* (Parrizas *et al.*, 2001), these data provide strong evidence that HNF1α contributes to the transcriptional regulation of these genes in the liver.

Finally, interindividual variances in UGT2B7 expression levels have been correlated with HNF1α expression in human liver samples (Toide *et al.*, 2002). In contrast, UGT2B15 mRNA levels were not associated with HNF1α, highlighting that while many *UGT* genes possess HNF1 sites in their proximal promoters, the exact function of these sites varies.

Functional HNF1-Binding Sites Are Part of a Larger
  Regulatory Network

In the chromosomal setting, a single HNF1 site is not sufficient to drive targeted expression of a gene. Rather, each gene is regulated by a transcription factor network that is specific both to that gene and to each cell type in which it is active. Promoters that are under the control of HNF1 generally have additional binding sites nearby for other transcription factors that participate in the overall activation of transcription. Regulatory regions may also have multiple HNF1 sites, particularly those belonging to genes expressed in the liver (Costa *et al.*, 2003; Frain *et al.*, 1990; Schrem *et al.*, 2002; Song *et al.*, 1998). The relationship between HNF1 proteins and UGT expression is therefore, like many other genes, complicated by interactions with other transcription factors. Of the *UGT* genes that possess functional HNF1 sites, two groups of genes are already known to rely on additional transcription factor sites to modify HNF1α activation *in vitro*.

One case in point is the *UGT2B17* gene. In the *UGT2B17* promoter the HNF1 site is situated immediately adjacent to an element that binds complexes of Pbx2 and Prep1. Binding of Pbx2-Prep1 to this site attenuates HNF1α-mediated activation of the promoter by restricting the access of HNF1α dimers to the DNA in HepG2 cells. This combination of elements is found in some, but not all of the other human *UGT2B* promoters (Fig. 1A) and may correlate with expression levels of UGT2B proteins in HNF1-containing tissues such as the liver (Congiu *et al.*, 2002; Gregory and Mackenzie, 2002).

A second example is the increased response of the *UGT1A8*, *UGT1A9*, and *UGT1A10* promoters to HNF1α in the presence of caudal-related homeoprotein 2 (Cdx2). Even though all three genes have very similar HNF1 and Cdx2 sites (see Fig. 1B), the contribution of each element to promoter activation *in vitro* varies. Cdx2 works independently to activate *UGT1A8* and *UGT1A10*, albeit weakly for the latter, but its overexpression has no detectable effect on *UGT1A9*. However, in the presence of an intact HNF1 site, Cdx2 can increase HNF1α-mediated activation of all three genes. This synergy is dependent on the HNF1α site and is also moderately reduced if the Cdx2 site in *UGT1A8* or *UGT1A10* is mutated. This effect is particularly marked in the case of *UGT1A8*. Therefore, Cdx2 appears to act as both transcription factor and coactivator on the *UGT1A8* and *UGT1A10* promoters. In contrast, because Cdx2 does not activate the *UGT1A9* promoter unless cotransfected with HNF1α, it appears that it contributes only coactivator function in this instance (Gregory *et al.*, 2004).

Other examples where HNF1 regulation of UGT expression is not straightforward include coactivation of *UGT2B7* by Oct-1 as discussed earlier (Ishii *et al.*, 2000) and the transcriptional control of *UGT1A1* and rat *UGT1A6*. *UGT1A1* has an HNF1 site that may be bound competitively by HNF3 family members. However, the functional consequences of this observation have not been investigated (Bernard *et al.*, 1999). The human *UGT1A1* HNF1 site has also been implicated in the *UGT1A1* promoter response to dexamethasone, but again the mechanism has yet to be elucidated (Kanou *et al.*, 2004). In the case of rat UGT1A6 expression, two alternative transcripts have been described. These differ only by their 5'-untranslated sequence, but have distinct expression patterns, particularly in terms of relative abundance (Auyeung *et al.*, 2001). HNF1α appears to be involved in the initiation of only one of the two transcripts generated from the rat *UGT1A6* gene (Auyeung *et al.*, 2003). Thus, from these and the instances given earlier, it is evident that even once a potential HNF1 site has been shown to have functional significance it is but the beginning of a long journey to understand the context in which it operates.

Materials and Methods

Screening Genes for Potential HNF1-Binding Elements

There are several approaches, both experimental and *in silico*, for identifying genes that are likely to be under the transcriptional control of HNF1. Target genes may be identified by means such as computer prediction of regulatory elements, comparative microarray analysis of expression profiles between knockout animals and their wild-type counterparts (Lockwood and Frayling, 2003; Shih *et al.*, 2001), or *in vitro* experiments using promoter–reporter systems (see later). Cross-species comparison of orthologous genes can also help highlight conserved regulatory elements that are likely to serve an indispensable purpose (Bernard *et al.*, 1999; Eskinazi *et al.*, 1999; Lenhard *et al.*, 2003; Metz *et al.*, 2000; Sandelin *et al.*, 2004; The ENCODE Project Consortium, 2004; The Rat Genome Sequencing Project Consortium, 2004).

Some examples of freely accessible software (to nonprofit organizations) that can assist in identifying transcription factor-binding sites include MatInspector (Quandt *et al.*, 1995), Match (Matys *et al.*, 2003), TESS (Schug and Overton, 1997), TFBind (Tsunoda and Takagi, 1999), MAPPER (Marinescu *et al.*, 2005), PROMO (Farre *et al.*, 2003), and ConSite (Sandelin *et al.*, 2004). These utilize the HNF1-binding site matrices of one or more databases such as TRANSFAC (Matys *et al.*, 2003), IMD (Chen *et al.*, 1995), and JASPAR (Sandelin *et al.*, 2004), with the HNF1 consensus statistics being derived from various compilations of experimental data (Frech *et al.*, 1993; Tronche and Yaniv, 1992; Tronche *et al.*, 1997). However, low sensitivity or accuracy can often limit the usefulness of such computer programs as they tend to be biased toward easily recognized, strong binding sites (Locker *et al.*, 2002). This has indeed been the case in our hands when searching for potential HNF1 sites that deviate too far from consensus. Many genes shown to contain functional HNF1-binding sites are poorly predicted by these tools. This seems to be a consequence of the fact that, in the right context, only one-half of the palindrome needs to be relatively well conserved for HNF1-mediated activation, at least *in vitro* (Auyeung *et al.*, 2003; Bernard *et al.*, 1999; Tronche and Yaniv, 1992).

While computer-assisted analysis and microarray technology can result in vast quantities of potential targets for HNF1, the more tedious experimental approach of simply testing an undefined stretch of DNA in an

HNF1 reporter assay can be quite revealing. If the number of promoters under scrutiny is small and possible HNF1 sites are poor consensus matches, promoter studies can be one of the more efficient means of determining whether HNF1 proteins play a potential role in gene expression.

Luciferase Reporter Assays

Whether aiming to verify the functional nature of a predicted HNF1 site or to assess a particular promoter for potential HNF1-binding elements, the same approaches can be used. Generally, requirements for testing the responsiveness of a promoter region to HNF1α or HNF1β in a luciferase-based reporter assay are as follow: the promoter in question, or portions thereof, cloned into a suitable reporter construct; plasmids that will overexpress the HNF1 proteins when transiently transfected into mammalian cell culture; the host cell line or lines for transfection; and a plasmid encoding a second reporter gene under constitutive noninducible control, to serve as an internal reference for transfection efficiency.

The promoter region to be investigated must first be incorporated into the multiple cloning site of a plasmid intended for reporter assays. pGL3-basic (Promega, Madison, WI), which is designed to express firefly luciferase under the direction of any user-inserted elements, is one such example. If a putative HNF1 site (or sites) has already been identified in the test sequence, it is prudent to create several promoter constructs of varying lengths that include or exclude the suspected element(s). Unless suitable restriction sites are available within the promoter sequence to be cloned, this is achieved easily by polymerase chain reaction amplification of the desired regions utilizing primers with restriction sites engineered into their 5′ ends.

To test the influence of HNF1α or HNF1β on a cloned promoter, one or more expression plasmids can be cotransfected with the reporter constructs as appropriate. Ideal plasmids for overexpression of transcription factors are those controlled by a promoter that gives high-level constitutive transcription of the encoded cDNA. The murine HNF1α and HNF1β expression plasmids used in our published work were the kind gift of Dr. Gerald Crabtree (Stanford University, Stanford, CA) (Kuo *et al.*, 1990; Mendel *et al.*, 1991).

Transfection of most cells and cell lines can be achieved with cationic lipid reagents, although each system must be optimized individually, as some reagents are toxic or ineffective on certain cell types. The seeding cell density must also be determined empirically for each system to achieve the maximum transfection efficiency and reproducible promoter activity at the point of harvest. The cell lines used in our models have included HepG2, Caco-2, HEK293, and LNCaP cells, all of which can be transfected with Lipofectamine2000 (Invitrogen, Carlsbad, CA) as detailed later.

Tissue culture 24-well plates (Nunc, Roskilde, Denmark) are seeded with the appropriate number of cells for the cell type (2×10^5, 7.5×10^4, and 1×10^5 for HepG2, Caco-2, and HEK293 cells, respectively) and left for 20 to 24 h to allow the cells to reattach. LNCaP cells are plated similarly at a passage ratio of 1:2 and allowed to reach 90% confluence before use. Triplicate wells are then transfected with either 0.5 μg of pGL3-basic or the pGL3-promoter constructs using 2–4 μl of Lipofectamine2000, according to the supplied protocol. Twenty-five nanograms of pRL-Null plasmid (Promega), which expresses low constitutive levels of the *Renilla* luciferase gene, is also included in every transfection to serve as the internal reference. Other members of the pRL vector family (Promega) are also suitable for use as internal control reporters; the suitability of each will depend on the cell lines and experimental conditions chosen. If cotransfection with expression plasmids is required, these are included in the transfection at 0.25 μg each, and all other samples are adjusted to match the highest DNA concentration by the addition of appropriate quantities of empty expression plasmid. Large-scale plasmid preparations of suitable quality for transfection are produced with Qiagen plasmid midi or maxi kits (Qiagen, Clifton Hill, Australia).

Twenty-four hours posttransfection the medium is replaced on all cells, except in the case of LNCaP cells, which are likely to detach from the plate if handled at this stage. The transfected cells are harvested 48 h after transfection by washing each well with approximately 500 μl of phosphate-buffered saline (PBS; 137 mM NaCl, 2.7 mM KCl, 10 mM Na$_2$PO$_4$) (again, skip this step if using LNCaP cells) and the addition of 100 μl of $1\times$ passive lysis buffer (Promega), followed by gentle rocking for 15 min at room temperature. The lysate (20 μl) can then be analyzed directly in 96-well format for luciferase activity using the dual luciferase reporter system from Promega as instructed. One exception to this last step can arise when HEK293 cells have been used as the host cells. We have found it necessary to dilute HEK293 lysates 1:20 in $1\times$ lysis buffer before performing the luciferase assay. This is required in order to obtain light emissions that are sufficiently reduced to be read accurately by our luminometer (Packard TopCount luminescence and scintillation counter; Mt. Waverly, Australia).

Site-Directed Mutagenesis of Promoter–Reporter Constructs

One method of confirming the contribution of a particular element to promoter activity is to introduce specific mutations within the binding site, hopefully leaving the function of remaining sequences unperturbed. This has an advantage over deletion of nucleotides, as it does not change the relative distance of upstream sequences from the transcription start site and other elements. Mutation of HNF1 sites is achieved using the Quik-Change site-directed mutagenesis kit (Invitrogen), which requires the design of a complementary pair of oligonucleotides containing the desired changes in the center of each. The design of the mutagenic primers is an important consideration when mutating HNF1 sites, as HNF1 elements vary widely in their structure. Strong binding sites typically consist of a palindrome that closely resembles the consensus along its entire length. These sites will usually require mutation in both halves of the palindrome to completely abolish HNF1 binding in EMSA experiments and promoter assays. However, many weaker binding sites are elements that match consensus far more strongly in one-half of the palindrome than the other, and mutations directed to the more perfect region may be sufficient to eliminate binding.

Choosing Cell Lines for Transfection Assays and EMSA

It is important to consider the effect of endogenous HNF1 proteins when interpreting results gained from *in vitro* luciferase reporter studies, as discussed later. Published data regarding HNF1 expression for many cell lines are now available, but depending on the cell line chosen, it may still be necessary to determine this experimentally. The relative abundance of the HNF1α protein expressed by various cell lines can be assessed by a Western blot using commercially available antibodies as described later.

Among the human tissue culture lines commonly used for HNF1 transfection studies are HepG2, Caco-2, HEK293, and HeLa cells. HepG2 is a hepatocellular carcinoma cell line that is considered one of the more highly differentiated models of the hepatocyte available (Ishiyama *et al.*, 2003). These cells express abundant levels of HNF1α, but, unlike the adult liver, also contain significant levels of HNF1β protein (Auyeung *et al.*, 2003; Ktistaki and Talianidis, 1997; Song *et al.*, 1998; see also later). Caco-2 cells are derived from a colorectal adenocarcinoma and also express both HNF1 proteins. These cells differentiate on confluence, at which point HNF1α levels rise considerably from preconfluence concentrations, becoming the predominant HNF1 isoform (Hu and Perlmutter, 1999; Mouchel *et al.*, 2004). HEK293 (transformed human embryonic kidney) and HeLa (cervical adenocarcinoma) cells have also been widely utilized in transfection experiments because they have undetectable levels of either of the HNF1 transcription factors (Barbacci *et al.*, 2004; Bernard *et al.*, 1999; Hu and Perlmutter, 1999).

Because cofactors and other transcription factors also vary between cell lines and can result in alternate outcomes from transfection experiments, further care should be taken to choose an appropriate host. If basal promoter activity is to be studied, ideally the transfected cell line should express the gene of interest and share as many characteristics as possible with the tissue from which it was derived, but in the case of UGTs this is not always possible. Primary cells are usually considered the best model, but are not obtained easily by many laboratories.

Detection of HNF1α Expression by Western Blot

Twenty-five micrograms of nuclear extract (see later) is subjected to SDS–PAGE electrophoresis on a 10% gel and transferred to a polyvinylidene difluoride membrane (Bio-Rad, Hercules, CA). After transfer, the membrane is blocked for 2 h in TBST [Tris-buffered saline (10 mM Tris, 150 mM NaCl) plus 0.05% Tween 20] containing 5% skim milk powder. Once blocking is complete the membrane is incubated with 1 μg/ml anti-HNF1α (sc-6547; Santa Cruz Biotechnology, Santa Cruz, CA) in blocking buffer for 2 h and then given three washes in TBST of 5 min each. This is followed by incubation of the membrane for a further hour in blocking buffer containing a 1:2500 dilution of rabbit antigoat horseradish peroxidase-conjugated secondary antibody (Southern Biotechnology Associates Inc., Birmingham, AL). A second series of washes as just described then precedes a single 5-min wash in TBS, before the membrane is treated with ECL Western blotting reagents (Amersham Biosciences, Piscataway, NJ) according to the manufacturer's instructions and exposed to X-Omat Blue XB-1 autoradiographic film (Eastman-Kodak, Rochester, NY). All blocking, washing, and antibody incubations steps are performed at room temperature with gentle rocking.

Preparation of Nuclear Extracts

Nuclear extracts are prepared by the following method adapted from Schreiber *et al.* (1989). Tissue culture cells are grown to confluence in 175-cm^2 flasks (Nunc) and harvested by scraping into 10 ml ice-cold PBS. The pellet from each flask is collected by centrifugation at 1500g for 5 min, resuspended in 1 ml of ice-cold PBS, and transferred to a 1.5-ml microcentrifuge tube. The cells are collected again in a microcentrifuge at 10,000g, 4° for 1 min and resuspended in 800 μl cold buffer A [10 mM Tris-HCl, pH 7.9, 10 mM KCl, 1 mM dithiothreitol (DTT), 1.5 mM MgCl$_2$ containing 1× complete protease inhibitor cocktail (Roche Diagnostics, Mannheim, Germany)] and 0.5% Nonidet P-40. After 15 min on ice the nuclear fraction is pelleted at 10,000g for 1 min (4°), washed briefly by resuspending in 800 μl cold buffer A, and collected by centrifugation as described previously. The pellets from four flasks are pooled and resuspended in

400 μl ice-cold buffer B (50 mM Tris-HCl, pH 7.9, 500 mM KCl, 2 mM DTT, 5 mM MgCl$_2$, 0.1 mM EDTA, 10% sucrose, 20% glycerol containing 1× complete protease inhibitor cocktail) in a 1.5-ml microcentrifuge tube. This tube is then laid between two layers of ice in a small beaker and shaken vigorously for an hour to facilitate lysis of the nuclei. The nuclear extract supernatant is separated from the remaining debris by centrifugation for 15 min at 10,000g at 4° and transferred to a Slide-A-Lyzer dialysis cassette (Pierce Biotechnology, Rockford, IL). Dialysis is performed for at least 2 h against 200 ml of buffer TM-1 (25 mM Tris-HCl, pH 7.6, 100 mM KCl, 0.5 mM DTT, 5 mM MgCl$_2$, 0.5 mM EDTA, and 20% glycerol).

EMSA and Supershift Experiments

Once an HNF1 element has been identified, direct interaction with HNF1 proteins can be confirmed through EMSA experiments. Five micrograms each of complementary oligonucleotides covering the region of interest is denatured at 95° for 2 min in 100 μl of annealing buffer (40 mM Tris-HCl, pH 7.5, 20 mM MgCl$_2$, and 50 mM NaCl) and allowed to cool to room temperature over the course of several hours. One hundred nanograms of the annealed probes is then end labeled with 5 units of T4 polynucleotide kinase (New England Biolabs, Beverly, MA) using [r-^{32}P] ATP (PerkinElmer, Rowville, Australia) for 1 h at 37° and purified through G25 columns (Amersham Biosciences).

HNF1 EMSA reactions require 5 μg of nuclear extract from cells known to express sufficient levels of one or both of the HNF1 proteins, such as HepG2, Caco-2, and LNCaP cells. Alternatively, cells that have been manipulated to express either transcription factor from introduced plasmids can be substituted (Mendel *et al.*, 1991). The nuclear extract is added to 1 μg of poly(dI-dC) (Sigma Chemical Co., St. Louis, MO), made up to 15 μl with TM-1 buffer, and incubated on ice for 10 min. Fifty thousand counts per minute of labeled probe are then added and the reaction is incubated on ice for a further 30 min. DNA–protein complexes are resolved on 4% nondenaturing polyacrylamide gels that are then dried and exposed to X-Omat Blue XB-1 film (Eastman-Kodak) with the aid of intensifying screens.

To confirm the presence of HNF1 proteins in any observed complexes, 2 μg of antibody specific to either HNF1α or HNF1β (sc-6547 and sc-7411, respectively; Santa Cruz Biotechnology) can be added directly after addition of the labeled oligonucleotide. This will cause a decrease in mobility of complexes containing the relevant protein when subjected to PAGE. Excess (up to 500×) unlabeled probes (wild type and mutated) can also be added to EMSA reactions to show that protein binding is specific and sequence dependent. If required, these are added before the initial 10-min incubation on ice.

Results

Exogenous HNF1α and HNF1β in Luciferase Reporter Assays

Results obtained from transient transfection of tissue culture cells with an HNF1-responsive promoter will naturally depend on the promoter sequence and the cell type chosen. If HNF1 expression plasmids are co-transfected with the reporter construct, the outcome will also depend on the particular HNF1 isoform used. Generally, HNF1β has a lower capacity to stimulate *UGT* promoters than HNF1α (Auyeung *et al.*, 2003; Gregory *et al.*, 2000, 2004; Hansen *et al.*, 1997), with the exception of the upstream regions of the mouse and human *UGT1A1* and rat *UGT1A7* genes, which respond equally well to both (Bernard *et al.*, 1999; Metz *et al.*, 2000).

Deletion Constructs Can Indicate the Position of Regulatory Elements

One method of verifying the functional capacity of a potential HNF1-binding site is to transfect a cell line known to express HNF1α or HNF1β with a series of reporter constructs containing progressively shorter portions of promoter. Deletion of an active HNF1 site can lead to a decrease in basal promoter activity (Fig. 2A). A published example of this approach is found in Ishii *et al.* (2000) where the activity of the *UGT2B7* −44/+57 reporter construct is reduced to background levels when nucleotides −44 to −28 are not included.

Use of HNF1-Negative Cell Lines to Determine Relative Importance of HNF1 Elements

Promoters that absolutely require HNF1 factors for basal activity are silent when transfected into HNF1-deficient cells such as HEK293. Examples of *UGT* promoters that are inactive in HEK293 cells include those from the human and mouse *UGT1A1* genes (Bernard *et al.*, 1999), as well as from rat *UGT1A7* (Metz *et al.*, 2000). In these cases HNF1 seems to be necessary for recruitment of the general transcription machinery, and upstream regulatory elements cannot compensate for the lack of HNF1α or HNF1β. In contrast, we have observed that *UGT1A8* promoter constructs in HEK293 cells retain about half of the activity seen in HNF1-expressing cells, suggesting that the initiator-like region in this TATA-less promoter (Gregory *et al.*, 2003) can function independently of HNF1 to instigate transcription. When HNF1α is transiently overexpressed in HEK293 cells, promoters from genes such as human *UGT1A1* are highly upregulated (Bernard *et al.*, 1999), whereas the *UGT1A8* promoter is increased less than twofold (Fig. 2B).

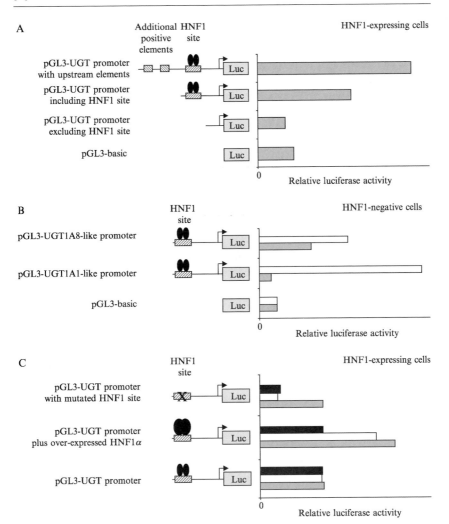

FIG. 2. Theoretical outcomes of UGT promoter transfections and cotransfections with HNF1α. (A) Typical result from UGT2B7-like promoter constructs transfected into HNF1-expressing cell types. The consequences of excluding the proximal HNF1 site on the basal activity of this promoter type are illustrated. (B) Representative results from UGT1A1-like and UGT1A8-like promoters in HNF1-negative cell lines. Filled bars represent basal activity of each type of promoter, and open bars indicate their relative responses to overexpressed HNF1α. (C) Some of the possible responses of UGT promoters transfected into HNF1α-expressing cells. UGT2B17-like promoters (gray bars) can be upregulated by overexpressed HNF1α, but basal activity is unaffected by mutation of the HNF1 element. UGT2B7-like promoters (white bars) can also be upregulated by exogenous HNF1α, but mutation of the HNF1 site decreases basal promoter activity. Rat UGT1A7-like promoters (black bars) cannot be increased by overexpressed HNF1α, yet mutation of the HNF1 site also decreases basal promoter activity.

Use of HNF1-Positive Cell Lines and Promoter Mutants to Demonstrate Functionality of HNF1-Binding Sites

Cotransfection of cells that already express HNF1 transcription factors, such as HepG2 or Caco-2, with *UGT* promoter–reporter constructs and exogenous HNF1α or HNF1β can lead to a myriad of different outcomes, including those illustrated in Fig. 2C. For example, the *UGT2B17* promoter can be upregulated by overexpressed HNF1α in HepG2 cells, and mutation of the HNF1 element prevents this increase. However, the HNF1 site mutation does not diminish the basal activity of the *UGT2B17* promoter in HepG2 cells (Gregory and Mackenzie, 2002). A similar phenomenon is seen for promoters of the *UGT1A8-10* cluster in Caco-2 cells (Gregory *et al.*, 2004). In contrast, mutation of the HNF1-binding element in the *UGT2B7* promoter not only abolishes HNF1-mediated upregulation of the transfected promoter in HepG2 cells, but also decimates basal activity of the reporter construct (Ishii *et al.*, 2000). The rat *UGT1A7* promoter demonstrates another variation in response, where transfection of HNF1α into HepG2 cells or primary hepatocytes does not result in an increase in promoter activity. However, mutation of the HNF1 site decreases rat *UGT1A7* promoter basal activity in these cell types. Because this promoter is HNF1 responsive in HEK293 cells, the lack of upregulation in HepG2 cells and hepatocytes by transiently transfected HNF1α is presumably due to sufficient expression of endogenous HNF1 factors to saturate this particular reporter system (Metz *et al.*, 2000). The rat *UGT1A6* promoter is similar, with a reduced or nonexistent response to introduced HNF1α or HNF1β in HepG2 cells and primary rat hepatocytes, respectively, compared to similarly treated HNF1-negative cells (Auyeung *et al.*, 2003).

Variation between Reporter Systems and Other Issues

As discussed earlier, the response of promoters to overexpressed HNF1α or HNF1β can depend on whether there is already sufficient HNF1 protein expression in the recipient cell to reach maximal activity. Likewise, the basal activity of HNF1-responsive promoters may vary as a consequence of changes in HNF1 protein expression. HNF1α and HNF1β levels in a given cell line are not necessarily static and may rely on the tissue culture conditions used. The change in HNF1α expression in Caco-2 cells depending on confluence is one such example. We have also noted that the HNF1 content of HepG2 cells seems to fluctuate considerably between laboratories (Auyeung *et al.*, 2003; Gregory *et al.*, 2004; Jover *et al.*, 1998; Metz *et al.*, 2000). Western blot analysis, such as the one illustrated in Fig. 3, can give some idea of the relative HNF1 content of a cell line.

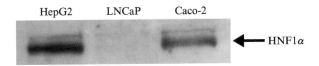

FIG. 3. HNF1α Western blot. HepG2, LNCaP, and Caco-2 nuclear abstracts were probed for the presence of HNF1α protein as described in the text.

The species from which overexpressed HNF1 proteins are cloned may also affect the results of cotransfection experiments. Our laboratory achieved greater than 50-fold upregulation of the rat *UGT2B1* promoter with murine HNF1α in HepG2 cells (Hansen *et al.*, 1997), but this was not replicated by Metz and co-workers (2000) when using rat HNF1α. In addition, we have observed that mouse and human HNF1α isoforms are not always equal in their ability to upregulate our UGT promoter constructs either independently or in concert with other transcription factors such as Cdx2 (unpublished observations).

One final complication that also needs to be kept in mind can occur when performing cotransfection experiments with HNF1-expressing vectors and reporter constructs containing very short promoter elements. We have found that empty pGL3 basic vector or pGL3 plasmids with small inserts are stimulated by HNF1α in the absence of an introduced HNF1 site. Bernard *et al.* (1999) have also reported this issue with the pBLCAT3 vector. To avoid spurious upregulation of reporter genes due to interaction of HNF1 factors with vector sequences, mutation of an HNF1-site in the context of a longer promoter is recommended over deletion constructs, particularly when the targeted element is very close to the transcription start site.

Electrophoretic Mobility-Shift Assays

Figure 4 shows an example of an HNF1 EMSA generated with HepG2 nuclear extracts. These cells express both HNF1α and HNF1β; therefore HNF1 probes incubated with this extract will typically form three complexes that migrate as distinct bands on a PAGE gel (Rey-Campos *et al.*, 1991). Because of its truncated activation domain, the HNF1β protein is sufficiently smaller than HNF1α to allow separation of HNF1α homodimers, HNF1α/HNF1β heterodimers, and HNF1β homodimers based on differences in their mobility. Monoclonal HNF1α antibodies will complex with either the HNF1α homodimers or the HNF1α/HNF1β heterodimers, further slowing their migration during electrophoresis. Similarly, antibodies to HNF1β are able to supershift the HNF1α/HNF1β heterodimers and HNF1β homodimers, as shown by Fig. 4.

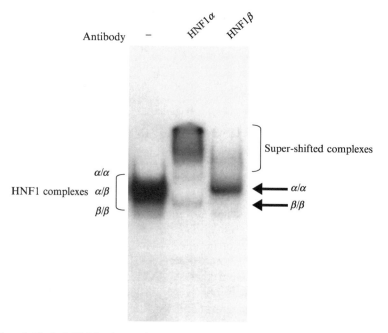

FIG. 4. Typical HNF1 electrophoretic mobility-shift assay with HNF1α and HNF1β supershifts. HNF1α and HNF1β proteins bind equally well to HNF1 elements. When both are present in a nuclear extract they form three distinct complexes with the DNA probe, which have differing mobility when subjected to PAGE analysis. HNF1α has a greater molecular weight than HNF1β, thus HNF1α homodimers migrate through the gel more slowly than complexes containing the smaller HNF1β. Antibodies against HNF1α and HNF1β each specifically supershift two of the three complexes, such that the only remaining band consists of homodimers of the opposite isoform.

If additional transcription factors or coactivators complex with the bound HNF1 proteins or compete for binding on the HNF1 site, it may also be possible to confirm this using EMSA. A *UGT2B7* HNF1 probe used by Ishii and co-workers (2000) forms a HNF1 complex with HepG2 nuclear proteins that also contains Oct-1. This was demonstrated by supershift of the entire complex with an antibody raised to Oct-1 (Ishii *et al.*, 2000). Competitive binding between HNF1 and other factors is exemplified by a *UGT2B17* probe that exclusively binds either HNF1α or Pbx2 complexes. Mutations that abolish binding of either factor allow increased binding of the other (Gregory and Mackenzie, 2002). However, it should also be noted that not all interactions that occur in the cellular environment can be replicated in an EMSA. In some cases, DNA sequences beyond the limits of the probe may influence protein binding, whereas in others the protein–protein bonds formed may simply not be stable enough to survive electrophoresis. For example, although physical association of Cdx2 and HNF1α can be demonstrated by GST pull-down experiments, EMSA conditions appear to preclude observation of this interaction (Mitchelmore *et al.*, 2000).

Final Remarks

Evidence that HNF1 transcription factors play a regulatory role in the expression of many human UGT isoforms, as well as those of several rodent species, is becoming increasingly strong. However, there is still plenty of scope for research that addresses the nature and importance of these roles. Many unanswered questions remain, not only regarding the molecular mechanisms of *UGT* regulation by HNF1, but also concerning clinical relevance. Do individuals with inheritable HNF1 diseases, such as maturity-onset diabetes of the young (MODY3 and MODY5), have perturbed glucuronidation profiles? If so, can these patients benefit clinically from taking this into account? Can other transcription factors compensate for insufficient HNF1 expression in MODY and how? Do non-MODY variants of HNF1α or HNF1β affect UGT expression? Are there individuals with mutations in the HNF1 sites of any of the HNF1-responsive *UGT* genes and, if so, what are the consequences? Do changes in HNF1α expression during inflammatory liver disease (Geier *et al.*, 2003) contribute to the associated repression of UGT mRNA transcription reported by Congiu *et al.* (2002)?

While the techniques described in this chapter will continue to yield valuable insights into the relationship among HNF1α, HNF1β, and *UGT* regulation, they are likely to add little to our understanding of the chromatin-dependent functions of the HNF1 transcription factors and their mutants (Hiesberger *et al.*, 2005) on these genes. Models that mimic more closely the chromosomally integrated *UGT* gene will eventually provide the answers to the aforementioned questions, and many others besides.

Acknowledgment

This work was supported by grants from the National Health and Research Council of Australia. PIM is a NHMRC Senior Principal Research Fellow.

References

Akiyama, T. E., and Gonzalez, F. J. (2003). Regulation of P450 genes by liver-enriched transcription factors and nuclear receptors. *Biochim. Biophys. Acta* **1619**, 223–234.
Archer, T. K., Lefebvre, P., Wolford, R. G., and Hager, G. L. (1992). Transcription factor loading on the MMTV promoter: A bimodal mechanism for promoter activation. *Science* **255**, 1573–1576.

Auyeung, D. J., Kessler, F. K., and Ritter, J. K. (2001). An alternative promoter contributes to tissue- and inducer-specific expression of the rat UDP-glucuronosyltransferase 1A6 gene. *Toxicol. Appl. Pharmacol.* **174**, 60–68.

Auyeung, D. J., Kessler, F. K., and Ritter, J. K. (2003). Differential regulation of alternate UDP-glucuronosyltransferase 1A6 gene promoters by hepatic nuclear factor-1. *Toxicol. Appl. Pharmacol.* **191**, 156–166.

Bach, I., Mattei, M. G., Cereghini, S., and Yaniv, M. (1991). Two members of an HNF1 homeoprotein family are expressed in human liver. *Nucleic Acids Res.* **19**, 3553–3559.

Bach, I., Pontoglio, M., and Yaniv, M. (1992). Structure of the gene encoding hepatocyte nuclear factor 1 (HNF1). *Nucleic Acids Res.* **20**, 4199–4204.

Barbacci, E., Chalkiadaki, A., Masdeu, C., Haumaitre, C., Lokmane, L., Loirat, C., Cloarec, S., Talianidis, I., Bellanne-Chantelot, C., and Cereghini, S. (2004). HNF1beta/TCF2 mutations impair transactivation potential through altered co-regulator recruitment. *Hum. Mol. Genet.* **13**, 3139–3149.

Bernard, P., Goudonnet, H., Artur, Y., Desvergne, B., and Wahli, W. (1999). Activation of the mouse TATA-less and human TATA-containing UDP-glucuronosyltransferase 1A1 promoters by hepatocyte nuclear factor 1. *Mol. Pharmacol.* **56**, 526–536.

Boudreau, F., Zhu, Y., and Traber, P. G. (2001). Sucrase-isomaltase gene transcription requires the hepatocyte nuclear factor-1 (HNF-1) regulatory element and is regulated by the ratio of HNF-1 alpha to HNF-1 beta. *J. Biol. Chem.* **276**, 32122–32128.

Chen, Q. K., Hertz, G. Z., and Stormo, G. D. (1995). MATRIX SEARCH 1.0: A computer program that scans DNA sequences for transcriptional elements using a database of weight matrices. *Comput. Appl. Biosci.* **11**, 563–566.

Congiu, M., Mashford, M. L., Slavin, J. L., and Desmond, P. V. (2002). UDP glucuronosyltransferase mRNA levels in human liver disease. *Drug Metab. Dispos.* **30**, 129–134.

Costa, R. H., Kalinichenko, V. V., Holterman, A. X., and Wang, X. (2003). Transcription factors in liver development, differentiation, and regeneration. *Hepatology* **38**, 1331–1347.

Dallas, P. B., Yaciuk, P., and Moran, E. (1997). Characterization of monoclonal antibodies raised against p300: Both p300 and CBP are present in intracellular TBP complexes. *J. Virol.* **71**, 1726–1731.

De Simone, V., De Magistris, L., Lazzaro, D., Gerstner, J., Monaci, P., Nicosia, A., and Cortese, R. (1991). LFB3, a heterodimer-forming homeoprotein of the LFB1 family, is expressed in specialized epithelia. *EMBO J.* **10**, 1435–1443.

Dohda, T., Kaneoka, H., Inayoshi, Y., Kamihira, M., Miyake, K., and Iijima, S. (2004). Transcriptional coactivators CBP and p300 cooperatively enhance HNF-1alpha-mediated expression of the albumin gene in hepatocytes. *J. Biochem. (Tokyo)* **136**, 313–319.

Eskinazi, R., Thony, B., Svoboda, M., Robberecht, P., Dassesse, D., Heizmann, C. W., Van Laethem, J. L., and Resibois, A. (1999). Overexpression of pterin-4a-carbinolamine dehydratase/dimerization cofactor of hepatocyte nuclear factor 1 in human colon cancer. *Am. J. Pathol.* **155**, 1105–1113.

Farre, D., Roset, R., Huerta, M., Adsuara, J. E., Rosello, L., Alba, M. M., and Messeguer, X. (2003). Identification of patterns in biological sequences at the ALGGEN server: PROMO and MALGEN. *Nucleic Acids Res.* **31**, 3651–3653.

Frain, M., Hardon, E., Ciliberto, G., and Sala-Trepat, J. M. (1990). Binding of a liver-specific factor to the human albumin gene promoter and enhancer. *Mol. Cell. Biol.* **10**, 991–999.

Frech, K., Herrmann, G., and Werner, T. (1993). Computer-assisted prediction, classification, and delimitation of protein binding sites in nucleic acids. *Nucleic Acids Res.* **21**, 1655–1664.

Geier, A., Dietrich, C. G., Voigt, S., Kim, S. K., Gerloff, T., Kullak-Ublick, G. A., Lorenzen, J., Matern, S., and Gartung, C. (2003). Effects of proinflammatory cytokines on rat organic anion transporters during toxic liver injury and cholestasis. *Hepatology* **38**, 345–354.

Gregori, C., Porteu, A., Lopez, S., Kahn, A., and Pichard, A. L. (1998). Characterization of the aldolase B intronic enhancer. *J. Biol. Chem.* **273**, 25237–25243.

Gregori, C., Porteu, A., Mitchell, C., Kahn, A., and Pichard, A. L. (2002). In vivo functional characterization of the aldolase B gene enhancer. *J. Biol. Chem.* **277**, 28618–28623.

Gregory, P. A., Gardner-Stephen, D. A., Lewinsky, R. H., Duncliffe, K. N., and Mackenzie, P. I. (2003). Cloning and characterization of the human UDP-glucuronosyltransferase 1A8, 1A9, and 1A10 gene promoters: Differential regulation through an initiator-like region. *J. Biol. Chem.* **278**, 36107–36114.

Gregory, P. A., Hansen, A. J., and Mackenzie, P. I. (2000). Tissue-specific differences in the regulation of the UDP glucuronosyltransferase 2B17 gene promoter. *Pharmacogenetics* **10**, 809–820.

Gregory, P. A., Lewinsky, R. H., Gardner-Stephen, D. A., and Mackenzie, P. I. (2004). Coordinate regulation of the human UDP-glucuronosyltransferase 1A8, 1A9, and 1A10 genes by hepatocyte nuclear factor 1alpha and the caudal-related homeodomain protein 2. *Mol. Pharmacol.* **65**, 953–963.

Gregory, P. A., and Mackenzie, P. I. (2002). The homeodomain Pbx2-Prep1 complex modulates hepatocyte nuclear factor 1alpha-mediated activation of the UDP-glucuronosyltransferase 2B17 gene. *Mol. Pharmacol.* **62**, 154–161.

Hansen, A. J., Lee, Y. H., Gonzalez, F. J., and Mackenzie, P. I. (1997). HNF1 alpha activates the rat UDP glucuronosyltransferase UGT2B1 gene promoter. *DNA Cell Biol.* **16**, 207–214.

Hiesberger, T., Shao, X., Gourley, E., Reimann, A., Pontoglio, M., and Igarashi, P. (2005). Role of the hepatocyte nuclear factor-beta (HNF-beta) C-terminal domain in Pkhd1 (ARPKD) gene transcription and renal cystogenesis. *J. Biol. Chem.* **280**, 10578–10586.

Hu, C., and Perlmutter, D. H. (1999). Regulation of alpha1-antitrypsin gene expression in human intestinal epithelial cell line caco-2 by HNF-1alpha and HNF-4. *Am. J. Physiol.* **276**, G1181–G1194.

Ishii, Y., Hansen, A. J., and Mackenzie, P. I. (2000). Octamer transcription factor-1 enhances hepatic nuclear factor-1alpha-mediated activation of the human UDP glucuronosyltransferase 2B7 promoter. *Mol. Pharmacol.* **57**, 940–947.

Ishiyama, T., Kano, J., Minami, Y., Iijima, T., Morishita, Y., and Noguchi, M. (2003). Expression of HNFs and C/EBP alpha is correlated with immunocytochemical differentiation of cell lines derived from human hepatocellular carcinomas, hepatoblastomas and immortalized hepatocytes. *Cancer Sci.* **94**, 757–763.

Jover, R., Bort, R., Gomez-Lechon, M. J., and Castell, J. V. (1998). Re-expression of C/EBP alpha induces CYP2B6, CYP2C9 and CYP2D6 genes in HepG2 cells. *FEBS Lett.* **431**, 227–230.

Kanou, M., Usui, T., Ueyama, H., Sato, H., Ohkubo, I., and Mizutani, T. (2004). Stimulation of transcriptional expression of human UDP-glucuronosyltransferase 1A1 by dexamethasone. *Mol. Biol. Rep.* **31**, 151–158.

Ktistaki, E., and Talianidis, I. (1997). Modulation of hepatic gene expression by hepatocyte nuclear factor 1. *Science* **277**, 109–112.

Kuo, C. J., Conley, P. B., Hsieh, C. L., Francke, U., and Crabtree, G. R. (1990). Molecular cloning, functional expression, and chromosomal localization of mouse hepatocyte nuclear factor 1. *Proc. Natl. Acad. Sci. USA* **87**, 9838–9842.

Lee, Y. H., Sauer, B., and Gonzalez, F. J. (1998). Laron dwarfism and non-insulin-dependent diabetes mellitus in the Hnf-1alpha knockout mouse. *Mol. Cell. Biol.* **18**, 3059–3068.

Lenhard, B., Sandelin, A., Mendoza, L., Engstrom, P., Jareborg, N., and Wasserman, W. W. (2003). Identification of conserved regulatory elements by comparative genome analysis. *J. Biol.* **2**, 13.

Liu, S. Y., and Gonzalez, F. J. (1995). Role of the liver-enriched transcription factor HNF-1 alpha in expression of the CYP2E1 gene. *DNA Cell Biol.* **14**, 285–293.

Locker, J., Ghosh, D., Luc, P. V., and Zheng, J. (2002). Definition and prediction of the full range of transcription factor binding sites: The hepatocyte nuclear factor 1 dimeric site. *Nucleic Acids Res.* **30**, 3809–3817.

Lockwood, C. R., and Frayling, T. M. (2003). Combining genome and mouse knockout expression data to highlight binding sites for the transcription factor HNF1alpha. *In Silico Biol.* **3**, 57–70.

Marinescu, V. D., Kohane, I. S., and Riva, A. (2005). The MAPPER database: A multi-genome catalog of putative transcription factor binding sites. *Nucleic Acids Res.* **33**, D91–D97.

Matys, V., Fricke, E., Geffers, R., Gossling, E., Haubrock, M., Hehl, R., Hornischer, K., Karas, D., Kel, A. E., Kel-Margoulis, O. V., Kloos, D. U., Land, S., Lewicki-Potapov, B., Michael, H., Munch, R., Reuter, I., Rotert, S., Saxel, H., Scheer, M., Thiele, S., and Wingender, E. (2003). TRANSFAC: Transcriptional regulation, from patterns to profiles. *Nucleic Acids Res.* **31**, 374–378.

Mendel, D. B., Hansen, L. P., Graves, M. K., Conley, P. B., and Crabtree, G. R. (1991). HNF-1 alpha and HNF-1 beta (vHNF-1) share dimerization and homeo domains, but not activation domains, and form heterodimers in vitro. *Genes Dev.* **5**, 1042–1056.

Mendel, D. B., Khavari, P. A., Conley, P. B., Graves, M. K., Hansen, L. P., Admon, A., and Crabtree, G. R. (1991). Characterization of a cofactor that regulates dimerization of a mammalian homeodomain protein. *Science* **254**, 1762–1767.

Metz, R. P., Auyeung, D. J., Kessler, F. K., and Ritter, J. K. (2000). Involvement of hepatocyte nuclear factor 1 in the regulation of the UDP-glucuronosyltransferase 1A7 (UGT1A7) gene in rat hepatocytes. *Mol. Pharmacol.* **58**, 319–327.

Mitchelmore, C., Troelsen, J. T., Spodsberg, N., Sjostrom, H., and Noren, O. (2000). Interaction between the homeodomain proteins Cdx2 and HNF1alpha mediates expression of the lactase-phlorizin hydrolase gene. *Biochem. J.* **346**, 529–535.

Mouchel, N., Henstra, S. A., McCarthy, V. A., Williams, S. H., Phylactides, M., and Harris, A. (2004). HNF1alpha is involved in tissue-specific regulation of CFTR gene expression. *Biochem. J.* **378**, 909–918.

Nakshatri, H., Nakshatri, P., and Currie, R. A. (1995). Interaction of Oct-1 with TFIIB: Implications for a novel response elicited through the proximal octamer site of the lipoprotein lipase promoter. *J. Biol. Chem.* **270**, 19613–19623.

Odom, D. T., Zizlsperger, N., Gordon, D. B., Bell, G. W., Rinaldi, N. J., Murray, H. L., Volkert, T. L., Schreiber, J., Rolfe, P. A., Gifford, D. K., Fraenkel, E., Bell, G. I., and Young, R. A. (2004). Control of pancreas and liver gene expression by HNF transcription factors. *Science* **303**, 1378–1381.

Parrizas, M., Maestro, M. A., Boj, S. F., Paniagua, A., Casamitjana, R., Gomis, R., Rivera, F., and Ferrer, J. (2001). Hepatic nuclear factor 1-alpha directs nucleosomal hyperacetylation to its tissue-specific transcriptional targets. *Mol. Cell. Biol.* **21**, 3234–3243.

Pogge von Strandmann, E., Senkel, S., and Ryffel, G. U. (1998). The bifunctional protein DCoH/PCD, a transcription factor with a cytoplasmic enzymatic activity, is a maternal factor in the rat egg and expressed tissue specifically during embryogenesis. *Int. J. Dev. Biol.* **42**, 53–59.

Pontoglio, M., Barra, J., Hadchouel, M., Doyen, A., Kress, C., Bach, J. P., Babinet, C., and Yaniv, M. (1996). Hepatocyte nuclear factor 1 inactivation results in hepatic dysfunction, phenylketonuria, and renal Fanconi syndrome. *Cell* **84**, 575–585.

Pontoglio, M., Faust, D. M., Doyen, A., Yaniv, M., and Weiss, M. C. (1997). Hepatocyte nuclear factor 1alpha gene inactivation impairs chromatin remodeling and demethylation of the phenylalanine hydroxylase gene. *Mol. Cell. Biol.* **17,** 4948–4956.

Quandt, K., Frech, K., Karas, H., Wingender, E., and Werner, T. (1995). MatInd and MatInspector: New fast and versatile tools for detection of consensus matches in nucleotide sequence data. *Nucleic Acids Res.* **23,** 4878–4884.

Rey-Campos, J., Chouard, T., Yaniv, M., and Cereghini, S. (1991). vHNF1 is a homeo-protein that activates transcription and forms heterodimers with HNF1. *EMBO J.* **10,** 1445–1457.

Rhee, K. H., Stier, G., Becker, P. B., Suck, D., and Sandaltzopoulos, R. (1997). The bifunctional protein DCoH modulates interactions of the homeodomain transcription factor HNF1 with nucleic acids. *J. Mol. Biol.* **265,** 20–29.

Rollini, P., Xu, L., and Fournier, R. E. (1999). Partial activation of gene activity and chromatin remodeling of the human 14q32.1 serpin gene cluster by HNF-1 alpha and HNF-4 in fibroblast microcell hybrids. *Somat. Cell Mol. Genet.* **25,** 207–221.

Ryffel, G. U. (2001). Mutations in the human genes encoding the transcription factors of the hepatocyte nuclear factor (HNF)1 and HNF4 families: Functional and pathological consequences. *J. Mol. Endocrinol.* **27,** 11–29.

Sandelin, A., Alkema, W., Engstrom, P., Wasserman, W. W., and Lenhard, B. (2004). JASPAR: An open-access database for eukaryotic transcription factor binding profiles. *Nucleic Acids Res.* **32,** D91–D94.

Sandelin, A., Wasserman, W. W., and Lenhard, B. (2004). ConSite: Web-based prediction of regulatory elements using cross-species comparison. *Nucleic Acids Res.* **32,** W249–W252.

Schreiber, E., Matthias, P., Muller, M. M., and Schaffner, W. (1989). Rapid detection of octamer binding proteins with 'mini-extracts,' prepared from a small number of cells. *Nucleic Acids Res.* **17,** 6419.

Schrem, H., Klempnauer, J., and Borlak, J. (2002). Liver-enriched transcription factors in liver function and development.I. The hepatocyte nuclear factor network and liver-specific gene expression. *Pharmacol. Rev.* **54,** 129–158.

Schug, J., and Overton, G. C. (1997). TESS: Transcription Element Search Software on the WWW. Technical Report CBIL-TR-1997–1001-v0.0..

Shih, D. Q., Bussen, M., Sehayek, E., Ananthanarayanan, M., Shneider, B. L., Suchy, F. J., Shefer, S., Bollileni, J. S., Gonzalez, F. J., Breslow, J. L., and Stoffel, M. (2001). Hepatocyte nuclear factor-1alpha is an essential regulator of bile acid and plasma cholesterol metabolism. *Nature Genet.* **27,** 375–382.

Song, Y. H., Ray, K., Liebhaber, S. A., and Cooke, N. E. (1998). Vitamin D-binding protein gene transcription is regulated by the relative abundance of hepatocyte nuclear factors 1alpha and 1beta. *J. Biol. Chem.* **273,** 28408–28418.

Soutoglou, E., Papafotiou, G., Katrakili, N., and Talianidis, I. (2000). Transcriptional activation by hepatocyte nuclear factor-1 requires synergism between multiple coactivator proteins. *J. Biol. Chem.* **275,** 12515–12520.

Soutoglou, E., Viollet, B., Vaxillaire, M., Yaniv, M., Pontoglio, M., and Talianidis, I. (2001). Transcription factor-dependent regulation of CBP and P/CAF histone acetyltransferase activity. *EMBO J.* **20,** 1984–1992.

The ENCODE Project Consortium (2004). The ENCODE (ENCyclopedia Of DNA Elements) project. *Science* **306,** 636–640.

The Rat Genome Sequencing Project Consortium (2004). Genome sequence of the brown Norway rat yields insights into mammalian evolution. *Nature* **428,** 493–521.

Toide, K., Takahashi, Y., Yamazaki, H., Terauchi, Y., Fujii, T., Parkinson, A., and Kamataki, T. (2002). Hepatocyte nuclear factor-1alpha is a causal factor responsible for interindividual differences in the expression of UDP-glucuronosyltransferase 2B7 mRNA in human livers. *Drug Metab. Dispos.* **30,** 613–615.

Tronche, F., Ringeisen, F., Blumenfeld, M., Yaniv, M., and Pontoglio, M. (1997). Analysis of the distribution of binding sites for a tissue-specific transcription factor in the vertebrate genome. *J. Mol. Biol.* **266,** 231–245.

Tronche, F., and Yaniv, M. (1992). HNF1, a homeoprotein member of the hepatic transcription regulatory network. *Bioessays* **14,** 579–587.

Tsunoda, T., and Takagi, T. (1999). Estimating transcription factor bindability on DNA. *Bioinformatics* **15,** 622–630.

Tukey, R. H., and Strassburg, C. P. (2001). Genetic multiplicity of the human UDP-glucuronosyltransferases and regulation in the gastrointestinal tract. *Mol. Pharmacol.* **59,** 405–414.

Vorachek, W. R., Steppan, C. M., Lima, M., Black, H., Bhattacharya, R., Wen, P., Kajiyama, Y., and Locker, J. (2000). Distant enhancers stimulate the albumin promoter through complex proximal binding sites. *J. Biol. Chem.* **275,** 29031–29041.

[3]   Substrate Specificity of Human Hepatic Udp-Glucuronosyltransferases

By Brian Burchell, David J. Lockley, Adam Staines, Yoshihiro Uesawa, and Michael W. H. Coughtrie

Abstract

Five human hepatic UDP-glucuronosyltransferases (UGTs) catalyze the facilitated excretion of more than 90% of drugs eliminated by glucuronidation. The substrate specificity of these UGTs has been examined using cloned expressed enzymes and liquid chromatography–mass spectrometry assays to determine the intrinsic clearance of drug glucuronidation *in vitro*. Specific substrates for the five individual UGTs have been identified. These five probe substrates could be used to predict the drug clearance catalyzed by individual UGTs *in vivo*.

Introduction

UDP-glucuronosyltransferases (UGTs) are a major family of hepatic microsomal enzymes responsible for catalyzing the conjugation of glucuronic acid with a wide range of xeno- and endobiotics, including many drugs (Dutton, 1980). These UGTs are expressed from multiple genes that are specifically regulated by a battery of transcriptional factors (see other chapters in this volume).

Determination of the substrate specificity of these enzymes has been an important task since the existence of multiple UGT isoforms was proposed

METHODS IN ENZYMOLOGY, VOL. 400   0076-6879/05 $35.00

(Mulder, 1971) and established through enzyme purification more than 30 years ago (Sanchez and Tephly, 1974). Purification of UGTs from tissue microsomal fractions was (and remains) exceedingly difficult, and the poor retention of functional activity and incomplete separation of the different isoforms prevented a clear description of their substrate specificities (Burchell, 1977; Coughtrie et al., 1987; Weatherill and Burchell, 1980).

Cloning and functional expression of individual UGTs in mammalian cell lines some 20 years ago provided the essential systems to define substrate specificity (Fournel-Gigleux et al., 1991; Jackson et al., 1985). Even so, the variety of analytical methodologies used in different laboratories has caused extensive debate about the true substrate specificity of individual UGTs; this is probably only defined for a few isoforms accepting endogenous substrates (Burchell et al., 1997). The use of xenobiotics to define the specificity of glucuronidation often confuses the investigator, although some UGTs cannot be characterized easily with an endogenous compound. Compiling data from many laboratories without consideration of enzyme kinetics and analytical technique may produce a very confusing outcome (Tukey and Strassburg, 2000).

UGTs are key enzymes in the metabolism of many drugs, carcinogens, and other xenobiotics, and therefore it is important to try to predict metabolic clearance of drugs by glucuronidation to avoid toxicity in the development of new chemical entities (Burchell, 2003; Burchell et al., 2000; Ethell et al., 2001). We have attempted to predict the specificity of human drug clearance by glucuronidation using the protocol outlined in Fig. 1.

Genetic variation and polymorphisms of UGTs have potential to result in adverse drug reactions or increased risk of cancer (Burchell, 2003). The most familiar example is the TA repeat polymorphism in the UGT1A1 TATA box where the variant allele UGT1A1*28 causes the mild hyperbilirubinemia of Gilbert's disease (Bosma et al., 1995; Monaghan et al., 1996). Certain drugs in development, such as the antiallergic and angiogenesis inhibitor Tranilast, may cause increased bilirubin levels in a percentage of the population identified with the UGT1A1*28 mutation (TA7/TA7) (Danoff et al., 2004). However, toxicity from drug–drug and drug–endogenous compound interactions due to inhibition of glucuronidation is relatively rare in vivo. The low affinity of xenobiotic substrates and drug inhibitors for UGTs and the existence of multiple forms of UGT explain the relatively small impact on the AUCi/AUC ratio as a result of inhibition of glucuronidation (Williams et al., 2004).

There are two subfamilies of human UGTs involved in drug glucuronidation: UGT1 and UGT2 comprising a total of 18 isoenzymes (see Owens et al., 2005). Individual isoforms in each subfamily are differentially expressed in different tissues; for example, UGTs 1A8 and 1A10 are

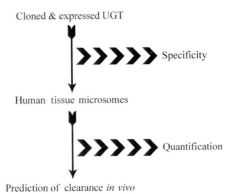

FIG. 1. Prediction of drug clearance by glucuronidation.

preferentially expressed in intestine, and UGTs 1A1 and 1A4 are *not* expressed in kidney (Soars *et al.*, 2001b). This chapter focuses on the determination of substrate specificity of the major drug-glucuronidating UGTs expressed in human liver.

Studies of the glucuronidation of drugs by humans reviewed by Williams *et al.* (2004) have demonstrated that seven hepatic UGT enzymes are responsible for drug conjugation. Figure 2 shows that 40% of drugs are glucuronidated by UGT2B7, with UGTs 1A1, 1A4, and 1A9 equally responsible for a further 47% of hepatic drug glucuronidation. Therefore, we focused our attention on five major human hepatic UGTs.

Methods

UGT Preparations Used for Enzyme Assays

Microsomes are prepared from frozen human tissue (UK Human Tissue Bank, Leicester, UK) by standard methods (Coughtrie *et al.*, 1991). Cell lines (V79 Chinese hamster lung fibroblasts) expressing human UGTs 1A1, 1A6, 1A9, and 2B7 have been created previously in this laboratory (Fournel-Gigleux *et al.*, 1989, 1991; Harding *et al.*, 1988; Wooster *et al.*, 1991). UGT1A4 expressed in the baculovirus/Sf9 insect cell system was kindly donated by Robert Tukey (La Jolla, CA). UGT-expressing cells are grown, and cell lysates prepared, as described previously (Fournel-Gigleux *et al.*, 1991), and cell membranes are disrupted by sonication. Frozen cell pellets (originally from two 15-cm-diameter tissue culture plates) are thawed and resuspended in 200 μl phosphate-buffered saline (PBS) at pH 7.4. The 200 μl suspension is sonicated for 5×5 s (MSE Soniprep 150,

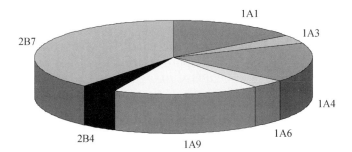

FIG. 2. Relative contributions to the glucuronidation of drugs by human liver UGTs. Based on data reported by Williams *et al.* (2004), but specifically modified to focus on hepatic glucuronidation. (See color insert.)

Sanyo Gallenkamp) on ice, with 1 min cooling on ice between bursts. Tissue microsomes are diluted in PBS to give 5–10 mg/ml and sonicated as described earlier. Protein concentrations are determined postsonication by the method of Lowry *et al.* (1951).

Drug Glucuronidation Assays

The substrate (5 m*M* final, from stock solution in dimethyl sulfoxide) is added to the assay mixture containing 350 µg sonicated microsomes (or 200 µg sonicated cell lysate) in 100 m*M* Tris/maleate buffer (pH 7.5) containing 5 m*M* MgCl$_2$. The reaction is initiated by the addition of UDPGA (10 m*M*, from 100 m*M* stock in PBS) to give a final reaction volume of 100 µl. Samples (assayed in triplicate) are incubated between 30 min and 3 h at 37° depending on substrate and UGT enzyme; a time course experiment with human liver microsomes indicated that the reaction is linear up to 3 h, depending on the drug substrate used (data not shown). The enzyme reactions are terminated by the addition of 100 µl cold (−20°) acetonitrile to the incubation mixtures. Samples are frozen at −20° for 20 min, thawed, and centrifuged at 14,000*g* for 5 min. The supernatant is removed and stored at −20° until analysis for the presence of glucuronide conjugate.

Gradient HPLC Assay of UGTs Using [^{14}C]UPDGA and Radiochemical Detection

To ease quantification when no glucuronide standard is available and when high activities are observed (e.g., for probe substrates), it may be preferable to use HPLC with radiochemical detection. Each frozen cell pellet is resuspended in approximately 200 µl of PBS, followed by homogenization in the presence of alamethicin (50 µg/mg protein). UGT assays with the substrates listed in Table I are performed as described previously (Ethell *et al.*, 1998). Briefly, 100 mM Tris/maleate buffer, pH 7.4, containing 5 mM magnesium chloride, 10 mM saccharic acid 1,4-lactone, 1 mM concentration of each substrate, approximately 1 mg of cellular homogenate protein, and 2 mM UDPGA (0.1 Ci of [^{14}C]UPDGA/assay) are combined in a total volume of 100 μl. Assays are prewarmed at 37° for 5 min, and the reaction is started with the addition of [^{14}C]UDPGA. Reactions containing cell homogenates of each recombinant UGT isoform are incubated at 37° for 2 h, and then the reaction is terminated by the addition of 100 μl of methanol that has been prechilled to −20°. The mixture is centrifuged at 14,000g for 10 min at room temperature. The resulting supernatant is then transferred to an HPLC vial, and 150 μl is injected directly onto the HPLC column. [^{14}C]UDPGA and glucuronide are detected using Reeve 9701 radioactivity monitors (Reeve Analytical, Glasgow, UK) fitted with a 200-μl solid scintillant flow cell packed with silanized cerium-activated lithium glass.

Sample Analysis by LC/MS-MS

Samples are analyzed by LC/MS-MS using a HP1100 LC system (Agilent Technologies, Stockport, UK) connected to a Micromass LC Quattro mass spectrometer (Waters, Manchester, UK) with a 10-μl injection volume per run. The LC separation uses a mobile phase of 0.1% (v/v) aqueous

TABLE I
ANALYSIS OF SUBSTRATE INTRINSIC CLEARANCE BY UGTs EXPRESSED IN BACULOVIRUS-INFECTED Sf9 CELL LYSATE[a]

Substrate	Intrinsic clearance—V_{max}/K_m (ml min^{-1} mg^{-1} × 10^{-3})				
	UGT1A1	UGT1A4	UGT1A6	UGT1A9	UGT2B7
Bilirubin	**5.5**	ND[b]	ND	ND	ND
17α-Ethinylestradiol	**1.2 (66)**	ND	ND	1.1 (179)	0.4 (488)
2-Aminobiphenyl	ND	**1.5 (67)**	ND	1.6 (171)	ND
Amitryptyline	ND	**0.7 (88)**	ND	ND	0.7 (230)
1-Naphthol	0.9 (209)	ND	**11.8 (88)**	3.5 (524)	3.0 (329)
Paracetamol	0.01 (10)	ND	**0.1 (2.3)**	0.06 (25)	ND
Propofol	ND	ND	ND	**18.9**	ND
Tolcapone	ND	ND	ND	**9.1 (72)**	0.8 (331)
Hyodeoxycholic acid	ND	ND	ND	ND	**2.5**
Morphine	ND	ND	ND	ND	**0.4**

[a] Values in parentheses indicate K_m (μM) except for paracetamol (mM).
[b] Below the level of detection of the assay method (approximately 8 pmol/min/mg).

formic acid (buffer A) and acetonitrile containing 0.1% formic acid (v/v) (buffer B). LC separation and elution are achieved using a 1-min isocratic segment at 5% buffer B followed by a gradient of 5–100% buffer B over 4 min. This is followed by a 2-min wash phase at 95% buffer B and a 3-min reequilibration step at 5% buffer B. Separations are performed with a Waters Spherisorb (ODS2) 2 μm, 2.1 × 150-mm column at a flow rate of 0.3 ml/min with a 2-cm Hypersil (ODS) guard column.

Mass spectral analysis is performed by direct infusion into the electrospray source, with column diversion during the first 2.5 min to protect the source from excessive salt. The glucuronide peak from the LC column is analyzed using a multiple reaction monitoring (MRM) method in positive ion mode.

Drug Glucuronide Quantification

Drug glucuronide standards are not usually available and have to be synthesized to facilitate enzyme assay and quantification. Here we use the assay of carbamazepine (CBZ) glucuronide to illustrate the process. The carbamazepine glucuronide standard was not available and, due to lack of sensitivity, we could not quantify it using a parallel incubation performed in the presence of [^{14}C]UDPGA as described previously (Staines *et al.*, 2004a). In order to quantitate the glucuronide produced, two identical CBZ glucuronide standards are generated by parallel incubation of CBZ with rabbit liver microsomes and UDPGA, using the conditions described earlier, in a final volume of 500 μl. CBZ is extracted from the samples by addition of an equal volume of chloroform, followed by mixing for 10 s. The chloroform layer containing the CBZ is removed and this process is repeated twice. For one sample, the aqueous layer, containing the carbamazepine glucuronide, is subjected to alkaline lysis at 50° for 1 h by the addition of 250 μl of 30% (v/v) ammonia, followed by neutralization with 130 μl formic acid. The parallel control sample has 380 μl of 1 *M* ammonium acetate added. This experiment is performed in triplicate, and samples are prepared for LC-MS/MS analysis as before.

Quantification is achieved by analysis of the aforementioned samples for both carbamazepine and carbamazepine glucuronide. This is performed using the same LC conditions as described previously. Detection is with a dual MRM method in positive ion mode with dual transitions at 413.3 > 237.0 (cone voltage 25 eV, capillary 3.0 eV, collision energy 15 eV) and at 237.3 > 194.1 (cone voltage 40 eV, capillary 3.0 eV, collision energy 20 eV) for CBZ glucuronide and CBZ, respectively; collision gas is at 3 mbar. The relative decrease in the CBZ glucuronide between hydrolyzed and nonhydrolyzed samples is analyzed with respect to the consequent increase in carbamazepine levels, which is used to calculate the difference in ionization potential between them. A standard curve for each experiment is subsequently generated using a CBZ standard from 0.1 to 200 μM. Levels of CBZ glucuronide are then extrapolated from the standard curve using the relative difference in ionization potential between the CBZ glucuronide and the CBZ (Staines *et al.*, 2004a).

Determination of Kinetic Parameters

Kinetic parameters for drug glucuronide formation are determined using the standard assay described earlier, performed in duplicate, with variations in the concentration of drug (final concentrations 0, 5, 10, 20, 50, 75, 100, 200, 350, 500, 750, 1000, 5000, 10,000 μM). Standard Michaelis–Menten curves are used to calculate the kinetic parameters (Kaleidagraph, Synergy Software, Reading, PA).

Results

Determination of Substrate Specificity of UGTs

Analysis of substrate specificity *in vitro* by the methods described in the previous section demands a complete kinetic analysis of UGT activity using saturating concentration of the cosubstrate UDP-glucuronic acid (10 mM). Determination of K_m and intrinsic clearance *in vitro* of cloned expressed UGTs provides the best indication of aglycone substrate specificity in the absence of purified enzymes. K_{cat} would be of better value, but the difficulty of purification, lipid environment, and stability of UGTs prevents this analysis.

Table I demonstrates the determination of *in vitro* intrinsic clearance for 10 substrates using five cloned and expressed human hepatic UGTs. These data indicate that UGT1A1 primarily catalyzes the glucuronidation of bilirubin and 17α-ethinylestradiol; UGT1A4 is the major isoform using 2-aminobiphenyl; UGT1A6 has a significant role in the glucuronidation of 1-naphthol at low substrate concentration; UGT1A9 specifically catalyzes the glucuronidation of propofol and tolcapone; and UGT2B7 specifically glucuronidates morphine and hyodeoxycholic acid.

Our analysis of the substrate specificity of drug glucuronidation examined four drugs of varied structure (see Fig. 3). The specificity of glucuronidation of these drugs was determined using *in vitro* intrinsic clearance measurements with seven cloned and expressed human hepatic UGTs. Data in Table II clearly indicate that UGT2B7 was solely responsible for

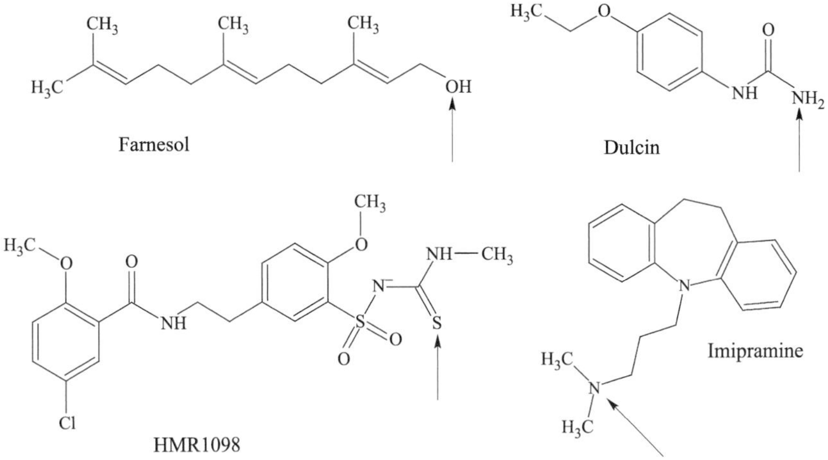

FIG. 3. Chemical structures of four substrates for human hepatic UGTs.

TABLE II
IDENTIFICATION OF INDIVIDUAL UGTs IN DRUG GLUCURONIDATION

UGT isoform	In vitro clearance (nl/mg/min)			
	Carbamazepine	Dulcin	HMR1098	Farnesol
1A1	0	10.3	853	710
1A3	–	0	–	23
1A4	0	0	0	207
1A6	0	0	0	44
1A9	0	15.5	189	2600
2B4	0	0	0	6509
2B7	3.7	0	0	19,156
Human liver microsomes	15	90	545	84,747

the catalysis of hepatic glucuronidation of carbamazepine (Staines *et al.*, 2004a) and is a major UGT involved in the glucuronidation of farnesol (Staines *et al.*, 2004b). However, UGT1A1 is primarily responsible for the *S*-glucuronidation of HMR1098 (Ethell *et al.*, 2003), whereas UGT1A9 and UGT1A1 are both significant contributors to the catalysis of the glucuronidation of the artificial sweetener dulcin (Uesawa *et al.*, 2004).

Discussion

Comparison of tissue microsomal drug glucuronidation catalyzed by a variety of species shows significant differences (Soars *et al.*, 2001a). The predictive value of animal drug glucuronidation for drug development is hence debatable. The identification of individual UGTs responsible for drug glucuronidation using *in vitro* intrinsic clearance is therefore essential in facilitating the prediction of *in vivo* drug clearance by glucuronidation. More recently, key roles for particular UGTs in the synthesis of drug glucuronides have been identified using LC-MS-MS assay technology. Thus, individual drug glucuronidation can be directly linked to individual UGTs. This process has enabled the identification of probe substrates to support the application of UGT specificity to support drug development (see Court, 2005).

Table III shows some probe substrates for five major hepatic UGTs. Serotonin has been determined to be a specific substrate for UGT1A6 (Krishnaswamy *et al.*, 2003), and morphine-6-glucuronidation is highly selective for UGT2B7 (Stone *et al.*, 2003). Other selective probe substrates identified are etoposide for UGT1A1 (Watanabe *et al.*, 2003) and tamoxifen *N*-glucuronide for UGT1A4 (Kaku *et al.*, 2004). Obviously, not all of these substrates are applicable to the *in vivo* investigation of UGT specificity, but all are important for characterization and normalization of the function of individual UGTs.

Knowledge of individual UGTs and quantitation of these enzymes are required to predict drug clearance by glucuronidation *in vivo* (see Fig. 1). Quantitation of UGTs in human tissues poses a significant technical problem due to the lack of highly specific antibodies and purified UGT

TABLE III
UGT PROBE SUBSTRATES

UGT isoform	Probe substrate
UGT1A1	Bilirubin
UGT1A4	Amitriptyline
UGT1A6	Serotonin
UGT1A9	Propofol
UGT2B7	Morphine, carbamazepine

standards. We have developed a novel technique to quantitate human UGTs by preparing monospecific anti-UGT antibodies using multiple antigenic peptides (Milne *et al.*, 2004a) and S-tag fusion proteins (Milne *et al.*, 2004b). Quantitation of UGT1A1 in human liver microsomes showed a 26-fold interindividual variation in the amount of protein and level of enzyme activity. However, there is a good correlation between the quantity of immune reactive UGT1A1 and bilirubin UGT activity (Milne *et al.*, 2004b). These predictions of drug clearance by glucuronidation *in vivo* are possible using the process described in Fig. 1.

Acknowledgment

This work was supported in part by the Wellcome Trust.

References

Bosma, P. J., Roy Chowdhury, J., Bakker, C., Gantla, S., De Boer, A., Oostra, B. A., Lindhout, D., Tytgat, G. N. J., Jansen, P. L. M., Oude Elferink, R. P. J., and Roy Chowdhury, N. (1995). The genetic basis of the reduced expression of bilirubin UDP-glucuronoysltransferase in Gilbert's syndrome. *N. Eng. J. Med.* **333,** 1171–1175.

Burchell, B. (1977). Studies on the purification of rat liver uridine diphosphate glucuronyltransferase. *Biochem. J.* **161,** 543–549.

Burchell, B. (2003). Genetic variation of human UDP-glucuronosyltransferase: Implications in disease and drug glucuronidation. *Am. J. Pharmacogen.* **3,** 37–52.

Burchell, B., McGurk, K., Brierley, C. H., and Clarke, D.J (1997). UDP-glucuronosyltransferases. *In* "Comprehensive Toxicology: Biotransformation" (F. P. Guengerich, ed.). Pergamon/Elsevier Science, Amsterdam.

Burchell, B., Soars, M., Monaghan, G., Cassidy, A., Smith, D., and Ethell, B. (2000). Drug-mediated toxicity caused by genetic deficiency of UDP-glucuronosyltransferases. *Toxicol. Lett.* **112–113,** 333–340.

Coughtrie, M. W. H., Blair, J. N. R., Hume, R., and Burchell, A. (1991). Improved preparation of hepatic microsomes for *in vitro* diagnosis of inherited disorders of the glucose-6-phosphatase system. *Clin. Chem.* **37,** 739–742.

Coughtrie, M. W. H., Blair, J. N. R., Hume, R., and Burchell, A. (1987). Purification and properties of rat kidney UDP-glucuronosyltransferase. *Biochem. Pharmacol.* **36,** 245–251.

Court, M. H. (2005). Isoform-selective probe substrates for *in vitro* studies of human UDP-glucuronosyltransferases. *Methods Enzymol* **400**[7] this volume.

Danoff, T. M., Campbell, D. A., McCarthy, L. C., Lewis, K. F., Repasch, M. H., Saunders, A. M., Spurr, N. K., Purvis, I. J., Roses, A. D., and Xu, C. F. (2004). A Gilbert's syndrome UGT1A1 variant confers susceptibility to tranilast-induced hyperbilirubinemia. *Pharmacogen. J.* **4,** 49–53.

Dutton, G. J. (1980). "Glucuronidation of Drugs and Other Compounds." CRC Press, Boca Raton, FL.

Ethell, B. T., Anderson, G. D., Beaumont, K., Rance, D. J., and Burchell, B. (1998). A universal radiochemical high-performance liquid chromatographic assay for the determination of UDP-glucuronosyltransferase activity. *Anal. Biochem.* **255,** 142–147.

Ethell, B. T., Beaumont, K., Rance, D. J., and Burchell, B. (2001). Use of cloned and expressed human UDP-glucuronosyltransferases for the assessment of human drug conjugation and identification of potential drug interactions. *Drug Metab. Dispos.* **29**, 48–53.

Ethell, B. T., Riedel, J., Englert, H., Jantz, H., Oekonomopulos, R., and Burchell, B. (2003). Glucuronidation of HMR1098 in human microsomes: Evidence for the involvement of UGT1A1 in the formation of *S*-glucuronides. *Drug Metab. Dispos.* **31**, 1027–1034.

Fournel-Gigleux, S., Jackson, M. R., Wooster, R., and Burchell, B. (1989). Expression of a human liver cDNA encoding a UDP-glucuronosyltransferase catalysing the glucuronidation of hyodeoxycholic acid in cell culture. *FEBS Lett.* **243**, 119–122.

Fournel-Gigleux, S., Sutherland, L., Sabolovic, N., Burchell, B., and Siest, G. (1991). Stable expression of two human UDP-glucuronosyltransferase cDNAs in V79 cell cultures. *Mol. Pharmacol.* **39**, 177–183.

Harding, D., Fournel-Gigleux, S., Jackson, M. R., and Burchell, B. (1988). Cloning and substrate specificity of a human phenol UDP-glucuronosytransferase expressed in COS-7 cells. *Proc. Natl. Acad. Sci. USA* **85**, 8381–8385.

Jackson, M. R., McCarthy, L. R., Corser, R. B., Barr, G. C., and Burchell, B. (1985). Cloning of cDNAs coding for rat hepatic microsomal UDP-glucuronyltransferases. *Gene* **34**, 147–153.

Kaku, T., Ogura, K., Nishiyama, T., Ohnuma, T., Muro, K., and Hiratsuka, A. (2004). Quaternary ammonium-linked glucuronidation of tamoxifen by human liver microsomes and UDP-glucuronosyltransferase 1A4. *Biochem. Pharmacol.* **67**, 2093–2102.

Krishnaswamy, S., Duan, S. X., von Moltke, L. L., Greenblatt, D. J., and Court, M. H. (2003). Validation of serotonin (5-hydroxtryptamine) as an *in vitro* substrate probe for human UDP-glucuronosyltransferase (UGT) 1A6. *Drug Metab. Dispos.* **31**, 133–139.

Lowry, O. H., Rosebrough, N. J., Farr, A. L., and Randall, R. J. (1951). Protein measurement with the Folin phenol reagent. *J. Biol. Chem.* **193**, 265–275.

Milne, A. M., Burchell, B., and Coughtrie, M. W. H. (2004a). Design of isoform-specific anti-UGT antibodies. *Drug Metab. Rev.* **36**(Suppl. 1), 130.

Milne, A. M., Burchell, B., and Coughtrie, M. W. H. (2004b). Quantification of UDP-glucuronosyltransferase expression in human tissues. *Drug Metab. Rev.* **36**(Suppl. 1), 130.

Monaghan, G., Ryan, M., Seddon, R., Hume, R., and Burchell, B. (1996). Genetic variation in bilirubin UPD-glucuronosyltransferase gene promoter and Gilbert's syndrome. *Lancet* **347**, 578–581.

Mulder, G. J. (1971). The heterogeneity of uridine diphosphate glucuronyltransferase from rat liver. *Biochem. J.* **125**, 9–15.

Owens, I. S., Basu, N. B., and Banerjee, R. (2005). Gene structures of *UGT1* and *UGT2* families. *Methods Enzymol.* **400**[1] this volume.

Sanchez, E., and Tephly, T. R. (1974). Evidence for separate enzymes in the glucuronidation of morphine and p-nitrophenol by rat hepatic microsomes. *Drug Metab. Dispos.* **2**, 247–253.

Soars, M. G., Riley, R. J., and Burchell, B. (2001a). Evaluation of the marmoset as a model species for drug glucuronidation. *Xenobiotica* **31**, 849–860.

Soars, M. G., Riley, R. J., Findlay, K. A. B., Coffey, M. J., and Burchell, B. (2001b). Evidence for significant differences in microsomal drug glucuronidation by canine and human liver and kidney. *Drug Metab. Dispos.* **29**, 121–126.

Staines, A. G., Coughtrie, M. W. H., and Burchell, B. (2004a). *N*-glucuronidation of carbamazepine in human tissues is mediated by UGT2B7. *J. Pharmacol. Exp. Ther.* **311**, 1131–1137.

Staines, A. G., Sindelar, P., Coughtrie, M. W. H., and Burchell, B. (2004b). Farnesol is glucuronidated in human liver, kidney and intestine *in vitro*, and is a novel substrate for UGT2B7 and UGT1A1. *Biochem. J.* **384,** 637–645.

Stone, A. N., Mackenzie, P. I., Galetin, A., Houston, J. B., and Miners, J. O. (2003). Isoform selectivity and kinetics of morphine 3- and 6-glucuronidation by human UDP-glucuronosyltransferases: Evidence for atypical glucuronidation kinetics by UGT2B7. *Drug Metab. Dispos.* **31,** 1086–1089.

Tukey, R. H., and Strassburg, C. P. (2000). Human UDP-glucuronosyltransferases: Metabolism, expression and disease. *Annu. Rev. Pharmacol. Toxicol.* **40,** 581–616.

Uesawa, Y., Staines, A. G., O'Sullivan, A., Mohri, K., and Burchell, B. (2004). Identification of the rabbit liver UDP-glucuronosyltransferase catalyzing the glucuronidation of 4-ethoxyphenylurea (dulcin). *Drug Metab. Dispos.* **32,** 1476–1481.

Watanabe, Y., Nakajima, M., Ohashi, N., Kume, T., and Yokoi, T. (2003). Glucuronidation of etoposide in human liver microsomes is specifically catalyzed by UDP-glucuronosyltransferase 1A1. *Drug Metab. Dispos.* **31,** 589–595.

Weatherill, P. J., and Burchell, B. (1980). The separation and purification of rat liver UDP glucuronyltransferase activities towards testosterone and oestrone. *Biochem. J.* **189,** 377–380.

Williams, J. A., Hyland, R., Jones, B. C., Smith, D. A., Hurst, S., Goosen, T. C., Peterkin, V., Koup, J. R., and Ball, S. E. (2004). Drug-drug interactions for UDP-glucuronosyltransferase substrates: A pharmacokinetic explanation for typically observed low exposure (AUCi/AUC) ratios. *Drug Metab. Dispos.* **32,** 1201–1208.

Wooster, R., Sutherland, L., Ebner, T., Clarke, D., Da Cruz e Silva, O., and Burchell, B. (1991). Cloning and stable expression of a new member of the human liver phenol/bilirubin: UDP-glucuronosyltransferase cDNA family. *Biochem. J.* **278,** 465–469.

[4] UDP-Glucuronosyltransferase 1A6: Structural, Functional, and Regulatory Aspects

By Karl Walter Bock and Christoph Köhle

Abstract

Glucuronidation, catalyzed by two families of UDP-glucuronosyltransferases (UGTs), represents a major phase II reaction of endo- and xenobiotic biotransformation. UGT1A6 is the founding member of the rat and human UGT1 family. It is expressed in liver and extrahepatic tissues, such as intestine, kidney, testis, and brain, and conjugates planar phenols and arylamines. Serotonin has been identified as a selective endogenous substrate of the human enzyme. UGT1A6 is also involved in conjugation of the drug paracetamol (acetaminophen) and of phenolic metabolites of benzo[a]pyrene (together with rat UGT1A7 and human UGT1A9). High interindividual variability of human liver protein levels is due to a number of influences, including genetic, tissue-specific, and environmental factors.

METHODS IN ENZYMOLOGY, VOL. 400   0076-6879/05 $35.00
DOI: 10.1016/S0076-6879(05)00004-2

Evidence shows that homo- and heterozygotic expression of UGT1A6 alleles markedly affects enzyme activity. HNF1 may be responsible for tissue-specific UGT1A6 expression. Multiple environmental factors controlling UGT1A6 expression have been identified, including the pregnane X receptor, the constitutive androstane receptor, the aryl hydrocarbon receptor, and Nrf2, a bZIP transcription factor mediating stress responses. However, marked differences have been noted in the expression of rat and human UGT1A6. Regulatory factors have been studied in detail in the human Caco-2 colon adenocarcinoma cell model.

Introduction

Glucuronidation represents a major route of detoxification and elimination for many xenobiotics and endobiotics. The conjugation reaction is catalyzed by a superfamily of UDP-glucuronosyltransferases (UGTs) (>16 human UGTs), which evolved as two enzyme families in mammals. UGT1A6 is the founding member of the rat and human UGT1A family (Harding et al., 1988; Iyanagi et al., 1986; Mackenzie et al., 1997; Tukey and Strassburg, 2000). It is the most studied of these enzymes and may serve as a paradigm. The isoform is known as phenol UGT, conjugating planar phenols (4-nitrophenol, 1-naphthol, 4-methylumbelliferone, etc.), and is expressed ubiquitously in tissues, in contrast to isoforms UGT1A7–UGT1A10, which conjugate bulky + planar phenols and are restricted in tissue-dependent expression (Tukey and Strassburg, 2000). This chapter discusses structural aspects only briefly, as comprehensive reviews have been published elsewhere (Ouzzine et al., 2003; Radominska-Pandya et al., 1999). Functions of UGT1A6 are discussed using paracetamol, serotonin, and environmental carcinogens as examples. Regulatory aspects are described in more detail, including genetic and endogenous factors responsible for tissue-dependent expression. In addition, UGT1A6 expression is modulated by hormones, drugs, and other xenobiotics that serve as ligands for multiple sensors, including the pregnane X receptor (PXR), the constitutive androstane receptor (CAR) , the aryl hydrocarbon (Ah) receptor, and transcription factor Nrf2, which responds to oxidative/electrophile stress.

Structural Aspects

The domain structure of UGTs revealed that these type I transmembrane proteins reside mostly, including the active site, at the luminal side of the endoplasmic reticulum (ER). This compartmentation explains the ''latency'' of UGT activity, i.e., the observation that microsomal enzyme activity can be activated within seconds by the addition of detergents or pore-forming agents such as alamethicin, allowing unrestricted access of the cofactor UDP-glucuronic acid to the active site (Ouzzine *et al.*, 2003; Radominska-Pandya *et al.*, 1999). In ER membranes, specific transporters for UDP-glucuronic acid control access of the cofactor to the active site. Accumulating evidence suggests that lipophilic aglycones, which are concentrated in ER membranes, have direct access to the active site (Bock and Lilienblum, 1994; Hauser *et al.*, 1988). In addition, evidence shows that UGTs act as dimers, thereby affecting their substrate specificity (Ishii *et al.*, 2001; Meech and Mackenzie, 1997). In the case of diphenol diglucuronide formation, a tetrameric structure may be required, as discussed later (Gschaidmeier and Bock, 1994; Peters *et al.*, 1984). More work is needed to understand the influence of multimeric structures on the substrate specificity of UGTs, particularly human UGT1A6 on cooperative, non-Michaelis–Menten enzyme kinetics of substrates, such as 1-naphthol (Uchaipichat *et al.*, 2004).

Functions of UGT1A6

Pharmacological and physiological aspects are discussed first, exemplified by paracetamol and serotonin. Then, roles of UGT1A6 in detoxification of carcinogens is summarized using arylamines and aryl hydrocarbons as examples.

Paracetamol Glucuronidation

Due to its wide use as an analgesic drug and its intensive pharmacokinetic characterization, paracetamol (acetaminophen) was, in early studies, selected as a probe drug to screen for the interindividual variation of the hepatic glucuronidation capacity of patients under the influence of inducing agents. Using this "paracetamol test," glucuronidation was found to be enhanced in patients treated with phenobarbital-type inducers, such as phenytoin and rifampicin, and in heavy smokers (individuals exposed to Ah-type inducers) (Bock *et al.*, 1987), confirming previous clinical pharmacological observations. A strong correlation was found in male heavy smokers between CYP1A2-mediated caffeine oxidation and paracetamol glucuronidation, suggesting the existence of Ah receptor-induced UGTs in humans (Bock *et al.*, 1987).

Using human UGT1A6-expressing cells, UGT1A6 was identified as the relatively high-affinity enzyme paracetamol glucuronidation (K_m 2 mM) and UGT1A9 (the major phenol UGT conjugating both bulky + planar phenols in human liver) as a low-affinity enzyme (K_m 50 mM) (Bock *et al.*, 1993).

UGT1A1 has been identified as a third major paracetamol-conjugating enzyme in human liver (Court *et al.*, 2001). It represents an intermediate affinity enzyme (K_m 9 mM). Comparing the three isoforms UGT1A6 is the most active at low paracetamol concentrations (<50 μM, a concentration reached in serum after the ingestion of 1000 mg paracetamol = two tablets). Under these conditions the relative contribution of the three major UGT enzymes involved in paracetamol conjugation was estimated to be 61% (UGT1A9), 29% (UGT1A6), and 10% (UGT1A1) (Court *et al.*, 2001). It is surprising that UGT1A9 as a low-affinity enzyme appears to be mainly responsible for paracetamol glucuronidation. This observation may be better explained once the relative levels of the three enzymes (UGT1A1, UGT1A6, and UGT1A9) in liver microsomes are known and cooperative enzyme kinetics in oligomeric enzyme complexes are better understood.

Serotonin as an Endogenous Substrate of UGT1A6

Astonishingly, serotonin was identified as a highly selective substrate of UGT1A6 in human liver microsomes (Krishnaswamy *et al.*, 2003). Serotonin glucuronidation could not be detected in cat liver (Leakey, 1978), in a species that expresses UGT1A6 as a pseudogene (Court and Greenblatt, 2000). Serotonin is known as an important neurotransmitter in brain. However, the largest quantity of serotonin in the body (ca. 95%) is found in the gastrointestinal tract where serotonin is stored in enterochromaffin cells (Erspamer, 1966). Evidence shows that the intestinal peristaltic reflex leading to ascending contraction and descending relaxation is initiated by serotonin released from enterochromaffin cells during the passage of intestinal contents (Grider, 2003). Early findings suggested that serotonin glucuronidation (in addition to oxidation by monoamine oxidase) represents an important compensatory catabolic pathway of this neurotransmitter in mice: While normally 30% of injected serotonin was excreted in urine as serotonin glucuronide, the amount increased to 70% when monoamine oxidase was inhibited by iproniazid (Weissbach *et al.*, 1961). To better understand a possible role of UGT1A6 in intestinal serotonin homeostasis, we studied serotonin disposition in Caco-2 cell monolayers, which differentiate to enterocyte-like cells during culture and are widely used as a model of the human intestinal epithelium (Hidalgo *et al.*, 1989). Disposition of serotonin in the model of Caco-2 cell monolayers was studied similar to investigations of 4-methylumbelliferone disposition (Bock-Hennig *et al.*, 2002). Evidence was obtained that serotonin was mostly taken up and secreted as serotonin glucuronide at the basolateral surface of Caco-2 cell monolayers; within 24 h, 35% of the absorbed serotonin was secreted as serotonin glucuronide (Köhle *et al.*, 2005). In support of a role of UGT1A6 in serotonin homeostasis in brain, UGT1A6 expression was detected in human and rat brain, mostly in the hippocampus and cerebellum (Brands *et al.*, 2000; King *et al.*, 1999; Martinasevic *et al.*, 1998). These observations in intestinal and brain cells hint at a homeostatic function of human UGT1A6 in serotonin catabolism.

Roles of UGT1A6 in Detoxification of Carcinogenic Arylamines and Aryl Hydrocarbons

Arylamines such as 2-naphthylamine and 4-aminobiphenyl are long known carcinogens (Miller and Miller, 1981). Using cell-expressed rat and human UGT1A6, it could be demonstrated that *N*-hydroxy derivatives of these carcinogens are better substrates of UGT1A6 than free amines (Orzechowsky *et al.*, 1994), possibly protecting the liver from these reactive metabolites. However, glucuronidation provides transport forms of these ultimate bladder carcinogens (Miller and Miller, 1981). Similarly, glucuronidation by UGT1A6 of phenol and hydroquinone (major myelotoxic metabolites of the carcinogen benzene) could be demonstrated (Schrenk *et al.*, 1996). Preferential glucuronidation of these benzene metabolites compared to sulfation in mice has been suggested to explain in part the relative resistance of the rat to benzene toxicity.

Metabolism and carcinogenicity of benzo[a]pyrene (BaP), the prototype of environmental carcinogenic aryl hydrocarbons , has been studied intensely (Conney, 1982). It was demonstrated that BaP-7,8-diol-9,10-epoxide represents a major ultimate carcinogen forming stable DNA adducts. Several UGTs, including rat and human UGT1A7 and UGT1A9 (but not UGT1A6), have been shown to detoxify BaP-7,8-dihydrodiol, an intermediate in the metabolic pathway to the ultimate carcinogen (Zheng *et al.*, 2002). However, UGT1A6 is effective in the detoxification of major BaP metabolites, such as phenols, quinones, and quinols, the latter formed readily from the corresponding quinones by NAD(P)H quinone oxidoreductase = NQO1 (Fig. 1) (Lilienblum *et al.*, 1985). Quinones undergo quinone–quinol redox cycles with the generation of semiquinones and

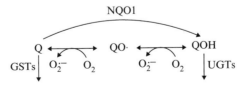

FIG. 1. Detoxification role of NQO1 together with UGTs in quinone–quinol redox cycles. Q, quinone; NQO1, NAD(P)H quinone oxidoreductase 1; GSTs, glutathione *S*-transferases.

reactive oxygen species. The action of NQO1, which transfers two electrons, bypasses the semiquinone step (Lorentzen and Ts'o, 1977) and provides diphenolic substrates of UGT1A6. Quinones may be mainly responsible for the cytotoxic, tumor-promoting actions of BaP (Burdick *et al.*, 2003). With rat liver microsomes and cell-expressed UGT1A6, BaP-3,6-diphenol is converted readily to mono- and diglucuronides by Ah-inducible UGT1A6 (Bock *et al.*, 1992; Gschaidmeier *et al.*, 1995). With human UGT1A6, only monoglucuronides are formed, which, however, can be converted to diglucuronides by UGT1A9 (Gschaidmeier *et al.*, 1995). Diglucuronides of BaP-3,6-quinol have been detected in rat bile (Bevan and Sadler, 1992) and urine (Yang *et al.*, 1999). Furthermore, the mutagenicity of BaP-3,6-quinone and of BaP has been shown to be reduced markedly by the addition of UDP-glucuronic acid to the "Ames test" (Bock *et al.*, 1990a).

BaP-3,6-quinol Diglucuronide Formation as Model to Study Oligomeric Interaction of UGT1A6

Using radiation-inactivation analysis of BaP-3,6-quinol glucuronidation, it was found that monoglucuronides are formed by UGT dimers, whereas diglucuronides are formed by tetramers (Gschaidmeier and Bock, 1994). The oligomeric structure of UGTs in ER membranes of hepatocytes may provide a microenvironment facilitating efficient diglucuronide formation. Microsomal enzyme kinetic studies of diglucuronide formation revealed much lower apparent K_m values (10–20 μM) when diglucuronides are formed from the quinol than when formed from synthesized monoglucuronides (>70 μM), suggesting that monoglucuronides appear to reach high concentrations at the active site under the former condition (Hartung, 1992). Hence, it is tempting to speculate that homo- and heterodimers of rat and human UGT1A6, together with rat UGT1A7 or human UGT1A9, may facilitate the detoxification of Ah diphenols in liver, providing diglucuronides for efficient biliary excretion by export transporters.

Regulation of UGT1A6 Expression

Factors responsible for regulation of UGT expression are beginning to be understood (Mackenzie *et al.*, 2003). Based in part on a high constitutive expression in rat kidney and on a low basal expression of UGT1A6 in rat liver with high inducibility by Ah-type inducers (similar to expression of CYP1A1 in many tissues) (Münzel *et al.*, 1994), two modes of UGT1A6 expression have been proposed: one associated with a high constitutive/lower inducible expression and the other with a low constitutive/high inducible expression (Münzel *et al.*, 1996, 1998). Tissue-specific, high constitutive expression is probably regulated by a number of endogenous factors, including HNF1 and HNF4 (Mackenzie *et al.*, 2003). In human liver, UGT1A6 is constitutively expressed with high interindividual variability in protein levels (Court *et al.*, 2001; Nagar *et al.*, 2004). Gender is one factor affecting the hepatic UGT1A6 level; it is about 50% lower in females (Court *et al.*, 2001), in line with lower metabolic ratios in females in the "paracetamol test" (Bock *et al.*, 1994).

Genetic Factors Regulating UGT1A6 Levels

Allelic variants of the coding region of human UGT1A6 have been described leading to over 85- and 25-fold differences in immunoreactive protein and in enzyme activity, respectively (Nagar *et al.*, 2004). Homodimerization of different alleles may affect UGT activity, similar to observations with UGT1A1 variants (Gosh *et al.*, 2001). In addition, polymorphisms in the promoter region of UGT1A6 are likely to exist and to affect the UGT1A6 protein level, similar to findings with UGT1A9 (Girard *et al.*, 2004). Not only polymorphisms of *cis*-acting response elements, which bind transcription factors, but also polymorphisms of the *trans*-acting transcription factors themselves may be responsible for variable UGT1A6 protein levels. In this way genetic factors may be responsible for polymorphisms in the tissue-dependent expression of UGTs (as discussed for HNF1 in the subsequent paragraph); for example, the number of individuals expressing UGT1A6 was 4/5 in duodenum, 1/5 in jejunum, 6/8 in ileum, and 16/16 in liver (Strassburg *et al.*, 2000).

Homeodomain Protein HNF1 Is, in Part, Responsible for Tissue-Dependent UGT1A6 Expression

Human UGT1A6 is widely expressed in tissues, including intestine, kidney, testis, and brain, whereas UGT1A9 is mainly expressed in liver (Table I). Brain and testis are discussed here in more detail. Based on strong UGT1A6 expression at the hepato-gastrointestinal barrier, it was assumed that the isoform was also present at internal barriers such as the choroid plexus, the blood–brain barrier, and the blood–testis barrier. However, using selective anti-UGT1A6 antibodies and *in situ* hybridization studies, UGT1A6 was detected in hippocampal pyramidal cells and in Purkinje cells, as well as in spermatogonia, primary spermatocytes, and Sertoli cells of rat testis (Brands *et al.*, 2000; King *et al.*, 1999; Martinasevic *et al.*, 1998). In the rat, two phenol UGTs (UGT1A6 and UGT1A7) appear to be largely coregulated in liver and intestine (Metz and Ritter, 1998), whereas in humans, UGT1A7 is not expressed in liver, but in the upper

TABLE I

Tissue Distribution of Human and Rat UGT1A6 in Comparison with Human UGT1A9 and Rat UGT1A7[a]

Tissue	Human		Rat	
	hUGT1A6	hUGT1A9	rUGT1A6	rUGT1A7
Liver	+	+	+	+
Larynx/pharynx	+	−	nd[b]	nd
Esophagus	−	−	+	+
Stomach	+	−	+	+
Duodenum	+	−	+	+
Jejunum/ileum	+	−	+	+
Colon	+	+	+	+
Kidney	+	+	+	+
Lung	+	−	+	nd
Brain	+	−	+	nd
Testis	nd	nd	+	nd
Ovary	nd	nd	+	nd

[a] Tissue-dependent human and rat UGT1A6 expression is described in Tukey and Strassburg (2000) and Grams et al. (2000), respectively. Expression of the brain and testis enzyme is described in the text.
[b] Not determined.

gastrointestinal tract, including the oral cavity, esophagus, stomach, and pancreas (Tukey and Strassburg, 2000).

As already mentioned, factors responsible for tissue-dependent regulation are just beginning to be understood (Mackenzie et al., 2003). Interestingly, alternative promoters (P1 and P2) may contribute to the tissue- and inducer-specific expression of rat UGT1A6 (no second promoter has been found in humans) (Auyeung et al., 2001). P2 contains a HNF1 domain close to the TATA box, similar to the HNF1 domain in the human UGT1A6 gene (Fig. 2). P1 is positioned 3.7 kb upstream of the coding first exon of UGT1A6 and contains one xenobiotic response element (XRE) domain responsible for Ah receptor binding (discussed later) and HNF4 sites (Emi et al., 1996). Despite expressing the same enzyme, the two mRNAs transcribed from P1 and P2 can be distinguished by their unique 5′ sequences (Auyeung et al., 2001): the class 1 transcript from P1 was found to be preferentially expressed in spleen, lung, and ovary, whereas the class 2 transcript from P2 was mostly found in kidney, liver, and intestine. However, Ah-type inducers enhanced both transcripts. In the latter three tissues, transcription factor HNF1 may be involved in tissue-dependent expression, because HNF1 is enriched in these tissues. However, it remains unclear why tissues expressing class 1 transcripts are unresponsive to

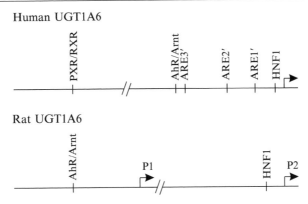

Fɪɢ. 2. Schematic diagram comparing binding sites of transcription factors and ARE-like motifs in the promoter/enhancer domains of human and rat UGT1A6 described in the text. AhR, Ah receptor; Arnt, Ah receptor nuclear translocator; PXR, pregnane X receptor; RXR, retinoid X receptor; HNF1, hepatic nuclear factor 1; ARE′, antioxidant-like response element; P1 and P2, promoters 1 and 2 (Auyeung et al., 2001).

Ah-type inducers. In discussing the roles of HNF1, it is noteworthy that factors such as HNF1 are under the control of a hierarchy of transcription factors, including the key regulator of liver development and differentiation HNF4 (Kamiya et al., 2003; Watt et al., 2003), a member of the steroid receptor supergene family. The role of HNF1 in the hepatic expression of UGT1A6 is supported by the observation that extrahepatic human UGT1A7, UGT1A8, and UGT1A10 do not contain HNF1-binding domains (Mackenzie et al., 2003).

Control of UGT1A6 by Nuclear Receptors PXR and CAR

Evidence was obtained in primary cultures of hepatocytes (the most accepted tool of investigating the potential of an agent to cause enzyme induction in humans) that UGT1A6 is under the control of phenobarbital-type inducers such as carbamazepine, as suggested from a 1.7-fold increase of 1-naphthol glucuronidation (Soars et al., 2004). Using transgenic mice, it was shown that UGT1A6 (and UGT1A1) is under transcriptional control by PXR/retinoid X receptor (RXR) and CAR/RXR heterodimers (Xie et al., 2003). Both PXR and CAR are active as heterodimers with RXR. These nuclear receptors regulate both phase I and II enzymes of drug metabolism, as well as glucuronide transporters in a coordinate manner (Bock and Köhle, 2004; Xie et al., 2003). Transgenic mice expressing a constitutively active human PXR, which had been shown previously to upregulate CYP2B and CYP3A, were used to explore the regulation of UGTs by PXR (Xie et al., 2003). Using Northern blot analysis it was shown that UGT1A6 mRNA was elevated compared to nontransgenic litter-mates. *In vitro*-translated PXR/RXR and CAR/RXR bound to a DR-3 NR domain (Table II) in the promoter region of UGT1A6. Furthermore, exposure of transfected UGT1A6 reporter genes containing the DR-3 NR domain led to over 10-fold induction by a variety of inducers such as St. John's wort and clotrimazole (Xie *et al.*, 2003). Using the information theory, a PXR-binding site was suggested at −9.2 kb of the human UGT1A6 promoter (Vyhlidal *et al.*, 2004).

Ah Receptor Control of UGT1A6

A search for Ah-inducible UGTs led to sequencing of rat UGT1A6, the founding member of the UGT1 family, and to characterization of one functional Ah receptor-binding XRE in its P1 promoter region (see earlier discussion) (Emi *et al.*, 1996; Iyanagi *et al.*, 1986). One functional XRE was also identified in human UGT1A6 using the Caco-2 cell model (discussed later; Fig. 2 and Table II) (Münzel *et al.*, 1998, 2003). The Ah receptor (AhR) represents a ligand-activated transcription factor that acts as a heterodimer with its partner protein Arnt. Both are members of the basic helix–loop–helix/Per-Arnt-Sim family of transcription factors (Gu *et al.*, 2000). In Ah receptor-null mice, UGT1A6 inducibility and, interestingly, its basal expression were abolished (Fernandez-Salguero *et al.*, 1995). As already mentioned, Ah-type induction of UGT1A6 in human hepatocyte cultures appears to be moderate and lower than induction by PXR or CAR (Soars *et al.*, 2004). These observations indicate marked species differences between humans and rats in the regulation of UGT1A6 by Ah-type inducers.

TABLE II

PXR-, Ah Receptor-, and Nrf2-Mediated Induction of UGT1A6: Sequences of DR-3 NR Responsible Element, XRE, and ARE-like Elements[a]

Rat PXR, DR-3 NR	
CYP3A23:	tagac**AGTTCA**tga**AGTTCA**tctaa
UGT1A6:	actgt**AGGTCA**taa**AGTTCA**catgg
Human XRE and ARE-like elements	
ARE1' (forward):	cagaagctcaggtgaagc**TGAC**acg**GC**catagt
ARE2' (reverse):	ttacccacaacttctgtc**TGAC**ttg**GC**aaaaat
XRE + ARE3':	aac**T**c**GCGTG**ccagccaggtgtgca**TGAC**t-a**GC**tctggg
ARE consensus	**TGACnnnGC**
XRE consensus	**TnGCGTG**

[a] Sequences for rat DR-3 (direct hexanucleotide repeat separated by three nucleotide spacer) NR and for human XRE and ARE-like elements were taken from Xie *et al.* (2003) and Münzel *et al.* (2003), respectively. Consensus sequences are shown in bold.

*Key Role of Nrf2 in Oxidative/Electrophile Stress-Mediated
UGT Induction*

Early studies in rats suggested induction of UGT activities by antioxidants (Bock *et al.*, 1980). The induction of human UGT1A6 (and of other UGTs such as UGT1A9 and UGT2B7) has been demonstrated in human colon adenocarcinoma Caco-2 cells (Münzel *et al.*, 1999). We used a high passage cell clone of Caco-2 cells (TC-7) (Caro *et al.*, 1995). Caco-2 cells have been generated from a male patient who was a homozygous carrier of the NQO1*2 allele (Bonnesen *et al.*, 2001), leading to loss of NQO1 activity (Siegel *et al.*, 2001). The homozygous variant is present in 4% of Caucasians and in 20% of the Chinese population (Kelsey *et al.*, 1997). In addition, the donor of Caco-2 cells was a homozygous carrier of a frequent haplotype consisting of three UGT1A variants (UGT1A1*28, responsible for Gilbert's syndrome; UGT1A6*2; and UGT1A7*3, leading to an increased risk of various Ah-induced cancers) (Köhle *et al.*, 2003). Studies demonstrated high levels of UGT1A6 mRNA and of UGT activity in Caco-2/TC-7 cells. Quantitative real-time polymerase chain reaction (RT-PCR) demonstrated that UGT1A6 mRNA was induced about twofold by treatment with the prototypical oxidative/electrophile stress inducer *t*-butylhydroquinone (tBHQ) (Köhle *et al.*, 2005). It was enhanced fourfold by the Ah receptor ligand 2,3,7,8-tetrachlorodibenzo-*p*-dioxin (TCDD) and by the mixed inducer *β*-naphthoflavone (BNF) (Fig. 3). BNF is an Ah receptor ligand that produces electrophile stress following CYP1A1 induction, leading to efficient BNF metabolism to stress-generating electrophiles. tBHQ is assumed to undergo quinone–quinol redox cycles with the production of semiquinones and reactive oxygen species (Fig. 1); the semiquinone is mainly responsible for triggering the stress response (Yu *et al.*, 1997). Compared to UGT1A6, UGT1A1 is poorly expressed in Caco-2/TC7 cells (Fig. 3). As expected, CYP1A1 mRNA is preferentially enhanced by TCDD (ca. 1000-fold), and by tBHQ (ca. 10-fold).

Nuclear factor E2-related factor 2, a bZIP transcription factor (Nrf2), is a key transcription factor mediating stress responses (Chan and Kan, 1999; Ramos-Gomez *et al.*, 2001). Nrf2 is sequestered in the cytoplasm by Keap-1, a sulfhydryl-rich chaperone protein. Oxidation of sulfhydryl groups by oxidative/electrophile stress disrupts the cytoplasmic complex, leading to nuclear translocation of Nrf2, to association with small Maf or other proteins, binding of the heterodimer to antioxidant response elements (AREs), and transmission of the signal to the transcription machinery of phase II enzymes of detoxification (Kwak *et al.*, 2003; Thimmulappa *et al.*, 2002). Nrf2 is a redox-sensitive transcription factor whereby glutathione regulates nuclear translocation and thioredoxin regulates binding to the ARE

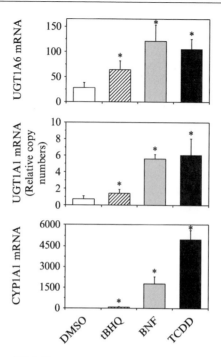

FIG. 3. Quantitative RT-PCR analysis of human UGT1A6 induction by Ah receptor agonists and oxidative/electrophile stress in Caco-2 cells. Cells were grown on 100×20-mm Falcon dishes (Becton Dickinson) in Dulbecco's modified Eagle's medium supplemented with 20% fetal bovine serum (heat inactivated at 56° for 30 min), 25 mM glucose, and 1% nonessential amino acids (Life Technologies). The medium was changed daily. Preconfluent cells were treated with 10 nM TCDD, 50 μM BNF, or 80 μM tBHQ, dissolved in 0.1% dimethyl sulfoxide (DMSO). Solvent controls contained 0.1% DMSO. Exposure time was 24 h. Thereafter, total RNA was isolated and RT-PCR analysis of UGT1A6, UGT1A1, CYP1A1, and cyclophilin B was carried out using the LightCycler (Roche) as described (Köhle et al., 2005). Cyclophilin B was examined as the internal standard. For all quantitative assays, an external calibration curve was used, based on synthesized standards. Relative copy numbers are given per 1000 copies of cyclophilin B. tBHQ, t-butylhydroquinone; BNF, β-naphthoflavone; TCDD, 2,3,7,8-tetrachlorodibenzo-p-dioxin. *$p < 0.05$, compared to DMSO controls.

(Hansen et al., 2004). Selective phase II enzyme induction by Nrf2 is in contrast to Ah receptor-mediated induction, which induces both phase I and II enzymes. In Nrf2-null mice, this nonreceptor-mediated stress response (including UGT1A6 induction by tBHQ) was diminished (Chan and Kan, 1999; Ramos-Gomez et al., 2001). Induction of Nrf2-induced genes has gained much attention because it may influence the susceptibility to carcinogenesis and degenerative diseases dramatically (Chan and Kan, 1999; Kwak et al., 2003; Ramos-Gomez et al., 2001; Thimmulappa et al., 2002).

ARE-like elements (ARE1' and ARE2') have been found in the promoter region of human UGT1A6 (Münzel et al., 2003). However, their functionality has not been proven. In addition, a truncated ARE (ARE3') is present close to the Ah receptor-binding XRE of human UGT1A6 (Table II). Site-directed mutagenesis of this ARE3' abolished the XRE response, suggesting a linkage between ARE and XRE responses. Evidence for linkage between Ah receptor and stress responses has been obtained in 5L cells, a rat hepatoma cell line. In 5L cells, UGT activity of both UGT1A6 and UGT1A7 is inducible by TCDD and tBHQ. However, in a 5L-derived cell line lacking the Ah receptor (BP8), both TCDD and tBHQ induction is lost, whereas after stable transfection of the Ah receptor into BP8 cells, both TCDD and tBHQ induction is regained (Münzel et al., 2003).

The essential role of Nrf2-mediated induction of UGTs, including UGT1A6, has been substantiated in studies on the chemopreventive efficacy of oltipraz (4-methyl-5-[2-pyrazinyl]-1,2-dithiole-3-thione) against urinary bladder carcinogenesis (Iida et al., 2004). The tumor incidence of n-nitrosobutyl(4-hydroxybutyl)amine (BBN) was enhanced significantly in Nrf2-null mice. The antischistosomal drug oltipraz induced BBN glucuronidation significantly, the major detoxification pathway, as well as the expression of UGT1A6 and other UGT1A enzymes. Importantly, BBN was found to suppress the expression of UGT1A enzymes, especially in the urinary bladder. Oltipraz counteracted this suppression and may thus diminish BBN-induced tumors by enhancing detoxification of this carcinogen in liver and the urinary bladder.

Changes of UGT1A6 Expression at Early Stages of Carcinogenesis

As discussed in the previous section, the bladder carcinogen BBN suppresses UGT1A6 and UGT1A7 expression in the urinary bladder (Iida et al., 2004). Similarly, downregulation of some UGTs (with the exception of UGT1A6) has been described in human hepatocarcinogenesis (Strassburg et al., 1997). This is in contrast to a markedly increased expression of UGT1A6 at early stages of rat hepatocarcinogenesis (Bock et al., 1990b; Fischer et al., 1985). However, mechanisms responsible for changes of UGTs and GSTs in hepatocarcinogenesis are unknown.

Conclusions

UGT1A6 can be considered as a paradigm for structural, functional, and regulatory aspects of multiple UGTs. The founding member of the UGT1A family and most-studied UGT enzyme is expressed in liver and most extrahepatic tissues. Glucuronidation of serotonin, a highly selective substrate of human hepatic UGT1A6, in the intestine and brain hints at a homeostatic function of this isoform, in addition to its role in detoxification. The transmembrane protein forms dimers and tetramers in endoplasmic reticulum membranes, with the active site within the lumen. This oligomeric structure may facilitate efficient detoxification of carcinogenic aryl hydrocarbons and other toxins. Multiple factors are responsible for the large interindividual differences in the tissue-dependent expression of UGT1A6, including genetic factors and endogenous factors regulating tissue-specific expression, such as the homeodomain protein HNF1. In addition, UGT1A6 expression is modulated by hormones, drugs, and other xenobiotics, which serve as ligands for multiple sensors, including PXR, CAR, Ah receptor, and transcription factor Nrf2, which responds to oxidative/electrophile stress. Marked species differences in UGT1A6 regulation between humans and rats were discussed.

References

Auyeung, D. J., Kessler, F. K., and Ritter, J. K. (2001). An alternative promoter contributes to tissue- and inducer-specific expression of the rat UDP-glucuronosyltransferase 1A6 gene. *Toxicol. Appl. Pharmacol.* **174,** 60–68.

Bevan, D. R., and Sadler, V. M. (1992). Quinol diglucuronides are predominant conjugated metabolites found in bile of rats following intratracheal instillation of benzo(a)pyrene. *Carcinogenesis* **13,** 403–407.

Bock, K. W., Forster, A., Gschaidmeier, H., Brück, M., Münzel, P., Scharek, W., Fournel-Gigleux, S., and Burchell, B. (1993). Paracetamol glucuronidation by recombinant rat and human phenol UDP-glucuronosyltransferases. *Biochem. Pharmacol.* **45,** 1809–1814.

Bock, K. W., Gschaidmeier, H., Seidel, A., Baird, S., and Burchell, B. (1992). Mono- and diglucuronide formation from chrysene and benzo(a)pyrene phenols by 3-methylcholanthrene-inducible phenol UDP-glucuronosyltransferase (UGT1A1). *Mol. Pharmacol.* **42,** 613–618.

Bock, K. W., Kahl, R., and Lilienblum, W. (1980). Induction of rat hepatic UDP-glucuronosyltransferases by dietary ethoxyquin. *Naunyn-Schmiedeberg's Arch. Pharmacol.* **310,** 249–252.

Bock, K. W., and Köhle, C. (2004). Coordinate regulation of drug metabolism by xenobiotic nuclear receptors: UGTs acting together with CYPs and glucuronide transporters. *Drug Metab. Rev.* **36,** 595–615.

Bock, K. W., and Lilienblum, W. (1994). Roles of uridine diphosphate glucuronosyltransferases in chemical carcinogenesis. *Handbook Exp. Pharmacol.* **112,** 391–428.

Bock, K. W., Lipp, H. P., and Bock-Hennig, B. S. (1990a). Induction of drug metabolizing enzymes by xenobiotics. *Xenobiotica* **20,** 1101–1111.

Bock, K. W., Münzel, P. A., Röhrdanz, E., Schrenk, D., and Eriksson, L. C. (1990b). Persistently increased expression of a 3-methylcholanthrene-inducible phenol uridine diphosphate-glucuronosyltransferase in rat hepatocyte nodules and hepatocellular carcinomas. *Cancer Res.* **50,** 3569–3573.

Bock, K. W., Schrenk, D., Forster, A., Griese, E. U., Mörike, K., Brockmeier, D., and Eichelbaum, M. (1994). The influence of environmental and genetic factors on CYP2D6,

CYP1A2 and UDP-glucuronosyltransferases in man using sparteine, caffeine, and paracetamol as probes. *Pharmacogenetics* **4**, 209–218.

Bock, K. W., Wildfang, J., Blume, R., Ullrich, D., and Bircher, J. (1987). Paracetamol as a test drug to determine glucuronide formation in man: Effects of inducers and smoking. *Eur. J. Clin. Pharmacol.* **31**, 677–683.

Bock-Hennig, B. S., Köhle, C., Nill, K., and Bock, K. W. (2002). Influence of t-butylhydroquinone and ß-naphthoflavone on formation and transport of 4-methylumbelliberone glucuronide in Caco-2/TC-7 cell monolayers. *Biochem. Pharmacol.* **63**, 123–128.

Bonnesen, C., Eggleston, I. M., and Hayes, J. D. (2001). Dietary indoles and isothiocyanates that are generated from cruciferous vegetables can both stimulate apoptosis and confer protection against DNA damage in human colon cell lines. *Cancer Res.* **61**, 6120–6130.

Brands, A., Münzel, P. A., and Bock, K. W. (2000). *In situ* hybridization studies of UDP-glucuronosyltransferase UGT1A6 expression in rat testis and brain. *Biochem. Pharmacol.* **59**, 1441–1444.

Burdick, A. D., Davis, J. W., Liu, K. J., Hudson, L. G., Shi, H., Monske, M. L., and Burchiel, S. W. (2003). Benzo(a)pyrene quinones increase cell proliferation, generate reactive oxygen species, and transactivate the epithelial growth factor receptor in breast epithelial cells. *Cancer Res.* **63**, 7825–7833.

Caro, I, Boulenc, X., Rousste, M., Meunier, V., Bourrie, M., Joyeux, H., Roques, C., Berger, Y., Zweibaum, A., and Fabre, G. (1995). Characterization of a newly isolated Caco-2 clone (TC-7), as a model of transport processes and biotransformation of drugs. *Int. J. Pharm.* **116**, 147–158.

Chan, K., and Kan, Y. W. (1999). Nrf2 is essential for protection against acute pulmonary injury in mice. *Proc. Natl. Acad. Sci. USA* **96**, 12731–12736.

Conney, A. H. (1982). Induction of microsomal enzymes by foreign chemicals and carcinogenesis by polycyclic aromatic hydrocarbons: G.H.A. Clowes memorial lecture. *Cancer Res.* **42**, 4875–4917.

Court, M. C., Duan, S. X., von Moltke, L. L., Greenblatt, D. J., Patten, C. J., Miners, J. O., and Mackenzie, P. I. (2001). Interindividual variability in acetaminophen glucuronidation by human liver microsomes: Identification of relevant acetaminophen UDP-glucuronosyltransferase isoforms. *J. Pharmacol. Exp. Ther.* **299**, 998–1006.

Court, M. H., and Greenblatt, D. J. (2000). Molecular genetic basis for deficient paracetamol glucuronidation in cats: UGT1A6 is a pseudogene and evidence for reduced diversity of expressed hepatic UGT1A isoforms. *Pharmacogenetics* **10**, 355–369.

Emi, Y., Ikushiro, S., and Iyanagi, T. (1996). Xenobiotic responsive element-mediated transcriptional activation in the UDP-glucuronosyltransferase family 1 gene complex. *J. Biol. Chem.* **271**, 3952–3958.

Erspamer, V. (1966). Occurrence of indolealkylamines in nature. *Handbook Exp. Pharmacol.* **19**, 132–181.

Fernandez-Salguero, P., Pineau, T., Hilbert, D. M., McPhail, T., Lee, S. S. T., Kimura, S., Nebert, D. W., Rudikoff, S., Ward, J. M., and Gonzalez, F. J. (1995). Immune system impairment and hepatic fibrosis in mice lacking the dioxin-binding Ah receptor. *Science* **268**, 722–726.

Fischer, G., Ullrich, D., and Bock, K. W. (1985). Effects of N-nitrosomorpholine and phenobarbital on UDP-glucuronosyltransferase in putative preneoplastic foci of rat liver. *Carcinogenesis* **6**, 605–609.

Girard, H., Court, M. H., Bernard, O., Fortier, L. C., Villeneuve, L., Hao, Q, Greenblatt, D. J., von Moltke, L. L., Perussed, I., and Guillemette, C. (2004). Identification of common polymorphisms in the promoter of the UGT1A9 gene: Evidence that UGT1A9 protein and activity levels are strongly genetically controlled in the liver. *Pharmacogenetics* **14,** 501–515.

Gosh, S. S., Sappal, B. S., Kalpana, G. V., Lee, S. W., Roy Chowdhury, J. R., and Roy Chowdhury, N. (2001). Homodimerization of human bilirubin-UDP-uridine-diphosphoglucuronate glucuronosyltransferase-1 (UGT1A1) and its functional implications. *J. Biol. Chem.* **276,** 42108–42115.

Grams, B., Harms, A., Braun, S., Strassburg, C. P., Manns, M. P., and Obermayer-Straub, P. (2000). Distribution and inducibility by 3-methylcholanthrene of family 1 UDP-glucuronosyltransferases in the rat gastrointestinal tract. *Arch. Biochem. Biophys.* **377,** 255–265.

Grider, J. R. (2003). Neurotransmitters mediating the intestinal peristaltic reflex in the mouse. *J. Pharmacol. Exp. Ther.* **307,** 460–467.

Gschaidmeier, H., and Bock, K. W. (1994). Radiation inactivation analysis of microsomal UDP-glucuronosyltransferases catalysing mono- and diglucuronide formation of 3,6-dihydroxybenzo(a)pyrene and 3,6-dihydroxychrysene. *Biochem. Pharmacol.* **48,** 1545–1549.

Gschaidmeier, H., Seidel, A., Burchell, B., and Bock, K. W. (1995). Formation of mono- and diglucuronides and other glycosides of benzo(a)pyrene-3,6-quinol by V79 cell-expressed human phenol UDP-glucuronosyltransferases of the UGT1 gene complex. *Biochem. Pharmacol.* **49,** 1601–1606.

Gu, Y. Z., Hogenesch, J. B., and Bradfield, C. A. (2000). The PAS superfamily: Sensors of environmental and developmental signals. *Annu. Rev. Pharmacol. Toxicol.* **40,** 519–561.

Hansen, J. M., Watson, W. H., and Jones, D. P. (2004). Compartmentation of Nrf2 redox control: Regulation of cytoplasmic activation by glutathione and DNA binding by thioredoxin-1. *Toxicol. Sci.* **82,** 308–317.

Harding, D., Fournel-Gigleux, S., Jackson, M. R., and Burchell, B (1988). Cloning and substrate specificity of a human phenol UDP-glucuronosyltransferase expressed in COS-7 cells. *Proc. Natl. Acad. Sci. USA* **85,** 8381–8385.

Hartung, T. (1992). "Mono- und Diglucuronid-Bildung von Benzpyren-3,6-chinol durch Lebermikrosomen von Ratte und Mensch." Thesis, Med. Faculty University of Tübingen.

Hauser, S. C., Ziurys, J. C., and Gollan, J. L. (1988). A membrane transporter mediates access of uridine 5′-diphosphoglucuronic acid from the cytosol into the endoplasmic reticulum of rat hepatocytes: Implications for glucuronidation reactions. *Biochim. Biophys. Acta* **967,** 149–157.

Hidalgo, I. J., Raub, T. J., and Borchardt, R. T. (1989). Characterization of the human colon carcinoma cell line (Caco-2) as a model system for intestinal permeability. *Gastroenterology* **96,** 736–749.

Iida, K., Itoh, K., Kumagai, Y., Oyasu, R., Hattori, K., Kawai, K., Shimazui, T., Akaza, H., and Yamamoto, M. (2004). Nrf2 is essential for the chemopreventive efficacy of oltipraz against urinary bladder carcinogenesis. *Cancer Res.* **64,** 6424–6431.

Ishii, Y., Mioshi, A., Watanabe, R., Tsuruda, K., Tsuda, M., Yamagushi-Nagamatsu, Y., Yoshisue, K., Tanaka, M., Maji, D., Ohgiya, S., and Oguri, K. (2001). Simultaneous expression of guinea pig UDP-glucuronosyltransferase 2B21 and 2B22 in COS-7 cells enhances UDP-glucuronosyltransferase 2B21-catalyzed morphine-6-glucuronide formation. *Mol. Pharmacol.* **60,** 1040–1048.

Iyanagi, T., Haniu, M., Sogawa, K., Fujii-Kuriyama, Y., Watanabe, S., Shively, J. E., and Anan, K. F. (1986). Cloning and characterization of cDNA encoding 3-methylcholanthrene inducible rat mRNA for UDP-glucuroronosyltransferase. *J. Biol. Chem.* **261,** 15607–15614.

Kamiya, A., Inoue, Y., and Gonzalez, F. J. (2003). Role of the hepatocyte nuclear factor 4α in control of the pregnane X receptor during fetal liver development. *Hepatology* **37,** 1375–1384.

Kelsey, K. T., Wiencke, J. K., Christiani, D. C., Zuo, Z., Spitz, M. R., Xu, X., Lee, B. K., Schwatz, B. S., Traver, R. D., and Ross, D. (1997). Ethnic variation in the prevalence of a common NAD(P)H:Quinone oxidoreductase polymorphism and its implications for anticancer chemotherapy. *Br. J. Cancer* **76**, 852–854.

King, C. D., Rios, G. R., Assouline, J. A., and Tephly, T. R. (1999). Expression of UDP-glucuronosyltransferases (UGTs) 2B7 and 1A6 in the human brain and identification of 5-hydroxytryptamine as a substrate. *Arch. Biochem. Biophys.* **365**, 156–162.

Köhle, C., Badary, O. A., Nill, K., Bock-Hennig, B. S., and Bock, K. W. (2005). Serotonin glucuronidation by Ah receptor- and oxidative stress-inducible human UDP-glucurono-syltransferase (UGT) 1A6 in Caco-2 cells. *Biochem. Pharmacol.* **69**, 1397–1402.

Köhle, C., Möhrle, B., Münzel, P. A., Schwab, M., Wernet, D., Badary, O. A., and Bock, K. W. (2003). Frequent co-occurrence of the TATA box mutation associated with Gilbert's syndrome (UGT1A1*28) with other polymorphisms of the UDP-glucuronosyl-transferase-1 locus (UGT1A6*2 and UGT1A7*3) in Caucasians and Egyptians. *Biochem. Pharmacol.* **65**, 1521–1527.

Krishnaswamy, S., Duan, S. X., von Moltke, L. L., Greenblatt, D. J., and Court, M. H. (2003). Validation of serotonin (5-hydroxytryptamine) as an *in vitro* substrate probe for human UDP-glucuronosyltransferase (UGT) 1A6. *Drug Metab. Disp.* **31**, 133–139.

Kwak, M. K., Wakabayashi, N., Itoh, K., Motohashi, H., Yamamoto, M., and Kensler, T. W. (2003). Modulation of gene expression by cancer chemopreventive dithiolthiones through the keap1-Nrf2 pathway. *J. Biol. Chem.* **278**, 8135–8145.

Leakey, J. (1978). An improved assays technique for UDP-glucuronosyltransferase activity towards 5-hydroxytryptamine and some properties of the enzyme. *Biochem. J.* **175**, 1119–1124.

Lilienblum, W., Bock-Hennig, B. S., and Bock, K. W. (1985). Protection against toxic redox cycles between benzo(a)pyrene-3,6-quinone and its quinol by 3-methylcholanthrene-inducible formation of the quinol mono- and diglucuronide. *Mol. Pharmacol.* **27**, 451–458.

Lorentzen, L. J., and Ts'o, P. O. P. (1977). Benzo(a)pyrenedione/benzo(a)pyrenediol oxidation-reduction couples and the generation of reduced molecular oxygen. *Biochemistry* **16**, 1467–1473.

Mackenzie, P. I., Gregory, P. A., Garner-Stephen, D. A., Lewinsky, R. H., Jorgensen, B. R., Nishiyama, T., Xie, W., and Radominska-Pandya, A. (2003). Regulation of UDP glucuronosyltransferase genes. *Curr. Drug Metab.* **4**, 249–257.

Mackenzie, P. I., Owens, I. S., Burchell, B., Bock, K. W., Bairoch, A., Belanger, A., Fournel-Gigleux, S., Green, M., Hum, D. W., Iyanagi, T., Lancet, D., Louisot, P., Magdalou, J., Roy Chowdhury, J., Ritter, J. K., Schachter, H., Tephly, T. R., Tipton, F. K., and Nebert, D. W. (1997). The UDP glycosyltransferase superfamily: Recommended nomenclature update based on evolutionary divergence. *Pharmacogenetics* **7**, 255–269.

Martinasevic, M. K., King, C. D., Rios, G. R., and Tephly, T. R. (1998). Immunohistochemical localization of UDP-glucuronosyltransferases in rat brain during early development. *Drug Metab. Disp.* **26**, 1039–1041.

Meech, R., and Mackenzie, P. I. (1997). UDP-glucuronosyltransferase, the role of the amino terminus in dimerization. *J. Biol. Chem.* **272**, 26913–26917.

Metz, R. P., and Ritter, J. K. (1998). Transcriptional activation of the UDP-glucuronosyl-transferase 1A7 gene in rat liver by aryl hydrocarbon ligands and oltipraz. *J. Biol. Chem.* **273**, 5607–5614.

Miller, E. C., and Miller, J. A. (1981). Searches for ultimate chemical carcinogens and their reactions with cellular macromolecules. *Cancer* **47**, 2327–2345.

Münzel, P. A., Bookjans, G., Mehner, G., Lehmköster, T., and Bock, K. W. (1996). Tissue-specific 2,3,7,8-tetrachlorodibenzo-p-dioxin-inducible expression of human UDP-glucur-onosyltransferase UGT1A6. *Arch. Biochem. Biophys.* **335**, 205–210.

Münzel, P. A., Brück, M., and Bock, K. W. (1994). Tissue-specific constitutive and inducible expression of rat phenol UDP-glucuronosyltransferase. *Biochem. Pharmacol.* **47,** 1445–1448.

Münzel, P. A., Lehmköster, T., Brück, M., Ritter, J. K., and Bock, K. W. (1998). Aryl hydrocarbon receptor-inducible or constitutive expression of human UDP glucuronosyltransferase UGT1A6. *Arch. Biochem. Biophys.* **350,** 72–78.

Münzel, P. A., Schmohl, S., Buckler, F., Jaehrling, J., Raschko, F. T., Köhle, C., and Bock, K. W. (2003). Contribution of the Ah receptor to the phenolic antioxidant-mediated expression of human and rat UDP-glucuronosyltransferase UGT1A6 in Caco-2 and rat hepatoma 5L cells. *Biochem. Pharmacol.* **66,** 841–847.

Münzel, P. A., Schmohl, S., Heel, H., Kälberer, K., Bock-Hennig, B. S., and Bock, K. W. (1999). Induction of human UDP glucuronosyltransferases (UGT1A6, UGT1A9, and UGT2B7) by t-butylhydroquinone and 2,3,7,8-tetrachlorodibenzo-p-dioxin in Caco-2 cells. *Drug Metab. Disp.* **27,** 569–573.

Nagar, S., Zalatoris, J. J., and Blanchard, R. L. (2004). Human UGT1A6 pharmacogenetics: Identification of a novel SNP, characterization of allele frequencies and functional analysis of recombinant allozymes in human liver tissue and in cultured cells. *Pharmacogenetics* **14,** 487–499.

Orzechowsky, A., Schrenk, D., Bock-Hennig, B. S., and Bock, K. W. (1994). Glucuronidation of carcinogenic arylamines and their N-hydroxy derivatives by rat and human phenol UDP-glucuronosyltransferases of the UGT1 gene complex. *Carcinogenesis* **15,** 1549–1553.

Ouzzine, M., Barre, L., Netter, P., Magdalou, J., and Fournel-Gigleux, S. (2003). The human UDP-glucuronosyltransferases: Structural aspects and drug glucuronidation. *Drug Metab. Rev.* **35,** 287–303.

Peters, W. H.M., Jansen, P. L. M., and Nauta, N. (1984). The molecular weights of UDP-glucuronosyltransferase determined with radiation-inactivation analysis: A molecular model of bilirubin UDP-glucuronosyltransferase. *J. Biol. Chem.* **259,** 11701–11705.

Radominska-Pandya, A., Czernik, P. J., Little, J. M., Battaglia, E., and Mackenzie, P. I. (1999). Structural and functional studies of UDP-glucuronosyltransferases. *Drug Metab. Rev.* **31,** 817–899.

Ramos-Gomez, M., Kwak, M. K., Dolan, P. M., Itoh, K., Yamamoto, M., Talalay, P., and Kensler, T. W. (2001). Sensitivity to carcinogenesis is increased and chemoprotective efficacy of enzyme inducers is lost in nrf2 transcription factor-deficient mice. *Proc. Natl. Acad. Sci. USA* **98,** 3410–3415.

Schrenk, D., Orzechowsky, A., Schwarz, L. R., Snyder, R., Burchell, B., Ingelman-Sundberg, M., and Bock, K. W. (1996). Phase II metabolism of benzene. *Environ. Health Perp.* **104** (Suppl. 6), 1183–1188.

Siegel, D., Anwar, A., Winski, S. L., Kepa, J. K., Zolman, K. L., and Ross, D. (2001). Rapid polyubiquitination and proteasomal degradation of a mutant form of NAD(P)H:Quinone oxidoreductase 1. *Mol. Pharmacol.* **59,** 263–268.

Soars, M. G., Petullo, D. M., Eckstein, J. A., Kasper, S. C., and Wrighton, S. A. (2004). An assessment of UDP-glucuronosyltransferase induction using primary human hepatocytes. *Drug Metab. Disp.* **32,** 140–148.

Strassburg, C. P., Kneip, S., Topp, J., Obermayer-Straub, P., Barut, A., Tukey, R. H., and Manns, M. P. (2000). Polymorphic gene regulation and interindividual variation of UDP-glucuronosyltransferase activity in human small intestine. *J. Biol. Chem.* **275,** 36164–36171.

Strassburg, C. P., Manns, M. P., and Tukey, R. T. (1997). Differential down-regulation of the UDP-glucuronosyltransferase 1A locus is an early event in human liver and biliary cancer. *Cancer Res.* **57,** 2979–2985.

Thimmulappa, R. K., Mai, K. H., Srisuma, S., Kensler, T. W., Yamamato, M., and Biswal, S. (2002). Identification of Nrf2-regulated genes induced by the chemopreventive agent sulforaphane by oligonucleotide microarray. *Cancer Res.* **62,** 5196–5203.

Tukey, R. H., and Strassburg, C. P. (2000). Human UDP-glucuronosyltransferases: Metabolism, expression, and disease. *Annu. Rev. Pharmacol. Toxicol.* **40,** 581–616.

Uchaipichat, V., Mackenzie, P. I., Guo, X. H., Gardner-Stephen, D., Galetin, A., Houston, J. B., and Miners, J. O. (2004). Human UDP-glucuronosyltransferases: Isoform selectivity and kinetics of 4-methylumbelliferone and 1-naphthol glucuronidation, effects of organic solvents, and inhibition by diclofenac and probenecid. *Drug Metab. Disp.* **32,** 413–423.

Vyhlidal, C. A., Rogan, P. K., and Leeder, J. S. (2004). Development and refinement of pregnane X receptor (PXR) DNA binding site model using information theory. *J. Biol. Chem.* **279,** 46779–46786.

Watt, A. J., Garrison, W. D., and Duncan, S. A. (2003). HNF4: A central regulator of hepatocyte differentiation and function. *Hepatology* **37,** 1249–1253.

Weissbach, H., Lovenberg, W., Redfield, B. G., and Udenfriend, S. (1961). *In vivo* metabolism of serotonin and tryptamine: Effect of monoamine oxidase inhibition. *J. Pharmacol. Exp. Ther.* **131,** 26–30.

Xie, W., Yeuh, M. F., Radominska-Pandya, A., Saini, S. P. S., Negishi, Y., Bottroff, B. S., Cabrera, G. Y., Tukey, R. H., and Evans, R. M. (2003). Control of steroid, heme, and carcinogen metabolism by nuclear pregnane X receptor and constitutive androstane receptor. *Proc. Natl. Acad. Sci. USA* **100,** 4150–4155.

Yang, Y., Griffiths, W. J., Midtvedt, T., Sjövall, J., Rafter, J., and Gustafsson, J. A. (1999). Characterization of conjugated metabolites of benzo(a)pyrene in germ-free rat urine by liquid chromatography/electrospray tandem mass spectrometry. *Chem. Res. Toxicol.* **12,** 1182–1189.

Yu, R., Tan, T. H., and Kong, A. N. T. (1997). Butylated hydroxyanisole and its metabolite tert-butylhydroquinone differentially regulate mitogen-activated protein kinases. *J. Biol. Chem.* **272,** 28962–28970.

Zheng, Z., Fang, J. L., and Lazarus, P. (2002). Glucuronidation: An important mechanism for detoxification of benzo(a)pyrene metabolites in aerodigestive tract tissues. *Drug Metab. Disp.* **30,** 397–403.

[5] The Role of Ah Receptor in Induction of Human UDP-Glucuronosyltransferase 1A1

By Mei-Fei Yueh, Jessica A. Bonzo, and Robert H. Tukey

Abstract

UDP-glucuronosyltransferases (UGTs) catalyze a major metabolic pathway initiating the transfer of glucuronic acid from uridine 5′-diphosphoglucuronic acid to endogenous and exogenous substances. Endogenous substances include bile acids, steroids, phenolic neurotransmitters, and bilirubin. Xenobiotic substances include dietary substances,

METHODS IN ENZYMOLOGY, VOL. 400
Copyright 2005, Elsevier Inc. All rights reserved.
0076-6879/05 $35.00
DOI: 10.1016/S0076-6879(05)00005-4

therapeutics, and environmental compounds. The versatility in the selection of substrates for glucuronidation results from the multiplicity of the UGTs in addition to the ability of these genes to be regulated. UDP-glucuronosyltransferase 1A1 (UGT1A1), responsible for the glucuronidation of bilirubin, is controlled in a tissue-specific manner and can be regulated following environmental exposure. This chapter describes materials and methods for the examination of molecular interactions that control UGT1A1 expression and induction in response to 2,3,7,8-tetrachlorodibenzo-*p*-dioxin (TCDD). Using an *in vitro* cell culture system, we mapped a regulatory sequence that contains a xenobiotic response element core sequence in the enhancer region of the *UGT1A1* gene. Similar to regulation of *CYP1A1*, the transcriptional activation of *UGT1A1* he aryl hydrocarbon receptor.

Introduction

Induction of drug-metabolizing enzymes is one important way of enhancing or attenuating the *in vivo* effects of endogenous or xenobiotic compounds that are substrates for these enzymes. Aryl hydrocarbon (Ah) receptor-mediated induction of *CYP1* genes (i.e., *CYP1A1, CYP1A2,* and *CYP1B1*) represents a classical mechanism of upregulation of drug-metabolizing enzymes. Without ligand stimulation, the Ah receptor is present in the cytosol in an inactive complex containing Hsp90, p23, and XAP2 (Kazlauskas *et al.*, 2000). Many structurally diverse compounds, including polycyclic aromatic hydrocarbons (PAHs) and halogenated aromatic hydrocarbons (HAHs), are capable of binding to and activating the Ah receptor. The activated Ah receptor subsequently translocates to the nucleus and heterodimerizes with the nuclear protein Ah receptor nuclear translocator (Arnt) to form an Ah receptor–Arnt complex that acts as a transcription factor to turn on a battery of target genes (Nebert *et al.*, 2000). The nuclear receptors pregnane xenobiotic receptor (PXR), constitutive active receptor (CAR), and peroxisome proliferator-activated receptor α (PPARα) have also been linked to the regulation of drug-metabolizing enzymes (Waxman, 1999).

Glucuronidation, catalyzed by the superfamily of UDP-glucuronosyltransferase (UGT) enzymes, is a major metabolic pathway for xenobiotics and steroids. Previously, 2,3,7,8-tetrachlorodibenzo-*p*-dioxin (TCDD) and PAHs have been reported to induce UGT activity in rodents (Emi *et al.*, 1996; Kessler and Ritter, 1997; Malik and Owens, 1981) and in human cell lines (Bock *et al.*, 1999), but the involvement of specific UGT isoforms and the underlying mechanisms have not been elucidated clearly. This study was initiated by the observation in our laboratory that UGT1A1 was regulated in HepG2 cells in response to the Ah receptor ligands, TCDD and β-naphthoflavone (BNF), as monitored by increases in protein levels and catalytic activity, as well as *UGT1A1* mRNA. Considering the ability of these conjugation enzymes to metabolize endogenous bilirubin and therapeutics, the alteration of UGT1A1 levels might have a profound impact on physiological and pathological consequences. Normally, gene function is influenced by a combination of *cis*-acting elements and *trans*-acting factors. Advances in the understanding of promoter–enhancer sequences and external transcription regulatory proteins involved in the control of gene expression continue to evolve using methods in molecular biology. In the systems explored here, cell culture provides a convenient *in vitro* model to dissect and analyze regulatory regions of a gene, to define functional enhancer sequences, and to monitor gene expression levels in the presence or absence of chemical inducers.

Materials

1-Naphthol, 17α-ethynylestradiol, and BNF are from Sigma. TCDD, benzo[a]pyrene (B[a]P) and its derivatives are from the National Cancer Institute, National Institutes of Health, Chemical Carcinogen Reference Standard Repository (Kansas City, MO). The Bradford assay for protein concentration analysis is from Bio-Rad. Restriction enzymes and T4 DNA ligase are from New England Biolabs (Beverly, MA). *Taq* polymerase, dual luciferase reporter assay system, and reporter plasmids—pGL3-basic vector, pGL3-promoter vector, and pRL-TK vector—are from Promega (Madison, WI). Lipofectamine 2000 for transfection and medium for cell culture are from Invitrogen. Custom oligonucleotides used in polymerase chain reaction (PCR) cloning, subcloning, DNA sequencing, and gel shift assay are from Genbase (San Diego, CA). Thin-layer chromatography (TLC) plates for glucuronidation activity are from Whatman (Clifton, NJ).

Methods and Results

Cell Culture and Treatment

Human hepatoma-derived HepG2 cells, obtained from ATCC, are grown in Dulbecco's modified Eagle's medium (DMEM) supplemented with 10% fetal bovine serum (FBS). TV101 cells are HepG2 cells stably integrated with a reporter gene that carries the *CYP1A1* regulatory region, containing multiple XREs and luciferase reporter (Postlind *et al.*, 1993). Luciferase expression is driven by the promoter activity of *CYP1A1*.

TV101 cells are grown in the same medium with HepG2 in the presence of neomycin (800 μg/ml). Cell lines are trypsinized, and 2×10^6 cells are seeded in P100 plates. Chemicals are dissolved in DMSO, and the DMSO concentration in media never exceeds 0.1% (v/v). Cells are treated for 24–72 h with TCDD (10 nM), BNF (20 μM), or DMSO. For transient transfection experiments, 1×10^5 cells/well are seeded on 24-well plates a day before transfection followed by chemical treatment for 48 h a day after transfection. Fresh media and chemical treatments are changed every 24 h.

Detection of UGT1A1 Levels

Glucuronidation Activity by TLC Assay. UGT1A1 activities are determined using 1-naphthol and 17α-ethynylestradiol as substrates by TLC assay according to the method of Bansal and Gessner (1980) with modification.

1. HepG2 cells are grown to ~70% confluency in DMEM supplemented with 10% FBS.
2. Whole cell lysates are prepared by washing and scraping with ice-cold phosphate-buffered saline (PBS) after 24–72 h of treatment. Cell pellets are then dissolved in a fivefold volume of lysis buffer [50 mM Tris-HCl (pH 7.6), 10 mM MgCl$_2$].
3. Each UGT assay contains 50 mM Tris-HCl (pH 7.6), 10 mM MgCl$_2$, 100 μM UDPGlcA, 0.04 μCi [^{14}C]UDPGlcA (0.14 nmol), 8.5 mM sacchrolactone, 100 μM substrate, and 100 μg of protein from HepG2 cell extracts in a total volume of 100 μl.
4. Reactions are performed at 37$°$ in a shaking water bath for 90 min.
5. At the end of the incubation, 100 μl of ethanol is added to the sample, centrifuged briefly to spin down cell lysates, and 100 μl of the supernatant applied to the TLC plate.
6. Glucuronides in the TLC plate are visualized with a Molecular Dynamics Storm 820 phosphorimager.
7. Resident glucuronides are then removed by scraping and are quantitated by liquid scintillation counting.

The small phenolic compound 1-naphthol is used as a substrate to examine UGT activity in HepG2 cells. Simple phenols have been shown to be glucuronidated by most of the UGT1A proteins (Tukey and Strassburg, 2000), with a preference for UGT1A1, UGT1A6, UGT1A8, and UGT1A9. Treatment of HepG2 cells with TCDD (10 nM) or BNF (20 μM) for 72 h leads to a time-dependent increase in 1-naphthol UGT activity of 3- and 4.5 fold, respectively (Fig. 1). Glucuronidation of 17 α-ethynylestradiol, a

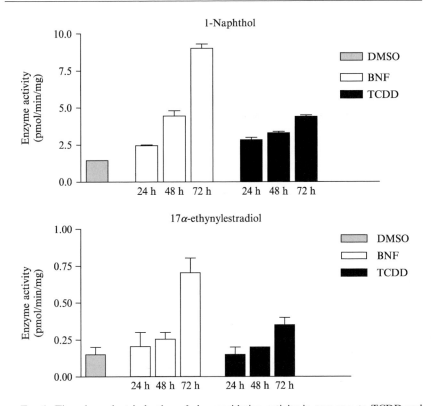

FIG. 1. Time-dependent induction of glucuronidation activity in response to TCDD and BNF. HepG2 cells were treated for 24, 48, and 72 h with DMSO, 10 nM TCDD, or 20 μM BNF and whole cell lysates were collected. Assays for UGT activity toward 1-naphthol and 17α-ethynylestradiol were performed using 100 μg protein. Glucuronides formed were scraped from TLC plates and counted. Each reaction was performed in triplicate and average activity displayed.

substrate that is preferentially glucuronidated by UGT1A1, is increased 2.5- to 5-fold in TCDD- or BNF-treated cells (Fig. 1).

UGT1A1 RNA Transcripts. Quantization of UGT1A1 mRNA is conducted by Northern blot analysis with a gene-specific cDNA probe.

1. Total RNA is isolated from TCDD-, BNF-, or DMSO-treated HepG2 cells using Trizol (Invitrogen).
2. Total RNA (15 μg) is separated through 1% formaldehyde agarose gels. RNA is subsequently blotted onto a GeneScreen membrane (PerkinElmer) by capillary transfer.

3. After transfer, the blot is stained with 0.01% methylene blue in 0.5 M sodium acetate (pH 5.2) to visualize RNA for loading and transfer efficiency.

4. For cDNA probe labeling, a 423-bp fragment recovered by digesting the *UGT1A1* cDNA (GenBank NM_000463, bases 299 to 722) with *Ava*I/*Eco*RI is ^{32}P labeled by random priming (Invitrogen). After labeling, the cDNA probe is purified using a nucleotide removal kit (Qiagen).

5. Total RNA is fixed onto the membranes with a cross-linker (Stratagene). The blot is prehybridized with the hybridization solution (Stratagene) at 68° for 30 min and then with denatured hybridization solution containing the *UGT1A1* cDNA probe at 68° for 2 h. The membrane is then washed in 0.1× SSC (0.15 M NaCl and 0.015 M sodium citrate) and 0.1% SDS at 60° for 30 min.

6. RNA is visualized on each membrane by a phosphorimager.

Results of Northern blot analysis demonstrate that both TCDD- and BNF-treated HepG2 cells lead to a time-dependent increase in *UGT1A1* mRNA (Fig. 2).

Analysis of UGT1A1 Protein by Western Blot

1. After treatment, HepG2 cells are collected by scraping, washed in cold PBS, and resuspended in 5 volumes of PBS.

2. Cell suspensions are homogenized 20 times with a Kontes Potter–Elvehjem tissue grinder, and homogenates are centrifuged at 5000g in a Sorvall RT 6000B refrigerated centrifuge.

3. The supernatant is collected and centrifuged at 150,000g for 1 h in a Beckman TL100 tabletop ultracentrifuge. The microsomal pellet is resuspended in 500 μl of 50 mM Tris-HCl, pH 7.4, 10 mM MgCl$_2$, and 1 mM phenylmethylsulfonyl fluoride (PMSF).

4. Protein (10 μg) is loaded onto Nupage Bis-Tris 10% polyacrylamide gels (Invitrogen). Electrophoresis is conducted at 200 V for 50 min, and the protein is transferred at 30 V for 1 h to nitrocellulose membranes (Millipore).

5. The membranes are blocked with 5% nonfat dry milk in Tris-buffered saline (10 mM Tris, pH 8.0; 150 mM NaCl, 0.05% Tween 20) for 1 h followed by incubation with anti-human UGT1A1 (1:1000) (Ritter *et al.*, 1999) or anti-human CYP1A1 (1:5000) (Soucek *et al.*, 1995) prepared in Tris-buffered saline with 5% bovine serum albumin (Sigma) overnight at 4°.

6. The membranes are washed and then treated with horseradish peroxidase-conjugated antimouse (Cell Signaling) (1:5000) for UGT1A1 or antirabbit (1:5000) for CYP1A1 antibody for 1 h.

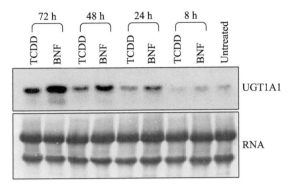

FIG. 2. UGT1A1 transcript levels after TCDD and BNF treatment. Northern blot of UGT1A1 RNA in HepG2 cells after treatment with DMSO (untreated), 10 nM TCDD, and 20 μM BNF for 8, 24, 48, and 72 h. RNA (15 μg) was separated on a formaldehyde agarose gel, transferred to a nitrocellulose membrane, and incubated with a 423-bp probe from *UGT1A1* cDNA. RNA was visualized by a phosphoimager. (See color insert.)

7. The membranes are washed again with Tris-buffered saline. Protein is visualized using the renaissance Western blot chemiluminescence reagent according to the manufacturer's instructions (PerkinElmer Life Science) followed by exposure to X-ray film.

In HepG2 cells, TCDD and BNF induce UGT1A1 as shown by increased levels of UGT1A1 protein in Western blot analysis (Fig. 3). In addition, TCDD and BNF are capable of inducing CYP1A1. Because induction of CYP1A1 by TCDD and BNF has been linked to activation of the Ah receptor, this result indicates that induction of UGT1A1 might be an Ah receptor-dependent event.

Identification and Genotyping of UGT1A Locus and Amplification of UGT1A1 Regulatory Fragments from a BAC Clone

Characterization of the *UGT1A* genomic cluster revealed that the locus consists of multiple first exons encoding isoform-specific sequences that are located at intervals of 3–15 kb apart and followed by a single set of common exons (exon 2, 3, 4, and 5) encoding the sequence that is identical in all UGT1A proteins (Gong *et al.*, 2001). The expression of each enzyme is regulated independently by splicing of the first exon to the common exons. An 11-kb region of the *UGT1A1* promoter is amplified by PCR from a BAC clone encoding the *UGT1A* locus with primers corresponding to sites on the promoter sequence, as published in NCBI GenBank accession number AF297093 (Gong *et al.*, 2001). The transcription start site is designated as +1. The PCR products for the −3712/−7 *UGT1A1* promoter

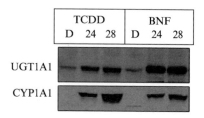

FIG. 3. Protein analysis of UGT1A1 and CYP1A1. Microsomal fractions were collected from HepG2 cells treated with DMSO (D), 10 nM TCDD, or 20 μM BNF for 48 and 72 h. Western blots were performed using 10 μg protein and blotted using anti-UGT1A1 or anti-CYP1A1 antibodies. (See color insert.)

and individual enhancer sequences containing bases $-10998/-8134$ (enhancer 1, E1), $-8533/-4738$ (enhancer 2, E2), and $-3712/-2081$ (enhancer 3, E3) are amplified using primers with sequences as follows: *UGT1A1* promoter, 5′-tttag*gagctc*TCAGACAAAAGGAA-3′ and 5′-tcctg-*ctcgag*GTTCGCCCTCTCCT-3′;   E1,   5′-atat*ggagctc*AAAGAAGAG-AACT-3′ and 5′-atct*actcga*GGGAATGATCCTTT-3′; E2, 5′-atatt*gagctc* TTGCTTGCTGC-3′ and 5′-aattt*ctcgag*ACCATGGCTGGTT-3′; and E3, 5′-tttag*gagctc*TCAGACAAAAGGAA-3′   and   5′-ttacactcgagAACCAC-TACTAAGC-3′. The restriction enzyme sites *Sac*I and *Xho*I are incorporated at the 5′ end of sense and antisense primers (lower-case), respectively, and PCR products are subcloned into the *Sac*I/*Xho*I-digested pGL3 vector.

Transient Transfections with Reporter Constructs for Promoter Function Assay

To determine the ability of the 5′-flanking region of the *UGT1A1* gene to confer transcriptional activity in response to TCDD, luciferase reporter constructs containing promoter or enhancer sequences of *UGT1A1* are transiently transfected into HepG2 cells, and the expression of luciferase activity is determined after treatment of cells for 48 h with TCDD, BNF, or DMSO.

1. Cells are seeded in a 24-well plate (2×10^5 cells/well) and grown to 70% confluency by the following day.
2. DNA and Lipofectamine 2000 (Invitrogen) are diluted separately with OptiMEM (Invitrogen). Luciferase DNA constructs (100 ng) are combined with 50 ng pRL-TK (internal control) and diluted with OptiMEM (50 μl).

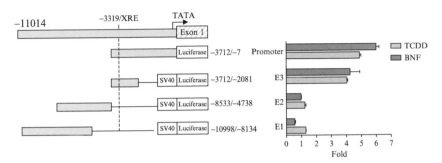

FIG. 4. Promoter activity of *UGT1A1* in response to TCDD and BNF by luciferase reporter assay. Four DNA fragments that cover the entire *UGT1A1* promoter and enhancer (11 kb) were cloned from a BAC library and subcloned into the pGL3 reporter plasmid. HepG2 cells were transiently transfected with the reporter plasmids, and firefly luciferase activity was measured in the cytosolic fraction after 48 h of treatment. Values were normalized to renilla luciferase activity and shown as fold over DMSO treatment. (See color insert.)

3. Lipofectamine 2000 (2 μl) is diluted with 50 μl OptiMEM and added to the DNA mixture. The DNA–Lipofectamine complex is incubated for 30 min at room temperature.

4. Medium containing FBS is removed from cells prior to transfection. HepG2 cells are transfected by adding 100 μl OptiMEM and 100 μl DNA–Lipofectamine complex to each well.

5. After overnight incubation, transfection medium is removed. Transfected HepG2 cells are treated with chemical inducers or DMSO for 48 h.

6. Cells are harvested with 1× permissive lysis buffer (Promega), and the supernatant is collected by brief centrifugation.

7. Promoter activity is determined by expression of firefly luciferase and is normalized to the renilla luciferase levels by using a dual luciferase reporter assay (Promega).

The *UGT1A1* −3712/−7 promoter-luciferase fragment is induced after treatment with TCDD and BNF (Fig. 4). An enhancer sequence from −3712 to −2081 (E3) relative to the transcriptional start site is also responsive. Enhancer sequences E2 and E1, which cover the region from −10998 to −4738, are refractory to both TCDD and BNF.

Generation of UGT1A1-Luciferase MH1A1L Cells

1. The neomycin gene is subcloned into the *Sal*I site of the −3712/−7 *UGT1A1* promoter-pGL3 basic vector to generate the pLUGT1A1-neo plasmid (Postlind *et al.*, 1993; Yueh *et al.*, 2003).

2. HepG2 cells are trypsinized, and 5×10^6 cells are seeded in P150 plates 1 day before transfection.
3. The pLUGT1A1neo plasmid is transfected into HepG2 cells using Lipofectamine 2000 according to the manufacturer's instruction.
4. After 48 h, neomycin (800 μg/ml) selection is initiated.
5. Stable transformants are obtained at approximately 3 weeks after neomycin selection. Positive transformants containing pLUGT1A1 are chosen by detecting luciferase activity.

Previously, HepG2 cells were stably integrated with the *CYP1A1* promoter that contains multiple xenobiotic response elements (XREs) to create TV101 cells that have been utilized as a convenient *in vitro* biomarker for detecting Ah receptor ligands (Allen *et al.*, 2001). TCDD and BNF are capable of inducing *CYP1A1*-luciferase in TV101 cells as shown in Fig. 5A indicating their ability to activate the Ah receptor.

Induction of the *UGT1A1* $-3712/-7$ promoter-luciferase construct with TCDD indicates that transcriptional activation may occur through an Ah receptor-dependent mechanism. MH1A1L cells carrying the *UGT1A1* promoter-luciferase demonstrate that classic PAHs are capable of inducing *UGT1A1* promoter-driven luciferase (Fig. 5B). Along with TCDD induction, B[a]P metabolites including the 1-, 2-, 3-, 4-, 6-, 8-, 9-, and 10-hydroxylated B[a]P isomers and B[a]P *cis*- and *trans*-4,5-dihydrodiol increase luciferase activity two- to fivefold. The 3- and 9-hydroxy B[a]P and *trans*-4,5-dihydrodiol B[a]P serve as the most efficient inducers. UGT1A1 reporter gene assays with PAHs further suggest that these agents might elicit transcriptional activation of *UGT1A1* through an Ah receptor-dependent pathway.

Localization of cis-Acting Response Element by 5'/3' Deletion Analysis and Site-Directed Mutation Assay

To localize the region on the *UGT1A1* gene that controls TCDD- and BNF-mediated induction, an additional series of expression plasmids, E4 ($-3529/-3143$), E5 ($-3430/-3285$), and E6 ($-3430/-3337$) are generated by progressive truncation of the *UGT1A1* $-3712/-2081$ fragment. The sequences of the primers used for these enhancers are as follow: E4, 5'-tccttgagctcTTTTTGACACTGGA-3' and 5'-aaattctcgagCTCATTCCTCC-TCT-3'; E5, 5'-aaagggagctcTAACGGTTCATAAA-3' and 5'-aaattctcgag CTTACTATGACTG-3'; and E6, 5'-aaagggagctcTAACGGTTCATAAA-3' and 5'-aatggctcgagGTTATGTAACTAGA-3'. Each of these amplified inserts is digested with *Sac*I and *Xho*I and subcloned into the *Sac*I/*Xho*I-digested pGL3-promoter vector. Each plasmid is transiently transfected

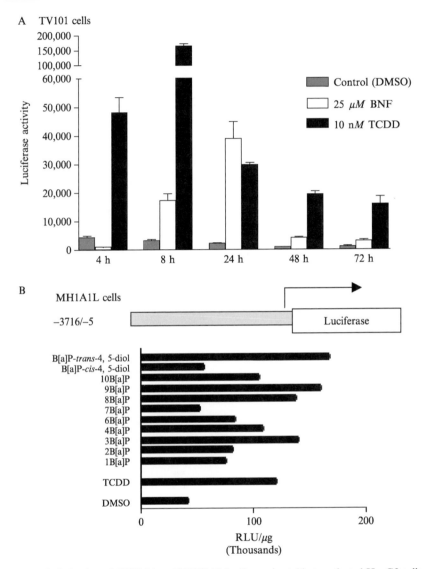

FIG. 5. Induction of CYP1A1-and UGT1A1-luciferase in stably transfected HepG2 cells. (A) The *CYP1A1* promoter luciferase plasmid that contains multiple XREs was used to establish TV101 cells. Luciferase activity was measured at various times after treatments. Activity is expressed as relative light units (RLU)/μg of protein. (B) HepG2 cells were stably transfected with a 3.7-kb *UGT1A1* promoter luciferase plasmid to generate the MH1A1L cell line. Treatment of cells was carried out for 48 h with 5 μM of each B[a]P metabolite. Luciferase activity is expressed as RLU/μg of protein. (See color insert.)

into HepG2 cells, and expression of luciferase activity is determined after treatment of transfected cells for 48 h.

Mutational analysis on the E4 clone demonstrates that a sharp drop in induction is observed between bases −3337 and −3285 (Fig. 6A). Sequence analysis in this region reveals the presence of a single copy of the Ah receptor XRE motif (CACGCA) starting at position −3309.

Using DNA fragments spanning −3529 to −3143, site-directed mutagenesis is carried out on the conserved *UGT1A1* XRE sequence, altering *CA*CGCA to *AC*CGCA. The reporter plasmid containing either wild type or the mutated *UGT1A1*-XRE is inserted into the pGL3-promoter vector and used in transient transfections. CA to AC mutation within the XRE results in a lack of TCDD-dependent induction of transcriptional activity (Fig. 6B), confirming a role for the Ah receptor in control and expression of *UGT1A1*.

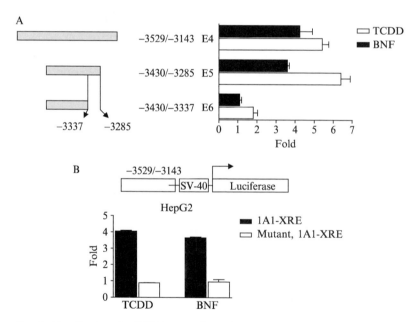

FIG. 6. Identification of a TCDD-responsive region in the *UGT1A1* promoter. (A) Using plasmid E3 as a DNA template, E4, E5, and E6 were generated by progressive truncation of the 5′ and 3′ ends. Transient transfections were conducted with these truncated plasmids, and promoter responsiveness to 48 h treatment with TCDD was determined by luciferase expression. (B) The core sequence of the XRE, CACGCA, was mutated to ACCGCA by site-directional mutagenesis PCR. The luciferase reporter plasmid containing either wild-type or mutated *UGT1A1*-XRE was transiently transfected into HepG2 cells. Cells were treated for 48 h with 10 n*M* TCDD and 20 μ*M* BNF. Results shown are fold activity over DMSO. (See color insert.)

Analysis of Ah Receptor/Arnt Binding to UGT1A1 XRE Sequence
by Electrophoretic Mobility-Shift Assay (EMSA) and Antibody
Competition Assay

EMSA is used to analyze TCDD-inducible protein–DNA interactions to confirm that the *UGT1A1* CACGCA motif is a binding site for the Ah receptor. In addition, to determine whether the induced nuclear proteins are the Ah receptor/Arnt complex, binding reactions are carried out in the presence of antibodies generated against the human Ah receptor (Anti-AhR) or Arnt (Anti-Arnt).

1. HepG2 cells are treated with TCDD or DMSO for 24 h. Nuclear extracts from HepG2 cells are isolated as outlined (Chen and Tukey, 1996).

2. Cells are washed twice with 10 mM HEPES buffer, pH 7.5, collected by scraping into MDH buffer [3 mM MgCl$_2$, 1 mM dithiothreitol (DTT), 25 mM HEPES, pH 7.5, 10 μg/ml aprotinin, 10 μg/ml leupeptin, 0.2 mM PMSF] and homogenized.

3. The homogenates are centrifuged at 5000g for 5 min, and the pellet is washed with MDHK buffer (3 mM MgCl$_2$, 1 mM DTT, 25 mM HEPES, pH 7.5, 0.1 M KCl, 10 μg/ml aprotinin, 10 μg/ml leupeptin, 0.2 mM PMSF). This procedure is repeated three times.

4. The cell pellet is then lysed in HDK buffer (25 mM HEPES, pH 7.5, 1 mM DTT, 0.4 M KCl, 10 μg/ml aprotinin, 10 μg/ml leupeptin, 0.2 mM PMSF) and centrifuged at 105,000g for 60 min, and the supernatant is designated as the nuclear extract.

5. A complementary pair of oligonucleotides containing the consensus core sequence of the *UGT1A1* XRE (underlined) is synthesized (5'-GCT AGGCACTTGGTAAG<u>CACGCA</u>ATGAACATGCA-3' and 5'-GCTA TGACTGTTCAT<u>TGCGTG</u>CTTACCAAGTGCC-3'). In addition, the human *CYP1A1* XRE oligonucleotides (5'-GATCCGGCTCTTGT<u>CAC GCA</u>ACTCCGAGCTCA-3' and 5'-GATCTGAGCTCGGAG<u>TGCG-TG</u>AGAAGAGCCG-3') are used as described previously (Chen and Tukey, 1996).

6. Double-stranded oligonucleotides are assembled by annealing at 80° equal concentrations of the sense and antisense XRE strands in a total of 20 μl sterile water. Double-stranded oligonucleotides are labeled for 20 min at room temperature in a 20-μl reaction containing 2 μl double-stranded oligonucleotides, 25 μM dATP, 25 μM dGTP, 25 μM dTTP, 5 μl 10 mCi/ml [α-^{32}P]dCTP, and 1 μl Klenow fragment. The reaction is terminated by the addition of NaCl, and double-stranded oligonucleotides are purified using the Qiagen nucleotide removal kit. The activity of double-stranded oligonucleotides is determined by liquid scintillation counting.

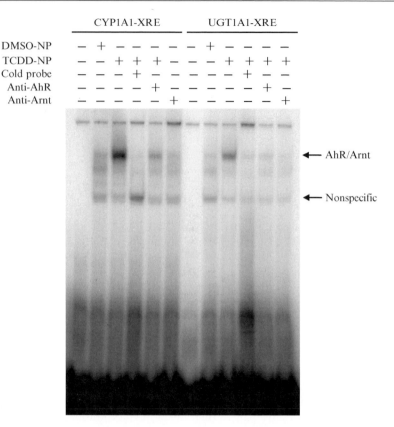

FIG. 7. Confirmation of Ah receptor/Arnt-binding complex on *UGT1A1* XRE. HepG2 cells were treated for 24 h with DMSO or TCDD, and nuclear extracts were collected (DMSO-NP, TCDD-NP). Ten micrograms of protein was used in binding reactions with α-^{32}P-dCTP-labeled oligonucleotides. Specificity of binding was confirmed using 200-fold excess unlabeled probe. One hundred nanograms of antibodies against the Ah receptor and Arnt was used for competition experiments.

7. Binding assays containing HEDG (25 mM HEPES, pH 7.5, 1.5 mM EDTA, 1 mM DTT, 10% glycerol), 10 μg of nuclear extract, 2 μg of poly dI-dC), and 1 μg of salmon sperm DNA in a final reaction volume of 20–25 μl are carried out on ice for 15 min. The binding reaction is further continued for 20 min at room temperature with the addition of 1×10^6 cpm of labeled oligonucleotide.

8. Competition assays are performed by adding a 200-fold excess of unlabeled *CYP1A1* XRE or *UGT1A1* XRE oligonucleotide to the binding reaction.

9. The specificity of the Ah receptor/Arnt complex binding is examined by adding 100 ng of anti-Ah receptor or anti-Arnt antibody (a generous gift from Christopher Bradfield) to the binding reaction.

10. Protein–DNA complexes are separated on a 6% nondenaturing polyacrylamide gel for ~2 h at 180 V using 45 mM Tris-borate, 10 mM EDTA (1× TBE) as running buffer.

11. The gels are separated from glass plates, rinsed with 1× TBE, placed on a stack of Whatman paper, wrapped in plastic wrap, and exposed to a phosphorimager plate for 2 h. Protein–DNA complexes are visualized by a Molecular Dynamics Storm 840 phosphorimager.

When a nuclear extract prepared from TCDD-treated HepG2 cells is incubated with a ^{32}P-labeled *UGT1A1*-XRE probe, an induced DNA–protein complex is detected (Fig. 7). Excess unlabeled *UGT1A1*-XRE competes efficiently for the labeled XRE, suggesting that TCDD induction of nuclear protein specifically binds to the *UGT1A1*-XRE. Competition experiments with Ah receptor and Arnt antibodies confirmed involvement of the Ah receptor/Arnt complex. Control experiments with *CYP1A1*-XRE as a probe show similar results to that of the *UGT1A1*-XRE, further confirming the presence of a classically functional XRE in the *UGT1A1* promoter.

Summary

Several recent findings confirm that UGTs are capable of undergoing differential regulation by environmental influences, resulting in an enhanced glucuronidation capacity. In primary human hepatocytes, treatments with phenobarbital, oltipraz, and 3-methylcholanthrene led to the induction of *UGT1A1* mRNA and protein (Ritter *et al.*, 1999). In addition, exposure of HepG2 and Caco-2 cells to flavonoids induces UGT1A1 (Galijatovic *et al.*, 2001; Walle and Walle, 2002; Walle *et al.*, 2000). The *UGT1A1* gene has been shown to be inducible by the nuclear receptors PXR and the CAR. This chapter described the procedures, reagents, and DNA constructs used to characterize Ah receptor-mediated UGT1A1 induction. The *UGT1A* locus was first characterized by screening a human BAC library. DNA constructs covering the entire enhancer and promoter segments that were amplified and established in a luciferase reporter gene were analyzed in transient transfection assays in the presence or absence of inducers. Results identified the −3712/−7 promoter-luciferase construct as inducible with TCDD and BNF treatments. A series of deletion mutants then generated from the original reporter construct confirmed that region–3529/–3142 of the *UGT1A1* gene is essential for

TCDD-induced luciferase activity. Nucleotide sequence analysis revealed axenobiotic response element core sequence CACGCA starting at position -3309. When the XRE core sequence was deleted or mutated, inducibility with the Ah receptor ligand was lost, suggesting that the XRE exists as a functional response element. EMSA and competition experiments with Ah receptor and Arnt antibodies demonstrated that the TCDD-activated nuclear protein specifically binds the UGT1A1-XRE and that the Ah receptor/Arnt complex is responsible for this increase in binding. HepG2 cells exposed to TCDD and BNF induce CYP1A1 as shown by Western blot analysis. Reporter assay of a *CYP1A1*-luciferase promoter, as well as EMSA assay with the *CYP1A1* XRE as a probe, confirmed this induction. These control experiments suggest that the Ah receptor is functional in these cells and that Ah receptor ligands may regulate UGT1A1 in a manner comparable with CYP1A1. Combined, these experiments reveal that the Ah receptor is involved in human UGT1A1 induction.

References

Allen, S. W., Mueller, L., Williams, S. N., Quattrochi, L. C., and Raucy, J. (2001). The use of a high-volume screening procedure to assess the effects of dietary flavonoids on human Cyp1a1 expression. *Drug Metab. Disp.* **29,** 1074–1079.

Bansal, S. K., and Gessner, T. (1980). A unified method for the assay of uridine diphosphoglucuronyltransferase activities toward various aglycones using uridine diphospho[U-14C]glucuronic acid. *Anal. Biochem.* **109,** 321–329.

Bock, K. W., Gschaidmeier, H., Heel, H., Lehmkoster, T., Munzel, P. A., and Bock-Hennig, B. S. (1999). Functions and transcriptional regulation of PAH-inducible human UDP-glucuronosyltransferases. *Drug Metab. Rev.* **31,** 411–422.

Chen, Y.-H., and Tukey, R. H. (1996). Protein kinase C modulates regulation of the CYP1A1 gene by the Ah receptor. *J. Biol. Chem.* **271,** 26261–26266.

Emi, Y., Ikushiro, S., and Iyanagi, T. (1996). Xenobiotic responsive element-mediated transcriptional activation in the UDP-glucuronosyltransferase family 1 gene complex. *J. Biol. Chem.* **271,** 3952–3958.

Galijatovic, A., Otake, Y., Walle, U. K., and Walle, T. (2001). Induction of UDP-glucuronosyltransferase UGT1A1 by the flavonoid chrysin in Caco-2 cells: Potential role in carcinogen bioinactivation. *Pharm. Res.* **18,** 374–379.

Gong, Q. H., Cho, J. W., Huang, T., Potter, C., Gholami, N., Basu, N. K., Kubota, S., Carvalho, S., Pennington, M. W., Owens, I. S., and Popescu, N. C. (2001). Thirteen UDPglucuronosyltransferase genes are encoded at the human UGT1 gene complex locus. *Pharmacogenetics* **11,** 357–368.

Kazlauskas, A., Poellinger, L., and Pongratz, I. (2000). The immunophilin-like protein XAP2 regulates ubiquitination and subcellular localization of the dioxin receptor. *J. Biol. Chem.* **275,** 41317–41324.

Kessler, F. K., and Ritter, J. K. (1997). Induction of a rat liver benzo[a]pyrene-trans-7, 8-dihydrodiol glucuronidating activity by oltipraz and beta-naphthoflavone. *Carcinogenesis* **18,** 107–114.

Malik, N., and Owens, I. S. (1981). Genetic regulation of bilirubin-UDP-glucuronosyltrans-ferase induction by polycyclic aromatic compounds and phenobarbital in mice association with aryl hydrocarbon (benzo[a]pyrene) hydroxylase induction. *J. Biol. Chem.* **256,** 9599–9604.

Nebert, D. W., Roe, A. L., Dieter, M. Z., Solis, W. A., Yang, Y., and Dalton, T. P. (2000). Role of the aromatic hydrocarbon receptor and [Ah] gene battery in the oxidative stress response, cell cycle control, and apoptosis. *Biochem. Pharmacol.* **59,** 65–85.

Postlind, H., Vu, T. P., Tukey, R. H., and Quattrochi, L. C. (1993). Response of human *CYP1*-luciferase plasmids to 2,3,7,8-tetrachlorodibenzo-*p*-dioxin and polycyclic aromatic hydro-carbons. *Toxicol. Appl. Pharmacol.* **118,** 255–262.

Ritter, J. K., Kessler, F. K., Thompson, E. T., Grove, A. D., Auyeung, D. J., and Fisher, R. A. (1999). Expression and inducibility of the human bilirubin UDP-glucuronosyltransferase UGTLA1 in liver and cultured primary hepatocytes: Evidence for both genetic and environmental influences. *Hepatology* **30,** 476–484.

Soucek, P., Martin, M. V., Ueng, Y., and Guengerich, F. P. (1995). Identification of a common cytochrome P450 epitope near the conserved heme-binding peptide with antibodies raised against recombinant cytochrome P450 family 2 proteins. *Biochemistry* **34,** 16013–16021.

Tukey, R. H., and Strassburg, C. P. (2000). Human UDP-glucuronosyltransferases: Metabolism, expression, and disease. *Annu. Rev. Pharmacol. Toxicol.* 581–618.

Walle, T., Otake, Y., Galijatovic, A., Ritter, J. K., and Walle, U. K. (2000). Induction of UDP-glucuronosyltransferase UGT1A1 by the flavonoid chrysin in the human hepatoma cell line Hep G2. *Drug Metab. Disp.* **28,** 1077–1082.

Walle, U. K., and Walle, T. (2002). Induction of human UDP-glucuronosyltransferase UGT1A1 by flavonoids: Structural requirements. *Drug Metab. Disp.* **30,** 564–569.

Waxman, D. J. (1999). P450 Gene induction by structurally diverse xenochemicals: Central role of nuclear receptors CAR, PXR, and PPAR. *Arch. Biochem. Biophys.* **369,** 11–23.

Yueh, M. F., Huang, Y. H., Hiller, A., Chen, S. J., Nguyen, N., and Tukey, R. H. (2003). Involvement of the xenobiotic response element (XRE) in Ah receptor-mediated induction of human UDP-glucuronosyltransferase 1A1. *J. Biol. Chem.* **278,** 15001–15006.

[6]   Regulation of the Human *UGT1A1* Gene by Nuclear Receptors Constitutive Active/Androstane Receptor, Pregnane X Receptor, and Glucocorticoid Receptor

By Junko Sugatani, Tatsuya Sueyoshi, Masahiko Negishi, and Masao Miwa

Abstract

Human UDP-glucuronosyltransferase (UGT) 1A1 is the enzyme that detoxifies neurotoxic bilirubin by conjugating it with glucuronic acid. In addition to bilirubin, UGT1A1 conjugates various endogenous and exogenous lipophilic compounds such as estrogens and the active metabolite of the anticancer drug irinotecan SN-38. Thus, activation by specific inducers of the *UGT1A1* gene is critical in treating patients with unconjugated hyperbilirubinemia and in preventing side effects of drug treatment such as SN-38-induced toxicity. This chapter describes the experimental processes used to identify the 290-bp distal enhancer module at $-3499/-3210$ of the *UGT1A1* gene and to characterize its regulation by nuclear receptors: constitutive active/androstane receptor, pregnane X receptor, and glucocorticoid receptor.

Introduction

Human UDP-glucuronosyltransferase (UGT) 1A1 plays a critical role in the detoxification of potentially neurotoxic bilirubin by excreting conjugated bilirubin into the bile duct (Iyanagi *et al.*, 1998; King *et al.*, 1996; Ostrow and Murphy, 1970; Tukey and Strassburg, 2000). A reduced level of UGT1A1 activity is associated with unconjugated hyperbilirubinemia (Crigler–Najjar syndrome and Gilbert's syndrome) (Mackenzie *et al.*, 1997) and with decreased glucuronidation of SN-38, leading to an increased risk for the development of severe irinotecan-associated toxicity (Gagne *et al.*, 2002; Tukey *et al.*, 2002). Phenobarbital (PB) is used as a therapeutic drug for treating patients with the Crigler–Najjar-type II syndrome because it increases the expression of bilirubin glucuronosyltransferase (UGT1A1) and markedly reduces the incidence of unconjugated hyperbilirubinemia (Sueyoshi *et al.*, 2001; Yaffe *et al.*, 2003). These observations prompted us to investigate PB response enhancer activity in the 5'-flanking region of the human *UGT1A1* gene and to delineate it to the 290-bp PB-responsive enhancer module (gtPBREM) located at $-3499/-3210$. Moreover, we have also identified the nuclear receptor constitutive active/androstane receptor

METHODS IN ENZYMOLOGY, VOL. 400
0076-6879/05 $35.00
DOI: 10.1016/S0076-6879(05)00006-6

(CAR) as the transcription factor that regulates gtPBREM in response to PB treatment (Sugatani *et al.*, 2001), pregnane X receptor (PXR) as the nuclear receptor responsible for the rifampicin (Rif)-induced activation of gtPBREM (Sugatani *et al.*, 2004), and glucocorticoid receptor (GR) as the nuclear hormone receptor capable of activating gtPBREM by dexamethasone (DEX) at its physiological concentration (Sugatani *et al.*, 2005). Thus, the 290-bp distal enhancer module is characterized as a composite regulatory element containing the multiple binding sites for these nuclear receptors, DR4, gtNR1, DR3 elements (Sugatani *et al.*, 2001), and GR response elements (GRE1 and GRE2) (Sugatani *et al.*, 2005) (Fig. 1). In addition to these nuclear receptor-binding sites, an aryl hydrocarbon receptor response element (XRE) is also localized within the 290-bp gtPBREM (Yueh *et al.*, 2003).

Understanding the molecular mechanism of the induction of human UGT1A1 may provide a clinically important tool for the prevention and treatment of unconjugated hyperbilirubinemia and for effective drug treatments. The purpose of this chapter is to describe the investigation methods used for the induction mechanism of the human *UGT1A1* gene and to define the effects of various agents on the gene expression by focusing on the assay system used to examine various nuclear receptor activities within the 290-bp distal enhancer module.

```
                                 DR4
 -3499  TACACTAGTAAAGGTCACTCAATTCCAAGGGGAAAATGATTAACCAA
                   G CG C

                                PXRE
        AGAACATTCTAACGGTTCATAAAGGGTATTAGGTGTAATGAGGATGT
                     TT

         GRE1                              gtNR1
        GTTATCTCACCAGAACAAAACTTCTGAGTTTATATAACCTCTAGTTACA
                   GG G        G C CC

                                          XRE
        TAACCTGAAACCCGGACTTGGCACTTGGTAAGCACGCAATGAACAGT
                                       AC

                                DR3
        CATAGTAAGCTGGCCAAGGGTAGAGTTCAGTTTGAACAAAGCAATTT
                               T C

          GRE2
        GAGAACATCAAAGGAAGTTTGGGGAACAGCAAGGGATCCAGAATGG
                             TCG G

        CTAGAGGG  -3210
```

Fig. 1. Nucleotide sequence analysis of 290-bp DNA. The nucleotide sequence of 290-bp DNA (−3499/−3210) is shown. Sequences of the binding sites of nuclear receptors (CAR, PXR, and GR) and receptor-type transcription factor (AhR) are in bold. The seven binding sites are designated DR4, PXRE, GRE1, gtNR1, XRE, DR3, and GRE2, respectively.

Methods

Determination of mRNA and Protein Levels

mRNA Levels. To measure UGT1A1 mRNA levels in HepG2 cells and g2car-3 cells, a HepG2-derived cell line expressing constitutive mCAR, the cells (1×10^5 cells) are plated onto a 24-well plate at 1×10^5 cells/ml in Dulbecco's modified Eagle's medium. The medium is replaced 48 h later, and the cells are treated with various agents. After an additional 24 or 48 h of culture, total RNA is extracted using Trisol reagent by Invitrogen (Carlsbad). cDNA prepared from total cellular RNA (100 ng) is subjected to quantitative real-time polymerase chain reaction (PCR) with ABI GeneAmp 5700 (PE Applied Biosystems) using 5'-GGCCCATCATGCC-CAATAT-3' and 5'-TTCAAATTCCTGGGATAGTGGATT-3' as primers and 6FAM-TTTTTGTTGGTGGAATCAACTGCCTTCAC-TAMRA for detection (Sugatani *et al.*, 2001). The mRNA levels are normalized against β-actin mRNA determined by predeveloped TaqMan assay reagents for human β-actin (PE Applied Biosystem).

While the UGT1A1 mRNA level in g2car-3 cells is 2.70 ± 0.1-fold higher than that in HepG2 cells, pretreatment with a CAR repressor, androstenol (8×10^{-6} M), decreases the UGT1A1 mRNA levels to 1.63 ± 0.24-fold. The most potent PB-type inducer, TCPOBOP (2.5×10^{-7} M), increases the suppressed mRNA levels to 2.50 ± 0.28-fold that in HepG2 cells. PXR activators, Rif (5×10^{-6} M) and clotrimazole (5×10^{-6} M), induce UGT1A1 mRNA in HepG2 cells (2.28 ± 0.81- and 3.15 ± 0.56-fold that of the vehicle-treated control, respectively). DEX at physiological concentration (10^{-7} M) not only causes UGT1A1 mRNA induction in HepG2 cells (2.15 ± 0.19-fold that of the vehicle-treated control), but also synergistically enhances Rif (5×10^{-6} M)- and clotrimazole (5×10^{-6} M)-inducible expression of UGT1A1 mRNA in HepG2 cells (3.97 ± 0.61- and 6.09 ± 0.82-fold that of the vehicle-treated control), while there is no significant change in CAR and PXR mRNA levels in the DEX-treated cells.

Immunoblot Analysis. For Western blot analysis, we prepare a microsomal fraction from cells as follows: HepG2 cells cultured in a 75T flask and treated with various agents are collected by centrifugation at 200g for 2 min and homogenized with a motorized Teflon/glass homogenizer (10 strokes) in homogenized buffer [10 mM KH_2PO_4 (pH 7.4), 250 mM sucrose, 1 mM EDTA, and 0.04 mg/ml phenylmethylsulfonyl fluoride]. After 10-min low speed (10,000g) centrifugation at 4°, the supernatant is removed and the microsomes are pelleted by high-speed centrifugation

(100,000g, 60 min) and stored at $-80°$ (Chen *et al.*, 2003). The microsomal protein concentration is determined with a bicinchoninic-acid protein assay kit (Pierce Chemicals) using bovine serum albumin as the standard. Microsomal proteins (20 μg) are resolved on a sodium dodecyl sulfate–12.5% polyacrylamide gel, electroblotted onto a polyvinylidene difluoride membrane (Millipore), and incubated with antihuman UGT1A1 antibody (BD Biosciences). After incubation with horseradish peroxidase-conjugated goat antirabbit IgG (Jackson Immuno Research Laboratories) to detect UGT1A1, the resultant immunoproducts are visualized with an enhanced chemiluminescence system (Amersham Biosciences).

Assessment of UGT1A1 Induction with the Reporter Gene

Because only a 290-bp fragment ($-3499/-3210$) in the 11-kbp 5′-flanking region of the *UGT1A1* gene displays transcriptional activation by CAR, PXR, AhR, and GR, we can assess UGT1A1 induction by various agents utilizing transactivation experiments with the 290-bp reporter gene.

Plasmids. The 290-bp fragment is amplified from human white blood cell genomic DNA using primers, 5′-gtttccgctagcTACACTAGTAAAGG-TCACTC-3′ and 5′-gtttaactcgagCCCTCTAGCCATTCTGGATC-3′, and then cloned into the pGL3-tk-firefly luciferase reporter plasmid at the *Nhe*I and *Xho*I sites (Sugatani *et al.*, 2001). Bases in lowercase letters are added to digest the oligonucleotides with restriction enzymes, *Nhe*I and *Xho*I. We identified a polymorphism that T normally presents at nucleotide -3279 in the 290-bp enhancer module of *UGT1A1* substituted with G as in HepG2 cell genomic DNA, thereby significantly decreasing transcriptional activity as indicated by the luciferase-reporter assay (Sugatani *et al.*, 2002). We therefore used the wild-type 290-bp enhancer module from white blood cell DNA for the reporter assay. Vectors for the expression of (1) human CAR (Honkakoski *et al.*, 1998), (2) human PXR (Maglich *et al.*, 2002), and (3) human GR (Cidlowski *et al.*, 1990; Hollenberg *et al.*, 1985) can be constructed by inserting the corresponding cDNAs, which were amplified by PCR with the primers: (1) 5′-gtttccggatccATGGCCAGTAGGGAA-GATGAGCT-3′ and 5′-gtttaactcgagTCAGCTGCAGATCTCCTGG-3′, (2) 5′-gccggatccgcaaacATGGAGGTGAGACCCAAA-3′ and 5′-gcgctc-gagTCAGCTACCTGTGATGCC-3′, and (3) 5′-gtttccggatccATGGACT-CCAAAGAATCATTAACTCCTG-3′ and 5′-gtttaactcgagTCACTTTTG-ATGAAACAGAAGTTTTTTG-3′, respectively, into the *Bam*HI and *Xho*I sites of pCR3, respectively. Bases in lowercase letters are added to allow digestion of the oligonucleotides with the restriction enzymes *Bam*HI and *Xho*I.

Transient Transfection and Luciferase Assays. Transient transfection assays are performed using human hepatoma cell line HepG2 cells (Riken Cell Bank) and pig proximal tube-like cell line LLC-PK1 cells (American Type Culture Collection). HepG2 cells and LLC-PK1 cells are plated at 1×10^5 and 2.5×10^4 cells/ml, respectively, 24 h before transfection with the UGT1A1 290-bp luciferase reporter gene plasmid (0.2 μg), the pCR3-human nuclear receptor (CAR, PXR or GR) expression plasmid (0.2 μg) or the control pCR3 vector (0.2 μg), and pRL-SV40 plasmid (0.2 μg) using a calcium phosphate coprecipitation method (Amersham Pharmacia). The medium is replaced with growth medium after 12 h. When the effects of xenobiotics are investigated, the cells are subsequently traced with xenobiotics. The cells are harvested after an additional 24 h of culture. Luciferase activity, measured by the pGL3-tk basic reporter gene in vehicle-treated HepG2 cells, is calculated as one.

CAR is a ligand-independent receptor and the 290-bp reporter gene is induced by exogenously expressed CAR with no ligands in HepG2 cells (Choi *et al.*, 1997; Zelko and Negishi, 2000). In contrast, PXR is a ligand-dependent nuclear receptor (Lehmann *et al.*, 1998; Watkins *et al.*, 2003), and the PXR activator Rif requires exogenously expressed PXR in HepG2 cells for the transcriptional activation of UGT1A1 290-bp DNA (Sugatani *et al.*, 2004). The GR activator DEX not only activates UGT1A1 290-bp DNA but also enhances CAR- and PXR-induced transactivation more strongly in the presence of exogenously expressed GR than in its absence in HepG2 cells (Sugatani *et al.*, 2005). While the induction of UGT1A1 expression by glucocorticoids is rather low (about twofold that of the control), they can be a pathophysiological factor regulating UGT1A1 activity: glucocorticoids may signal an increased expression of UGT1A1 in response to the sudden interruption of maternal glucose in newborn infants to suppress neonatal jaundice or may maintain the blood bilirubin level within the physiological range. GR enhancement of CAR- and PXR-mediated transcriptional activation almost reaches a plateau after 24 h of the transfection, similar to CAR- and PXR-dependent enhancer activity and peaks at the physiological concentration (10^{-7} M) of DEX. Results obtained by the reporter gene assay are consistent with those of UGT1A1 gene expression. In contrast, AhR activators, benzo[a]pyrene and 3-methylcholanthrene, do not require exogenously expressed AhR for transcriptional activation. In this regard, chrysin and baicalein activity for the 290-bp DNA element found in HepG2 seems to depend on the Ah receptor, as the two flavones activate the reporter without exogenously expressed nuclear receptors (Sugatani *et al.*, 2004) and mutation of the functional Ah receptor-binding site, XRE, in the 290-bp DNA abrogates the induction.

Characterization of the UGT1A1 *290-bp Distal Enhancer Module by Site-Directed Mutagenesis, Deletion Assays, and Electrophoretic Mobility Shift Assays*

Site-Directed Mutagenesis. Mutations of the DR4 element, gtNR1 and DR3 element (Sugatani *et al.*, 2001), PXRE (Xie *et al.*, 2003), XRE (Yueh *et al.*, 2003), and GRE1 and GRE2 (Sugatani *et al.*, 2005) in the 290-bp distal enhancer module are performed using a QuickChange site-directed mutagenesis kit (Stratagene). The primers used are 5'-TACACTAGTAA-*GGCGCC*CTCAATTCCAAGG-3', 5'-AGAACAAACTTC*GGCGCC*T ATATAACCTC-3', 5'-TGGCCAAGGGTAGA*TT*CCAGTTTGAAC AAAG-3', 5'-ACATTCTAAC*TTTT*CATAAAGGGTATTAGG-3', 5'-C TTGGTAAGA*CC*GCAATGAAC-3', 5'-TGTTATCTCACCAG*GGCG* AACTTCTGAGTT-3', 5'-AAAGGAAGTTTGGG*TCGCG*GCAAGGG ATC-3', and the complements of the DR4 element, gtNR1, DR3 element, PXRE, XRE, GRE1, and GRE2, respectively. Italicized characters show the bases of the mutation.

In reporter gene assays with 290-bp DNA, gtNR1 mutation result in a 100% decrease of both CAR- and PXR-dependent enhancer activities in HepG2 cells, whereas mutations of the DR4 element, PXRE, and DR3 element retain 56–76% in CAR-dependent enhancer activity and about 60% in Rif-induced enhancer activity (Fig. 2A). These observations suggest that gtNR1 (-3382/-3367) in 290-bp DNA plays a central role in *UGT1A1* gene expression by PB and Rif. Mutation of the GRE1 and GRE2 sites results in 75.1 ± 4.7 and 77.7 ± 4.1% decreases of 10^{-7} *M* DEX-induced enhancer activity, respectively, in HepG2 cells transfected with the GR expression plasmid (Fig. 2B). The CAR-dependent enhancer activity of GRE1- and GRE2-mutated constructs retains 85.2 ± 10.2 and 92.7 ± 3.0% of the control, respectively, and the PXR-dependent enhancer activity of the mutants retains 97.7 ± 17.8 and 90.4 ± 7.4% of the control, respectively (Fig. 2B). These results indicate that GRE1 (−3404/−3389) and GRE2 (−3251/−3236) contribute to the UGT1A1 gene expression via GR by DEX.

Deletion Assays. To delineate the minimal sequence activated by nuclear receptors, we prepared six subfragments located at nucleotide numbers −3483/−3434, −3433/−3384, −3392/−3333, −3343/−3284, −3293/−3234, and −3243/−3194 by amplifying with the following 12 primers and cloned them into pGL3-tk-firefly luciferase reporter plasmid at the *Nhe*I and *Xho*I sites (Sugatani *et al.*, 2004). The sequences of the primers are (1) 5'-gtttccgctagcTACACTAGTAAAGGTCACT-3', (2) 5'-gtttccgctagcACA TTCTAACGGTTCATA-3', (3) 5'-gtttccgctagcACATTCTAACGGTTC ATA-3', (4) 5'-gtttaactcgagGATAACACATCCTCATTA-3', (5) 5'-gtttc

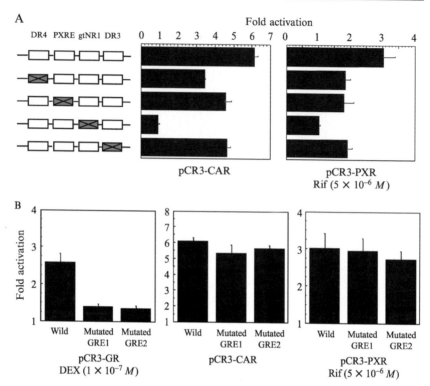

FIG. 2. The role of nuclear receptor-binding elements in the transcriptional activation of UGT1A1 290-bp DNA mediated by CAR, PXR, and GR (Sugatani *et al.*, 2005). Each mutated DNA (indicated by closed boxes with X) (0.2 μg) was cotransfected into HepG2 cells with pCR3-CAR, pCR3-PXR, or pCR3-GR expression vector or the control vector pCR3 (0.2 μg), together with pRL-SV40 (0.2 μg) as described in the text. Transfected cells were incubated for 24 h with the vehicle (dimethyl sulfoxide) alone or DEX (10^{-7} M) for GR-expressing cells, without the ligand for CAR-expressing cells, or with the vehicle (dimethyl sulfoxide) alone or Rif (5×10^{-6} M) for PXR-expressing cells, then harvested, and assayed for luciferase activity. Fold activation was calculated by dividing the activity with the inducer (DEX or Rif) by that without the inducer for GR- and PXR-expressing cells, respectively, or by dividing the activity in the presence of CAR by that in the absence of CAR for CAR-expressing cells. Data presented are the average of four independent experiments ± SD. DEX, dexamethasone; Rif, rifampicin.

cgctagcTGTGTTATCTCACCAGAA-3', (6) 5'-gtttaactcgagTTCAGGT-TATGTAACTAG-3', (7) 5'-gtttccgctagcCATAACCTGAAACCCGGA-3', (8) 5'-gtttaactcgagAGCTTACTATGACTGTTC-3', (9) 5'-gtttccgctagcA-TAGTAAGCTGGCCAAGG-3', (10) 5'-gtttaactcgagCCTTTGATGT TCTCAAAT-3', (11) 5'-gtttccgctagcACATCAAAGGAAGTTTGG-3', and

(12) 5′-gtttaactcgagCCCTCTAGCCATTCTGGA-3′. Bases in lowercase letters are added to digest the oligonucleotides with the restriction enzymes *Nhe*I and *Xho*I.

When we tested the enhancer activities of the six constructs (−3499/−3450, −3499/−3400, −3400/−3349, −3359/−3300, −3309/−3250, and −3259/−3210), activation of the reporter gene by Rif is abolished in all six fragments, even in the presence of exogenously expressed PXR. In contrast, in the presence of exogenously expressed GR, two fragments (−3406/−3349 and −3259/−3210) containing GRE1 and GRE2, respectively, produce full activation by 10^{-7} M DEX. In the presence of exogenously expressed CAR, three fragments (−3499/−3450, −3406/−3349, and −3309/−3250) containing the DR4 element, gtNR1, and DR3 element, respectively, are activated, but the extent of transcriptional activation is decreased to 8.1 ± 0.4, 13.6 ± 1.2, and 6.6 ± 0.4% of that of the 290-bp enhancer module, respectively. Thus, it appears that DR4 and DR3 need to confer full enhancer activity by CAR/PXR to 290-bp DNA, in addition to gtNR1.

Electrophoretic Mobility Shift Assays. Recombinant full-length human CAR or PXR is produced in *Escherichia coli* BL21[DE3(pLys)], to which pGEX4T.3-CAR, pGEX4T.3-PXR, or pGEX4T.3-GR is transformed, as N-terminal glutathione *S*-transferase (GST) fusion protein, and purified by binding to glutathione-Sepharose (Amersham Biosciences). The vectors for the expression of (1) human CAR, (2) human PXR, and (3) human GR can be constructed by inserting the corresponding cDNAs, which are amplified by PCR with the primers described earlier, into the *Bam*HI and *Xho*I sites of pGEX4T.3. The *in vitro* transcription/translation TNT rabbit reticulocyte system (Promega) is used to synthesize human RXRα and GR. The following sets of complementary oligonucleotides are synthesized and annealed for electrophoretic mobility shift assays. gtNR1, 5′-gatcTTCT-GAGTTTATATAACCTCTA-3′ and 5′-gatcTAGAGGTTATATAAAC TCAGAA-3′; DR4, 5′-gatcTAAAGGTCACTCAATTCCAAGG-3′ and 5′-gatcCCTTGGAATTGAGTGACCTTTA-3′; DR3, 5′-gatcTAGAGTT-CAGTTTGAACAAAG-3′ and 5′-gatcCTTTGTTCAAACTGAACTC-TA-3′; PXRE, 5′-gatcCTTTGTTCAAACTGAACTCTA-3′ and 5′-gatcT AATACCCTTTATGAACCGTT-3′; GRE1, 5′-gatcTGTTATCTCACCA GAACAAAC-3′ and 5′-gatcGTTTGTTCTGGTGAGATAACA-3′; GRE1 3′, 5′-gatcACCAGAACAAACTTCTGAGTTTAT-3′ and 5′-gatcATAAA CTCAGAAGTTTGTTCTGGT-3′; GRE2, 5′-gatcAAGGAAGTTTGGG-GAACAGCA-3′ and 5′-gatcTGCTGTTCCCCAAACTTCCTT-3′; GRE2 3′, 5′-gatcTGGGGAACAGCAAGGGATCCA-3′ and 5′-gatcTGGATCCC TTGCTGTTCCCCA-3′; and MMTV GRE (Malkoski and Dorin, 1999), 5′-gttggGTTACAAACTGTTCT-3′ and 5′-tggttAGAACAGTTTGTAAC-3′. Bases in lowercase letters are added to allow labeling of the oligonucleotides with Klenow fragment and $[\alpha\text{-}^{32}P]dCTP$. Binding assays are performed using GST-CAR (1 μg) and RXRα (1 μl), GST-PXR (1 μg) and RXRα (1 μl), or GR (1 μl) in 10 μl of binding buffer [10 mM Tris-HCl buffer (pH 7.5), 50 mM NaCl, 0.5 mM dithiothreitol, 1.5 μg poly(dI-dC), 10% glycerol, 0.05% NP40, and radiolabeled probe]. After incubation at room temperature for 15 min, the protein/DNA complexes are analyzed by electrophoresis on a 4% polyacrylamide gel with running buffer (7 mM Tris, 3 mM sodium acetate, and 1 mM EDTA, pH 7.5). For competition, a 100-fold concentration of unlabeled oligonucleotides is added before the radioactive probes.

In the previous study, we compared the binding affinity of CAR to nuclear receptor-binding sites, DR4, gtNR1, and DR3. Unfortunately, DR3, from HepG2 genomic DNA, has a T-3279G mutation (Sugatani *et al.*, 2002). Thus, we compared the binding affinities of CAR and PXR to wild-type DR4, gtNR1, and DR3, together with PXRE. Consistent with the mutation assays described earlier, a mixture of GST-CAR and *in vitro*-translated RXRα bound to gtNR1 most strongly (Fig. 3A), whereas GST-CAR alone did not bind to gtNR1 (Sugatani *et al.*, 2001). Binding of the CAR:RXRα heterodimer to DR4 and PXRE was at trace level. Furthermore, a mixture of GST-PXR and *in vitro*-translated RXRα also bound to gtNR1 most strongly among four nuclear receptor-binding sites, DR4, PXRE, gtNR1, and DR3 (Fig. 3B). The order of the binding affinity of PXR:RXRα heterodimer to nuclear receptor-binding sites was gtNR1 > DR3 > DR4. In addition to gtNR1, DR3 appeared to be the binding site of the PXR:RXRα heterodimer. These bands were specifically eliminated by the addition of 100-fold molar excesses of unlabeled probes. Thus, gtNR1 appeared to be the strongest binding site of both CAR:RXRα and PXR:RXRα heterodimers, indicating that gtNR1 plays the most important role in CAR- and PXR-mediated transcriptional activation. In addition, GR bound more strongly to GRE1 and GRE2 probes than GRE1 3′ and GRE2 3′ probes, producing protein–DNA complexes consistent with that observed on the binding of the dimeric GR-mouse mammary tumor virus (MMTV) GRE (Fig. 3C). GR showed decreased binding to GRE1 and GRE2 probes mutated at either the 5′ half-site or the 3′ half-site (Sugatani *et al.*, 2005). These bands were specifically eliminated by the addition of 100-fold molar excesses of unlabeled probes. Collectively, the regions of −3404/−3389 and −3251/−3236 are UGT1A1 GREs.

Concluding Remarks

Xenobiotics that induce human UGT1A1 include CAR activators, PB and methoxychlor (Sueyoshi *et al.*, 1999), PXR activators, Rif and clotrimazole (Lehmann *et al.*, 1998), and GR activators, DEX and hydrocortisone

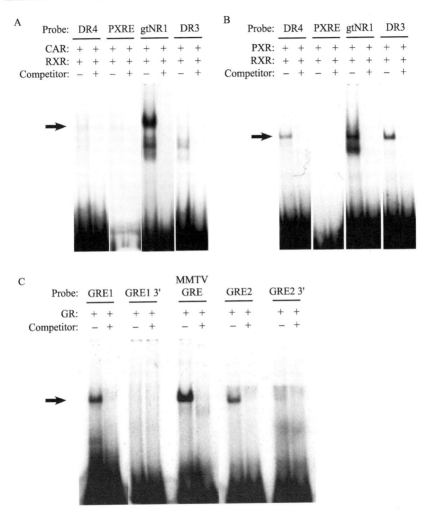

Fig. 3. Electrophoretic mobility shift assays for the binding of CAR, PXR, and GR to each nuclear receptor-binding site. The bacterially expressed GST-CAR and GST-PXR and *in vitro*-translated RXRα and GRα were prepared, and electrophoretic mobility shift assays were performed as described in the text.

(Hollenberg *et al.*, 1985). The 290-bp distal enhancer module gtPBREM fully accounted for the CAR-, PXR-, and GR-mediated gene activation of the *UGT1A1* gene (Sugatani *et al.*, 2001, 2004, 2005). The gtNR1 (−3382/−3367) within gtPBREM plays a central role in UGT1A1 induction mediated by both CAR and PXR, which is supported by the following: the site-directed mutagenesis of gtNR1 completely abolished CAR- and PXR-mediated transcriptional activities in the reporter gene assays, and both CAR:RXRα and PXR:RXRα heterodimers bound most strongly to gtNR1 in electrophoretic mobility shift assays. The so-called PXRE (Xie et al., 2003) neither bound to PXR nor contributed to the PXR activation of the 290-bp enhancer module. We also identified the GR-regulated enhancer elements located at $-3404/-3389$ and $-3251/-3236$ within the gtPBREM. Including XRE, the 290-bp gtPBREM can respond to numerous therapeutic drugs, xenobiotics, and steroid hormones to activate the transcription of the *UGT1A1* gene through the ability to cross talk among CAR, PXR, GR, and AhR, providing a useful system to test drug candidates for UGT1A1 induction.

Acknowledgments

We gratefully acknowledge the assistance of Kasumi Yamakawa, Shinichi Nishitani, and Eri Tonda. This work was supported in part by a Grant-in-Aid for Scientific Research (14572057 and 16590056) and The 21st Century COE Program from the Ministry of Education, Culture, Sports, Science and Technology and a Goto Research Grant from University of Shizuoka.

References

Cidlowski, J. A., Bellingham, D. L., Powell-Oliver, F. E., Lubahn, D. B., and Sar, M. (1990). Novel antipeptide antibodies to the human glucocorticoid receptor: Recognition of multiple receptor forms *in vitro* and distinct localization of cytoplasmic and nuclear receptors. *Mol. Endocrinol.* **4,** 1427–1437.

Chen, J., Lin, H., and Hu, M. (2003). Metabolism of flavonoids via enteric recycling: Role of intestinal disposition. *J. Pharmacol. Exp. Ther.* **304,** 1228–1235.

Choi, H.-S., Chung, M., Tzameli, I., Simha, D., Lee, Y.-K., Seol, W., and Moore, D. D. (1997). Differential transactivation by two isoforms of the orphan nuclear hormone receptor CAR. *J. Biol. Chem.* **272,** 23565–23571.

Gagne, J. F., Montminy, V., Belanger, P., Journault, K., Gaucher, G., and Guillemette, C. (2002). Common human UGT1A polymorphisms and the altered metabolism of irinotecan active metabolite 7-ethyl-10-hydroxycamptothecin (SN-38). *Mol. Pharmacol.* **62,** 608–617.

Hollenberg, S. M., Weinberger, C., Ong, E. S., Cerelli, G., Oro, A., Lebo, R., Thompson, E. B., Rosenfeld, M. G., and Evans, R. M. (1985). Primary structure and expression of a functional human glucocorticoid receptor cDNA. *Nature* **318,** 635–641.

Honkakoski, P., Moore, R., Washburn, K. A., and Negishi, M. (1998). Activation by diverse xenochemicals of the 51-base pair phenobarbital-responsive enhancer module in the CYP2B10 gene. *Mol. Pharmacol.* **53,** 597–601.

Iyanagi, T., Emi, Y., and Ikushiro, S. (1998). Biochemical and molecular aspects of genetic disorders of bilirubin metabolism. *Biochim. Biophys. Acta* **1407,** 173–184.

King, C. D., Green, M. D., Rios, G. R., Coffman, B. L., Owens, I. S., Bishop, W. P., and Tephly, T. R. (1996). The glucuronidation of exogenous and endogenous compounds by stably expressed rat and human UDP-glucuronosyltransferase 1.1. *Arch. Biochem. Biophys.* **332**, 92–100.

Lehmann, J. M., McKee, D. D., Watson, M. A., Willson, T. M., Moore, J. T., and Kliewer, S. A. (1998). The human orphan nuclear receptor PXR is activated by compounds that regulate *CYP3A4* gene expression and cause drug interactions. *J. Clin. Invest.* **102**, 1016–1023.

Mackenzie, P. I., Owens, I. S., Burchell, B., Bock, K. W., Bairoch, A., Belanger, A., Fournel-Gigleux, S., Green, M., Hum, D. W., Iyanagi, T., Lancet, D., Louisot, P., Magdalou, J., Chowdhury, J. R., Ritter, J. K., Schachter, H., Tephly, T. R., Tipton, K. F., and Nebert, D. W. (1997). The UDP glycosyltransferase gene superfamily: Recommended nomenclature update based on evolutionary divergence. *Pharmacogenetics* **7**, 255–269.

Maglich, J. M., Stoltz, C. M., Goodwin, B., Hawkins-Brown, D., Moore, J. T., and Kliewer, S. A. (2002). Nuclear pregnane x receptor and constitutive androstane receptor regulate overlapping but distinct sets of genes involved in xenobiotic detoxification. *Mol. Pharmacol.* **62**, 638–646.

Malkoski, S. P., and Dorin, R. I. (1999). Composite glucocorticoid regulation at a functionally defined negative glucocorticoid response element of the human corticotropin-releasing hormone gene. *Mol. Endocrinol.* **13**, 1629–1644.

Ostrow, J. D., and Murphy, M. H. (1970). Isolation and properties of conjugated bilirubin from bile. *Biochem. J.* **120**, 311–327.

Sueyoshi, T., Kawamoto, T., Zelko, I., Honkakoski, P., and Negishi, M. (1999). The repressed nuclear receptor CAR responds to phenobarbital in activating the human *CYP2B6* gene. *J. Biol. Chem.* **274**, 6043–6046.

Sueyoshi, T., and Negishi, M. (2001). Phenobarbital response elements of cytochrome P450 genes and nuclear receptors. *Annu. Rev. Pharmacol. Toxicol.* **41**, 123–143.

Sugatani, J., Kojima, H., Ueda, A., Kakizaki, S., Yoshinari, K., Gong, Q. H., Owens, I. S., Negishi, M., and Sueyoshi, T. (2001). The phenobarbital response enhancer module in the human bilirubin UDP-glucuronosyltransferase *UGT1A1* gene and regulation by the nuclear receptor CAR. *Hepatology* **33**, 1232–1238.

Sugatani, J., Yamakawa, K., Yoshinari, K., Machida, T., Takagi, H., Mori, M., Kakizaki, S., Sueyoshi, T., Negishi, M., and Miwa, M. (2002). Identification of a defect in the UGT1A1 gene promoter and its association with hyperbilirubinemia. *Biochem. Biophys. Res. Commun.* **292**, 492–497.

Sugatani, J., Yamakawa, K., Tonda, E., Nishitani, S., Yoshinari, K., Degawa, M., Abe, I., Noguchi, H., and Miwa, M. (2004). The induction of human UDP-glucuronosyltransferase 1A1 mediated through a distal enhancer module by flavonoids and xenobiotics. *Biohem. Pharmacol.* **67**, 989–1000.

Sugatani, J., Nishitani, S., Yamakawa, K., Yoshinari, K., Sueyoshi, T., Negishi, M., and Miwa, M. (2005). Transcriptional regulation of human UGT1A1 gene expression: Activated GR enhances CAR/PXR-mediated UGT1A1 regulation with GRIP1. *Mol. Pharmacol.* **67**, 845–855.

Tukey, R. H., and Strassburg, C. P. (2000). Human UDP-glucuronosyltransferases: Metabolism, expression, and disease. *Annu. Rev. Pharmacol. Toxicol.* **40**, 581–616.

Tukey, R. H., Strassburg, C. P., and Mackenzie, P. I. (2002). Pharmacogenomics of human UDP-glucuronosyltransferases and irinotecan toxicity. *Mol. Pharmacol.* **62**, 446–450.

Watkins, R. E., Davis-Searles, P. R., Lambert, M. H., and Redinbo, M. R. (2003). Coactivator binding promotes the specific interaction between ligand and the pregnane X receptor. *J. Mol. Biol.* **331,** 815–828.

Xie, W., Yeuh, M.-F., Radminska-Pandye, A., Saini, S. P. S., Negishi, Y., Bottroff, B. S., Cabrera, G. Y., Tukey, R. H., and Evans, R. M. (2003). Control of steroid, heme, and carcinogen metabolism by nuclear pregnane X receptor and constitutive androstane receptor. *Proc. Natl. Acad. Sci. USA* **100,** 4150–4155.

Yaffe, S. J., Levy, G., Matsuzawa, T., and Baliah, T. (2003). Enhancement of glucuronide-conjugating capacity in a hyperbilirubinemic infant due to apparent enzyme induction by phenobarbital. *N. Engl. J. Med.* **275,** 1461–1466.

Yueh, M.-F., Huang, Y.-H., Chen, S., Nguyen, N., and Tukey, R. H. (2003). Involvement of the xenobotic response element (XRE) in Ah-receptor mediated induction of human UDP-glucuronosyltransferase 1A1. *J. Biol. Chem.* **278,** 15001–15006.

Zelko, I., and Negishi, M. (2000). Phenobarbital-elicited activation of nuclear receptor CAR in induction of cytochrome P450 genes. *Biochem. Biophys. Res. Commun.* **277,** 1–6.

[7] Isoform-Selective Probe Substrates for *In Vitro* Studies of Human UDP-Glucuronosyltransferases

By Michael H. Court

Abstract

The majority of UDP-glucuronosyltransferases (UGT), like other drug-metabolizing enzymes, display broad and often overlapping substrate specificities, complicating evaluation of the function of individual UGT isoforms within human tissues. Despite this, there have been recent advances in identifying UGT-selective probes—UGT substrates that are primarily glucuronidated by a single isoform. Such probes can be used to (1) facilitate identification of UGT isoforms mediating a particular glucuronidation activity in human liver through activity correlation analysis; (2) evaluate the role of particular UGTs in drug–drug interactions through either enzyme induction or inhibition; and (3) elucidate the functional significance of genetic polymorphisms associated with the gene encoding the UGT of interest. UGT-selective probes currently being used in our laboratory for the evaluation of glucuronidation activities in human liver tissues include estradiol (3OH-glucuronidation; UGT1A1), trifluoperazine (UGT1A4) serotonin (UGT1A6), propofol (UGT1A9), 3'-azidothymidine (UGT2B7), and S-oxazepam (UGT2B15). *In vitro* incubation protocols and the HPLC analysis methods used to determine each of these activities are described in detail. Future work is needed to elucidate more highly selective probes

METHODS IN ENZYMOLOGY, VOL. 400
0076-6879/05 $35.00
DOI: 10.1016/S0076-6879(05)00007-8

than those in current usage, as well as probes for the extrahepatic UGT isoforms.

Introduction

Glucuronidation represents a major metabolic pathway that facilitates efficient detoxification and elimination of potentially harmful xenobiotics (drugs, toxins, carcinogens) and endobiotics (bilirubin, bile acids, hormones, inflammatory mediators) (for a review, see Fisher *et al.*, 2001). This reaction is catalyzed by UDP-glucuronosyltransferases, a superfamily of enzymes known in humans to contain at least 17 distinct isoforms. UGTs are expressed primarily in the liver, although extrahepatic tissues (particularly kidney and aerodigestive tracts) are also rich sources, reflecting the critical role for these enzymes in protecting against potentially harmful xenobiotics. The majority of UGTs, like other xenobiotic-metabolizing enzymes, display broad and overlapping substrate specificities. While this provides an effective protective barrier against xenobiotic exposure, it greatly complicates study of the function of individual isoforms within human tissues. Despite this, substantial progress has been made in identifying substrates that are selectively glucuronidated by individual UGT isoforms (see Table I).

Uses for UGT-Selective Probes

UGT-selective probes have a number of important applications in the study of drug glucuronidation. First, they are used to substantiate the identification of UGTs mediating a particular glucuronidation reaction in human tissues via activity correlation analysis. For example, we have verified the role of UGT2B15 in S-lorazepam glucuronidation by human liver by showing significant correlation between this activity and S-oxazepam glucuronidation activities (a UGT2B15 probe) measured in the same bank of human liver microsomes (HLMs) (Fig. 1A). Second, UGT-selective probes can be used to evaluate the role of particular UGTs in drug–drug interactions through either enzyme induction or inhibition. Using a human intestinal cell line (LS-180) we have substantiated UGT1A4 as a target gene for pregnane X receptor-mediated induction of lamotrigine glucuronidation by rifampin (Fig. 1B), confirming the observed enhancement of the *in vivo* clearance of lamotrigine by rifampin (Ebert *et al.*, 2000). Furthermore, using this same model system, we have identified ritonavir as a potential inhibitor of glucuronidation by UGT1A4 (Fig. 1B). Finally, glucuronidation activities measured using a UGT-selective probe in tissues from different individuals can be used as a phenotype measure to elucidate

TABLE I
POTENTIALLY USEFUL ISOFORM-SELECTIVE PROBE SUBSTRATES FOR UGTs EXPRESSED IN HUMAN LIVER

UGT isoform	Probe substrate	Evidence for selectivity			Effect of polymorphism		Reference
		Other rUGT activities		HLMs vs rUGT K_m			
		Hepatic[a]	Extrahepatic		HLMs	In vivo	
UGT1A1	Bilirubin	None	None	=	UGT1A1*28	UGT1A1*28	Patten et al. (2001)
	Estradiol (3-OH)	UGT1A3	UGT1A8, 10	=	UGT1A1*28	ND[d]	Patten et al. (2001)
	SN-38	UGT1A6, 9	UGT1A7, 8, 10	=	UGT1A1*28	UGT1A1*28	Gagne et al. (2002)
	Etoposide	None	None	=	ND	ND	Watanabe et al. (2003)
UGT1A3	R-lorazepam	None[c]	None[c]	ND	ND	ND	Unpublished
	(F)6-1,23,25 (OH)3 vitamin D3	UGT1A4, 2B4, 7	None	=(high affinity)	ND	ND	Kasai et al. (2005)
UGT1A4	Trifluoperazine	None	None	=	ND	ND	Dehal et al. (2001)
	Imipramine	None	None	=(low affinity)	ND	ND	Nakajima et al. (2002)
UGT1A6	Lamotrigine	None	None	ND	ND	ND	Unpublished
	Serotonin	None	None	=	No UGT1A6*2-*5	ND	Krishnaswamy et al. (2003)
	5-Hydroxy-tryptophol	UGT1A9 [high]	None	=	No UGT1A6*2-*5	ND	Krishnaswamy et al. (2004)
UGT1A9	Propofol	None	UGT1A7, 8, 10	HLMs > rUGT (4×)	UGT1A9*22	ND	Unpublished
	Mycophenolic acid (7-OH)	UGT1A1 [low]	UGT1A7, 8, 10	HLMs > rUGT (2×)	UGT1A9*22	ND	Bernard et al. (2004)

UGT2B4	Codeine	UGT2B7[b]	None	=	ND	ND	Court et al. (2003)
UGT2B7	AZT	UGT2B4, 17	None	HLMs > rUGT (2×)	No UGT2B7*2	ND	Court et al. (2003)
	Morphine (3-OH)	UGT1A3, 9, 2B4, 15,17	UGT1A10	HLMs > rUGT(6×)	No UGT2B7*2	No UGT2B7*2	Court et al. (2003)
	Morphine (6-OH)	UGT1A1,3	None	HLMs > rUGT (3×)	No UGT2B7*2	No UGT2B7*2	Court et al. (2003)
	Epirubicin	None	None	HLMs > rUGT (4×)	No UGT2B7*2	ND	Innocenti et al. (2001)
UGT2B15	S-oxazepam	UGT1A1, 6, 2B7	None	=	UGT2B15*2	ND	Court et al. (2002)
	S-lorazepam	None[c]	None[c]	ND	UGT2B15*2	UGT2B15*2	Unpublished
	E-OH-tamoxifen	UGT1A1, 4, 9, 2B7	None	=	ND	ND	Nishiyama et al. (2002)
	5-OH-rofecoxib	UGT1A9, 2B7	None	HLMs > rUGT (3×)	ND	ND	Zhang et al. (2003)
UGT2B17	Dihydrotestosterone	UGT2B7, 15	ND	ND	ND	ND	Turgeon et al. (2001)

[a] Activities less that 20% that of the primary UGT (except for codeine).
[b] Note that although preliminary evidence indicates UGT2B4 may be the primary enzyme, there is also significant contribution from UGT2B7.
[c] Apparent lack of other rUGT activities could be a consequence of limitations in assay sensitivity.
[d] ND = not determined.

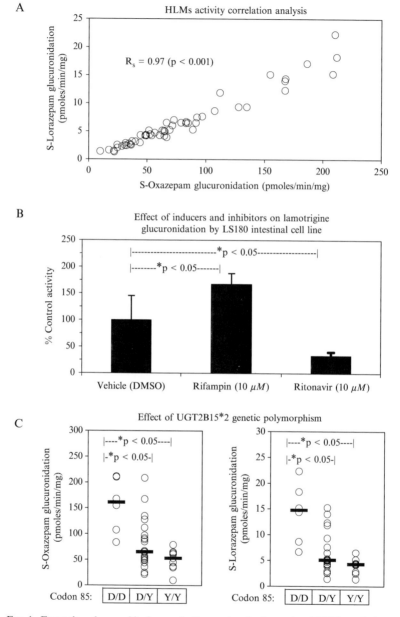

FIG. 1. Examples of uses of isoform-selective probes in the study of UGT regulation and function in human tissues and cell lines. (A) A highly significant correlation (R_s: Spearman correlation coefficient) between S-oxazepam (UGT2B15 probe) and S-lorazepam glucuronidation activities measured in the same set of human liver microsomes ($n = 54$) is consistent

the functional significance of genetic polymorphisms associated with the gene encoding the UGT of interest. Using a bank of HLMs and recombinant UGTs (rUGTs), we have shown that the UGT2B15*2 (D85Y) polymorphism is associated with over 60% lower oxazepam (Court *et al.*, 2004) and lorazepam (unpublished data) glucuronidation activities compared with wild-type UGT2B15*1 (Fig. 1C).

Although potentially any class of UGT substrate (including carcinogens and toxins) may serve as an *in vitro* probe, we have focused primarily on identifying drugs or endogenous substrates, as these compounds could also have utility as UGT-selective probes for *in vivo* glucuronidation studies. We have also concentrated on evaluating probes for the hepatic UGT isoforms given the relative importance of this organ in drug glucuronidation and availability of this tissue for study.

Evaluation of UGT Probe Selectivity

UGT substrates that are either being used as UGT-selective probes or show potential for this purpose based on experimental data from the literature (Bernard and Guillemette, 2004; Court *et al.*, 2002, 2003; Dehal *et al.*, 2001; Gagne *et al.*, 2002; Innocenti *et al.*, 2001; Kasai *et al.*, 2005; Krishnaswamy *et al.*, 2003, 2004; Nakajima *et al.*, 2002; Nishiyama *et al.*, 2002; Patten *et al.*, 2001; Turgeon *et al.*, 2001; Watanabe *et al.*, 2003; Zhang *et al.*, 2003) or unpublished work from this laboratory are listed in Table I. Three criteria that we have found useful in evaluating the potential selectivity of these substrates include comparative rUGT activities, human tissue versus rUGT K_m value, and effect of genetic polymorphism on glucuronidation activity. Ideally, only one rUGT (i.e., the target UGT) shows glucuronidation activity toward the UGT-selective probe, and the K_m value for this rUGT is essentially identical to that of human tissues (HLMs), as has been found for serotonin glucuronidation by UGT1A6 (Krishnaswamy *et al.*, 2003) and trifluoperazine glucuronidation by UGT1A4 (Dehal *et al.*, 2001). However, in many instances other rUGTs are found to have some activity and the K_m value of human tissue tends to be higher than the K_m value of the target rUGT. Consequently, it is often

---

with a major role for UGT2B15 in S-lorazepam glucuronidation. (B) Lamotrigine (UGT1A4 probe) glucuronidation by LS180 human intestinal cell line homogenates is enhanced by a 72-h treatment with 10 μM rifampin, suggesting regulation of UGT1A4 gene expression by the pregnane X receptor (PXR). In contrast, 10 μM ritonavir treatment for 72 h was associated with decreased lamotrigine glucuronidation. (C) Carriers of UGT2B15*2 polymorphism (85D/Y and 85Y/Y) show lower glucuronidation of both S-oxazepam (UGT2B15 probe) and S-lorazepam measured in a bank of human liver microsomes ($n = 38$ male donors).

necessary to account for the abundance of off-target UGTs in the tissue being studied and to use probe substrate concentrations that approximate the K_m value for the target UGT. For example, despite significant glucuronidation by extrahepatic forms UGTs 1A7, 8, and 10, propofol may be an appropriate probe for UGT1A9 in human liver when used at a substrate concentration of 100 μM. Finally, genetic polymorphisms that have been shown to influence (decrease or increase) the abundance or intrinsic activity of a particular UGT should have a similar effect on the probe substrate activity for that UGT. Examples include the effects of UGT1A1*28 on bilirubin, SN-38, and estradiol (3OH) glucuronidation (Fisher *et al.*, 2000), UGT1A9*22 on propofol and mycophenolic acid glucuronidation (Girard *et al.*, 2004), and UGT2B15*2 on S-oxazepam and S-lorazepam glucuronidation (Court *et al.*, 2004; unpublished data).

The primary UGT-selective probes that are currently being used in our laboratory for the evaluation of human liver glucuronidation activities include estradiol (3OH-glucuronide; UGT1A1), trifluoperazine (UGT1A4), serotonin (UGT1A6), propofol (UGT1A9), 3′-azidothymidine (AZT; UGT2B7), and S-oxazepam (UGT2B15). The methods used to determine glucuronidation activities using these probes are described in the following sections.

Measurement of Probe Glucuronidation Activities

Incubation Protocol

Details regarding incubation conditions specific to each substrate are given in Table II. The following description assumes a 100-μl final incubation volume, although it can be scaled to other volumes. Substrates, glucuronide standards, and internal standards are conveniently prepared by dissolving in methanol (some additional water may be needed to dissolve glucuronides) and stored at $-20°$. Substrate is added to each incubation tube (0.5- or 1.5-ml polypropylene microcentrifuge) and dried down in a vacuum oven (set at 45°) or Speed-Vac microcentrifuge. An exception to this is propofol (UGT1A9 substrate), which is prepared in dimethyl sulfoxide (DMSO) and added to each tube (100 μM propofol and 1% DMSO final concentrations). Tubes are then placed on ice, and tissue microsomes, alamethicin (2.5 μg/μl methanol; 50 μg alamethicin/mg microsomal protein), and aqueous buffer added to a volume of 50 μl. Some laboratories routinely add saccharolactone (5 mM) to incubations to inhibit endogenous β-glucuronidase activity. We have not found this necessary for these assays and would caution that saccharolactone can be inhibitory for some glucuronidation activities (Alkharfy and Frye, 2001). We normally use 50 mM

TABLE II
DETAILS OF *In Vitro* INCUBATION METHODS USED FOR DETERMINING UGT-SELECTIVE GLUCURONIDATION ACTIVITIES

Substrate[a]	Substrate concentration	Protein concentration	Incubation time (min)	Internal standard	Metabolite[a] assayed
Estradiol	100 μM	0.25 mg/ml	30	Phenacetin	Estradiol-3-glucuronide
Trifluoperazine	200 μM	0.25 mg/ml	30	Acetaminophen	Trifluoperazine-glucuronide[b]
Serotonin	4 mM	0.05 mg/ml	30	Acetaminophen	Serotonin-glucuronide[c]
Propofol	100 μM	0.25 mg/ml	30	Thymol	Propofol-glucuronide[b]
3'-azidothymidine (AZT)	500 μM	0.5 mg/ml	120	3-Acetamidophenol	AZT-glucuronide[d]
Oxazepam	100 μM	0.5 mg/ml	120	Phenacetin	S-oxazepam-glucuronide[b]

[a] Unless otherwise indicated, compounds were from Sigma-Aldrich (St. Louis, MO).
[b] No commercial source of these compounds available as yet.
[c] Available from SRI International (Menlo Park, CA) through contract to the NIMH Chemical Synthesis Program.
[d] Available from Toronto Research Chemicals (North York, Ontario, Canada).

potassium phosphate in water (pH 7.5) as the primary buffer, although Tris buffer (50–100 mM in water) works equally well. A UDP-glucuronic acid (UDPGA) cofactor solution is also prepared on ice containing UDPGA (5 mM final concentration) and MgCl$_2$ (5 mM final concentration) in buffer.

Incubation tubes are preincubated for 5 min in a water bath set at 37°, and reactions are initiated by the addition of 50 μl of the UDPGA cofactor solution to each tube, vortexing briefly, and returning to the water bath. Reactions are terminated by the addition of 100 μl of ice-cold methanol (or acetonitrile) containing the appropriate internal standard. Tubes are then centrifuged at 14,000g for 10 min, and the supernatant is transferred to glass HPLC tubes, dried down by a vacuum oven, reconstituted with 100 μl of water, and analyzed for glucuronide concentration by HPLC.

HPLC Analysis

The HPLC methods used to measure glucuronide concentrations are given in Table III. The minimum HPLC equipment setup would include a solvent pump capable of variable mixing (for mobile phase gradient formation), variable wavelength UV absorbance detector, and C$_{18}$ HPLC column (4.6 × 250 mm). A fluorescence detector is optional but provides higher sensitivity and ready detection of serotonin glucuronide. HPLC column packing materials capable of working under highly aqueous conditions, such as Synergi Hydro-RP (Phenomenex, Torrance, CA), are recommended for the analysis of polar glucuronides to avoid problems related to stationary phase collapse and subsequent loss of column performance.

Normally, analysis of 10 to 50 μl of sample is sufficient for most assays. Analyte peaks are identified by comparison of peak retention times to that of injected pure reference standards (if available). The identity of glucuronide peaks can also be verified by showing absence in negative control samples (no incubation; no UDPGA) and following treatment of positive samples with β-glucuronidase. The chromatogram "overlay" capability of modern HPLC systems is particularly useful for the purpose of comparing peaks in positive and negative control samples. Glucuronide peak identification is also simplified greatly through use of a HPLC/mass spectrometry detector, if available.

Data Analysis

Standard curves are generated by preparing a series of known amounts of pure glucuronide in incubation buffer. In instances where a glucuronide

TABLE III
DETAILS OF HPLC METHODS USED TO ASSAY FOR GLUCURONIDES GENERATED BY UGT-SELECTIVE GLUCURONIDATION ACTIVITIES

Glucuronidation activity	HPLC conditions[a]	UV absorbance wavelength	Glucuronide retention time (RT)	Substrate RT	Internal standard RT
Estradiol-3-glucuronidation	20–30% A over 15 min; balance with B	280 nm	9 min — E-3-glu; 10 min — E-17-glu	19 min	13 min
Trifluoperazine glucuronidation	10–70% A over 20 min; balance with C	254 nm	14 min	15 min	7 min
Serotonin glucuronidation	5% A for 8 min, 5–50% A over next 9 min; balance with B	270 nm(225 nm EX/330 nm EM[b])	5 min	8 min	14 min
Propofol glucuronidation	20–100% A over 20 min; balance with B	214 nm	7 min	16 min	14 min
AZT glucuronidation	15% A for 15 min, 15–50% A over next 10 min; balance with D	266 nm	7 min	11 min	8 min
S-oxazepam glucuronidation	25% mobile phase A for 15 min; 25–60% mobile phase A over 10 min Balance with mobile phase B	214 nm	8 min — R-oxazepam-glucuronide; 9 min — S-oxazepam-glucuronide	25 min	17 min

[a] Flow rate is 1 ml/min, column is 4.6 × 250 mm Synergy Hydro-RP (Phenomenex) ; mobile phase A (A): acetonitrile; mobile phase B (B): 20 mM potassium phosphate buffer in water, pH 4.5; mobile phase C (C): 0.1% trifluoroacetic acid in water; mobile phase D (D): 20 mM potassium phosphate buffer in water, pH 2.2.

[b] Use of an additional fluorescence detector set at these excitation (EX) and emission (EM) wavelengths is optional but provides higher sensitivity and ready identification of serotonin glucuronide. Note that a serially connected UV detector is still needed for quantitation of the internal standard.

standard is not available, it is possible to obtain activity estimates (expressed as "glucuronide equivalents") by use of a standard curve generated with known amounts of substrate, assuming similar UV absorbance of substrate and glucuronide, as in most cases this assumption is correct. Use of an internal standard is highly recommended to enhance assay precision. For each standard, the ratio of the glucuronide peak area to the internal standard peak area is determined and plotted against the absolute amount of added standard. Linear regression is then used to obtain a standard curve slope (usually forced through the origin), which is in turn used to calculate the amount of glucuronide within unknown samples. Sufficient standard curve points should be used to bracket all unknown points.

Glucuronidation activities are calculated by dividing the amount of glucuronide formed by incubation time and protein concentration. For reference purposes, the ranges of glucuronidation activities for HLMs determined previously are:

Estradiol-3-glucuronidation: 0.04–1.2 nmol/min/mg protein
Trifluoperazine glucuronidation: 0.03–1.5 nmol/min/mg protein
Serotonin glucuronidation: 1.9–33 nmol /min/mg protein
Propofol glucuronidation: 0.4–5.6 nmol/min/mg protein
AZT glucuronidation: 0.03–0.65 nmol/min/mg protein
S-oxazepam glucuronidation: 0.008–0.23 nmol/min/mg protein

Concluding Remarks

Finally, it should be pointed out that while the particular probes described here have proven of substantial utility in studies in our laboratory (and elsewhere), based on the evidence provided in Table I not all of these probes are considered ideal in terms of isoform selectivity. Moreover, isoform-selective substrates have not yet been identified for the important extrahepatic UGT isoforms, such as UGTs 1A7, 8, and 10. Given the growing number of drugs and other compounds that are being recognized as UGT substrates, it is likely that novel and more highly selective probes than those in current usage will be identified within the near future.

Acknowledgments

This work was supported in part by Grant GM-61834 from the National Institutes of Health (Bethesda, MD) and by Pfizer Global Research and Development (Ann Arbor, MI).

References

Alkharfy, K. M., and Frye, R. F. (2001). High-performance liquid chromatographic assay for acetaminophen glucuronide in human liver microsomes. *J. Chromatogr. B Biomed. Sci. Appl.* **753**, 303–308.

Bernard, O., and Guillemette, C. (2004). The main role of UGT1A9 in the hepatic metabolism of mycophenolic acid and the effects of naturally occurring variants. *Drug Metab. Dispos.* **32**, 775–778.

Court, M. H., Duan, S. X., Guillemette, C., Journault, K., Krishnaswamy, S., von Moltke, L. L., and Greenblatt, D. J. (2002). Stereoselective conjugation of oxazepam by human UDP-glucuronosyltransferases (UGTs): S-oxazepam is glucuronidated by UGT2B15, while R-oxazepam is glucuronidated by UGT2B7 and UGT1A9. *Drug Metab. Dispos.* **30**, 1257–1265.

Court, M. H., Krishnaswamy, S., Hao, Q., Duan, S. X., Patten, C. J., von Moltke, L. L., and Greenblatt, D. J. (2003). Evaluation of 3'-azido-3'-deoxythymidine, morphine, and codeine as probe substrates for UDP-glucuronosyltransferase 2B7 (UGT2B7) in human liver microsomes: Specificity and influence of the UGT2B7*2 polymorphism. *Drug Metab. Dispos.* **31**, 1125–1133.

Court, M. H., Hao, Q., Krishnaswamy, S., Bekaii-Saab, T., Al-Rohaimi, A., Von Moltke, L. L., and Greenblatt, D. J. (2004). UDP-Glucuronosyltransferase (UGT) 2B15 pharmacogenetics: UGT2B15 D85Y genotype and gender are major determinants of oxazepam glucuronidation by human liver. *J. Pharmacol. Exp. Ther.* **310**, 656–665.

Dehal, S. S., Gange, P. V., Crespi, C. L., and Patten, C. J. (2001). Characterization of a probe substrate and an inhibitor of UDP-glucuronosyltransferase 1A4 activity in human liver microsomes and cDNA-expressed UGT-enzymes. *Drug Metab. Rev.* **33**, 162. [Abstract].

Ebert, U., Thong, N. Q., Oertel, R., and Kirch, W. (2000). Effects of rifampicin and cimetidine on pharmacokinetics and pharmacodynamics of lamotrigine in healthy subjects. *Eur. J. Clin. Pharmacol.* **56**, 299–304.

Fisher, M. B., Paine, M. F., Strelevitz, T. J., and Wrighton, S. A. (2001). The role of hepatic and extrahepatic UDP-glucuronosyltransferases in human drug metabolism. *Drug Metab. Rev.* **33**, 273–297.

Fisher, M. B., Vandenbranden, M., Findlay, K., Burchell, B., Thummel, K. E., Hall, S. D., and Wrighton, S. A. (2000). Tissue distribution and interindividual variation in human UDP-glucuronosyltransferase activity: Relationship between UGT1A1 promoter genotype and variability in a liver bank. *Pharmacogenetics* **10**, 727–739.

Gagne, J. F., Montminy, V., Belanger, P., Journault, K., Gaucher, G., and Guillemette, C. (2002). Common human UGT1A polymorphisms and the altered metabolism of irinotecan active metabolite 7-ethyl-10-hydroxycamptothecin (SN-38). *Mol. Pharmacol.* **62**, 608–617.

Girard, H., Court, M. H., Bernard, O., Fortier, L. C., Villeneuve, L., Hao, Q., Greenblatt, D. J., von Moltke, L. L., Perussed, L., and Guillemette, C. (2004). Identification of common polymorphisms in the promoter of the UGT1A9 gene: Evidence that UGT1A9 protein and activity levels are strongly genetically controlled in the liver. *Pharmacogenetics* **14**, 501–515.

Innocenti, F., Iyer, L., Ramirez, J., Green, M. D., and Ratain, M. J. (2001). Epirubicin glucuronidation is catalyzed by human UDP-glucuronosyltransferase 2B7. *Drug Metab. Dispos.* **29**, 686–692.

Kasai, N., Sakaki, T., Shinkyo, R., Ikushiro, S., Iyanagi, T., Ohta, M., and Inouye, K. (2005). Metabolism of 26,26,26,27,27,27-F(6)-1alpha,23S,25-trihydroxyvitamin D(3) by human UDP-glucuronosyltransferase 1A3. *Drug Metab. Dispos.* **33**, 102–107.

Krishnaswamy, S., Duan, S. X., von Moltke, L. L., Greenblatt, D. J., and Court, M. H. (2003). Validation of serotonin (5-hydroxtryptamine) as an *in vitro* substrate probe for human UDP-glucuronosyltransferase (UGT) 1A6. *Drug Metab. Dispos.* **31,** 133–139.

Krishnaswamy, S., Hao, Q., von Moltke, L. L., Greenblatt, D. J., and Court, M. H. (2004). Evaluation of 5-hydroxytryptophol and other endogenous serotonin (5-hydroxytryptamine) analogs as substrates for udp-glucuronosyltransferase 1a6. *Drug Metab. Dispos.* **32,** 862–869.

Nakajima, M., Tanaka, E., Kobayashi, T., Ohashi, N., Kume, T., and Yokoi, T. (2002). Imipramine N-glucuronidation in human liver microsomes: Biphasic kinetics and characterization of UDP-glucuronosyltransferase isoforms. *Drug Metab. Dispos.* **30,** 636–642.

Nishiyama, T., Ogura, K., Nakano, H., Ohnuma, T., Kaku, T., Hiratsuka, A., Muro, K., and Watabe, T. (2002). Reverse geometrical selectivity in glucuronidation and sulfation of cis- and trans-4-hydroxytamoxifens by human liver UDP-glucuronosyltransferases and sulfotransferases. *Biochem. Pharmacol.* **63,** 1817–1830.

Patten, C. J., Code, E. L., Dehal, S. S., Gange, P. V., and Crespi, C. L. (2001). Analysis of UGT enzyme levels in human liver microsomes using form specific anti-peptide antibodies, probe substrate activities and recombinant UGT enzymes. *Drug Metab. Rev.* **33,** 165.[Abstract].

Turgeon, D., Carrier, J. S., Levesque, E., Hum, D. W., and Belanger, A. (2001). Relative enzymatic activity, protein stability, and tissue distribution of human steroid-metabolizing UGT2B subfamily members. *Endocrinology* **142,** 778–787.

Watanabe, Y., Nakajima, M., Ohashi, N., Kume, T., and Yokoi, T. (2003). Glucuronidation of etoposide in human liver microsomes is specifically catalyzed by UDP-glucuronosyltransferase 1A1. *Drug Metab. Dispos.* **31,** 589–595.

Zhang, J. Y., Zhan, J., Cook, C. S., Ings, R. M., and Breau, A. P. (2003). Involvement of human UGT2B7 and 2B15 in rofecoxib metabolism. *Drug Metab. Dispos.* **31,** 652–658.

[8]   Structure of UDP-Glucuronosyltransferases in Membranes

By Anna Radominska-Pandya, Mohamed Ouzzine,
Sylvie Fournel-Gigleux, and Jacques Magdalou

Abstract

This chapter presents the most recent experimental approaches to the investigation of UDP-glucuronosyltransferase (UGTs) in membranes. The first topic described is the subcellular localization of UGTs with special emphasis on the association of these proteins with the endoplasmic reticulum (ER). Experimental methods include subfractionation of tissue for microsome preparation, evaluation of the purity of the membrane fraction obtained, and measurement of UGT activity in the presence of detergents. Next, the recently demonstrated formation of UGT homo- and

METHODS IN ENZYMOLOGY, VOL. 400   0076-6879/05 $35.00

heterodimer formation and its functional relevance is discussed and the appropriate methods used to characterize such interactions are given (radiation inactivation, size exclusion chromatography, immunopurification, cross-linking, two-hybrid system). The structural determinants of UGTs in relation to membrane association, residency, and enzymatic activity are the next topic, supplemented by a description of the appropriate methods, including the design and expression of chimeric proteins, membrane insertion, and subcellular localization by immunofluorescence. Also presented is new information on the structure and function of UGTs obtained by molecular modeling, bioinformatics (sequence alignment), and comparison with selected crystallized glycosyltransferases. Finally, we discuss the important, and still not fully developed, issue of UGT active site architecture and organization within the ER. This is addressed from two perspectives: (1) chemical modification of UGT active sites by amino acid-specific probes and (2) photoaffinity labeling of UGTs. The detailed synthesis of a photoaffinity probe for an aglycon-binding site is provided and the use of this probe and direct photoaffinity labeling with retinoids is discussed. The application of proteomics techniques, including proteolytic digestion and protein sequencing by liquid chromatography/tandem mass spectrometry and *matrix-assisted laser desorption ionization/time of flight*, to the identification of crucial amino acids of the active sites, and subsequent site-directed mutagenesis of identified amino acids, is discussed in detail.

Introduction: Characterization of Membrane-Bound UDP-Glucuronosyltransferase (UGT) Proteins

UDP-Glucuronosyltransferases are a family of membrane glycoproteins located in the endoplasmic reticulum (ER) that can be conceptually divided into two parts: the amino (N-) and carboxy (C-) terminal domains. UGTs are synthesized as precursors of about 530 residues that contain an N-terminal signal peptide that mediates the integration of the polypeptide chain into the ER (Mackenzie and Owens, 1984; Ouzzine *et al.*, 1999a,b). The mature protein of about 505 residues is classified as a type I ER transmembrane protein (Meech and Mackenzie, 1998) with a lumenal domain consisting of about 95% of the polypeptide chain and a cytoplasmic domain of only 20 or so residues (Mackenzie, 1986b, 1987). A conserved region of 17 hydrophobic residues between a N-terminal aspartate and a C-terminal lysine connects the two domains through the ER lipid bilayer (Iyanagi *et al.*, 1986; Mackenzie, 1986b).

UGTs have been a subject of intense research during the last several decades. Even though these enzymes have been investigated from the perspectives of regulation, toxicology, oncology, endocrinology, and drug development, few studies have examined the structural properties of UGTs. A UGT crystal structure is not available; there are only a few reports of computer-aided molecular modeling being applied to this system (Coffman *et al.*, 2001, 2003) and site-directed mutagenesis studies are limited (Iwano *et al.*, 1997, 1999; Senay *et al.*, 1997). However, several novel studies related to the characterization of membrane-bound UGTs have been published by Magdalou and colleagues (Barre *et al.*, 2005; Ouzzine *et al.*, 1999a,b) and others (Kurkela *et al.*, 2004), which are addressed in detail in this work. Additionally, progress has been made in the identification of UGT-binding sites and will supplement previous structural studies that have resulted in the prediction of the organization of UGT active site structures by utilizing selective inhibitors, amino acid-specific chemical modification reagents, and amino acid alignments. Specifically, new developments in photoaffinity labeling have been applied to the characterization of UGT active sites with the goal of identifying critical motifs, which are discussed here.

Molecular Mechanism of Membrane Targeting and Translocation of UGTs

Subcellular Localization

Determination of the localization of UGTs within the different subcellular compartments is an important issue with regard to the toxicity, mutagenicity, and metabolism of drugs and foreign compounds. Such subfractionation is also necessary to isolate UGT-enriched preparations, allowing functional studies of these enzymes and their purification for structural investigations.

Methodological Approach. The subcellular localization of UGTs has been mainly investigated in the liver, which is the main organ involved in drug glucuronidation. Subfractionation of hepatic tissue is classically performed using homogenates of liver tissue prepared at $4°$ in 100 mM Tris-HCl, 0.25 M sucrose, 1 mM EDTA, pH 7.8, in a Potter–Elvehjem (Teflon pestle) homogenizer in ice. This crude homogenate is centrifuged at $4°$ for 10 min at 1000g, and the pellet, consisting of cellular debris and nuclei, is discarded. The supernatant is subsequently centrifuged at $4°$ for 20 min at 10,000g. The resulting pellet contains mainly mitochondria. The supernatant is centrifuged at $4°$ for 60 min at 105,000g to give the microsomal fraction as the pellet and a soluble fraction (cytosol) in the supernatant. The microsomal pellet is then carefully homogenized by hand, on ice, with a Dounce B glass/glass homogenizer in the same buffer and centrifugation is repeated. The resulting washed microsomal pellet is suspended in the same buffer containing 20% (v/v) glycerol and 1 mM dithiothreitol. At this point, the microsomal suspension can be aliquoted and stored at $-80°$ with no significant loss of UGT activity for several months.

More sophisticated procedures than the aforementioned method have been described for analytical subcellular fractionation, especially those using isopycnic (density gradient) ultracentrifugation of homogenates prepared with highly precise conditions (speed of the homogenizer, number of strokes, choice of the rotor, etc). Such a procedure is required for the isolation of subcellular components present in low quantity, such as Golgi apparatus, plasma or nuclear membranes, and peroxysomes and is also necessary for the isolation of highly purified subcellular fractions not contaminated with other subcellular components.

The quality of the fractions obtained, as judged by contamination by other organelles, is determined from the activity of marker enzymes known to be predominantly associated with a given subcellular compartment and/or by electron microscopy (e.g., see Wibo *et al.*, 1981).

UGTs Are Associated with ER Membranes

Subfractionation experiments clearly showed that UGT activity was preferentially associated with microsomes, which are mainly composed of ER membranes. In contrast to plants, in which soluble UGTs involved in the formation of pigments have been reported, no cytosolic form of UGT has been described in mammals. Additionally, the presence of UGT in mitochondria or plasma membranes has been reported, but these data probably reflect contamination with the microsomal fraction during the subcellular fractionation. The association of UGTs with the Golgi apparatus is more controversial (Roy Chowdhury *et al.*, 1985). However, the existence in this compartment of UGT forms that catalyze the biosynthesis of oligosaccharides such as heparan or chondroitin sulfates has been definitively confirmed (Lindt *et al.*, 1993).

Subcellular localization of UGTs has been investigated further by taking advantage of the availability of specific antibodies raised against individual isoforms and sophisticated imaging analysis. Using immunofluorescence microscopy combined with immunogold electron microscopy and densitometry, Radominska-Pandya *et al.* (2002) could unambiguously demonstrate the presence of UGT1A6 and UGT2B7 in the inner and outer nuclear membranes of human liver, thus emphasizing a possible role of these UGTs as a protective barrier against genotoxic substances.

UGTs are considered to be integral ER membrane proteins and detergents are required to solubilize them. Indeed, the activity of the enzyme is strongly dependent upon interactions with membrane phospholipids. This situation has for a long time impaired the characterization of these proteins due to the difficulty of purifying the proteins in active form for structural studies or the development of antibodies. Additionally, efforts to crystallize UGTs have so far been unsuccessful. In that context, structural information at the molecular level on this large group of enzymes is very limited. An additional property of the enzyme is latency, resulting from its tight association with microsomal membranes. The enzyme activity in sealed, intact microsomes is very low. Two hypotheses have been raised to account for UGT latency: (1) it is difficult for the hydrophilic donor substrate, UDP-glucuronic acid (UDP-GlcUA), which is synthesized in the cytosol, to reach the active site located on the lumenal surface of the microsomes, in contrast to the hydrophobic acceptor substrate that can diffuse through the membrane easily, and (2) the enzyme is constrained in a low activity form as a consequence of interaction with the surrounding phospholipids. There is now considerable evidence favoring the view that the catalytic-binding site of the UGTs is located on the lumenal surface of the ER (Drake *et al.*, 1992; Mackenzie, 1986a; Shepherd *et al.*, 1989; Vanstapel and Blanckaert, 1988; Yokota *et al.*, 1992). As a result, glucuronidation activity in microsomes can only be measured after the addition of various types of detergents (nonionic detergents, Triton X-100, digitonin; ionic detergents, cholate, deoxycholate; zwitterionic detergents, CHAPS) or pore-forming reagents (alamethicin) in order to fully remove latency. Experimental conditions for achieving optimal UGT activation and getting reproducible activity data can be determined by incubating microsomes on ice with increasing concentrations of detergent or permeabilizing agent and for various periods of time. Typically, the enzymatic activity increases to a maximum as a function of detergent concentration and time of incubation, after which it decreases at higher concentrations due to progressive disruption of the membrane organization of the UGT. Establishment of conditions for optimal activation will vary for each detergent, substrate, and microsomal source used and should, therefore, be determined each time one of these parameters changes.

The mode of homogenization of tissues and cells strongly affects the integrity of the final microsomal fraction and, therefore, the UGT activity. Homogenization with a Potter–Elvejhem homogenizer with a Teflon pestle leads to microsomal vesicles exhibiting maximal latency, whereas disruption of cells expressing recombinant UGTs (V79 or HEK cells) by sonication does not allow the formation of sealed microsomes but results in filament-like membrane preparations, in which UGT latency is fully removed. In this situation, the addition of detergent either has no effect or inhibits enzymatic activity. In contrast, when using the yeast *Pichia pastoris* expressing recombinant UGT, homogenization with glass beads allows intact microsomes to be obtained that have latency and can be treated similarly to tissue microsomes.

Functional Association of UGTs in the ER

UGTs have been found to form macromolecular structures by the association of several identical or different isoforms (homo/heterodimers). However, they can also interact with other microsomal drug-metabolizing enzymes, such as cytochromes P450 and epoxide hydrolase (mEH). Such a cooperation favored by the fluidity state of the phospholipid bilayer may be required for optimal, efficient drug biotransformation.

Homo/heterooligomerization of UGTs. The question of whether UGTs exist as homo- or heterodimers and the implications of such organization in their function was raised in the mid-1980s. Initial experiments providing evidence that UGTs may exist as dimers or tetramers were based on radiation inactivation, a method for determining molecular masses of membrane-bound enzymes *in situ*. When a membrane preparation is exposed to ionizing radiation, the degree of inactivation is directly related to the radiation dose and is dependent on the target size, or molecular mass, of the enzyme. Using a ^{60}Co source for γ irradiation, Peters *et al.* (1984) suggested that UGTs may be oligomers composed of one to four subunits with similar molecular masses. Gschaidmeier and Bock (1994) have also examined microsomal UGTs by radiation inactivation, concluding that monoglucuronidation of phenols may be catalyzed by a dimeric form of UGT whereas diglucuronidation is catalyzed by a tetramer.

The most direct estimation of the oligomeric state of UGTs could be determination of the molecular weight of the purified enzyme by size exclusion chromatography. However, such evaluation is often unreliable when applied to membrane-bound proteins, mainly due to the contribution of bound detergent micelles. Nevertheless, gel permeation chromatography of solubilized rat liver microsomal UGTs with activity toward chenodeoxycholic acid (Matern *et al.*, 1982) or phenols (Gschaidmeier and Bock, 1994) suggested that they exist at larger than monomeric size. Similar experiments have been performed on microsomes of Gunn rat fibroblasts expressing recombinant UGT1A1, solubilized with *n*-octylglucoside kept at a concentration below its critical micellar concentration to avoid incorporation into large micellar particles (Ghosh *et al.*, 2001). Results suggested that human recombinant UGT1A1 was organized as a dimer. These data may be of functional significance in the occurrence of Crigler–Najjar type II syndrome, as discussed later.

Early purification studies using conventional chromatography methods (chromatofocusing, column isoelectric focusing, UDP-hexanolamine

Sepharose 4B affinity chromatography) have failed to separate some UGT isoforms, such as the UGTs glucuronidating testosterone and androsterone. This result can be interpreted as an indication of heterooligomer formation or may result from the artifactual formation of aggregates (Matsui and Nagai, 1986). More reliable evidence for UGT heterooligomers has been obtained by immunopurification studies using isoform-specific antipeptide-antibodies. UGT1A isoforms and UGT2B1 from rat liver microsomes were readily copurified using specific Sepharose-conjugated antibodies directed against either UGT1A isoforms or UGT2B1. Crosslinking experiments using 1,6-bis(maleimido)hexane, which generates disulfide bonds between "nearest-neighbor" proteins, confirmed a possible heterooligomerization between UGT1A isoforms and UGT2B1 (Ikushiro et al., 1997).

Epitope tagging methods, now made possible by the availability of a battery of vectors allowing fusion of proteins with peptide tags (poly-His/hemagglutinin (HA)/FLAG/Myc) and the corresponding specific antibodies, offer new possibilities for investigating UGT organization. His-tagged proteins have been expressed in baculovirus-infected insect cells and purified by immobilized metal-chelating chromatography following detergent extraction (Kurkela et al., 2003). Recombinant UGT1A9 in which the His tag was replaced by a HA tag was also prepared. When cotransfected with His-tagged UGT1A9, a portion of the HA-tagged protein was bound to the column and coeluted with the His-tagged protein. This was not observed when membranes of cells expressing these constructs separately were mixed prior to being subjected to the same procedure. Taken together, these results support the idea that the formation of stable dimers occurs during the membrane assembly process.

The two-hybrid systems offer attractive methods for studying protein–protein interactions. These approaches are based on the coexpression of a "bait" vector containing a DNA-binding domain and the protein of interest and a "prey" vector containing a transactivating domain fused to another, different protein, or the same protein in the case of the study of UGT–UGT interactions. In the yeast system, human UGT1A1 was expressed as a fusion protein with the Gal4 transactivator protein and, in the second construct, as a fusion with the bacterial Lex DNA-binding domain (Ghosh et al., 2001). In the appropriate yeast strain, the LexA operator is engineered into the flanking region of the Escherichia coli galactosidase reporter gene (lacZ). Interactions of the proteins expressed by the "bait" and "prey" constructs bring the Gal4 activation domain into proximity with LexA, resulting in the transcription of lacZ and the appearance of blue colonies. Upon coexpression of "prey" and "bait" vectors containing UGT1A1, the lacZ reporter gene was transcribed, clearly demonstrating intermolecular interactions between these proteins. In the mammalian two-hybrid system, the wild-type UGT1A1 has been expressed in the "bait" vector as a fusion protein with the DNA-binding domain of the yeast protein Gal4. The "prey" vector consisted of a series of test constructs containing wild-type UGT1A1 or UGT1A6 and mutated forms of UGT1A1 expressed as fusion proteins with the transcriptional activation domain of NFκB. Expression of luciferase activity after cotransfection of the "bait" and "prey" vectors into COS-7 cells indicated a strong interaction between wild-type UGT1A1 molecules but no binding with UGT1A6. Further constructs indicated that the formation of homodimers of UGT1A1 probably involved the N-terminal domain of the protein.

Functional Relevance of UGT–UGT Interactions. A most important issue regarding the organization of UGTs is whether homo- or hetero-oligomer formation impacts the function of these enzymes. Meech and Mackenzie (1997) showed that rat recombinant UGT2B1 formed stable homooligomers when analyzed by SDS–PAGE under reducing conditions and investigated the functional relevance of this observation. When two inactive forms of UGT2B1 with different functional defects were coexpressed in cell culture, catalytic activity toward testosterone was detected, indicating that a functional interaction must occur between the proteins that permits partial compensation for each defect and restoration of activity. These results favor the idea that UGT2B1 functions as a dimer.

Another example of the relationship between a UGT oligomer and its function has been reported for UGT1A1. Clinical observations of the bilirubin glucuronidation deficiency syndrome Crigler–Najjar type II have suggested that interaction of some mutant forms of human UGT1A1 with the wild-type enzyme may be dominant negative (Koiwai *et al.*, 1996). Experimental studies described earlier (Ghosh *et al.*, 2001) support the idea that homodimerization of recombinant human UGT1A1 may explain this dominant-negative effect. The functional consequences of single nucleotide polymorphisms are also interesting to consider with regard to oligomerization of different allozymes. When the allelic variant UGT1A6*2 was expressed homozygously in HEK293 cells, it exhibited almost twofold greater activity toward 4-nitrophenol than that of wild-type 1A6*1. However, simultaneous expression of 1A6*1 and UGT1A6*2 was shown to produce low enzymatic activity. This observation suggests that the 1A6*1/A6*2 oligomer has a lower activity than the corresponding homodimers (Nagar *et al.*, 2004). Together, these results suggest that genetic deficiencies or polymorphisms may impact clinical efficacy or toxicity as a result of dimerization.

A most puzzling question is whether the formation of heterooligomers may lead to differences in substrate specificity. Evidence for a functional heterooligomer between guinea pig UGT2B21 and UGT2B22 has been provided by studies on the glucuronidation of morphine. Ishii *et al.* (2001) reported that COS-7 cells transfected with UGT2B21 cDNA catalyzed mainly formation of morphine-3-glucuronide, although morphine-6-glucuronide was also formed to some extent, whereas UGT2B22 did not show any significant activity toward morphine. When UGT2B21 and UGT2B22 were expressed simultaneously, extensive morphine-6-glucuronide formation was observed. This was the first report suggesting that UGT heterooligomer formation leads to altered substrate specificity. This work has been extended to the study of a range of substrates, and results suggest that chloramphenicol glucuronidation was also enhanced by coexpression of UGT2B21 and UGT2B22 (Ishii *et al.*, 2004).

UGT–Drug-Metabolizing Enzyme Interactions. UGTs have also been described as interacting within the ER with the P450 monooxygenase complex and mEH. The unique concentration of drug-metabolizing enzymes in the ER makes microsomes the most important organelle for the sequential biotransformation of these substances. Protein–protein interactions among UGTs, P450s, and mEH have been characterized by coexpression and copurification, cross-linking procedures, and immunoenzymology studies (Ishii *et al.*, 2005). Such association and cooperation may facilitate the multistep metabolic conversion of drugs via hydroxylation/epoxidation, epoxide hydration, and the subsequent glucuronidation process.

Structural Determinants of UGT for Membrane Association, Residency, and Activity

Biogenesis of membrane proteins involves several events, such as targeting to the ER, translocation of certain domains into the ER lumen, and integration of transmembrane domains (TMDs) into the lipid bilayer. Protein targeting and translocation to the ER are mostly cotranslational, i.e., protein translocation and membrane insertion are coupled to protein synthesis. Most of the proteins destined for the ER contain a signal sequence at or near the N terminus, which triggers transport of the nascent chain from the cytosol into the ER with the assistance of the signal recognition particle (SRP) and its membrane receptor (Corsi and Schekman, 1996). The polypeptide is then translocated into the ER by multiprotein assembly machinery termed translocon (Johnson and van Waes, 1999). This protein complex provides an aqueous protein-conducting channel spanning the membrane bilayer. The translocon is able to bind ribosomes with high affinity, recognize functional signal sequences, and allow the lateral partitioning of TMDs into the lipid bilayer. The orientation and integration of membrane proteins determine protein topology and are coupled to protein folding.

Many prediction algorithms have been developed in order to determine the topology of integral membrane proteins (Ott and Lingappa, 2002). Protein topology prediction programs generally attempt to predict whether the protein is likely to be an integral membrane protein, how many membrane-spanning domains the protein contains, the orientations, and the length of the transmembrane domains. Limited information is available about eukaryotic membrane proteins because they are generally hard to crystallize.

UGTs are predicted to be type I ER membrane proteins containing a cleavable N-terminal signal peptide, a glycosylated lumenal domain, and a short cytoplasmic tail. Therefore, the majority of the protein is located in the lumen of the ER and attached to the membrane via a C-terminal TMD.

The transmembrane orientation of UGTs has been a subject of controversy for several years (Dutton, 1980). The activity of these enzymes in sealed hepatic microsomes is latent and can be increased several times by the disruption of microsomal vesicles by detergents. This suggested that disruption of the membrane barrier would allow free access of the donor substrate, UDP-GlcUA, a hydrophilic and charged substrate, to the active site of the enzyme, reflecting a lumenal orientation of the catalytic center of the enzyme. To gain insight into the orientation of UGT proteins in the ER, different methods, including protease digestion and immunoblot analysis or the use of photoaffinity probes, as well as *in vitro* transcription/translation of full length and truncated forms of UGTs, have been used. Based on protease digestion and immunoblot analyses, Shepherd et al. (1989) showed that treatment of sealed hepatic microsomes with proteases resulted in an increase in the mobility of UGT1A1, indicating the removal of a fragment with a molecular mass of approximately 2 kDa. In contrast, immunoblot analysis of protease-treated hepatic microsomes solubilized previously by detergent showed extensive destruction of the UGT1A1 protein. These analyses indicated that only a small portion of the UGT1A1 protein is exposed on the cytoplasmic surface of the ER.

Membrane protein topology prediction programs suggested that UGTs contain only one membrane-spanning domain near the C terminus. However, expression of the truncated form of UGT2B1 and UGT1A6, from which the cytosolic tail and the TMD have been removed, in mammalian cells and in yeast did not abrogate ER retention, as shown by sensitivity of these proteins to endoglycosidase H (Meech *et al.*, 1996; Ouzzine *et al.*, 1999b), indicating that they have not moved forward to the distal organelles in the secretory pathway. Despite the lack of their unique predicted transmembrane domain, UGT2B1 and UGT1A6 were still retained in the ER membranes. Based on studies of the membrane binding of chimeric proteins created with the N-terminal half of UGT2B1 and the C-terminal half of the soluble glycosyltransferase ecdysteroid glucosyltransferase, it has been shown that the N-terminal part of UGT2B1 was able to retain the chimera in the ER, as demonstrated by the sensitivity of the protein to endoglycosidase H. This N-terminal part also contains a region of strong interaction with the membrane, as the chimera was resistant to extraction by alkali treatment and low concentration of detergents. Investigation of UGT1A6 determinants involved in membrane association of the polypeptide in the absence of the C-terminal TMD was carried out by expressing truncated forms of UGT1A6 created by successive deletions from the C-terminal TMD, using the rabbit reticulocyte lysate transcription/translation system in the presence of microsomal membranes. Membrane insertion of the resulting polypeptides, translocation of lumenal domains, and membrane orientation were characterized by glycosylation pattern and resistance to alkaline treatment and protease digestion. Analyses showed that deletion of the N-terminal end did not affect the glycosylation pattern or membrane targeting and retention until residues 140–240 were deleted, indicating that this N-terminal region contains a topogenic element strongly interacting with ER membranes. Protease digestions revealed that the 140–240 region was not spanning the membrane (Ouzzine et al., 1999b). Such an internal, hydrophobic peptide may be critical for the function of the enzyme. Indeed, it must be emphasized that UGT substrates are generally lipophilic molecules that reach their binding site(s) in the N-terminal half of the proteins by passive diffusion through the lipid bilayer. An attractive hypothesis is that the membrane interaction conferred by this region may provide a hydrophobic path from the membrane interior to the catalytic site.

Proteins are maintained in the ER by two means, static retention or dynamic retention by continuous retrieval of the escaped protein from post-ER compartments. In type I membrane proteins such as UGTs, the retrieval signal has been defined as two lysine residues at positions −3 and −5 from the C terminus exposed on the cytosolic side of the ER membrane. Expression of chimeric proteins in which the TMD and cytoplasmic tail domain of a plasma membrane protein were replaced by the same domain of an ER resident protein is widely used to identify the determinant involved in ER retention. When the ectodomain of the plasma membrane protein CD4 was appended with the TMD and cytoplasmic tail domain of UGT1A, this cell surface protein colocalized with the ER marker, calnexin, in recombinant HeLa cells and was sensitive to endoglycosidase H. Immunofluorescence and colocalization studies and glycosylation pattern analyses showed that mutation of lysine residues positioned at 3 and 5 in the cytoplasmic tail resulted in leakage of CD4 chimeric protein from ER to the Golgi and plasma membrane and the chimera became resistant to endoglycosidase H. However, a major part of the protein was still retained in ER, indicating that the lysine residues were not the only determinant ensuring ER retention (Barre et al., 2005). In agreement with this, Jackson et al. (1990) reported that mutation of the dilysine motif of UGT1A6 resulted in only slow, if any, migration of the protein out of the ER, and Meech and Mackenzie (1998) showed that the cytosolic dilysine motif may be redundant for residency of UGT in the ER, as truncation of this motif did not abrogate ER residency of UGT2B1. Based on these observations, the role of the TMD (without the cytosolic tail) in ER retention was investigated by expressing the CD4 ectodomain fused to the TMD of UGT1A. Immunofluorescence and glycosylation pattern studies showed that UGT1A TMD was capable of retaining the CD4 plasma membrane reporter protein in the ER of recombinant HeLa cells (Barre et al., 2005). These observations suggested that, in addition to the dilysine motif of the cytosolic tail, the TMD of UGT1A acts as an ER retention domain. Together, these observations suggest that ER residency conferred by the TMD and the cytoplasmic tail domain involves at least two determinants: the TMD, probably acting by static ER retention, and the KSKTH dilysine motif on the cytoplasmic tail for the retrieval of escaped proteins from the post-ER compartment.

Architecture and Organization of the Active Site within the Membrane

In order to better predict the glucuronidation of drugs in human and their possible associated toxicity, it is necessary to investigate the molecular mechanisms that account for substrate recognition and catalysis.

UGTs are inverting glycosyltransferases, which catalyze the transfer of glucuronic acid from the high-energy donor, UDP-GlcUA, to a variety of functional groups, e.g., -OH, -COOH, $-NH_2$, -SH, and C-C, of endogenous compounds, xenobiotics and drugs, leading to the formation of β-D-glucuronides. UGT activity is strongly dependent on the presence of magnesium. The reaction is believed to proceed through a S_N2 mechanism with an oxocarbanium ion as the transition state. The nucleophilic acceptor substrate is activated by deprotonation by a general base, which can thereafter attack carbon 1 of GlcUA, with concomitant formation of a β-glycosidic bond and the release of UDP, facilitated by the presence of magnesium. Identification of the general base catalyst and of the other amino acids that participate in the glucuronidation reaction has been investigated.

Methodological Approach. In the absence of UGT crystal structures, identification of crucial amino acids in the active site(s) on the lumenal surface of microsomal membranes is performed by a mutidisciplinary approach combining modification of the protein by electrophilic chemicals, multiple sequence alignments carried out using bioinformatic software, site-directed mutagenesis, and, finally, modeling by homology with glycosyltransferases (GTs) whose three-dimensional structure has been established by X-ray diffraction.

In the first step, a cDNA encoding a UGT isoform is expressed in active form in heterologous cells (V79 fibroblasts, HEK cells, or the methyltrophic yeast, *Pichia pastoris*). Yeast, especially, is a very convenient host cell for expressing wild-type and mutant UGTs in sufficient amounts for biochemical investigations. Microsomes prepared from recombinant cells by differential centrifugation are incubated with amino acid-directed reagents able to interact covalently and preferentially with side chains of polar (serine, cysteine, asparagine) or prototrophic (aspartic, glutamic acid, lysine, arginine, histidine) residues. These residues are known to potentially play a key role in catalysis or protein–substrate interactions. Even if the specificity of the chemical reaction is not always strict, this method allows the determination of the class(es) of amino acid residues important for enzymatic activity. When a critical amino acid is modified, irreversible inhibition occurs with a characteristic decrease in activity as a function of time and the concentration of the chemical that allows the corresponding kinetic parameters (pseudo-first order, and second-order inactivation rate constants, order of the reaction) to be determined (Battaglia *et al.*, 1994). In a second step, the importance of this residue is confirmed by site-directed mutagenesis. Mutants generated by conventional polymerase chain reaction (PCR) methods using sense and antisense primers introducing the chosen mutations are expressed in *P. pastoris* and the corresponding UGT proteins are subjected to kinetic analysis.

Crucial Amino Acids of UGTs. Chemical modification of the human recombinant UGT1A6 revealed that the isoform was irreversibly inactivated by the histidyl-directed reagent, diethylpyrocarbonate, carboxyl-directed reagents, carbodiimide, guanidinium-directed reagents, 2,3-butanedione, and thiol-directed reagents, such as maleimides. These experiments suggested the crucial importance of four main groups of amino acids: histidine, carboxyl amino acids (aspartic, glutamic acid), arginine, and cysteine residues, respectively. Their importance was further confirmed by site-directed mutagenesis after selection of the amino acid by sequence alignment of multiple UGTs. Arg52, located in a highly conserved amino-terminal hydrophobic region (LX2-*R52*-G-H54-X4-V-L), was critical for the function and structural integrity of UGT1A6, although it was not involved directly in catalysis or substrate binding (Senay et al., 1997). However, Cys126 was also an essential residue and was not involved in the intra- or interdisulfide bond (Senay et al., 2002). Although informative, these results did not provide evidence on the nature of the catalytic base. However, extensive studies performed with dicyclohexylcarbodiimide and diethylpyrocarbonate strongly indicated that histidyl and aspartic/glutamic acid residues were likely candidates as base catalysts. Therefore, five highly conserved histidine residues (H38, H54, H361, H370, H485) in the sequences of rat and human UGTs were mutated systematically. Results indicated that H38 and 485 were critical for activity, as mutation led to an inactive enzyme. H361 was found to interact with UDP-glucuronic acid, whereas H54 was required for optimal enzyme efficiency. Finally, H370 was found to be the residue reacting with diethylpyrocarbonate, and therefore could play a role in catalysis (Ouzzine et al., 2000) However, this series of experiments failed to determine the carboxyl amino acid, which would also be involved in catalysis.

To gain additional information on the nature of such residues and on the structural organization of UGT, we considered the large GT family as a potential unexplored mine of data that could contribute new insights into their structure and function. UGTs belong to the GT family of enzymes that is present in mammals, bacteria, fungi, and plants. They all catalyze glycosidic bond formation according to a retaining or inverting mechanism. More than 7000 sequences have been characterized to date and are classified into more than 65 distinct families. The sequences are indexed in the CAZY database (http://afmb.cnrs-mrs.fr/CAZY/index.html) (Coutinho et al., 2003). Interestingly, 15 GTs, some of them membrane bound, have been crystallized and their three-dimensional structure solved. Surprisingly, although the sequence homology is very low among these GTs, fold-recognition methods led to the conclusion that they all belong to only two structural superfamiles, referred to as GT-A and GT-B. Their characteristics are described in Fig. 1. The basic motif for the two structures is the presence of two $\alpha/\beta/\alpha$ subunits or Rossmann folds, which are often present in nucleotide-binding proteins. In GT-A proteins, these two domains interact with the donor and acceptor substrates and are tightly associated, whereas in GT-B they are more loosely associated. Another difference concerns activation of the GT by metal ions. The metal ion requirement is strict for GT-A, but metal ions are only required for optimal activity of GT-B. Finally, in both GT structures, an aspartic or glutamic acid residue has been identified as the base catalyst.

A careful comparison of the structure and properties of GT-A and GT-B with those known for UGTs revealed that UGTs have three lines of

A

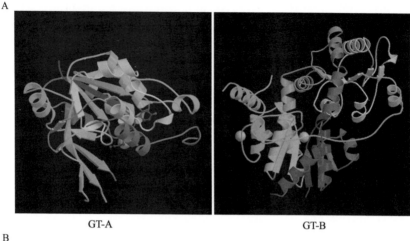

GT-A   GT-B

B

GT representative member	• SpsA (*Bacillus substilis*), production of mature spore coat • GlcAT-I (man), glycosaminoglycan synthesis	• GtfB (bacteria), antibiotics synthesis • MurG (bacteria), peptidoglycan synthesis
Fold	2 Rossmann-like $\alpha/\beta/\alpha$ domains tightly associated (N- and C-terminal domains, which bind the donor and the acceptor substrate, respectively)	2 Rossmann-like $\alpha/\beta/\alpha$ domains less tightly associated (N- and C-terminal independent domains, which bind the acceptor and the donor substrate, respectively)
Metal activation	Strict, with a DXD motif (Mn^{2+}, Mg^{2+})	Less strict (Mn^{2+}, Mg^{2+})
Catalytic base	Asp or Glu	Asp or Glu
PDB code	SpsA, 1QG8	GtfB, 1IIR

FIG. 1. Structural and functional comparison of crystallized glycosyltransferases of GT-A and GT-B families. (See color insert.)

homology with GT-B and, therefore, would adopt a three-dimensional structure similar to that of GT-B proteins.

UGTs are known to display two independent N- and C-terminal domains of similar size that are believed to interact with the acceptor and donor substrates, respectively.

UGT activity is optimal in the presence of Mg^{2+}, but is not strictly dependent on the presence of the metal ion.

UGTs have a membrane organization that is thought to be similar, in some aspects, to that of the bacterial MurG involved in peptidoglycan synthesis. This plasma membrane-associated GT catalyzes the transfer of *N*-acetylglucosamine from UDP-*N*-acetylglucosamine to lipophilic acceptor substrates that have crossed the membrane to the active site of MurG on the cytoplasmic side.

Sequence alignment comparisons of MurG from *E. coli* and homologs from seven other bacterial strains with UGT isoforms led to identification of the UGT sequence consensus region His/Arg-X7-Glu (Ha *et al.*, 2000). Interestingly, the conserved histidine residue in MurG, which has been proposed to be important in stabilizing the UDP group during catalysis, corresponded to the catalytic His370 of UGT1A6. It is therefore tempting to speculate that His370 plays a similar role in the glucuronidation reaction. Similarly, by analogy with the function of the conserved glutamic acid residue in MurG, Glu378 of UGT would be involved in recognition of the ribose moiety of UDP-GlcUA.

In conclusion, comparison with GTs whose three-dimensional architecture has been solved is a helpful approach in predicting the role and function of key amino acids and peptide domains of UGTs. Even if identification of the base catalyst residue has been unsuccessful, a careful search for homology between UGTs and MurG using powerful structural informatics programs will surely contribute to new structural information about these membrane-bound enzymes.

Application of Photoaffinity Labeling to the Characterization of Membrane-Bound and Purified Recombinant UGTs

Photoaffinity labeling has been applied to the characterization of UGT active sites with the goal of identifying critical amino acids within the active sites, as well as for the characterization of these membrane-associated proteins (Drake and Elbein, 1992; Radominska and Drake, 1994). This technique involves the use of a substrate or inhibitors that have been modified with a photolabile group. The photoactivation of this probe occurs *in situ* and results in its covalent binding to the active site. Azido group modifications, which, upon photoactivation, yield nitrenes capable of insertion into carbon–hydrogen, oxygen–hydrogen, and nitrogen–hydrogen bonds, are useful for labeling the substrate-binding site residues of UGTs, which are likely to have more than a single amino acid residue interacting with substrates. Direct photoaffinity labeling also represents a valuable variation of photoaffinity labeling. It has an important advantage over indirect labeling because the compounds used in this method possess an intrinsic photoreactivity and do not act as competitive inhibitors as many modified photoprobes do. [^{3}H]*all-trans* and [^{3}H]9-*cis* retinoic acid ([^{3}H]-*at*RA and [^{3}H]9*cis*RA) have been used as effective direct photoaffinity probes (Chen and Radominska-Pandya, 2000; Little and Radominska, 1997; Radominska-Pandya *et al.*, 2000, 2001).

Photoaffinity labeling allows membrane-bound UGTs to be labeled in their native state, and valuable information on the structure, architecture, and orientation of the active sites of UGTs has been obtained (Radominska-Pandya *et al.*, 1999). Specifically, the use of 5-azido-substituted nucleotide affinity analogs has resulted in substantial progress in the characterization of native UGTs (Drake and Elbein, 1992; Radominska and Drake, 1994; Rachmel *et al.*, 1985). The identification of novel proteins, effective monitoring of UGT purification processes, and evaluation of the effects of pH, inhibitors, detergents, competing nucleotides, and proteolytic digestion on enzyme activity can be evaluated, as is described in several papers (Drake *et al.*, 1989, 1991a,b; Radominska *et al.*, 1994a,b,c).

Photoaffinity labeling can generate the most important information when applied to the specific amino acid present in the active site of UGTs. However, a homogeneous protein is required for identification in this type of experiment. Recombinant UGTs purified to homogeneity have not been available until very recently. The major development in the identification of active site amino acids has been the generation of recombinant human UGTs containing a C-terminal His tag that are expressed in baculovirus-infected Sf9 insect cell (Kurkela *et al.*, 2003). This system results in high levels of expressed proteins that are modified posttranscriptionally by glycosylation. Therefore, UGT membrane fractions and purified proteins are now available for photolabeling by several photoaffinity probes.

Several photoaffinity probes have been developed for identification of the UGT cosubstrate, UDP-GlcUA, active site. Specifically, $[\beta-^{32}P]5$-azido-UDP-glucuronic acid ($[\beta-^{32}P]5N_3UDP$-GlcUA), $[\beta-^{32}P]5$-azido-UDP-glucose ($[\beta-^{32}P]5N_3UDP$-Glc) (radioactive photoaffinity probes), and periodate-oxidized $[\beta-^{32}P]UDP$-GlcUA (o-UDP-GlcUA; affinity probe) have been designed, synthesized, characterized, and studies published (Battaglia *et al.*, 1998; Drake *et al.*, 1991b, 1992; Radominska and Drake, 1994; Senay *et al.*, 1999). As far as the aglycon-binding site(s) is concerned, three major probes are available: 7-azido-4-methylcoumarin (AzMC; a fluorescent photoaffinity probe), $[^3H]at$RA, and $[^3H]9cis$RA (Chen and Radominska-Pandya, 2000; Little and Radominska, 1997; Radominska-Pandya *et al.*, 2000, 2001).

Now that sufficient amounts of catalytically active UGT protein are available (as described in the following sections), these probes are indispensable labels for the rigorous identification of both cosubstrate and substrate-binding domains of the UGTs.

Cloning and Expression of Human Recombinant UGTs

This section presents published methods for the preparation and purification of recombinant human His-tag UGTs.

Cloning. Human UGTs are cloned by RT-PCR using commercially available (Invitrogen) total human liver RNA (Kurkela *et al.*, 2003). To express UGTs with C-terminal His tags, derivatives of the shuttle vector pFastBac (Invitrogen) are prepared with the *Sal*1–*Hin*d3 segments of the multicloning site replaced by fragments (in different reading frames) that encode an enterokinase cleavage site followed by six histidine residues (6× His tag) and a stop codon. These modified shuttle vectors, pFBXHA, pFBXHB, and pFBXHC, differ from each other by a single nucleotide. To express His-tagged UGTs, a *Sal*1 restriction site is inserted immediately upstream of the authentic UGT stop codon, which also adds two or three amino acids between the end of the cloned human UGTs and the start of the enterokinase recognition site within the C-terminal tag. Insertion of the *Sal*1 site, leading to "bypassing" the authentic stop codon, is done so that that all the UGT1A isoforms are "in frame" with the His tag in the shuttle vector pFBXHC, whereas the UGT2B isoforms are "in frame" with pFBXHA (Kurkela *et al.*, 2003).

The putative extrahepatic UGTs, 1A7, 1A8, and 1A10, are amplified by PCR from human DNA (control Finnish sample) and a new restriction site for *Hin*d3 is generated by a silent mutation in the 3' end of the amplified DNAs. The same restriction site is inserted at the beginning of exon 2 of the human ugt1A gene (sample from the already cloned UGT1A9), and the amplified exons 1 are each joined to exons 2–5 by digestion (with *Hin*d3) and ligation (Kurkela *et al.*, 2003). The cloned genes are sequenced in both directions and, if mutations are detected, they are corrected by site-directed mutagenesis, followed by another round of DNA sequencing.

Expression. Recombinant baculovirus for the expression of human UGTs in insect cells is prepared using the "Bac-to-Bac" procedure (Invitrogen). The cloned gene, in frame with the C-terminal tag in the pFastBac derivative, is transferred to the baculovirus genome by transposition within the DH10Bac strain of *E. coli.* The viral DNA is subsequently isolated from the bacterial cells and, after verification that it carries the correct insert, is used to transfect insect cells. About 48 h later, intact viruses carrying the inserted gene are collected from the medium and the viral titer is increased by two or three rounds of infection. Recombinant protein expression optimization is carried out by infecting cultures of insect cells at 2×10^6 live cells/ml with various amounts of the virus stock. At 48 h postinfection, cells are collected and activity assays are carried out. In this way, a large amount of infected cells can be prepared, e.g., 200–500 ml containing 2×10^6 cells/ml. The infected cells are then collected by centrifugation (1000g, 10 min, 20°), washed with phosphate-buffered saline (PBS), and stored at −20° for up to 6 weeks.

Membrane Isolation. Membranes are prepared from frozen baculovirus-infected insect cells by mild disruption of the cells by osmotic shock, followed by centrifugation at 40,000g at 4° for 60 min, suspension of the pellets in cold distilled water, and a second centrifugation under the same conditions. The final membrane pellets are suspended in a small volume of buffer (25 mM Tris-HCl, 0.5 mM EDTA, pH 7.5), homogenized with a glass tissue homogenizer, divided into aliquots, and stored at −70° until use.

Immobilized Metal Affinity Chromatography (IMAC) Purification. Purification of recombinant His-tagged UGTs is done by IMAC at 4°. Membranes are thawed and suspended at a concentration of 1–2 mg of protein/ml in extraction medium (25 mM Tris, pH 7.5, 500 mM NaCl, 1% Triton X-100, or 0.8 mg Emulgen 911/mg protein). The suspension is mixed for 1 h at 4° and centrifuged at 40,000g for 1 h. The resulting supernatant is filtered through a syringe filter (0.45 μm) and loaded onto a nickel-charged His Hi-trap column (Amersham, Piscataway, NJ) preequilibrated with 25 mM Tris, pH 7.5, 500 mM NaCl, 0.05% Triton X-100, and 50 mM imidazole (buffer A). After extensive washing, the bound protein is eluted by a stepwise gradient of imidazole (50–400 mM) in buffer A. One-milliliter fractions are collected and examined by SDS–PAGE (Fig. 2) and Western blotting. Fractions containing purified protein are pooled and either used directly or concentrated by centrifugation through an ultrafilter Centriplus (YM30 Millipore).

Synthesis of Photoaffinity Probes for the Labeling of UGTs

5-Azido-UDP-glucuronic acid (5N$_3$UDP-GlcUA). The photoaffinity probe [β-^{32}P]5N$_3$UDP-Glc has proven to be a valuable biochemical tool in the study of nucleotide diphosphate sugar-utilizing enzymes, especially membrane-associated UDP-glucosyltransferases, and the synthesis has been described in detail previously (Drake and Elbein, 1992; Drake *et al.*, 1991b; Radominska and Drake, 1994). There are several advantages to using 5-azido-substituted nucleotide photoaffinity analogs. The photoaffinity probe is structurally very similar to the UDP-GlcUA: the two compounds differ only in the presence of the 5-azido group. In the absence of activating light, [β-^{32}P]5N$_3$UDP-GlcUA serves as a substrate for the glucuronidation reaction without inhibiting the enzyme (Battaglia *et al.*, 1997). Other advantages are that, upon exposure to low intensity UV light, the highly reactive nitrene intermediate has a short half-life. As a result, there

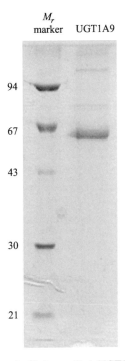

FIG. 2. SDS–PAGE analysis of affinity-purified UGT1A9 visualized using Coomassie blue.

is little or no damage to the protein and incorporation of ^{32}P is achieved readily. Finally, the photoprobe can be synthesized with high specific activity. The major disadvantage of photolabeling with azido-nucleotide sugar is the fact that only a small percentage (up to 15%) of the protein molecules are modified. Moreover, the photolabeling reaction does not follow the stoichiometry of the catalytic reaction.

5N$_3$UDP-GlcUA is synthesized enzymatically from 5N$_3$-UDP-glucose (Glc), which is prepared by coupling 5N$_3$UTP and Glc-1-phosphate enzymatically. These reactions depend on the chemical synthesis of 5N$_3$UMP and, subsequently, 5N$_3$UTP. For all procedures described here that require water, the purest, deionized water available should be used. It is recommended that UDP-glucose pyrophosphorylase from yeast (Sigma, St. Louis, MO) and ammonium bicarbonate from Fisher Scientific (Pittsburgh, PA) be used in the syntheses for best results. 5N$_3$-UDP-Glc and 5N$_3$-UDP-GlcUA photoaffinity analogs also can be purchased from Affinity Labeling Technologies (Lexington, KY). (More information is available at http://www.altcorp.com/.) Both probes can be purchased in radioactive form ($[\beta\text{-}^{32}\text{P}]5\text{N}_3\text{UDP-Glc}$ or -GlcUA), and labeled probes with specific activities of 2–5 mCi/μmol are available.

Synthesis of 7-Azido-4-methylcoumarin (AzMC). The most efficient probe for identification of the active site of UGTs that carry out the glucuronidation of phenols is AzMC. Because this probe is fluorescent, the efficiency of photoaffinity labeling can be monitored easily. In order to synthesize AzMC, all steps must be carried out in the dark on ice. Three grams (17 mmol) of 7-amino-4-methylcoumarin is dissolved in 125 ml 12 *N* HCl. This solution is diazotized with 5.5 ml of 25% (w/v) NaNO$_2$ (20 mmol) in water. After 10 min, 6.5 ml of 20% (w/v) NaN$_3$ (20 mmol) in water is added to the solution of diazonium salt. After additional stirring for 48 h, the acidic mixture is neutralized with a cold solution of 12 *N* NaOH and is extracted five times with 50 ml of chloroform. The extracts are pooled and dried under vacuum to give a yellow powder (96% yield; m.p. 110°). TLC on silica gel in chloroform-methanol (3:1, v/v) yields one spot with R_f 0.9. The molar extinction coefficient in chloroform is 13,348 $M^1\text{cm}^{-1}$ at 327 nm. The probe is then stored as a dry powder in the dark at −20°.

Retinoic Acid Probes [³H]atRA and [³H]9cisRA. Radioactive *at*RA and 9*cis*RA are available commercially and can be used for direct labeling of the UGT active site without additional modification. Their specific activity is very high so that labeling and characterization of the active site can be followed easily.

Methods for Photoaffinity Labeling

In order to describe labeling of UGTs with photoaffinity probes as a true photolabeling process, several criteria must be satisfied, as established by Haley (1991) for active site photolabeling. First, only proteins known to specifically bind the probe must be photolabeled. Second, photoinsertion of the probe into the protein must be saturable. Third, protection against photoinsertion of the probe by substrate analogs and/or various active site-directed inhibitors has to be observed. Finally, labeling into the protein has to be strictly dependent on UV irradiation, indicating that only specific photolabeling of the proteins occurs and excluding any unspecific enzymatic phosphorylation processes.

Photoaffinity Labeling with 5N₃UDP-GlcUA. Human liver microsomes (50 μg protein) or membrane fractions of cells expressing recombinant UGT protein (5–50 μg protein) are incubated for 10 min, on ice, in the presence of 0.05% Triton X-100, in 100 m*M* HEPES, pH 7.0, and 5.0 m*M* MgCl2 in a total volume of 20 μl. The samples are left at room temperature for 1 min, and [^{32}P]5N3UDP-GlcUA (200 μM; 2–5 mCi/mmol) is added to a final concentration of 40 μM (5 μl) to give a final volume of 25 μl and allowed to equilibrate for 20 s, followed by UV irradiation with a hand-held short wave (254 nm) UV lamp (UVP-11, Ultraviolet Products, Inc., San Gabriel, CA) for 90 s at room temperature. It is most convenient to prepare the samples in microcentrifuge tubes arranged in a rack in such a way that the UV lamp can be placed on top of several tubes at once. Control samples, which are not irradiated, are prepared at the same time. Reactions are terminated by the addition of 2 volumes of cold 10% TCA. Samples are centrifuged at 13,000g for 3 min, the supernatant is removed as completely as possible, and the pellets are dissolved in 10–50 μl of denaturing buffer (3.6 M urea, 100 mM Tris, 50 mM dithiothreitol, 4% SDS, pH 8.0, containing bromphenol blue as a tracking dye). Proteins are separated on 10% SDS–polyacrylamide gels (Laemmli, 1970), followed by autoradiography for 1–2 days at $-80°$. Autoradiographs are analyzed and quantitated by densitometry. For example, photolabeling of human recombinant UGT1A6 with [β-^{32}P]5-N$_3$UDP-GlcUA results in photoincorporation of [β-^{32}P]5-N$_3$UDP-GlcUA that is saturable and inhibited by preincubation with cold UDP-GlcUA (Fig. 3).

Photoaffinity Labeling with AzMC. For studies involving determination of inhibition of enzymatic activity by AzMC, photolabeling reactions are routinely performed on membrane fractions of cells expressing recombinant UGT protein (UGT1A6 or other phenol-binding protein) diluted to 8.9 mg/ml in 180 mM Tris-HCl, pH 7.4, 28 mM MgCl$_2$, and 2 mM AzMC dissolved in dimethyl sulfoxide (5%, v/v). Samples are transferred to microcentrifuge tubes and incubated in the dark for 1 min at 25° and are

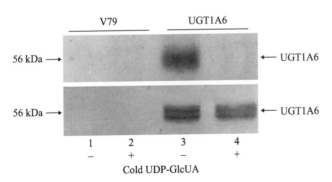

FIG. 3. Photolabeling of UGT1A6 expressed in V79 cells with [β-^{32}P]5-N$_3$UDP-GlcUA. (Top) Autoradiograph of photolabeled proteins. (Bottom) Corresponding Western blot. Lanes 1 and 2, V79 control membranes; lanes 3 and 4, V79 expressing human UGT1A6.

then exposed to a hand-held long-wave (366 nm) UV light (UVP-21, Ultraviolet Products) for 20 min on ice. After light activation, the reaction is stopped by a 60-fold dilution with 180 mM Tris-HCl, pH 7.4, 28 mM MgCl$_2$. In control experiments, no damage to UGT1A6 activity induced by light alone at this wavelength can be detected.

For SDS–PAGE of protein labeled with AzMC, membrane fractions expressing UGT (60 μg protein) are incubated with 25 μg bovine serum albumin (used as a scavenger) and 0.1 mM AzMC in 180 mM Tris-HCl, pH 7.4, and 28 mM MgCl$_2$ in a final volume of 50 μl for 5 min at room temperature prior to irradiation. After irradiation for 20 min, proteins are precipitated with 10% (w/v) ice-cold TCA and centrifuged at 10,000g for 5 min. The resulting pellet is washed with ethanol:ether (1:2, v/v), and proteins are separated on 10% SDS–PAGE gels. Proteins are fixed in a methanol/acetic acid/formaldehyde (50:12:0.05, v/v/v) for 1 h, and fluorescent protein adducts are detected at 312 nm with a UV transilluminator. An example of photolabeling of recombinant UGTs that carry out the glucuronidation of phenols is presented in Fig. 4.

Direct Photoaffinity Labeling with Retinoids. Retinoids can be used effectively as photoprobes without derivatization (direct photoaffinity labeling). The availability of high specific activity [^{3}H]at-RA and [^{3}H] 9-cis-RA makes it possible to characterize the substrate-binding sites of retinoid-binding proteins in both microsomal and recombinant preparations (Bernstein *et al.*, 1995; Chen and Radominska-Pandya, 2000; Little and Radominska, 1997; Radominska-Pandya and Chen, 2002; Radominska-Pandya *et al.*, 2001). An example of the direct labeling of cellular

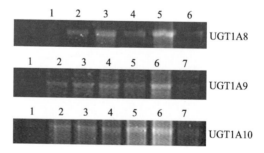

Fig. 4. Photolabeling of membrane fractions of recombinant UGT1A8, 1A9, and 1A10 with AzMC. UGT1A8. Lanes: 1, DMSO; 2–5, photolabeled with 0.1, 0.15, 0.25, and 1.0 mM AzMC; and 6, labeled is 0.15 mM AzMC in absence of UV. UGT1A9. Lanes: 1, DMSO; 2–6, photolabeled with 0.1, 0.15, 0.25, 0.75, and 2.0 mM AzMC; and 6, labeled is 0.15 mM AzMC in absence of UV._UGT1A10. Lanes: 1, DMSO; 2–6, photolabeled with 0.1, 0.15, 0.25, 0.75, and 2.0 mM AzMC; and 6, labeled is 0.15 mM AzMC in absence of UV.

retinoic acid-binding protein II (CRABP-II) with [³H]*at*RA is presented in Fig. 5.

Photolabeling of Hepatic Microsomes. All experiments involving retinoids are carried out under yellow light. Photolabeling with labeled retinoids is done using the method of Bernstein *et al.* (1995) modified as follows. For studies with detergent-activated microsomes, [³H]*at*-RA (solubilized in micellar form with Triton X-100) is added to microsomal protein (50 μg) in 100 m*M* HEPES, pH 7.0, containing 5 m*M* MgCl2 to a final concentration of 2 μ*M* [³H]*at*RA (2.0 μCi) and 0.05% Triton in a total volume of 25 μl. The reaction is incubated on ice for 10 min, followed by irradiation with a hand-held long-wave (366 nm) UV lamp for 15 min on ice. The reaction is stopped with 10% TCA (150 μl) and processed for SDS–PAGE on 10% gels (Laemmli, 1970) as described earlier for photolabeling with [³²P]5N₃UDP-GlcUA. Following electrophoresis, gels are stained with Coomassie blue, washed thoroughly in distilled water, treated with AutoFluor (National Diagnostics, Manville, NJ) according to the manufacturer's directions, dried, and subjected to autoradiography for 1–7 days at −80°.

Photoaffinity Labeling of Recombinant Proteins. Because optimum conditions for photolabeling need to be established for each specific

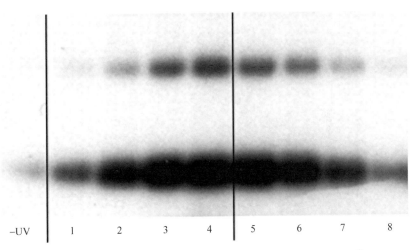

FIG. 5. Photoaffinity labeling of CRABP-II by differing concentrations of [³H]RA and protection by unlabeled atRA. Lanes: -UV, control; 1–4, concentration dependence of [³H] RA photolabeling (0.825, 1.65, 3.3, and 6.6 μ*M*); 5–8, protection with increasing concentrations of unlabeled atRA (0, 5, 25, and 125 μ*M*). The two bands in each lane represent the monomer and dimer forms of the enzyme.

recombinant protein, labeling of recombinant protein kinase C (PKC) isoforms (Radominska-Pandya *et al.*, 2000) is used as an example (Fig. 6). For screening PKC isoforms, 7 pmol each of PKCα, βI, βII, γ, δ, ε, and ζ in 20 m*M* HEPES, pH 7, is incubated for 2 min at room temperature with or without 30 μ*M* *at*RA. [³H]*at*RA (30 Ci/mmol) is added to a final concentration of 3.3 μ*M* (0.03 nmol, ~1 μCi, in a final volume of 20 μl). [³H]*at*RA is added in ethanol with a final concentration of 2% ethanol in all samples. Samples are incubated for 10 min on ice prior to photolabeling with a hand-held long-wave (366 nm) UV lamp for 15 min. Proteins are denatured by the addition of NuPAGE denaturing buffer (Invitrogen, Carlsbad, CA) followed by sonication, boiling for 1 min, and centrifugation. Proteins are separated on NuPAGE minigels (1.5 mm; Invitrogen). Following electrophoresis, gels are stained with Coomassie blue, washed thoroughly with water, treated with AutoFluor, and subjected to autoradiography for 2–5 days at −80°.

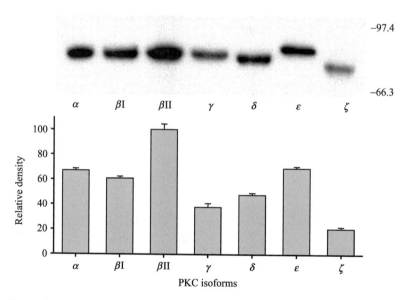

Fig. 6. Differential photoaffinity labeling of protein kinase C (PKC) isoforms by [³H]*at*RA. Seven different isoforms (α, βI, βII, γ, δ, ε, and ζ) of PKC were photolabeled with 3.3 μ*M* [³H]*at*RA. The relative photoincorporation of [³H]*at*RA into the PKC isoforms was determined by densitometry. PKCβII showed the highest photoincorporation of [³H]*at*RA; therefore, it was assigned the density of 100%, and all other densities are expressed as percentages of PKCβII activity.

Protein Digestion and Identification and Mutation of Amino Acids

Proteolytic Digestion. The next step in the identification of amino acids of the active site is proteolytic digestion of the conjugated proteins. Here we provide the example of digestion with trypsin, which is done routinely in the field of proteomics. Photolabeled and control irradiated samples are mixed with a trypsin solution (12.5 ng/μl) and incubated in an ice bath for approximately 45 min. The trypsin solution is removed and replaced with 40–50 μl of 50 mM ammonium bicarbonate, pH 8.0, and left overnight. The resulting peptides are extracted with 50% acetonitrile, 5% formic acid and dried under reduced pressure. The peptides are resuspended in 5% acetonitrile, 0.5% formic acid, 0.005% heptafluorobutyric acid, and 3 μl is applied to a 70-μm i.d., 15-cm Magic C18 reversed-phase capillary column. Peptides are eluted with a 5–80% acetonitrile gradient and analyzed on a ThermoFinnigan LCQ DECA XP ion trap tandem mass spectrometer. Sequence analysis is performed with Sequest using an indexed human subset database of the nonredundant protein database from NCBI.

Protein Identification by Liquid Chromatography/Tandem Mass Spectrometry (LC-MS/MS). Although several spectral methods are used in the field of proteomics to sequence conjugated polypeptides, the electrospray LC-MS/MS seems to be the method of choice and is described here. An electrospray tandem mass spectrometer coupled to a nanoscale liquid chromatography column (ESI LC-MS/MS; Thermo Finnigan, LCQ Deca) is utilized for solution sequence identification. Tryptic protein digests are loaded onto a C-18 12-cm × 0.075-mm capillary column, and peptides are eluted directly into the source by applying a linear gradient of 5–95% acetonitrile buffer at a final flow rate of ~200 nl/min. The LCQ runs in automatic collection mode with an instrument method composed of a single segment and four data-dependent scan events with a full MS scan. Strict relevance parameters are used to safeguard against false interpretation of data output from the tandem MS. Peptide lysates from nonphotocross-linked control samples are used to blank the instrument for subsequent analysis of the corresponding photocross-linked sample. Unique peptide peaks in the first MS run of the treated samples are selected, trapped, and collected for MS2/MS3 fragmentation and sequence identity. The TurboSEQUEST software provided by ThermoFinnigan automatically sorts and cross-correlates the peptides in the MS/MS profile to genomic and protein database information, providing the protein identity.

Protein Identification by Matrix-Assisted Laser Desorption Ionization (MALDI). MALDI is often used as a complementary approach to LC-MS/MS sequencing. Using a MALDI/time of flight (TOF) instrument with robotics and data analysis software (Zhang *et al.*, 1991), peptide digest maps of control and photolabeled UGTs are generated. Cross-linked peptide peaks of potential interest are subjected to high-energy laser power to induce fragmentation and the fragmented spectra are acquired in 12 segments (decrement ratio: 0.75). Selected metastable and collision-induced fragments of targeted peptides are then sent down the second TOF tube. Because the fragmentation pattern is greatly simplified and derived from one peptide, identification is made easier and more sensitive. Angiotensin is used as a calibration marker. Lists of the obtained peptide masses are used to search protein sequence databases (MASCOT) (Perkins *et al.*, 1999).

Site-Directed Mutagenesis. The final step in the identification of UGT active sites is investigation of the functionality of the identified amino acids by site-directed mutagenesis. QuikChange site-directed mutagenesis kits (Stratagene, La Jolla, CA) are used to generate mutations of specific amino acid residues and/or motifs from the wild-type UGT2B7 and UGT1A6 expression vectors. Primers containing the desired mutation are annealed to the target expression vectors and are incubated with dNTPs and Pfu DNA polymerase to extend the new mutant DNA strand. A restriction digestion with the *Dpn*I enzyme is used to digest the remaining methylated parental plasmid, leaving only the unmethylated mutated plasmid. The plasmid mixture is transformed into *E. coli* strain XL1-Blue for nick repair. The plasmid is amplified and isolated from liquid cultures of these transformants using a Qiagen miniprep kit (Qiagen). Positive clones are directly sequenced to verify the desired mutations.

Concluding Remarks

Despite important progress that has been made to date in characterizing the structure and function of UGTs, which is mainly due to efficient cloning and expression procedures in heterologous systems, as well as to the design of molecular tools and probes to investigate the active sites, elucidation of the three-dimensional structure of UGTs still remains a high-priority challenge. The original mode of integration of UGT in ER membranes and the absolute requirement of these proteins for membrane phospholipids result in a unique situation that makes the task difficult. Under these conditions, alternative strategies based on structure homology with crystallized GTs, analysis by high-field nuclear magnetic resonance of the two domains separately, or the design of soluble, active proteins will surely allow more structural data to be obtained about this family of enzymes. Such information should provide a better understanding of the molecular basis of glucuronidation, which can be taken into account for the development of safer drugs.

Acknowledgments

The studies reported here were supported by NIH Grants DK56226 and DK60109 and tobacco settlement funds (AR-P) and by Région Lorraine (JM). The authors thank Joanna Little for her assistance in preparing and editing the manuscript and Stacie Bratton for her assistance in preparing both the manuscript and the figures.

References

Barre, L., Magdalou, J., Netter, P., Fournel-Gigleux, S., and Ouzzine, M. (2005). The stop transfer syquence of the human UDP-glucuronosyltransferase 1A determines localization to the endoplasmic reticulum by both static retention and retrieval mechanisms. *FEBS J.* **272,** 1063–1071.

Battaglia, E., Nowell, S., Drake, R. R., Magdalou, J., Fournel-Gigleux, S., Senay, C., and Radominska, A. (1997). Photoaffinity labeling studies of the human recombinant UDP-glucuronosyltransferase, UGT1*6, with 5-azido-UDP-glucuronic acid. *Drug Metab. Dispos.* **25,** 406–411.

Battaglia, E., Pritchard, M., Ouzzine, M., Fournel-Gigleux, S., Radominska, A., Siest, G., and Magdalou, J. (1994). Chemical modification of human UDP-glucuronosyltransferase UGT1*6 by diethyl pyrocarbonate: Possible involvement of a histidine residue in the catalytic process. *Arch. Biochem. Biophys.* **309,** 266–272.

Battaglia, E., Terrier, N., Mizeracka, M., Senay, C., Magdalou, J., Fournel-Gigleux, S., and Radominska, A. (1998). Interaction of periodate-oxidized UDP-glucuronic acid with recombinant human liver UDP-glucuronosyltransferase UGT1A6. *Drug Metab. Dispos.* **26,** 812–817.

Bernstein, P. S., Choi, S. Y., Ho, Y. C., and Rando, R. R. (1995). Photoaffinity labeling of retinoic acid-binding proteins. *Proc. Natl. Acad. Sci. USA* **92,** 654–658.

Chen, G., and Radominska-Pandya, A. (2000). Direct photoaffinity labeling of cellular retinoic acid-binding protein I (CRABP-I) with all trans-retinoic acid: Identification of amino acids in the ligand binding site. *Biochemistry* **39,** 12568–12574.

Coffman, B. L., Kearney, W. R., Goldsmith, S., Knosp, B. M., and Tephly, T. R. (2003). Opioids bind to the amino acids 84 to 118 of UDP-glucuronosyltransferase UGT2B7. *Mol. Pharmacol.* **63,** 283–299.

Coffman, B. L., Kearney, W. R., Green, M. D., Lowery, R. G., and Tephly, T. R. (2001). Analysis of opioid binding to UDP-glucuronosyltransferase 2B7 fusion proteins using nuclear magnetic resonance spectroscopy. *Mol. Pharmacol.* **59,** 1464–1469.

Corsi, A. K., and Schekman, R. (1996). Mechanism of polypeptide translocation into the endoplasmic reticulum. *J. Biol. Chem.* **271,** 30299–30302.

Coutinho, P., Deleury, E., Davies, G. J., and Henrissat, B. (2003). An evolving hierarchical family classification for glycosyltransferases. *J. Mol. Biol.* **328,** 307–317.

Drake, R., Igari, I., Lester, R., Elbein, A., and Radominska, A. (1992). Application of 5-azido-UDP-glucose and 5-azido-UDP-glucuronic acid photoaffinity probes for the determination of the active site orientation of microsomal UDP-glucosyltransferases and UDP-glucuronosyltransferases. *J. Biol. Chem.* **267,** 11360–11365.

Drake, R. R., and Elbein, A. D. (1992). Photoaffinity labeling of glycosyltransferases. *Glycobiology* **2,** 279–284.

Drake, R. R., Evans, R. K., Wolf, M. J., and Haley, B. E. (1989). Synthesis and properties of 5-azido-UDP-glucose: Development of photoaffinity probes for nucleotide diphospate sugar binding sites. *J. Biol. Chem.* **264,** 11928–11933.

Drake, R. R., Kaushal, G. P., Pastuszak, I., and Elbein, A. D. (1991a). Purification, photo-affinity labeling and properties of plant UDP-glucose: Phosphoryldolichol glucosyltrans-ferase. *Plant Physiol.* **97,** 396–401.

Drake, R. R., Zimniak, P., Haley, B. E., Lester, R., Elbein, A. D., and Radominska, A. (1991b). Synthesis and characterization of 5-azido-UDP-glucuronic acid: A new photo-affinity probe for UDP-glucuronic acid-utilizing proteins. *J. Biol. Chem.* **266,** 23257–23260.

Dutton, G. J. (1980). "Glucuronidation of Drugs and Other Compounds." CRC Press, Boca Raton, FL.

Ghosh, S. S., Sappal, B. S., Kalpana, G. V., Lee, S. W., Roy Chowdhury, J., and Roy Chowdhury, N. (2001). Homodimerization of human bilirubin-uridine-diphoshoglucur-onate glucuronosyltransferase-1 (UGT1A1) and its functional implications. *J. Biol. Chem.* **276,** 42108–42115.

Gschaidmeier, H., and Bock, K. W. (1994). Radiation inactivation analysis of microsomal UDP-glucuronosyltransferases catalysing mono- and diglucuronide formation of 3,6-dihydroxybenzo(a)pyrene and 3,6-dihydroxychrysene. *Biochem. Pharmacol.* **48,** 1545–1549.

Ha, S., Walker, D., Shi, Y., and Walker, S. (2000). The 1.9 Å crystal structure of *Escherichia coli* MurG, a membrane-associated glycosyltransferase involved in peptidoglycan bio-synthesis. *Protein Sci.* **9,** 1045–1052.

Haley, B. E. (1991). Nucleotide photoaffinity labeling of protein kinase subunits. *Methods Enzymol.* **200,** 477–487.

Ikushiro, S., Emi, Y., and Iyanagi, T. (1997). Protein-protein interactions between UDP-glucuronosyltransferase isozymes in rat hepatic microsomes. *Biochemistry* **36,** 7154–7161.

Ishii, Y., Miyoshi, A., Maji, D., Yamada, H., and Oguri, K. (2004). Simulataneous expression of guinea pig UDP-glucuronosyltransferase 2B21 (UGT2B21) and 2B22 in COS-7 cells enhances UGT2B21-catalyzed chloramphenicol glucuronidation. *Drug Metab. Dispos.* **32,** 1057–1060.

Ishii, Y., Miyoshi, A., Watanabe, R., Tsuruda, K., Tsuda, M., Yamguchi-Nagamatsu, Y., Yoshisue, K., Tanaka, M., Maji, D., Ohgiya, S., and Oguri, K. (2001). Simultaneous expression of guinea pig UDP-glucuronosyltransferase 2B21 and 2B22 in COS-7 cells enhance UDP-glucuronosyltransferase 2B21-catalyzed morphine-6-glucuronide forma-tion. *Mol. Pharmacol.* **60,** 1040–1048.

Ishii, Y., Takeda, S., Yamada, H., and Oguri, K. (2005). Functional protein-protein interaction of drug metabolizing enzymes. *Front. Biosci.* **10,** 887–895.

Iwano, H., Yokota, H., Ohgiya, S., Yotumoto, N., and Yuasa, A. (1997). A critical amino acid residue, asp446, in UDP-glucuronosyltransferase. *Biochem. J.* **325,** 587–591.

Iwano, H., Yokota, H., Ohgiya, S., and Yuasa, A. (1999). The significance of amino acid residue Asp446 for enzymatic stability of rat UDP-glucuronosyltransferase UGT1A6. *Arch. Biochem. Biophys.* **363,** 116–120.

Iyanagi, T., Haniu, M., Sogawa, K., Fujii-Kuriyama, Y., Watanabe, S., Shively, J. E., and Anan, K. F. (1986). Cloning and characterization of cDNA encoding 3-methylcholan-threne inducible rat mRNA for UDP-glucuronosyltransferase. *J. Biol. Chem.* **261,** 15607–15614.

Jackson, M. R., Nilsson, T., and Peterson, P. A. (1990). Identification of a consensus motif for retention of transmembrane proteins in the endoplasmic reticulum. *EMBO J.* **9,** 3153–3162.

Johnson, A. E., and van Waes, M. A. (1999). The translocon: A dynamic gateway to the ER membrane. *Annu. Rev. Cell Dev. Biol.* **15,** 799–842.

Koiwai, O., Aono, S., Adachi, Y., Kamisako, T., Yasui, Y., Nishizawa, M., and Sato, H. (1996). Crigler-Najjar syndrome type II is inherited both as a dominant and as a recessive trait. *Hum. Mol. Genet.* **5,** 645–647.

Kurkela, M., García-Horsman, J. A., Luukkkanen, L., Mörsky, S., Taskinen, J., Baumann, M., Kostiainen, R., Hirvonen, J., and Finel, M. (2003). Expression and characterization of recombinant human UDP-glucuronosyltransferases (UGTs): UGT1A9 is more resistant to detergent inhibition than the other UGTs and was purified as an active dimeric enzyme. *J. Biol. Chem.* **278,** 3536–3544.

Kurkela, M., Hirvonen, J., Kostiainen, R., and Finel, M. (2004). The interactions between the N-terminal and C-terminal domains of the human UDP-glucuronosyltransferases are partly isoform-specific, and may involve both monomers. *Biochem. Pharmacol.* **68,** 2443–2450.

Laemmli, U. K. (1970). Cleavage of structural proteins during the assembly of the head of the bacteriophage T4. *Nature* **227,** 680–685.

Lindt, T., Lindahl, U., and Lidholt, K. (1993). Bioxynthesis of heparin/heparan sulfate: Identification of a 70 kDa protein catalyzing both D-glucuronyl- and the N-acetyl-D-glucosaminyltransferase reactions. *J. Biol. Chem.* **268,** 20705–20708.

Little, J. M., and Radominska, A. (1997). Application of photoaffinity labeling with [11,12-^{3}H] all *trans*-retinoic acid to characterization of rat liver microsomal UDP-glucuronosyltransferase(s) with activity toward retinoic acid. *Biochem. Biophys. Res. Commun.* **230,** 497–500.

Mackenzie, P. I. (1986a). Rat liver UDP-glucuronosyltransferase: cDNA sequence and expression of a form glucuronidating 3-hydroxyandrogens. *J. Biol. Chem.* **261,** 14112–14117.

Mackenzie, P. I. (1986b). Rat liver UDP-glucuronosyltransferase: Sequence and expression of a cDNA encoding a phenobarbital-inducible form. *J. Biol. Chem.* **261,** 6119–6125.

Mackenzie, P. I. (1987). Rat liver UDP-glucuronosyltransferase: Identification of cDNAs encoding two enzymes which glucuronidate testosterone, dihydrotestosterone, and β-estradiol. *J. Biol. Chem.* **262,** 9744–9749.

Mackenzie, P. I., and Owens, I. S. (1984). Cleavage of nascent UDP-glucuronosyltransferase from rat liver by dog pancreatic microsomes. *Biochem. Biophys. Res. Commun.* **122,** 1441–1449.

Matern, H., Matern, S., and Gerok, W. (1982). Isolation and characterization of rat liver microsomal UDP-glucuronosyltransferase activity toward chenodeoxycholic acid and testosterone as a single form of enzyme. *J. Biol. Chem.* **257,** 7422–7429.

Matsui, M., and Nagai, F. (1986). Genetic deficiency of androsteron UDP-glucuronosyltransferase activity in Wistar rats is due to the loss of genzyme protein. *Biochem. J.* **234,** 139–144.

Meech, R., and Mackenzie, P. I. (1997). UDP-Glucuronosyltransferase, the role of the amino terminus in dimerization. *J. Biol. Chem.* **272,** 26913–26917.

Meech, R., and Mackenzie, P. I. (1998). Determinants of UDP glucuronosyltransferase membrane association and residency in the endoplasmic reticulum. *Arch. Biochem. Biophys.* **356,** 77–85.

Meech, R., Yogalingam, G., and Mackenzie, P. (1996). Mutational analysis of the carboxy-terminal region of UDP-glucuronosyltransferase 2B1. *DNA Cell Biol.* **15,** 489–494.

Nagar, S., Zalatoris, J. J., and Blanchard, R. L. (2004). Human UGT1A6 pharmacogenetics: Identification of a novel SNP, characterization of allele frequencies and functional analysis of recombinant allozymes in human liver tissue and in cultured cells. *Pharmacogenetics* **14,** 487–499.

Ott, C. M., and Lingappa, V. R. (2002). Integral membrane protein biosynthesis: Why topology is hard to predict. *J. Cell Sci.* **115,** 2003–2009.

Ouzzine, M., Antonio, L., Burchell, B., Netter, P., Fournel-Gigleux, S., and Magdalou, J. (2000). Importance of histidine residues for the function of the human liver

UDP-glucuronosyltransferase UGT1A6: Evidence for the catalytic role of histidine 370. *Mol. Pharmacol.* **58,** 1609–1615.

Ouzzine, M., Magdalou, J., Burchell, B., and Fournel-Gigleux, S. (1999a). Expression of a functionally active human hepatic UDP-glucuronosyltransferase (UGT1A6) lacking the N-terminal signal sequence in the endoplasmic reticulum. *FEBS Lett.* **454,** 187–191.

Ouzzine, M., Magdalou, J., Burchell, B., and Fournel-Gigleux, S. (1999b). An internal signal sequence mediates the targeting and retention of the human UDP-glucuronosyltransferase 1A6 to the endoplasmic reticulum. *J. Biol. Chem.* **274,** 31401–31409.

Perkins, D. N., Pappin, D. J., Creasy, D. M., and Cottrell, J. S. (1999). Probability-based protein identification by searching sequence database using mass spectrometry data. *Electrophoresis* **20,** 3551–3567.

Peters, W. H., Jansen, P. L., and Nauta, H. (1984). The molecular weights of UDP-glucuronyltransferase determined with radiation-inactivation analysis: A molecular model of bilirubin UDP-glucuronyltransferase. *J. Biol. Chem.* **259,** 11701–11705.

Rachmel, A., Hazelton, G. A., Yergey, A. L., and Liberato, D. J. (1985). Furosemide 1-O-acyl glucuronide: *In vitro* biosynthesis and pH-dependent isomerization to β-glucuronidase-resistant forms. *Drug Metab. Dispos.* **13,** 705–710.

Radominska, A., Berg, C., Treat, S., Little, J. M., Gollan, J., Lester, R., and Drake, R. (1994a). Characterization of UDP-glucuronic acid transport in rat liver microsomal vesicles with photoaffinity analogs. *Biochim. Biophys. Acta* **1195,** 63–70.

Radominska, A., and Drake, R. (1994). Synthesis and uses of azido-nucleoside diphosphate sugar photoaffinity analogs. *Methods Enzymol.* **230,** 330–339.

Radominska, A., Little, J. M., Lester, R., and Mackenzie, P. I. (1994b). Bile acid glucuronidation by rat liver microsomes and cDNA-expressed UDP-glucuronosyltransferases. *Biochim. Biophys. Acta* **1205,** 75–82.

Radominska, A., Paul, P., Treat, S., Towbin, H., Pratt, C., Little, J., Magdalou, J., Lester, R., and Drake, R. R. (1994c). Photoaffinity labeling for evaluation of uridinyl analogs as specific inhibitors of rat liver microsomal UDP-glucuronosyltransferases. *Biochim. Biophys. Acta* **1205,** 336–345.

Radominska-Pandya, A., and Chen, G. (2002). Photoaffinity labeling of human retinoid X receptor-β (RXR-β) with 9-*cis*-retinoic acid: Identification of phytanic acid, docosahexaenoic acid and lithocholic acid as ligands for RXR-β. *Biochemistry* **41,** 4883–4890.

Radominska-Pandya, A., Chen, G., Czernik, P. J., Little, J. M., Samokyszyn, V. M., Carter, C. A., and Nowak, G. (2000). Direct interaction of all *trans*-retinoic acid with protein kinase C: Implications for PKC signaling and cancer therapy. *J. Biol. Chem.* **275,** 22324–22330.

Radominska-Pandya, A., Chen, G., Samokyszyn, V. M., Little, J. M., Gall, W. E., Zawada, G., Terrier, N., Magdalou, J., and Czernik, P. (2001). Application of photoaffinity labeling with [3H]all *trans*- and 9-*cis*-retinoic acids for characterization of cellular retinoic acid binding proteins I and II. *Protein Sci.* **10,** 200–211.

Radominska-Pandya, A., Czernik, P., Little, J. M., Battaglia, E., and Mackenzie, P. I. (1999). Structural and functional studies of UDP-glucuronsyltransferases. *Drug Metab. Rev.* **31,** 817–900.

Radominska-Pandya, A., Pokrovskaya, I. D., Xu, J., Little, J. M., Jude, A. R., and Czernik, P. J. (2002). Nuclear UDP-glucuronosyltransferases: Identification of UGT2B7 and UGT1A6 in human liver nuclear membranes. *Arch. Biochem. Biophys.* **399,** 37–48.

Roy Chowdhury, J., Novikoff, P. M., Roy Chowdhury, N., and Novikoff, A. B. (1985). Distribution of UDP-glucuronosyltransferase in rat tissue. *Proc. Natl. Acad. Sci. USA* **82,** 2990–2994.

Senay, C., Battaglia, E., Chen, G., Breton, R., Fournel-Gigleux, S., Magdalou, J., and Radominska-Pandya, A. (1999). Photoaffinity labeling of the aglycon binding site of the recombinant human liver UDP-glucuronosyltransferase UGT1A6 with 7-azido-4-methyl-coumarin. *Arch. Biochem. Biophys.* **368,** 75–84.

Senay, C., Jedlitschky, G., Terrier, N., Burchell, B., Magdalou, J., and Fournel-Gigleux, S. (2002). The importance of cysteine 126 in the human liver UDP-glucuronosyltransferase UGT1A6. *Biochim. Biophys. Acta* **1597,** 90–96.

Senay, C., Ouzzine, M., Battaglia, E., Pless, D., Cano, V., Burchell, B., Radominska, A., Magdalou, J., and Fournel-Gigleux, S. (1997). Arginine 52 and histidine 54 located in a conserved N-terminal hydrophobic region (LX2-R52-G-H54-X3-V-L) are important amino acids for the functional and structural integrity of the human liver UDP-glucuronosyltransferase UGT1*6. *Mol. Pharmacol.* **51,** 406–413.

Shepherd, S. R. P., Baird, S. J., Hallinan, T., and Burchell, B. (1989). An investigation of the transverse topology of bilirubin UDP-glucuronosyltransferase in rat hepatic endoplasmic reticulum. *Biochem. J.* **259,** 617–620.

Vanstapel, F., and Blanckaert, N. (1988). Topology and regulation of bilirubin UDP-glucuronosyltransferase in sealed native microsomes from rat liver. *Arch. Biochem. Biophys.* **263,** 216–225.

Wibo, M., Thines-Sempoux, D., Amar-Costesec, A., Beaufay, H., and Godeaine, D. (1981). Analytical study of microsomes and isolated subcellular membranes from rat liver VIII: Subfractionation of preparations enriched with plasma membranes outer mitochondrial membranes or Golgi complex membranes. *J. Cell Biol.* **89,** 456–474.

Yokota, H., Yuasa, A., and Sato, R. (1992). Topological disposition of UDP-glucuronyl-transferase in rat liver microsomes. *J. Biochem.* **112,** 192–196.

Zhang, P., Graminski, G. F., and Armstrong, R. N. (1991). Are the histidine residues of glutathione S-transferase important in catalysis? An assessment by 13C NMR spectros-copy and site-specific mutagenesis. *J. Biol. Chem.* **266,** 19475–19479.

[9] Human *SULT1A* Genes: Cloning and Activity Assays of the *SULT1A* Promoters

By Nadine Hempel, Masahiko Negishi, and Michael E. McManus

Abstract

The three human SULT1A sulfotransferase enzymes are closely related in amino acid sequence (>90%), yet differ in their substrate preference and tissue distribution. SULT1A1 has a broad tissue distribution and metabo-lizes a range of xenobiotics as well as endogenous substrates such as estrogens and iodothyronines. While the localization of SULT1A2 is poor-ly understood, it has been shown to metabolize a number of aromatic amines. SULT1A3 is the major catecholamine sulfonating form, which is consistent with it being expressed principally in the gastrointestinal tract.

METHODS IN ENZYMOLOGY, VOL. 400
0076-6879/05 $35.00
DOI: 10.1016/S0076-6879(05)00009-1

SULT1A proteins are encoded by three separate genes, located in close proximity to each other on chromosome 16. The presence of differential 5′-untranslated regions identified upon cloning of the SULT1A cDNAs suggested the utilization of differential transcriptional start sites and/or differential splicing. This chapter describes the methods utilized by our laboratory to clone and assay the activity of the promoters flanking these different untranslated regions found on SULT1A genes. These techniques will assist investigators in further elucidating the differential mechanisms that control regulation of the human SULT1A genes. They will also help reveal how different cellular environments and polymorphisms affect the activity of SULT1A gene promoters.

Introduction

The human SULT1A subfamily of cytosolic sulfotransferases is unique as it contains more than one isoform (SULT1A1, SULT1A2, and SUL-T1A3), compared to a solitary SULT1A1 member identified in all other species to date. The human SULT1A enzymes were first classified as the aryl/phenol sulfotransferases according to their substrate preference for small phenolic compounds (Hart et al., 1979). Originally, two isoforms were differentiated based on their substrate preference and thermostability (Veronese et al., 1994). The more thermostable (TS-PST) and phenol-preferring (P-ST) form is now referred to as SULT1A1 and the thermo-labile (TL-PST) and monoamine-preferring (M-ST) isoform as SULT1A3 (Blanchard et al., 2004; Wilborn et al., 1993; Zhu et al., 1993a,b). Later, a second thermostable isoforms was cloned and named SULT1A2 (Ozawa et al., 1995; Zhu et al., 1996).

The three SULT1A enzymes exhibit relative molecular masses on SDS–PAGE between 32 and 34 kDa and each form is composed of 295 amino acids, which share >90% sequence identity (Fig. 1). Although the three SULT1A enzymes share high amino acid sequence identity, their

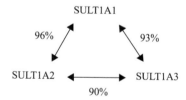

FIG. 1. Human SULT1A isoform amino acid homology.

tissue-specific expression and substrate profiles vary. Not surprisingly, the greatest variability in amino acid sequence is found in the areas of substrate binding. Site-directed mutagenesis studies and elucidation of the human SULT1A1 and SULT1A3 protein crystal structures have given insight into the mechanisms determining substrate specificity differences between these two enzymes (Barnett et al., 2004; Bidwell et al., 1999; Brix et al., 1999; Dajani et al., 1999; Gamage et al., 2003). The negatively charged glutamic acid residue at position 146 has been shown to play an important role in the binding of positively charged biogenic amines such as dopamine by SULT1A3 (Barnett et al., 2004; Brix et al., 1999). SULT1A1, which lacks a charge at this position, has less affinity for these substrates (Brix et al., 1999). However, compared to SULT1A3, the hydrophobic substrate-binding pocket of SULT1A1 has greater affinity for small phenolic compounds such as the model substrate p-nitrophenol and the drugs acetaminophen and minoxidil (Brix et al., 1999; Falany, 1997; Gamage et al., 2003). SULT1A1 also has activity in sulfonating estrogens and thyroid hormone, and crystallographic studies have shown that the hydrophobic-binding pocket is large enough to accommodate two substrate molecules, explaining the substrate inhibition kinetic profile observed at high substrate concentrations (Gamage et al., 2003).

Studies investigating the mechanisms underlying the differential tissue expression profiles of the SULT1A enzymes have only recently been undertaken. One of the most striking differences in expression occurs in the adult liver, where SULT1A1 is expressed in high amounts, yet SULT1A3 at negligible levels (Richard et al., 2001; Windmill et al., 1998). Our laboratory has reported on a differential mechanism of transcriptional regulation between SULT1A1 and SULT1A3 genes, which is a first step in elucidating the tissue-specific patterns of expression of these two closely related enzymes (Hempel et al., 2004). This chapter describes the methods employed by our laboratory to investigate the properties of the human SULT1A genes, with specific emphasis on the approach used to clone and study the properties of the SULT1A promoters in mammalian and Drosophila S2 cell lines.

Properties of Human SULT1A Genes

Members of the human SULT1A subfamily are encoded by three separate genes, which are found in close proximity to each other on chromosome 16 (16p11.2–12.1). Their relative position on chromosome 16 of the Homo sapiens genome can be attained using the NCBI Entrez Map View browser (http://www.ncbi.nlm.nih.gov/mapview/) with the search terms

"SULT1A1," "SULT1A2," or "SULT1A3." A schematic representation with links to sequence information can be gained by clicking the "Map Element" Link. The gene symbol leads to a summary data page with information on the gene, including links to references and NCBI curated nucleotide and protein sequence records. For reference, the curated nucleotide records for SULT1A1, SULT1A2, and SULT1A3 are NM_001055, NM_001054, and NM_003166, respectively. Within the Map Viewer page the area spanning between the three *SULT1A* genes can be displayed by entering the two genes of interest (e.g., SULT1A1 and SULT1A3) in the search field "Region shown" on the left-hand side of the page. With current data from the annotated genome sequence, *SULT1A2* is positioned the most telomeric of the three *SULT1A* genes. The *SULT1A2* and *SULT1A1* genes are positioned head to tail, approximately 9.75 kb apart from the stop codon of *SULT1A1* to the ATG codon of *SULT1A2*. These two genes share more than 93% sequence identity and no other gene has been positioned between the two (Aksoy and Weinshilboum, 1995; Bernier *et al.*, 1996; Her *et al.*, 1996). Their close proximity and high homology suggest that these genes have arisen recently in evolution due to a gene duplication event. *SULT1A3* is located most centromeric and at a distance of approximately 1.58 Mb from the position of *SULT1A1* and *SULT1A2*, with several other genes occupying the sequence between them. *SULT1A3* is currently positioned in the opposite direction to the other *SULT1A* genes. The *SULT1A3* gene differs mainly in its 5′ promoter sequence and shares about 70% sequence identity with the other two *SULT1A* genes (Aksoy and Weinshilboum, 1995; Bernier *et al.*, 1994a; Dooley *et al.*, 1994).

Structurally the three *SULT1A* genes are similar. All contain seven coding exons and have been shown to have alternatively transcribed 5′-untranslated regions (5′UTRs). Figure 2 shows the 5′UTRs thus far identified in the SULT1A cDNA species isolated to date. Why these alternate 5′UTRs are transcribed is currently unknown. Due to their isolation from different tissue libraries, it was suggested that transcription of these may represent a tissue-specific mechanism (Zhu *et al.*, 1993a,b). The endogenous consequences for this phenomenon are unclear, as all SULT1A cDNAs contain the same open reading frame. It has not been investigated whether this compromises mRNA translation or stability as a way to control SULT1A expression posttranscriptionally. Alternate splicing and/or the utilization of alternate promoters in front of these sequences has been proposed as the mechanism for the occurrence of these 5′UTRs. Our laboratory and others have shown that sequences on the SULT1A genes flanking these alternate 5′UTRs have different promoter activities.

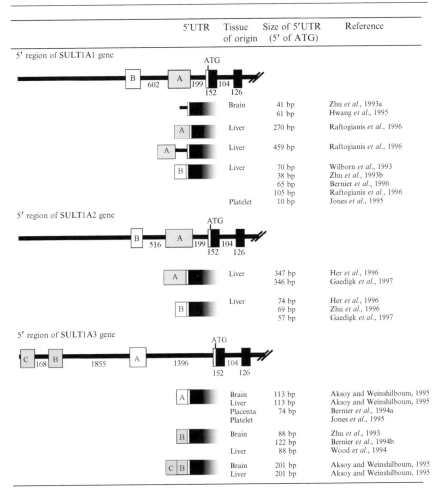

FIG. 2. 5'-Untranslated regions of SULT1A cDNAs. Black boxes represent coding exons; white and gray boxes represent alternate 5'UTRs. Sizes of introns are given below on the 5' region of the genes (Aksoy and Weinshilboum, 1995; Her *et al.*, 1996; Raftogianis *et al.*, 1996).

Cloning of Human *SULT1A* Promoters

The *SULT1A* promoters can be amplified easily from human genomic DNA. Primers were originally designed by our laboratory based on the SULT1A3 gene sequence submitted to GenBank prior to the release of human genome data (Accession No.: U20499; Aksoy and Weinshilboum,

1995). The sense primer 5'-<u>ACGCGTGCTAGC</u>GAGCTGTGAGGAA-GTTCAGGTC-3', containing *Mlu*I and *Nhe*I restriction sites (underlined), and the antisense primer 5'-<u>AGATCTCTCGAG</u>GATCAGCTCCATG-TTCCTGCATC-3', containing *Bgl*II and *Xho*I restriction sites (underlined), are located 4023 bp upstream and 12 bp downstream of the ATG start codon of *SULT1A3*, respectively.

The *Taq*Plus Long polymerase system by Stratagene, containing a mixture of Taq2000 and the proofreading polymerase Pfu, allows for efficient and more accurate amplification of long polymerase chain reaction (PCR) products. To a 300-ng genomic DNA template the following components are added in a final volume of 50 μl: 1× low-salt *Taq*Plus Long polymerase buffer, 800 μM dNTP mix, 1 μM of each sense and antisense primer, and 2 U of *Taq*Plus Long polymerase. The PCR cycling parameters are as follows: an initial 2-min denaturing step at 94° is followed by 30 cycles of denaturing at 92° for 30 s, annealing of primers at 66° for 1 min, and extension at 72° for 5 min. This is followed by an additional 5 cycles, with an increased extension time of 15 min. The PCR product is electrophoresed on a 1% agarose gel containing ethidium bromide, excised, and purified. Approximately 150 ng of the PCR product is used for cloning into the TOPO TA pCR2.1 vector (Invitrogen) according to the manufacturer's instructions, followed by transformation into TOP10 chemically competent cells (Invitrogen). Sequences are subcloned into the pGL3Basic luciferase reporter gene vector using restriction sites *Mlu*I and *Bgl*II at the 3' and 5' ends of the insert, respectively. The pGL3Basic reporter gene vector encodes the firefly (*Photinus pyralis*) luciferase protein when a functional promoter is cloned upstream of its transcriptional start site. This vector has no enhancer or intrinsic promoters, and therefore will only transcribe the luciferase gene based on the activity of the cloned promoter. The high sequence similarity of the three *SULT1A* genes in areas near the ATG start codon and 4 kb upstream means that the sequences of all three *SULT1A* genes are amplified in the aforementioned PCR. The restriction enzymes *Nsi*I, *Cla*I, and *Kpn*I, which specifically digest the *SULT1A1*, *SULT1A2*, and *SULT1A3* sequences, respectively, can be used to differentiate between the amplified sequences. Figure 3 shows the restriction digest of three plasmid preparations containing the *SULT1A1*, *SULT1A2*, or *SULT1A3* 5' regions cloned into the 4818-bp pGL3Basic vector.

To amplify the 6-kb sequence flanking the 5'UTR C of *SULT1A3* the sense primer 5'-GTGAGAGACCTGGCAGGAACAGG-3' and the antisense primer 5'-AGATCACATGGGCCCTTAGC-3', which are 9689 and 3691 bp upstream of the *SULT1A3* ATG start codon, respectively, are used. The following components are added to 300 ng of human genomic DNA in a final volume of 50 μl: 1× Expand Long PCR buffer 1 (Roche),

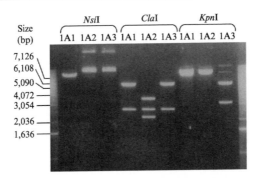

Fig. 3. Differential restriction digestion of 4-kb upstream *SULT1A* gene sequences cloned into pGL3Basic. Four kilobases of the *SULT1A* sequences flanking the ATG start codon was isolated from human genomic DNA by PCR and cloned into the pGL3Basic vector as described. The constructs (1 μg) were digested with 2 U of *Nsi*I, *Cla*I, or *Kpn*I restriction enzymes for 4 h at 37°. Digested DNA was electrophoresed on a 1% agarose gel containing ethidium bromide.

1.75 mM MgCl$_2$, 4 mM dNTP mix, 300 ng of each primer, and 0.75 μl of Expand Long PCR enzyme mix (Roche). The cycling parameters are as follows: initial denaturing at 94° for 5 min, followed by 10 cycles of 30 sec denaturing at 94°, 30 s annealing at 62°, and 7 min extension at 68°. This is followed by 25 cycles of the same temperature parameters, where the extension time is increased in length by 20 s at each cycle. A final extension at 68° for 10 min concludes the PCR. The product is electrophoresed, gel extracted, and cloned into the TOPO TA pCR2.1 vector. For subcloning into the pGL3Basic vector the sequence is excised from pCR2.1 using *Xho*I and *Hin*dIII restriction sites of the vector. A 2564-bp deletion construct of the SULT1A3 promoter flanking 5'UTR C can be created using the restriction enzyme *Nhe*I.

The cloned *SULT1A* 5' gene sequences are used as templates for amplification of the different promoters flanking the alternate 5'UTRs and promoter deletion constructs using the primers listed in Table I. To allow for subcloning of sequences into pGL3Basic, sense primers are designed to incorporate 5' *Mlu*I or *Nhe*I restriction enzyme recognition sites and antisense primers 5' *Bgl*II or *Xho*I sites. The PCR is carried out with 2.5 U *Taq* DNA polymerase, 200 ng of each primer, 50 ng of template DNA, 1.5 mM MgCl$_2$, and polymerase buffer in a final volume of 50 μl. Cycling is carried out under the following conditions: an initial denaturing step at 94° for 2 min, followed by 25 cycles at 94° for 30 s, annealing at 64° for 30 s, and extension at 72° for 1 min per kilobase PCR product length. The cycling is concluded with a final extension at 72° for 5 min. The PCR products are

TABLE I
OLIGONUCLEOTIDE PRIMERS FOR *SULT1A* PROMOTER PCR

Name	Location	S/AS	Sequence[a]
SULT1A promoter deletions			
Promoter flanking first coding exon I			
1A1/2/3 RPExI	−1 bp of ATG	AS	5'-AGATCTCTCGAGTGTTCCTGCGTCAGGGGCCAGAGC-3'
Promoter flanking 5'UTR A 1A1/1A2			
1A1/2 FP1A	5'UTR B	S	5'-ACGCGGTTCACCCTGCTCAGCTTGTGGCTC-3'
1A1/2/3 RP1	5'UTR A 1A1/2	AS	5'-AGATCTCTCGAGTGTCTCACCATTTCCTGCTGG-3'
Promoter flanking 5'UTR B 1A1/1A2 and 5'UTR A 1A3	Location from transcriptional start site		
1A1/2/3 RP2	5'UTR B 1A1/2 5'UTR A 1A3	AS	5'-AGATCTCTCGAGACCTGAGCTCTTGGGAACCTG-3'
1A1/2 FP3	−2325 1A1 −2258 1A2	S	5'-ACGCGTGCTAGCGGTAGCTGTGAGGCGTCACTGCTTTGG-3'
1A2 FP4	−1663	S	5'-ACGCGTGCTAGCGGCTCTTGGCACCTTAGCCAGA-3'
1A3 FP4	−1554	S	5'-ACGCGTGCTAGCGCCTCGAGCTCATGCAATTCTTGG-3'
1A1/2 FP5	−1217 1A1 −1156 1A2	S	5'-ACGCGTGCTAGCCACATTCTCGCCCTCTTTTCTGTGTCA-3'
1A3 FP5	−1074	S	ACGCGTGCTAGCGCCGCCATCATGCCCAGCTAA

1A1/2 FP6	−542 1A1 −488 1A2	S	5'-ACGCGTGCTAGCGGCGGCCTTGTGTGGTCAGAGCCTGGA-3'
1A3 FP6	−560	S	ACGCGTGCTAGCCATGGCAAAACCCGTCTCTACTAAA
1A1/2 FP7	−232 1A1 −221 1A2	S	5'-ACGCGTGCTAGCCCTTTCCCCTTTCATTCTTCTGTTTTC-3'
1A3 FP7	−265	S	5'-ACGCGTGCTAGCCCAAATACCAATGTTGGCCCCTTTT-3'
1A1/2/3 FP8	−144 1A1 −132 1A2 −157 1A3	S	5'-ACGCGTTAAGGAGGGTAATGGAGAAGCT-3'
1A1/3 FP9	−112 1A1 −125 1A3	S	5'-ACGCGTCAACCCCACCCTTCCTTCC-3'
1A2 FP9	−100	S	5'-ACGCGTCAACCCTACCCCTTCCTTCC-3'
1A1 FP9b	−89	S	5'-GGACGCGTGTAGCAAATCTAAGTCCAGCCCCG-3'
1A1/2 FP10	−68 1A1 −56 1A2	S	5'-ACGCGTGGCTCCAGATCCCTCCCACA-3'
1A3 FP10	−89	S	5'-ACGCGTGACTTTAGATCCTCCCACACTG-3'
1A1 FP11	−1	S	5'-ACGCGTCACAGCACCACAATCAGCCACT-3'
1A2 FP11	+10	S	5'-ACGCGTCACAACACCACAACTCAGCCACT-3'
1A3 FP11	−22	S	5'-ACGCGTCACCACCACCATACTCAGCCCCT-3'
Promoter flanking 5'UTR B 1A3	5'UTR B		
1A3 utrB RP		AS	5'-CTCGAGAGATCTCAGTGTCCATCTGGCACAGCCA-3'

[a] Underlined bases represent restriction enzyme recognition sites.

electrophoresed on 1–2% agarose gels and purified. After initial TA cloning into pCR2.1 the sequences are subcloned into the pGL3Basic vector.

Cell-Based Transfection of *SULT1A* Promoter Constructs

A range of cell lines are useful tools in studying the activity of the human *SULT1A* promoters. In the past we have utilized mammalian cell lines such as the human hepatocarcinoma HepG2 and Hep3B cells as representatives of a liver-like environment, human Caco2 cells as a representative of the gastrointestinal tract, and the human breast cancer cell line MCF-7 as a representative of mammary tissue. Further, Schneider's *Drosophila melanogaster* 2 cells (S2) lack a large variety of mammalian transcription factors (Galvagni *et al.*, 2001) and are a useful system to study the regulation of promoters by ubiquitous or highly expressed mammalian transcription factors using cotransfection. We have used this method to show an induction of the human SULT1A1 promoter by Ets and Sp1 transcription factors (Hempel *et al.*, 2004).

Mammalian cells are grown in MEM containing 10% fetal bovine serum and penicillin (50 U/ml)/streptomycin (50 μg/ml) at 37° in a humidified incubator with 5% CO_2. For transfection of luciferase reporter plasmids, cells are seeded into 24-well plates at approximately 5×10^4 cells per well in 500 μl medium. Transfection is carried out at 70–80% confluence using a modified protocol of the calcium phosphate transfection method of the CellPhect transfection kit (Amersham). Each *SULT1A* promoter luciferase reporter construct or the empty pGL3-Basic vector (0.1 μg/well) is transfected together with the transfection control reporter vector pRL-SV40 (0.05 μg/well). For comparison of different promoter constructs that vary significantly in length, the DNA amounts are adjusted to bring samples to equimolar concentrations of plasmid. In cotransfection experiments, mammalian expression vectors containing transcription factor cDNAs (0.1–1 μg/well) can be included in the transfection mix. Empty vectors are used to bring transfection samples to equal DNA concentrations between wells. Each sample is transfected in triplicate. For transfection into 3 wells the DNA is mixed with MilliQ H_2O to a final volume of 37.5 μl, to which an equal volume of buffer A (CellPhect kit) is added. After brief vortexing and incubation at room temperature for 10 min, 75 μl of buffer B is added, followed by immediate vortexing and incubation at room temperature for 15 min. The amount of DNA and the volumes of buffers A and B are scaled up or down according to the number of wells transfected. The DNA mix (50 μl per well) is added to the cells and the medium is replaced 24 h posttransfection. After an additional 24 h in culture the medium is aspirated and the cells are washed once with phos-phate-buffered saline, followed by the addition of $1\times$ passive lysis buffer (100 μl/well) of the dual luciferase assay (Promega). Cells are lysed by gentle rocking at room temperature.

S2 cells are grown in the dark in Schneider's *Drosophila* medium (Invitrogen), supplemented with 10% heat-inactivated fetal bovine serum and penicillin/streptomycin at room temperature. Cells are transferred to 12-well plates at a cell density of 0.5×10^6 cells per well in 1 ml media and transfected after 24 h in culture using the calcium phosphate method. Each well is transfected with 2 μg *SULT1A* promoter reporter constructs and 2–5 μg of the desired transcription factor, cloned into a *Drosophila* expression vector such as pAC5.1 (Invitrogen). The DNA is mixed with 2 M CaCl$_2$ and the sample is brought to a final volume of 105 μl with sterile MilliQ H$_2$O. An equal volume of $2\times$ HEPES-buffered saline (50 mM HEPES, 1.5 mM Na$_2$HPO$_4$, 280 mM NaCl; pH 7) is added drop wise while mixing the sample gently. The mix is incubated at room temperature for 30 min and 200 μl is added to each well containing the S2 cells while swirling the plates gently. The medium is changed 24 h posttransfection by collecting the cells using centrifugation at $100g$ for 10 min in 1.7-ml tubes, resuspending them in 1 ml fresh medium, and returning cells to the wells. After an additional 24 h in culture, cells are collected by centrifugation at $1000g$ for 10 min in 1.7-ml tubes. The medium is aspirated, 100 μl of $1\times$ passive lysis buffer is added, and the cells are shaken vigorously for 10 min. After aiding lysis by freeze/thaw at $-80°$, the cells are centrifuged at $3000g$ for 10 min.

SULT1A Promoter Luciferase Activity Assays

The luciferase reporter assay is a quantitative gene reporter system used to assess the ability of sequences to act as promoters by driving transcription of the firefly luciferase enzyme. The transfection standard vector pRLSV-40 encodes a second *Renilla reniformis* luciferase enzyme. The activities of both enzymes are assessed by a sequential chemilumines-cence assay (dual luciferase assay system, Promega), according to the manufacturer's specifications. Briefly, after lysis of cells with passive lysis buffer, 50 μl of the firefly luciferase enzyme substrate (luciferase assay reagent II) is added to 2–10 μl of total lysed mammalian cells or 20 μl cleared lysate of the S2 *Drosophila* cells. Chemiluminescence is measured for 10 s after an initial 2-s premeasurement delay on either a single sample or a 96-multiwell injection luminometer. The reaction is quenched by the addition of 50 μl *Renilla* luciferase substrate (Stop & Glo) and *Renilla* chemiluminescence is measured for 10 s. Firefly luciferase activity

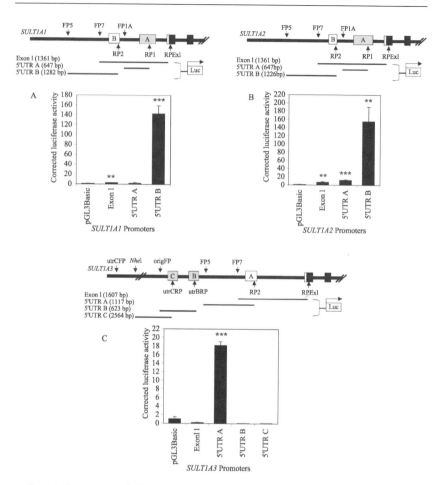

FIG. 4. Promoter activities of sequences flanking the 5′UTRs of *SULT1A* genes. Sequences flanking the different 5′UTRs of the *SULT1A1* (A), *SULT1A2* (B), and *SULT1A3* (C) genes were created by PCR and cloned into the pGL3Basic vector. The location and size of the constructs and the position of primers (arrows) used to amplify these are displayed schematically. Constructs were transfected into HepG2 cells and lysed cells were assayed for luciferase activity as described. Results represent firefly luciferase activity corrected for *Renilla* luciferase activity of the transfection control and are expressed relative to the activity of the empty pGL3Basic vector. Each bar represents the mean result from three transfected wells of cells ± SD. Asterisks indicate significant differences to the activity of the empty pGL3Basic basic vector (**$p < 0.01$; ***$p < 0.001$, Student t test). (Adapted with permission from Hempel *et al.*, 2004.)

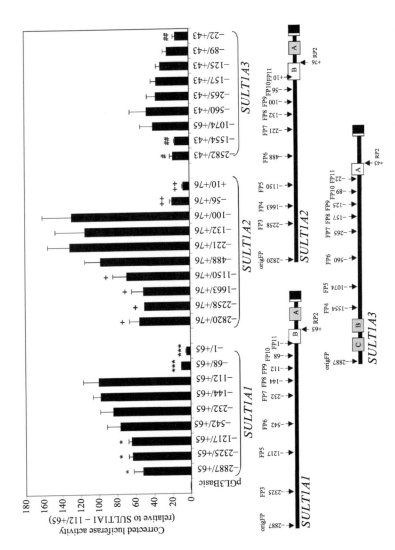

FIG. 5. *SULT1A* promoter deletion construct activities in HepG2 cells. Promoter deletion constructs of the sequences flanking 5′UTR B of *SULT1A1* and *SULT1A2* and 5′UTR A of *SULT1A3* were created by PCR and cloned into the pGL3Basic vector. Positions of the sense primers used to create these are represented schematically by arrows above the gene and are numbered according to their position relative to the

is generally normalized against *Renilla* luciferase activity. Alternatively, in S2 cells, firefly luciferase activity is normalized against the protein concentration of the cleared lysate.

Activity of *SULT1A* Promoters

Using the aforementioned methods, we cloned and assessed the activity of sequences flanking the different 5'UTRs of the *SULT1A* genes in HepG2 cells (Hempel *et al.*, 2004). Sequences flanking the first coding exon of *SULT1A1* and *SULT1A2* increase luciferase expression by 2- and 6-fold, respectively, compared to the empty vector (exon I; Fig. 4A and B). This is not observed for the sequence flanking exon I of *SULT1A3* (Fig. 4C). The *SULT1A1* sequence flanking 5'UTR A, which is a 5'UTR so far only observed in SULT1A1 and SULT1A2 cDNA species, has no statistically significant promoter activity (Fig. 4A). However, the highly homologous *SULT1A2* sequence displays a statistically significant 11-fold higher luciferase activity than the empty pGL3Basic vector (Fig. 4B). Of the regions flanking the different SULT1A 5'UTRs identified in the literature, sequences upstream of *SULT1A1* and *SULT1A2* 5'UTR B have the highest promoter activity (Fig. 4; Hempel *et al.*, 2004). Similarly, the homologous sequence upstream of *SULT1A3* 5'UTR A also efficiently drives luciferase transcription.

Figure 5 shows the activity of promoter deletion constructs of the *SULT1A* promoters flanking 5'UTR B of *SULT1A1* and *SULT1A2* and the homologous *SULT1A3* 5'UTR A. Upon deletion of 44 bp from construct $-112/+65$ and $-100/+76$ of the *SULT1A1* and *SULT1A2* promoters, respectively, a significant decrease in promoter activity is observed, indicating the presence of a crucial regulatory element (Fig. 5; Hempel *et al.*, 2004). Our laboratory has reported on the importance of the Ets and Sp1 transcription factor response elements in this region and their role

---

transcriptional start site ($+1$). The antisense primer is marked below the gene by an arrow. Constructs were transfected into HepG2 cells and lysed cells were assayed for luciferase activity as described. Results represent firefly luciferase activity corrected for *Renilla* luciferase activity of the transfection standard and are expressed relative to the activity of the *SULT1A1* $-112/+65$ construct. Each bar represents the mean result from three transfected wells of cells $\pm$ SD. Asterisks indicate significant differences to the activity of the *SULT1A1* $-112/+65$ construct (*$p<0.05$; ***$p<0.001$, Student t test); plus signs indicate significant differences to the activity of the *SULT1A2* $-100/+76$ construct ($^+p < 0.05$; $^{++}p < 0.01$, Student t test); and hash marks indicate significant differences to the activity of the *SULT1A3* $-125/+43$ construct ($^\#p < 0.05$; $^{\#\#}p < 0.01$, Student t test). (Adapted with permission from Hempel *et al.*, 2004.)

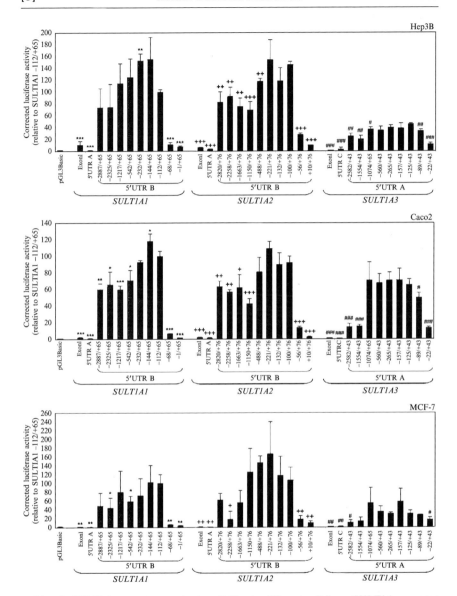

FIG. 6. *SULT1A* promoter constructs activities in different cell lines. *SULT1A* promoter deletion constructs were created by PCR and cloned into the pGL3 Basic vector. Promoter constructs were transfected into Hep3B, Caco2, and MCF-7 cells and lysed cells were assayed for luciferase activity as described. Results represent firefly luciferase activity corrected for *Renilla* luciferase activity of the transfection standard and are expressed relative to the activity of the *SULT1A1* promoter deletion construct −112/+65. Each bar represents the mean result from three transfected wells of cells ± SD. Asterisks indicate significant differences to the

in regulating the *SULT1A1* promoter (Hempel *et al.*, 2004). In HepG2 cells the *SULT1A3* promoter constructs display approximately 70% less activity than the *SULT1A1* and *SULT1A2* promoters. The promoter activities of *SULT1A3* constructs $-2582/+43$ and $-1554/+43$ are approximately half that observed for the $-1074/+43$ construct, which may indicate the presence of an inhibitory transcription factor-binding site. This is particularly apparent when Caco2 cells are used as the experimental model (Fig. 6).

The relative activities of the *SULT1A* promoters in different cellular environments are shown in Fig. 6, where constructs are transfected into another hepatocarcinoma cell line, Hep3B, the colon carcinoma cell line Caco2, and the human breast cancer cell line MCF-7. Although no direct comparison among cell lines can be made due to differential transfection efficiencies, differences in activities can be assessed by comparison with the highly active *SULT1A1* $-112/+65$ promoter construct (Fig. 6). The general pattern of the promoter activities in the different cell lines tested is similar to that observed in HepG2 cells. In all cell lines the sequence flanking *SULT1A1* and *SULT1A2* 5′UTR B is the most highly active promoter (Fig. 6). Similarly, the promoter flanking 5′UTR A of *SULT1A3* has the highest activity of the *SULT1A3* constructs (Fig. 6). As observed for the transfection in HepG2 cells, the deletion of 44 bp from -122 to -68 of the *SULT1A1* promoter and -100 to -56 of the *SULT1A2* promoter results in a greater than 90% drop in activity, suggesting that the regulatory element present in this region controls the activity of these two promoters in all the cells tested. One interesting observation is the activity of the *SULT1A3* promoter in Caco2 cells. Relative to the *SULT1A1* and *SULT1A2* promoters, *SULT1A3* promoter activity in the colon carcinoma cell line is higher than in the other cell lines tested. It cannot be concluded whether this indicates that *SULT1A1* and *SULT1A2* promoters display a decreased activity level in Caco2 cells or whether there is an increase in *SULT1A3* promoter activity in this cell line. However, tissue distribution profiles match this difference in observed promoter activity (Windmill *et al.*, 1998). Similar to its promoter activity, SULT1A3 protein levels are lower in the liver and higher in the gastrointestinal tract (Windmill *et al.*, 1998).

---

activity of the *SULT1A1* $-112/+65$ construct ($*p < 0.05$; $**p < 0.01$; $***p < 0.001$, Student t test); plus signs indicate significant differences to the activity of the *SULT1A2* $-100/+76$ construct ($^+p < 0.05$; $^{++}p < 0.01$; $^{+++}p < 0.001$, Student t test); and hash marks indicate significant differences to the activity of the *SULT1A3* $-125/+43$ construct ($^\#p < 0.05$; $^{\#\#}p < 0.01$; $^{\#\#\#}p < 0.001$, Student t test).

Conclusion

Using the aforementioned methods, we have studied the mechanisms of the *SULT1A* promoters and have identified a crucial regulatory region in the proximal *SULT1A1* promoter containing Ets and Sp1 transcription factor-binding sites (Hempel *et al.*, 2004). Further, we were able to show that a lack of the full Ets-binding site in the *SULT1A3* promoter compromised its activity in hepatocarcinoma cells due to an inability of Ets and Sp1 transcription factors to act in synergy in activating this promoter (Hempel *et al.*, 2004). The described techniques in this chapter will allow further investigations into the promoter properties of the human *SULT1A* genes, including the influence of polymorphisms on the activity of the promoters. Interindividual differences in SULT1A expression levels have been reported in numerous studies (Abenhaim *et al.*, 1981; Iida *et al.*, 2001) and thus far a mechanism controlling this variation has not been fully explained. Single nucleotide polymorphisms (SNPs) have been described for all three *SULT1A* genes. One study of a Japanese population reported 13 and 7 SNPs in the 3-kb region upstream of the ATG start codon of the *SULT1A1* and *SULT1A2* genes, respectively (Iida *et al.*, 2001). Additionally, SNPs on the SULT1A genes can be visualized using the NCBI SNP site (http://www.ncbi.nlm.nih.gov/SNP/). Whether these nucleotide changes influence promoter activities of the genes by altering transcription factor recognition sequences has not been investigated.

References

Abenhaim, L., Romain, Y., and Kuchel, O. (1981). Platelet phenolsulfotransferase and catecholamines: Physiological and pathological variations in humans. *Can. J. Physiol. Pharmacol.* **59**, 300–306.

Aksoy, I. A., and Weinshilboum, R. M. (1995). Human thermolabile phenol sulfotransferase gene (STM) molecular cloning and structural characterization. *Biochem. Biophys. Res. Commun.* **208**, 786–795.

Barnett, A. C., Tsvetanov, S., Gamage, N., Martin, J. L., Duggleby, R. G., and McManus, M. E. (2004). Active site mutations and substrate inhibition in human sulfotransferase 1A1 and 1A3. *J. Biol. Chem.* **279**, 18799–18805.

Bernier, F., Leblanc, G., Labrie, F., and Luu-The, V. (1994a). Structure of human estrogen and aryl sulfotransferase gene: Two mRNA species issued from a single gene. *J. Biol. Chem.* **269**, 28200–28205.

Bernier, F., Soucy, P., and Luu-The, V. (1996). Human phenol sulfotransferase gene contains two alternative promoters: Structure and expression of the gene. *DNA Cell Biol.* **15**, 367–375.

Bidwell, L. M., McManus, M. E., Gaedigk, A., Kakuta, Y., Negishi, M., Pedersen, L., and Martin, J. L. (1999). Structure of catecholamine sulfotransferase (SULT1A3) and comparison with estrogen and heparin sulfotransferases. *Nature Struct. Biol.* **293**, 521–530.

Blanchard, R. L., Freimuth, R. R., Buck, J., Weinshilboum, R. M., and Coughtrie, M. W. H. (2004). A proposed nomenclature system for the cytosolic sulfotransferase (SULT) superfamily. *Pharmacogenetics* **14**, 199–211.

Brix, L. A., Barnett, A., Duggleby, R. G., and McManus, M. E. (1999). Analysis of the catalytic specificity of human sulfotransferases SULT1A1 and SULT1A3: Site-directed mutagenesis and kinetic studies. *Biochemistry* **38,** 10474–10479.

Dajani, R., Cleasby, A., Neu, M., Wonacott, A. J., Jhoti, H., Hood, A. M., Modi, S., Hersey, A., Taskinen, J., Cooke, R. M., Manchee, G. R., and Coughtrie, M. W. H. (1999). X-ray crystal structure of human dopamine sulfotransferase, SULT1A3. *J. Biol. Chem.* **274,** 37862–37868.

Dooley, T. P., Mitchison, H. M., Munroe, P. B., Probst, P., Neal, M., Siciliano, M. J., Deng, Z., Doggett, N. A., Callen, D. F., Gardiner, R. M., and Mole, S. E. (1994). Mapping of two phenol sulphotransferase genes, STP and STM, to 16p: Candidate genes for Batten disease. *Biochem. Biophys. Res. Commun.* **205,** 482–489.

Falany, C. N. (1997). Enzymology of human cytosolic sulfotransferases. *FASEB J.* **11,** 206–216.

Galvagni, F., Capo, S., and Oliviero, S. (2001). Sp1 and Sp3 physically interact and co-operate with GABP for the activation of the utrophin promoter. *J. Mol. Biol.* **306,** 985–996.

Gamage, N. U., Duggleby, R. G., Barnett, A. C., Tresillian, M., Latham, C. F., Liyou, N. E., McManus, M. E., and Martin, J. L. (2003). Structure of a human carcinogen-converting enzyme, SULT1A1: Structural and kinetic implications of substrate inhibition. *J. Biol. Chem.* **278,** 7655–7662.

Hart, R. F., Renskers, K. J., Nelson, E. B., and Roth, J. A. (1979). Localization and characterization of phenol sulfotransferase in human platelets. *Life Sci.* **24,** 125–130.

Hempel, N., Wang, H., Le Cluyse, E. L., McManus, M. E., and Negishi, M. (2004). The human sulfotransferase SULT1A1 gene is regulated in a synergistic manner by Sp1 and GA binding protein. *Mol. Pharmacol.* **66,** 1690–1701.

Her, C., Raftogianis, R., and Weinshilboum, R. (1996). Human phenol sulfotransferase *STP2* gene: Molecular cloning, structural characterization, and chromosomal localization. *Genomics* **33,** 409–420.

Iida, A., Sekine, A., Saito, S., Kitamura, Y., Kitamoto, T., Osawa, S., Mishima, C., and Nakamura, Y. (2001). Catalog of 320 single nucleotide polymorphisms (SNPs) in 20 quinone oxidoreductase and sulfotransferase genes. *J. Hum. Genet.* **46,** 225–240.

Ozawa, S., Nagata, K., Shimada, M., Ueda, M., Tsuzuki, T., Yamazoe, Y., and Kato, R. (1995). Primary structures and properties of two related forms of aryl sulfotransferases in human liver. *Pharmacogenetics* **5,** S135–S140.

Raftogianis, R. B., Her, C., and Weinshilboum, M. (1996). Human phenol sulfotransferase pharmacogenetics: *STP1* * gene cloning and structural characterization. *Pharmacogenetics* **6,** 473–487.

Richard, K., Hume, R., Kaptein, E., Stanley, E. L., Visser, T. J., and Coughtrie, M. W. (2001). Sulfation of thyroid hormone and dopamine during human development: Ontogeny of phenol sulfotransferases and arylsulfatase in liver, lung, and brain. *J. Clin. Endocrinol. Metab.* **86,** 2734–2742.

Veronese, M. E., Burgess, W., Zhu, X., and McManus, M. E. (1994). Functional characterization of two human sulfotransferase cDNAs that encode monoamine- and phenol-sulphating forms of phenol sulphotransferase: Substrate kinetics, thermal-stability and inhibitor-sensitivity studies. *Biochem. J.* **302,** 497–502.

Wilborn, T. W., Comer, K. A., Dooley, T. P., Reardon, I. M., Heinrikson, R. L., and Falany, C. N. (1993). Sequence analysis and expression of the cDNA for the phenol-sulfating form of human liver phenol sulfotransferase. *Mol. Pharmacol.* **43,** 70–77.

Windmill, K. F., Christiansen, A., Teusner, J. T., Bhasker, C. R., Birkett, D. J., Zhu, X., and McManus, M. E. (1998). Localisation of aryl sulfotransferase expression in human tissues using hybridisation histochemistry and immunohistochemistry. *Chem. Biol. Interact.* **109,** 341–346.

Zhu, X., Veronese, M. E., Bernard, C. C., Sansom, L. N., and McManus, M. E. (1993a). Identification of two human brain aryl sulfotransferase cDNAs. *Biochem. Biophys. Res. Commun.* **195**, 120–127.

Zhu, X., Veronese, M. E., Iocco, P., and McManus, M. E. (1996). cDNA cloning and expression of a new form of human aryl sulfotransferase. *Int. J. Biochem. Cell Biol.* **28**, 565–571.

Zhu, X., Veronese, M. E., Sansom, L. N., and McManus, M. E. (1993b). Molecular characterisation of a human aryl sulfotransferase cDNA. *Biochem. Biophys. Res. Commun.* **192**, 671–676.

[10] Vitamin D Receptor Regulation of the Steroid/Bile Acid Sulfotransferase SULT2A1

By Bandana Chatterjee, Ibtissam Echchgadda, and Chung Seog Song

Abstract

SULT2A1 is a sulfo-conjugating phase II enzyme expressed at very high levels in the liver and intestine, the two major first-pass metabolic tissues, and in the steroidogenic adrenal tissue. SULT2A1 acts preferentially on the hydroxysteroids dehydroepiandrosterone, testosterone/dihydrotestosterone, and pregnenolone and on cholesterol-derived amphipathic sterol bile acids. Several therapeutic drugs and other xenobiotics, which include xenoestrogens, are also sulfonated by this cytosolic steroid/bile acid sulfotransferase. Nonsteroid nuclear receptors with key roles in the metabolism and detoxification of endobiotics and xenobiotics, such as bile acid-activated farnesoid X receptor, xenobiotic-activated pregnane X receptor and constitutive androstane receptor, and lipid-activated peroxisome proliferator-activated receptor-α, mediate transcription induction of SULT2A1 in the enterohepatic system. The ligand-activated vitamin D receptor (VDR) is another nuclear receptor that stimulates SULT2A1 transcription, and the regulatory elements in human, mouse, and rat promoters directing this induction have been characterized. Given that bile acid sulfonation is catalyzed exclusively by SULT2A1 and that the 3α-sulfate of the highly toxic lithocholic acid is a major excretory metabolite in humans, we speculate that a role for the VDR pathway in SULT2A1 expression may have emerged to shield first-pass tissues from the cytotoxic effects of a bile acid overload arising from disrupted sterol homeostasis triggered by endogenous and exogenous factors.

METHODS IN ENZYMOLOGY, VOL. 400   0076-6879/05 $35.00
Copyright 2005, Elsevier Inc. All rights reserved.   DOI: 10.1016/S0076-6879(05)00010-8

Introduction

An elaborate network of regulated detoxification systems has evolved in humans and other mammals as a defense against the harmful overload of endobiotics and diverse xenobiotics that arise from food products to pesticides and other pollutants to clinically active drugs. The phase I metabolic products of cytochrome P450 (CYP) mixed function oxidases become more water soluble after polar groups (such as sulfate) are conjugated to the functionalized metabolites by phase II transferases. Common phase II reactions include sulfonation, glucuronidation, glutathionation, amino acylation (glycine, taurine), and alkylation (methylation, N-acetylation). Phase II biotransformation also leads to inactivation of the reactive oxidation products of CYP-catalyzed reactions so that damage to DNA, RNA, proteins, and other cellular entities may be averted. Indeed, induced phase I events and/or impaired phase II activities are associated with heightened risks for diseases, including cancer and neurodegenerative maladies (Bandmann et al.., 1997; Liska, 1998). Cytosolically localized sulfotransferases (SULTs), glutathione S-transferases (GSTs), N-acetyltransferases (NATs), and membrane-bound UDP-glucuronosyltransferases (UGTs) are among the phase II enzymes that are well characterized regarding the genomic organization, genetic polymorphism, structural and functional diversity, and regulation by endocrine and intracrine (metabolic) factors. Phase III transporters mediating the uptake and efflux of ions and metabolites are also integral to the detoxification network; their disrupted/impaired activity is linked to the etiology of specific human diseases (Dietrich et al.., 2003).

The key roles of nuclear receptors (NRs) in the hormone-, metabolite-, and xenobiotic-mediated regulation of phase I/phase II enzymes and phase III transporters have been the subject of extensive investigations in recent years. The NRs that regulate one or more of these enzymes/transporters include the bile acid-activated receptor farnesoid X receptor (FXR), xenobiotic-activated receptors pregnane X receptor (PXR) and constitutive androstane receptor (CAR), lipid-sensing receptor peroxisome proliferator-activated receptor-α (PPAR-α), and the anti-inflammatory stress hormone receptor glucocorticoid receptor (GR). Vitamin D receptor (VDR)-mediated signaling is yet another NR pathway that influences steroid and xenobiotic metabolism by inducing CYP2 and CYP3 enzymes (Adachi et al., 2005; Drocourt et al., 2002; Makishima et al., 2002). The VDR-binding regulatory sequences in promoters of the vitamin D responsive $CYP2$ and $CYP3$ genes have been characterized. The involvement of the VDR pathway extends to phase II activity, as the ligand-activated VDR robustly induces gene transcription for SULT2A1 in liver and intestinal cells (Echchgadda et al., 2004a). SULT2A1 is a hydroxysteroid sulfotransferase with potent sulfonation activity for the steroids dehydro-epiandrosterone (DHEA), testosterone, dihydrotestosterone, and preg-nenolone and for amphipathic sterols bile acids, which are the catabolic end products of cholesterol metabolism.

This chapter provides a brief background on SULT2A1 and other mammalian sulfotransferases and reviews SULT2A1 regulation by several members of the NR family of transcription factors. The VDR-mediated induction of SULT2A1 in mice and in human cell lines is described in detail, and the methods optimized in our laboratory for the investigation of SULT2A1 enzymology and gene induction and for the characterization of VDR-responsive regulatory elements are presented.

Sulfotransferases and the SULT2 Family

SULT2 enzymes catalyze sulfonation of specific endogenous hydroxy-steroids, including the prototypical substrate DHEA. The mammalian cyto-solic sulfotransferases encompassing at least 47 different members are grouped as five distinct families, with further classification as subfamilies, based on primary structure homology and on preferential activity toward specific substrates (Blanchard et al., 2004). SULT enzymes catalyze the sulfate group transfer from the cofactor 3'-phosphoadenosine-5'-phosphosulfate (PAPS) to small molecule endobiotics and xenobiotics at the hydroxyl (OH) or amino ($-NH_2$; NH) functional group, producing sulfated or sulfa-mated derivatives, respectively. The five SULT1 subfamilies (SULT1A→1E) use phenolic OH as the target, while the SULT2 family converts the alcohol OH to a sulfated ester. SULT3 targets heterocyclic amines to produce sulfa-mates. The target substrates for SULT4 and SULT5 are not known; never-theless, the deduced amino acid sequences of the corresponding cDNAs show that SULT4 and SULT5 have significant amino acid homology with other SULTs and contain the conserved PAPS-binding domain.

Members of the SULT2 family are distinguished as SULT2A1 or SULT2B1 based on primary structures, tissue-specific expression, and sub-strate predilection. Two different genes, each containing six exons, encode SULT2A1 and SULT2B1 (Her et al., 1998; Otterness et al., 1995; Shimizu et al., 2003). Human SULT2A1 is found as a single form; rodents express multiple Sult2A1 isoforms (Kong and Fei, 1994; Ogura et al., 1994). SULT2A1 expression is abundant in the first-pass liver and intestinal tissues and in the steroidogenic cortical tissue of the adrenal gland (Chatterjee et al., 1987; Falany, 1997; Saner et al., 2005; Weinshilboum et al., 1997). SULT2A1 preferentially sulfonates DHEA, androgenic steroids and primary and sec-ondary bile acids. SULT2A1 substrates also include certain drugs, including the breast cancer drug hydroxytamoxifen, the anti-inflammatory drug budesonide, and various environmental estrogens (Meloche *et al.*, 2002; Pai *et al.*, 2002; Shibutani *et al.*, 1998). The catalytic site, the substrate-binding site, and the cofactor-binding site of SULT2A1 have been defined by X-ray crystallography (Pedersen *et al.*, 2000; Rehse *et al.*, 2002).

SULT 2B1 exists as two isoforms (SULT2B1a and SULT2B1b), which arise from a single gene (*SULT2B1*) by alternative splicing. These two isoforms exhibit differential substrate preferences and tissue expression (Fuda *et al.*, 2002; Geese and Raftogianis, 2001; He *et al.*, 2004; Shimizu *et al.*, 2003). SULT2B1b is expressed abundantly in the skin, prostate, placenta, lung, and colon; SULT2B1a shows high expression in the brain and spinal cord. Neither isoform is detected to any significant extent in SULT2A1-expressing tissues. SUL2B1b is preferentially a cholesterol sulfotransferase; it also sulfonates DHEA with a relatively high efficiency. SULT2B1a is a pregnenolone sulfotransferase with no appreciable activity toward cholesterol and DHEA. It appears that SULT2A1 is involved primarily in detoxification and in the metabolism of therapeutic drugs and endogenous steroids. The SULT2B1 isoforms appear to have tissue-specific roles, serving as a regulator of neurosteroid biosynthesis (due to pregnenolone sulfonation activity in the brain), a regulator of skin development (due to a role in cholesterol sulfate synthesis in keratinocytes), and a regulator of steroid metabolism (due to the cholesterol/steroid sulfonation activity in the prostate, placenta, and several other tissues).

SULT2A1 Enzymology

The substrate specificity of SULT2A1 was examined in our laboratory using the baculovirus-expressed recombinant enzyme. We have developed a quick and convenient protocol for autoradiographic visualization of sulfonated substrates produced by incubation of the substrate with ^{35}S-labeled PAPS and the recombinant enzyme. The ^{35}S-labeled products are separated from free PAPS by thin-layer chromatography (TLC). An example of the TLC autoradiogram for sulfonated steroids is shown in Fig. 1. In addition to DHEA and pregnenolone, SULT2A1 sulfonates 17β-estradiol and estrone, albeit less efficiently (Fig. 1A). The sulfonation of estriol was minimal. The activity of SULT2A1 to sulfonate estrogens, which usually are the preferred substrates for SULT1E (Blanchard *et al.*, 2004; Demyan *et al.*, 1992), is in agreement with the general consensus that SULTs are broad-specificity enzymes with partially overlapping substrate profiles. Corticosterone was sulfonated by SULT2A1 only weakly (barely above the background level) and progesterone sulfonation was completely absent. Figure 1B is an example of sulfonation of androgens (5α-dihydrotestosterone and the synthetic androgen R1881) and the secondary bile

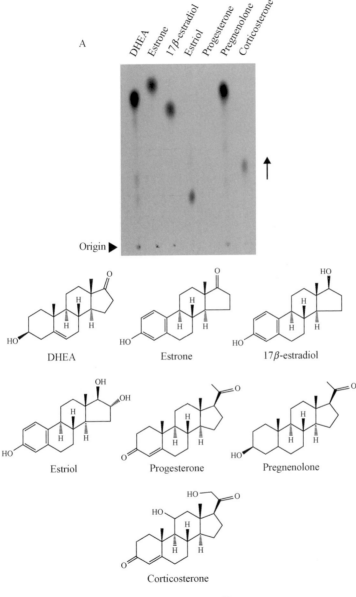

FIG. 1. *(continued)*

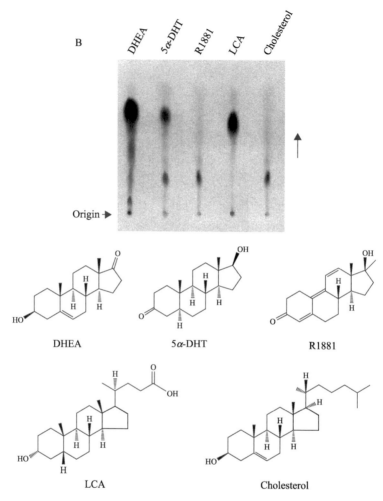

FIG. 1. Sulfonation of various steroids and secondary bile acid lithocholic acid by recombinant SULT2A1. The structure for each substrate is shown. The upward arrow indicates the direction of solvent migration. The TLC plates were exposed to X-ray films overnight. (A) Sulfonation of DHEA and non-androgenic steroids, namely estrone, 17β-estradiol and estriol, which are the estrogenic steroids; progesterone; pregnenolone and corticosterone. (B) Sulfonation of DHEA, androgens (5α-DHT and R1881); the secondary bile acid lithocholic acid (LCA) and cholesterol.

acid lithocholic acid (LCA) by SULT2A1. The minor spot closer to the point of origin in each lane appears to be due to a contaminant. Primary and secondary bile acids are sulfonated exclusively by SULT2A1 (Radominska *et al.*, 1990), and SULT2A1 can play a protective role against bile acid toxicity (Kitada *et al.*, 2003). Importantly, unlike SULT2B1b (a preferentially cholesterol sulfotransferase), SULT2A1 has no activity toward cholesterol (Chatterjee *et al.*, 1994; Fig. 1B).

Production of Recombinant SULT2A1 and Assay for Its Activity

Expression of Recombinant Enzyme. Active SULT2A1 has been produced in stably transfected mammalian cells, in eukaryotic Sf9 insect cells through the baculovirus vector, and in *Escherichia coli* through bacterial expression vectors (Czich *et al.*, 1994; Rehse *et al.*, 2002; Song *et al.*, 2001). The insect and the bacterial systems are suitable for the large-scale production of recombinant SULT2A1; the mammalian expression in stable cells generates a limited amount of the enzyme and thus is not practical for many experiments, including detailed studies on SULT2A1 enzymology. The bacterial expression is quick and less tedious than the baculovirus expression. This ease of procedure and the new generation of bacterial expression vectors containing convenient cloning sites make the bacterial system the method of choice for producing SULT2A1 and other sulfotransferases.

In earlier studies we produced active recombinant Sult2A1 in Sf9 cells from the AcNPV (*Autographa californica* nuclear polyhedrosis virus) vector in which the viral polyhedrin gene was replaced with the full-length rat Sult2A1 cDNA by homologous recombination *in vivo* in Sf9 cells cotransfected with the wild-type (AcNPV) vector DNA and the recombinant transfer vector pVL1393 containing the Sult2A1 cDNA (Chatterjee *et al.*, 1994; Song *et al.*, 2001). Infected Sf9 cells were grown at 27° for 4 to 5 days and progeny virus particles released into the culture medium were used to infect new Sf9 cells. The monolayer culture of infected Sf9 cells was overlaid with agarose and plaques were grown in a humid chamber by incubating the plates at 27° for ~4 days. Recombinant plaques were identified from the absence of the occlusion body, as polyhedrin crystals are expressed only from the wild-type baculovirus. The occlusion body-negative recombinant plaques were purified through two rounds of screening and the high-titer recombinant viral stock was used to infect Sf9 cells. After 3 days of culture the cytosol from the harvested Sf9 cells was used as the source of recombinant SULT2A1.

A vastly improved and more facile version of the baculovirus system containing new generations of the viral vector (BaculoGold) and transfer vectors is now available (BD Biosciences-Pharmingen). Baculovirus-produced human SULT2A1 is available commercially (Invitrogen Corp.). Despite being more labor intensive, for certain studies, SULT2A1 expression in insect cells may be preferable to the bacterial expression, as the Sf9-expressed SULT2A1 is potentially similar to the mammalian cell-expressed native enzyme with regard to posttranslational modification (correct folding, disulfide bond formation, oligomerization, glycosylation, etc.) and thus to biological function.

Bacteria-expressed SULT2B1 isoforms have been used to compare the kinetic parameters and substrate specificity of the two enzymes (Pai *et al.*, 2002; Shimuzi *et al.*, 2001). Using the pGEX-4T1 vector (Pharmacia), we have produced SULT2B1b at high yield in *E. coli* as a GST-fusion product showing robust activity.

Assay for SULT2A1. A quick and quantitative assay for SULT2A1 activity, as developed by us, involves incubation of the test substrate with the radioactive cofactor ^{35}S-labeled PAPS in the presence of the recombinant enzyme, separation of the ^{35}S-labeled sulfated steroids/sterols from free PAPS by ethyl acetate extraction, and parallel analysis of the recovered radiolabeled steroid sulfate by direct scintillation counting and by performing TLC on a glass fiber plate impregnated with silica gel (Demyan *et al.*, 1992; Song *et al.*, 2001).

Procedure

1. A 50-μl reaction mixture contains 2–4 μg recombinant SULT2A1, 0.2 μM DHEA (or another test substrate), 0.1 μl (45 pmol) ^{35}S-labeled PAPS (0.865 mCi/ml, NEN), and an optimized buffer (30 mM Tris-HCl, pH 7.6; 3 mM MgCl$_2$). Incubation is at 37° for 1 h (or 10 min for studies on the kinetic parameters of the enzyme activity). The reaction is stopped with 100 μl 4 M ammonium sulfate.
2. Ethyl acetate (300 μl) is added to the aforementioned reaction and mixed by vortexing vigorously for 1 min. After a spin down in a microfuge (full speed, 5 min, room temp), the upper layer is transferred to a fresh 500-μl microfuge tube, without touching the lower aqueous phase.
3. Approximately 10 μl is used for liquid scinitillation counting.
4. The rest of the sample is air dried (30 min inside a hood to dry).
5. The dried sample is dissolved in 20 μl ethyl acetate (vortex, spin down), and the sample is applied to the TLC plate (Gelman Sciences) at the point of origin.
6. The TLC is run with a developing solvent (chloroform, 200 ml; acetone, 70 ml; acetic acid, 12 ml).
7. The run is terminated when the solvent front reaches the top of the plate.
8. The air-dried plate is exposed to an X-ray film for autoradiography.

Critical Points

The TLC chamber should be kept saturated with the solvent vapor by soaking a filter paper (Whatman 3MM) in the TLC solvent and placing it upright inside the chamber to ensure that the inside space of the chamber remains filled with the solvent vapor at all times during the TLC run. The chamber is filled with the developing solvent to ~1 in. from the bottom and is kept closed with a lid (at least 2 h) before TLC is initiated. The slots holding the plate should be clean to ensure uniform migration of the solvent front.

Nuclear Receptors as Regulators of SULT2A1

Nuclear receptors are ligand-inducible transcription factors with a high degree of structural and functional conservation preserved through evolution (Escriva *et al.*, 2004). Human NRs, of which there are at least 48 members, are activated by hormones (steroids, vitamin D, retinoids, thyroid hormone) and by metabolites (cholesterol derivatives, bile acids, lipids). In addition, diverse xenobiotic molecules from pollutants to herbal products to pharmaceuticals can activate the xenobiotic-sensing NRs PXR and CAR (Handschin *et al.*, 2004). Ligand-independent, constitutive activation for CAR, and several other orphan nuclear receptors are also known. The majority of nonsteroid NRs, including bile acid-activated FXR, xenobiotic-activated PXR, CAR, and vitamin D activated-VDR, bind to the hormone response elements of target genes as a heterodimer with obligate partner RXR-α, the receptor for 9-*cis*-retinoic acid (Mangelsdorf and Evans, 1985). Several NRs regulate SULT2A1 expression by activating human, mouse, and rat SULT2A1 promoters.

The androgen receptor (AR) was the first NR shown to be a regulator of SULT2A1 in the authors' laboratory (Song *et al.*, 1998). A negative androgen response region residing between -235 and -310 positions of the promoter imparted inhibition of the rat Sult2A1 promoter in transfected hepatoma cells by the ligand-activated AR. This negative regulatory region lacks a direct AR-binding site; however, its deletion abolished androgen-dependent repression of the transfected Sult2A1 promoter. Androgenic repression of the *SULT2A1* gene accounts for its high expression in the androgen-insensitive rodent liver when the AR expression is very low or completely lost, as in females, in prepubertal males, in senescent males, and in testicular feminized (Tfm) males (Chatterjee *et al.*, 1987, 1996; Demyan *et al.*, 1992; Echchgadda *et al.*, 2004b).

Unlike the negative regulation by AR, the *Sult2A1* gene is induced in the liver, in the intestine, and in cells derived from these tissues by bile acid-activated FXR, xenobiotic-activated PXR and CAR, and lipid-activated PPAR-α (Assem *et al.*, 2004; Echchgadda *et al.*, 2004b; Fang *et al.*, 2005; Runge-Morris *et al.*, 1999; Song *et al.*, 2001; Sonoda *et al.*, 2002). Induction of SULT2A1 expression by vitamin D/VDR signaling has been demonstrated (Echchgadda *et al.*, 2004a). The role of VDR as a regulator of calcium and phosphate homeostasis and as a regulator of cellular differentiation and proliferation is well recognized (Lin and White, 2004). An additional role of VDR in steroid and xenobiotic metabolism has become evident with the finding that gene transcription of phase I CYP enzymes is induced by the activated VDR (Adachi *et al.*, 2005; Drocourt *et al.*, 2002; Makishima *et al.*, 2002; Thummel *et al.*, 2001). Our finding that a phase II enzyme such as SULT2A1 is also induced by vitamin D implies that VDR has a broad role in the detoxification network. Phase III transporters are likely to be regulated similarly, although no published report to this end has yet come out. The following section describes the induction of SULT2A1 by VDR and the regulatory elements directing this induction and, in addition, considers the implication of this regulation in the protection of the liver and gastrointestinal system against bile acid-induced cytotoxicity. We also briefly discuss our finding that VDR mediates SULT2B1b induction in the prostate.

VDR Regulation of SULT2A1

Endogenous SULT2A1 mRNA and protein expression is induced by vitamin D (Echchgadda *et al.*, 2004a). Figure 2 shows an example of this induction in liver cells and in the mouse liver. Treatment of HepG2 hepatoma cells with vitamin D (1α,25-dihydroxy vitamin D_3) for 24 h induced SULT2A1 mRNAs $\sim$6-fold in a RT-PCR assay (Fig. 2A); in mice injected with vitamin D, Sult2A1 mRNAs in the liver increased $\sim$3-fold in a Northern blot assay (Fig. 2B). This induction reflected activation of the corresponding promoter (Fig. 3). In Caco-2 intestinal cells, vitamin D induced the human promoter (from -1102 to 10) $\sim$6-fold and the mouse promoter (from -292 to 42) $\sim$4.5-fold; similar induction of the promoter

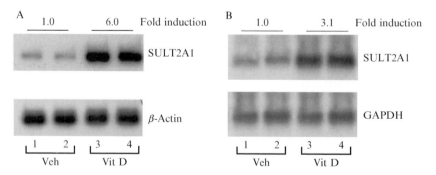

FIG. 2. Vitamin D-mediated induction of endogenous SULT2A1 mRNAs. (A) Induction in HepG2 hepatoma cells assayed by RT-PCR. (B) Induction in the mouse liver assayed by Northern blot. β-Actin and GAPDH mRNA expression is used as the invariant control.

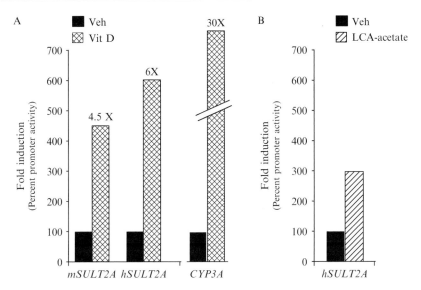

FIG. 3. Induction of the SULT2A1 promoter by ligand-activated VDR in transfected Caco-2 intestinal cells. (A) Induction by vitamin D-mediated VDR and (B) induction by LCA-acetate-activated VDR. VDR activation of the human CYP3A4 promoter is also shown (A).

has been observed in HepG2 hepatoma cells. Vitamin D induction of the human *CYP3A4* promoter (from the *CYP3A*-Luc construct) is also shown (Fig. 3A). This construct contains a proximal (at −173) and a distal (at −7715) VDR-binding element (Drocourt *et al.*, 2002; Makishima *et al.*, 2002) in the *CYP3A4* promoter.

Secondary bile acid lithocholic acid (LCA) and its derivatives (LCA-acetate, LCA-formate, 3-keto LCA) are known to serve as sensitive VDR agonists and to induce VDR target genes (Adachi *et al.*, 2005; Makishima *et al.*, 2002). Figure 3B shows induction of the *SULT2A1* promoter by LCA-acetate. The LCA-mediated induction of SULT2A1 may be a key defense response of the intestinal and hepatobiliary tissues against the toxic/carcinogenic effects of bile acids.

The cholesterol/steroid sulfotransferase SULT2B1b in prostate cells is also induced by VDR (Seo *et al.*, 2005). An example of this induction (Fig. 4) shows that within 6 h of vitamin D treatment, SULT2B1b mRNAs were induced in CWR22R human prostate cancer cells (RT-PCR assay). Sequencing of the RT-PCR product verified the identity of the induced mRNA. Given that SULT2B1b is expressed in normal prostate, BPH prostate, human prostate tumors, and human prostate cancer cell lines

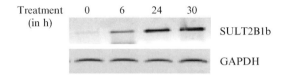

FIG. 4. Induction of SULT2B1b mRNAs in vitamin D-treated CWR22R human prostate cancer cells. The induced mRNAs at different time points after vitamin D treatment are shown in the RT-PCR assay. The induced band was confirmed by sequencing.

(He *et al.*, 2004) and that human prostate cancer cell growth is inhibited by vitamin D (Krishnan *et al.*, 2003), the induction of SULT2B1b in vitamin D-treated cells may signify a functional link of cholesterol/androgen metabolism to cell proliferation in the prostate through SULT2B1b activity.

Methods Used to Study SULT Regulation in Cell Lines and in Mice

Hormone Treatment and Endogenous SULT2A1 Expression. Cell lines from enterohepatic tissues should be used to study SULT2A1 induction. Not all cell lines express the regulatory factors necessary to support a robust induction of the gene. Thus, the optimal cell line should be assessed first. Although primary hepatocytes are ideal for regulatory studies, isolation of hepatocytes is time-consuming and transfection in primary hepatocytes is technically challenging (although not insurmountable). HepG2 hepatoma and Caco-2 intestinal adenocarcinoma lines (available from ATCC) are suitable for investigating the transcription regulation of SULT2A1 by VDR and other NRs. The cells are cultured for 3 days in a medium containing 5% charcoal-stripped serum before hormone treatment. The hormonally active vitamin D metabolite (1α,25-dihydroxy vitamin D_3) is dissolved in 50% ethanol and added (at 10 nM) to the medium. Hormone and control (vehicle) treatments are for 24 h. To study induction in mice, 1.5 μg of 1α,25-dihydroxy vitamin D_3 or the vitamin D agonist EB1089 [dissolved in an 80:20 (v/v) mixture of propylene glycol and phosphate-buffered saline, pH 7.5] is injected intraperitoneally into C57BL/6J mice daily for 3 days.

The cells and tissues are processed for total RNAs using Trizol (Invitrogen) and for total lysates by homogenizing cells/tissue in lysis buffer (2% SDS, 5% β-mercaptoethanol, 125 mM Tris-HCl, pH 6.8, 20% glycerol). SULT2A1 mRNAs are analyzed by RT-PCR or Northern blot using standard protocols. An optimal RT-PCR primer set for the human SULT2A1 mRNA is 5'-GTATACAGCACTCAGTGA (forward primer) and 5'-CCCAGGAATTGACAGATC (reverse primer). This primer set does not amplify mRNAs of other SULTs (not even the closely related SULT2B1 forms). The RT-PCR product should always be sequence confirmed.

For Northern analysis, the membrane-bound RNAs are hybridized at 52° with the ^{32}P-labeled cDNA probe, specific to an N-terminal segment of SULT2A1, in a formamide-containing hybridization buffer. The probe spans the SULT2A1 mRNA from 52 to 420 and is generated by PCR. The probe is prepared to high specific activity ($>10^9$ cpm/μg) by random primer labeling (RediprimeII kit, Amersham-Bioscience) using [α-^{32}P] dCTP (3000 Ci/mmol). The membrane is washed twice (20 min each) with 2× SSC, 0.1% SDS at room temperature and twice (30 min each) with 0.1× SSC, 0.1% SDS at 55° to clear out the background.

For Western analysis, proteins are transferred onto a PVDF membrane (Amersham-Biosciences), prehybridized in a blocking buffer (same as the hybridization buffer), and hybridized with the primary antibody in TBS buffer (50 mM Tris HCl, pH 7.4; 200 mM NaCl) containing 3% nonfat dry milk (Bio-Rad) and 0.3% bovine serum albumin (BSA). The membrane is washed in the TBS buffer containing 0.1% Tween–20. Washing of the membrane, hybridization with the secondary antibody solution followed by a second step of washing, and development of the Western blot signal with enhanced chemiluminescence reagents (Perkin-Elmer-NEN) are performed using standard conditions. The antibody to human SULT2A1 is available commercially (Oxford Biomedical Research). All standard conditions are from a Web site for molecular biology-related protocols (www.protocol-online.org/prot/molecular_biology/).

Critical Points

1. *Northern blot.* In order to get a high signal-to-noise ratio, the probe must have high specific activity ($>10^9$ cpm/μg). To generate a high-specific activity probe, the DNA fragment should be completely free of ethidium bromide to avoid inhibition of the labeling reaction and the final probe preparation should be cleaned by repeated ethanol precipitations. Approximately 30 μg of total RNAs and 2 to 5 μg poly A-enriched RNAs in each lane are generally sufficient for strong Northern signals.

2. *Western blot.* Optimal dilutions for the primary and secondary antibodies need to be tested out. Approximately 30 μg total protein in each lane gives good signals for proteins expressed at low to moderately low abundance. A back-and-forth motion (as opposed to a circulatory motion) of the membrane during blocking, hybridizing, and washing ensures a clear background.

Promoter Isolation, Promoter–Reporter Constructs, and Mutagenesis. Promoters from the rodent or human genomic DNA are isolated by PCR. Multiple primers selected on the basis of the sequence information in the database may be tested to find the optimum primer set. Long promoters (up to ~20 to 30 kbp) can be isolated using a high-fidelity TaKaRa LA *Taq*-thermostable polymerase (TAKARA Bio Inc.). The PCR product is gel purified and cloned upstream of a promoter-less reporter vector such as pGL3b (Promega).

Genomic fragments cloned into the bacterial artificial chromosome (BAC) vector are useful for isolating very long promoters (>50 kbp) if the goal is to analyze regulatory regions located very far away from the proximal promoter. BAC clones are also used to isolate an entire gene, including all exons and introns. BAC libraries for human and mouse genomic DNAs (average insert 100 to 300 kpb) are available commercially (Invitrogen). The BAC clone containing a region of interest is used to isolate the target sequence by restriction enzyme cleavage or by long-range PCR (using reagents from TAKARA Bio Inc.). Because human SULT2A1 is localized in chromosome 19, a BAC clone of chromosome 19 that includes the SULT2A1 region is used to isolate a very long SULT2A1-specific gene/promoter sequence.

Commercial mutagenesis kits (e.g., QuikChange XL site-directed muta-genesis kit from Stratagene) work well, but are expensive. A cost-saving method that we use routinely involves PCR-mediated splicing after overlap extension, developed by Ho *et al.* (1989) and presented schematically in Fig. 5. Initially, separate PCR amplifications are performed to produce two DNA products, both spanning the site selected for mutagenesis. One amplified DNA extends to the 5' end from the target site (5' product) and the other extends to the 3' end (3' product). DNAs are generated from the original template using a mutated reverse oligonucleotide primer (for the 5' DNA) and a mutated forward primer (for the 3'DNA). The gel-purified DNAs are subjected to PCR in a single reaction using the forward primer (primer #1) of the 5' DNA and reverse primer (primer #4) of the 3' DNA. Annealing of the overlapping complementary strands and subsequent PCR result in the spliced product containing the desired mutation. Wild-type DNA is replaced with the mutated DNA in the promoter–reporter construct via suitable cloning sites.

Assay for Promoter Activity in Transfected Cells. Cells, maintained in a medium containing the charcoal-stripped serum, are seeded in 24-well flasks at ~100,000 cells/well overnight before transfection. The reporter construct (300 ng) and the VDR-expressing plasmid (50 ng) are transferred to cells using Fugene (Roche-BMB) or lipofectamine (Invitrogen) reagents. Approximately 12 h after DNA transfer, hormonally active vitamin D (at 20–50 n*M*) or ethanol (<0.005%) is added to the medium and cells are harvested at ~36 h after transfection. Transfection results are normalized either to constant expression of the cotransfected (2 ng) tk-*Renilla* luciferase (Promega) or to constant protein amounts (see comments under critical points). Total DNAs are kept constant using an empty vector. Firefly and *Renilla* luciferase activities are assayed using the dual luciferase kit (Promega).

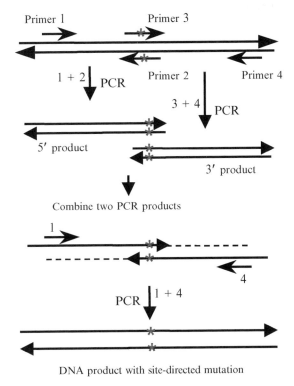

Combine two PCR products

PCR ↓ 1 + 4

DNA product with site-directed mutation

Fig. 5. Schema for site-directed mutagenesis using PCR. The 5′ PCR product is amplified by primers 1 and 2, and the 3′ PCR product results from primers 3 and 4. Primers 2 (reverse) and 3 (forward) carry the target mutations. Primers 1 and 4 are used to generate the spliced-mutated DNA. Modified from Ho *et al.* (1989), with permission.

Critical Points

1. Depending on experimental conditions (cell passage number, the type of cell line, the type of ligand, etc.) cotransfection of cells with an RXR-α plasmid (10 ng) may result in improved fold induction.

2. Normalization against *Renilla* luciferase expressed from a constitutive promoter can be problematic depending on the cell type and test promoter, as the constitutive *Renilla* luciferase expression is reported to be promoter and cell type dependent (Mulholland *et al.*, 2004). We have confirmed this limitation and found that depending on the cell passage number, the activity of *Renilla* luciferase expressed from constitutive promoters in Caco-2 and HepG2 cells can differ substantially for two test conditions (control vs experimental) in the context of the transfected human SULT2A1 promoter. In this case, normalization is done against constant protein amounts.

Regulatory Elements Directing VDR Regulation

DNase I footprinting is an informative approach for the initial mapping of the vitamin D responsive region within a long promoter fragment. Figure 6 shows the DNase I-footprinted region from -170 to -190 positions in the mouse *Sult2A1* promoter produced by recombinant VDR and RXR-α (A: lanes 3 and 4). Incubation with BSA alone or with each receptor alone rendered no protection (A, lanes 1, 2, and 5). This region was also protected by mouse liver nuclear extracts (B, lane 2 versus lanes 1 and 3). The -170 to

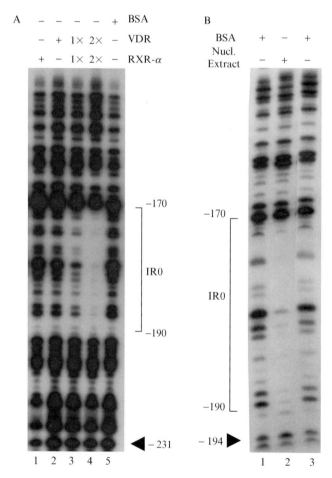

Fig. 6. DNase I footprinting of the mouse *Sult2A1* promoter. Footprinting was conducted with recombinant VDR and/or RXR-α (A) and with mouse liver nuclear extract (B). VDR and RXR-α amounts in lane 4 were twice as much as those in lane 3.

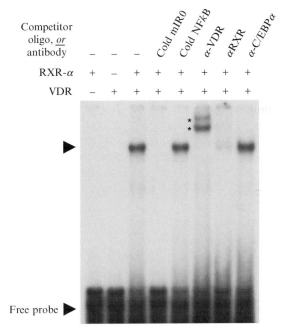

FIG. 7. EMSA for VDR/RXR-α binding to the −170 to −190 footprint. Competition with homologous and heterologous double-stranded oligonucleotides and immunoreactivity of the EMSA complex with antibodies to VDR and RXR-α are shown. The double-stranded DNA spanning −170 to −190 positions was used as the ^{32}P-labeled probe.

−190 footprint contains an IR0 sequence (GGGTCATGAACT) that conforms to an imperfect inverted repeat (without any spacer nucleotide) of the (A/G)G(G/T)TCA sequence, which is a consensus half-site for nonsteroid NRs (Mangelsdorf and Evans, 1985). In the electrophoretic gel mobility shift assay (EMSA), VDR/RXR-α specifically bound to IR0 to produce a gel-shifted complex (marked as an arrowhead; Fig. 7). The complex was competed out by the cold homologous competitor oligonucleotide, not by the heterologous NF-κB oligonucleotide. The anti-VDR antibody supershifted the complex (the asterisk positions) and the anti-RXR-α antibody inhibited the complex. The unrelated C/EBP-α antibody had no immunoreactivity to the complex. A detailed characterization showed that point mutations within the IR0 motif prevented formation of the EMSA complex; mutations outside IR0 had no effect. Furthermore, IR0 can direct VDR-mediated induction of the thymidine kinase (tk) promoter, and inactivating mutations within IR0 in the natural promoter (−292mSult2A1mt-Luc) or in the IR0 linked to the tk promoter [(IR0mt)$_3$-tk–Luc construct]

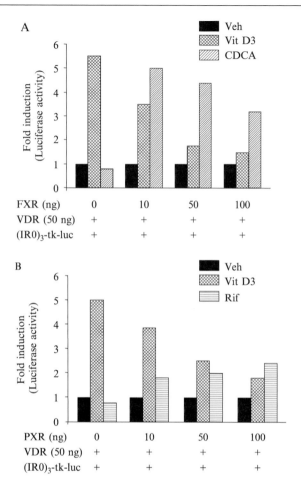

FIG. 8. Competition of FXR and PXR with VDR for the common IR0 element in the mouse Sult2A1 promoter. HepG2 cells (seeded in a 24-well flask) were cotransfected with VDR (50 ng) and increasing amounts of rat FXR (A) or human PXR (B). The IR0 containing heterologous reporter construct, $(IR0)_3$-tk-Luc, was transfected at 300 ng. Cells were treated with vitamin D_3 (50 nM), primary bile acid CDCA (25 μM) as the ligand for FXR, or rifampicin (10 μM) as the ligand for PXR. Redrawn from Echchgadda et al. (2004a).

abrogated VDR induction (Echchgadda et al., 2004a). The IR0 in the rat Sult2A1 promoter responded similarly.

This IR0 also binds to FXR/RXR-α and PXR/RXR-α heterodimers and regulates FXR- and PXR-directed induction of the mouse and rat Sult2A1 promoters (Echchgadda et al., 2004a,b; Song et al., 2001; Sonoda *et al.*, 2002). Competitive functional interactions among VDR, FXR, and PXR at the common *cis*-acting binding sequence are shown in Fig. 8. The induction of IR0 by vitamin D decreased progressively with increasingly higher amounts of the cotransfected FXR or PXR plasmid. Competition between VDR and CAR for the induction of Sult2A1 can also be expected because IR0 in the rodent Sult2A1 promoter acts as a CAR responsive element (Sonoda *et al.*, 2002). Competitive interaction of VDR with PXR and CAR at common response elements in the *CYP3A4, CYP2B6,* and *CYP2C9* genes has been observed (Drocourt *et al.*, 2002). Thus it is likely that the specific hormonal and metabolic milieu of cells would determine which of the competing NR pathways would play a dominant role in the induction of SULT2A1 and CYP genes at the converging regulatory site.

The VDR-binding element (VDRE) in the human SULT2A1 promoter is an IR2 arrangement of the (A/G)G(G/T)TCA half-site. VDR responsiveness of the human promoter requires functional interaction of the IR2 with a second element that binds to the CAAT/enhancer binding protein (C/EBP, a liver-enriched transcription factor) (Song *et al.*, 2004). The essential role of C/EBP in this case is proven by functional analysis of the promoters carrying inactivating point mutations at the VDRE or C/EBP site alone (single mutant) or concurrently (double mutant).

Several repeat motifs (DR3, DR4, DR6, IR9, ER6) of the consensus half-site are known to direct vitamin D-mediated regulation of target gene transcription (Toell *et al.*, 2000). The SULT2A1 system further expands the choices of the VDRE motif to IR0 (in mouse and rat Sult2A1 promoters) and to IR2 (human SULT2A1 promoter). Additionally, a DR7 motif serves as the VDR element directing vitamin D responsiveness of the SULT2B1b promoter in prostate cells (Seo *et al.*, 2005). A homology scan using the NUBIScan program (Podvinec *et al.*, 2002) has been useful in searching for potential NR-binding sites. The program is created as an algorithm for preferred nucleotide distribution matrices and uses combined analysis of both half-sites involved in the binding of NR dimers. Diversity in the sequence organization of a response element in the promoter for any given ligand is thought to arise from the conformational flexibility of NR-bound DNAs and from the likelihood that DNA-bound NRs may assume multiple conformations (Droucourt *et al.*, 2002).

Methods Used to Identify and Characterize Regulatory Elements

DNase I Footprinting. The following protocol uses optimized conditions to prepare an unnicked DNA probe and to ensure specific binding of the probe to proteins. The probe is generated by PCR amplification of the promoter with a ^{32}P-end-labeled sense primer and an unlabeled antisense primer.

Procedure

1. *End labeling of the primer and probe preparation.*The primer (~23 base primer; 1 pmol/μl) is end labeled by incubating with [γ-^{32}P]ATP (3000 Ci/mmol, NEN) and T4 polynucleotide kinase in a kinase buffer [50 mM Tris-HCl, pH 7.6, 10 mM MgCl$_2$, 5 mM dithiothreitol (DTT)]. After incubation (30 min at 37°), 5 μg yeast tRNA, 100 μl water, 20 μl 3 M sodium acetate, and 500 μl 100% ethanol are added. The alcohol-precipitated DNA is kept at −20° (at least 10 min), spun down in an Eppendorf centrifuge at 12,000g (5 min; room temperature), rinsed twice with 100% ethanol (200 μl each time; invert and then spin down), and air dried. The pellet is dissolved in sterile water (mixing by gentle tapping and a quick spin down). The labeled primer is ready for use in the PCR step.

2. *PCR amplification of the promoter.*

    1 μl 50 ng/μl plasmid template

    2 μl unlabeled primer (1 pmol/μl)

    3 μl 10× buffer (100 mM Tris-HCl, pH 8.3, 500 mM KCl, 15 mM MgCl$_2$, 1% Triton X-100)

    4 μl 250 μM dNTP mixture

    1 μl *Taq* polymerase (any commercial source)

    5 μl radiolabeled primer

    H$_2$O to a total volume of 30 μl

    PCR is performed for 30 cycles at 94° for 30 s; 54° for 2 min; and 72° for 1 min.

3. *Probe recovery and probe purification.*

    a. Run the PCR product on a 1.2 to 1.5% agarose gel containing ethidium bromide (1 μg in 100 ml agarose solution). Visualize the gel by a quick exposure to UV light. A single sharp band should be visible. *Note*: A long UV exposure to visualize the radiolabeled probe should be avoided to prevent nicking.

    b. Excise the ethidium bromide-stained DNA band from the gel and gel extract the DNA using the Zymoclean Gel DNA recovery kit (Zymo Research).

    c. Elute the probe with ~50 μl autoclaved water from the Zymoclean column.

    d. Count the probe for radioactivity. To get a strong signal in the footprint pattern, the probe should be at least 50,000 cpm/μl.

4. *DNA binding and DNase I-mediated cleavage.*

    a. Binding reaction: 10 μl 5× binding buffer [50 mM Tris-HCl, pH 7.5; 0.25 M NaCl; 5 mM EDTA; 25 % glycerol (v/v); 5 mM DTT], 2 μg poly(dI-dC) (1 μg/μl), 50–100 μg nuclear extract (or 50–100 ng recombinant receptor), and H_2O to 50 μl. Incubate at room temperature for 5 min.

b. Add the DNA probe at ~50,000–100,000 cpm to each reaction tube. Continue incubation in ice for 30 min.

c. DNase I cleavage: 50 μl binding reaction from step a and 50 μl Ca^{2+}/Mg^{2+} solution (2 mM $CaCl_2$ 10 mM $MgCl_2$). Vortex and spin down (incubation at room temperature for 1 min).

d. Add 1 to 5 μl DNase I at 20 ng/μl to 1× binding buffer. Continue incubation for 30 s to 2 min. Time of incubation should be determined empirically on a case-by-case basis.

e. Add 100 μl stop solution (50 mM Tris, pH 8.0, 20 mM EDTA, 2% SDS, 0.1 mg/ml tRNA).

f. Add 100 μl phenol–chloroform mixture (at 1:1, v/v).

g. Mix by vortexing (1 min, high speed) and centrifuge (12,000g, 5 min, room temperature).

h. Transfer the aqueous phase to a new tube (without touching the interphase).

i. Add 20 μl 3 M sodium acetate and 400 μl ethanol, mixing; keep at −20° for 10 min.

j. Spin down cleaved DNAs, rinsing with 100 % ethanol (twice) and air drying DNAs.

k. Add 5 μl sequencing dye, vortex, collect sample by spin down,and heat sample at 95° for 1 min; chill in ice for 5 min.

l. Electrophorese on a 6 to 8% sequencing gel (polyacrylamide–urea) at 1800 V using the sequencing gel running buffer (0.5× TBE).

m. Dry the gel; expose to X-ray film for autoradiography or to a phosphorimager cassette for image analysis. *Note*: Solutions used to prepare the gels and the running buffer are as recommended in the following Web site: www.protocol-online.org/protein/molecular_biology.

Critical Points

1. An unnicked, high-specific activity probe is critical in producing a clearly defined footprint pattern. The probe should not be used if it produces a smear in the gel.

2. With nuclear extract in the binding reaction, a longer incubation with higher amounts of DNase I is required than that with recombinant proteins or BSA.

3. The optimal gel concentration for running a footprinting gel depends on the location of the footprint with respect to the

end-labeled sequence. The closer the footprint is to the probe, the higher the gel percentage.

4. Clean glass plates with Gel-Plate Clean Solutions (Gene Technology, Inc.). The Gel Slick Solution (Cambrex Biosciences) is applied to one plate only.

EMSA and Antibody Supershift. EMSA is performed with the recombinant NRs (VDR and RXR-α) or liver nuclear extracts isolated from mice. Recombinant VDR and RXR-α are produced as GST-fusion proteins in *E. coli*. Alternately, the recombinant NR can be generated by *in vitro* transcription and translation of the receptor-encoding cDNA using the rabbit reticulocyte TNTsystem (Promega).

Procedure

1. *Probe preparation.* To prevent formation of a single-stranded labeled oligonucleotide during electrophoresis it is preferable to label one oligonucleotide strand and then anneal the complementary unlabeled strand at fivefold molar excess. This ensures that the labeled probe remains in a double-stranded state. Label the oligonucleotide with $[\gamma\text{-}^{32}P]ATP$ and T4 polynucleotide kinase using standard reaction conditions.

2. Binding reaction under probe excess: 1 μl poly(dI-dC)(1 μg/μl in TE), 4 μl 5× binding buffer [50 mM Tris-HCl, pH 7.5; 250 mM NaCl; 5 mM EDTA; 25% glycerol (v/v); 5 mM DTT], 1 μl cold double-stranded DNA competitor (if needed), and 1 μl nuclear extract (~5 μg) or recombinant proteins (20–50 ng). Add H_2O to 20 μl. Incubate for 5 min at room temperature. Add the antibody at this time in the supershift assay.

3. Add 1 μl (~20,000 cpm) radiolabeled probe. Incubate for an additional 30 min at room temperature.

4. Perform electrophoresis on 5% nondenaturing polyacrylamide gel. Apply 5 μl of the sequencing dye (without any sample) to a corner slot. Prerun the gel at 150 V for 30 min (0.5× TBE as running buffer) and run at 180 V. Continue the gel run until the dye front for bromphenol blue migrates approximately four-fifths of the gel length.

5. Autoradiograph the dried gel.

Critical Points

1. The concentration for poly(dI-dC) should be adjusted for an optimal signal-to-noise ratio.
2. Sharp bands are produced with a higher (5 mM) DTT level in the binding buffer.

3. Not all antibodies (polyclonal or monoclonal) recognize corresponding antigens within the EMSA complex in the supershift. Thus multiple antibodies may have to be tested in the supershift assay.
4. Depending on the conformation of the DNA-bound receptors, an antibody immunoreactive to the EMSA complex either upshifts a complex or the complex formation is inhibited.
5. The TNT-produced recombinant proteins may not always be optimal for EMSA. In this case the quality of EMSA will improve when recombinant proteins expressed in bacteria or in insect (Sf9) cells are used.

Conclusion

Vitamin D_3 induces endogenous SULT2A1 *in vivo* at 10–50 nM and considering that this concentration range falls within the normal spread of the circulating vitamin D_3 level in human (Feldman *et al.*, 1997), it is highly likely that VDR also promotes basal SULT2A1 expression in enterohepatic tissues. The VDR induction of SULT2A1 may be most significant regarding the metabolic disposition of the toxic secondary bile acids, especially lithocholic acid (LCA), which is produced from chenodeoxycholic acid by the dehydratases in the intestinal flora. An LCA overload in humans could result from specific metabolic disorders (endogenous cause) or from a high-fat diet (exogenous factor), which would stimulate bile acid secretions and would elevate the levels of LCA and other bile acids in the fecal excreta (Nagengast *et al.*, 1995; Reddy *et al.*, 1977). In experimental systems, LCA is known to cause DNA damage and promote cellular hyperplasia, and epidemiological data suggest that LCA may have a role in the etiology of colon cancer (Bernstein *et al.*, 2005; Nagengast *et al.*, 1995). The primary LCA metabolite in humans is its 3α-sulfate form, which, after efflux into the intestinal lumen, is poorly reabsorbed and thus is cleared out readily through the fecal route (Hofmann, 2004). Because sulfoconjugation of LCA and other bile acids is mediated exclusively by SULT2A1, it is expected that SULT2A1 would protect hepatic, biliary, and intestinal tissues from LCA toxicity. Direct evidence for the protective role of SULT2A1 in LCA-induced liver injury comes from a study in mice (Kitada *et al.*, 2003).

LCA and LCA derivatives are agonist ligands of VDR and are able to induce CYP2, CYP3, and several other VDR target genes at micromolar concentrations (Adachi *et al.*, 2005). The SULT2A1 promoter is also induced by LCA-acetate (Fig. 3B). We predict that VDR takes a central role in LCA-mediated induction of the SULT2A1 gene, even though PXR and FXR, the two NRs that are also activated by LCA, can induce SULT2A1 (Song *et al.*, 2001; Sonoda *et al.*, 2002). This prediction is made on the ground that LCA is an FXR antagonist (Yu *et al.*, 2002) and that the functional LCA concentration for PXR activation is about 10-fold higher than that needed to activate VDR (Makishima *et al.*, 2002). In conclusion, a role for the VDR pathway in the basal and induced expression of SULT2A1, a phase II transferase, may have emerged to protect cells of the first-pass tissues from the adverse effects of disrupted bile acid/sterol homeostasis caused by endogenous or exogenous factors.

Acknowledgments

This work was supported by NIH Grants R01-AG–10486; R01-AG–19660, a Merit Review grant from the Department of Veterans Affairs (VA), and a grant from the Philip Morris USA. B.C. is a Senior Career Scientist in the VA. We thank Dr. Rommel Tirona (Vanderbilt University) for the *Cyp3A*-Luc construct; Dr. Ronald Evans (Salk Institute) for the RXR-α and PXR plasmids; and Dr. David Mangelsdorf (Southwestern Medical School) for the FXR plasmid. We thank Drs. Taesung Oh and Young-kyo Seo (for SULT2B1b and protein–DNA interaction studies), Dr. Sunghwan Cho (for Northern blot), and Gilbert Torralva for graphics.

References

Adachi, R., Honma, Y., Masuno, H., Kawana, K., Shimomura, I., Yamada, S., and Makishima, M. (2005). Selective activation of vitamin D receptor by lithocholic acid acetate, a bile acid derivative. *J. Lipid Res.* **46**(1), 46–57.

Assem, M., Schuetz, E. G., Leggas, M., Sun, D., Yasuda, K., Reid, G., Zelcer, N., Adachi, M., Strom, S., Evans, R. M., Moore, D. D., Borst, P., and Schuetz, J. D. (2004). Interactions between hepatic Mrp4 and Sult2a as revealed by the constitutive androstane receptor and Mrp4 knockout mice. *J. Biol. Chem.* **279**(21), 22250–22257.

Bandmann, O., Vaughan, J., Holmans, P., Marsden, C. D., and Wood, N. W. (1997). Association of slow acetylator genotype for N-acetyltransferase 2 with familial Parkinson's disease. *Lancet* **350**(9085), 1136–1139.

Bernstein, H., Bernstein, C., Payne, C. M., Dvorakova, K., and Garewal, H. (2005). Bile acids as carcinogens in human gastrointestinal cancers. *Mutat. Res.* **589**(1), 47–65.

Blanchard, R. L., Freimuth, R. R., Buck, J., Weinshilboum, R. M., and Coughtrie, M. W. (2004). A proposed nomenclature system for the cytosolic sulfotransferase (SULT) superfamily. *Pharmacogenetics* **14**(3), 199–211.

Chatterjee, B., Majumdar, D., Ozbilen, O., Murty, C. V., and Roy, A. K. (1987). Molecular cloning and characterization of cDNA for androgen-repressible rat liver protein, SMP-2. *J. Biol. Chem.* **262**(2), 822–825.

Chatterjee, B., Song, C. S., Jung, M. H., Chen, S., Walter, C. A., Herbert, D. C., Weaker, F. J., Mancini, M. A., and Roy, A. K. (1996). Targeted overexpression of androgen receptor with a liver-specific promoter in transgenic mice. *Proc. Natl. Acad. Sci. USA* **93**(2), 728–733.

Chatterjee, B., Song, C. S., Kim, J. M., and Roy, A. K. (1994). Androgen and estrogen sulfotransferases of the rat liver: Physiological function, molecular cloning, and *in vitro* expression. *Chem. Biol. Interact.* **92**(1–3), 273–279.

Czich, A., Bartsch, I., Dogra, S., Hornhardt, S., and Glatt, H. R. (1994). Stable heterologous expression of hydroxysteroid sulphotransferase in Chinese hamster V79 cells and their use for toxicological investigations. *Chem. Biol. Interact.* **92**(1–3), 119–128.

Demyan, W. F., Song, C. S., Kim, D. S., Her, S., Gallwitz, W., Rao, T. R., Slomczynska, M., Chatterjee, B., and Roy, A. K. (1992). Estrogen sulfotransferase of the rat liver: Complementary DNA cloning and age- and sex-specific regulation of messenger RNA. *Mol. Endocrinol.* **6**(4), 589–597.

Dietrich, C. G., Geier, A., and Elferink, R. P. J. O. (2003). ABC of oral bioavailability: Transporters as gatekeepers in the gut. *Gut* **52,** 1788–1795.

Drocourt, L., Ourlin, J. C., Pascussi, J. M., Maurel, P., and Vilarem, M. J. (2002). Expression of CYP3A4, CYP2B6, and CYP2C9 is regulated by the vitamin D receptor pathway in primary human hepatocytes. *J. Biol. Chem.* **277,** 25125–25132.

Echchgadda, I., Song, C. S., Oh, T. S., Cho, S. H., Rivera, O. J., and Chatterjee, B. (2004b). Gene regulation for the senescence marker protein DHEA-sulfotransferase by the xenobiotic-activated nuclear pregnane X receptor (PXR). *Mech. Aging Dev.* **125**(10–11), 733–745.

Echchgadda, I., Song, C. S., Roy, A. K., and Chatterjee, B. (2004a). Dehydroepiandrosterone sulfotransferase is a target for transcriptional induction by the vitamin D receptor. *Mol. Pharmacol.* **65**(3), 720–729.

Escriva, H., Bertrand, S., and Laudet, V. (2004). The evolution of the nuclear receptor superfamily. *Essays Biochem.* **40,** 11–26.

Falany, C. N. (1997). Enzymology of human cytosolic sulfotransferases. *FASEB J.* **11,** 206–216.

Fang, H. L., Strom, S. C., Cai, H., Falany, C. N., Kocarek, T. A., and Runge-Morris, M. (2005). Regulation of human hepatic hydroxysteroid sulfotransferase gene expression by the peroxisome proliferator activated receptor alpha transcription factor. *Mol. Pharmacol.* **67,** 1257–1267.

Feldman, D., Glorieux, F. H., and Pike, J. W. (eds.) (1997). *In* "Vitamin D." Academic Press, New York.

Fuda, H., Lee, Y. C., Shimizu, C., Javitt, N. B., and Strott, C. A. (2002). Mutational analysis of human hydroxysteroid sulfotransferase SULT2B1 isoforms reveals that exon 1B of the SULT2B1 gene produces cholesterol sulfotransferase, whereas exon 1A yields pregnenolone sulfotransferase. *J. Biol. Chem.* **277**(39), 36161–36166.

Geese, W. J., and Raftogianis, R. B. (2001). Biochemical characterization and tissue distribution of human SULT2B1. *Biochem. Biophys. Res. Commun.* **288**(1), 280–289.

Handschin, C., Blattler, S., Roth, A., Looser, R., Oscarson, M., Kaufmann, M. R., Podvinec, M., Gnerre, C., and Meyer, U. A. (2004). The evolution of drug-activated nuclear receptors: One ancestral gene diverged into two xenosensor genes in mammals. *Nuclear Recept.* **2**(1), 7.

He, D., Meloche, C. A., Dumas, N. A., Frost, A. R., and Falany, C. N. (2004). Different subcellular localization of sulphotransferase 2B1b in human placenta and prostate. *Biochem. J.* **379**(Pt. 3), 533–540.

Her, C., Wood, T. C., Eichler, E. E., Mohrenweiser, H. W., Ramagli, L. S., Siciliano, M. J., and Weinshilboum, R. M. (1998). Human hydroxysteroid sulfotransferase SULT2B1: Two enzymes encoded by a single chromosome 19 gene. *Genomics* **53**(3), 284–295.

Ho, S. N., Hunt, H. D., Horton, R. M., Pullen, J. K., and Pease, L. R. (1989). Site-directed mutagenesis by overlap extension using the polymerase chain reaction. *Gene* **77**(1), 51–59.

Hofmann, A. F. (2004). Detoxification of lithocholic acid, a toxic bile acid: Relevance to drug hepatotoxicity. *Drug Metab. Rev.* **36**(3–4), 703–722.

Kitada, H., Miyata, M., Nakamura, T., Tozawa, A., Honma, W., Shimada, M., Nagata, K., Sinal, C. J., Guo, G. L., Gonzalez, F. J., and Yamazoe, Y. (2003). Protective role of hydroxysteroid sulfotransferase in lithocholic acid-induced liver toxicity. *J. Biol. Chem.* **278**, 17838–17844.

Kong, A. N., and Fei, P. (1994). Molecular cloning of three sulfotransferase cDNAs from mouse liver. *Chem. Biol. Int.* **92**, 161–168.

Krishnan, A. V., Peehl, D. M., and Feldman, D. (2003). The role of vitamin D in prostate cancer. *Recent Results Cancer Res.* **164**, 205–221.

Lin, R., and White, J. H. (2004). The pleiotropic actions of vitamin D. *Bioessays* **26**(1), 21–28.

Liska, D. J. (1998). The detoxification enzyme systems. *Altern. Med. Rev.* **3**(3), 187–198.

Makishima, M., Lu, T. T., Xie, W., Whitfield, G. K., Domoto, H., Evans, R. M., Haussler, M. R., and Mangelsdorf, D. J. (2002). Vitamin D receptor as an intestinal bile acid sensor. *Science* **296**, 1313–1316.

Mangelsdorf, D. J., and Evans, R. M. (1985). The RXR heterodimers and orphan receptors. *Cell* **83**(6), 841–850.

Meloche, C. A., Sharma, V., Swedmark, S., Andersson, P., and Falany, C. N. (2002). Sulfation of budesonide by human cytosolic sulfotransferase, dehydroepiandrosterone-sulfotransferase (DHEA-ST). *Drug Metab. Dispos.* **30**, 582–585.

Mulholland, D. J., Cox, M., Read, J., Rennie, P., and Nelson, C. (2004). Androgen responsiveness of Renilla luciferase reporter vectors is promoter, transgene, and cell line dependent. *Prostate* **59**, 115–119.

Nagengast, F. M., Grubben, M. J., and van Munster, I. P. (1995). Role of bile acids in colorectal carcinogenesis. *Eur. J. Cancer* **31A**(7–8), 1067–1070.

Ogura, K., Satsukawa, M., Okuda, H., Hiratsuka, A., and Watabe, T. (1994). Major hydroxysteroid sulfotransferase STa in rat liver cytosol may consist of two microheterogeneous subunits. *Chem. Biol. Interact.* **92**(1–3), 129–144.

Otterness, D. M., Her, C., Aksoy, S., Kimura, S., Wieben, E. D., and Weinshilboum, R. M. (1995). Human dehydroepiandrosterone sulfotransferase gene: Molecular cloning and structural characterization. *DNA Cell Biol.* **14**(4), 331–341.

Pai, T. G., Sugahara, T., Suiko, M., Sakakibara, Y., Xu, F., and Liu, M. C. (2002). Differential xenoestrogen-sulfating activities of the human cytosolic sulfotransferases: Molecular cloning, expression, and purification of human SULT2B1a and SULT2B1b sulfotransferases. *Biochim. Biophys. Acta* **1573**, 165–170.

Pedersen, L. C., Petrotchenko, E. V., and Negishi, M. (2000). Crystal structure of SULT2A3, human hydroxysteroid sulfotransferase. *FEBS Lett.* **475**(1), 61–64.

Radominska, A., Comer, K. A., Zimniak, P., Falany, J., Iscan, M., and Falany, C. N. (1990). Human liver steroid sulphotransferase sulphates bile acids. *Biochem. J.* **272**(3), 597–604.

Reddy, B. S., Mangat, S., Sheinfil, A., Weisburger, J. H., and Wynder, E. L. (1977). Effect of type and amount of dietary fat and 1,2-dimethylhydrazine on biliary bile acids, fecal bile acids, and neutral sterols in rats. *Cancer Res.* **37**(7 Pt. 1), 2132–2137.

Rehse, P. H., Zhou, M., and Lin, S. X. (2002). Crystal structure of human dehydroepiandrosterone sulphotransferase in complex with substrate. *Biochem. J.* **364**(Pt. 1), 165–171.

Runge-Morris, M., Wu, W., and Kocarek, T. A. (1999). Regulation of rat hepatic hydroxysteroid sulfotransferase (SULT2–40/41) gene expression by glucocorticoids: Evidence for a dual mechanism of transcriptional control. *Mol. Pharmacol.* **56**, 1198–1206.

Saner, K. J., Suzuki, T., Sasano, H., Pizzey, J., Ho, C., Strauss, J. F., 3rd, Carr, B. R., and Rainey, W. E. (2005). Steroid sulfotransferase 2A1 gene transcription is regulated by steroidogenic factor 1 and GATA-6 in the human adrenal. *Mol. Endocrinol.* **19**(1), 184–197.

Seo, Y.-K., Oh, T.-S., Cho, S.-H., Shi, L.-H., Ko, S.-Y., Kim, S.-A., Song, C. S., and Chatterjee, B. (2005). Vitamin D Receptor Regulated Expression of the Human Cholesterol and Steroid Sulfotransferase (SULT2B1) in the Prostate. 86th US Endocrine Society Meeting, San Diego, CA.

Shibutani, S., Shaw, P. M., Suzuki, N., Dasaradhi, L., Duffel, M. W., and Terashima, I. (1998). Sulfation of alpha-hydroxytamoxifen catalyzed by human hydroxysteroid sulfotransferase results in tamoxifen-DNA adducts. *Carcinogenesis* **19**(11), 2007–2011.

Shimizu, C., Fuda, H., Yanai, H., and Strott, C. A. (2003). Conservation of the hydroxysteroid sulfotransferase SULT2B1 gene structure in the mouse: Pre- and postnatal expression, kinetic analysis of isoforms, and comparison with prototypical SULT2A1. *Endocrinology* **144**(4), 1186–1193.

Song, C., S., Echchgadda, I., Baek, B. S., Ahn, S. C., Oh, T., Roy, A. K., and Chatterjee, B. (2001). Dehydroepiandrosterone sulfotransferase gene induction by bile acid activated farnesoid X receptor. *J. Biol. Chem.* **276**, 42549–42556.

Song, C. S., Echchgadda, I., Oh, T., Shi, I. H., Kim, S. A., Kim, S. Y., Chatterjee, B. (2004). Vitamin D Induced DHEA-sulfotransferase Gene Transcription Requires Functional Interaction between the Vitamin D Receptor (VDR) and CAAT/Enhancer-Binding Protein (C/EBP). 85th US Endocrine Society Meeting, New Orleans, LA.

Song, C. S., Jung, M. H., Kim, S. C., Hassan, T., Roy, A. K., and Chatterjee, B. (1998). Tissue-specific and androgen-repressible regulation of the rat dehydroepiandrosterone sulfo-transferase gene promoter. *J. Biol. Chem.* **273**, 21856–21866.

Sonoda, J., Xie, W., Rosenfeld, J. M., Barwick, J. L., Guzelian, P. S., and Evans, R. M. (2002). Regulation of a xenobiotic sulfonation cascade by nuclear pregnane X receptor (PXR). *Proc. Natl. Acad. Sci. USA* **99**(21), 13801–13816.

Thummel, K. E., Brimer, C., Yasuda, K., Thottassery, J., Senn, T., Lin, Y., Ishizuka, H., Kharasch, E., Schuetz, J., and Schuetz, E. (2001). Transcriptional control of intestinal cytochrome P-4503A by 1alpha,25-dihydroxy vitamin D3. *Mol. Pharmacol.* **60**, 1399–1406.

Toell, A., Polly, P., and Carlberg, C. (2000). All natural DR3-type vitamin D response elements show a similar functionality *in vitro*. *Biochem. J.* **352**, 301–309.

Weinshilboum, R. M., Otterness, D. M., Aksoy, I. A., Wood, T. C., Her, C., and Raftogianis, R. B. (1997). Sulfation and sulfotransferases 1: Sulfotransferase molecular biology: cDNAs and genes. *FASEB J.* **11**, 3–14.

Yu, J., Lo, J. L., Huang, L., Zhao, A., Metzger, E., Adams, A., Meinke, T., Wright, S. D., and Cui, J. (2002). Lithocholic acid decreases expression of bile salt export pump through farnesoid X receptor antagonist activity. *J. Biol. Chem.* **277**, 31441–31447.

[11] Screening and Characterizing Human NAT2 Variants

By MIHAELA R. SAVULESCU, ADEEL MUSHTAQ, and P. DAVID JOSEPHY

Abstract

Acetyl CoA:arylamine N-acetyltransferase (NAT; E.C. 2.3.1.5) enzymes play a key role in the metabolic activation of aromatic amine and nitroaromatic mutagens to electrophilic reactive intermediates. We have developed a system in which the activation of mutagens by recombinant human NAT2, expressed in *Escherichia coli*, can be detected by the appearance of Lac$^+$ revertants. The mutagenesis assay is based on the reversion of an *E. coli lacZ* frameshift allele; the host strain for the assay is devoid of endogenous NAT activity and a plasmid vector is used for expression of human NAT2. A high-throughput version of the assay facilitates rapid screening of pools of NAT2 variants generated (for example) by random mutagenesis. Along with the methods for these assays, we present selected results of a screening effort in which mutations along the length of the NAT2 sequence have been examined. Homology modeling and simulated annealing have been used to analyze the potential effects of these mutations on structural integrity and substrate binding.

Introduction

NAT Enzymes

Acetyl CoA:arylamine N-acetyltransferase (NATs; E.C. 2.3.1.5) enzymes catalyze the N-acetylation (detoxication) of arylamine carcinogens and the O-acetylation activation of aryl hydroxylamines (Hein, 2000; Saito *et al.*, 1986) (Fig. 1). Humans express two NAT enzymes, NAT1 and NAT2, both of which are highly polymorphic (Butcher *et al.*, 2000; Blum *et al.*, 1990). To date, 26 NAT1 and 29 NAT2 alleles have been identified in the human population (see listing at www.louisville.edu/medschool/pharmacology/NAT.html). The NAT1 and NAT2 enzymes have distinct, although overlapping, substrate specificities and tissue distributions (Grant *et al.*, 1991). NAT2 is found mainly in the liver and the gut (Hein *et al.*, 2000a,b; 2002; Hickman *et al.*, 1998; Ilett *et al.*, 1994) whereas NAT1 is found in leukocytes (Cribb *et al.*, 1991), erythrocytes (Risch *et al.*, 1996), and placenta (Smelt *et al.*, 1998, 2000), as well as the liver and gut.

METHODS IN ENZYMOLOGY, VOL. 400
0076-6879/05 $35.00
DOI: 10.1016/S0076-6879(05)00011-X

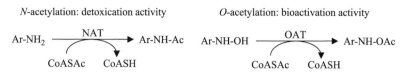

FIG. 1. Reactions catalyzed by the NAT and OAT activities of acetyl CoA:arylamine *N*-acetyltransferase enzymes. Ar, aromatic or heterocyclic ring.

NAT Structures

The three-dimensional structures of *Salmonella typhimurium* and *Mycobacterium smegmatis* NAT enzymes have been solved by X-ray crystallography (Pompeo *et al.*, 2002; Sinclair *et al.*, 2000) and reveal the presence of an active site catalytic triad, aspartate-histidine-cysteine, similar to that of cysteine proteases (Turk *et al.*, 1997). Site-directed mutagenesis experiments with mammalian NATs have shown that the cysteine (Dupret and Grant, 1992) and aspartate (Wang *et al.*, 2004a) residues are essential for catalysis.

Comparison of the sequences of human NAT1 and NAT2 shows that the amino acid differences, which must determine substrate specificity differences, are greatest in the C-terminal portion. No three-dimensional structure of any eukaryotic NAT protein is yet available. The high-level expression and purification of eukaryotic NATs are difficult (Sinclair and Sim, 1997; Wagner *et al.*, 1996; Wang *et al.*, 2005). An alternative approach is modeling of the structures, based on the known structures of prokaryotic NATs. A previous model of the human NAT1 protein included only a 102 residue N-terminal portion (Rodrigues-Lima and Dupret, 2002; Rodrigues-Lima *et al.*, 2001). However, the C-terminal portion of the enzyme plays a crucial role in prokaryotic NATs; futile hydrolysis of acetyl CoA was observed in truncated proteins (Mushtaq *et al.*, 2002).

NAT and Bioactivation of Mutagens

Aryl hydroxlamines are converted to DNA-reactive *N*-acetoxy esters (Bosold and Boche, 1990; Humphreys *et al.*, 1992; Miller *et al.*, 1966) by the acetyl CoA:arylhydroxylamine *O*-acetyltransferase (OAT) activity of NAT enzymes (Josephy, 1997). NAT activity is critical for detection of the mutagenicity of aromatic amines and nitroaromatic compounds in short-term bacterial tests. Cytochrome P450 catalyzes the oxidation of aromatic amines to aryl hydroxylamines (Guengerich, 2002), and aryl hydroxylamines can also be formed by the enzymatic reduction of nitroaromatic compounds (Johansson *et al.*, 2003; Nokhbeh *et al.*, 2002).

Salmonella typhimurium NAT enzyme activity was first identified by virtue of the resistance of NAT-minus mutant derivatives of Ames tester strains (McCoy *et al.*, 1983) to the mutagenicity and toxicity of nitropyrenes (Rosenkranz and Mermelstein, 1983). As in *S. typhimurium*, *Escherichia coli* strains bearing a null mutation in the gene encoding NAT are also highly resistant to the mutagenicity of aromatic amines and nitro compounds (Josephy *et al.*, 2002). Overexpression of prokaryotic or eukaryotic (human) NAT enzymes greatly enhances the mutagenicity of these compounds in *S. typhimurium* (Ames test) and *E. coli* (*lacI* and *lacZ*) mutagenicity assay systems (Grant *et al.*, 1992; Josephy, 2000; Kosakarn *et al.*, 1993; Suzuki *et al.*, 1998; Watanabe *et al.*, 1987; Wild *et al.*, 1995).

The lacZ *Mutagenicity Assay*

Recent studies have relied on the *E. coli lacZ* reversion assay. Miller and colleagues developed this assay as a short-term test for the detection of specific classes of mutations (Cupples and Miller, 1989; Cupples *et al.*, 1990). The *lacZ* gene encodes the enzyme β-galactosidase, which catalyzes the hydrolysis of lactose to glucose and galactose. A functional β-galactosidase enzyme is required for lactose to be used as a carbon source; selection for revertants is carried out by plating onto minimal lactose medium. The *lacZ* assay utilizes a series of *E. coli* strains, each possessing a different mutation of the *lacZ* gene and therefore detecting a different reversion mutation. The target *lacZ* allele in strain CC109 has a $+2$ base pair frameshift at a GC-rich site that is a hot spot for reversion mutations. Attractive features of the *lacZ* assay (compared to the *S. typhimurium* Ames test) include the ease with which plasmids can be introduced to, and recovered from, *E. coli* and the flexibility that results from having the *lacZ* target carried on an episome, which can be moved between strains by conjugation.

Screening for Variants of the Human NAT2 Sequence with Altered Activity

Because NAT activity determines the mutagenicity of aromatic amines and nitro compounds in bacterial assays, the mutagenic response to these compounds provides a simple "readout" of NAT enzyme activity, and variant enzymes can be screened on this basis. Our approach is as follows. A pool of variant DNAs is generated by random mutagenesis of the NAT2 coding sequence (Summerscales and Josephy, 2004). The mutagenized open reading frame (ORF) is ligated into the expression vector to give a pool of variant plasmids. The plasmid pool is used to transform a host mutagenicity tester strain of *E. coli*, resulting in a pool of variant strains.

The variant strains are then screened on the basis of phenotype, namely the induction of $lacZ^+$ revertants following exposure to a nitroaromatic mutagen. Clones for which the mutagenic response differs from that of the wild type are candidates for further characterization. To screen the clones with a standard mutagenicity assay protocol would require the use of one petri dish per clone, making large-scale screening cumbersome and expensive. However, the screening process can be scaled down. Variant clones are cultured in multiwell plates and small aliquots of culture are applied to a petri dish in arrays using a multichannel pipetter or a robot. We call this approach "DAVERAMA" (detection of active variant enzymes by reversion assay/ mutagen activation) (Josephy, 2002).

The DAVERAMA strategy can be used for any enzyme catalyzing mutagen bioactivation in a short-term assay. In addition to the NAT2 studies reported previously (Summerscales and Josephy, 2004) and in this chapter, the method has also been applied to the analysis of P450 1A2 variants (Kim and Guengerich, 2004; Parikh *et al.*, 1999; Zhou *et al.*, 2004). Kim and Guengerich (2004) have reported the largest-scale application of the method, using a robotic colony-picking system and 384-well plates for culture growth to screen tens of thousands of sequence variants of cytochrome P450 1A2 and identifying variants with enhanced catalytic activities.

Our screening approach can be employed to identify enzymes with activities either higher or lower than that of the wild type; in either case, the variants are evident from an altered mutagenic response. Once variants are identified [e.g., from a pool of plasmids constructed by PCR mutagenesis (Summerscales and Josephy, 2004)], they are submitted for DNA sequencing to identify the amino acid substitution that they bear. In this way, we can pinpoint specific amino acid residues that govern catalytic activity in NAT enzymes. In our first application of this approach to human NAT2 (Summerscales and Josephy, 2004), polymerase chain reaction (PCR) random mutagenesis along the entire NAT2 coding sequence was employed, and 18 mutants with altered enzymatic activity were identified and characterized.

Here, we have expanded our study of human NAT2, focusing on variants exhibiting low but nonzero bioactivation capacity. NO_2-IQ (3-methyl-2-nitroimidazo[4,5-*f*]quinoline), a mutagen that is highly dependent on NAT activation (Dirr and Wild, 1988), is used in screening the variants. We present the methods used for the DAVERAMA screening protocol and discuss the characteristics of several of the NAT2 variants that we have isolated. In addition, we applied computational methods to predict the locations of these variants and to identify residues that may mediate binding of xenobiotic substrates to the NAT2 enzyme.

Materials and Methods

E. coli *Strains*

DJ2002 (Josephy *et al.*, 2002) is used as the background strain for screening the recombinant human NAT2 variants. This *E. coli* strain bears the F-prime episome from strain CC109 (see earlier discussion). This episome carries a *lacZ* allele designed to detect frameshift mutations (Cupples *et al.*, 1990); reversion occurs by a -2-bp deletion in the sequence GCGCGC. *E. coli* strains carrying this episome have been used previously in our laboratory (Marwood *et al.*, 1995) to detect the mutagenicity of aromatic and heterocyclic amines. The endogenous *E. coli nat* gene is disrupted in strain DJ2002. In the absence of endogenous NAT activity, nitro-compound-induced reversion is entirely dependent on the expression of active recombinant NAT, thereby improving the dynamic range of the screening process.

Mutagenicity Assays

Chemicals

Sources of chemicals are as follows: NO_2-IQ, Toronto Research Chemicals (Toronto, ON); NF and other specialty chemicals are from Aldrich and Sigma Chemical Co. (St. Louis, MO).

Solutions and Materials

Minimal lactose (ML) plates: In a 2-liter Erlenmeyer flask, dissolve 15 g Difco agar (granulated) in 900 ml water and add 20 ml 10% lactose (w/v), 100 ml 10× A salts (see later), 1 ml $MgSO_4$ (1 *M*), and 0.5 ml 1% thiamine. After autoclaving, dispense the solution into 100-mm petri dishes.

10× A salts: In 1 liter water, combine 105 g K_2HPO_4, 45 g KH_2PO_4, 10 g $(NH_4)_2SO_4$, and 5 g sodium citrate; autoclave in 100-ml aliquots.

LB broth: Combine two capsules of LB medium (Q-BioGene Inc., Carlsbad, CA) in 80 ml water.

Isopropyl-ß-D-thiogalactopyranoside (IPTG): Dissolve IPTG, 200 mg, in 1 ml water and filter sterilize (0.2-μm Acrodisc syringe filter, Fisher Scientific, Pittsburgh, PA).

Ampicillin: Dissolve 100 mg ampicillin (Sigma Chemicals, St. Louis, MO) in 1 ml water and filter sterilize as for IPTG.

Top agar: In a 250-ml Erlenmeyer flask, combine 0.6 g Difco agar (granulated), 0.5 g NaCl, and 100 ml water; autoclave.

Nutrient broth: Combine 25 g Oxoid nutrient broth No. 2 powder (Basingstoke, Hampshire, England) and 1 liter water; autoclave in 75-ml aliquots.

lacZ assay preincubation buffer: 0.1 M sodium phosphate buffer, pH 7.4, + 60 mM KCl.

lacZ *Assay Protocol I: Initial Screening*

Construction of the NAT2 variant library by PCR mutagenesis of the NAT2 open reading frame and ligation into vector pKK223-3 has been described previously (Summerscales and Josephy, 2004). Initial screening of presumptive NAT2 variants is done using the DAVERAMA method (Josephy, 2000). The pooled DNA of the NAT2 variant library is transformed into the DJ2002 background and transformants are plated onto minimal glucose medium supplemented with thiamine and ampicillin. Individual colonies picked from these plates are inoculated into 96-well Cluster Tube racks (Fisher Scientific, Toronto, ON) filled with LB expression medium (LB + 100 μg/ml ampicillin + 1 mM IPTG; 0.5 ml per well). Positive (DJ2002 expressing wild-type NAT2; vector pKK223–3) and negative (DJ2002 + pKK223–3) control strains are also cultured in each ClusterTube rack. Aliquots (5 μl) from each well of the 96-well tube rack (including positive and negative controls) are plated onto mutagenicity test plates.

Mutagenicity test plates are prepared by combining specific concentrations of mutagen (e.g., 0, 10, or 30 pmol NO$_2$-IQ, in DMSO; final volume 10 μl) and lacZ assay preincubation buffer (see earlier discussion), 0.5 ml, in 5-ml snap-cap tubes (Sarstedt, Montreal, QC). Melted top agar (2 ml) supplemented with nutrient broth (0.8 ml) is added to each tube and the mixture is overlaid onto ML plates.

Cultures are grown (37°, 175 rpm, 24 h) in a 96-well cluster tube to OD$_{600}$ = 0.8. A multichannel micropipette is used to transfer aliquots (5 μl) of each culture onto the mutagenicity assay test plates. In addition, aliquots of each culture are also plated onto master plates [minimal glucose + ampicillin (50 μg/ml) + thiamine]. After incubation (48 h, 37°), small revertant "microcolonies" can be seen on the background lawn of each culture spot. Using a magnifier, the small colonies are counted manually. Each determination is done in triplicate; values are averaged and used to construct NO$_2$-IQ dose–response curves. This screening procedure is used to find clones with mutagen bioactivation activities substantially lower than the wild type. Only these clones are studied further. Permanent stocks of each variant to be characterized further are prepared by combining overnight culture (1 ml) with DMSO (90 μl), freezing on dry ice, and storing at −70°.

lacZ *Assay Protocol II: Characterization of Variants*

Escherichia coli cultures are grown in 5 ml LB + IPTG (5 μl × 200 mg/ml) + ampicillin (5 μl ×100 mg/ml) at 37°, 250 rpm shaking, for about 10 h, to $OD_{600} = 0.8$. Mutagen (30 pmol NO_2-IQ in DMSO; 10 μl) and *lacZ* assay preincubation buffer (see earlier discussion; 0.5 ml) are combined in 5-ml snap-cap tubes to which an aliquot of culture (0.1 ml) is added. Tubes are incubated at 37° for 30 min. Top agar (2 ml) supplemented with nutrient broth (0.8 ml) is then added to each tube and the mixture is overlaid onto ML plates. Plates are incubated at 37° for 48 h and colonies are counted with the aid of a video analysis system.

Some clones that pass the first stage of screening are discarded at this stage because their responses are judged to be similar to the wild-type control. Variants exhibiting consistent low responses are selected and characterized further. DNA minipreps are performed and both strands of the entire NAT2 ORF are sequenced (Molecular Supercentre, University of Guelph) using primers located on the 5' and 3' sides of the insert.

pKK5: 5' GGATAACAATTTCACACAGG 3'
pKK3: 5' AATCTTCTCTCATCCGCCAA 3'

In almost all cases, putative variants that pass the phenotypic screening tests are found to have a single amino acid substitution in the NAT2 coding sequence. (We have not yet screened heavily mutagenized pools in which multiple mutations would be common.) We also tested the response of the variants to NF in standard *lacZ* assays.

Enzyme Activity Measurements

NAT activities (substrate, 2-aminofluorene; AF) are measured in crude *E. coli* extracts by quantifying the disappearance of the AF substrate. AF (but not the acetylated product) forms a Schiff base with DMAB, measured by absorbance at 450 nm (Smith *et al.*, 1992).

Solutions and Materials

10% TCA: Dissolve 10 g TCA in 100 ml water. Colorimetric reagent: Combine equal volumes of *p*-dimethylaminobenzaldehyde [DMAB, 1% (w/v) in ethanol] and 1 M sodium acetate-HCl buffer, pH 1.4.

Sonication buffer: To 0.975 ml buffer (0.066 M sodium phosphate, pH 7.2, + 1 mM EDTA), add dithiothreitol (DTT, 20 μl × 0.1 M) and phenylmethylsulfonyl fluoride (PMSF, 5 μl × 0.1 M).

Assay Protocol

Overnight cultures (40 ml, $A_{600} = 1.0$) are centrifuged ($4000g$, 10 min, 4°); resuspend the cells in sonication buffer (see earlier discussion). Add lysozyme (0.1 ml × 10 mg/ml) and incubate on ice for 1 h. Transfer suspensions to 1.5-ml microcentrifuge tubes and disrupt with a "Microson" sonicator (Misonix Inc., Farmingdale, NY) at 70% power level; 1 × 30 s and 2 × 15 s, on ice, with 15-s intervals between bursts. Centrifuge the crude extracts (10 min, 13,000 rpm, 4°) and assay the resulting supernatant for NAT activity (samples may be frozen in aliquots at −20°). Determine the protein concentration of each extract using the Bradford assay (Bradford, 1976).

Prior to each assay, extracts are diluted appropriately, according to the anticipated activity, e.g., the wild type is diluted 10-fold. Assay premix contains diluted crude extract in sonication buffer [23 μl lysate (diluted if necessary) + 250 μl sonication buffer] and AF (21 μl × 4 mM in DMSO) per time point per strain sample. Aliquots (20 μl of premix) are dispensed into wells of the 96-well plate and chilled on ice. To initiate the reaction, acetyl-CoA (6 μl per time point per sample, 10 mM, dissolved in water immediately prior to assay) is added to three wells and water (6 μl) to the fourth (control). Final substrate concentrations are 0.22 mM AF and 2.3 mM acetyl-CoA. The plate is transferred into a 37° water bath for the desired incubation period. To terminate the reaction, the plate is placed on ice and TCA [10% (w/v); 15 μl] is added to each well. The colorimetric solution, *p*-DMAB (0.3 ml) is added to each well, and color is allowed to develop for a minimum of 5 min at room temperature. The absorbance is read at 450 nm on a plate reader.

One absorbance unit at 450 nm corresponds to 8.40 nmol AF, as determined by a standard curve. Using the difference in absorbance between the samples (average of three values) and the negative control, the amount of AF acetylated is calculated and expressed as the *N*-acetylation activity per minute per milligram total protein.

Modeling the Three-Dimensional Structures of Human NAT Enzymes

The amino acid sequences of human NAT1 and 2 were aligned to the amino acid sequences of NATs from *S. typhimurium* and *M. smegmatis* using the program ClustalW (Thompson *et al.*, 1994), with gap creation penalty = 2 and gap extension penalty = 0.1. NAT1 and NAT2 structures were then modeled using the MODELLER (Sali and Blundell, 1993) suite of programs, using the crystal structures of *S. typhimurium* and *M. smegmatis* (PDB accession codes 1E2T and 1GX3, respectively) as templates.

The effects of point mutations on the structure of human NAT2 were analyzed by changing the appropriate amino acid residue in the initial alignment, followed by repeating the modeling process. Final models were assigned secondary structure elements with the Program DSSP and structural integrity was checked with the program PROCHECK (Laskowski *et al.*, 1993).

Simulated "In Silico" Annealing (Docking) of Substrates

All nonpolar hydrogens and terminal oxygen groups were attached and Gasteiger charges were assigned to the three-dimensional modeled structures of NAT1 and NAT2 with the programs SYBYL 6.5 (www.tripos. com). Exact three-dimensional structures of the amine mutagens IQ (3-methyl-2-aminoimidazo[4,5-*f*]quinoline) and AF were created using DSViewer Pro (Accelrys, Inc., San Diego, CA). Rotatable bonds and fixed rings were assigned using the program AUTOTORS (Morris *et al.*, 1996). The two substrate structures were annealed independently to the protein structure using the AUTODOCK suite of programs (Morris *et al.*, 1998). The protein–ligand interactions of the best-fitting lowest energy model were then analyzed with the program LIGPLOT (Wallace *et al.*, 1995).

Results

Overview

Thirty two NAT2 sequence variants with altered activities have been identified in our laboratory: 18 in our previously published study (Summerscales and Josephy, 2004) and 14 in the present report. Each variant bears a single amino acid change in the NAT2 sequence. (A few variants have additional silent base pair substitutions.) The positions of the mutations are distributed throughout the NAT2 ORF. Figure 2 shows the human NAT2 sequence aligned with four other NAT sequences, and the positions of the 14 new mutations are marked.

For each of these variants, the standard *lacZ* assay was repeated in triplicate (three plates per dose) for both NO_2-IQ and NF. Figure 3 summarizes these results; for clarity, the response at only one dose of each mutagen is shown. Figure 4 shows the correlation between the potencies of the two mutagens. Variants exhibiting shifts in substrate specificity between the two mutagens tested would be expected to lie away from the line connecting the negative control and the wild-type enzyme.

Five of the variants identified in this study have been tested for NAT (AF acetylation) enzyme activity of crude extracts. [Variations in activity

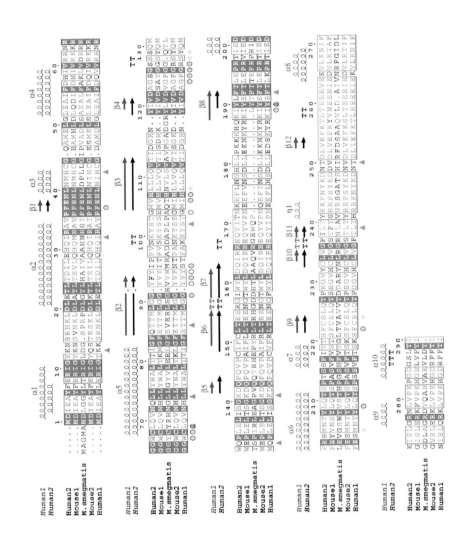

could reflect either reduced expression of the variant proteins or diminished enzyme activity, as none of the variants has yet been purified. In our previous study (Summerscales and Josephy, 2004), expression levels were similar among all variants examined, as judged by immunoblotting.] Figure 5 shows the activities of the five variants examined. Generally, AF NAT enzyme activity correlates closely with mutagen bioactivation activity in the *lacZ* assay.

Specific Variants

Our long-term goal is to understand how specific residues in the NAT sequence contribute to the catalytic activity of the enzyme. This research is motivated by several general observations about NAT enzymes. First, we still do not understand their biological role. In contrast to cytochrome P450 enzymes, glutathione transferases, and many other enzymes of xenobiotic biotransformation, few (if any) physiologically relevant endogenous substrates for NATs have been identified (Kawamura *et al.*, 2005). The structural similarity of prokaryotic (and, presumably, eukaryotic) NATs to cysteine proteases is remarkable, but we do not know whether this represents convergent or divergent evolution. Second, the human NAT genes/enzymes are highly polymorphic. Single amino acid changes can reduce NAT activity drastically: this is the basis of the human slow acetylator phenotype, a classic example of pharmacogenetic variation that has been known for almost half a century (Hein *et al.*, 2000b). However, characterization of the phenotypic consequences of human NAT single nucleotide polymorphisms (functional genomics) is far from complete (Fretland *et al.*, 2001a,b). Third, in the absence of an experimental structure for any eukaryotic NAT enzyme, analysis of the properties of sequence-variant enzymes can provide valuable clues to NAT protein structure.

Human NAT1 and NAT2 are 87% identical in amino acid sequence. The two human enzymes have many common substrates (Hein *et al.*, 1993), but they also have unique activities. Both NAT1 and NAT2 were modeled successfully (Fig. 6) with initial G-factor values of −0.41, which were improved to −0.21 after several rounds of refinement within the Modeller. The RMSD value from the template structures was 0.8 Å and the RMSD between NAT1 and NAT2 was 0.5 Å. The modeled structures closely

---

FIG. 2. Amino acid sequence alignment of NATs, indicating the secondary structural elements from the models. Numbering follows the human NAT2 sequence. The sequences of NAT from *M. smegmatis* and mouse (NAT1 and NAT2) are shown for comparison. Sites of the mutations in the 14 variants investigated in this study are indicated with purple triangles. Residues calculated to be interacting (within 6Å) of the docked ligands are shown with green stars (2-AF) and black circles (IQ). (See color insert.)

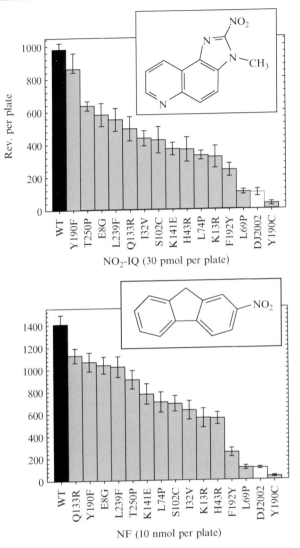

FIG. 3. Mutagenicity of nitro compounds in each of the 14 NAT2 variants. Compounds tested: NO₂-IQ, 3-methyl-2-nitroimidazo[4,5-ƒ]quinoline; NF, 2-nitrofluorene. Strains tested were wild-type NAT2 (black bars), variants (light gray bars), and DJ2002 (negative control, white bars). Variants were tested at multiple doses, but, for simplicity of presentation, only responses at a single dose are shown. Data points represent three independent triplicate experiments. Error bars represent standard errors of counts.

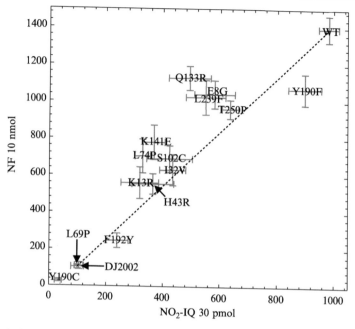

Fig. 4. Scatter plot illustrating correlation of mutagenic responses to NO_2-IQ and NF. Results from assays of all 14 variants are shown, including positive (NAT2, WT) and negative (DJ2002) controls. Mutagenic potencies were measured at 30 pmol NO_2-IQ and 10 nmol NF revertants per plate. Error bars represent standard errors of counts from at least 12 plates.

resembled those of the template (prokaryotic NAT) structures. The catalytic triad, for both NAT1 and NAT2 enzymes, was found to be in an almost identical position to that of the NATs from *S. typhimurium* and *M. smegmatis*. Major differences were found, as anticipated, in the interdomain (residues 170–190 in NAT2) and third domain regions (beyond residue 190 in NAT2). The interdomain region, which does not exist in prokaryotic NATs, could not be modeled reliably, as indicated by the high modeler violation values (>0.7) at the beginning of this region (Fig. 7). A comparison between modeled structures of NAT1 and NAT2 also points to this interdomain region as the major point of difference between them. There was a large variation in RMSD (>2 Å) in this region, most likely due to poor modeling.

The interdomain region is unlikely to play any role in the catalytic mechanism. However, in the modeled structures, the region is close enough to the active site cleft that it might loop over the opening to it, thereby influencing substrate specificity. The interdomain loop could also control

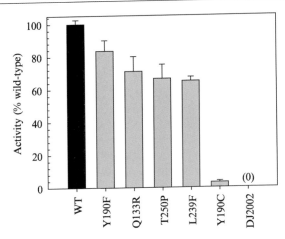

FIG. 5. 2-Aminofluorene *N*-acetylation activities of crude extracts from *E. coli* strains expressing five NAT2 variants, normalized to 100% activity for the wild-type enzyme. DJ2002 (negative control) shows zero activity. Data represent three independent triplicate experiments.

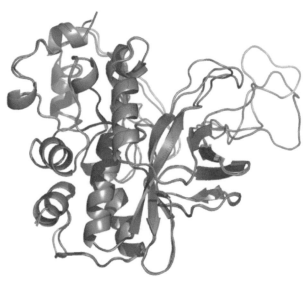

FIG. 6. Modeled structures of human NAT1 (gray) and NAT2 (magenta) are overlaid. The structures of the human NATs were modeled using the methods described in the text. (See color insert.)

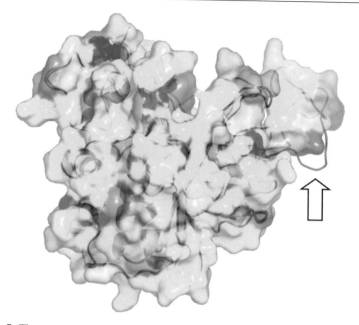

Fig. 7. The same structures shown in Fig. 6, with addition of a transparent surface colored according to Modeller violation values (RGB, blue ∼0, green ∼0.5, and red >1). The structures are similar, except in the vicinity of the interdomain region (residues 190–210), seen as a patch of red (arrow). (See color insert.)

protein–protein interactions or may represent a domain-swapping type of interface (Newcomer, 2002). The interdomain region points away from the core of the protein and into the medium, as is commonly the case in interaction interfaces (Xenarios *et al.*, 2002).

Residue 127 (arginine in NAT1 and serine in NAT2), suggested to represent a key difference between the two enzymes, was found to be in a position identical to that reported previously (Rodrigues-Lima and Dupret, 2002; Rodrigues-Lima *et al.*, 2001).

Docking studies (Figs. 8 and 9) with amine forms of the mutagens showed the positions of the aromatic moieties of AF and IQ to be nearly identical to those observed for aromatic substrates of prokaryotic NATs (Brooke *et al.*, 2003b; Mushtaq *et al.*, 2002). All 14 point mutations investigated in this study were modeled. None caused a significant perturbation of the NAT2 structure, with RMSD deviations less than 0.1 Å in all cases. Nevertheless, many of the mutated residues appear to be within 6 Å of the docked ligand and might have a "knock-on" effect on the ligand-binding

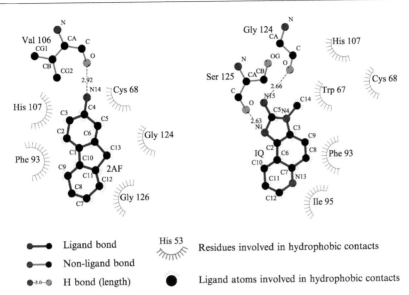

FIG. 8. Predicted nonbonded interactions of AF (left) and IQ (right) with NAT2, determined using AutoDock. The aromatic rings from both ligands form hydrophobic interactions with the phe37, phe93, and tyr190 residues, with each of these residues playing a differing role when binding IQ or AF. The amino groups of both substrates face toward the active site cysteine (indicated by orange lettering). (See color insert.)

pocket. We have chosen to discuss 5 of the 14 NAT2 variants in greater detail.

The human NAT2 model structure possesses three structural domains, as is observed for both known NAT crystal structures. The third domain (residues 190 and beyond) forms a "cap" in the *S. typhimurium* and *M. smegmatis* enzymes (Fig. 10). The sequence of this domain varies markedly among NATs from different species and it may be the region contributing most strongly to substrate profile determination (Payton *et al.*, 2001). Two of our mutants, L239F and T250P, are at residues in this "cap" domain. Neither is close to the active site and neither inactivates the enzyme.

Mutation L239F replaces one hydrophobic residue by another but introduces a substantial steric change. Position 239 shows some variability among eukaryotic NATs and, indeed, the corresponding residue in mouse NAT1 (the murine enzyme most closely resembling human NAT2, in terms of substrate specificity) is phenylalanine. Residue 239 lies on a sheet directly above the aromatic cleft within which the substrate is believed to

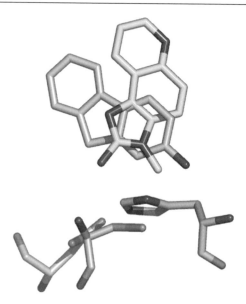

Fɪɢ. 9. The docked positions of AF (gray) and IQ (green) are overlaid. Catalytic triad residues are shown. (See color insert.)

bind. Modeling predicts that phenylalanine can be accommodated (as was expected, because of the flexible nature of the NAT core) but cannot predict the effect that such a change might have on substrate specificity.

Mutation T250P is highly nonconservative in terms of side chain functionality. Site 250 is not conserved, with valine (murine NAT1 and NAT2), isoleucine (rat NAT2), and methionine (chicken NAT1 and NAT2), although not proline, found at this site in other reported eukaryotic NAT sequences. This residue is also predicted to lie just above the active site cleft, on the lip of domain three.

Of the variants analyzed in this study, Q133R shows the most marked shift in mutagen substrate selectivity, from NO_2-IQ toward NF (Fig. 4). Glutamine 133 is not a well-conserved residue; the corresponding residue is glutamate in all reported rodent NATs. The Q133R variant retains more than half of the wild-type NAT2 enzyme activity. Thus, at position 133, acidic, neutral, and basic side chains are all compatible with enzyme activity. In the core of the protein, such charge changes might be highly disruptive to tertiary structure, but position 133 is predicted to lie on the surface, directly behind the catalytic triad and away from the substrate cleft. The aspartate residue of the NAT2 catalytic triad is at position 122, and residues such as glycine 124, serine 125, and glycine 126 are predicted

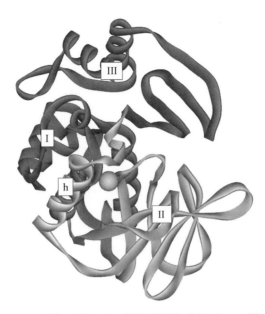

FIG. 10. The structure of *S. typhimurium* NAT (1E2T.pdb) is shown. The catalytic cysteine sulfur atom is represented as a ball. The three domains are indicated by Roman numerals, and the helix connecting domains I and II is represented by the letter "h". (See color insert.)

to interact with NAT2 substrates, as has been clearly observed for other NAT enzymes (Mushtaq *et al.*, 2002). An earlier site-directed mutagenesis investigation focused on the serine residues at NAT2 positions 125, 127, and 129 as important determinants of substrate selectivity (Goodfellow *et al.*, 2000). Perhaps substitutions at position 133 exert their effects by perturbing the positions of these preceding residues in the chain.

Residue tyrosine 190 is conserved among all known NAT sequences. Hanna and colleagues have suggested that this tyrosine interacts with the histidine residue of the catalytic triad (Wang *et al.*, 2004b). However, other studies have suggested that this residue forms part of an aromatic "greasy hat" system, helping hold the substrate in the correct orientation with respect to the catalytic triad (Brooke *et al.*, 2003a; Mushtaq *et al.*, 2002). These "greasy hat" residues are also present in the human proteins (F37/F93/Y190); F93 may have an anchoring role, via direct π-π interactions with the substrate. We identified two variants at Y190. The highly non-conservative variant Y190C had drastically reduced, although detectable, enzyme activity and its mutagen activation capacity was less than observed for the negative control; it was, in fact, one of the least active variants that has "slipped through" our screening, as we generally avoid picking variants with zero activity. However, variant Y190F had enzyme and mutagen activation activities only slightly reduced from that of the wild-type enzyme. This suggests that the role of tyrosine 190 is primarily to provide a hydrophobic site rather than specifically to provide a phenolic hydroxyl group, e.g., for H bonding. This variant is also the only one in our study that shows a marked shift on the correlation plot (Fig. 4) in the direction of NO_2-IQ.

Conclusions

Using a high-throughput screening system based on the *E. coli lacZ* mutagenicity assay, we have identified 14 new enzyme variants exhibiting low to intermediate activity, compared to wild-type recombinant human NAT2. All mutants correspond to single amino acid substitutions and the variants are distributed rather uniformly along the NAT2 sequence, including both positions near the active site and distant from it (both in the primary structure and, apparently, in the tertiary structure).

Many investigators have noted that amino acid residues distant from the active site of an enzyme (Raman *et al.*, 2004), including surface residues (el Hawrani *et al.*, 1994), can markedly influence the activity and substrate specificity of an enzyme. As has been noted, "Residues distant from an active site may be important in holding the catalytic residues in their required orientations. ... It has been difficult to demonstrate by protein engineering the importance of residues that are remote from the active site because the number of these residues is large and the contribution of each residue may be modest" (Oue *et al.*, 1999). Testing the roles of individual amino acid residues in protein function by site-directed mutagenesis experiments is laborious and expensive. High-throughput screening strategies (Cohen *et al.*, 2001) and *in vitro* evolution of enzymes by artificial selection (Jestin and Kaminski, 2004) provide powerful alternatives to one position at a time site-directed mutagenesis methods. Our screening strategy, based on measuring the response to chemical mutagens that are substrates for recombinant bioactivation enzymes, can identify residues whose properties exert a subtle influence on catalytic activity and can highlight the functional importance of residues that might be overlooked in planning site-directed mutagenesis studies.

Acknowledgment

This research was supported by the Natural Sciences and Engineering Research Council of Canada.

References

Bosold, F., and Boche, G. (1990). The ultimate carcinogen O-acetyl-N-(2-fluorenyl)hydroxylamine (N-acetoxy-2-aminofluorene) and its *in vitro* reaction to give 2-[N-(deoxyguanosin-8-yl)amino]fluorene. *Angew. Chem.* **102**, 99–100.

Bradford, M. M. (1976). A rapid and sensitive method for the quantitation of microgram quantities of protein utilizing the principle of protein-dye binding. *Anal. Biochem.* **72**, 248–254.

Brooke, E. W., Davies, S. G., Mulvaney, A. W., Okada, M., Pompeo, F., Sim, E., Vickers, R. J., and Westwood, I. M. (2003a). Synthesis and *in vitro* evaluation of novel small molecule inhibitors of bacterial arylamine N-acetyltransferases (NATs). *Bioorg. Med. Chem. Lett.* **13**, 2527–2530.

Brooke, E. W., Davies, S. G., Mulvaney, A. W., Pompeo, F., Sim, E., and Vickers, R. J. (2003b). An approach to identifying novel substrates of bacterial arylamine N-acetyltransferases. *Bioorg. Med. Chem.* **11**, 1227–1234.

Cohen, N., Abramov, S., Dror, Y., and Freeman, A. (2001). *In vitro* enzyme evolution: The screening challenge of isolating the one in a million. *Trends Biotechnol.* **19**, 507–510.

Cribb, A. E., Grant, D. M., Miller, M. A., and Spielberg, S. P. (1991). Expression of monomorphic arylamine N-acetyltransferase (NAT1) in human leukocytes. *J. Pharmacol. Exp. Ther.* **259**, 1241–1246.

Cupples, C. G., Cabrera, M., Cruz, C., and Miller, J. H. (1990). A set of *lacZ* mutations in *Escherichia coli* that allow rapid detection of specific frameshift mutations. *Genetics* **125**, 275–280.

Cupples, C. G., and Miller, J. H. (1989). A set of *lacZ* mutations in *Escherichia coli* that allow rapid detection of each of the six base substitutions. *Proc. Natl. Acad. Sci. USA* **86**, 5345–5349.

Dirr, A., and Wild, D. (1988). Synthesis and mutagenic activity of nitro-imidazoarenes: A study on the mechanism of the genotoxicity of heterocyclic arylamines and nitroarenes. *Mutagenesis* **3**, 147–152.

Dupret, J. M., and Grant, D. M. (1992). Site-directed mutagenesis of recombinant human arylamine N-acetyltransferase expressed in *Escherichia coli*: Evidence for direct involvement of Cys 68 in the catalytic mechanism of polymorphic human NAT2. *J. Biol. Chem.* **267**, 7381–7385.

el Hawrani, A. S., Moreton, K. M., Sessions, R. B., Clarke, A. R., and Holbrook, J. J. (1994). Engineering surface loops of proteins: A preferred strategy for obtaining new enzyme function. *Trends Biotechnol.* **12**, 207–211.

Fretland, A. J., Doll, M. A., Leff, M. A., and Hein, D. W. (2001a). Functional characterization of nucleotide polymorphisms in the coding region of N-acetyltransferase 1. *Pharmacogenetics* **11**, 511–520.

Fretland, A. J., Leff, M. A., Doll, M. A., and Hein, D. W. (2001b). Functional characterization of human N-acetyltransferase 2 (NAT2) single nucleotide polymorphisms. *Pharmacogenetics* **11**, 207–215.

Goodfellow, G. H., Dupret, J. M., and Grant, D. M. (2000). Identification of amino acids imparting acceptor substrate selectivity to human arylamine acetyltransferases NAT1 and NAT2. *Biochem. J.* **348**, Pt 1, 159–166.

Grant, D. M., Blum, M., Beer, M., and Meyer, U. A. (1991). Monomorphic and polymorphic human arylamine N-acetyltransferases: A comparison of liver isozymes and expressed products of two cloned genes. *Mol. Pharmacol.* **39**, 184–191.

Grant, D. M., Josephy, P. D., Lord, H. L., and Morrison, L. D. (1992). *Salmonella typhimurium* strains expressing human arylamine N-acetyltransferases: Metabolism and mutagenic activation of aromatic amines. *Cancer Res.* **52,** 3961–3964.

Guengerich, F. P. (2002). N-hydroxyarylamines. *Drug Metab Rev.* **34,** 607–623.

Hein, D. W. (2000). N-Acetyltransferase genetics and their role in predisposition to aromatic and heterocyclic amine-induced carcinogenesis. *Toxicol. Lett.* **112–113,** 349–356.

Hein, D. W. (2002). Molecular genetics and function of NAT1 and NAT2: Role in aromatic amine metabolism and carcinogenesis. *Mutat. Res. Fundam. Mol. Mech. Mutagen.* **506,** 65–77.

Hein, D. W., Doll, M. A., Fretland, A. J., Leff, M. A., Webb, S. J., Xiao, G. H., Devanaboyina, U. S., Nangju, N. A., and Feng, Y. (2000a). Molecular genetics and epidemiology of the NAT1 and NAT2 acetylation polymorphisms. *Cancer Epidemiol. Biomark. Prev.* **9,** 29–42.

Hein, D. W., Doll, M. A., Rustan, T. D., Gray, K., Feng, Y., Ferguson, R. J., and Grant, D. M. (1993). Metabolic activation and deactivation of arylamine carcinogens by recombinant human NAT1 and polymorphic NAT2 acetyltransferases. *Carcinogenesis* **14,** 1633–1638.

Hein, D. W., McQueen, C. A., Grant, D. M., Goodfellow, G. H., Kadlubar, F. F., and Weber, W. W. (2000b). Pharmacogenetics of the arylamine N-acetyltransferases: A symposium in honor of Wendell W. Weber. *Drug Metab. Dispos.* **28,** 1425–1432.

Hickman, D., Pope, J., Patil, S. D., Fakis, G., Smelt, V., Stanley, L. A., Payton, M., Unadkat, J. D., and Sim, E. (1998). Expression of arylamine N-acetyltransferase in human intestine. *Gut* **42,** 402–409.

Humphreys, W. G., Kadlubar, F. F., and Guengerich, F. P. (1992). Mechanism of C8 alkylation of guanine residues by activated arylamines: Evidence for initial adduct formation at the N7 position. *Proc. Natl. Acad. Sci. USA* **89,** 8278–8282.

Ilett, K. F., Ingram, D. M., Carpenter, D. S., Teitel, C. H., Lang, N. P., Kadlubar, F. F., and Minchin, R. F. (1994). Expression of monomorphic and polymorphic N-acetyltransferases in human colon. *Biochem. Pharmacol.* **47,** 914–917.

Jestin, J. L., and Kaminski, P. A. (2004). Directed enzyme evolution and selections for catalysis based on product formation. *J. Biotechnol.* **113,** 85–103.

Johansson, E., Parkinson, G. N., Denny, W. A., and Neidle, S. (2003). Studies on the nitroreductase prodrug-activating system: Crystal structures of complexes with the inhibitor dicoumarol and dinitrobenzamide prodrugs and of the enzyme active form. *J. Med. Chem.* **46,** 4009–4020.

Josephy, P. D. (1997). Aromatic amines and other N-aryl compounds. *In* "Molecular Toxicology," pp. 352–363. Oxford Univ. Press, New York.

Josephy, P. D. (2000). The *Escherichia coli lacZ* reversion mutagenicity assay. *Mutat. Res.* **455,** 71–80.

Josephy, P. D. (2002). Genetically-engineered bacteria expressing human enzymes and their use in the study of mutagens and mutagenesis. *Toxicology* **181–182,** 255–260.

Josephy, P. D., Summerscales, J., De Bruin, L. S., Schlaeger, C., and Ho, J. (2002). *N*-hydroxyarylamine *O*-acetyltransferase-deficient *Escherichia coli* strains are resistant to the mutagenicity of nitro compounds. *Biol. Chem.* **383,** 977–982.

Kawamura, A., Graham, J., Mushtaq, A., Tsiftsoglou, S. A., Vath, G. M., Hanna, P. E., Wagner, C. R., and Sim, E. (2005). Eukaryotic arylamine N-acetyltransferase: Investigation of substrate specificity by high-throughput screening. *Biochem. Pharmacol.* **69,** 347–359.

Kim, D., and Guengerich, F. P. (2004). Selection of human cytochrome P450 1A2 mutants with enhanced catalytic activity for heterocyclic amine N-hydroxylation. *Biochemistry* **43,** 981–988.

Kosakarn, P., Halliday, J. A., Glickman, B. W., and Josephy, P. D. (1993). Mutational specificity of 2-nitro-3,4-dimethylimidazo[4,5-f]quinoline in the lacI gene of *Escherichia coli. Carcinogenesis* **14**, 511–517.

Laskowski, R. A., Mac Arthur, M. W., Moss, D. S., and Thornton, J. M. (1993). Procheck: A program to check the stereochemical quality of protein structures. *J. Appl. Crystallogr.* **26**, 283–291.

Marwood, T. M., Meyer, D., and Josephy, P. D. (1995). *Escherichia coli lacZ* strains engineered for detection of frameshift mutations induced by aromatic amines and nitroaromatic compounds. *Carcinogenesis* **16**, 2037–2043.

McCoy, E. C., Anders, M., and Rosenkranz, H. S. (1983). The basis of the insensitivity of *Salmonella typhimurium* strain TA98/1,8-DNP$_6$ to the mutagenic action of nitroarenes. *Mutat. Res.* **121**, 17–23.

Miller, E. C., Juhl, U., and Miller, J. A. (1966). Nucleic acid guanine: Reaction with the carcinogen N-acetoxy-2-acetylaminofluorene. *Science* **153**, 1125–1127.

Morris, G. M., Goodsell, D. S., Halliday, R. S., Huey, R., Hart, W. E., Belew, R. K., and Olson, A. J. (1998). Automated docking using a Lamarckian genetic algorithm and an empirical binding free energy function. *J. Comp. Chem.* **19**, 1639–1662.

Morris, G. M., Goodsell, D. S., Huey, R., and Olson, A. J. (1996). Distributed automated docking of flexible ligands to proteins: Parallel applications of AutoDock 2.4. *J. Comput. Aided Mol. Des* **10**, 293–304.

Mushtaq, A., Payton, M., and Sim, E. (2002). The COOH terminus of arylamine N-acetyltransferase from *Salmonella typhimurium* controls enzymic activity. *J. Biol. Chem.* **277**, 12175–12181.

Newcomer, M. E. (2002). Protein folding and three-dimensional domain swapping: A strained relationship? *Curr. Opin. Struct. Biol.* **12**, 48–53.

Nokhbeh, M. R., Boroumandi, S., Pokorny, N., Koziarz, P., Paterson, E. S., and Lambert, I. B. (2002). Identification and characterization of SnrA, an inducible oxygen-insensitive nitroreductase in Salmonella enterica serovar Typhimurium TA1535. *Mutat. Res.* **508**, 59–70.

Oue, S., Okamoto, A., Yano, T., and Kagamiyama, H. (1999). Redesigning the substrate specificity of an enzyme by cumulative effects of the mutations of non-active site residues. *J. Biol. Chem.* **274**, 2344–2349.

Parikh, A., Josephy, P. D., and Guengerich, F. P. (1999). Selection and characterization of human cytochrome P450 1A2 mutants with altered catalytic properties. *Biochemistry* **38**, 5283–5289.

Payton, M., Mushtaq, A., Yu, T. W., Wu, L. J., Sinclair, J., and Sim, E. (2001). Eubacterial arylamine N-acetyltransferases: Identification and comparison of 18 members of the protein family with conserved active site cysteine, histidine and aspartate residues. *Microbiology* **147**, 1137–1147.

Pompeo, F., Brooke, E., Kawamura, A., Mushtaq, A., and Sim, E. (2002). The pharmacogenetics of NAT: Structural aspects. *Pharmacogenomics* **3**, 19–30.

Raman, J., Sumathy, K., Anand, R. P., and Balaram, H. (2004). A non-active site mutation in human hypoxanthine guanine phosphoribosyltransferase expands substrate specificity. *Arch. Biochem. Biophys.* **427**, 116–122.

Risch, A., Smelt, V., Lane, D., Stanley, L., van der, S. W., Ward, A., and Sim, E. (1996). Arylamine N-acetyltransferase in erythrocytes of cystic fibrosis patients. *Pharmacol. Toxicol.* **78**, 235–240.

Rodrigues-Lima, F., Delomenie, C., Goodfellow, G. H., Grant, D. M., and Dupret, J. M. (2001). Homology modelling and structural analysis of human arylamine N-acetyltransferase

NAT1: Evidence for the conservation of a cysteine protease catalytic domain and an active-site loop. *Biochem. J.* **356**, 327–334.

Rodrigues-Lima, F., and Dupret, J. M. (2002). 3D model of human arylamine N-acetyltransferase 2: Structural basis of the slow acetylator phenotype of the R64Q variant and analysis of the active-site loop. *Biochem. Biophys. Res. Commun.* **291**, 116–123.

Rosenkranz, H. S., and Mermelstein, R. (1983). Mutagenicity and genotoxicity of nitroarenes: All nitro-containing chemicals were not created equal. *Mutat. Res.* **114**, 217–267.

Saito, K., Shinohara, A., Kamataki, T., and Kato, R. (1986). N-hydroxyarylamine O-acetyltransferase in hamster liver: Identity with arylhydroxamic acid N,O-acetyltransferase and arylamine N-acetyltransferase. *J. Biochem. (Tokyo)* **99**, 1689–1697.

Sali, A., and Blundell, T. L. (1993). Comparative protein modelling by satisfaction of spatial restraints. *J. Mol. Biol.* **234**, 779–815.

Sinclair, J. C., Sandy, J., Delgoda, R., Sim, E., and Noble, M. E. (2000). Structure of arylamine N-acetyltransferase reveals a catalytic triad. *Nat. Struct. Biol.* **7**, 560–564.

Smelt, V. A., Mardon, H. J., and Sim, E. (1998). Placental expression of arylamine N-acetyltransferases: Evidence for linkage disequilibrium between NAT1*10 and NAT2*4 alleles of the two human arylamine N-acetyltransferase loci NAT1 and NAT2. *Pharmacol. Toxicol.* **83**, 149–157.

Smelt, V. A., Upton, A., Adjaye, J., Payton, M. A., Boukouvala, S., Johnson, N., Mardon, H. J., and Sim, E. (2000). Expression of arylamine N-acetyltransferases in pre-term placentas and in human pre-implantation embryos. *Hum. Mol. Genet.* **9**, 1101–1107.

Smith, B. J., De Bruin, L., Josephy, P. D., and Eling, T. E. (1992). Mutagenic activation of benzidine requires prior bacterial acetylation and subsequent conversion by prostaglandin H synthase to 4-nitro-4′-(acetylamino)biphenyl. *Chem. Res. Toxicol.* **5**, 431–439.

Summerscales, J. E., and Josephy, P. D. (2004). Human acetyl CoA:arylamine N-acetyltransferase variants generated by random mutagenesis. *Mol. Pharmacol.* **65**, 220–226.

Suzuki, A., Kushida, H., Iwata, H., Watanabe, M., Nohmi, T., Fujita, K., Gonzalez, F. J., and Kamataki, T. (1998). Establishment of a *Salmonella* tester strain highly sensitive to mutagenic heterocyclic amines. *Cancer Res.* **58**, 1833–1838.

Thompson, J. D., Higgins, D. G., and Gibson, T. J. (1994). CLUSTAL W: Improving the sensitivity of progressive multiple sequence alignment through sequence weighting, position-specific gap penalties and weight matrix choice. *Nucleic Acids Res.* **22**, 4673–4680.

Turk, B., Turk, V., and Turk, D. (1997). Structural and functional aspects of papain-like cysteine proteinases and their protein inhibitors. *Biol. Chem.* **378**, 141–150.

Wallace, A. C., Laskowski, R. A., and Thornton, J. M. (1995). LIGPLOT: A program to generate schematic diagrams of protein-ligand interactions. *Protein Eng.* **8**, 127–134.

Wang, H., Vath, G. M., Gleason, K. J., Hanna, P. E., and Wagner, C. R. (2004a). Probing the mechanism of hamster arylamine N-acetyltransferase 2 acetylation by active site modification, site-directed mutagenesis, and pre-steady state and steady state kinetic studies. *Biochemistry* **43**, 8234–8246.

Wang, H., Vath, G. M., Gleason, K. J., Liu, L., Hanna, P. E., and Wagner, C. R. (2004b). Kinetic studies of hamster arylamine N-acetyltransferase 2: A unique catalytic mechanism facilitates N-acetylation of arylamine substrates in an aqueous environment. 3rd Intl. Workshop on Arylamine N-Acetyltransferases, Vancouver, BC, Canada, Aug. 2004.

Watanabe, M., Nohmi, T., and Ishidate, M. J. (1987). New tester strains of *Salmonella typhimurium* highly sensitive to mutagenic nitroarenes. *Biochem. Biophys. Res. Commun.* **147**, 974–979.

Wild, D., Feser, W., Michel, S., Lord, H. L., and Josephy, P. D. (1995). Metabolic activation of heterocyclic aromatic amines catalyzed by human arylamine N-acetyltransferase isozymes (NAT1 and NAT2) expressed in *Salmonella typhimurium. Carcinogenesis* **16**, 643–648.

Xenarios, I., Salwinski, L., Duan, X. J., Higney, P., Kim, S. M., and Eisenberg, D. (2002). DIP, the Database of Interacting Proteins: A research tool for studying cellular networks of protein interactions. *Nucleic Acids Res.* **30**, 303–305.

Zhou, H., Josephy, P. D., Kim, D., and Guengerich, F. P. (2004). Functional characterization of four allelic variants of human cytochrome P450 1A2. *Arch. Biochem. Biophys.* **422**, 23–30.

[12] Inactivation of Human Arylamine N-Acetyltransferase 1 by Hydrogen Peroxide and Peroxynitrite

By Jean-Marie Dupret, Julien Dairou, Noureddine Atmane, and Fernando Rodrigues-Lima

Abstract

Arylamine N-acetyltransferases (NAT) are xenobiotic-metabolizing enzymes responsible for the acetylation of many arylamine and heterocyclic amines. They therefore play an important role in the detoxification and activation of numerous drugs and carcinogens. Two closely related isoforms (NAT1 and NAT2) have been described in humans. NAT2 is present mainly in the liver and intestine, whereas NAT1 is found in a wide range of tissues. Interindividual variations in NAT genes have been shown to be a potential source of pharmacological and/or pathological susceptibility. Evidence now shows that redox conditions may also contribute to overall NAT activity. This chapter summarizes current knowledge on human NAT1 regulation by reactive oxygen and nitrogen species.

Introduction

Arylamine N-acetyltransferases (NATs) are xenobiotic-metabolizing enzymes (XME) that catalyze the transfer of an acetyl moiety from acetyl-CoA to the nitrogen atom of primary arylamines and hydrazines. They are also responsible for the O-acetylation of N-hydroxyarylamines (Pompeo *et al.*, 2002). NATs therefore participate in the detoxification and/or activation of a variety of drugs and carcinogens (Badawi *et al.*, 1995; Hein, 2002).

NATs have been identified in several species, ranging from bacteria to mammals (Deloménie *et al.*, 2001; Payton *et al.*, 2001; Rodrigues-Lima and

METHODS IN ENZYMOLOGY, VOL. 400
0076-6879/05 $35.00
DOI: 10.1016/S0076-6879(05)00012-1

Dupret, 2002b). Two functional isoforms of NAT (NAT1 and NAT2) have been described in humans (Matas *et al.*, 1997). NAT1 is found in a wide range of tissues and organs (Rodrigues-Lima *et al.*, 2003). NAT2 is present mainly in the liver and intestine (Hickman *et al.*, 1998). However, it is likely that NAT2 expression is more widespread (Dairou *et al.*, 2005; Dupret and Rodrigues-Lima, 2005).

Genetically determined interindividual variation in NAT2 content and activity is the basis of the well-known isoniazid acetylation polymorphism. Polymorphism associated with the NAT1 isoform may also be a source of pharmacological susceptibility (Hein *et al.*, 2000). To date, at least 26 *NAT1* alleles and 36 *NAT2* alleles have been identified, resulting from numerous single-nucleotide polymorphisms (Dupret *et al.*, 2004). A detailed description of *NAT* alleles is available at http://www.louisville.edu/medschool/pharmacology/NAT.html. Although the relationships between polymorphic substitutions and the activity of variant enzymes have been investigated thoroughly (Fretland *et al.*, 2001a,b), further studies are required to determine the effects of nucleotide variations outside the NAT coding regions and to characterize the enzyme activity of allelic variants more fully.

Studies have shown that other mechanisms, such as substrate-dependent downregulation (Butcher *et al.*, 2000, 2004), the existence of splice variants (Husain *et al.*, 2004), and posttranslational inhibition of NAT activity (Atmane *et al.*, 2003; Dairou *et al.*, 2005), may also contribute to the overall activity of NATs. This chapter summarizes current knowledge on NAT inhibition by reactive oxygen or nitrogen species, consistent with the fundamental role in NAT-mediated catalysis of a conserved active-site residue.

NAT-Mediated Catalysis

Results of early steady-state kinetic studies were consistent with a simple two-step substituted enzyme ("ping-pong") kinetic mechanism for the NAT reaction (Riddle and Jencks, 1971). Subsequent site-directed mutagenesis and functional studies have shown that an acetylcysteinyl enzyme is involved in the catalytic process (Dupret and Grant, 1992). Finally, the crystallographic determination of the structure of NATs from prokaryotic species (*Salmonella typhimurium, Mycobacterium smegmatis, Pseudomonas aeruginosa, Mesorhizobium loti*) (Holton *et al.*, 2005; Sandy *et al.*, 2002; Sinclair *et al.*, 2000; Westwood *et al.*, 2005) and the construction of homology models of human NAT1 and NAT2 (Rodrigues-Lima and Dupret, 2002a; Rodrigues-Lima *et al.*, 2001) have revealed structural similarities with cysteine proteases. These studies revealed the existence of a protease-like catalytic triad (residues Cys 68-His 107-Asp 122 in humans) (Fig. 1). The strict conservation of these structural features suggests that

FIG. 1. Ribbon-and-stick representation of the human NAT1 catalytic triad. This figure was created with Deep-View 3.7b2 (Guex and Peitsch, 1997) using the homology model of the N-terminal domain of human NAT1 (Rodrigues-Lima *et al.*, 2001). The hydrogen bonds among cysteine, histidine, and aspartate residues are shown as dotted lines.

NATs have adapted a catalytic mechanism commonly found in cysteine proteases for use in the acetyl-transfer reaction.

Although the exact catalytic mechanism of NAT enzymes is not fully understood, it has been shown to depend on the formation of a thiolate-imidazolium ion pair between the triad cysteine and histidine (Wang *et al.*, 2004). The use of chemical modification procedures to probe the reactivity of the active-site cysteine revealed that it undergoes alkylation upon NAT1 labeling with *N*-arylbromoacetamido reagents (Guo *et al.*, 2003). This type of covalent modification can also occur upon NAT1-catalyzed bioactivation of *N*-hydroxy-2-acetylaminofluorene and subsequent sulfinamide adduct formation (Guo *et al.*, 2004). In addition to its importance as the primary site of covalent modification in NAT1, Cys 68 has been shown to be the molecular target of reactive oxygen or nitrogen species (Atmane *et al.*, 2003; Dairou *et al.*, 2003, 2004).

Oxidative Inhibition of NATs

Several enzymes with a reactive catalytic cysteine residue have been shown to be regulated by the reactive oxygen or nitrogen species (ROS and RNS, respectively) generated during oxidative stress (Halliwell and Gutteridge, 1999). Hydrogen peroxide (H_2O_2), one of the major cellular oxidants, inactivates phosphatases both *in vitro* and *in vivo* (Caselli *et al.*, 1998; Lee *et al.*, 1998, 2002). Generally, *in vivo* and *in vitro*, H_2O_2 can oxidize the cysteine residue to give cysteine sulfenic acid (cys-SOH) or disulfide, which can be reduced back to cysteine by cellular reductants (Halliwell and Gutteridge, 1999). *S*-nitrosothiols (RSNO) are reactive nitrogen species known to inactivate cysteine-containing enzymes, such as papain (Xian *et al.*, 2000a), phosphatases (Xian *et al.*, 2000b), and transglutaminases (Bernassola *et al.*, 1999), either through *S*-nitrosylation (formation of an S-NO) or through formation of a mixed disulfide bond. Both modifications can be reversed by reducing conditions. Peroxynitrite ($ONOO^-$) also affects protein function by modifying essential reactive thiols or tyrosine residues (Groves, 1999; Radi *et al.*, 1991). It irreversibly inactivates several enzymes, including creatine kinase (Konorev *et al.*, 1998), caspases (Mohr *et al.*, 1997), and phosphatases (Takakura *et al.*, 1999). XMEs such as cytochrome P450 (Lin *et al.*, 2003) and glutathione *S*-transferases (Wong *et al.*, 2001) are also irreversibly inactivated by peroxynitrite.

We assessed the ability of reactive oxygen and nitrogen species to inactivate human NAT1. We also investigated the possible reactivation of oxidized NAT1 by reducing agents. Our observations led us to propose that H_2O_2 and RSNO reversibly inhibit human NAT1, whereas peroxynitrite irreversibly inhibits the enzyme (Rodrigues-Lima and Dupret, 2004) (Fig. 2).

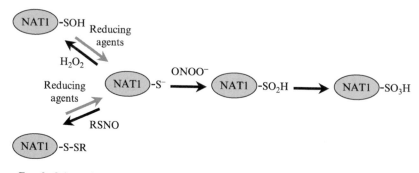

Fig. 2. Schematic representation of the regulation of human NAT1 by oxidation. The catalytic cysteine residue of the reduced active form of NAT1 is present predominantly as a thiolate anion. This enhances its nucleophilic properties but renders NAT1 susceptible to oxidation. Low levels of H_2O_2 lead to the reversible oxidation of the catalytic cysteine residue to a sulfenic acid (SOH) with concomitant reversible inactivation of NAT1 activity. When NAT1 is oxidized by *S*-nitrosothiols (RSNO), a mixed disulfide (NAT1-S-SR) is formed at the catalytic cysteine residue with reversible inactivation of NAT1. Oxidation by low concentrations of peroxynitrite leads to the irreversible inactivation of NAT1 via oxidation of the catalytic cysteine to a sulfinic (SO_2H) or sulfonic acid (SO_3H).

Here, we describe *in vitro* and *ex vivo* approaches that allow the study of the effect of ROS or RNS on the activity of human NAT1. These reactive species are produced in response to a wide variety of stimuli, such as exposure to chemicals, hormones, and ultraviolet, and following exposure to deleterious physiological/physiopathological conditions, such as apoptosis and inflammation (Groves, 1999; Halliwell and Gutteridge, 1999; Liu and Stamler, 1999; Tannickal and Fanburg, 2000). Excessive production of ROS and RNS is known to participate in the pathological processes of several diseases such as cancer and cataract (Halliwell and Gutteridge, 1999). In addition, these oxidants may affect XME-dependent biotransformation (Boelsterli, 2003; Pagano, 2002). Thus, the protocols described herein may be used to study the sensitivity of NATs toward oxidative stress induced by various types of stimulus.

General Methods

Expression and Purification of Recombinant Human NAT1

To study the effect of oxidants on recombinant human NAT1, the enzyme is expressed (PET28A vector from Novagen) as a polyhistidine-tagged fusion protein (His-NAT1) in *Escherichia coli* BL21(DE3) grown for 4 h at 37° in the presence of 0.5 mM isopropyl-1-thio-β-D-galactopyranoside. To purify recombinant NAT1, lysates are loaded onto nickel-agarose affinity chromatography columns. Purified His-NAT1 is reduced by treating with 10 mM dithiothreitol (DTT) (final concentration) for 10 min at 4° prior to dialysis against 25 mM Tris-HCl, pH 7.5, 1 mM EDTA. SDS–PAGE analysis is carried out at each stage of purification, and protein concentrations are determined using a standard Bradford assay.

Cells Expressing NAT1

The effects of oxidants on endogenous cellular NAT1 have been studied using MCF7 (human mammary carcinoma) cells, which are known to express NAT1 (Adam *et al.*, 2003). Other cells have also been used to study the effects of oxidants on endogenous NAT1 (Dairou *et al.*, 2005). Given that the NAT1 gene appears to be expressed ubiquitously, it is likely that most cell types could be used.

NAT1 Assays

NAT1 activity can be detected easily using spectrophotometric assays. HPLC assays may also be used (Dupret and Grant, 1992). Only spectrophotometric assays are described here. These assays can be done easily in the wells of microtiter plates.

First Method. NAT1 enzyme activity is determined spectrophotometrically (410 nm) using p-nitrophenylacetate (PNPA) as the acetyl donor and a NAT1-specific arylamine substrate such as p-aminosalicylic acid (PAS) (Musthaq *et al.*, 2002). Briefly, oxidized and unoxidized forms of NAT1 (10–50 μl) are assayed in a reaction mixture containing 500 μM PAS (final concentration) in 25 mM Tris-HCl, pH 7.5, 1 mM EDTA. Reactions are started by adding 125 μM PNPA (final concentration). Total reaction volume is 1 ml and final enzyme concentration is usually 10–20 nM. After 10 min at 37°, reactions are quenched by SDS (1% final concentration). p-Nitrophenol, generated through hydrolysis of PNPA by NAT1 in presence of PAS, is quantified by measuring absorbance at 410 nm with an ELISA plate analyzer. One unit of p-nitrophenol is defined as the amount of enzyme that gives an absorbance at 410 nm of 0.5 per 10 min/ml. Controls are included without enzyme, PNPA, or PAS. All enzymatic reactions are performed in quadruplicate under conditions where the initial rates are linear.

Second Method. The rate of hydrolysis of acetyl-CoA (first substrate) by NAT1 in the presence of the NAT1-specific arylamine substrate (PAS) can be determined by detecting the free CoA thiol with 5,5′-dithio-bis 2-nitrobenzoic acid (DTNB or Ellman's reagent) as described previously by Brooke *et al.* (2003). PAS (500 μM final) and samples containing purified NAT1 ($\approx$0.25 μg enzyme) in 25 mM Tris-HCl, pH 7.5, 1 mM EDTA buffer are mixed and preincubated (37°, 5 min) in a 96-well ELISA plate. Acetyl-CoA is added to a final concentration of 400 μM to start the reaction in a total volume of 100 μl. The reaction is quenched with 25 μl of 5 mM DTNB in guanidine hydrochloride solution (6.4 M guanidine-HCl, 0.1 M Tris-HCl, pH 7.3). The absorbance at 405 nm is measured using an ELISA plate analyzer. The reaction rate is such that the hydrolysis of acetyl-CoA is within the linear range. Controls are carried out in absence of enzyme, acetyl-CoA, or PAS. The amount of CoA produced in the reaction is determined by comparison with a standard curve obtained with DTNB.

Third Method. N-Acetylation of arylamines (PAS) can also be measured by the colorimetric detection of remaining arylamine with 4-dimethylaminobenzaldehyde (DMAB) in 96-well plates as described by Coroneos *et al.* (1991). This method can be used to measure the activity of recombinant NAT1 and to measure endogenous NAT1 activity in cell extracts. Briefly, NAT1 activity in cell extracts is detected as described previously (Sinclair and Sim, 1997) in a total volume of 100 μl. Cell extracts (50 μl, obtained from cells exposed or not to oxidants) and PAS (200 μM final concentration) in assay buffer (20 mM Tris-HCl, 1 mM EDTA, pH 7.5) are preincubated at 37° for 5 min. AcCoA (400 μM final) is added to start the reaction and the samples are incubated at 37° for different times (up to 30 min). The reaction is quenched with 100 μl of ice-cold aqueous TCA (20%, w/v) and proteins are pelleted by centrifugation for 5 min at 12,000g. DMAB [800 μl, 5% (w/v) in 9:1 acetonitrile:water] is then added and absorbance is measured in 10-mm path length cuvettes at 450 nm. The amount of remaining arylamine (not acetylated) is determined from a standard curve. All assays are performed in triplicate, in conditions such that the initial rates are linear. Enzyme activities are normalized according to the protein concentration of cell extracts. Controls are carried out without extract, PAS, or AcCoA.

Effect of H_2O_2 on Recombinant NAT1

As stated earlier, H_2O_2 is one of the major cellular oxidants. It regulates several cell functions by oxidizing active cysteine residues in proteins (Lee *et al.*, 2002). The physiological/pathophysiological concentration of H_2O_2 can reach the micromolar range (Halliwell and Gutteridge, 1999), although concentrations above 500 μM have been detected (Spector and Garner, 1981). In most studies on the effect of H_2O_2 on enzyme activities, H_2O_2 is added as a bolus to the enzyme (either purified or in cell extracts). After an incubation period, residual activity is measured (generally excess H_2O_2 is removed by treatment with catalase) (Borutaite and Brown, 2001; Caselli *et al.*, 1998; Lee *et al.*, 1998). Using this approach, we showed that recombinant NAT1 is inactivated by H_2O_2 in a dose-dependent manner (Atmane *et al.*, 2003). Inactivated enzyme can be reactivated by adding an excess of reducing agents such as DTT or glutathione (GSH) (Atmane *et al.*, 2003). Although simple, the bolus addition of H_2O_2 to purified enzymes or cell extracts is far from being physiological. Therefore to assess the effect of H_2O_2 on recombinant NAT1 activity in more realistic and physiological conditions, we used an enzyme system that continuously generates physiological levels of H_2O_2. The glucose/glucose oxidase or xanthine/xanthine oxidase systems can be used to this end (Barbouti *et al.*, 2002; Lee *et al.*, 2002; Mueller *et al.*, 1997; Ravid *et al.*, 2002). We used the glucose/glucose oxidase system (Fig. 3A). We assessed the effect of continuous generation of H_2O_2 by incubating purified NAT1 (1.5 μM) with glucose oxidase (0.15 units/ml, from *Aspergillus niger*, Sigma) and glucose (5 mM) in 25 mM Tris-HCl, pH 7.5, 1 mM EDTA (total volume of 15 μl) at 37°. At various time intervals, catalase (300 units/ml, bovine from Sigma) is added and residual NAT1 activity is measured. In controls, catalase (300 unit/ml) is added directly to the glucose/glucose oxidase system. We showed that H_2O_2 is produced at a constant rate of $\approx$6 μM H_2O_2/min in these experiments,

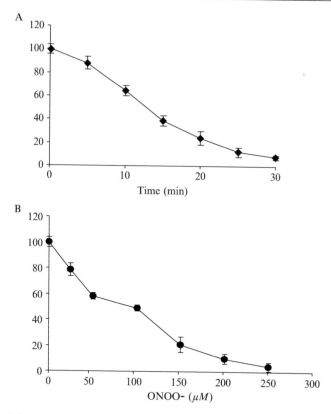

Fig. 3. (A) Inactivation of human recombinant NAT1 by continuous generation of H_2O_2 via the glucose/glucose oxidase enzyme system. NAT1 was incubated with glucose (5 mM) and glucose oxidase (0.15 unit/ml) in 25 mM Tris-HCl, 1 mM EDTA, pH 7.5, at 37° for the indicated times. After the addition of catalase (300 units/ml), residual NAT1 activity was measured. NAT1 was inactivated in a time-dependent manner. Treatment of NAT1 with the glucose/glucose oxidase system in the presence of catalase did not result in inactivation of the enzyme, showing that NAT1 is indeed inactivated by the generated H_2O_2. (B) Inactivation of endogenous NAT1 in MCF7 cells upon exposure to peroxynitrite (PN). MCF7 cells in culture dishes were exposed to the specified final concentrations of PN in 10 ml of PBS for 10 min at 37°. After washing with PBS, cells were scraped into lysis buffer, a total extract was made, and NAT1 activity was measured. Endogenous NAT1 was inactivated by PN in a dose-dependent manner, showing that NAT1 is a target of PN in cells.

which corresponds to physiological levels (Ravid *et al.*, 2002). The amount of H_2O_2 generated by 0.15 units/ml of glucose oxidase is determined by measuring the absorbance at 240 nm and/or with the peroxidase/orthophenylene-diamine assay as described previously (Barbouti *et al.*, 2002; Panayiotidis *et al.*, 1999).

NAT1 is inactivated by the continuous generation of H_2O_2 in a time-dependent manner (Fig. 3A). When exposed to the glucose/glucose oxidase system in the presence of catalase, NAT1 is not inactivated (not shown), demonstrating that the oxidative modification of NAT1 by H_2O_2 leads to its inactivation. H_2O_2-dependent inactivation of NAT1 is reversible as the enzyme is fully active when its oxidized form is incubated with reducing agents (Atmane et al., 2003). The continuous generation of H_2O_2 by the glucose/glucose oxidase system can also be used to expose cells to H_2O_2 and to study the subsequent effect of this oxidant on enzymes. We performed such an experiment with cultured lens epithelial cells in petri dishes and showed that endogenous NAT1 and NAT2 (measured in cell extracts) are impaired by continuous exposure to H_2O_2 (Dairou et al., 2005). A similar protocol using cultured cells and peroxynitrite has been used and is described in more detail here.

Effect of Peroxynitrite (PN) on Cellular NAT1

Peroxynitrite is a one of the major biologically relevant oxidants. PN is a highly reactive nitrogen species that exerts many of its biological effects through its capacities to alter protein structure and to function via cysteine oxidation and tyrosine nitration (Groves, 1999). PN is formed during the diffusion rate-limited bimolecular reaction between nitric oxide and super-oxide ion. PN formation and reactions are thought to contribute to the pathogenesis of several physiopathological processes, such as chronic inflammation, sepsis, ischemia-reperfusion, and cancer (Radi et al., 2001). Physiological concentrations of PN in vivo have been estimated to be around 50 μM; however, concentrations of around 500 μM have been detected within phagolysosomes of activated macrophages (Denicola et al., 1993). PN has been demonstrated to inhibit enzymatic processes in vitro through either nitration or oxidation of critical amino acids or cofactors (Crow et al., 1995; Mihm and Bauer, 2002; Stachowiak et al., 1998).

We assessed the sensitivity of human NAT1 to PN-dependent inactivation by exposing MCF7 cells (a human breast carcinoma cell line known to express NAT1) to physiological concentrations of PN (Fig. 3B) and assaying residual activity (Dairou et al., 2004). MCF7 cells are cultured as monolayers in 100-mm petri dishes at 37° in Dulbecco's modified Eagles medium supplemented with 20% (v/v) fetal bovine serum and penicillin/streptomycin. At ≈90% confluence, cell monolayers are washed with phosphate-buffered saline (PBS) (Ca^{2+}/Mg^{2+}). Cell monolayers are exposed to different concentrations of PN (bolus addition of peroxynitrite obtained from Calbiochem-Novabiochem) in 10 ml of PBS and kept for 10 min at 37°. Controls consist of decomposed PN (obtained by allowing decomposition at room temperature in the dark for 48 h) or PBS only. After treatment, monolayers are washed with PBS, scraped in 1 ml of lysis buffer (20 mM Tris-HCl, 1 mM EDTA, pH 7.5, 0.2 % Triton X-100, protease inhibitors), and centrifuged for 45 min at 100,000 g. The Bradford method is used to determine the protein concentration of supernatants (total cell extract). Cell extracts are all adjusted to the same protein concentration by adding 20 mM Tris-HCl, 1 mM EDTA, pH 7.5, and then subjected to the enzymatic assays described earlier. Exposure of cells to PN for 10 min leads to the concentration-dependent inactivation of endogenous NAT1 (Fig. 3B). Given that PN may be highly toxic for cells, cell viability must be checked using standard protocols (e.g., Trypan blue) to adjust PN concentrations and time of exposure. To mimic physiological PN generation, we have also used 3-morpholinosydnonimine N-ethylcarbamide (SIN1), a PN donor. This compound releases both superoxide and nitric oxide at a constant rate, leading to the formation of PN (Groves, 1999; Takakura et al., 1999). SIN1, which is more stable than PN itself, attacks many biological targets in the same manner as PN and has thus been used widely as a source of PN in studies using cultured cells (Singh et al., 1999; Takakura et al., 1999). It is important to remember that the amount of PN generated by SIN1 is lower than the amount of the parent PN donor (Hosker et al., 2001; Percival et al., 1999). Thus, higher concentrations of SIN1 are generally used. Treatment of MCF7 cells with SIN1 gave similar results to those obtained with PN. Decomposed SIN1 may be used to ensure that the observed effect is due to the generation of PN by SIN1. More details about the chemistry of SIN1 can be found in Singh et al. (1999).

The inactivation of endogenous NAT1 in cells exposed to PN or SIN1 showed that the reducing intracellular environment of MCF7 cells does not protect endogenous NAT1 sufficiently from PN-dependent inactivation (Dairou et al., 2004), suggesting that NAT1 is likely to be a target of PN in vivo. Similar results have been obtained with human lens epithelial cells, which are known to have high levels of protective antioxidant systems (Dairou et al., 2005). These results are also supported by data obtained in vitro using recombinant NAT1. Indeed, using the kinetics approach published by Radi et al. (1991), we have shown that NAT1 is inactivated rapidly by PN with a second-order rate constant (k_{inact}) of 5×10^4 $M^{-1}s^{-1}$. In addition, unlike the H_2O_2-dependent inactivation of NAT1, the PN-dependent inactivation of NAT1 is not reversed by reducing agents (Dairou et al., 2004). As stated earlier, PN can oxidize cysteine and nitrate tyrosine residues. Chemical modification of recombinant NAT1 by acetylimidazole (a compound that modifies tyrosine side chains) did not inactivate NAT1. Substrate protection assays with AcCoA and CoA were used to demonstrate that the oxidative modification of the catalytic cysteine residue of NAT1 by H_2O_2 or PN was responsible for the inactivation of the enzyme. The active site cysteine residue of NAT is known to form a covalent acetylcysteinyl enzyme with the acetyl group of AcCoA, the physiological acetyl donor substrate of NAT, but not with CoA (the product resulting from hydrolysis of AcCoA) (Pompeo et al., 2002; Riddle and Jencks, 1971). Thus, recombinant NAT1 was preincubated with increasing concentrations of AcCoA and CoA (from 100 μM to 5 mM final) and further incubated with a given concentration of PN (or SIN1) that inhibits 90% of NAT1 activity. AcCoA protected NAT1 activity from inactivation by PN in a dose-dependent manner. Indeed, 73% of the enzyme activity was protected with 5 mM AcCoA, compared to 9% with 5 mM CoA. Although we cannot rule out the possibility that other amino acids of NAT1 are modified by PN, it appears that it is the specific oxidation of the catalytic cysteine residue that leads to enzyme inactivation (Dairou et al., 2004).

Overall, these studies show that human NAT1 and, more broadly, NAT enzymes may be the target of biological oxidants. Oxidative-dependent inactivation of these enzymes could affect the metabolic pathway of numerous xenobiotics. Further studies are needed to assess the real impact of such a regulation in terms of pharmacological susceptibility.

Acknowledgments

The authors' laboratory work was supported by grants from ARC (Association pour la Recherche sur le Cancer), AFM (Association Française contre les Myopathies) and Rétina-France. J.D. holds a postdoctoral fellowship from AFM. N.A. holds a Ph.D. fellowship from le Ministère de la Jeunesse, de l'Education Nationale et de la Recherche.

References

Adam, P. J., Berry, J., Loader, J. A., Tyson, K. L., Craggs, G., Smith, P., De Belin, J., Steers, G., Pezzella, F., Sachsenmeir, K. F., Stamps, A. C., Herath, A., Sim, E., O'Hare, M. J., Harris, A. L., and Terrett, J. A. (2003). Arylamine N-acetyltransferase-1 is highly expressed in breast cancers and conveys enhanced growth and resistance to etoposide in vitro. Mol. Cancer Res. 1, 826–835.

Atmane, N., Dairou, J., Paul, A., Dupret, J. M., and Rodrigues-Lima, F. (2003). Redox regulation of the human xenobiotic metabolizing enzyme arylamine N-acetyltransferase 1 (NAT1): Reversible inactivation by hydrogen peroxide. J. Biol. Chem. 278, 35086–35092.

Badawi, A. F., Hirvonen, A., Bell, D. A., Lang, N. P., and Kadlubar, F. F. (1995). Role of aromatic amine acetyltransferases, NAT1 and NAT2, in carcinogen-DNA adduct formation in the human urinary bladder. Cancer Res. 55, 5230–5237.

Barbouti, A., Paschalis-Thomas, D., Lambros, N., Tenopoulou, M., and Galaris, D. (2002). DNA damage and apoptosis in hydrogen peroxide-exposed jurkat cells: Bolus addition versus continuous generation of H_2O_2. *Free Radic. Biol. Med.* **33**, 691–702.

Bernassola, F., Rossi, A., and Melino, G. (1999). Regulation of transglutaminase by nitric oxide. *Ann. N.Y. Acad. Sci.* **887**, 83–91.

Boelsterli, U. (2003). "Mechanistic Toxicology: The Molecular Basis of How Chemicals Disrupt Biological Targets." Taylor & Francis, London.

Borutaite, V., and Brown, G. (2001). Caspases are reversibly inactivated by hydrogen peroxide. *FEBS Lett.* **500**, 114–118.

Brooke, E. W., Davies, S. G., Mulvaney, A. W., Pompeo, F., Sim, E., and Vickers, R. J. (2003). An approach to identifying novel substrates of bacterial arylamine N-acetyltransferases. *Bioorg. Med. Chem.* **11**, 1227–1234.

Butcher, N. J., Arulpragasam, A., and Minchin, R. F. (2004). Proteasomal degradation of N-acetyltransferase 1 is prevented by acetylation of the active site cysteine: A mechanism for the slow acetylator phenotype and substrate-dependent down-regulation. *J. Biol. Chem.* **279**, 22131–22137.

Butcher, N. J., Ilett, K. F., and Minchin, R. F. (2000). Substrate-dependent regulation of human arylamine N-acetyltransferase-1 in cultured cells. *Mol. Pharmacol.* **57**, 468–473.

Caselli, A., Marzocchini, R., Camici, G., Manao, G., Moneti, G., Pieraccini, G., and Ramponi, G. (1998). The inactivation mechanism of low molecular weight phosphotyrosine-protein phosphatase by H_2O_2. *J. Biol. Chem.* **273**, 32554–32560.

Coroneos, E., Gordon, J. W., Kelly, S. L., Wang, P. D., and Sim, E. (1991). Drug metabolising N-acetyltransferase activity in human cell lines. *Biochim. Biophys. Acta* **1073**, 593–599.

Crow, J., McCord, J., and Beckman, J. (1995). Sensitivity of the essential zinc-thiolate moiety of yeast alcohol dehydrogenase to hypochlorite and peroxynitrite. *Biochemistry* **34**, 3544–3552.

Dairou, J., Atmane, N., Dupret, J. M., and Rodrigues-Lima, F. (2003). Reversible inhibition of the human xenobiotic-metabolizing enzyme arylamine N-acetyltransferase 1 by S-nitrosothiols. *Biochem. Biophys. Res. Commun.* **307**, 1059–1065.

Dairou, J., Atmane, N., Rodrigues-Lima, F., and Dupret, J. M. (2004). Peroxynitrite irreversibly inactivates the human xenobiotic-metabolizing enzyme arylamine N-acetyl-transferase 1 (NAT1) in human breast cancer cells: A cellular and mechanistic study. *J. Biol. Chem.* **279**, 7708–7714.

Dairou, J., Malecaze, F., Dupret, J. M., and Rodrigues-Lima, F. (2005). The xenobiotic-metabolizing enzymes arylamine N-acetyltransferases (NAT) in human lens epi-thelial cells: Inactivation by cellular oxidants and UVB-induced oxidative stress. *Mol. Pharmacol.* **67**, 1299–1306.

Deloménie, C., Fouix, S., Longuemaux, S., Brahimi, N., Bizet, C., Picard, B., Denamur, E., and Dupret, J. M. (2001). Identification and functional characterization of arylamine n-acetyltransferases in eubacteria: Evidence for highly selective acetylation of 5-aminosa-licylic acid. *J. Bacteriol.* **183**, 3417–3427.

Denicola, A., Rubbo, H., Rodriguez, D., and Radi, R. (1993). Peroxynitrite-mediated cytotoxicity to *Trypanosoma cruzi*. *Arch. Biochem. Biophys.* **304**, 279–286.

Dupret, J. M., Dairou, J., Atmane, N., and Rodrigues-Lima, F. (2004). Pharmacogenetics, regulation and structural properties of the drug-metabolizing enzymes arylamine N-acetyltransferases. *Curr. Pharmacogenomics* **2**, 333–338.

Dupret, J. M., and Grant, D. M. (1992). Site-directed mutagenesis of recombinant human arylamine N-acetyltransferase expressed in *Escherichia coli*: Evidence for direct involvement of Cys68 in the catalytic mechanism of polymorphic human NAT2. *J. Biol. Chem.* **267**, 7381–7385.

Dupret, J. M., and Rodrigues-Lima, F. (2005). Structure and regulation of the drug-metabolizing enzymes arylamine N-acetyltransferases. *Curr. Med. Chem.* **12**, 763–771.

Fretland, A. J., Doll, M. A., Leff, M. A., and Hein, D. W. (2001a). Functional characterization of nucleotide polymorphisms in the coding region of N-acetyltransferase 1. *Pharmacogenetics* **11**, 511–520.

Fretland, A. J., Leff, M. A., Doll, M. A., and Hein, D. W. (2001b). Functional characterization of human N-acetyltransferase 2 (NAT2) single nucleotide polymorphisms. *Pharmacogenetics* **11**, 207–215.

Groves, J. (1999). Peroxynitrite: Reactive, invasive and enigmatic. *Curr. Opin. Chem. Biol.* **3**, 226–235.

Guex, N., and Peitsch, M. C. (1997). SWISS-MODEL and the Swiss-PdbViewer: An environment for comparative protein modeling. *Electrophoresis* **18**, 2714–2723.

Guo, Z., Vath, G. M., Wagner, C. R., and Hanna, P. E. (2003). Arylamine N-acetyltransferases: Covalent modification and inactivation of hamster NAT1 by bromoacetamido derivatives of aniline and 2-aminofluorene. *J. Protein Chem.* **22**, 631–642.

Guo, Z., Wagner, C. R., and Hanna, P. E. (2004). Mass spectrometric investigation of the mechanism of inactivation of hamster arylamine N-acetyltransferase 1 by N-hydroxy-2-acetylaminofluorene. *Chem. Res. Toxicol.* **17**, 275–286.

Halliwell, B., and Gutteridge, J. M. C. (1999). "Free Radicals in Biology and Medicine." Oxford Univ. Press, Oxford.

Hein, D. W. (2002). Molecular genetics and function of NAT1 and NAT2: Role in aromatic amine metabolism and carcinogenesis. *Mutat. Res.* **65**, 506–507.

Hein, D. W., Doll, M. A., Fretland, A. J., Leff, M. A., Webb, S. J., Xiao, G. H., Devanaboyina, U. S., Nangju, N. A., and Feng, Y. (2000). Molecular genetics and epidemiology of the NAT1 and NAT2 acetylation polymorphisms. *Cancer Epidemiol. Biomark. Prev.* **9**, 29–42.

Hickman, D., Pope, J., Patil, S. D., Fakis, G., Smelt, V., Stanley, L. A., Payton, M., Unadkat, J. D., and Sim, E. (1998). Expression of arylamine N-acetyltransferase in human intestine. *Gut* **42**, 402–409.

Holton, S. J., Dairou, J., Sandy, J., Rodrigues-Lima, F., Dupret, J. M., Noble, M. E. M., and Sim, E. (2005). Structure of Mesorhizobium loti arylamine N-acetyltransferase 1. *Acta Cryst.* **F61**, 14–16.

Hosker, B., Bilton, R., Lowe, G., and Loweth, A. (2001). Quantification of peroxynitrite to mixed radical and nitric oxide mediated cell death. *Diabetologia* **44**. [Abstract.]

Husain, A., Barker, D. F., States, J. C., Doll, M. A., and Hein, D. W. (2004). Identification of the major promoter and non-coding exons of the human arylamine N-acetyltransferase 1 gene (NAT1). *Pharmacogenetics* **14**, 397–406.

Konorev, E. A., Hogg, N., and Kalyanaraman, B. (1998). Rapid and irreversible inhibition of creatine kinase by peroxynitrite. *FEBS Lett.* **427**, 171–174.

Lee, S., Kwon, K., Kim, S., and Rhee, S. (1998). Reversible inactivation of protein-tyrosine phosphatase 1B in A431 cells stimulated with epidermal growth factor. *J. Biol. Chem.* **273**, 15366–15372.

Lee, S., Yang, K., Kwon, J., Lee, C., Jeong, W., and Rhee, S. (2002). Reversible inactivation of the tumor suppressor PTEN by H2O2. *J. Biol. Chem.* **277**, 20336–20342.

Lin, H., Kent, U., Zhang, H., Waskell, L., and Hollenberg, P. (2003). Mutation of tyrosine 190 to alanine eliminates the inactivation of cytochrome P450 2B1 by peroxynitrite. *Chem. Res. Toxicol.* **16**, 129–136.

Liu, L., and Stamler, J. S. (1999). NO: An inhibitor of cell death. *Cell Death Differ.* **6**, 937–942.

Matas, N., Thygesen, P., Stacey, M., Risch, A., and Sim, E. (1997). Mapping AAC1, AAC2 and AACP, the genes for arylamine N-acetyltransferases, carcinogen metabolising enzymes on human chromosome 8p22, a region frequently deleted in tumours. *Cytogenet. Cell Genet.* **77,** 290–295.

Mihm, M., and Bauer, J. (2002). Peroxynitrite-induced inhibition and nitration of cardiac myofibrillar creatine kinase. *Biochimie* **84,** 1013–1019.

Mohr, S., Zech, B., Lapetina, E. G., and Brune, B. (1997). Inhibition of caspase-3 by S-nitrosation and oxidation caused by nitric oxide. *Biochem. Biophys. Res. Commun.* **238,** 387–391.

Mueller, S., Riedel, H., and Stremmel, W. (1997). Determination of catalase activity at physiological hydrogen peroxide concentrations. *Anal. Biochem.* **245,** 55–60.

Pagano, G. (2002). Redox-modulated xenobiotic action and ROS formation: A mirror or a window? *Hum. Exp. Toxicol.* **21,** 77–81.

Panayiotidis, M., Tsolas, O., and Galaris, D. (1999). Glucose oxidase-produced H2O2 induces Ca^{2+}-dependent DNA damage in human peripheral blood lymphocytes. *Free Radic. Biol. Med.* **26,** 548–556.

Payton, M., Mushtaq, A., Yu, T. W., Wu, L. J., Sinclair, J., and Sim, E. (2001). Eubacterial arylamine N-acetyltransferases: Identification and comparison of 18 members of the protein family with conserved active site cysteine, histidine and aspartate residues. *Microbiology* **147,** 1137–1147.

Percival, M., Ouellet, M., Campagnolo, C., Claveau, D., and Li, C. (1999). Inhibition of cathepsin K by nitric oxide donors: Evidence for the formation of mixed disulfides and a sulfenic acid. *Biochemistry* **38,** 13574–13583.

Pompeo, F., Brooke, E., Akane, K., Mushtaq, A., and Sim, E. (2002). The pharmacogenetics of NAT: Structural aspects. *Pharmacogenomics* **3,** 19–30.

Radi, R., Beckman, J., Bush, K., and Freeman, B. (1991). Peroxynitrite oxidation of sulfhydryls. *J. Biol. Chem.* **266,** 4244–4250.

Radi, R., Peluffo, G., Alvarez, M. N., Naviliat, M., and Cayota, A. (2001). Unraveling peroxynitrite formation in biological systems. *Free Radic. Biol. Med.* **30,** 463–488.

Ravid, T., Sweeney, C., Gee, P., Carraway, K., III, and Goldkorn, T. (2002). Epidermal growth factor receptor activation under oxidative stress fails to promote c-Cbl mediated down-regulation. *J. Biol. Chem.* **277,** 31214–31219.

Riddle, B., and Jencks, W. P. (1971). Acetyl-coenzyme A: Arylamine N-acetyltransferase: Role of the acetyl-enzyme intermediate and the effects of substituents on the rate. *J. Biol. Chem.* **246,** 3250–3258.

Rodrigues-Lima, F., Cooper, R., Goudeau, B., Atmane, N., Chamagne, A., Butler-Browne, G., Sim, E., Vicart, P., and Dupret, J. (2003). Skeletal muscles express the xenobiotic-metabolizing enzyme arylamine N-acetyltransferase. *J. Histochem. Cytochem.* **51,** 789–796.

Rodrigues-Lima, F., Deloménie, C., Goodfellow, G. H., Grant, D. M., and Dupret, J. M. (2001). Homology modelling and structural analysis of human arylamine N-acetyltransferase NAT1: Evidence for the conservation of a cysteine protease catalytic domain and an active-site loop. *Biochem. J.* **356,** 327–334.

Rodrigues-Lima, F., and Dupret, J. M. (2002a). 3D model of human arylamine N-acetyltransferase 2: Structural basis of the slow acetylator phenotype of the R64Q variant and analysis of the active site loop. *Biochem. Biophys. Res. Commun.* **291,** 116–123.

Rodrigues-Lima, F., and Dupret, J. M. (2002b). In silico sequence analysis of arylamine N-acetyltransferases: Evidence for an absence of lateral gene transfer from bacteria to vertebrates and first description of paralogs in bacteria. *Biochem. Biophys. Res. Commun.* **293,** 783–792.

Rodrigues-Lima, F., and Dupret, J. M. (2004). Regulation of the activity of the human drug metabolizing enzyme arylamine N-acetyltransferase 1: Role of genetic and non genetic factors. *Curr. Pharm. Des.* **10,** 2519–2524.

Sandy, J., Mushtaq, A., Kawamura, A., Sinclair, J., Noble, M., and Sim, E. (2002). The structure of arylamine N-acetyltransferase from Mycobacterium smegmatis: An enzyme which inactivates the anti-tubercular drug, isoniazid. *J. Mol. Biol.* **318,** 1071–1083.

Sinclair, J., and Sim, E. (1997). A fragment consisting of the first 204 amino-terminal amino acids of human arylamine N-acetyltransferase one (NAT1) and the first transacetylation step of catalysis. *Biochem. Pharmacol.* **53,** 11–16.

Sinclair, J. C., Sandy, J., Delgoda, R., Sim, E., and Noble, M. E. (2000). Structure of arylamine N-acetyltransferase reveals a catalytic triad. *Nature Struct. Biol.* **7,** 560–564.

Singh, R. J., Hogg, N., Joseph, J., Konorev, E., and Kalyanaraman, B. (1999). The peroxynitrite generator, SIN-1, becomes a nitric oxide donor in the presence of electron acceptors. *Arch. Biochem. Biophys.* **361,** 331–339.

Spector, A., and Garner, W. H. (1981). Hydrogen peroxide and human cataract. *Exp. Eye. Res.* **33,** 673–681.

Stachowiak, O., Dolder, M., Wallimann, T., and Richter, C. (1998). Mitochondrial creatine kinase is a prime target of peroxynitrite-induced modification and inactivation. *J. Biol. Chem.* **273,** 16694–16699.

Takakura, K., Beckman, J., MacMillan-Crow, L., and Crow, J. (1999). Rapid and irreversible inactivation of protein tyrosine phosphatases PTP1B, CD45, and LAR by peroxynitrite. *Arch. Biochem. Biophys.* **369,** 197–207.

Tannickal, V., and Fanburg, B. (2000). Reactive oxygen species in cell signaling. *Am. J. Physiol. Lung Cell Mol. Physiol.* **279,** 1005–1028.

Wang, H., Vath, G. M., Gleason, K. J., Hanna, P. E., and Wagner, C. R. (2004). Probing the mechanism of hamster arylamine N-acetyltransferase 2 acetylation by active site modification, site-directed mutagenesis and pre-steady state and steady state kinetic studies. *Biochemistry* **43,** 8234–8246.

Westwood, I. M., Holton, S. J., Rodrigues-Lima, F., Dupret, J.-M., Bhakta, S., Noble, M. E. M., and Sim, E. (2005). Expression, purification, characterisation and structure of *Pseudomonas aeruginosa* arylamine N-acetyltransferase. *Biochem. J.* **385,** 605–612.

Wong, P., Eiserich, J., Reddy, S., Lopez, C., Cross, C., and van der Vliet, A. (2001). Inactivation of glutathione S-transferase by nitric oxide-derived oxidants: Exploring a role for tyrosine nitration. *Arch. Biochem. Biophys.* **394,** 216–228.

Xian, M., Chen, X., Liu, Z., Wang, K., and Wang, P. (2000a). Inhibition of papain by S-nitrosothiols. *J. Biol. Chem.* **275,** 20467–20473.

Xian, M., Wang, K., Chen, X., Hou, Y., McGill, A., Chen, X., Zhou, B., Zhang, Z., Chen, J., and Wang, P. (2000b). Inhibition of protein tyrosine phosphatases by low-molecular-weight S-nitrosothiols and S-nitrosylated human serum albumin. *Biochem. Biophys. Res. Commun.* **268,** 310–314.

[13] Sulfotransferases and Acetyltransferases in Mutagenicity Testing: Technical Aspects

By HANSRUEDI GLATT and WALTER MEINL

Abstract

Sulfotransferases (SULTs) and *N*-acetyltransferases (NATs) mediate the terminal activation step of various mutagens and carcinogens. Target cells of standard *in vitro* mutagenicity tests do not express any endogenous SULTs. NATs are expressed in some cells, but may not reflect the substrate specificity of human NATs. External activating systems usually lack the cofactors for these enzymes. Upon addition of the cofactor, the ultimate mutagen may be formed, but especially sulfo conjugates—anions—may not reliably penetrate into the target cells. This chapter presents methods used to incorporate these enzyme systems into *in vitro* mutagenicity test systems and to identify the critical human forms. The method of choice is direct expression of the enzymes in target cells. We present procedures on how this can be reached in bacteria and in mammalian cell lines in culture. Furthermore, genetically manipulated mouse models are a very promising perspective for answering open questions.

Introduction

Mutagenicity is an important toxicological effect, as it can lead to inheritable damage to the progeny, the initiation and progression of neoplasias, and degenerative changes in tissues. Because the structure and processing of DNA are highly conserved, mutagenicity studies conducted in any biological systems may be useful for detecting possible genotoxic hazards to human. Indeed, *in vitro* investigations using bacterial or mammalian target cells form the dominating part in standard programs for studying the genotoxicity of new compounds.

Chemical damage to DNA is by far the most common mechanism underlying chemical mutagenesis. It involves reactive molecules, which usually are formed within the organism, either as side products of endogenous metabolism or during the biotransformation of xenobiotics. Biotransformation can vary enormously among species, genotypes, tissues, and physiological states. This variation tremendously complicates the design of appropriate genotoxicity studies *in vitro* and *in vivo* and the assessment of risks for humans. It is often assumed that phase I metabolism is particularly prone

METHODS IN ENZYMOLOGY, VOL. 400
0076-6879/05 $35.00
DOI: 10.1016/S0076-6879(05)00013-3

to the formation of reactive intermediates. Therefore, hepatic microsomal or postmitochondrial ("S9") fractions, supplemented with NADPH (the cofactor of cytochromes P450) or an NADPH-generating system, are used frequently in *in vitro* genotoxicity assays. Cofactors for phase II enzymes are normally not added to activating systems, as many toxicologists associate phase II metabolism primarily with detoxification rather than toxification. We are convinced that this view is one-sided and that involvements of phase II enzymes in toxification are rather frequent. However, we recognize another reason for omitting cofactors for phase II enzymes in activating systems. The primary function of typical phase II enzymes is the introduction of an anionic group into the acceptor molecule to prevent the passive penetration of cell membranes, the property needed for its vectorial transport and effective excretion. Therefore, many reactive (and often short-lived) phase II metabolites formed externally may not reach the DNA of the target cell. For this reason, other strategies are required for the reliable detection of genotoxicants formed by phase II enzymes.

A phase II metabolite, 2-acetylaminofluorene *N*-sulfate, was the first ultimate carcinogen, formed from a procarcinogen, to be discovered (DeBaun *et al.*, 1968; King and Phillips, 1968). Subsequently, it was found that sulfotransferases (SULTs) are also involved in the bioactivation of other carcinogens, e.g., many aromatic amines (Miller, 1994). Using SULT-proficient target cells we then demonstrated that a large number of compounds belonging to several chemical classes could be activated to mutagens by SULTs (Glatt, 2000, 2005). The high chemical reactivity of many sulfuric acid esters, formed via sulfo conjugation of hydroxylated acceptor molecules, can be explained by the fact that sulfate is a good leaving group in certain chemical linkages. For example, heterolytic cleavage is facilitated if the resulting cation is resonance stabilized, as is the case with sulfuric acid esters derived from aromatic amines (Fig. 1). *O*-Acetylation is another metabolic reaction creating a potential leaving group, acetate, in xenobiotics. The enzymes catalyzing this reaction in vertebrates are termed *N*-acetyltransferases (NAT) because they appear to transfer the acetyl group exclusively to nitrogen atoms and *N*-centered hydroxyl groups. Therefore, NAT-mediated activation has been observed primarily with aromatic hydroxylamines (usually formed from amino- and nitroarenes). Various aromatic hydroxylamines may be activated by SULTs and NATs via analogous mechanisms (Fig. 1). In both cases, the same nitrenium/carbonium ion is transferred to the DNA (and other nucleophilic acceptors). However, in agreement with the fact that sulfuric acid (pK_a -3 and 1.92 for the first and second proton, respectively) is a more potent acid than acetic acid (pK_a 4.74), sulfuric acid esters are usually more reactive than the corresponding acetic acid esters. Moreover, acetylation is an

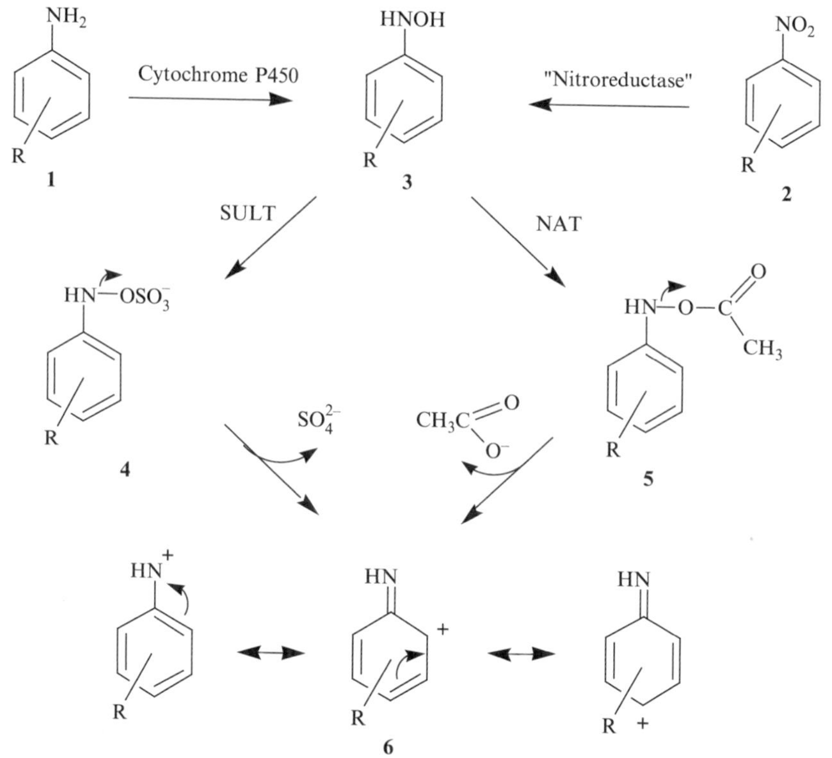

FIG. 1. Activation of aromatic amines (1) and nitro compounds (2) via hydroxylamines (3) to reactive esters (4,5), which may react spontaneously with DNA and other nucleophiles via a resonance-stabilized nitrenium/carbonium ion (6). Sulfuric acid esters (4) differ from acetic acid esters (5) by their negative charge and higher reactivity.

atypical conjugation reaction, as it involves the transfer of a neutral group, which may not constrict the penetration of membranes. Finally, the substrate specificity of the enzymes may be another factor determining which pathway is important for a given substrate. In the mammalian species investigated, usually 2 members of the NAT superfamily (three in the mouse) and approximately 12 members of the SULT superfamily have been detected. The activation of various promutagens by NATs (Hein, 2000) and SULTs (Glatt, 2000, 2005) has been reviewed elsewhere. This chapter focuses on methodological aspects—how to detect the involvement of these enzyme classes in the activation of a compound and how to identify the enzyme forms involved. The dominating part deals with the expression of these enzymes in standard target cells of mutagenicity assays, using gene-technical approaches, and with the usage of the systems.

Natural Expression of SULTs and NATs in Target Cells of Test Systems

SULTs are expressed *in vivo* with high tissue and cell-type selectivity (Glatt, 2002). In human, some SULTs are highly expressed in liver, whereas other forms are absent or very low in this tissue, but high at specific extrahepatic sites. In rats and mice, the expression of most SULTs is much more focused to the liver than in human. The expression of the mRNA of SULTs is completely lost (or is reduced for SULT1A1) in primary cultures of rat hepatocytes within a few hours (Liu *et al.*, 1996). Likewise, SULT activity levels are much lower in hepatic and other epithelial cell lines in culture than in liver tissue (Glatt *et al.*, 1994). In general, fast-growing mesenchymal cell lines, such as Chinese hamster V79 and CHO and mouse L5178Y cells, are used for mutagenicity experiments. Using numerous substrates and antisera (cross-reacting with rat, mouse, and human SULTs), we have not detected any SULT expression in V79 cells, the cell line widely employed in our laboratory for genetic engineering and genotoxicity studies (Glatt *et al.*, 1994; and many results that have not been published systematically). Sporadic investigations in CHO and L5178Y cells consistently gave negative results. Furthermore, we have not detected SULT activity toward any xenobiotics in Ames's *Salmonella typhimurium* strains (Glatt *et al.*, 1994; and many results that have not been published systematically), the most widely used target cells in mutagenicity research.

The V79 cells maintained in our laboratory for many years (termed V79-MZ) do not express any NAT proteins or activities (Glatt *et al.*, 2004). The situation may be different in V79 sublines used in other laboratories. In particular, the subline V79-NH demonstrated some endogenous NAT activity (Perchermeier *et al.*, 1994). However, Chinese hamster NATs have not yet been characterized on a molecular level, and thus it is not known which forms are expressed in V79-NH cells and whether they are similar to human forms. Ames's *S. typhimurium* strains express an endogenous acetyltransferase. Knockout of this enzyme (in strains marked with the suffix "DNP" or "1,8-DNP") leads to a drastic decrease in the mutagenicity and cytotoxicity of many nitro- and aminoarenes, such as 1,8-dinitropyrene (1,8-DNP) (Rosenkranz and Mermelstein, 1983). This bacterial enzyme differs in its substrate specificity from mammalian enzymes, despite some overlaps. For example, mammalian NATs, unlike *Salmonella* acetyltransferase, show *N*-hydroxyacetamide *N,O*-transacylase activity, and thus can activate the N-hydroxy metabolites of various aromatic amides (Ando *et al.*, 1993).

SULT and NAT Activity in Subcellular Activating Systems

Xenobiotic-metabolizing SULTs and NATs are soluble proteins normally localized in the cytoplasm. However, some SULTs were detected in the nuclei rather than the cytoplasm in some tissues (He *et al.*, 2004). This localization may require a modification of some protocols for subcellular fractionation. Furthermore, each SULT and NAT form is not expressed in each tissue and ontogenetic stage. In adult rat liver, several SULT forms are only present in females or males (reviewed by Glatt, 2002). Apart from these complications, a role of SULTs or NATs in a biotransformation (e.g., activation) reaction in a subcellular system can be demonstrated readily by conducting incubations in the presence and absence of their characteristic cofactors, 3'-phosphoadenosine-5'-phosphosulfate (PAPS) and acetyl coenzyme A (acetyl-CoA), respectively. Identification of the critical form may be strived for by comparing the effect of preparations from tissues differing in the forms expressed or by using selective inhibitors. Likewise, individual cDNA-expressed and/or purified enzymes may be employed (Glatt *et al.*, 1995).

Tissue levels of PAPS and acetyl-CoA are low. PAPS levels of 3.6 to 76.8 nmol per gram tissue have been reported for various tissues from different species (Klaassen and Boles, 1997). In liver of the guinea pig and the rat, acetyl-CoA was detected at levels of 3 and 14 nmol per gram, respectively (Erfle and Sauer, 1967; Kato, 1978). Thus, their endogenous concentration is negligible in the strongly diluted subcellular preparations commonly used in biotransformation studies, and normally they have not to be removed (e.g., by dialysis) for the negative controls. For the addition of the cofactor, the following information may be useful.

Apparent K_m values of SULTs for PAPS can range from <1 to 155 μM depending on the enzyme form and experimental conditions used (reviewed by Glatt, 2002). PAPS, unlike its desulfonated form, PAP, does not inhibit SULTs when its concentration is increased. Therefore, we recommend a generous supply of PAPS, especially if the sulfo acceptor substrate is used at a high concentration and a high turnover is strived for.

Commercially available PAPS usually contains high levels of PAP, which inhibits SULTs. Fortunately, complete cytosolic fractions from tissues and cells contain PAP-degrading enzymes and thus reduce this inhibition. When accurate quantification of reaction parameters is aimed at or when a purified SULT is studied, it is, however, advised to use high-quality PAPS (synthesized in some laboratories working with SULTs).

PAPS (which is relatively expensive) may be replaced in some situations by a PAPS-generating system. The physiological synthesis of PAPS is conducted by PAPS synthetases from inorganic sulfate and ATP and requires the presence of Mg^{2+}. PAPS synthetase is found in the cytosolic fraction of most tissues, including liver. Therefore, it is sufficient to supplement such preparations with ATP (e.g., 10 mM) and $MgSO_4$ (e.g., 10 mM). An alternative is the "transfer assay," which can lead to particularly high sulfo conjugation rates and involves the transfer of the sulfo group from a conjugate (usually 4-nitrophenylsulfate, used at a high concentration, e.g., 5 mM) to SULT-bound PAP and from there to the actual acceptor substrate (Frame et al., 2000; Duffel et al., 2001). Although this approach can be very efficient, it should only been used with well-characterized and appropriate enzyme systems. In particular, it is necessary that a SULT is present that accepts 4-nitrophenol/4-nitrophenylsulfate as a substrate, and ideally the same enzyme mediates the sulfo conjugation of the final substrate. Primarily, human SULT1A1 is appropriate for this reaction (Frame et al., 2000). This enzyme shows particularly broad substrate tolerance toward promutagens (Glatt, 2000, 2005). However, the use of PAPS-generating systems is not possible for determining accurate kinetic data. In general, we prefer the use of highly pure PAPS in subcellular enzyme systems.

Some SULTs are strongly inhibited at excessive substrate concentrations. For example, human liver thermostable phenol sulfotransferase (SULT1A1) showed an apparent K_m of 0.94 μM for the standard substrate 4-nitrophenol; the conjugation rate was maximal at 4 μM and then decreased rapidly with increasing substrate concentrations (Campbell et al., 1987). This effect was also observed with various other substrates and has to be taken into account when studying possible bioactivation reactions. The use of high substrate concentrations may lead to an underestimation of the activation potential. We address this point here, as conventional activity measurements with subcellular preparations are usually conducted under enzyme-limiting, substrate-saturated conditions, whereas toxicological experiments with cells are normally done under substrate-limiting conditions in the presence of high enzyme levels. This difference may be the reason for various conflicting results in the literature.

A concentration of 200 μM acetyl-CoA is appropriate for studying NAT activities; it is recommended to add acetyl phosphate (5 mM) and phosphotransacetylase (4.5 U/ml) to recycle CoA, which is an inhibitor of NATs (Andres et al., 1985). Others have used 2 mM of acetyl-CoA in the absence of a regenerating system (Arlt et al., 2005).

Subcellular preparations have been incorporated in cell-free genotoxicity test systems using the formation of adducts to DNA and other biomolecules as the end point. This was how the SULT-dependent activation of

2-acetylaminofluorene in the presence of hepatic preparations was detected (DeBaun *et al.*, 1968; King and Phillips, 1968). The approach has also been employed with cDNA-expressed SULTs and NATs (e.g., Arlt *et al.*, 2005). More often, subcellular preparations are combined with cellular genotoxicity test systems, such as the NADPH-fortified hepatic S9 fraction in standard *Salmonella* and mammalian cell mutagenicity tests (Bradley *et al.*, 1981; Maron and Ames, 1983). It appears that the cofactor acetyl-CoA has not been fortified in such studies, probably because the most popular target cells (*S. typhimurium* and *Escherichia coli*) express their own acetyltransferase and produce the necessary cofactor. PAPS-fortified activating systems have been used with varying success. The major problem involves penetration of the active metabolite into the target cells, which depends on the individual sulfo conjugate, the target cell, and the incubation conditions. We noticed that some benzylic sulfuric acid esters readily exchange the anionic leaving-group sulfate against a neutral leaving group (such as chloride) in a medium containing the corresponding component and that the resulting secondary reactive species can readily penetrate into the target cells (Enders *et al.*, 1993). Thus, we have used the bioactivation of 1-hydroxymethylpyrene to a bacterial mutagen as an assay for monitoring the active fractions in the purification of a SULT (Czich *et al.*, 1994). Other classes of reactive sulfuric acid esters, including those derived from aromatic amines, are usually not detected as mutagens when generated extracellularly. For example, chemically synthesized 2-acetylaminofluorene *N*-sulfate showed negligible mutagenicity to *S. typhimurium* (Smith *et al.*, 1986). Likewise, the addition of PAPS to the hepatic-metabolizing system decreased the mutagenic activity of *N*-hydroxy-2-acetylaminofluorene (Mulder *et al.*, 1977), whereas intracellular (cDNA-expressed) SULT strongly enhanced its effect in isogenic bacterial strains (next section).

Recombinant Systems

Historical Background

Watanabe *et al.* (1990) cloned the acetyltransferase of *S. typhimurium* into the pBR322 vector. This plasmid was introduced into the standard *S. typhimurium* strains TA98 and TA100, yielding strains YG1024 and YG1029, respectively. This manipulation led to an approximately 100-fold increase in acetyltransferase activity and to strongly enhanced mutagenic activity of various *N*-hydroxylamino- and nitroarenes (tested directly) and aminoarenes (in the presence of a hepatic S9 preparation). Grant *et al.* (1992) were the first to express human enzymes directly in target cells of a bacterial mutagenicity test system. They cloned human NAT1 and NAT2 into the phagemid vector pKEN2 and then transformed strain TA1538/1,8-DNP (deficient in endogenous acetyltransferase). The resulting strains, DJ400 and DJ460, respectively, were much more sensitive to the mutagenic action of 2-aminofluorene, benzidine, and 2-amino-3,4-dimethylimidazo-4,5-f]quinoline than the parental strain TA1538/1,8-DNP. We constructed analogous strains (named TA1538-DNP-hNAT1 and -hNAT2) using the pKK223–3 vector (Table I) (Muckel et al., 2002). The responsiveness of these strains was similar to those of strains DJ400 and DJ460, respectively. E. coli and S. typhimurium strains coexpressing a human cytochrome P450, together with high levels of Salmonella acetyltransferase, have also been constructed for detection of the mutagenicity of various aromatic amines without the need of external activating systems (Josephy et al., 1998; Suzuki et al., 1998). However, analogous strains with human NATs are not yet available. We have used vector pKK233-2 (Table I) for expressing a total

TABLE I

Vectors Used for the Expression of SULTs and NATs in Target Cells of
Mutagenicity Tests

Vector	Unique restriction sites within multiple cloning site	Promoter	Expression mode	Selection marker
	Expression in *S. typhimurium*			
pKK233-2	*Nco*I, *Pst*I, *Hind*III	*trc*	Constitutive	Ampicillin
pKKnew	*Nco*I, *Sac*I, *Sma*I, *Xba*I, *Pst*I, *Hind*III	*trc*	Constitutive	Ampicillin
pKK223-3	*Eco*RI, *Sma*I, Pst I, *Hind*III	*tac*	Constitutive	Ampicillin
pKNeo	see main text	*trc*	Constitutive	Neomycin
pCWmodI	*Nde*I, *Xba*I, *Sal*I, *Hind*III	Two adjacent *tac* motifs	Inducible[a]	Ampicillin
	Expression in V79 cells			
pMPSVEH	*Eco*RI, *Sal*I, *Bam*HI, *Hind*III	Long terminal repeat of myeloproliferative sarcoma virus	Constitutive	None[b]
pSI	*Eco*RI, *Xba*I, *Sal*I, *Sma*I, *Not*I	SV40 early promoter	Constitutive	None[b]

[a] Vector contains lac I[q] repressor (absent in the S. typhimurium LT2 genome), induction by isopropylthio-β-d-1-galactopyranoside.
[b] Usually cotransfected with a separate vector (pBSpacΔp) conferring resistance to puromycin.

of 40 different mammalian SULTs in *S. typhimurium* TA1538 (e.g., Glatt *et al.*, 1996; Meinl *et al.*, 2002).

Doehmer *et al.* (1988) were the first to employ gene transfer techniques for expressing a xenobiotic-metabolizing enzyme, a cytochrome P450, in mammalian target cells (V79) to be used in gene mutation tests. We adopted this technology for expressing SULTs and NATs in V79 cells (Czich *et al.*, 1994; Teubner *et al.*, 2002). However, we selected other vectors and resistance markers to make possible the expression of these conjugating enzymes in cells already engineered for a cytochrome P450 (Glatt *et al.*, 1996, 2004).

The following sections present technical aspects of our experience with the expression of SULTs and NATs.

Vectors

cDNAs are normally cloned in *E. coli*, e.g., in strain XL blue-1 (Stratagene, Heidelberg, Germany). Plasmids are adapted to the restriction enzymes of Ames's *S. typhimurium* strains (derivatives of LT2) by passaging through the restriction-deficient, but methylation-proficient *S. typhimurium* strain LB5000 (Bullas and Ryu, 1983). All bacterial expression vectors used contain a *trc* or *tac* promoter (Table I), leading to repression in common *E. coli* strains (containing an endogenous lac I^q) in the absence of an inducer (such as isopropylthio-β-D-1-galactopyranoside). However, these promoters lead to constitutive expression in *S. typhimurium* due to the absence of endogenous lac I^q unless such a repressor is contained on the plasmid.

For the expression of SULTs in bacteria, we generally use pKK233-2 (Amann and Brosius, 1985) (GenBank accession number X70478; the vector was obtained from Pharmacia, Freiburg, Germany, but is now delisted). The *Nco*I restriction site used for inserting the 5' end of the cDNA contains an ATG translation initiation codon followed by a G. Fortunately, the second codon of most SULTs starts with this base. In other cases (primarily SULT2A forms), the base was belatedly corrected by site-directed mutagenesis. To enhance the practicality of this vector, we introduced additional unique restriction sites within the multiple coding region (Meinl *et al.*, 2002); in the meantime we named this vector pKKnew. These pKK vectors contain an ampicillin resistance marker. Such a resistance factor is already present in standard *S. typhimurium* strains containing plasmid pKM101 (e.g., strains TA98 and TA100). For the transformation of these strains, we exchanged the ampicillin resistance marker in pKK233-2 against a neomycin resistance marker to give vector pKNeo (Glatt and Meinl, 2004). This vector contains additional *Nco*I restriction sites. Therefore, cDNAs were first cloned in pKK233-2 or pKKnew and then transferred into pKNeo using *Sal*I (pKKnew) or *Sal*I/*Hin*dIII fragments. Similar levels of SULT proteins were expressed from vectors with the different resistance markers in strains TA1538 or TA1537 (not containing pKM101). Nevertheless, the activities of SULT-dependent and -independent mutagens were stronger (in general by a factor of nearly 2) with the original vector than with the pKNeo vector, indicating additional influences of the resistance markers.

Vector pKK223-3 (M77749, from Pharmacia, delisted) differs primarily from pKK233-2 in the restriction sites in the multiple cloning region (Table I). It was more appropriate than the latter vector for the insertion of NATs.

Vector pCWmodI (a generous gift of Dr. T. Friedberg, University of Dundee, Scotland) was generated from pCW (Gegner and Dahlquist, 1991) by deleting the *CheW* gene. The *Nde*I restriction site contains the ATG translation initiation codon (but no additional bases reaching into the next codon). pCWmodI encodes a lac I^q repressor sequence.

For the stable expression in V79 cells, we initially used pMPSV (Artelt *et al.*, 1988), carrying the long terminal repeat promoter of the myeloproliferative sarcoma virus (Czich *et al.*, 1994; Teubner *et al.*, 2002). Later, we also explored other vectors. In particular, pSI (U47121, purchased from Promega, Mannheim, Germany) appeared to give somewhat higher expression levels than pMPSV, although a robust comparison is difficult, as the expression in each clone is affected by chance (resulting from the number of copies integrated and the integration sites). Expression from pSI is driven by the SV40 early promoter.

Characteristics and Levels of Expressed Proteins

The heterologous enzyme proteins in recombinant cells are analyzed by electrophoresis in polyacrylamide gels under denaturing conditions with subsequent immunoblotting. With most SULT and NAT expression vectors, a single polypeptide band is observed, which comigrates with the corresponding protein naturally expressed in mammalian tissues. The only exception among the 13 human SULTs and NATs (reference sequences) and their allelic variants was found with SULT2B1b. It produced a single band in *E. coli* and V79 cells, but an additional, weaker band, representing a smaller polypeptide, in *S. typhimurium*. Site-directed mutagenesis of an internal ATG abolished the occurrence of this second band, suggesting that this ATG was used as an additional translation initiation codon. Two double bands, which were just separated by electrophoresis, were also observed with mouse and rat Sult1e1 (containing ATG as codon 4) expressed in *S. typhimurium*. Dog SULT1D1 (Tsoi *et al.*, 2001) cDNA produced two immunoreactive bands differing by nearly 7 kDa in *E. coli*, with the larger band comigrating with the authentic protein. *S. typhimurium* only exhibited the shortened polypeptide. We have not studied its origin, but SULT1D1 contains an ATG codon in position 60. If used for starting the translation it would create a polypeptide with the approximate size of the shortened product found in bacteria.

For the estimation of expression levels, purified enzyme protein or— nearly homogeneous, but enzymatically inactive—inclusion bodies were used as standards in immunoblot analyses. Expressed from the natural cDNA sequence in pKK and pKNeo vectors in *S. typhimurium*, the level of SULTs usually was in the range of 0.5 to 10% of the soluble protein fraction (Meinl *et al.*, 2005). Human SULT2B1b was the most notable exception, with a very low expression under these conditions (see later). Using the same approach for quantification, we found SULT1A1 levels of 0.07 to 0.3% in the cytosolic fraction of three liver samples concurrently tested, the highest value being one-third of that observed in the standard *S. typhimurium* strain TA1538-hSULT1A1 (Meinl *et al.*, 2005). Expression levels for NATs were lower in *S. typhimurium* in absolute terms than for SULTs, but exceeded hepatic levels by factors of ≥100.

In some situations, one would like to vary the level of the activating enzyme in a mutagenicity study. This is simple with subcellular preparations (Glatt *et al.*, 1994, 1995), but difficult with enzymes expressed within the target cells. For example, when we compared the activation potential between alloenzymes of human SULT1A2, we noticed that they were expressed at different levels in the *S. typhimurium* strains constructed (Meinl *et al.*, 2002). We then analyzed several separately transformed clones, but only found negligible variation in expression levels and promutagen activation. We also varied the concentration of ampicillin, the selection marker, from zero to a high level (100 μg/ml) when growing the strains; this modification did not affect the activation of promutagens either. In an attempt to reduce the level of an enzyme, we introduced synonymous low-usage codons in the 5′-terminal region of its cDNA. To our surprise, this modification enhanced, rather than decreased, the expression (Meinl *et al.*, 2002). Consequentially, we then employed this method with various poorly expressed cDNA—consistently with success. For example, the level of SULT2B1b protein in *E. coli* and *S. typhimurium* was enhanced from barely detectable with the natural cDNA sequence to nearly 3% of the soluble protein after the introduction of several synonymous low-usage codons (M. Osterloh-Quiroz, W. Meinl, and H. R. Glatt, manuscript in preparation). We have not yet clarified whether the low-usage codons as such were decisive or accidentally associated changes in the secondary structure of the mRNA.

The other approach for modulating expression levels in *S. typhimurium* involved the use of an inducible promoter (pCWmodI, Table I). However, the repression of this vector in *S. typhimurium* is incomplete, as shown for the activation of *N*-hydroxy-2-amino-3-methylimidazo[4,5-*f*]quinoline in a strain engineered for expression of rat NAT1 (Fig. 2). The activation was only enhanced two- to threefold when the inducer, isopropylthio-β-D-1-galactopyranoside, was added to the growth medium. Effects of other test compounds and with other NAT forms were enhanced to similar extents.

Finally, it was possible to modify the expression of various NATs and SULTs by changing the temperature at which the cultures were grown. However, this modification may also alter the expression of endogenous factors involved in the handling of the test compound or the primary damage. Thus, the method is not appropriate for elucidating the influence of varying levels of a heterologous enzyme on the mutagenicity of a chemical.

Unlike in the bacterial model, the transfection of expression vectors into V79 cells created transformed cell clones that differed strongly in the level and stability of the heterologous protein. This is due to the fact that

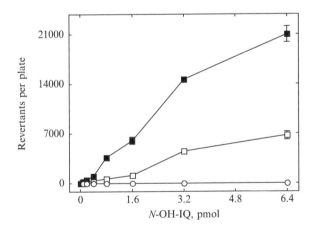

FIG. 2. Mutagenicity (reversion to histidine prototrophy) of *N*-hydroxy-2-amino-3-methylimidazo[4,5-*f*]quinoline (*N*-OH-IQ) in *S. typhimurium* strain TA1538-1,8-DNP-rNAT1 grown in the presence (■) or absence (□) of isopropylthio-β-D-1-galactopyranoside (1 m*M*, added to the overnight culture for the last 2 h before bacteria were used in the mutagenicity tests). This strain contains rat NAT1 cDNA inserted into vector pCWmodI. The NAT-deficient recipient strain TA1538–1,8-DNP (○) was used as a control. The mutagenicity experiment was conducted as described by Muckel *et al.* (2002). Values are means and SE of three plates. The initial slopes of the dose–response curves amounted to 3000, 1200, and 10 revertants per picomole test compound, respectively. Unpublished data from E. Muckel and H. R. Glatt.

the transfected DNA has to be integrated into the host genome. The number of copies integrated and the integration sites are accidental with the vectors used. With SULT, we usually selected strains with expression levels in the high natural range in tissues, e.g., liver for human SULT1A1 and SULT2A1 and female liver for rat hydroxysteroid sulfotransferase a (a member of the SULT2A subfamily) (Glatt *et al.*, 1998; Meinl *et al.*, 2005). Expression of NATs in our recombinant cells lines is markedly higher than in tissues such as liver (Glatt *et al.*, 2004). Various transformed V79 clones showed stable expression levels for 100 or more population-doubling times, even in the absence of a selecting agent, which we only use for the initial picking of transfectants (Czich *et al.*, 1994; Teubner *et al.*, 2002).

Cofactor Supply

The cofactors PAPS and acetyl-CoA can be added directly to subcellular-metabolizing systems, as described in a preceding section. However, intact cells would not take up these cofactors and thus have to synthesize them. In practice, the synthesis has to occur during the incubation period, as the tissue and cell levels of these cofactors are low. The regeneration of acetyl-CoA requires common intermediates from the endogenous metabolism and thus appears unproblematic—changes in the exposure media did not materially affect the responses with NAT-dependent mutagens in *S. typhimurium* and V79 cells. The situation is different for PAPS, which is synthesized from inorganic sulfate (and ATP). Mulder and Jakoby (1990) observed apparent K_m values of 0.3 to 0.5 mM for sulfate in the sulfonation of various substrates in isolated hepatocytes. Most cell culture media contain approximately 0.8 mM sulfate. We increased this concentration up to 5 mM when studying SULT-dependent mutagenicity in recombinant V79 cells, but did not find any influence of this modification. In these experiments, we conducted the exposure in regular medium (Dulbecco's modified Eagle's minimum essential medium). It is also possible, and not uncommon, to use phosphate-buffered saline for the exposure to the mutagen; it would be imperative to supplement sulfate when using this protocol with SULT-dependent mutagens.

Salmonella typhimurium does not appear to express any endogenous SULTs for transforming xenobiotics, but it produces PAPS for endogenous reactions. This is demonstrated by the fact that we could activate hundreds of chemicals to mutagens by mammalian SULTs expressed in these bacteria. There are two common exposure protocols for the Ames test: the plate incorporation assay (bacteria and test compound are mixed with soft agar and then immediately poured on agar plates) and the liquid preincubation assay (involving incubation of bacteria and test compound at 37° for some time before adding the top agar) (Maron and Ames, 1983). The second method is much more efficient than the former one for SULT-dependent mutagens. Omission of sulfate in the exposure medium completely abrogates SULT-dependent mutagenicity. We recommend using a concentration of at least 10 mM MgSO$_4$ and a liquid preincubation time of at least 20 min. In general we now expose the bacteria to the test compound in 100 mM MgSO$_4$ for 60 min (Meinl *et al.*, 2002), based on our findings from various optimization experiments.

Positive Controls

The expression of SULTs and NATs usually had little effect on the number of spontaneous revertants in *S. typhimurium* TA1538 (and other strains without plasmid pKM101). We frequently use SULT-independent, in addition to SULT-dependent, mutagens as positive controls. The mutagenic activity of benzo[*a*]pyrene 4,5-oxide (nearly 15,000 revertants per nanomole under our conditions) was usually unchanged after transformation with SULT- or NAT-encoding vectors. However, during this work we noticed that our stock culture of TA1538 contained a subpopulation of cells with reduced response toward this compound (usually by a factor of four). Since some previous recombinant strains were derived from this population, we made new strains for the corresponding SULTs and observed enhanced mutagenicity not only toward benzo[*a*]pyrene 4,5-oxide, but also some SULT-dependent mutagens. Transformation of strain TA100 (containing pKM101) frequently led to a decrease in the number of spontaneous and benzo[*a*]pyrene 4,5-oxide-induced revertants. The effect was enhanced when the same protein was expressed at a higher level (by synonymous base exchange in the cDNA).

1-Hydroxymethylpyrene (available from Aldrich, Taufkirchen, Germany) is particularly useful as a SULT-dependent positive control compound because it is not mutagenic in SULT-deficient recipient strains, but reverts most standard strains (TA1538, TA1537, TA98, and TA100, but not TA1535) after sulfo conjugation. The parental compound shows low bacteriotoxicity and thus can be used at high dose levels (up to 500 nmol per plate). It was activated to a mutagen by many different SULT forms. However, the appropriate dose levels varied substantially between different SULT forms, as shown in Fig. 3 for human SULTs. Activation of 1-hydroxymethylpyrene was weak (but unambiguous) with human SULT1C3 (Fig. 3, lower right) and absent (or marginal) with human SULT1C1, 2B1a, 2B1b, and 4A1. 6-Hydroxymethyanthathrene is a more potent positive control compound to SULT1C3, but shows some mutagenic activity in

Revertants per plate

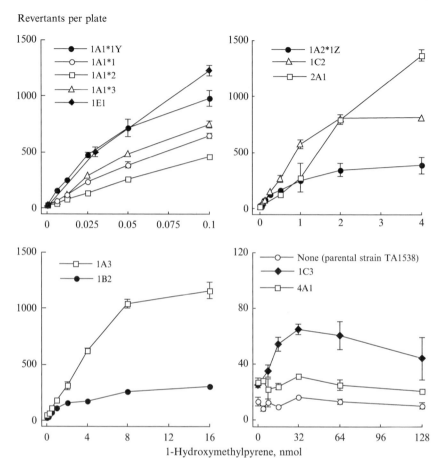

FIG. 3. Mutagenicity (reversion to histidine prototrophy) of 1-hydroxymethylpyrene to *S. typhimurium* TA1538 strains expressing the indicated human SULT. Strains TA1538-SULT1A1*1, *2, and *3 express allelic variants (involving amino acid substitutions) of SULT1A1 from their natural cDNA sequences. Expression levels were similar for all three alloenzymes. Synonymous codon exchanges were used to enhance the expression of SULT1A1 in strain TA1538-SULT1A1Y. The mutagenicity experiment was conducted as described by Meinl *et al.* (2002). Values are means and SE of three plates.

the parental strain (10% of the activity observed with the SULT1C3-expressing strain) (Fig. 4, left). We have not yet found satisfactory positive control compounds activated by human SULT1C1, 2B1a, and 2B1b. However, cytosolic preparations of these strains demonstrated SULT activity

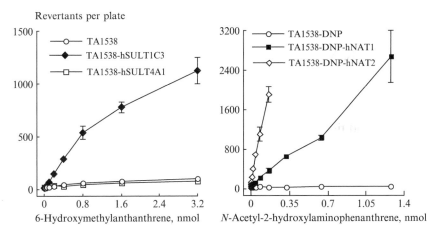

Fig. 4. Positive control compounds activated to mutagens by human SULT1C3 in TA1538-derived strains (left) and by human NAT1 and NAT2 (reference-type alleles 1*4 and 2*4, respectively) in TA1538-DNP-derived strains (right). The mutagenicity experiment was conducted as described by Meinl *et al.* (2002). Values are means and SE of three plates.

toward conventional substrates. Although SULT4A1 is a member of the SULT family based on its amino acid sequence, no substrates have been found for this form up to date (Falany *et al.*, 2000). We now use this strain as a (negative) vector control instead of, or in addition to, the parental strains (e.g., Figs. 3 and 4). Human and rat NATs, expressed in *Salmonella*, activate numerous aromatic hydroxylamines to mutagens. However, most of these compounds also show substantial mutagenic activity in acetyltransferase-deficient strains when used at high dose levels. This background activity is particularly low with *N*-acetyl-2-hydroxylaminophenanthrene, making it a convenient positive control compound (Fig. 4, right).

Positive control compounds in recombinant V79 cells have to be adjusted to the needs of individual cell lines. 1-Sulfooxymethylpyrene is mutagenic in these cells, but its metabolic precursor, 1-hydroxymethylpyrene, is cytotoxic at relatively low concentrations even in SULT-deficient control cells, limiting the concentrations that can be used. Its mutagenicity can be demonstrated in V79-derived cells expressing human SULT1A1 or SULT1E1, forms that are particularly active in the *Salmonella* system. 2-Nitropropane (used at a high concentration, 1 to 10 m*M*) is a convenient alternative for cell lines expressing human SULT1A1. Other positive control compounds are found in publications on the individual cell lines (Czich *et al.*, 1994; Glatt *et al.*, 1996, 2004; Teubner *et al.*, 2002).

Perspectives

Genotoxicity studies with recombinant bacteria and mammalian cell lines have demonstrated marked differences in the substrate specificity toward promutagens between orthologous SULT and NAT forms from human and rodent species normally used in carcinogenicity studies. Moreover, it is known that the tissue distribution of SULTs can vary strongly between different species. Based on these findings, one might suspect enhanced or reduced susceptibility of human toward standard rat and mouse models. Moreover, whereas SULT-dependent carcinogenicity in these models usually was targeted to the liver, one might expect a higher level of extrahepatic effects in human. To test these hypotheses, animal models are required that express enzymes with human-like substrate specificity and tissue distribution. Initial results from our laboratory indicate that this is possible for some SULTs by introducing human genomic sequences (rather than cDNAs) with long flanking regions into mouse oocytes.

Acknowledgment

Our current work on SULTs is financially supported by European Union (FP6–506820) and Philip Morris Incorporated.

References

Amann, E., and Brosius, J. (1985). "ATG vectors" for regulated high-level expression of cloned genes in *Escherichia coli*. *Gene* **40**, 183–190.

Ando, M., Shindo, Y., Fujita, M., Ozawa, S., Yamazoe, Y., and Kato, R. (1993). A new *Salmonella* tester strain expressing a hamster acetyltransferase shows high sensitivity for arylamines. *Mutat. Res.* **292**, 155–163.

Andres, H. H., Klem, A. J., Szabo, S. M., and Weber, W. W. (1985). New spectrophotometric and radiochemical assays for acetyl-CoA:arylamine *N*-acetyltransferase applicable to a variety of arylamines. *Anal. Biochem.* **145**, 367–375.

Arlt, V. M., Stiborova, M., Henderson, C. J., Osborne, M. R., Bieler, C. A., Frei, E., Martinek, V., Sopko, B., Wolf, C. R., Schmeiser, H. H., and Phillips, D. H. (2005). Environmental pollutant and potent mutagen 3-nitrobenzanthrone forms DNA adducts after reduction by NAD(P)H:quinone oxidoreductase and conjugation by acetyltransferases and sulfotransferases in human hepatic cytosols. *Cancer Res.* **65**, 2644–2652.

Artelt, P., Morelle, C., Ausmeier, M., Fitzek, M., and Hauser, H. (1988). Vectors for efficient expression in mammalian fibroblastoid, myeloid and lymphoid cells via transfection or infection. *Gene* **68**, 213–219.

Bradley, M. O., Bhuyan, B., Francis, M. C., Langenbach, R., Peterson, A., and Huberman, E. (1981). Mutagenesis by chemical agents in V79 Chinese hamster cells: A review and analysis of the literature. *Mutat. Res.* **87**, 81–142.

Bullas, L. R., and Ryu, J. I. (1983). *Salmonella typhimurium* LT2 strains which are R⁻ M⁺ for all three chromosomally located systems of DNA restriction and modification. *J. Bacteriol.* **156,** 471–474.

Campbell, N. R., Van Loon, J. A., and Weinshilboum, R. M. (1987). Human liver phenol sulfotransferase: Assay conditions, biochemical properties and partial purification of isozymes of the thermostable form. *Biochem. Pharmacol.* **36,** 1435–1446.

Czich, A., Bartsch, I., Dogra, S., Hornhardt, S., and Glatt, H. R. (1994). Stable heterologous expression of hydroxysteroid sulphotransferase in Chinese hamster V79 cells and their use for toxicological investigations. *Chem.-Biol. Interact.* **92,** 119–128.

DeBaun, J. R., Rowley, J. Y., Miller, E. C., and Miller, J. A. (1968). Sulfotransferase activation of *N*-hydroxy-2-acetylaminofluorene in rodent livers susceptible and resistant to this carcinogen. *Proc. Soc. Exp. Biol. Med.* **129,** 268–273.

Doehmer, J., Dogra, S., Friedberg, T., Monier, S., Adesnik, M., Glatt, H. R., and Oesch, F. (1988). Stable expression of rat cytochrome P450IIB1 cDNA in Chinese hamster cells (V79) and mutagenicity of aflatoxin B₁. *Proc. Natl. Acad. Sci. USA* **85,** 5769–5773.

Duffel, M. W., Marshal, A. D., McPhie, P., Sharma, V., and Jakoby, W. B. (2001). Enzymatic aspects of the phenol (aryl) sulfotransferases. *Drug Metab. Rev.* **33,** 369–395.

Enders, N., Seidel, A., Monnerjahn, S., and Glatt, H. R. (1993). Synthesis of 11 benzylic sulfate esters, their bacterial mutagenicity and its modulation by chloride, bromide and acetate anions. *Polycyclic Aromat. Compds.* **3,** 887s–894s.

Erfle, J. D., and Sauer, F. (1967). Acetyl coenzyme A and acetylcarnitine concentration and turnover rates in muscle and liver of the ketotic rat and guinea pig. *J. Biol. Chem.* **242,** 1988–1996.

Falany, C. N., Xie, X., Wang, J., Ferrer, J., and Falany, J. L. (2000). Molecular cloning and expression of novel sulphotransferase-like cDNAs from human and rat brain. *Biochem. J.* **346,** 857–864.

Frame, L. T., Ozawa, S., Nowell, S. A., Chou, H. C., de Longchamp, R. R., Doerge, D. R., Lang, N. P., and Kadlubar, F. F. (2000). A simple colorimetric assay for phenotyping the major human thermostable phenol sulfotransferase (SULT1A1) using platelet cytosols. *Drug Metab. Dispos.* **28,** 1063–1068.

Gegner, J. A., and Dahlquist, F. W. (1991). Signal transduction in bacteria: CheW forms a reversible complex with the protein kinase CheA. *Proc. Natl. Acad. Sci. USA* **88,** 750–754.

Glatt, H. R. (2000). Sulfotransferases in the bioactivation of xenobiotics. *Chem.-Biol. Interact.* **129,** 141–170.

Glatt, H. R. (2002). Sulphotransferases. *In* "Handbook of Enzyme Systems That Metabolise Drugs and Other Xenobiotics" (C. Ioannides, ed.), pp. 353–439. Wiley, Sussex.

Glatt, H. R. (2005). Activation and inactivation of carcinogens by human sulfotransferases. *In* "Human Sulphotransferases" (G. M. Pacifici and M. W. H. Coughtrie, eds.), pp. 281–306. Taylor & Francis, London.

Glatt, H. R., Christoph, S., Czich, A., Pauly, K., Schwierzok, A., Seidel, A., Coughtrie, M. W. H., Doehmer, J., Falany, C. N., Phillips, D. H., Yamazoe, Y., and Bartsch, I. (1996). Rat and human sulfotransferases expressed in Ames's *Salmonella typhimurium* strains and Chinese hamster V79 cells for the activation of mutagens. *In* "Control Mechanisms of Carcinogenesis" (J. G. Hengstler and F. Oesch, eds.), pp. 98–115. Institut Toxikologie, D-55131 Mainz, Germany.

Glatt, H. R., Davis, W., Meinl, W., Hermersdörfer, H., Venitt, S., and Phillips, D. H. (1998). Rat, but not human, sulfotransferase activates a tamoxifen metabolite to produce DNA adducts and gene mutations in bacteria and mammalian cells in culture. *Carcinogenesis* **19,** 1709–1713.

Glatt, H. R., and Meinl, W. (2004). Use of genetically manipulated *Salmonella typhimurium* strains to evaluate the role of sulfotransferases and acetyltransferases in nitrofen mutagenicity. *Carcinogenesis* **25**, 779–786.

Glatt, H. R., Pabel, U., Meinl, W., Frederiksen, H., Frandsen, H., and Muckel, E. (2004). Bioactivation of the heterocyclic aromatic amine 2-amino-3-methyl-9*H*-pyrido [2,3-*b*] indole (MeAαC) in recombinant test systems expressing human xenobiotic-metabolizing enzymes. *Carcinogenesis* **25**, 801–807.

Glatt, H. R., Pauly, K., Czich, A., Falany, J. L., and Falany, C. N. (1995). Activation of benzylic alcohols to mutagens by rat and human sulfotransferases expressed in *Escherichia coli. Eur. J. Pharmacol.* **293**, 173–181.

Glatt, H. R., Pauly, K., Piée-Staffa, A., Seidel, A., Hornhardt, S., and Czich, A. (1994). Activation of promutagens by endogenous and heterologous sulfotransferases expressed in continuous cell cultures. *Toxicol. Lett.* **72**, 13–21.

Glatt, H. R., Seidel, A., Harvey, R. G., and Coughtrie, M. W. H. (1994). Activation of benzylic alcohols to mutagens by human hepatic sulphotransferases. *Mutagenesis* **9**, 553–557.

Grant, D. M., Josephy, P. D., Lord, H. L., and Morrison, L. D. (1992). *Salmonella typhimurium* strains expressing human arylamine *N*-acetyltransferases: Metabolism and mutagenic activation of aromatic amines. *Cancer Res.* **52**, 3961–3964.

He, D., Meloche, C. A., Dumas, N. A., Frost, A. R., and Falany, C. N. (2004). Different subcellular localization of sulphotransferase 2B1b in human placenta and prostate. *Biochem. J.* **379**, 533–540.

Hein, D. W. (2000). *N*-Acetyltransferase genetics and their role in predisposition to aromatic and heterocyclic amine-induced carcinogenesis. *Toxicol. Lett.* **112**, 349–356.

Josephy, P. D., Evans, D. H., Parikh, A., and Guengerich, F. P. (1998). Metabolic activation of aromatic amine mutagens by simultaneous expression of human cytochrome P450 1A2, NADPH-cytochrome P450 reductase, and *N*-acetyltransferase in *Escherichia coli. Chem. Res. Toxicol.* **11**, 70–74.

Kato, T. (1978). Ischemic effect on CoASH and acetyl-CoA concentration levels in cerebrum, cerebellum and liver of mice. *J. Neurochem.* **31**, 1545–1548.

King, C. M., and Phillips, B. (1968). Enzyme-catalyzed reactions of the carcinogen *N*-hydroxy-2-fluorenylacetamide with nucleic acid. *Science* **159**, 1351–1353.

Klaassen, C. D., and Boles, J. W. (1997). Sulfation and sulfotransferases. 5. The importance of 3′-phosphoadenosine 5′-phosphosulfate (PAPS) in the regulation of sulfation. *FASEB J.* **11**, 404–418.

Liu, L., Lecluyse, E. L., Liu, J., and Klaassen, C. D. (1996). Sulfotransferase gene expression in primary cultures of rat hepatocytes. *Biochem. Pharmacol.* **52**, 1621–1630.

Maron, D. M., and Ames, B. N. (1983). Revised methods for the *Salmonella* mutagenicity test. *Mutat. Res.* **113**, 173–215.

Meinl, W., Meerman, J. H., and Glatt, H. (2002). Differential activation of promutagens by alloenzymes of human sulfotransferase 1A2 expressed in *Salmonella typhimurium. Pharmacogenetics* **12**, 677–689.

Meinl, W., Pabel, U., Osterloh-Quiroz, M., Hengstler, J. G., and Glatt, H. R. (2005). Human sulfotransferases are involved in the activation of aristolochic acids and is expressed in renal target tissue. *Int. J. Cancer.*

Miller, J. A. (1994). Sulfonation in chemical carcinogenesis: History and present status. *Chem.-Biol. Interact.* **92**, 329–341.

Muckel, E., Frandsen, H., and Glatt, H. R. (2002). Heterologous expression of human *N*-acetyltransferases 1 and 2 and sulfotransferase 1A1 in *Salmonella typhimurium* for mutagenicity testing of heterocyclic amines. *Food Chem. Toxicol.* **40**, 1063–1068.

Mulder, G. J., Hinson, J. A., Nelson, W. L., and Thorgeirsson, S. N. (1977). Role of sulfotransferase from rat liver in the mutagenicity of N-hydroxy-2-acetylaminofluorene in *Salmonella typhimurium. Biochem. Pharmacol.* **26,** 1356–1358.

Mulder, G. J., and Jakoby, W. B. (1990). Sulfation. *In* "Conjugation Reactions in Drug Metabolism: An Integrated Approach" (G. J. Mulder, ed.), pp. 107–161. Taylor and Francis, London.

Perchermeier, M. M., Kiefer, F., and Wiebel, F. J. (1994). Toxicity of monocyclic and polycyclic nitroaromatic compounds in a panel of mammalian test cell lines. *Toxicol. Lett.* **72,** 53–57.

Rosenkranz, H. S., and Mermelstein, R. (1983). Mutagenicity and genotoxicity of nitroarenes: All nitro-containing chemicals were not created equal. *Mutat. Res.* **114,** 217–267.

Smith, B. A., Springfield, J. R., and Gutmann, H. R. (1986). Interaction of the synthetic ultimate carcinogens, N-sulfonoxy- and N-acetoxy-2-acetylaminofluorene, and of enzymatically activated N-hydroxy-2-acetylaminofluorene with nucleophiles. *Carcinogenesis* **7,** 405–411.

Suzuki, A., Kushida, H., Iwata, H., Watanabe, M., Nohmi, T., Fujita, K., Gonzalez, F. J., and Kamataki, T. (1998). Establishment of a *Salmonella* tester strain highly sensitive to mutagenic heterocyclic amines. *Cancer Res.* **58,** 1833–1838.

Teubner, W., Meinl, W., and Glatt, H. R. (2002). Stable expression of rat sulfotransferase 1B1 in V79 cells: Activation of benzylic alcohols to mutagens. *Carcinogenesis* **23,** 1877–1884.

Tsoi, C., Falany, C. N., Morgenstern, R., and Swedmark, S. (2001). Identification of a new subfamily of sulphotransferases: Cloning and characterization of canine SULT1D1. *Biochem. J.* **356,** 891–897.

Watanabe, M., Ishidate, M., and Nohmi, T. (1990). Sensitive method for the detection of mutagenic nitroarenes and aromatic amines: New derivatives of *Salmonella typhimurium* tester strains possessing elevated O-acetyltransferase levels. *Mutation Res.* **234,** 337–348.

[14] A Comparative Molecular Field Analysis-Based Approach to Prediction of Sulfotransferase Catalytic Specificity

By Vyas Sharma and Michael W. Duffel

Abstract

Understanding the catalytic function and substrate specificity of cytosolic sulfotransferases (SULTs) involved in drug metabolism is essential for predicting the metabolic outcomes of many xenobiotics. Although multiple isoforms of cytosolic SULTs have been identified and characterized in humans and other species, relatively little is known about the specific molecular interactions that govern their selectivity for substrates. The use of three-dimensional quantitative structure–activity relationship

METHODS IN ENZYMOLOGY, VOL. 400
0076-6879/05 $35.00
DOI: 10.1016/S0076-6879(05)00014-5

(3D-QSAR) techniques has emerged as a powerful tool for understanding the relationships among protein structure, catalytic function, and substrate specificity. We have found that a specific adaptation of a ligand-based 3D-QSAR method, comparative molecular field analysis (CoMFA), is particularly useful for prediction of the catalytic efficiencies of SULTs. This approach has been used to study the function of a prototypical rat hepatic phenol SULT and has now been extended to a member of the hydroxysteroid SULT family. Key aspects of this methodology incorporate strategies for finding the most meaningful bioactive conformation with respect to the protein structure, use of a model of an enzyme–substrate complex incorporating the mechanism of sulfuryl transfer, and the utilization of log (k_{cat}/K_m) as the parameter for correlation analysis. The success of this approach with members of two different families of cytosolic SULTs suggests that it may be of more general use in the study of other SULTs.

Introduction

Quantitative structure–activity relationships (QSARs) that correlate biological activity with structural features of molecules are based on the general premise that molecular properties that are characteristic of all active compounds must in some way be essential for activity. The basic goal of a QSAR study is to explain the observed variation in biological activities of a series of molecules in terms of the variations due to changes in molecular structure so that associations within the data set can be examined and predictive information can be extrapolated from the data set. One powerful and well-established technique for three-dimensional QSAR (3D-QSAR) is comparative molecular field analysis (CoMFA), which is sensitive to both steric and electrostatic characteristics within a set of molecules (Cramer et al., 1988). As reviewed by Masimirembwa et al. (2002) for the cytochrome P450 enzymes, interest in the use of 3D-QSAR to understand the properties of drug-metabolizing enzymes has increased considerably over the past several years concomitant with increased recognition of the potential to predict more accurately the metabolic fate of drugs and other xenobiotics and to synthesize drug molecules with specific desired metabolic properties.

Sulfotransferases (SULTs) are a superfamily of enzymes that contribute to phase II drug metabolism in humans and other species. SULTs catalyze the transfer of a sulfuryl group from the endogenous donor 3'-phosphoadenosine 5'-phosphosulfate (PAPS) to an acceptor molecule, producing adenosine 3',5'-diphosphate (PAP) and the sulfuric acid ester of the acceptor molecule (Duffel, 1997; Duffel et al., 2001; Falany, 1997; Glatt and Meinl, 2004; Mulder and Jakoby, 1990; Weinshilboum et al., 1997). The sulfotransferases involved in the metabolism of drugs and other xenobiotics generally aid in detoxication by producing excretable hydrophilic products; however, this sulfation can also lead to the bioactivation of certain chemicals that are rendered electrophilic and mutagenic (Glatt *et al.*, 2001; Surh and Miller, 1994). Two of the major families of mammalian SULTs are phenol SULTs (classified as members of family 1) and hydroxysteroid SULTs (family 2). Sulfation reactions catalyzed by these two SULT families represent a significant route of conjugation for a wide variety of xenobiotics and their hydroxylated metabolites as well as endogenous molecules.

While earlier studies explored the application of QSAR methods to sulfotransferases (Campbell *et al.*, 1987; Dajani *et al.*, 1999), the use of 3D-QSAR techniques to study these enzymes is a more recent development. In particular, there are two recently published studies utilizing CoMFA-based approaches (Sharma and Duffel, 2002; Sipilä *et al.*, 2003). As described in more detail later, we have combined homology modeling with aspects of the catalytic mechanism of the sulfotransferase-catalyzed reaction to provide a general alignment rule for the use of CoMFA in predicting k_{cat}/K_m values as a measure of catalytic efficiency. Sipilä and co-workers (2003) have also utilized CoMFA in the prediction of K_m values for a human catecholamine sulfotransferase, hSULT1A3. Although there are differences between these two studies in the methods used for determining alignment rules for the data sets and the specific kinetic parameters fitted, they each provide an example of the overall utility of CoMFA in predicting the substrate specificity of sulfotransferases.

Our approaches to the development of CoMFA-based methods for sulfotransferases have been focused on the isoforms in the rat that are representative of the two major families of sulfotransferases, namely the aryl sulfotransferase IV from family 1 and hydroxysteroid sulfotransferase STa from family 2. Based on current systematic nomenclature (Blanchard *et al.*, 2004), these enzymes are known as rSULT1A1 and rSULT2A3, respectively. The rSULT1A1 can be considered as a prototypical phenol sulfotransferase due to similarities in its structure and function to phenol sulfotransferases in humans (e.g., hSULT1A1, hSULT1A2, and hSULT1A3) and other species (Duffel *et al.*, 2001; Negishi *et al.*, 2001). Moreover, this isoform has been studied extensively due to the long-standing use of the rat as a model for studies on drug metabolism, toxicology, and chemical carcinogenesis. It has been characterized with respect to kinetic mechanism (Duffel and Jakoby, 1981; Duffel *et al.*, 2001; Marshall *et al.*, 2000), and the kinetic constants for a broad range of structurally diverse substrates are available (Binder and Duffel, 1988; Chen *et al.*, 1992; Duffel *et al.*, 1998; King *et al.*, 2000; Rao and Duffel, 1991). The kinetic constants for many structurally diverse substrates are also available for the major hepatic hydroxysteroid SULT in the rat, rSULT2A3 (i.e., hydroxysteroid sulfotransferase STa) (Banoglu and Duffel, 1997, 1999; Chen et al., 1996). This enzyme bears similarities in both structure and function to other SULTs from family 2 (e.g., hSULT2A1 in the human) (Nagata and Yamazoe, 2000; Negishi et al., 2001; Pedersen et al., 2000).

Application of CoMFA to Studies on Sulfotransferases

The success of a CoMFA method is highly dependent on both the choice of parameters to be correlated and the features of the "alignment rule" (i.e., the method for determining structural conformations of the molecules comprising the data set). In the case of enzymes that may exhibit nonproductive binding interactions with some substrates, a CoMFA study that aligns the structures in the data set based on analogy to a single substrate conformation and then correlates structural parameters with a kinetic constant that may be influenced by nonproductive binding (e.g., a constant such as K_m or V_{max}) may not always provide results that yield mechanistically meaningful interpretations. However, a highly relevant kinetic parameter for correlation of the structure of a substrate to its ability to serve as a substrate for a sulfotransferase is the k_{cat}/K_m value, or log (k_{cat}/K_m). As a measure of catalytic specificity, k_{cat}/K_m is independent of nonproductive binding contributions (Brot and Bender, 1969), and therefore a single conformation of the substrate would be able to address the factors that relate to it. Furthermore, k_{cat}/K_m describes the apparent second-order rate constant for the capture of a substrate into a catalytically competent enzyme complex that will ultimately produce a product(s) (Northrup, 1999).

In addition to the choice of kinetic constant for correlation, the conformations of molecules in the data set and their alignment are of critical importance. Although it might be expected that a relevant conformation for correlating structure with enzymatic specificity would be that of a substrate in the active site of a sulfotransferase poised to be sulfated, it is not generally feasible to determine crystal structures of each possible substrate–sulfotransferase combination. The high degree of three-dimensional structural similarity present within sulfotransferases (Kakuta et al., 1998a; Negishi et al., 2001), however, presents the opportunity to utilize molecular modeling for constructing models of catalytically competent forms of PAPS–substrate–sulfotransferase complexes by homology.

We have utilized the crystal structures of the PAP–estradiol–enzyme and PAP–vanadate–enzyme complexes of mouse estrogen sulfotransferase (PDB IDs: 1AQU and 1BO6, respectively) (Kakuta et al., 1997, 1998b) to devise an alignment procedure for CoMFA that has been applied to both a phenol sulfotransferase, rSULT1A1, and to an alcohol (hydroxysteroid) sulfotransferase, rSULT2A3. As described in detail previously (Sharma and Duffel, 2002), this alignment procedure takes advantage of our knowledge of the mechanism of sulfation where there is an in-line transfer of the sulfuryl group of PAPS to the oxygen of an acceptor molecule. Furthermore, the method utilizes coordinates of the PAP–estradiol–estrogen sulfotransferase and PAP–vanadate–estrogen sulfotransferase complexes in order to model the structure of a PAPS–acceptor–enzyme complex at a point intermediate in the in-line transfer of a sulfuryl group (Kakuta *et al.*, 1998). While this is not necessarily the transition state of the sulfuryl transfer reaction, it does provide a structural model for a catalytically competent complex upon which to base an alignment rule for CoMFA. As a further refinement of this alignment, the position of the hydrogen on the phenol or alcohol acceptor is modeled for proton abstraction with respect to the catalytic histidine. Once the position of the oxygen and hydrogen of the acceptor substrate is fixed relative to an orientation consistent with in-line sulfuryl transfer, the conformation of the acceptor substrate is optimized with respect to the amino acid residues of the enzyme, which are held as fixed, during an energy minimization step. In the final step, the energy-minimized conformation of the substrate molecule is extracted from the enzyme and placed in a data set that is to be used for CoMFA. This alignment methodology takes into account the abstraction of a proton, an in-line transfer of the sulfuryl group, and the influence of the structure of the enzyme. Thus this alignment procedure results in a conformation for each substrate that is representative of a catalytically competent conformation. A diagrammatic representation of this process is shown in Fig. 1. The structural manipulations, energy optimizations, and subsequent CoMFA model development were performed with SYBYL (Tripos Associates, St. Louis, MO) software as described previously (Sharma and Duffel, 2002).

CoMFA for rSULT1A1 Substrates

We have earlier outlined the development of a 3D-QSAR methodology aimed at understanding the substrate specificity of rSULT1A1 and the successful application of the CoMFA method in combination with the alignment rules described earlier (Sharma and Duffel, 2002). Based on the results from CoMFA of a structurally diverse data set of 35 compounds and an additional external validation data set of 6 compounds, a region where steric bulk was predicted to be preferred in the substrate was

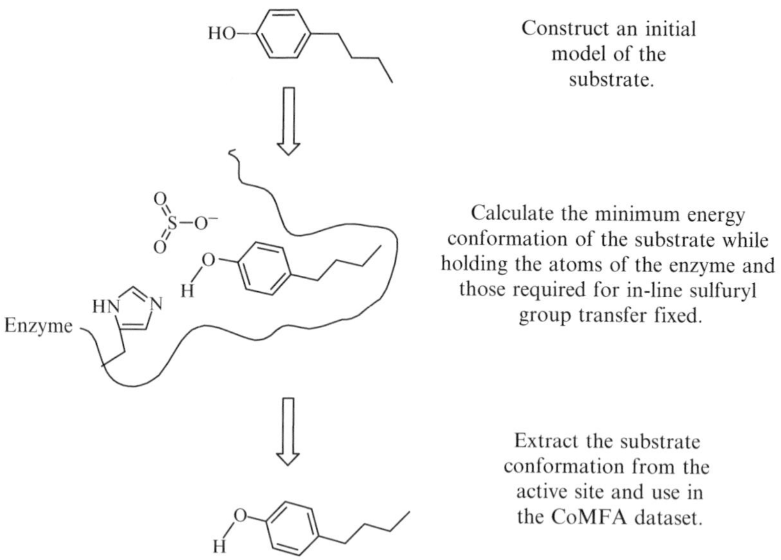

FIG. 1. Summary of the alignment procedure used for sulfotransferase substrates in CoMFA.

examined by testing a set of three additional compounds that were selected on the basis of having steric bulk at this position. The k_{cat}/K_m values for these three substrates were then determined experimentally as a further test of the predictive capacity of the model (Sharma and Duffel, 2002). A combination of all the molecules examined to date (i.e., initial data set, external validation set, and the additional set of three compounds) has been used to build the CoMFA model of rSULT1A1 that is presented in Fig. 2. The leave-one-out cross-validated q^2 value for this CoMFA model was 0.744 and the noncross-validated r^2 value for the model was 0.943. The quantitative result from this analysis is shown in Fig. 2A and the contour coefficient map is shown in Fig. 2B. When the contour coefficient map was examined by overlay with the homology model of the rSULT1A1, amino acids that corresponded to regions of unfavorable steric bulk for substrates were identified as Pro43, Lys44, Phe77, Phe80, Tyr135, Phe138, Tyr236, Met244, and Phe251 (Sharma and Duffel, 2002). Two of the residues that occupied the region where steric bulk in the substrate was not preferred, namely Phe77 and Phe138, have been shown to be responsible for the stereospecificity rSULT1A1 with enantiomers of 1,2,3,4-tetrahydro-1-naphthol (Sheng et al., 2004).

A   B

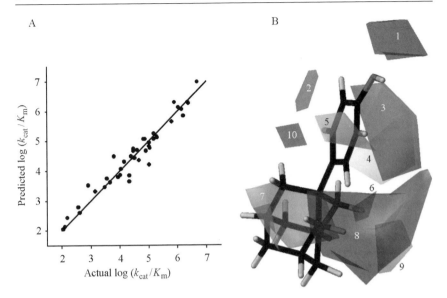

FIG. 2. CoMFA results for catalytic specificity of rSULT1A1. (A) Quantitative results for a total of 44 substrates are shown (data from Sharma and Duffel, 2002). The line indicates a perfect agreement between actual and predicted values and can be used to estimate the error in the predicted values; it is not a regression line. (B) The contour coefficient map from CoMFA of rSULT1A1 is overlaid on the structure of 4-(1-adamantyl)-phenol. Sterically favored regions for substrates (numbered 7 and 9) are shown in green, whereas sterically unfavorable regions (numbered 4 and 5) are shown in yellow. Regions favoring a positive charge on the substrate are shown in red (numbered 2, 3, and 6), and regions favoring a negative charge on the substrate are in blue (numbered 1, 8, and 10). (See color insert.)

Application of CoMFA Methodology to rSULT2A3

Homology Modeling of the Catalytically Competent Complex of rSULT2A3

The X-ray crystal structure of the PAP–enzyme complex of a human hydroxysteroid sulfotransferase, hSULT2A1 (known previously as DHEA-ST or human ST2A3), shows significant overall similarities in topology, PAP-binding site architecture, and catalytic core to the structures of other cytosolic sulfotransferases determined to date (Pedersen *et al.*, 2000). The homology model of rSULT2A3 was built using the coordinates for the crystal structure of human hydroxysteroid sulfotransferase, hSULT2A1 (PDB file 1EFH) (Pedersen *et al.*, 2000). As hSULT2A1 belongs to the same sulfotransferase gene family as rSULT2A3, it forms an excellent template for homology modeling of rSULT2A3. The construction and manipulation of the rSULT2A3 homology model and structures of the substrate molecules were performed using the molecular modeling software package SYBYL (version 6.5, Tripos Associates, St. Louis, MO). The initial homology model of rSULT2A3, produced with the COMPOSER option of SYBYL, afforded a high-quality alignment with hSULT2A1 showing sequence identity score, significance score, and native alignment score values of 60.8%, 40.6, and 332.6, respectively. The final model contained 278 amino acids (residues 4–281). Hydrogen atoms were added to the rSULT2A3 model, and it was minimized using the AMBER force field option in SYBYL. Because the starting template contained only the coordinates of PAP and no acceptor substrate, the positions of PAPS and a representative acceptor substrate were manually modeled based on the available information on the binding of estrogen in estrogen sulfotransferase (Kakuta et al., 1997, 1998b). The structure of PAP was modified to represent PAPS by extraction and modification of PAP–vanadate from PDB file 1BO6 for the PAP–vanadate–estrogen sulfotransferase complex. As the substrate for refinement of the homology model, structural data for dehydroepiandrosterone (DHEA) were extracted from PDB file 3DHE and merged into the rSULT2A3 homology model. Charges for the DHEA molecule and PAPS were calculated by the Gasteiger–Huckel method, whereas Kollman-all charges were used for the protein. Conformations of the amino acid side chains lining the DHEA-binding site were minimized iteratively. These iterations involved holding the amino acid backbone fixed along with the DHEA molecule and performing a minimization of the side chains, followed by holding the side chains fixed along with DHEA molecule and performing a set of minimizations for the amino acid backbone. This step was repeated three times. The final result was the energy-minimized rSULT2A3 homology model with PAPS and DHEA in its active site. The resulting homology model of rSULT2A3 had a high degree of structural similarity with the crystal structure of hSULT2A1, exhibiting a RMSD of 0.33 Å over the best-fit α carbon backbone. Most of the amino acids in the sulfuryl acceptor-binding site of both hSULT2A1 and rSULT2A3 were identical; however, conformational alterations of residues Trp77 and Phe133, above and below the DHEA molecule, were required to accommodate the substrate in the active site. Additionally, after the minimization of DHEA into the acceptor substrate-binding pocket of the homology model of rSULT2A3, conformational changes are also seen in the loop Tyr231-Gln244. These are the same regions that were identified earlier and suggested to flex in order to accommodate DHEA in the structure of hSULT2A1 (Pedersen et al., 2000).

Application of the Alignment Rule and Construction of the
CoMFA Model

The compounds shown in Table I were used for generation of the rSULT2A3 CoMFA, and compounds shown in Table II were used to evaluate its predictive capacity. These molecules range from simple primary alcohols to secondary aliphatic alcohols, *p*-alkyl substituted benzylic alcohols, chiral benzylic alcohols, and hydroxymethyl derivatives of polycyclic aromatic hydrocarbons. The alignment method described earlier for rSULT1A1 (Sharma and Duffel, 2002) was used for the conformational alignment of the substrates for rSULT2A3 that are shown in Table I. The structure of each rSULT2A3 substrate was constructed using SYBYL molecular modeling software. Molecular mechanics calculations were performed with the Powel force field using a minimum energy change of 0.01 kcal/mol as the convergence criterion and charges for the molecule were calculated using the Gasteiger–Hückel method. The resulting structure was placed into the substrate-binding pocket of rSULT2A3. The enzyme was given Kollman all-atom charges and the amino acid residues in the homology model were held fixed, as were the positions of the oxygen and hydrogen of the alcohol in relation to the sulfuryl group and catalytic histidine, respectively. The energy of the entire system was then minimized, and the conformation of the substrate was subsequently extracted from the enzyme and utilized in the data set for CoMFA studies. The leave-one-out cross-validated q^2 value for the resulting CoMFA model was 0.664 and the noncross-validated r^2 value was 0.948.

As the molecules in Table II were not used in the generation of this CoMFA model, they represented a set of molecules that challenged the predictive ability of the CoMFA technique. Thus, to further evaluate the CoMFA model of rSULT2A3, each molecule in Table II was constructed and aligned subject to the same alignment rules as the molecules from Table I. Their $\log(k_{cat}/K_m)$ values were predicted using the CoMFA model generated from Table I. Both the experimentally determined values and the predicted values are shown in Table II. It can be seen that 1-methylcyclohexanol had a predicted value matching the experimentally determined value, whereas the predicted value for *cis*-4-methylcyclohexylmethanol showed the largest difference from the experimental value (i.e., 0.75 log units). Other differences between predicted and experimental values for $\log k_{cat}/K_m$ were from 0.01 to 0.3.

A final CoMFA model was prepared using all data from Tables I and II, and both the quantitative comparison and the contour coefficient map are seen in Fig. 3. This CoMFA model had a leave-one-out cross-validated q^2 value of 0.719 and a noncross-validated r^2 value of 0.972. When the contour

TABLE I

Substrates and Kinetic Constants for rSULT2A3 Used in the Development of a CoMFA Model for Catalytic Specificity[a]

Compound	K_m (app)	V_{max}	log (k_{cat}/K_m)	Reference
Propan-1-ol	610	87.2	0.68	Chen et al. (1996)
Butan-1-ol	106	80.7	1.40	Chen et al. (1996)
Pentan-1-ol	19.2	70.5	2.09	Chen et al. (1996)
Heptan-1-ol	2.20	65.4	2.99	Chen et al. (1996)
Octan-1-ol	0.77	55.5	3.38	Chen et al. (1996)
Nonan-1-ol	0.37	51.0	3.66	Chen et al. (1996)
Decan-1-ol	0.31	50.2	3.73	Chen et al. (1996)
Undecan-1-ol	0.26	37.0	3.67	Chen et al. (1996)
Dodecan-1-ol	0.33	25.7	3.41	Chen et al. (1996)
Tridecan-1-ol	0.23	10.0	3.16	Chen et al. (1996)
Tetradecan-1-ol	0.23	5.70	2.91	Chen et al. (1996)
Phenylmethanol	25.8	96.4	2.09	Chen et al. (1996)
(4-Methyphenyl)methanol	2.19	74.9	3.05	Chen et al. (1996)
(4-Ethylphenyl)methanol	0.67	77.3	3.58	Chen et al. (1996)
(4-Butylphenyl)methanol	0.16	70.3	4.16	Chen et al. (1996)
(4-Pentylphenyl)methanol	0.12	63.1	4.24	Chen et al. (1996)
(4-Hexylphenyl)methanol	0.20	65.4	4.04	Chen et al. (1996)
(4-Heptylphenyl)methanol	0.19	67.4	4.07	Chen et al. (1996)
(4-Octylphenyl)methanol	0.29	58.2	3.82	Chen et al. (1996)
Cyclohexylmethanol	2.00	101	3.22	Chen et al. (1996)
trans-4-Methylcyclohexylmethanol	5.10	39.2	2.41	Chen et al. (1996)
(2R)-Heptan-2-ol	4.70	54.0	2.58	Chen et al. (1996)
(2S)-Heptan-2-ol	4.90	41.8	2.45	Chen et al. (1996)
Heptan-4-ol	7.20	43.4	2.30	Chen et al. (1996)
2-Methylhexan-2-ol	24.6	21.4	1.46	Chen et al. (1996)
3-Ethylpentan-3-ol	40.3	13.9	1.06	Chen et al. (1996)
(1R)-2-Methyl-1-phenylpropan-1-ol	2.80	27.0	2.50	Banoglu and Duffel (1997)
(1R)-1-Phenylbutan-1-ol	3.30	85.5	2.93	Banoglu and Duffel (1997)
(1R)-1-Phenylpentan-1-ol	1.40	154	3.56	Banoglu and Duffel (1997)
(1R)-1-Phenylhexan-1-ol	0.46	157	4.05	Banoglu and Duffel (1997)
(1R)-1-Phenylheptan-1-ol	0.38	94.3	3.92	Banoglu and Duffel (1997)
(R)-Cyclohexyl(phenyl)methanol	1.00	17.8	2.77	Banoglu and Duffel (1997)
(1S)-2-Methyl-1-phenylpropan-1-ol	3.00	22.8	2.40	Banoglu and Duffel (1997)
(1S)-1-Phenylbutan-1-ol	3.70	58.8	2.72	Banoglu and Duffel (1997)
(1S)-1-Phenylpentan-1-ol	1.90	70.3	3.09	Banoglu and Duffel (1997)
(1S)-1-Phenylhexan-1-ol	0.53	59.2	3.57	Banoglu and Duffel (1997)
(1S)-1-Phenylheptan-1-ol	0.42	37.3	3.47	Banoglu and Duffel (1997)
(1R)-1-(2-Naphthyl)ethanol	1.00	34.0	3.05	Banoglu and Duffel (1999)

(continued)

TABLE I (*continued*)

Compound	K_m (app)	V_{max}	log (k_{cat}/K_m)	Reference
(1R)-1-(9-Phenanthryl) ethanol	0.09	4.50	3.22	Banoglu and Duffel (1999)
(1R)-1,2-Dihydroacenaphthylen-1-ol	0.56	83.0	3.69	Banoglu and Duffel (1999)
(1S)-1-(2-Naphthyl)ethanol	1.20	28.5	2.90	Banoglu and Duffel (1999)
(1S)-1,2-Dihydroace-naphthylen-1-ol	0.40	121	4.00	Banoglu and Duffel (1999)

[a] Values for apparent K_m are expressed in mM, values for V_{max} are expressed in nanomoles $(min)^{-1}$ (mg rSULT2A3)$^{-1}$, and values of k_{cat}/K_m are expressed as min^{-1} M^{-1}.

TABLE II

EVALUATION OF KINETIC PARAMETERS FOR rSULT2A3 PREDICTED BY THE COMFA MODEL[a]

Compound	K_m(app)	V_{max}	log(k_{cat}/K_m) A	P
Hexan-1-ol	4.30	69.7	2.73	2.43
(4-Propylphenyl)methanol	0.40	74.7	3.79	3.78
cis-4-Methylcyclohexylmethanol	4.50	55.8	2.61	1.86
1-Methylcyclohexanol	26.8	30.6	1.58	1.58
(1R)-1-(1-Naphthyl)ethanol	0.80	52.0	3.33	3.63

[a] Values for apparent K_m are expressed in mM, and values for V_{max} are expressed in nanomoles min^{-1}(mg of rSULT2A3)$^{-1}$. Values of k_{cat}/K_m are expressed as min^{-1} M^{-1}. Column A represents the experimentally determined values, whereas column P represents the predicted value from the model. Experimentally determined values of K_m and V_{max} for (1R)-1-(1-naphthyl)ethanol were from Banoglu and Duffel (1999); all other experimental values were obtained from Chen et al. (1996).

coefficient map from this CoMFA was overlaid with the homology model of rSULT2A3, regions of unfavorable steric bulk for substrates were either near or overlapping with amino acid residues Phe133, Trp134, Tyr160, and Phe245. The region of favorable steric bulk for substrates corresponded to a cavity in the active site of the homology model that was lined by the hydrophobic amino acids Pro14-Gly17, Trp77-Thr80, Leu233, and Met234.

Conclusion

Although CoMFA is often utilized solely as a ligand-based approach for 3D-QSAR, the adaptation of this method to include structural and mechanistic features of sulfotransferases has led to the development of

A   B

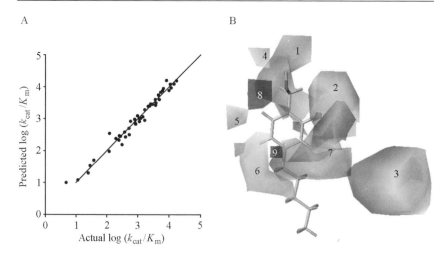

FIG. 3. CoMFA results for catalytic specificity of rSULT2A3. (A) Quantitative results
from CoMFA of substrates in Tables I and II. The line indicates a perfect agreement between
actual and predicted values and can be used to estimate the error in the predicted values; it is
not a regression line. (B) Contour coefficient map from CoMFA of rSULT2A3 substrates
from Tables I and II is shown overlaid on the structure of (4-pentylphenyl)methanol. A
sterically favored region for substrates (numbered 7) is shown in green, whereas sterically
unfavorable regions (numbered 1–6) are shown in yellow. Regions shown as favoring a
negative charge on the substrate (numbered 8 and 9) are in blue, and a single small region
favoring a positive charge is seen in red. (See color insert.)

highly predictive models for catalytic specificity of these enzymes with
their substrates. The most important features of this adaptation of the
CoMFA method are the use of k_{cat}/K_m values as the kinetic parameters
for correlation analysis and the coupling of this with the determination
of substrate conformations based on structural analysis of the sulfotransferase and relevant crystal structure models for the in-line sulfuryl
transfer. By utilizing coordinates obtained from the crystal structure of
PAP–vanadate–estrogen sulfotransferase (Kakuta et al., 1997, 1998b) as a
mimic of the sulfuryl transfer mechanism, it is possible to develop a
model of the position of a catalytically competent enzyme substrate complex within the structure of a sulfotransferase homology model, and the
resulting constraints on conformations for the substrate can be used to
guide determination of appropriate substrate conformations for use in
CoMFA.

    The CoMFA models that result from this enzyme- and mechanism-
guided approach to conformational assignments for substrates are consistent with sulfotransferase homology models and crystal structures and are

predictive with respect to new molecules that are not utilized in developing the models. Furthermore, the strong conservation of overall three-dimensional structure among those sulfotransferases thus far crystallized (Negishi *et al.*, 2001) suggests that the approaches outlined for rSULT1A1 and rSULT2A3 will be more generally applicable to a broad range of isoforms for both aryl (phenol) and alcohol (hydroxysteroid) sulfotransferases.

Acknowledgment

These studies were supported by the U.S. Public Health Service Grant CA38683 awarded by the National Cancer Institute, Department of Health and Human Services.

References

Banoglu, E., and Duffel, M. W. (1997). Studies on the interactions of chiral secondary alcohols with rat hydroxysteroid sulfotransferase STa. *Drug Metab. Dispos.* **25,** 1304–1310.

Banoglu, E., and Duffel, M. W. (1999). Importance of peri-interactions on the stereospecificity of rat hydroxysteroid sulfotransferase STa with 1-arylethanols. *Chem. Res. Toxicol.* **12,** 278–285.

Binder, T. P., and Duffel, M. W. (1988). Sulfation of benzylic alcohols catalyzed by aryl sulfotransferase IV. *Mol. Pharmacol.* **33,** 477–479.

Blanchard, R. L., Freimuth, R. R., Buck, J., Weinshilboum, R. M., and Coughtrie, M. W. (2004). A proposed nomenclature system for the cytosolic sulfotransferase (SULT) superfamily. *Pharmacogenetics* **14,** 199–211.

Brot, F. E., and Bender, M. L. (1969). Use of the specificity constant of α-chymotrypsin. *J. Am. Chem. Soc.* **91,** 7187–7191.

Campbell, N. R., Van Loon, J. A., Sundaram, R. S., Ames, M. M., Hansch, C., and Weinshilboum, R. (1987). Human and rat liver phenol sulfotransferase: Structure-activity relationships for phenolic substrates. *Mol. Pharmacol.* **32,** 813–819.

Chen, G., Banoglu, E., and Duffel, M. W. (1996). Influence of substrate structure on the catalytic efficiency of hydroxysteroid sulfotransferase STa in the sulfation of alcohols. *Chem. Res. Toxicol.* **9,** 67–74.

Chen, X., Yang, Y.-S., Zheng, Y., Martin, B. M., Duffel, M. W., and Jakoby, W. B. (1992). Tyrosine-ester sulfotransferase from rat liver: Bacterial expression and identification. *Protein Expression Purif.* **3,** 421–426.

Cramer, R. D., Patterson, D. E., and Bunce, J. D. (1988). Comparative molecular field analysis (CoMFA). 1. Effect of shape on binding of steroids to carrier proteins. *J. Am. Chem. Soc.* **110,** 5959–5967.

Dajani, R., Cleasby, A., Neu, M., Wonacott, A. J., Jhoti, H., Hood, A. M., Modi, S., Hersey, A., Taskinen, J., Cooke, R. M., Manchee, G. R., and Coughtrie, M. W. (1999). X-ray crystal structure of human dopamine sulfotransferase, SULT1A3: Molecular modeling and quantitative structure-activity relationship analysis demonstrate a molecular basis for sulfotransferase substrate specificity. *J. Biol. Chem.* **274,** 37862–37868.

Duffel, M. W. (1997). Sulfotransferases. *In* "Biotransformation" (F. P. Guengerich, ed.), Vol. 3, pp. 365–383. Elsevier, Oxford.

Duffel, M. W., Chen, G., and Sharma, V. (1998). Studies on an affinity label for the sulfuryl acceptor binding site in an aryl sulfotransferase. *Chem. Biol. Interact.* **109,** 81–92.

Duffel, M. W., and Jakoby, W. B. (1981). On the mechanism of aryl sulfotransferase. *J. Biol. Chem.* **256,** 11123–11127.

Duffel, M. W., Marshall, A. D., McPhie, P., Sharma, V., and Jakoby, W. B. (2001). Enzymatic aspects of the phenol (aryl) sulfotransferases. *Drug Metab. Rev.* **33,** 369–395.

Falany, C. N. (1997). Enzymology of human cytosolic sulfotransferases. *FASEB J.* **11,** 206–216.

Glatt, H., Boeing, H., Engelke, C. E., Ma, L., Kuhlow, A., Pabel, U., Pomplun, D., Teubner, W., and Meinl, W. (2001). Human cytosolic sulphotransferases: Genetics, characteristics, toxicological aspects. *Mutat. Res.* **482,** 27–40.

Glatt, H., and Meinl, W. (2004). Pharmacogenetics of soluble sulfotransferases (SULTs). *Naunyn. Schmiedebergs Arch. Pharmacol.* **369,** 55–68.

Kakuta, Y., Pedersen, L. G., Carter, C. W., Negishi, M., and Pedersen, L. C. (1997). Crystal structure of estrogen sulphotransferase. *Nature Struct. Biol.* **4,** 904–908.

Kakuta, Y., Pedersen, L. G., Pedersen, L. C., and Negishi, M. (1998a). Conserved structural motifs in the sulfotransferase family. *Trends Biochem. Sci.* **23,** 129–130.

Kakuta, Y., Petrotchenko, E. V., Pedersen, L. C., and Negishi, M. (1998b). The sulfuryl transfer mechanism: Crystal structure of a vanadate complex of estrogen sulfotransferase and mutational analysis. *J. Biol. Chem.* **273,** 27325–27330.

King, R. S., Sharma, V., Pedersen, L. C., Kakuta, Y., Negishi, M., and Duffel, M. W. (2000). Structure-function modeling of the interactions of N-alkyl-N-hydroxyanilines with rat hepatic aryl sulfotransferase IV. *Chem. Res. Toxicol.* **13,** 1251–1258.

Marshall, A. D., McPhie, P., and Jakoby, W. B. (2000). Redox control of aryl sulfotransferase specificity. *Arch. Biochem. Biophys.* **382,** 95–104.

Masimirembwa, C. M., Ridderstrom, M., Zamora, I., and Andersson, T. B. (2002). Combining pharmacophore and protein modeling to predict CYP450 inhibitors and substrates. *Methods Enzymol.* **357,** 133–144.

Mulder, G. J., and Jakoby, W. B. (1990). Sulfation. *In* "Conjugation Reactions in Drug Metabolism" (G. J. Mulder, ed.), pp. 107–161. Taylor & Francis, New York..

Nagata, K., and Yamazoe, Y. (2000). Pharmacogenetics of sulfotransferase. *Annu. Rev. Pharmacol. Toxicol.* **40,** 159–176.

Negishi, M., Pedersen, L. G., Petrotchenko, E., Shevtsov, S., Gorokhov, A., Kakuta, Y., and Pedersen, L. C. (2001). Structure and function of sulfotransferases. *Arch. Biochem. Biophys.* **390,** 149–157.

Northrup, D. B. (1999). Rethinking fundamentals of enzyme action. *Adv. Enzymol. Relat. Areas .Mol. Biol.* **73,** 25–55.

Pedersen, L. C., Petrotchenko, E. V., and Negishi, M. (2000). Crystal structure of SULT2A3, human hydroxysteroid sulfotransferase. *FEBS Lett.* **475,** 61–64.

Rao, S. I., and Duffel, M. W. (1991). Benzylic alcohols as stereospecific substrates and inhibitors for aryl sulfotransferase. *Chirality* **3,** 104–111.

Sharma, V., and Duffel, M. W. (2002). Comparative molecular field analysis of substrates for an aryl sulfotransferase based on catalytic mechanism and protein homology modeling. *J. Med. Chem.* **45,** 5514–5522.

Sheng, J., Saxena, A., and Duffel, M. W. (2004). Influence of phenylalanines 77 and 138 on the stereospecificity of aryl sulfotransferase IV. *Drug Metab. Dispos.* **32,** 559–565.

Sipilä, J., Hood, A. M., Coughtrie, M. W., and Taskinen, J. (2003). CoMFA modeling of enzyme kinetics: K(m) values for sulfation of diverse phenolic substrates by human catecholamine sulfotransferase SULT1A3. *J. Chem. Inf. Comput. Sci.* **43,** 1563–1569.

Surh, Y.-J., and Miller, J. A. (1994). Roles of electrophilic sulfuric acid ester metabolites in mutagenesis and carcinogenesis by some polynuclear aromatic hydrocarbons. *Chem. Biol. Interact.* **92,** 351–362.

Weinshilboum, R. M., Otterness, D. M., Aksoy, I. A., Wood, T. C., Her, C., and Raftogianis, R. B. (1997). Sulfation and sulfotransferases 1: Sulfotransferase molecular biology: cDNAs and genes. *FASEB J.* **11,** 3–14.

[15] Glucuronidase Deconjugation in Inflammation

By Kayoko Shimoi and Tsutomu Nakayama

Abstract

This chapter focuses on deglucuronidation by β-glucuronidase in inflammation. We investigated whether glucuronides were converted to free parent compounds by β-glucuronidase released from human-stimulated neutrophils in inflammation. β-Glucuronidase activity was assayed using 4-methylumbelliferyl-glucuronide and methanol extracts of rat plasma containing luteolin monoglucuronide as a substrate. The released 4-methylumbelliferone, a fluorescent molecule, was quantitated on a microplate fluorometer. Deglucuronidation of luteolin monoglucuronide was examined by high-performance liquid chromatography (HPLC) analysis. The β-glucuronidase activity in mouse plasma after iv injection of lipopolysaccharide (LPS) increased with time, as did the levels of inflammation marker, tumor necrosis factor-α, and soluble intercellular adhesion molecule-1. Four kinds of human cell (neutrophils, human umbilical vein endothelial cells, IMR-90, and Caco-2) possess β-glucuronidase activity. Among these, Caco-2 cells showed the highest level of β-glucuronidase activity. Supernatants obtained from neutrophils stimulated with cytochalasin B and ionomycin showed higher levels of β-glucuronidase activity than those of nonstimulated neutrophils. HPLC analyses also showed that supernatants obtained from stimulated neutrophils hydrolyzed luteolin monoglucuronide to free luteolin. As reported previously (Shimoi *et al.*, 1998), two main peaks (free luteolin and luteolin monoglucuronide) were observed in plasma of rats administered with luteolin. In LPS-treated rats, the peak of luteolin monoglucuronide decreased to about half and the ratio of luteolin to luteolin monoglucuronide increased. These results suggest that β-glucuronidase released from neutrophils or certain injured cells may hydrolyze glucuronide conjugates to free aglycones at the site of inflammation.

METHODS IN ENZYMOLOGY, VOL. 400   0076-6879/05 $35.00
Copyright 2005, Elsevier Inc. All rights reserved.   DOI: 10.1016/S0076-6879(05)00015-7

Introduction

Glucuronidation is catalyzed by UDP-glucuronosyltransferase and a metabolic pathway in the detoxification of exogenous compounds such as drugs, food components, and environmental pollutants. Many reports have indicated that glucuronidation plays an important role in the homeostasis of various active endobiotics, including steroid hormones, bilirubin, fatty acids, and biliary acids (Guillemette et al., 2004). Glucuronides are generally regarded as being pharmacologically inactive and excreted in urine and bile. Conversely, deglucuronidation has been reported to occur in the gut. Glucuronides such as bilirubin glucuronides excreted into bile can be hydrolyzed to produce free parent compounds by intestinal bacteria that have β-glucuronidase (Heneghan, 1988; Qin, 2002).

β-Glucuronidase is located intracellularly in lysosomes of mammalian cells (Sperker et al., 1997). Human β-glucuronidase (EC 3.2.1.31) is an acid hydrolase and its physiological role is glycosaminoglycan degradation in lysosomes (Paigen, 1989). Mürdter et al. (1997) reported that β-glucuronidase is present at higher levels in many tumors than in the surrounding normal tissue and that cleavage of the glucuronide prodrug by β-glucuronidase occurred at the tumor site. It was shown that active β-glucuronidase is present extracellularly in large amounts in necrotic areas of solid cancers and that cells present in necrosis with β-glucuronidase activity are mainly monocytes and granulocytes (Bosslet et al., 1998). In inflammation, β-glucuronidase is known to be released from granulocytes, including neutrophils (Marshall et al., 1988).

Flavonoids, which occur naturally in common plant foods such as vegetables, fruits, and tea, are absorbed and metabolized into glucuronide or sulfate conjugates. Many studies have demonstrated that conjugates circulate in the blood (Scalbert and Williamson, 2000). Quercetin glucuronides and sulfates have been detected in human plasma. However, free quercetin has never been detected (Day et al., 2001). It remains unclear whether these conjugates still retain the biological functions in vivo that their aglycones do. As the conjugates are polar and water soluble, an organic anion transporter or deconjugation is required for the uptake of the conjugates into cells. For this reason, we investigated whether luteolin monoglucuronide is converted to free luteolin during inflammation using human neutrophils stimulated with ionomycin/cytochalasin B and lipopolysaccharide (LPS)-treated rats. We reported that supernatants obtained from neutrophils stimulated with ionomycin/cytochalasin B hydrolyzed luteolin monoglucuronide to free luteolin and that β-glucuronidase activity in rat plasma increased after iv injection of LPS (Shimoi et al., 2000, 2001). These results suggested that β-glucuronidase was secreted from stimulated neutrophils at extracellular locations and then deglucuronidation of flavonoids occurred in inflammation. This chapter focuses on deconjugation, which makes inactive metabolites active, by β-glucuronidase in inflammation.

Inflammation and Plasma β-Glucuronidase Activity in Mice

Male ICR mice (6 weeks old, Japan, SLC, Inc., Hamamatsu) are housed in an air-conditioned room with free access to CE-2 commercial food pellets (Clea Japan, Tokyo). The diet is changed to a synthetic basic diet that consists of 38% corn starch, 25% casein, 10% α starch, 8% cellulose powder, 6% minerals, 5% sugar, 2% vitamins, and 6% lard (Oriental Yeast Co., Tokyo) 1 week before the experiment. Two or three mice are assigned to each experimental group. The animals are maintained and handled according to the guidelines for the regulation of animal experimentation committee of the University of Shizuoka. Mice fasted overnight receive an iv injection of LPS (*Escherichia coli* 0111:B4; Difco, Detroit, MI) dissolved in 0.9% NaCl at a dose of 3.75 mg/kg and are anesthetized with ethyl ether at the indicated times after injection. The mouse blood is withdrawn from the abdominal aorta into EDTA-treated tubes, and plasma is prepared by centrifugation for 15 min at 900g. The plasma levels of tumor necrosis factor-α (TNF-α) and soluble intercellular adhesion molecule 1(sICAM-1), which are regarded as inflammation markers, are determined by an ELISA kit(ENDOGEN, Cambridge, MA). β-Glucuronidase activity in each sample is assayed using the FluorAce β-glucuronidase reporter assay kit (Bio-Rad, Hercules, CA). Briefly, 50 μl of assay buffer and 10 μl of mouse plasma are added to each sample well of a 96-well microplate. The microtiter plate is incubated at 37° for 30 min. β-Glucuronidase hydrolyzes 4-methylumbelliferyl-glucuronide (4MUG), resulting in release of the fluorescent molecule 4-methylumbelliferone (4MU). The fluorescence of this molecule is then measured on a microplate fluorometer using an excitation wavelength of 360 nm and an emission wavelength of 460 nm.

Lipopolysaccharide is an outer membrane component of gram-negative bacteria that stimulates the release of various proinflammatory cytokines, such as TNF-α and interleukin (IL)-6. TNF-α induces cellular expression of ICAM-1, which is one of the adhesion molecules, and mediates the binding of leukocytes to endothelial cells and the entry of leukocytes into endothelium. This is a crucial process in the inflammatory response. As shown in Fig. 1, the plasma TNF-α levels reached their peak 1 h after injection of LPS and then decreased rapidly. The second peak level was observed 10 h after dosing. In contrast, the plasma levels of sICAM-1 increased in a time-dependent manner. It took much longer for sICAM-1 to reach a peak

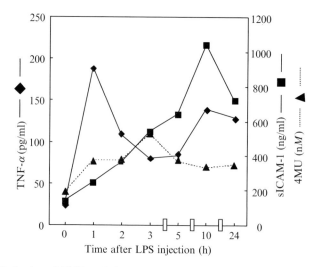

Fig. 1. Induction of TNF-α, β-glucuronidase, and sICAM-1 by treatment with LPS in mice. Male ICR mice were treated with LPS (3.75 mg/kg, iv). Blood was collected at the indicated times after LPS injection. The plasma levels of TNF-α and sICAM-1 were assayed by an ELISA kit (ENDOGEN). The β-glucuronidase activity of mouse plasma was assayed by liberation of 4MU from 4MUG using the FluoroAce β-glucuronidase reporter assay kit (Bio-Rad). Values are mean (n = 3 or 2).

than TNF-α. The β-glucuronidase activity in mouse plasma after LPS treatment also increased with time, as observed in the previous experiment using rats (Shimoi et al., 2001). We also reported that β-glucuronidase activity in human serum from patients on hemodialysis increased significantly compared with that from healthy volunteers (Shimoi et al., 2001). These results suggest that inflammation induces the release of β-glucuronidase.

β-Glucuronidase Activity in Various Human Cells

Human neutrophils are isolated from the venous blood of healthy volunteers (who gave informed consent before the experiment) using standard dextran sedimentation and gradient separation on Lymphoprep (NYCOMED, Oslo, Norway). Contaminating erythrocytes are lysed by 0.6 M KCl treatment. Neutrophils are >95% pure, and viability exceeds 98% as assessed by Trypan blue dye exclusion. HUVEC, IMR-90, and Caco-2 cells are cultured to the confluent in MCDB-104 (Nihon Pharmaceutical Co., Ltd.) supplemented with 10% fetal bovine serum (FBS) (Moregate, Australia and New Zealand), 100 ng/ml endothelial cell growth factor

(ECGF), 10 ng/ml epidermal growth factor (EGF, Collaborative Research Inc., Bedford, MA), 100 μg/ml heparin (Sigma, St. Louis, MO), Eagle's minimal essential medium (GIBCO BRL, Grand Island, NY) supplemented with 10% FBS, Dullbecco's modified Eagle's medium (GIBCO BRL) supplemented with 10% FBS, 10 ml/liter nonessential amino acid solution (GIBCO), and antibiotics (50,000 U/liter penicillin and 50 mg/liter streptomycin, GIBCO), respectively, at 37° in an incubator with 5% CO_2 and 95% air. The cell lysate is prepared as follows: The cells are rinsed with saline G and treated with lysis buffer (0.5% Triton X-100 in phosphate-buffered saline, Complete: protease inhibitor cocktail, Boehringer Mannheim, Mannheim, Germany) and kept on ice for 10 min so as to yield a concentration of 3×10^6 cells/ml lysis buffer. The solution is then centrifuged at 45,000g for 15 min at 4°, and the clear lysate is stored at –80°. The protein concentration of each lysate is determined by the BCA method. The β-glucuronidase activity of the cell lysate is assayed as described earlier.

As shown in Fig. 2, each human cell possesses β-glucuronidase. Among the four types of human cell examined, Caco-2 cells showed the highest level of β-glucuronidase activity. Caco-2 cells are derived from human colon adenocarcinoma and exhibit enterocyte-like characteristics. At the present time, the reason for the higher level of β-glucuronidase activity in Caco-2 cells is unknown. It may be possible that the expression level of β-glucuronidase is higher in cancerous cells such as Caco-2 cells and inflammatory cells such as neutrophils than in normal tissue cells.

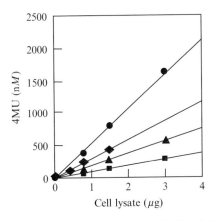

FIG. 2. β-Glucuronidase activity of various human cells. The β-glucuronidase activity in each cell lysate was assayed as described in the legend to Fig. 1. ◆, neutrophils; ■, HUVEC; ▲, IMR-90; ●, Caco-2.

Activity of β-Glucuronidase Released from Human Neutrophils

Human neutrophils are obtained as described earlier. Purified neutrophils (4.2 and 5.5 × 10^6 cells) are resuspended in saline G(200 μl). They are stimulated with 5 μg/ml of cytochalasin B (Sigma) for 5 min at 37° and further treated with 1 μM ionomycin (Sigma) for 20 min at 37°. After centrifugation for 10 min at 4° at 750g, the supernatant is obtained. The activity of β-glucuronidase released from human neutrophils is assayed by two methods using 4MUG and methanol extracts of plasma of rats orally administered with luteolin, which contain luteolin and luteolin monoglucuronide, as a substrate. The former method was described earlier. The latter assay is performed as follows: The supernatants are incubated with the methanol extract at 37° for 2 h. Methanol extracts are obtained by the method described (Shimoi et al., 1998). In brief, rat blood is collected from the abdominal aorta into heparinized tubes after the administration of luteolin (50 μmol/kg in propyleneglycol) by gastric intubation. The plasma (0.5 ml) is acidified with the same volume of 0.01 M oxalic acid. This solution is applied to a Sep-Pak C$_{18}$ cartridge. After washing the cartridge with 0.01 M oxalic acid, methanol/water/0.5 M oxalic acid (25:73:2, v/v/v), and distilled water, the methanol eluate is obtained. The eluate is evaporated to dryness, and the residue is dissolved in 100 μl of methanol. The reaction mixture of the supernatant and methanol extract is acidified with the same volume of 0.01 M oxalic acid. This solution is applied to a Sep-Pak C$_{18}$ cartridge, and then the methanol eluate obtained by the same method as mentioned earlier is evaporated to dryness. The residue dissolved in methanol is subjected to HPLC. HPLC analysis is carried out by the method described previously (Shimoi et al., 1998). Briefly, the samples are analyzed chromatographically by a JASCO HPLC system (Tokyo, Japan) using a Capcell Pak C$_{18}$-UG120 column (150 × 4.6 mm, Shiseido, Tokyo, Japan) and UV detection (349 nm). Elution is performed using methanol with 0.03% trifluoroacetic acid as solvent A and water with 0.01% trifluoroacetic acid as solvent B in gradient conditions at a flow rate of 0.7 ml/min.

The β-glucuronidase activity in the supernatant is assayed by the liberation of luteolin from luteolin monoglucuronide using the HPLC system. The area ratio of luteolin to luteolin monoglucuronide is calculated.

Supernatants obtained from neutrophils stimulated with cytochalasin B and ionomycin showed a higher level of β-glucuronidase activity than those of nonstimulated neutrophils (Fig. 3). The area ratio of luteolin to luteolin monoglucuronide in the methanol extracts after treatment with the supernatants from stimulated neutrophils increased from 0.42 to 3.72 and 3.69, respectively. These results indicate that luteolin monoglucuronide can be a substrate for β-glucuronidase released from neutrophils.

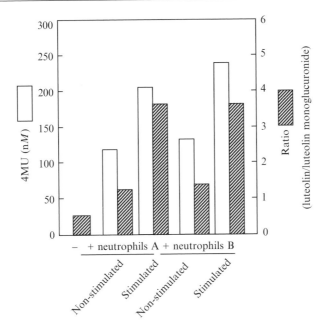

FIG. 3. Activity of β-glucuronidase released from stimulated human neutrophils. β-Glucuronidase activity was assayed with 4MUG as described in the legend to Fig. 1 and with rat plasma extract containing luteolin monoglucuronide. The area ratio of luteolin and luteolin monoglucuronide of the methanol extract of plasma from rats orally administered with luteolin after treatment with or without the supernatant which contain β-glucuronidase released from the stimulated human neutrophils. Neutrophils were isolated from the venous blood of two healthy volunteers (A: 4.2×10^6 cells, B: 5.5×10^6 cells). The reaction mixtures were analyzed by HPLC.

Inflammation and HPLC Plasma Profile of Luteolin-Administered Rats

Male SD rats (7–9 weeks old, Japan, SLC, Inc., Hamamatsu) weighing 180–200 g are housed in the same conditions as in the mice experiments. The rats fast overnight and receive an iv injection of LPS (*E. coli* 0111:B4; Difco) at a dose of 750 μg/kg. Luteolin (50 μmol/kg) is administered to the rats by gastric intubation 3.5 h after LPS injection and the plasma is prepared 1 h after oral administration of luteolin for HPLC analysis as described earlier. Rat blood is withdrawn from the abdominal aorta into EDTA-treated tubes and the plasma is prepared by centrifugation for 10 min at 900g.

As reported previously, two main peaks, free luteolin and luteolin monoglucuronide, were observed in the plasma of rats administered with

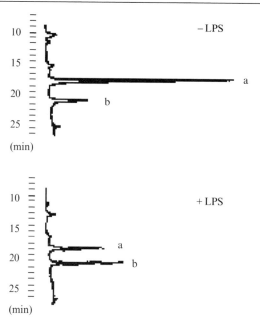

Fig. 4. HPLC chromatogram of the methanol extract of rat plasma with or without LPS treatment. Rats were treated with LPS (750 μg/kg), and luteolin (50 μmol/kg) was administered to rats by gastric intubation 3.5 h after LPS injection. The rat plasma was prepared 1 h after oral administration of luteolin for HPLC analysis. Peak a: luteolin monoglucuronide, peak b: luteolin.

luteolin (Shimoi et al., 1998). Treatment with LPS increased the peak of free luteolin. In contrast, the peak of luteolin monoglucuronide decreased (Fig. 4). In a previous paper, the ratio of luteolin to luteolin monoglucuronide was determined by the peak areas on HPLC chromatograms (Shimoi et al., 2001). The concentration of luteolin monoglucuronide decreased to about half in the LPS-treated rats. When inflammation occurs, β-glucuronidase for deglucuronidation may be dominant compared with UDP-glucuronyltransferase for glucuronidation in the liver. Further investigation of the expression of both enzymes in the liver is required. Conversely, the possibility exists that β-glucuronidase released from neutrophils or certain injured cells may hydrolyze glucuronide conjugates of flavonoids to free aglycones at the site of inflammation.

Discussion

A previous study (Shimoi et al., 2001) demonstrated that the β-glucuronidase activity of serum from patients on hemodialysis and plasma from LPS-treated rats is higher than that of healthy ones. This chapter also demonstrated a time-dependent increase of β-glucuronidase in the plasma of mice treated with LPS. Of the four types of cell investigated, the level of β-glucuronidase activity was found to be highest in Caco-2, followed by human neutrophils, IMR-90, and HUVEC cells in that order. β-Glucuronidase was shown to be released from human neutrophils stimulated with cytochalasin B and ionomycin. These results suggest that β-glucuronidase is released from stimulated neutrophils or injured cells and leaks into the circulation system when inflammation occurs. The content of β-glucuronidase may become high at the sites of inflammation, where the tissue fluid has a low pH (which is favorable for this hydrolyzing enzyme).

β-Glucuronidase hydrolyzes steroid/estrogen hormone glucuronides (Zhu *et al.*, 1996) and glycosaminoglycan (Paigen, 1989). We also showed that β-glucuronidase released from human neutrophils hydrolyzes glucuronide conjugates of flavonoids to free aglycones. Bosslet *et al.* (1998) reported that tumor-infiltrating leukocytes release β-glucuronidase extracellularly, thereby activating a nontoxic, glucuronyl spacer doxorubicin prodrug called HMR 1826. Sperker *et al.* (2000) also demonstrated a significant increase of β-glucuronidase activity and protein levels in pancreatic cancer and chronic pancreatitis. β-Glucuronidase is able to bioactivate prodrug HMR 1826. Therefore, the possibility exists that inactive glucuronides of exogenous compounds such as food components, drugs, and environmental toxic pollutants, which penetrate poorly into living cells, convert to active parent compounds at sites of inflammation.

Acknowledgment

This work was supported by Special Coordination Funds of the Ministry of Education, Culture, Sports, Science and Technology, the Japanese Government.

References

Bosslet, K., Straub, R., Blumrich, M., Czech, J., Gerken, M., Sperker, B., Kroemer, H. K., Gesson, J.-P., Koch, M., and Monneret, C. (1998). Elucidation of the mechanism enabling tumor selective prodrug monotherapy. *Cancer Res.* **58,** 1195–1201.

Day, A. J., Mellar, F., Barron, D., Savazin, G., Morgan, M. R. A., and Williamson, G. (2001). Human metabolism of dietary flavonoids: Identification of plasma metabolites of quercetin. *Free Radic. Res.* **35,** 941–952.

Guillemette, C., Bélanger, A., and Lépine, J. (2004). Metabolic inactivation of estrogens in breast tissue by UDP-glucuronosyltransferase enzymes: An overview. *Breast Cancer Res.* **6,** 246–254.

Heneghan, J. B. (1988). Alimentary tract physiology: Interaction between the host and its microbial flora. *In* "Role of the Gut Flora in Toxicity and Cancer" (I. R. Rowland, ed.), pp. 39–78. Academic Press, San Diego.

Marshall, T., Shult, P., and Busse, W. W. (1988). Release of lysosomal enzyme beta-glucuronidase from isolated human eosinophils. *J. Allergy Clin. Immunol.* **82,** 550–555.

Mürdter, T. E., Sperker, B., Kivistö, K. T., McClellan, M., Fritz, P., Friedel, G., Linder, A., Bosslet, K., Toomes, H., Dierkesmann, R., and Kroemer, H. K. (1997). Enhanced uptake of doxorubicin into bronchial carcinoma: β-Glucuronidase mediates release of doxorubicin from a glucuronide prodrug(HMR 1826) at the tumor site. *Cancer Res.* **57,** 2440–2445.

Paigen, K. (1989). Mammalian beta-glucuronidase: Genetics, molecular biology, and cell biology. *Prog. Nucleic. Acid Res. Mol. Biol.* **37,** 155–205.

Qin, X. F. (2002). Impaired inactivation of digestive proteases by deconjugated bilirubin: The possible mechanism for inflammatory bowel disease. *Med. Hypotheses* **59,** 159–163.

Scalbert, A., and Williamson, G. (2000). Dietary intake and bioavailability. *J. Nutr.* **130,** 2073S–2085S.

Shimoi, K., Okada, H., Furugori, M., Goda, T., Takase, S., Suzuki, M., Hara, Y., Yamamoto, H., and Kinae, N. (1998). Intestinal absorption of luteolin and luteolin 7-*o*-β-glucoside in rats and humans. *FEBS Lett.* **438,** 220–224.

Shimoi, K., Saka, N., Kaji, K., Nozawa, R., and Kinae, N. (2000). Metabolic fate of luteolin and its functional activity at focal site. *Biofactor* **12,** 181–186.

Shimoi, K., Saka, N., Nozawa, R., Sato, M., Amano, I., Nakayama, T., and Kinae, N. (2001). Deglucuronidation of a flavonoid, luteolin monoglucuronide, during inflammation. *Drug Metab. Dispos.* **29,** 1521–1524.

Sperker, B., Backman, J. T., and Kroemer, H. K. (1997). The role of β-glucuronidase in drug disposition and drug targeting in humans. *Clin. Pharmacokinet.* **33,** 18–31.

Sperker, B., Werner, U., Mürdter, T. E., Tekkaya, C., Fritz, P., Wacke, R., Adam, U., Gerken, M., Drewelow, B., and Kroemer, H. K. (2000). Expression and function of β-glucuronidase in pancreatic cancer: Potential role in drug targeting. *Naunyn-Schmiedeberg's Arch Pharmacol.* **362,** 110–115.

Zhu, B. T., Evaristus, E. N., Antoniak, S. K., Sarabia, S. F., Ricci, M. J., and Liehr, J. G. (1996). Metabolic deglucuronidation and demethylation of estrogen conjugates as a source of parent estrogens and catecholestrogen metabolites in Syrian hamster kidney, a target organ of estrogen-induced tumorigenesis. *Toxicol. Appl. Pharmacol.* **136,** 186–193.

[16] Three-Dimensional Structures of Sulfatases

By DEBASHIS GHOSH

Abstract

The sulfatase family of enzymes catalyzes the hydrolysis of sulfate ester bonds of a wide variety of substrates. Nine human sulfatase proteins and their genes have been identified, many of which are associated with genetic disorders leading to reduction or loss of function of the corresponding enzyme. A catalytic cysteine residue, strictly conserved in prokaryotic and eukaryotic sulfatases, is modified posttranslationally into a formylglycine. Hydroxylation of the formylglycine residue by a water molecule forming the activated hydroxylformylglycine (a formylglycine hydrate or a gem-diol) is a necessary step for sulfatase activity of the enzyme. Crystal structures of three human sulfatases, arylsulfatases A and B (ARSA and ARSB) and C, also known as steroid sulfatase or estrone/dehydroepiandrosterone sulfatase (ES), have been determined. In addition, the crystal structure of a homologous bacterial arylsulfatase from *Pseudomonas aeruginosa* (PAS) is also available. While ARSA, ARSB, and PAS are water-soluble enzymes, ES has a hydrophobic domain and is presumed to be bound to the endoplasmic reticulum membrane. This chapter compares and contrasts four sulfatase structures and revisits the proposed catalytic mechanism in light of available structural and functional data. Examination of the ES active site reveals substrate-specific interactions previously identified in another steroidogenic enzyme. Possible influence of the lipid bilayer in substrate capture and recognition by ES is described. Finally, mapping the genetic mutations into the ES structure provides an explanation for the loss of enzyme function in X-linked ichthyosis.

Introduction

The sulfatase family of enzymes catalyzes the hydrolysis of sulfate ester bonds of a wide variety of substrates ranging from sulfated proteoglycans to conjugated steroids and sulfate esters of small aromatics. Nine human sulfatase proteins and their genes have been identified (Parenti *et al.*, 1997). Several of them are associated with genetic disorders leading to reduction or loss of function of corresponding enzymes. The sequence homology

METHODS IN ENZYMOLOGY, VOL. 400
0076-6879/05 $35.00
DOI: 10.1016/S0076-6879(05)00016-9

among the members of the sulfatase family ranges between 20 and 60%. The residues of the active site, which is situated at the amino-terminal half of the polypeptide chain, are highly conserved, reflecting the common catalytic mechanism shared by the members of the family. One particular active site amino acid that is strictly conserved in prokaryotic and eukaryotic sulfatases is a cysteine, which is modified posttranslationally into a formylglycine (Schmidt et al., 1995). Hydroxylation of formylglycine by a water molecule forming the activated hydroxylformylglycine (a formylglycine hydrate or a gem-diol) is a necessary step for sulfatase activity of the enzyme (Recksiek et al., 1998; Schmidt et al., 1995).

Human estrone (E1)/dehydroepiandrosterone (DHEA) sulfatase (ES) is a microsomal enzyme and is known to be associated with the endoplasmic reticulum (ER) membrane (Hernandez-Guzman et al., 2001; Vaccaro et al., 1987). Human arylsulfatases, such as arylsulfatses A (ARSA) and B (ARSB), are lysosomal, optimally active at the acidic pH, and represent soluble forms of the enzyme. However, ES, also known as arylsulfatse C, is most active at or near the neutral pH and can be solubilized only in the presence of detergents (Hernandez-Guzman et al., 2001, 2003). Arylsulfatases D, E, and F are also supposedly localized in ER or golgi compartments (Parenti et al., 1997). Iduronate 2-sulfatase, galactose 6-sulfatase, and glucosamine sulfatases, other members of the human sulfatase family, are localized in lysosomes. The recently determined structure of the native, full-length human placental ES provides the first direct evidence of membrane integration of these enzymes, suggesting roles of lipid bilayer, possibly that of the edoplasmic reticulum membrane, in catalysis (Hernandez-Guzman et al., 2003). ES, which uses 3-hydroxysteroid sulfate esters of E1, DHEA, and cholesterol as substrates, is responsible for maintaining high levels of the active estrogen, 17β-estradiol (E2), in tumor cells. ES catalyzes the hydrolysis of E1-sulfate, which is subsequently reduced to E2 by 17β-hydroxysteroid dehydrogenase type1 (17HSD1) (Ghosh et al., 2002; Reed et al., 1997). The presence of ES in breast carcinomas and ES-dependent proliferation of breast cancer cells have been demonstrated (Billich et al., 2000; Pasqualini et al., 2002). Mutations in the ES gene and inactive enzyme have also been associated with X-linked ichthyosis, a disease related to scaling of the skin (Alperin et al., 1997; Hernandez-Martin et al., 1999). Localization of ES in the smooth and rough ER was demonstrated by immunohistochemical labeling (Noel et al., 1983). ES is synthesized in several tissues, including human placenta, skin fibroblasts, breasts, and fallopian tubes (Burns et al., 1983; Dibbelt et al., 1983; Hernandez-Guzman et al., 2001; Noel et al., 1983; Purohit et al., 1998; Stein et al., 1989; Suzuki et al., 1992; Vaccaro et al., 1987; Van der Loos et al., 1983; Yanaihara et al., 2001).

Crystal structures of three members of the human sulfatase family, human placental ES (Hernandez-Guzman *et al.*, 2003), ARSA (Lukatela *et al.*, 1998), ARSB (Bond *et al.*, 1997), and one homologous bacterial arylsulfatase from *Pseudomonas aeruginosa* (PAS) (Boltes *et al.*, 2001), have been determined. The overall three-dimensional structures of all three soluble sulfatases exhibit a high degree of homology for the domain of ES that scaffolds the catalytic residues. Furthermore, the spatial arrangement of amino acids responsible for hydrolysis of sulfate esters is virtually identical in all four sulfatases, demonstrating the high degree of similarity of their catalytic mechanism. Nonetheless, subtle differences in sequences of the substrate-binding cleft result in differences in the active site architecture that account for the variation in substrate specificity. Having a membrane-spanning domain bordering the lipid bilayer and partially contributing to the architecture of the active site makes ES uniquely different from the other three known structures of the sulfatase family. This chapter describes and discusses the commonality of the sulfatase structures in relation to their function, with special emphasis on the ES structure, probing the origin of its steroidal substrate specificity at the atomic level.

The cytoplasmic side of the rough ER is the site for biosynthesis for membrane and secreted proteins, which are then transported to the lumen side across the ER membrane by translocation signal peptides at amino termini. Many ER-resident steroidogenic enzymes are known to possess strong ER membrane association that presumably plays a role in their functionality. The proximity of the putative membrane-spanning domain and lipid-associated regions to the active site in ES suggests a direct influence of the lipid bilayer on the enzyme activity, thereby providing structural evidence for the functional significance of membrane association in this class of microsomal enzymes.

Overall Structure of Sulfatases

The three-dimensional (3-D) structure of the full-length ES from human placenta as determined by X-ray crystallography is shown in Fig. 1A (PDB ID code: 1P49). Figure 1B is a least-squares superposition of all four structures (PDB ID codes are ARSA, 1AUK; ARSB, 1FSU; and PAS, 1HDH). The topology of the overall fold of ES is shown in Fig. 2A. The tertiary structure consists of two domains: a globular (55 × 60 × 70 Å), polar domain containing the catalytic site and a putative transmembrane domain consisting of two antiparallel hydrophobic α helices. The major polar domain consists of two subdomains with the α/β sandwich fold (Fig. 2A). The secondary structure elements are described in Table I. Subdomain 1 (SD1), wound around a central 11-stranded (strands 1 and

A   B

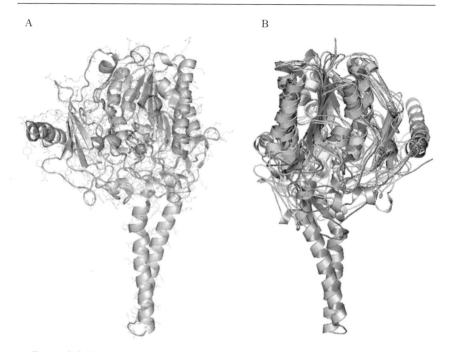

FIG. 1. (A) The crystal structure of ES showing the course of the polypeptide backbone through α helices, β sheets, and loop regions. Side chains are shown as thin bonds. Atoms are color coded: carbon, green; nitrogen, blue; oxygen, red; and sulfur, yellow/orange. The bivalent cation Ca^{2+} at the active site is shown as a grayish purple sphere. Side chain ligands to the metal atom are shown as thick bonds. (B) A least-square superposition of four crystal structures of sulfatases: ES, green; ARSA, cyan; ARSB, pink; and PAS, yellow. The view is roughly 180° rotated about a vertical axis with respect to the view in A. (See color insert.)

2, 4–11, and 17) mixed β sheet flanked by 13 α-helices/helical turns (helices 1–7, 10–15), contains the catalytic core. Subdomain 2 (SD2), consisting of roughly 110 C-terminal residues and wound around a 4-stranded antiparallel β sheet (strands 13–16) flanked by α helix 16, packs against turn and loop regions of the β sheet of subdomain 1. The two putative transmembrane helices protrude on one side of the nearly spherical polar domain, thereby giving the overall molecule a "mushroom-like" shape. In contrast, the topology (Fig. 2B) of the tertiary structure of ARSA, a representative structure of the three soluble sulfatases, has the transmembrane helix pair of ES missing. Nevertheless, the overall fold of the catalytic domain of ES, the location, and the composition of two β sheets of SD1 and SD2, as well as the locations, number, and lengths of the flanking helices, closely resemble those of the three soluble sulfatases.

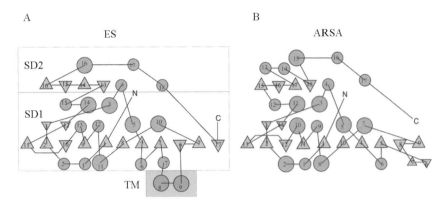

Fig. 2. Folding topologies of (A) ES and (B) ARSA representing the soluble sulfatases created with the topology cartoon server TOPS software (University of Glasgow, UK; http://www.tops.leeds.ac.uk). Helices are shown as circles and strands as triangles. Subdomains SD1 and SD2 and the putative transmembrane domain TM in ES are highlighted. (See color insert.)

TABLE I
LOCATIONS OF SECONDARY STRUCTURE ELEMENTS IN ES

α helices and 3_{10} turns	β strands
1: 43–45	1: 28–34
2: 53–56	2: 63–65
3: 75–84	3: 68–69
4: 88–91	4: 128–134
5: 118–124	5: 158–162
6: 151–153	6: 244–246
7: 168–170	7: 249–253
8: 179–206	8: 280–285
9: 213–241	9: 335–341
10: 260–273	10: 380–383
11: 307–328	11: 393–394
12: 332–334	12: 398–399
13: 373–376	13: 442–444
14: 400–410	14: 452–453
15: 427–430	15: 464–467
16: 523–537	16: 501–504
17: 550–553	17: 572–573
18: 557–559	

```
        5    10   15   20   25   30   35   40   45   50   55   60   65   70   75   80
     28      33      38      43     48      53     58      63      68      73     79     84     89     94      99     104
1  ASRPNIILVMADDLGIGDPGCYGNKTIRTPNIDRLASGGVKLTQHLAASPLXTPSRAAFMTGRYPVRSGMASWSRTGVFLFTA···
     23      28    33      38    43     48     53     58      63      68      73    78     83     88      93
2  ·RPPNIVLIFADDLGYGDLGCYGHPSSTTPNLDQLAAGGLRFTDFYVPVSLXTPSRAALLTGRLPVRMGMYPG·····VLVPS···
     47      52    57     62     67     72     77     82      87      93     98    103    108     113
3  ·RPPHLVFLLADDLGWNDVGFHGSR·IRTPHLDALAAGGVLLDNYYTOPL·XTPSRSOLLTGRYOIRTGLQHO·····IIWPC·
      7     12     17     22     27     32     37     42     47     52    57     62     67     72     77     82
4  ·KRPNFLVIVADDLGFSDIGAFGG·EIATPNLDALAIAGLRLTDFHT·ASTXSPTRSMLLTGTDHHIAGIGTMAEALTPELEGKP

        90   95   100  105  110  115  120  125  130  135  140  145  150  155  160  165
     109    114    119    124    129    134    139    144    149    154    159    164    169    174    179    184    189
1  ·SSGGLPTDEITFAKLLKDOGYSTALIGKWHLGMSCHSKTDFCHHPLHHGFNYFYGISLTNLRDCKPGEGSVFTTGFKRLVFLPL
      98    103    108    113    118    123    128    133    138    143    148    153    158    163    168
2  ·SRGGLPLEEVTVAEVLAARGYLTGMAGKWHLGVGPEGAFLPPHOGFHRFLGIPYSHDOGPCONLTCFPPATPCDGG·······
     118    123    128    133    138    143    148    153    158    163    168    173    178
3  ·OPSCVPLDEKLLPOLLKEAGYTTHMVGKWHLGMYRKECLPTRRGFDTYFGYLLGSEDYYSHERC·············
     87     92     97    102    107    112    117    122    127    132    137    142    147
4  GYEGHLNERVVALPELLREAGYQTLMAGKWHLGLKPEOTPHARGFERSFSLLPGAANHYGFEPPY·············

       175  180  185  190  195  200  205  210  215  220  225  230  235  240  245  250
     194    199    204    209    214    219    224    229    234    239    244    249    254    259    264    269    274
1  OIVGVTLLTLAALNCLGLLHVPLGVFFSLLFLAALILTLFLGFLHYFRPLNCFMMRNYEIIQQPMSYDNLTQRLTVEAAQFIQRN
                                            173    178    183    188    193    198    203    208    213
2  ··············CDQGLVPIPLLANLSVEAQPPWLPGLEARYMAFAHDLMADAQR
                                      183    188    193    198    203    208    213    218    223
3  ··············TLIDALNVTRCALDFRDGEEVATGYKNMYSTNIFTKRAIALITNHP
                                152    157    162    167    172    177    182    187    192
4  ··············DESTPRILKGTPALYVEDERYLDTLPEGFYSSDAFGDKLLQYLKERD

       260  265  270  275  280  285  290  295  300  305  310  315  320  325  330  335
     279    284    289    294    299    304
1  TETPFLLVLSYLHVHTALFSSKDFAGKSO································HG
     218    223    228    233    238    243
2  ODRPFFLYYASHHTHYPQFSGQSFAERSG································RG
   228    233    238    243    248    253    258                                        263
3  PEKPLFLYLALOSVHEPLQVPEEYLKPYDFI················ODKNRH
     197    202    207    212    217    222    227    232    237    242    247    252    257    262    267    272    277
4  OSRPFFAYLPFSAPHWPLQAPREIVEKYRGRYDAGPEALRQERLARLKELGLVEADVEAHPVLALTREWEALEDEERAKSARAME

       345  350  355  360  365  370  375  380  385  390  395  400  405  410  415  420
     309    314    319    324    329    334    339    344    349                                          354    359
1  VYGDAVEEMDWSVGQILNLLDELRLANDTLIYFTSDQGAHVEE····························VSSKGEIHGG
     248    253    258    263    268    273    278    283    288                                          293
2  PFGDSLMELDAAVGTLMTAIGDLGLLEETLVIFTADNGPETMR····························M····SRGG
     268    273    278    283    288    293    298    303                                                 308
3  HYAGMVSLMDEAVGNVTAALKSSGLWNNTVFIFSTDNGGQTLA····························GG
   282    287    292    297    302    307    312    317    322    327    332    337    342    347    352    357    362
4  VYAAMVERMDWNIGRVVDYLRRQGELDNTFVLFMSDNGAEGALLEAFPKFGPDLLGFLDRHYDNSLENIGRANSYVWYGPRWAQA

       430  435  440  445  450  455  460  465  470  475  480  485  490  495  500  505
     364    369    374    379    384    389    394    399    404    409    414                    419    424    429    434
1  SNGIYKGGKANNWEGGIRWPRVIQAGQKIDEPTSNMDIFPTVAKLAGAPLPED······RIIDGRDLMPLLEGKSQRS
     298    303    308    313    318    323    328    333    338    343    348                    353    358    363    368
2  CSGLLRCGKGTTYEGGVREPALAFWPGHI·APGVTHELASSLDLLPTLAALAGAPLPN·······VTLDGFDLSPLLLGTGKSP
     313    318    323    328    333    338    343    348    353    358    363    368             373    378    383
3  NNWPLRGRKWSLWEGGVRGVGFVASPLLKQKGVKNRELIHISDWLPTLVKLARGHTNGT·······KPLDGFDVWKTISEGSPSP
   367    372    377    382    387    392    397    402    407    412    417    422    427    432    437    442
4  ATAPSRLYKAFTTOGGIRVPALVRYPRLSRQGAISHAFATVMDVTPTLLDLAGVRHPGKRWRGREIAEPRGRSWLGWLSGETEAA

       515  520  525  530  535  540  545  550  555  560  565  570  575  580  585  590
     439    444    449                    454    459    464    469    474    483    488    493    498    503    508    513    518
1  DHEFLFHYCNAYL·····NAVRWHPONSTSIWKAFFFTPNFNPVCFATHVCFCFGSYVTHHDDPPLLFDISKDPRERNPLTPASEP
     373    378    383                    388    393    398    403    408    413    418    423    428    433    438    443
2  RQSLFFYPSYPDE·····VRGVFAVR·TGKYKAHFFTQGSAHSDTTADPACHASSSLTAHEPPLLYDLSKDPGENYNLLGATPE
   388    393    398    403    408    413    418    423    428    433    438    443    448    453    458    463    468    473    478    483    488
3  RIELLHNIDPNFVDSSPCSAFNTSVHAAIRHGNWKLLTGYPGCGYWFPPPSQYNVSEIPSSDPPTKTLWLFDIDRDPEERHDLSR
   452    457    462    467    472    477    482    487    492    497    502    507    512    517    522    527
4  HDENTVTGWELFGMRAIRQGDWKAVYLPAPVGPATWOLYDLARDPGEIHDLADSOPGKLAELIEHWKRYVSETGVV

       600  605  610  615  620  625  630  635  640  645  650  655  660  665  670  675
     523    528    533    538    543    548    553    558    563    568    573
1  RFYEILKVMOEAADRHTOTLPEVPDOFSWNNFLWKPWLQLCCPSTGLSCQCDRE
     452    457    462    467    472    477    482    487    492    497    502
2  VLOALKQLOLLKAQLDAAVTFGPSQVARGE··DPALQICCHPGCTPRPACCHCP
   480    495    500    505    510    515    520    525    530
3  EYPHIVTKLLSRLQFYHKHSVPVYFPAQDPRCDPKATGVWGPWM
4
```

A comparison of the four amino acid sequences is shown in Fig. 3. The sequence identities of ES with ARSA, ARSB, and PAS range between 20 and 32%. The superposition of the crystal structures of all four sulfatases shown in Fig.1b was achieved by a least-square fitting of the α-carbon atoms of the four conserved Asp/Asn/Gln side chains that are ligands to the bivalent cation, with a rmsd of 0.2 ?. Except for the transmembrane helices in ES, the major features of all four structures superpose quite well, as evidenced by Fig. 1B. Despite the overall similarity, however, each of the sulfatases has small but distinctive differences in its 3-D structure. The central β sheet in the SD1 of ES has 11 strands as opposed to 10 for other sulfatases. In the ARSB structure, the missing transmembrane helices 8 and 9 of ES are compensated by formation of a pair of antiparallel β strands. The ARSA structure also has a similar loop region, but in an entirely different conformation with only a hint of antiparallel strands. Additionally, the loop regions of the three human enzymes have significant differences, both in lengths and in conformations. Some of the loop regions in ES that have a proposed membrane association (Hernandez-Guzman et al., 2003), such as loops between α4 and α5 and between β9 and α13 that approach the lipid bilayer, have four to seven residue peptide insertions when compared with the structures of soluble ARSA and ARSB.

Two significant regions of differences between the bacterial PAS and the human enzymes worth noting are two long insertions in PAS as demonstrated by sequence alignment (Fig. 3). The first one is roughly a 52 residue insertion after a point in the 3-D structure corresponding to residue 304 of ES. The insertion results in additional helices G and H and a longer α11 (αI, according to PAS nomenclature; Boltes et al., 2001). The second insertion, about 30 residues long after residue 349 in ES (between β9 and α13), forms a loop region with two additional small helices, J and K. None of these structural changes, however, could possibly have any influence on catalysis as they are all on the surface of the molecule. Both ES and ARSA contain 12 cysteines and six disulfide bonds, whereas ARSB has eight cysteines and four disulfides. In contrast, PAS, being a cytoplasmic enzyme, has no disulfide bond and no cysteine. All of the cysteine residues in ES are distributed in two catalytic subdomains, with the Cys170-Cys242

FIG. 3. An alignment of the sequences of four sulfatases—1, ES; 2, ARSA; 3, ARSB; and 4, PAS—and their known secondary structures. Helical regions are marked in red lines, sheets in yellow, and loops in blue. Residues are color coded: acidic (DE), red; basic (RKH), blue; hydrophobic (AVILMFPW), green; and hydrophilic (GSTNQCY), cyan; X, Formylglycine. (See color insert.)

disulfide "zipper-lock" near the lipid–protein interface serving to stabilize the putative transmembrane helices. The presence of disulfides and four glycosylation sites in ES suggests that the catalytic domain is located in the lumen side of the ER.

Conservation of Catalytic Machinery

As a consequence of the location and orientation of the transmembrane helices, the opening to the active site buried deep in a cavity in the "gill" of the "mushroom" rests near the membrane surface. A close-up view of the catalytic residues is shown in Fig. 4A. The catalytic amino acid hydroxyl-formylglycine (FG) 75, created by posttranslational modification of Cys 75 (Schmidt et al., 1995), was found to be linked covalently to a sulfate moiety [i.e., as a sulfate ester of FG (FGS)]. The metal ion found at the center of the catalytic site near the FGS 75 side chain was interpreted to be a Ca^{2+} (Hernandez-Guzman et al., 2003). Metal ions in the catalytic sites of ARSB and PAS were also proposed to be Ca^{2+} ions (Boltes et al., 2001; Bond et al., 1997), but it was a Mg^{2+} in the wild-type ARSA (Lukatela et al., 1998), as well as in the C69A mutated complex of ARSA with p-nitrocatechol sulfate (von Bülow et al., 2001).

While the oxygen atoms of the Asp 35, Asp 36, Asp 342, Gln 343, and FGS 75 side chains serve as ligands for the bivalent cation (Ca^{2+} …O distances range between 2.2 and 2.8 Å), Lys 134, Lys 368, and Arg 79 are involved in neutralization of negative charges of the carboxylic moieties. The positively charged amino groups of the Lys 134 and Lys 368 side chains are also within contact distances (2.7–3.1 Å) of the sulfate oxygen atoms of FGS 75 (Fig. 4A). In addition, two sulfate oxygen atoms are within coordination distances of Ca^{2+} (2.6–2.7Å). Several histidine residues in the immediate vicinity may play important roles in catalysis as well. The imidazole ring of His 136 is situated within a hydrogen-bonding distance (2.6 Å) of the hydroxyl of FGS, and His 290 $N_{\varepsilon 2}$ is 2.6 Å away from a sulfate oxygen of FGS. Also, the His 346 side chain is linked to Lys 368 and the Thr 291 side chain via a bridging water molecule. The main chain NH groups of FGS 75 and Thr 76 point toward the sulfate-binding cavity and may thus be responsible for formation of an oxyanion hole. A few additional solvent molecules, the presence of which may be important for the completion of the hydrolysis, are located inside the substrate-binding cavity.

The catalytic end of the active site in ES is highly homologous to those in ARSA, ARSB, and PAS. Nine of the 10 catalytically important residues, namely Asp 35, Asp 36, FGS 75, Arg 79, Lys 134, His 136, His 290, Asp 342, and Lys 368, are strictly conserved in all four enzymes (Fig. 3). When these nine α-carbon atom positions were superimposed by least-square

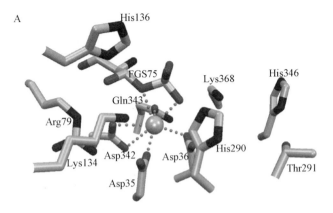

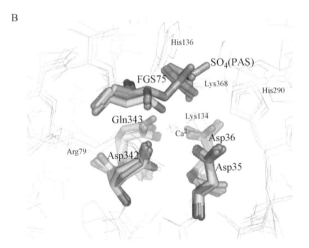

FIG. 4. (A) Active site catalytic residues in ES and the coordination of Ca^{2+} as described in the text. (B) A least-square superposition of four ligand side chains in crystal structures of sulfatases. The ligand side chains and formylglycine residues, along with their sulfate groups, are shown as thick bonds. The sulfate group in PAS not linked covalently to the side chain is labeled. Carbon atoms have the same colors as the colors used for the main chains (ES, green; ARSA, cyan; ARSB, pink; and PAS, yellow). Four bivalent cations shown as crosses for clarity have the same color codes as the main chains. Other side chains are shown as thin bonds. (See color insert.)

minimization, the rmsd was 0.4 Å. The 10th residue is Gln 343, a ligand to the cation, which is an asparagine in ARSA, ARSB, and PAS. Figure 4B is a close-up view of superimposed active sites of the four sulfatase structures by least-square fitting of the α carbon atoms of four side chains (Asp 35, Asp 36, Asp 342, and Gln 343 in ES) that provide ligand oxygen atoms to the bivalent cation (rmsd ~0.2 Å). The rmsd of the metal cation positions resulting from the superposition was 0.3 Å. The positions of the hydroxyl-formylglycine sulfate esters, including the sulfate moieties in ES and ARSB, are nearly identical, as shown in Fig. 4B. Although the position of the hydroxylformylglycine side chain in the bacterial PAS is also quite similar to that in ES, the sulfate moiety being bound noncovalently in PAS is 1.0 Å away from the superimposed sulfate position in ES and ARSB. In ARSA, the formylglycine residue is modeled as a glycine and the sulfate moiety as a chloride ion (Lukatela et al., 1998), the coordinates of which are not available. However, in two recently determined structures of human placental ARSA with covalently bound phosphate moieties (Chruszcz et al., 2003), the metal ion is proposed to be a bivalent Ca^{2+} instead of a Mg^{2+}, and the catalytic formylglycine 69 is found to be esterified with a phosphate group. It is possible that the sulfate moiety in ES is, in fact, a phosphate group, as phosphate buffer is used in the purification of ES and ammonium phosphate is used for crystallization (Hernandez-Guzman et al., 2001). ES, however, is not inhibited by an inorganic phosphate, as the enzyme maintains full activity during the purification and crystallization process. The experimental result derived by X-ray crystallography at the resolution and accuracy of the work described is unable to distinguish a phosphate group from a sulfate moiety.

Substrate-Binding Modes in ES

The substrate-binding end of the catalytic cavity, the end opposite to the metal center, has a very different architecture in each of these four enzymes, as expected. For ES, in particular, the cavity is considerably longer (~16 Å) than the other three (~10–12 Å) and extends to termini of the putative transmembrane helices, that is, the membrane surface (Figs. 1B and 5). An E1 sulfate (E1-SO4) molecule was docked in the active site with its sulfate moiety superimposed with the crystallographically observed sulfate of FGS 75. The energy of the system comprising the steroid mole-cule along with protein atoms lining the substrate-binding cleft was mini-mized using the Molecular Operating Environment package of software (Chemical Computing Group, Montreal, Canada). The resulting model of the covalently linked substrate that mimics an intermediate step of the proposed catalytic mechanism (see the following section) is shown in Fig. 5. Residues Leu 74, Arg 98, Thr 99, Val 101, Leu 103, Leu 167, Val 177, Phe 178, Thr 180, Gly 181, Thr 484, His 485, Val 486, and Phe 488 surround the steroid backbone, yielding mostly hydrophobic contacts and could, therefore, be involved in substrate recognition. Of particular interest here

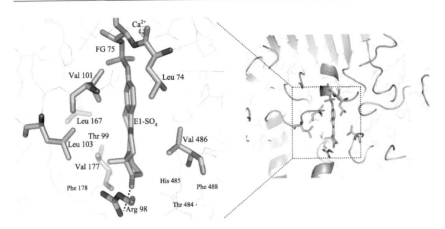

Fig. 5. A view of the active site of ES with a modeled E1 sulfate molecule linked covalently to FG 75 at an intermediate step, as described in the text. (Right) The boxed region showing a broader view of the active site cleft, including a turn of a transmembrane helix, is magnified (left). Important leucine and valine residues, as well as the Arg 98 side chain, are shown as thick bonds. Other side chains of interest are shown as thin bonds and are labeled. (See color insert.)

is the presence of several pairs of Val-Leu side chains that are closest to the substrate molecule, at or near van der Waals contact distances. As a sulfated steroid enters the catalytic cavity, several of these side chains, as well as others listed earlier, could play critical roles, from initial recognition and binding to the formation of reaction intermediates, and finally release of the unconjugated steroid from the catalytic cavity. The modeling result suggests that Val-Leu side chains along the substrate-binding cavity participate in these processes. The Arg 98 side chain closes in on the D ring of substrate during the energy minimization and forms a hydrogen bond with the 17-keto oxygen.

In an effort to determine the specific binding mode of a steroid molecule in the active site, we have collected 2.70-Å resolution synchrotron X-ray diffraction data at the Advanced Photon Source (Argonne, IL) on a crystal of ES soaked with 2 mM of E1. The difference map shows a weak electron density for E1, suggesting a partial occupancy of the E1 molecule and high thermal motion (D. Ghosh $et~al.$, unpublished results). Figure 6A illustrates the active site structure of the E1 complex of ES, showing that the A–B rings of E1 are sandwiched between a Leu-Val pair, Leu 103 and Val 486, while the 3-OH group makes a hydrogen bond with the sulfate oxygen of FGS 75 and the 17-keto oxygen accepts a proton from

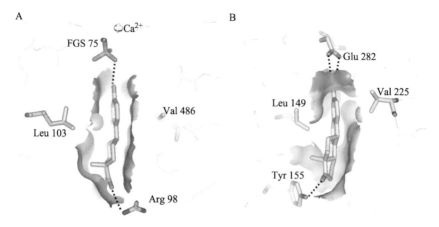

FIG. 6. Similarity of recognition of the estrogenic A ring by ES and 17HSD1: (A) E1 complex of ES as observed in a crystal structure and (B) crystal structure of the 17BHSD1–equilin complex. Residues of the Leu-Val sandwich are shown as thick bonds and are labeled. Molecular surfaces of the proteins are drawn in the same color code as the ligands. (See color insert.)

the guanidinium group of Arg 98. Interestingly, the active site of the human estrogenic enzyme 17HSD1 is known to have a similar recognition mechanism by which the estrogenic A ring forms a Leu-Val sandwich as observed in both E2 and equilin complexes of the enzyme (Sawicki *et al.*, 1999) (Fig. 6B). This binding mode of E1 in ES, however, possibly represents the location of the steroid backbone of a substrate (E1-SO4) molecule prior to its covalent linkage with F6 75, forming a reactive intermediate. It could also indicate the location of the product E1 in the active site prior to its release. Nonetheless, the E1 complex of ES elucidates steroid-specific recognition mechanism, which may be conserved across enzyme families. It is conceivable that the binding modes of DHEA and cholesterol sulfates are somewhat different from that of E1 sulfate.

The Catalytic Mechanism

As described earlier, 9 out of 10 catalytically important residues are identical in all four sulfatases of this family with known crystal structures. The 10th residue is Gln 343 in ES, as opposed to asparagines in ARSA, ARSB, and PAS. The bivalent cation also appears to be a Ca^{2+} in all four enzymes. Thus, the catalytic mechanism by which steroidal sulfates are hydrolyzed by ES into sulfate and unconjugated steroids is likely to be similar to those proposed for arylsulfatases (Boltes *et al.*, 2001). Results clearly demonstrate that catalytic formylglycine 75 is a hydroxylformylglycine and is linked covalently to a sulfate moiety. Interestingly, the catalytic formylglycine in ARSB was also found linked covalently to a sulfate moiety (Bond *et al.*, 1997). It appears that the catalytic residue as a sulfated hydroxylformylglycine could describe the resting state of the enzyme, as has been suggested for ARSB (Bond *et al.*, 1997).

The reaction mechanism of sulfatases resembles that of an alkaline phosphatase, which has a catalytic serine instead of a formylglycine (Sowadski *et al.*, 1985). The serine links covalently to the phosphate group during the first half-reaction at pH 5, which is then hydrolyzed by an activated water molecule at an alkaline pH (Recksiek *et al.*, 1998; Sowadski *et al.*, 1985). By substituting the catalytic residue in ARSA and ARSB by a serine, it was shown that the enzymes behaved like an alkaline phosphatase and were able to trap a sulfate by covalently linking the serine side chain to it (Recksiek *et al.*, 1998). Based on these data, it was proposed that the mechanistic difference between the two catalysis was due to the existence of the second, unesterified hydroxyl group of the gem-diol in sulfatases, which served as a nucleophile for rapid release of the sulfate moiety (Recksiek *et al.*, 1998). The proposed mechanism for ES, shown in Fig. 7, basically follows the same four-step scheme described for ARSA. Several positively charged side chains, Lys 134, Lys 368, and Arg 79, as well as His 136 and His 290, participate in catalysis, in addition to their role in charge neutralization inside the active site cavity. Step I is the activation of FG 75 by a water molecule, forming the gem-diol. In step II, a nucleophilic attack on the sulfur atom by one of the hydroxyls of the hydroxylformylglycine follows its activation by Ca^{2+}, while the other hydroxyl is deprotonated by His 136. This causes the sulfate moiety to link covalently with the formylglycine side chain and release of the unconjugated substrate. The free hydroxyl is involved in a nucleophilic attack on the ester bond in step III. In step IV, the HSO_4^- moiety is released, and the formylglycine side chain is regenerated.

The scheme, however, does not explain the observation of a sulfate linked covalently to hydroxylformylglycine in the crystal structures of both ARSB and ES. It is possible that the sulfate group is trapped within the crystal and is not hydrolyzed under the conditions of crystallization. Alternatively, the second hydroxyl of the gem-diol group may need the presence of an additional water molecule as a nucleophile to perform the deesterification reaction effectively. Water molecules can be transported to the catalytic cavity by solvated sulfate groups of substrates. In that case,

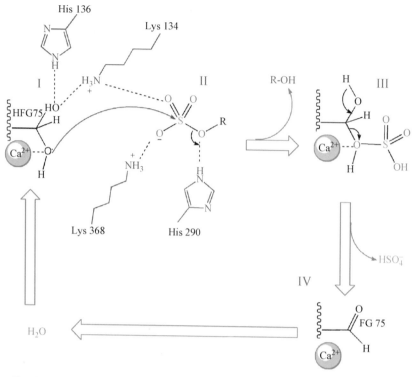

FIG. 7. The proposed catalytic mechanism for sulfatases. The catalytic side chain labeled as HFG75 is hydroxylformylglycine 75. See text for an explanation of the four steps of catalysis. (See color insert.)

sulfated hydroxylformylglycine would indeed be the resting state of the enzyme. More structural and biochemical data are necessary before this issue can be put to rest.

Does Membrane Anchoring Have a Functional Role?

What makes ES unique is its putative transmembrane association. Helices 8 and 9 in ES, each roughly 40 Å long (Figs. 1B and 8) and situated between residues 179 and 241, are presumed to traverse the membrane and anchor the functional domain to the membrane surface facing the ER lumen (Hernandez-Guzman et al., 2003). Due to the presence of several disulfide groups and four glycosylation sites, it is highly unlikely that the

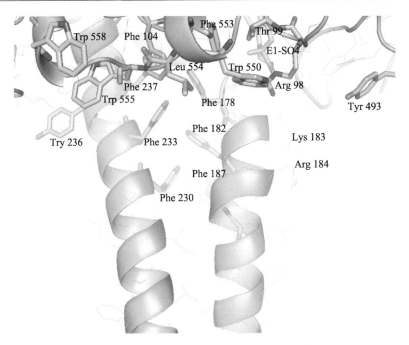

FIG. 8. Large hydrophobic residues at the putative protein–lipid interface in the crystal structure of ES. Some of these side chains are responsible for the hydrophobic "tunnel" described in the text. A bound E1 sulfate molecule at the active site and two transmembrane helices are partially visible. (See color insert.)

polar catalytic domain of ES can be situated in the reducing environment of the cytoplasm. In all likelihood, the polar catalytic domain rests on the lumen side of the lipid bilayer.

The opening to the active site cleft is embedded into what is presumed to be the lipid–protein interface. Near the opening, a constellation of large hydrophobic side chains, Phe 178, Phe 182, Phe 187, Phe 230, Phe 233, Tyr 236, and Phe 237 from the transmembrane domain and Phe 104, Tyr 493, Trp 550, Phe 553, Leu 554, Trp 555, and Trp 558 (Fig. 8) of the catalytic domain, line the surface of a hydrophobic "tunnel" leading to the active site (Hernandez-Guzman *et al.*, 2003). The Arg 98 and Thr 99 side chains position themselves as gatekeepers to the "tunnel." A bound steroid sulfate substrate, shown in Fig. 8, covers the entire length of the catalytic cleft up to the lipid interface. As has been proposed, in addition to helices 8 and 9 (residues 179–241), other loop and helical turn regions that the aforementioned residues belong to (residues 468–500, 550–559) associate with the lipid bilayer (Hernandez-Guzman *et al.*, 2003). Therefore, participation of the lipid bilayer in maintenance of the integrity of the active site and passage of the substrate and the product can be envisioned. One proposal is that steroidal substrates enter the active site through one of the three "swing gates" in the lumen side and the steroidal products are dispensed through the "tunnel" into the lipid bilayer (Hernandez-Guzman *et al.*, 2003). Alternatively, both substrates and products may use the "tunnel" and the lipid bilayer for entering into and exiting from the active site (Hernandez-Guzman *et al.*, 2003). Shielding of the negative charge on the sulfate moiety can be achieved through solvation. Thus, the crystal structure of ES is indicative of functional roles of the lipid bilayer with which it associates.

Oligomeric States of Sulfatases

Human enzyme ARSB is known to be a monomer. Both ARSA and ARSB crystal structures show one molecule per asymmetric unit (Bond *et al.*, 1997; Lukatela *et al.*, 1998). However, ARSA forms a crystallographic octamer within the crystal; the octameric form was also observed in solution at low pH (Lukatela *et al.*, 1998). Although the crystal structure of PAS has two molecules in the asymmetric unit, the contact is believed to be due to crystal packing and not maintained in solution (Boltes *et al.*, 2001). The ES crystal has one molecule per asymmetric unit, but it forms a trimer through a crystallographic threefold rotation symmetry (Fig. 9) (Hernandez-Guzman, 2002; Hernandez-Guzman *et al.*, 2003). Three transmembrane helix pairs (helices 8 and 9) from three symmetry-related molecules pack about the threefold rotation axis, thereby concealing one side of the hydrophobic surface, as shown in Fig. 9.

Does this oligomerization have any functional significance? Probably not, for two reasons. First, in this trimer the helix association is highly oblique, which does not allow for membrane spanning. Furthermore, in this packing, each active site opening is to the exterior of the trimeric association, opposite and away from the center, which is tightly packed with hydrophobic side chains (Fig. 9). Thus, oligomerization appears only to stabilize the helices and conceal hydrophobic surfaces from an aqueous environment while the three active sites are left to function independently. Second, it is believed that trimer formation takes place as the enzyme is extracted from microsomal lipids and solubulized with a detergent solution, which is consistent with the observation of a multimeric ES in solution during its purification (Hernandez-Guzman *et al.*, 2001). Formation of stable trimeric ES during its extraction could very well be the key to

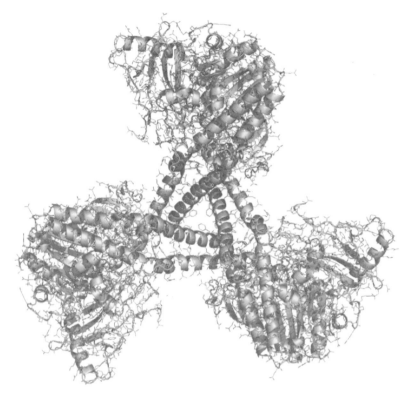

FIG. 9. A view of the observed crystallographic trimer of ES roughly along the threefold rotation axis. (See color insert.)

maintenance of the activity of the enzyme, its successful purification, and crystallization, but not necessarily for its function as a membrane-embedded sulfatase *in vivo*. Nevertheless, one cannot completely rule out the possibility of a rearranged trimer or another oligomer of ES as the functional form of the enzyme within the lipid bilayer of the endoplasmic reticulum.

Structural Explanation for Steroid Sulfatase Deficiency

X-linked ichthyosis is an inherited genetic disorder of the skin that results from ES or steroid sulfatase (STS) deficiency. While in a majority of the cases there are extensive deletions of the gene, several point mutations, each corresponding to a single base change, have been identified among patients with the disease. Six of the seven point mutations lead to amino acid substitutions and one to a premature termination of the polypeptide chain (Alperin *et al.*, 1997; Hernandez-Martin *et al.*, 1999). Six amino acid substitutions are Ser341Leu, Trp372Arg, Trp372Pro, His444Arg, Cys446Tyr, and Gln560Pro. The seventh mutation, a 19-bp insertion at the exon 8–intron 8 splice junction at nucleotide 1477, results in a shift in the open reading frame and termination of the chain at residue 427, 8 residues after the frameshift. The amino acid side chains that are mutated are shown on the crystal structure of ES in Fig. 10. Locations of the first five substitutions are all in the surrounding region of the active site cavity. Careful analysis suggests that each substitution leads to loss of integrity of the active site and/or interferes with the enzyme function (Ghosh, 2004). The Gln560Pro substitution, which is 22 Å away from the catalytic Ca^{2+}, may disrupt the precise orientation of the hydrophobic residues at the lipid interface. Premature termination of the polypeptide

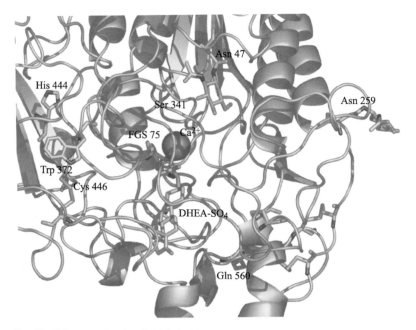

FIG. 10. Point mutation sites in X-linked ichthyosis displaying amino acids of the wild-type enzyme (Ser 341, Trp 372, His 444, Cys 446, and Gln 560). The catalytic residue FG 75, the bivalent cation Ca^{2+}, and a covalently linked modeled DHEA-sulfate are shown at the active site. Two observed glycosylation sites (Asn 47 and Asn 259), each with one molecule of *N*-acetyl glucosamine, are also shown. (See color insert.)

chain at residue 427, situated at the beginning of the C-terminal subdomain, completely wipes out SD2 and hence a section of the catalytic cavity, along with the membrane anchoring loops. Mapping of point mutations into the three-dimensional structure, thus, suggests severe destabilization of the active site architecture, thereby probably rendering the enzyme inactive or active at low levels.

Finally, ES has figured prominently in recent breast cancer literature as one of the major enzyme targets for the reduction of estrogen levels at or near the tumor site (Suzuki *et al.*, 2003; Thijssen, 2004). Our structural and modeling analyses have revealed evidence for steroid-specific selectivity in the substrate-binding cleft. However, analyses have also revealed flexibility and spaciousness of the cleft and the presence of several polar groups—all of which are important considerations for attempting rational design of inhibitors of ES.

Acknowledgment

This research was supported in part by National Institute of Health Grant GM62794.

References

Alperin, E. S., and Shapiro, L. J. (1997). Characterization of point mutations in patients with X-linked ichthyosis. *J. Biol. Chem.* **272**, 20756–20763.

Billich, A., Nussbaumer, P., and Lehr, P. (2000). Stimulation of MCF-7 breast cancer cell proliferation by estrone sulfate and dehydroepiandrosterone sulfate: Inhibition by novel non-steroidal steroid sulfatase inhibitors. *J. Steroid Biochem. Mol. Biol.* **73**, 225–235.

Boltes, I., Czapinska, H., Kahnert, A., von Bülow, R., Dierks, T., Schmidt, B., von Figura, K., Kertesz, M. A., and Usón, I. (2001). 1.3 Å structure of arylsulfatase from *P. aeruginosa* establishes the catalytic mechanism of sulfate ester cleavage in the sulfatase family. *Structure* **9**, 483–491.

Bond, C. S., Clements, P. R., Ashby, S. J., Collyer, C. A., Harrop, S. J., Hopwood, J. J., and Guss, J. M. (1997). Structure of human lysosomal sulfatase. *Structure* **5**, 277–289.

Bülow, R. V., Schmidt, B., Dierks, T., von Figura, K., and Usón, I. (2001). Crystal structure of an enzyme-substrate complex provides insight into the interaction between human arylsulfatase A and its substrates during catalysis. *J. Mol. Biol.* **305**, 269–277.

Burns, G. R. J. (1983). Purification and partial characterization of arylsulphatase C from human placental microsomes. *Biochim. Biophys. Acta* **759**, 199–204.

Chruszcz, M., Laidler, P., Monkiewicz, M., Ortlund, E., Lebioda, L., and Lewinski, K. (2003). Crystal structure of a covalent intermediate of endogenous human arylsulfatase A. *J. Inorg. Biochem.* **96**, 386–392.

Dibbelt, L., and Kuss, E. (1983). Human placental steryl-sulfatase: Enzyme purification, production of antisera, and immunoblotting reactions with normal and sulfatase-deficient placentas. *Biol. Chem. Hoppe-Seyler* **367**, 1223–1229.

Ghosh, D. (2004). Mutations in X-linked ichthyosis disrupt the active site structure of Estrone/DHEA sulfatase. *Biochim. Biophys. Acta* **1739**, 1–4.

Ghosh, D., and Vihko, P. (2002). Molecular mechanism of estrogen recognition and 17-keto reduction by 17β-hydroxysteroid dehydrogenase type 1. *Chem.-Biol. Interact.* **130–132**, 637–650.

Hernandez-Guzman, F. G. (2002). The State University of New York at Buffalo, Buffalo, New York, Ph.D. thesis.

Hernandez-Guzman, F. G., Higashiyama, T., Osawa, Y., and Ghosh, D. (2001). Purification, characterization and crystallization of human placental estrone/dehydroepiandrosterone sulfatase, a membrane-bound enzyme of the endoplasmic reticulum. *J. Steroid Biochem. Mol. Biol.* **78**, 441–450.

Hernandez-Guzman, F. G., Higashiyama, T., Pangborn, W., Osawa, Y., and Ghosh, D. (2003). Structure of human estrone sulfatase suggests functional roles of membrane association. *J. Biol. Chem.* **278**, 22989–22997.

Hernandez-Martin, A., Gonzalez-Sarmiento, R., and De, Unamuno P. (1999). X-linked ichthyosis: An update. *Br. J. Dermatol.* **141**, 617–627.

Lukatela, G., Krauss, N., Theis, K., Selmer, T., Gieselmann, V., von Figura, K., and Saenger, W. (1998). Crystal structure of human arylsulfatase A: The aldehyde function and the metal ion at the active site suggest a novel mechanism for sulfate ester hydrolysis. *Biochemistry* **37**, 3654–3664.

Noel, H., Plante, L., Bleau, G., Chapdelaine, A., and Roberts, K. D. (1983). Human placental steroid sulfatase: Purification and properties. *J. Steroid Biochem.* **19**, 1591–1598.

Parenti, G., Germana, M., and Ballabio, A. (1997). The sulfatase gene familty. *Curr. Opin. Genet. Dev.* **7**, 386–391.

Pasqualini, J. R., and Chetrite, G. S. (2002). The selective estrogen enzyme modulators in breast cancer. *In* "Breast Cancer: Prognosis, Treatment and Prevention" (J. R. Pasqualini, ed.), pp. 187–249. Dekker, New York..

Purohit, A., Potter, B. V. L., Parker, M. G., and Reed, M. J. (1998). Steroid sulfatase: Expression, isolation and inhibition for active site identification studies. *Chem. Biol. Interact.* **109**, 183–193.

Recksiek, M., Selmer, T., Dierks, T., Schmidt, B., and von Figura, K. (1998). Sulfatases, trapping of the sulfated enzyme intermediate by substituting the active site formylglycine. *J. Biol. Chem.* **273**, 6096–6103.

Reed, M. J., and Purohit, A. (1997). Breast cancer and the role of cytokines in the regulating estrogen synthesis: an emerging hypothesis. *Endocr. Rev.* **18**, 701–715.

Sawicki, M., Erman, M., Puranen, T., Vihko, P., and Ghosh, D. (1999). Structure of the ternary complex of human 17β-hydroxysteroid dehydrogenase type 1 with equilin and NADP$^+$. *Proc. Natl. Acad. Sci. USA* **96**, 840–845.

Schmidt, B., Selmer, T., Ingendoh, A., and von, Figura.K. (1995). A novel amino acid modification in sulfatases that is defective in multiple sulfatase deficiency. *Cell* **82**, 271–278.

Sowadski, J. M., Handschumacher, M. D., Kirshna Murthy, H. M., Foster, B. A., and Wyckoff, H. W. (1985). Refined structure of alkaline phosphatase from *E. coli* at 2.8 Å resolution. *J. Mol. Biol.* **186**, 417–433.

Suzuki, T., Hirato, K., Yanaihara, T., Kodofuku, T., Sato, T., Hoshino, M., and Yanaihara, N. (1992). Purification and properties of steroid sulfatase from human placenta,. *Endocrinol. Jpn.* **39**, 93–101.

Suzuki, T., Moriya, T., Ishida, T., Ohuchi, N., and Sasano, H. (2003). Intracrine mechanism of estrogen synthesis in breast cancer. *Biomed. Pharmacother.* **57**, 460–462.

Thijssen, J. H. H. (2004). Local biosynthesis and metabolism of oestrogens in the human breast. *Maturitas* **49,** 25–33.

Vaccaro, A. M., Salvioli, R., Muscillo, M., and Renola, L. (1987). Purification and properties of arylsulfatase C from human placenta. *Enzyme* **37,** 115–126.

Van der Loos, C. M., Van Breda, A. J., Van den Berg, F. M., Walboomers, J. M. M., and Jobsis, A. C. (1983). Human placental steroid sulfatase purification and monospecific antibody production in rabbits. *J. Inherit. Metab. Dis.* **7,** 97–103.

Yanaihara, A., Yanaihara, T., Toma, Y., Shimizu, Y., Saito, H., Okai, T., Higashiyama, T., and Osawa, Y. (2001). Localization and expression of steroid sulfatase in human fallopian tube. *Steroids* **66,** 87–91.

[17] Estrogen Sulfatase

By MASAO IWAMORI

Abstract

Estrogen sulfatase is a microsomal enzyme and is ubiquitously distributed in several mammalian tissues, among which the liver, placenta, and endocrine tissues exhibit relatively high activity. Because the major circulating precursors of estrogen are estrone 3-sulfate and dehydroepiandrosterone 3-sulfate, estrogen sulfatase plays an important role not only in their incorporation and metabolism, but also in the controls of estrogen activity by regulating the binding potential of estrogen as to its receptor through sulfoconjugation and desulfation reactions. Accordingly, an increase in sulfoconjugation through transfection of the sulfotransferase gene or inhibition of estrogen sulfatase by specific inhibitors has been successfully applied to abolish the estrogen activity in estrogen-dependent breast cancer- and uterine endometrial adenocarcinoma-derived cells. Inhibitors of estrogen sulfatase are expected to be developed as new drugs for estrogen-dependent cancer therapy, particularly in postmenopausal women.

Sulfoconjugates of Estrogens

Steroids are known to be converted into sulfoconjugates by cytosolic sulfotransferase, and the resultant steroid sulfates exhibit higher solubility in aqueous media than the original steroids, facilitating excretion from the body and transportation through the circulatory system to supply metabolic intermediates for steroidogenesis, as described by several workers (Hobkirk, 1985; Stein *et al.*, 1989). In addition, endocrine tissues with

METHODS IN ENZYMOLOGY, VOL. 400
0076-6879/05 $35.00
DOI: 10.1016/S0076-6879(05)00017-0

steroid receptors utilize sulfation and desulfation reactions to regulate the binding of steroids with the respective receptors in inhibitory and promotional manners, respectively (Kotov et al., 1999; Tanaka et al., 2003). Among steroids, regulation of estrogen activity at the levels of both precursor supplementation and ligand–receptor interaction by means of sulfoconjugation is particularly important for estrogen-mediated phenomena, including the progression of transformed cells. The sulfated derivatives of estrogens characterized so far in mammalian tissues and fluids are estrone 3-sulfate, estradiol 3-sulfate, estradiol 17-sulfate, estradiol 3,17-disulfate, estriol 3-sulfate, and estriol 3-sulfate 16-glucuronide. Among them, estrone 3-sulfate is the most abundant estrogen in human serum, where 95% of it is bound with albumin (Rosenthal et al., 1972) and provides a source of local bioactive estrogen. For example, Muir et al. (2004) demonstrated that estradiol, which is important in cellular maturation and homeostasis of human bone, is formed from circulating estrone 3-sulfate at higher metabolic rates than from androstenedione and testosterone and that estrone sulfatase plays a role in initiating the synthesis of estradiol in bone, indicating that the incorporation of estrone 3-sulfate from the bloodstream, followed by desulfation, is the major route of estrogen supply in bone for the control of estrogen-dependent cellular functions. However, in the case of pregnant women, circulating dehydroepiandrosterone sulfate is converted to estrogen in a high yield, as shown in early studies by Gant et al. (1971) and Grodin et al. (1973). When radioactive dehydroepiandrosterone sulfate was administered via the uterine artery, umbilical vein, or antecubital vein in pregnant women at the time of elective abortion, estriol was found to be the major phenolic steroid in the fetal–placental compartment. Because dehydroepiandrosterone sulfate does not cross from the maternal to the fetal side as the conjugated form, the placental steroid sulfatase first hydrolyzes the sulfoconjugate to permit the transfer of free dehydroepiandrosterone, which is then 16α-hydroxylated and aromatized to estriol. However, the double conjugate of estrogen estriol 3-sulfate 16-glucuronide is detected in the urine, amniotic fluid, and cord and maternal blood of humans and has been demonstrated to be formed through the sulfation of estriol 16-glucuronide (Touchstone et al., 1963). The same double conjugate of estrogen was included in the biliary metabolites 30 min after the perfusion of estrone (5 μmol) into rat liver, at the level of 0.7 μmol, together with 1.8 μmol of estrogen glucuronide and 0.07 μmol of estrogen sulfate, but not in the bile of guinea pig in the same experiments, due to the restricted ability of hydroxylation at the D ring of steroids in guinea pig (Roy et al., 1987). Thus, although there are several differences in the metabolic pathway for the sulfoconjugates of estrogens among animal species, derivatives sulfated at the C-3 positions of estrone, estradiol, and estriol are the key metabolites in mammalian tissues so far investigated in the past, such as human, rat, mouse, sheep, guinea pig, and cow (Glutek and Hobkirk, 1990; Hadd and Blickrstaff, 1969; Hoffman et al., 2001; Purinton et al., 1999).

Procedure for Determination of Estrogen Sulfatase Activity

Determination of estrogen sulfatase activity is generally performed with radioactive substrates, such as estrone 3-[^{35}S]sulfate, estradiol 3-[^{35}S] sulfate, and [6,7-^{3}H]estrone sulfate. Because decay of the radioactivity of ^{35}S occurs in rather a short period, in comparison to ^{3}H, and ^{3}H-labeled estrone sulfate is available commercially, [6,7-^{3}H]estrone sulfate (New England Nuclear, Boston, MA) is generally used as the substrate for estrogen sulfatase.

The optimum conditions for estrogen sulfatase were established with [6,7-^{3}H]estrone sulfate as the substrate and the microsomal fraction of human endometrial carcinoma-derived cells, HEC-108, as the enzyme source (Tanaka et al., 2003). The addition of Triton X-100 at 1 mg/ml stimulated the activity threefold, and other detergents, namely Nonidet P-40, Tween 20, Tween 80, sodium cholate, sodium deoxycholate, and sodium taurocholate, at the same concentration enhanced the activity to lesser extents, although it was unclear whether the detergents affected the solubilization of microsomal insoluble enzymes or micellar formation of substrates. Among the buffers examined, Tris-HCl buffer gave the highest activity, giving the optimum pH of 7.5 (Fig. 1A). Because phosphate, sulfate, and sulfite have been reported to be suppressive agents as to the activity of steroid sulfatases (Roy, 1987) and phosphate completely inhibits rat liver arylsulfatases A and B, but only moderately arylsulfatase C (Milson et al., 1972), the relatively lower activity of estrogen sulfatase in phosphate buffer (Fig. 1A) was thought to be due to the phosphate ion. Under the conditions in Tris-HCl, pH 7.5, containing Triton X-100, the apparent K_m for estrone sulfate was 41.7 μM (Figs. 1B and C), which was similar with those reported previously (Dolly et al., 1972; Huang et al., 1997). The standard procedure for determination of estrogen sulfatase is as follows:

100 mM Tris-HCl buffer, pH 7.5, 0.1% Triton-X-100, 80.6 μM[6,7-^{3}H]-estrone sulfate (10.7 nCi/nmol), and 50–100 μg enzyme protein in a total volume of 100 μl.

After incubation at 37° for 1 h, the reaction is terminated by the addition of 100 μl of chloroform/methanol (2:1, v), and then the lower

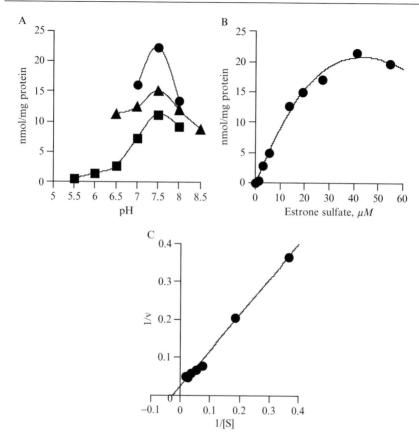

FIG. 1. Estrogen sulfatase activity as functions of pH (A) and substrate concentration (B), and a Lineweaver–Burk plot (C).For A, activity was determined with [6,7-³H]estrone sulfate (80.6 μM) as the substrate and the microsomal fraction of human endometrial adenocarcinoma cells as the enzyme source in various buffers, ●, Tris-HCl; ■, triethanolamine-NaOH; and ▲, phosphate containing 0.1% Triton X-100; and for B, activity was determined in 100 mM Tris-HCl, pH 7.5.

organic phase is spotted onto a plastic-coated thin-layer plate (Polygram Sil G; Macherey-Nagel, Duren, Germany), which is developed with benzene/ethyl acetate (2:1, v). The position of the product, estrone, on the plate is determined by development of standard estrone (Sigma, St. Louis, MO) on the same plate, followed by visualization with a cupric acetate-phosphoric acid reagent. The area corresponding to estrone is cut out and placed in a scintillation cocktail (ACS-II; Amersham, Piscataway, NJ), and then the radioactivity of estrone produced is determined with a liquid scintillation counter (TriCarb 1500; Packard/Perkin-Elmer, Foster City, CA).

Activity of Estrogen Sulfatase

Estrogen sulfatase is widely distributed among various mammalian tissues as well as in many molluscs, as described by Roy (1987) and Verde and Drucker (1972). In mammals, the liver exhibits the highest activity, and relatively high activity is also detected in the adrenals, kidneys, testes, prostate, and placenta (Table I), indicating that the endocrine organs and tissues related with estrogen-mediated hormonal stimuli require high estrogen sulfatase activity. However, estrone sulfate, as the substrate, is known to be hydrolyzed not only by estrogen sulfatase, but also by steroid sulfatase, which was usually determined with dehydroepiandrosterone sulfate or pregnenolone sulfate as the substrate, as demonstrated by Hernandez-Guzman et al. (2001), Purohit et al. (1998), Glutek and Hoblirk

TABLE I
SPECIFIC ACTIVITY OF ESTROGEN SULFATASE IN VARIOUS TISSUES[a]

	Rat (units/mg protein)		Mouse (nmol/mg protein/hr)	
	Male	Female	Male	Female
Liver	102.5	65.9	12.0	11.0
Adrenals	43.7	53.3	6.5	6.9
Kidneys	26.6	27.7	3.2	3.6
Lungs	25.5	12.6	1.7	1.9
Pancreas	22.1	17.4	2.4	2.7
Brain	20.8	11.1	2.8	2.3
Spleen	16.2	11.7	1.2	1.5
Heart	16.1	16.2	1.5	1.6
Small intestine	2.7	3.5	1.6	3.1
Large intestine	5.7	3.0	1.7	2.4
Testes	24.6	—	8.8	—
Prostate	—	—	4.5	—
Ovary	—	23.2	—	2.7
Uterus	—	14.0	—	1.0
Placenta (Early-trimester)	—	—	—	1.5
Placenta (Mid-trimester)	—	—	—	2.2
Placenta (Late-trimester)	—	—	—	7.1

[a] Estrogen sulfatase activities in the various rat tissues were cited from Dolly et al. (1972), whereas those in mouse tissues were determined with microsomal fractions according to the procedure described by Tanaka et al. (2003).

(1990), Dibbelt and Kuss (1986), Utsumi *et al.* (1999), Conary *et al.* (1986), and Huang *et al.* (1997). A difference in substrate specificity between estrogen sulfatase and steroid sulfatase was clearly revealed by the enzyme kinetics with quercetin, which inhibited the former and latter enzymes in competitive and uncompetitive manners, respectively (Huang *et al.*, 1997), and the tissue distribution of estrogen sulfatase was distinct from that of steroid sulfatase, with the former being found in the placenta, endometrium, decidua basalis, amnion, and chorion and the latter in the endometrium and decidua basalis, but not in the placenta, respectively (Glutek and Hobkirk, 1990). Consequently, pregnant women with recessive X-linked ichthyosis, who are deficient in steroid sulfatase activity in the placenta, exhibit the low urinary levels of estriol, which is a clinical marker of a healthy fetoplacental unit (Conary *et al.*, 1986; Horwitz *et al.*, 1986).

Estrogen Sulfatase and Hormonal Activity

Sulfoconjugation of estrogen by estrogen sulfotransferase is known to abolish the binding ability of estrogen as to its receptor. For example, as demonstrated by Tanaka *et al.* (2003), human uterine endometrial Ishikawa cells abolished the estrogen-stimulated proliferation potential through the increased activity of estrogen sulfotransferase after transfection of its gene. The same transfection experiment was performed on breast carcinoma-derived cells to prove the endocrine-disrupting nature of estrogen sulfotransferase (Falany *et al.*, 2002; Shimizu *et al.*, 2002). However, estrogen sulfatase was shown to enhance estrogen activity. For example, estrogen sulfatase in breast carcinoma-derived and uterine endometrial adenocarcinoma-derived cells, both of which exhibit the ability of estrogen-dependent cell growth, have been shown to exhibit significantly higher activity than estrogen-independent cells (Pasqualini *et al.*, 1992; Tanaka *et al.*, 2003), and either inhibition of estrogen sulfatase by structurally related steroids or increased sulfotransferase activity abolished the estrogen dependency of both types of cells, as described earlier. In the case of Ishikawa cells, the specific activities of estrogen sulfatase and estrogen sulfotransferase were 10 and 0.05 nmol/mg protein/h, respectively, and those in the cells that had lost the estrogen dependency through transfection of the sulfotransferase gene were 10 and 2 nmol/mg protein/h, respectively, indicating that the ratio of the specific activities of the two enzymes is important for regulating estrogen activity (Fig. 2). Tanaka *et al.* (2003) suggested that the reason why the lower specific activity of sulfotransferase than that of estrogen sulfatase was sufficient to eliminate the estrogen activity through sulfation

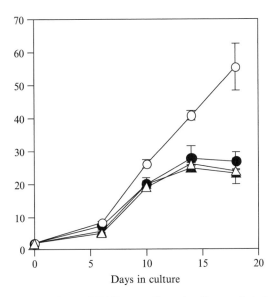

Fig. 2. Change in numbers of Ishikawa cells and cells transfected with the human estrogen sulfotransferase gene in medium with and without estradiol (Tanaka *et al.*, 2003). Human endometrial adenocarcinoma-derived Ishikawa cells were cultured in the presence (○) and absence (●) of estradiol, and Ishikawa cells transfected with the human estrogen sulfotransferase gene were cultured in the presence (△) and absence (▲) of estradiol (10^{-8} M).

was the different properties of the two enzymes. Compared with the insoluble nature of estrogen sulfatase on the membrane surface, the soluble sulfotransferase may be readily moved into the cytosol to achieve effective sulfation of estrogen, although the marginal ratio of the activities of sulfotransferase and sulfatase to maintain the estrogen-dependency of cell growth could not be explained in the transfection experiments by Tanaka *et al.* (2003), Falany *et al.* (2002), and Shimizu *et al.* (2002).

In addition, expression of steroid sulfatase mRNA in human breast carcinoma tissues has been shown to be closely related with the development and metastasis of cancer cells as a predictor of recurrence by Utsumi *et al.* (1999). In this connection, several inhibitors of estrogen sulfatase and steroid sulfatase have been evaluated as potential therapeutic agents for the treatment of estrogen-dependent cancers, that is, estrone formate (Schreiner and Billich, 2004), estrogen sulfamates (Chander *et al.*, 2004), chromenone sulfamates (Horvath *et al.*, 2004), 2-alkylsulfanyl estrogen (Leese *et al.*, 2004), 2-phenylindole sulfamates (Walter *et al.*, 2004), benzophenone-4,4-O,O-bis-sulfamate (Hejaz *et al.*, 2004), 2',4'-dicyanobiphenyl 4-O-sulfamate (Saito *et al.*, 2004), 2-difluoromethyloestrone 3-O-sulfamate (Reed *et al.*, 2004), flavonoids (Huang *et al.*, 1997), and daidzein sulfate (Wong and Keung, 1997). For one inhibitor, 2-difluoromethyloestrone 3-O-sulfamate showing the highest inhibitory activity among inhibitors reported, interaction between its 2-difluoromethyl group and hydrogen bond donor residues lining the catalytic site of steroid sulfatase was demonstrated, resulting in the irreversible inhibition with an IC$_{50}$ of 100 pM (Reed *et al.*, 2004). These drugs are expected to be applicable to therapy for estrogen-dependent breast cancers in postmenopausal women.

However, inhibition of estrogen sulfotransferase by several chlorinated phenol derivatives, such as 6,6-dichloro-4-nitrophenol and hydroxylated polychlorinated biphenol, has been shown to enhance the activity of endogenous estrogen and to be probably involved in human reproductive disorders and an increased incidence of breast carcinomas (Song, 2001). In addition, environmental xenoestrogens, such as bisphenol, 4-*n*-octylphenol, 4-*n*-nonylphenol, and 17α-ethynylestradiol, which have become ubiquitous in the environment and have been shown to cause a decrease in sperm quality and progression of estrogen-dependent cancers by interacting with the estrogen receptor, are converted to their sulfoconjugates by cytosolic sulfotransferase, resulting in detoxification of the chemicals through abolition of their binding ability as to the receptor and in facilitation of their removal from the body through an increase in water solubility (Suiko *et al.*, 2000; Wolff *et al.*, 1993). Thus, estrogen sulfatase is also involved in the regulation of artificial chemicals affecting estrogen activity.

References

Chander, S. K., Purohit, A., Woo, L. W., Potter, B. V., and Reed, M. J. (2004). The role of steroid sulphatase in regulating the oestrogenicity of oestrogen sulphamates. *Biochem. Biophys. Res. Commun.* **322,** 217–222.

Conary, J., Nauerth, A., Burns, G., Hasilik, A., and von Figura, K. (1986). Steroid sulfatase: Biosynthesis and processing in normal and mutant fibroblasts. *Eur. J. Biochem.* **158,** 71–76.

Dibbelt, L., and Kuss, E. (1986). Human placental steryl-sulfatase: Enzyme purification, production of antisera, and immunoblotting reactions with normal and sulfatase-deficient placentas. *Biol. Chem. Hoppe. Seyler* **367,** 1223–1229.

Dolly, J. O., Dodgson, K. S., and Rose, F. A. (1972). Studies on the oestrogen sulphatase and arylsulphatase C activities of rat liver. *Biochem. J.* **128,** 337–345.

Falany, J. L., Macrina, N., and Falany, C. N. (2002). Regulation of MCF-7 breast cancer cell growth by beta-estradiol sulfation. *Breast Cancer Res. Treat.* **74,** 167–176.

Gant, N. F., Hutchinson, H. T., Siiteri, P. K., and Mac Donald, P. C. (1971). Study of the metabolic clearance rate of dehydroisoandrosterone sulfate in pregnancy. *Am. J. Obstet. Gynecol.* **111,** 555–563.

Glutek, S. M., and Hobkirk, R. (1990). Estrogen sulfatase and steroid sulfatase activities in intrauterine tissues of the pregnant guinea pig. *J. Steroid Biochem. Mol. Biol.* **37,** 707–715.

Grodin, J. M., Siiteri, P. K., and Mac Donald, P. C. (1973). Source of estrogen production in postmenopausal women. *J. Clin. Endocrinol. Metab.* **36,** 207–214.

Hadd, H. E., and Blickrstaff, R. T. (1969). "Conjugates of Steroid Hormones." Academic Press, New York.

Hejaz, H. A., Woo, L. W., Purohit, A., Reed, M. J., and Potter, B. V. (2004). Synthesis, *in vitro* and *in vivo* activity of benzophenone-based inhibitors of steroid sulfatase. *Bioorg. Med. Chem.* **12,** 2759–2772.

Hernandez-Guzman, F. G., Higashiyama, T., Osawa, Y., and Ghosh, D. (2001). Purification, characterization and crystallization of human placental estrone/dehydroepiandrosterone sulfatase, a membrane-bound enzyme of the endoplasmic reticulum. *J. Steroid Biochem. Mol. Biol.* **78,** 441–450.

Hobkirk, R. (1985). Steroid sulfotransferases and steroid sulfate sulfatases: Characteristics and biological roles. *Can. J. Biochem. Cell Biol.* **63,** 1127–1144.

Hoffman, B., Falter, K., Vielemeier, A., Failing, K., and Schuler, G. (2001). Investigations on the activity of bovine placental oestrogen sulfotransferase and sulfatase from midgestation to parturition. *Exp. Clin. Endocrinol. Diabetes* **109,** 294–301.

Horvath, A., Nussbaumer, P., Wolff, B., and Billich, A. (2004). 2-(1-Adamantyl)-4-(thio) chromenone-6-carboxylic acids: Potent reversible inhibitors of human steroid sulfatase. *J. Med. Chem.* **47,** 4268–4276.

Horwitz, A. L., Warshawsky, L., King, J., and Burns, G. (1986). Rapid degradation of steroid sulfatase in multiple sulfatase deficiency. *Biochem. Biophys. Res. Commun.* **135,** 389–396.

Huang, Z., Fasco, M. J., and Kaminsky, L. S. (1997). Inhibition of estrone sulfatase in human liver microsomes by quercetin and other flavonoids. *J. Steroid Biochem. Mol. Biol.* **63,** 9–15.

Kotov, A., Falany, J. L., Wang, J., and Falany, C. N. (1999). Regulation of estrogen activity by sulfation in human Ishikawa endometrial adenocarcinoma cells. *J. Steroid Biochem. Mol. Biol.* **68,** 137–144.

Leese, M. P., Newman, S. P., Purohit, A., Reed, M. J., and Potter, B. V. (2004). 2-Alkylsulfanyl estrogen derivatives: Synthesis of a novel class of multi-targeted anti-tumour agents. *Bioorg. Med. Chem. Lett.* **14,** 3135–3138.

Milson, D. W., Rose, F. A., and Dodgson, K. S. (1972). The specific assay of arylsulphatase C, a rat liver microsomal marker enzyme. *Biochem. J.* **128,** 331–336.

Muir, M., Romalo, G., Wolf, L., Elger, W., and Schweikert, H. U. (2004). Estrone sulfate is a major source of local estrogen formation in human bone. *J. Clin. Endocrinol. Metab.* **89,** 4685–4692.

Pasqualini, J. R., Schatz, B., Varin, C., and Nguyen, B. L. (1992). Recent data on estrogen sulfatases and sulfotransferases activities in human breast cancer. *J. Steroid Biochem. Mol. Biol.* **41,** 323–329.

Purinton, S. C., Newman, H., Castro, M. I., and Wood, C. E. (1999). Ontogeny of estrogen sulfatase activity in ovine fetal hypothalamus, hippocampus, and brain stem. *Am. J. Physiol.* **276,** R1647–R1652.

Purohit, A., Potter, B. V., Parker, M. G., and Reed, M. J. (1998). Steroid sulphatase: Expression, isolation and inhibition for active-site identification studies. *Chem. Biol. Interact.* **109,** 183–193.

Reed, J. E., Woo, L. W., Robinson, J. J., Leblond, B., Leese, M. P., Purohit, A., Reed, M. J., and Potter, B. V. (2004). 2-Difluoromethyloestrone 3-O-sulphamate, a highly potent steroid sulphatase inhibitor. *Biochem. Biophys. Res. Commun.* **317,** 169–175.

Rosenthal, H. E., Pietrzak, E., Slaunwhite, W. R., Jr., and Sandberg, A. A. (1972). Binding of estrone sulfate in human plasma. *J. Clin. Endocrinol. Metab.* **34,** 805–813.

Roy, A. B. (1987). Evaluation of sulfate esters and glucuronides by enzymatic means. *Anal. Biochem.* **165,** 1–12.

Roy, A. B., Curtis, C. G., and Powell, G. M. (1987). The metabolism of oestrone and some other steroids in isolated perfused rat and guinea pig livers. *Xenobiotica* **17,** 1299–1313.

Saito, T., Kinoshita, S., Fujii, T., Bandoh, K., Fuse, S., Yamauchi, Y., Koizumi, N., and Horiuchi, T. (2004). Development of novel steroid sulfatase inhibitors. II. TZS-8478 potently inhibits the growth of breast tumors in postmenopausal breast cancer model rats. *J. Steroid Biochem. Mol. Biol.* **88,** 167–173.

Schreiner, E. P., and Billich, A. (2004). Estrone formate: A novel type of irreversible inhibitor of human steroid sulfatase. *Bioorg. Med. Chem. Lett.* **14,** 4999–5002.

Shimizu, M., Ohta, K., Matsumoto, Y., Fukuoka, M., Ohno, Y., and Ozawa, S. (2002). Sulfation of bisphenol A abolished its estrogenicity based on proliferation and gene expression in human breast cancer MCF-7 cells. *Toxicol. In Vitro* **16,** 549–556.

Song, W. C. (2001). Biochemistry and reproductive endocrinology of estrogen sulfotransferase. *Ann. N. Y. Acad. Sci.* **948,** 43–50.

Stein, C., Hille, A., Seidel, J., Rijnbout, S., Waheed, A., Schmidt, B., Geuze, H., and von Figura, K. (1989). Cloning and expression of human steroid-sulfatase: Membrane topology, glycosylation, and subcellular distribution in BHK-21 cells. *J. Biol. Chem.* **264,** 13865–13872.

Suiko, M., Sakakibara, Y., and Liu, M. C. (2000). Sulfation of environmental estrogen-like chemicals by human cytosolic sulfotransferases. *Biochem. Biophys. Res. Commun.* **267,** 80–84.

Tanaka, K., Kubushiro, K., Iwamori, Y., Okairi, Y., Kiguchi, K., Ishiwata, I., Tsukazaki, K., Nozawa, S., and Iwamori, M. (2003). Estrogen sulfotransferase and sulfatase: Roles in the regulation of estrogen activity in human uterine endometrial carcinomas. *Cancer Sci.* **94,** 871–876.

Touchstone, J. C., Greene, J. W., Jr., Mcelroy, R. C., and Murawec, T. (1963). Blood estriol conjugation during human pregnancy. *Biochemistry* **128,** 653–657.

Utsumi, T., Yoshimura, N., Takeuchi, S., Ando, J., Maruta, M., Maeda, K., and Harada, N. (1999). Steroid sulfatase expression is an independent predictor of recurrence in human breast cancer. *Cancer Res.* **59,** 377–381.

Verde, A., and Drucker, W. D. (1972). Distribution of dehydroepiandrosterone sulfate sulfatase in rat tissue. *Endocrinology* **90,** 138–143.

Walter, G., Liebl, R., and von Angerer, E. (2004). 2-Phenylindole sulfamates: Inhibitors of steroid sulfatase with antiproliferative activity in MCF-7 breast cancer cells. *J. Steroid Biochem. Mol. Biol.* **88,** 409–420.

Wolff, M. S., Toniolo, P. G., Lee, E. W., Rivera, M., and Dubin, N. (1993). Blood levels of organochlorine residues and risk of breast cancer. *J. Natl. Cancer Inst.* **85,** 648–652.

Wong, C. K., and Keung, W. M. (1997). Daidzein sulfoconjugates are potent inhibitors of sterol sulfatase (EC 3.1.6.2). *Biochem. Biophys. Res. Commun.* **233,** 579–583.

[18] Analysis for Localization of Steroid Sulfatase in Human Tissues

By Takashi Suzuki, Yasuhiro Miki, Tsuyoshi Fukuda, Taisuke Nakata, Takuya Moriya, and Hironobu Sasano

Abstract

Human steroid sulfatase (STS) is an enzyme that hydrolyzes several sulfated steroids, such as estrone sulfate, dehydroepiandrosterone sulfate, and cholesterol sulfate, and results in the production of active substances. STS has been demonstrated in human breast cancer tissues and is considered to be involved in intratumoral estrogen production. It is very important to analyze the cellular distribution of STS with accuracy in human tissues in order to obtain a better understanding of the biological significance of STS. Therefore, this chapter describes several morphological approaches used to study the localization of STS, including immunohistochemistry, mRNA *in situ* hybridization, and laser capture microdissection/reverse transcription–polymerase chain reaction, in human tissues.

Introduction

Steroid sulfatase (STS) is a single enzyme that hydrolyzes several sulfated steroids such as estrone sulfate, dehydroepiandrosterone sulfate, and cholesterol sulfate (Burns, 1983; Dibbelt and Kuss, 1986). In human, STS expression has been detected in the placenta (Miki *et al.*, 2002) and in a majority of breast cancer tissues (Evans *et al.*, 1994; Santner *et al.*, 1993; Utsumi *et al.*, 1999). In particular, STS is considered to mainly catalyze estrone sulfate to estrone in breast cancer tissues, which contributes to intratumoral or *in situ* estrogen production (Pasqualini and Chetrite, 1999). Results have demonstrated the importance of *in situ* estrogen production in the development of breast carcinomas, and the inhibition of this pathway could be clinically useful for reducing the progression of estrogen-dependent breast tumors (Santen and Harvey, 1999). Therefore, STS may become an important therapeutic target as an endocrine therapy for breast cancer patients, and STS inhibitors are currently being developed by several groups (Nussbaumer and Billich, 2004; Reed *et al.*, 2004).

It then becomes very important to analyze the expression of STS with accuracy in order to obtain a better understanding of the clinical and/or biological significance of STS in human tissues, including breast cancers.

METHODS IN ENZYMOLOGY, VOL. 400   0076-6879/05 $35.00
                                                          DOI: 10.1016/S0076-6879(05)00018-2

STS expression has been analyzed predominantly by enzymatic assay or reverse transcription–polymerase chain reaction (RT-PCR) in human tissues (Evans *et al.*, 1994; Santner *et al.*, 1993; Utsumi *et al.*, 1999). However, it is true that these quantitative results are influenced by the relative ratio of target cell population in the tissues examined because whole specimens were homogenized in these analyses. Therefore, an examination of STS localization provides very important information that cannot be obtained from these biochemical analyses. Therefore, this chapter summarizes several morphological approaches used to study the distribution of STS in human tissues, including immunohistochemistry, mRNA *in situ* hybridization, and laser capture microdissection (LCM)/RT-PCR.

Immunohistochemistry

Immunohistochemistry is a method used to detect a specific protein in tissues. Immunohistochemical analysis of STS has been reported previously in various human tissues, and STS immunoreactivity was detected in syncytiotrophoblasts of the placenta (Miki *et al.*, 2002) and carcinoma cells of 75–90% of the breast carcinoma (Saeki *et al.*, 1999; Suzuki *et al.*, 2003). However, it is also true that the results are subject to some technical artifacts, such as the antibody, tissue processing, and evaluation system employed.

It is well known that the sensitivity and/or specificity of immunohistochemistry is most dependent on which antibody is used in the study. Therefore, a reliable antibody is essential in the immunohistochemical study of STS. The STS antibody used in our study is an affinity-purified monoclonal antibody (KM1049), which was raised against the STS enzyme purified from human placenta and recognized the peptide corresponding to amino acids 420–428 (Saeki *et al.*, 1999). This antibody could demonstrate STS immunoreactivity in specimens fixed with 10% formalin and embedded in paraffin wax. However, the antibody for human STS is not available commercially at this juncture, to our knowledge, which makes it difficult to perform immunohistochemistry for STS in a great majority of laboratories. A Histofine kit (Nichirei, Tokyo, Japan), which employs the streptavidin–biotin amplification method, is used for immunostaining of STS, and no antigen retrieval is required (Suzuki *et al.*, 2003). Detailed immunohistochemical procedures for STS are summarized in Table I, and results are demonstrated in Fig. 1.

Tissue processing is also considered an important factor in detecting significant immunoreactivity (Suzuki *et al.*, 1994). STS immunoreactivity was markedly detected in human placental tissues embedded in paraffin wax after fixation in 10% formalin buffer, as shown in Fig. 1A. However,

TABLE I
PROCEDURES OF STS IMMUNOHISTOCHEMISTRY IN HUMAN TISSUES[a]

Step	Time	Temperature[b]
1. Deparaffinization		
2. Blocking of endogenous peroxidase activity		
a. 100% methanol with 0.3% hydrogen peroxidase	30 min	RT
b. Rinse in 0.01 M phosphate-buffered saline (PBS)	3×	RT
3. Blocking of nonspecific protein binding		
a. 1% normal rabbit serum[c]	30 min	RT
b. Rinse in 0.01 M PBS	3×	RT
4. Incubation with primary antibody		
a. Primary antibody for STS (KM1049) [1/9000 (0.37 (μg/ml) dilution]	O/N	4°
b. Rinse in 0.01 M PBS	3×	RT
5. Incubation with secondary antibody		
a. Biotinylated antimouse immunoglobulin (Ig)[c]	30 min	RT
b. Rinse in 0.01 M PBS	3×	RT
6. Streptavidin–biotin amplification		
a. Peroxidase-conjugated streptavidin[c]	30 min	RT
b. Rinse in 0.01 M PBS	3×	RT
7. Visualization of antigen–antibody complex		
a. Immersion in 3.3′-diaminobenzidine (DAB) solution [1 mM DAB, 50 mM Tris-HCl buffer (pH 7.6), and 0.006% hydrogen peroxidase]	3 min	RT
b. Immersion in tap water	5 min	RT
c. Immersion in 0.01 M PBS	10 min	RT
8. Counterstaining Hematoxylin	30 s	RT
9. Mounting		

[a] Specimens were fixed with 10% formalin and embedded in paraffin wax. These were sectioned into 3-μm slices and placed on regular clean glass slides (Matsunami, Tokyo, Japan) prior to immunohistochemistry.

[b] RT, room temperature; O/N, overnight.

[c] A Histofine kit (Nichirei, Tokyo, Japan).

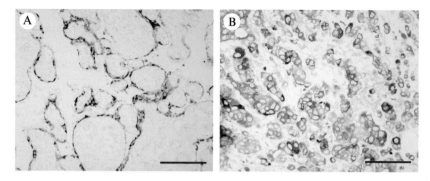

Fig. 1. Immunohistochemistry for STS in human tissues. STS immunoreactivity was detected in syncytiotrophoblasts of the placenta (A) or carcinoma cells of the breast carcinoma (B). Specimens were fixed with 10% formalin and embedded in paraffin wax. Bar: 100 μm.

immunoreactivity of STS was much weaker in those after fixed in paraformaldehyde–lysine–periodate buffer. In addition, the relative immunointensity of STS was influenced by the duration of fixation, and overfixation time such as 1 week markedly decreased the immunoreactivity in the tissues, regardless of the types of fixatives. Therefore, modes of tissue processing should be considered when evaluating STS immunohistochemistry in specimens retrieved from pathology archives.

Positive and negative controls are required in order to interpret the immunoreactivity with accuracy. Positive control tissues are generally used to confirm the expected immunoreactivity of the experiment, and placental tissues are generally considered suitable as a positive control of STS immunohistochemistry. A negative control is also important in evaluating specific immunoreactivity. The immunohistochemical preabsorption test is a gold standard as a negative control of immunohistochemistry when using a polyclonal antibody. However, in immunohistochemistry using a mouse monoclonal antibody, such as the STS antibody used in our study, normal mouse IgG is usually used instead of the primary antibody. No specific immunoreactivity was detected in these sections.

The evaluation system of STS immunoreactivity is also important in order to improve the reliability of immunohistochemical results. Immunoreactivity is generally evaluated by observers according to the immunointensity and/or immunostaining area. Immunointensity may reflect the amount of protein (Suzuki *et al.*, 1994), although immunohistochemistry is by no means a quantitative analysis. For instance, overexpression of HER2 protein is generally evaluated according to HER2 immunointensity in breast cancer tissues (HercepTest; DAKO, Carpintaria, CA). However, it is also true that immunointensity is markedly affected by modes of tissue processing and/or immunohistochemical procedures employed. Therefore, an accurate determination of immunoreactivity is not necessarily easy among different immunostained slides. Because of these factors, some pathologists are reluctant to use immunointensity as an evaluative instrument of immunohistochemistry (Tsuda et al., 2002). In our immunohistochemical study of STS in breast cancers, we evaluated STS immunoreactivity according to the immunostaining area, regardless of the intensity (i.e., ++, more than 50% positive carcinoma cells; +, 1–50% positive carcinoma cells; and -, no immunoreactivity) in breast carcinoma tissues (Suzuki et al., 2003). When three pathologists independently evaluated the immunoreactivity of STS in 113 breast cancer tissues using this system, interobserver differences were 1.8% and discordant results turned out to be mainly due to differences in the evaluation of weak immunopositive or –negative staining (background) (Suzuki et al., 2003). Comparison of the results to representative slides in each group could improve the reliability of immunohistochemical evaluation.

mRNA In Situ Hybridization

The mRNA in situ hybridization method can demonstrate the cellular localization of mRNA in tissue sections. Antibodies or full sequences of the substances are not required if specific oligonucleotides are employed as probes for mRNA hybridization. Therefore, this method has been utilized extensively in various studies. It is true that mRNA in situ hybridization is relatively cumbersome and requires technical skills. However, the development of computer-assisted control of mRNA in situ hybridization contributes greatly to the improvement of reliability and reproducibility of the method. This section describes mRNA in situ hybridization analysis for STS using a Discovery automated slide-processing system (Ventana Medical Systems, Tucson, AZ) (Nitta et al., 2003; Saruta et al., 2004). Detailed procedures are described in Table II, and representative results are shown in Fig. 2.

Tissue processing is the most important factor in the mRNA in situ hybridization method because mRNA is degradated easily by RNase in the specimens. No reliable hybridization signals could be obtained without proper fixation of the specimens, even using the best procedure available. The most suitable fixatives are generally considered 1% glutalaldehyde in 4% paraformaldehyde (PFA) adjusted to pH 7.4 (Sasano et al., 2001), but 10% formalin may be used as a fixative for specimens of mRNA in situ hybridization for STS (Fig. 2). Regardless of the fixative employed, specimens should be fixed promptly to prevent the degradation of mRNA targets in cells or distorted architecture of specimens. Appropriate dura-

TABLE II
PROCEDURES OF mRNA *IN SITU* HYBRIDIZATION FOR STS IN HUMAN TISSUES[a]

Step	Time	Temperature
1. Deparaffinization		
2. First fixation with RiboPrep[b] (formalin based)	30 min	37°
3. Acid treatment with RiboClear[b] (hydrochloride based)	10 min	37°
4. Protease digestion with Protease 2[b]	2 min	37°
5. Hybridization		
a. Denaturation of probe solution (200 ng/slide)	10 min	70°
b. Hybridization with probe solution	6 h	42°
6. Stringency wash using 0.1× RiboWash[b]	3×, 6 min	65°
7. Second fixation with RiboPrep[b]	20 min	37°
8. Incubation with biotin-labeled antidigoxigenin antibody (Sigma-Aldrich, St. Louis, MO)	30 min	37°
9. Incubation with streptavidin–alkaline phosphotase conjugate (SA-Alk Phos[b])	16 min	37°
10. Signal detection using a Blue Map NBT / BCIP[b] substrate kit[b]	6 h	37°
11. Counter staining by fast red	3 min	RT
12. Mounting		

[a] This protocol summarizes the procedures of mRNA *in situ* hybridization using Discovery (Ventana Medical Systems, Tucson, AZ). Specimens were fixed with 10% formalin and embedded in paraffin wax. These were sectioned into 3-μm slices and placed on silan-coated glass slides (Matsunami, Tokyo, Japan).
[b] Reagents were from Ventana Medical Systems.

tion of fixation time is 6 h to 1 day, and fixation for a long duration makes it difficult to detect mRNA hybridization signals. Paraffin-embedded sections are suitable mRNA *in situ* hybridization compared to frozen tissue sections, which provide good morphological details.

Oligonucleotide probes employed mRNA *in situ* hybridization for STS mRNA in our laboratory were TCT GGC AGG GTC TGG GTG TGT CTG TCC GCA GCT TCC TGC ATG ACT TTG AG [cDNA position (accession No. M16505); 1799–1848]. Satisfactory results are obtained when using the oligonucleotide probes with 50 mers in length and approximately 60% of GC content in the Discovery system in our study, although oligonucleotide probes with a higher GC content are considered to hybridize more stably. The oligonucleotide probes were designed using computer-assisted search engines and were subsequently labeled by a digoxigenin-RNA labeling kit (Roche Diagnostics GmbH, Mannheim,

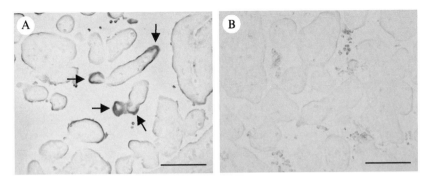

FIG. 2. mRNA *in situ* hybridization for STS in the human placental tissue. mRNA signals for STS were detected in the syncytiotrophoblasts (arrows) (A), but not in a negative control section using the sense probe (B). Specimens were fixed with 10% formalin and embedded in paraffin wax. A Discovery automated slide-processing system (Ventana Medical systems) was used. Bar: 100 μm.

Germany). It may be possible to use a mixture of several oligonucleotide probes that recognize different locations of STS mRNA to increase the sensitivity of hybridization reactions.

Controls are essential to appropriately evaluate the findings of mRNA *in situ* hybridization. Properly processed human placental tissues should be run simultaneously in each experiment as a positive control tissue for STS mRNA *in situ* hybridization, and positive hybridization reactions are confirmed in these tissue sections. In addition, a positive control probe such as the poly(T) oligonucleotide probe may be recommended in the method of mRNA *in situ* hybridization, as degradation of mRNA in tissues easily results in false-negative findings. This probe is labeled in the same manner as the probes employed in the experiment, and its hybridization signals should be detected in all tissue specimens examined. Hybridization reactions for the positive control probe represent presentation of mRNA in the section, which is useful in interpreting the negative reaction with the probes for STS. As a negative control, oligo probes of sense orientation should be employed in order to confirm the specificity of the hybridization signals (Fig. 2B).

Simultaneous mRNA *in situ* hybridization and immunohistochemistry may be performed in the same tissue section. This technique can provide detailed information as to the precise cellular localization at both mRNA and protein levels in the same cells (Suzuki *et al.*, 1992), which may be useful in analyzing abnormal expression or discrepancy between mRNA and protein expression of STS in human tissues. In this method, immunohistochemistry is generally performed prior to mRNA *in situ* hybridization.

RNase contamination through the immunohistochemical procedure may decrease the sensitivity of mRNA *in situ* hybridization. However, great care should be taken for the problem, as immunoreactivity is occasionally diminished by protease treatment that is necessary for the procedure of mRNA *in situ* hybridization.

Laser Capture Microdissection/Reverse Transcription–Polymerase Chain Reaction

RT-PCR analysis is a very useful method for examining mRNA expression in human tissues, and it is also possible to quantify the amount of PCR products using recently developed methods such as real-time PCR (Dumoulin *et al.*, 2000). However, RT-PCR basically requires homogenization of whole specimens; therefore, the results are influenced by the components of cell population. This fact is generally overlooked by many investigators, but it often results in serious problems in the interpretation of findings in tissues such as breast carcinoma. In order to overcome this problem, the microdissection technique was introduced as a tool for obtaining specific cell populations from tissue sections (Going and Lamb, 1996; Whetsell *et al.*, 1992). In the traditional method of this technique, cells were dissected manually from tissues under a microscope using a needle or micromanipulation. However, this manual method requires technical skill and is time-consuming. In addition, it is not necessarily easy to collect a sufficient quantity of RNA for RT-PCR analyses.

The development of laser-based dissection technologies such as LCM or laser microbeam microdissection with laser pressure catapulting has provided researchers with the ability to accurately procure near pure populations of target cells from specific microscopic regions of tissue sections (Emmert-Buck *et al.*, 1996; Kolble, 2000; Schutze and Lahr, 1998). LCM has also resulted in a breakthrough in terms of speed and ease of use from the standpoints of investigators, and cells isolated by LCM are generally considered suitable mRNA sources for RT-PCR analyses. In our laboratory, LCM is conducted using the Laser Scissors CRI-337 (Cell Robotics Inc., Albuquerque, NM) (Fig. 3A–D). The procedures of LCM/RT-PCR for STS in human tissues and the results obtained are summarized in Table III and Fig. 3E, respectively.

Tissue processing is the most important step in the LCM/RT-PCR method, as with mRNA *in situ* hybridization, because mRNA is degraded easily by RNase in specimens. Frozen tissue sections embedded in Tissue-Tek OTC compound (Sakura Finetechnical Co., Tokyo, Japan) are used in LCM/RT-PCR analysis in our laboratory. However, formalin-fixed and

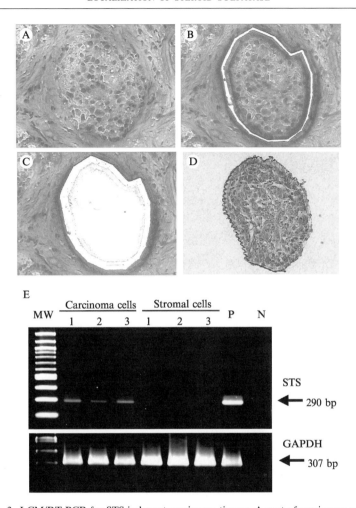

FIG. 3. LCM/RT-PCR for STS in breast carcinoma tissues. A nest of carcinoma cells was dissected from breast cancer tissue by LCM (A–C) and was collected separately (D). LCM was conducted using the Laser Scissors CRI-337 (Cell Robotics Inc). (E) RT-PCR analysis for STS followed LCM in human breast carcinoma tissues. mRNA expression for STS was detected as a specific single band (290 bp) in carcinoma cells, but not in intratumoral stromal cells. PCR was performed for 40 cycles using a light cycler system (Roche Diagnostics). P, positive control (human placental syncytiotrophoblasts); N, negative control (no cDNA substrate). Three cases are represented in these agarose gel photos.

paraffin-embedded tissues have also been reported to be employed for LCM/RT-PCR analysis (Schutze and Lahr, 1998; Specht *et al.*, 2000). Specimens are sectioned into 8- 10-μm slices, fixed with 100% methanol, and subsequently stained with toluidine blue or hematoxylin–eosin to

TABLE III
PROCEDURES OF LCM/RT-PCR FOR STS IN HUMAN TISSUES

Step	Time	Temperature
1. Preparation of tissue sections		
a. Sectioning of frozen tissues at 8 μm		$-20°$
(Cell Robotics, Inc., Albuquerque, NM)		
b. Mounting on membrane-attached glass slides		
(Cell Robotics Inc., Albuquerque, NM)		
c. Fixation with 100% methanol	5 min	RT
d. Staining with toluidine blue	10 s	RT
e. Immersion in RNAse-free water	5 min	RT
f. Rinse in 100% methanol	3×	RT
g. Air-dry		
2. LCM		
Collection of cells from tissues using LCM		RT
3. RNA extraction[a]		
a. Addition of RNA denaturing buffer (200 μl) containing	2 h	$4°$
guanidium thiocyanate, sodium N-lauroyl sarcosinate and		
β-mercaptoethanol to dissolve dissected tissues		
b. Addition of 2M sodium acetate, pH 4.0 (20 μl), phenol		
(220 μl), and chloroform-isoamyl alcohol (60 μl)		
c. Centrifugation (15,000 rpm)	30 min	$4°$
d. Transfer of upper aqueous layer		
e. Addition of glycogen (1 μg) and isopropanol (200 μl)	O/N	$-80°$
f. Precipitation of RNA and wash with 70% ethanol		
g. Resuspension of pellets in RNase-free water (9 μl)		
4. RT		
Reverse transcription in a reaction mixture containing	60 min	$50°$
50 mM Tris acetate, pH 8.4, 75 mM potassium acetate,		
8 mM magnesium acetate, 0.01 M dithiothreitol, 2 mM		
dNTP, 25 μM oligo(dT)12–18 primer, 25 μg/μl random		
hexamer oligonucleotides, and SuperScript II RNase		
H-Reverse Transcriptase (InVitrogen Life Technologies)		
5. PCR[b]		
a. Initial denaturation	1 min	$95°$
b. Denaturation	1 s	$95°$
c. Annealing	15 s	$60°$
d. Elongation 40 cycles in total (5–2, 5–3, and 5–4)	15 s	$72°$
e. Resolution on a 2% agarose ethidium bromide gel		

[a] RNA microisolation protocol is based on the reports by Emmert-Buck et al. (1999) and Niino et al. (2001).
[b] PCR was carried out with the Light Cycler System (Roche Diagnostics GmbH, Mannheim, Germany).

identify the morphological features and to isolate the specific cells under the microscopy of the LCM system. The application of immunohistochemistry prior to LCM has been reported in order to collect more specific cell populations according to phenotypes or functions (Fend *et al.*, 2000). In contrast to immunohistochemistry on paraffin sections, immunohistochemistry on frozen sections always results in a reduction of RNA recovery. Therefore, a brief immunostaining time (less than 15 min) is one of the most important keys in obtaining successful results of immuno-LCM (Fend *et al.*, 1999; Murakami *et al.*, 2000).

It is important to establish minimum amounts of cells required for molecular analyses because it is not necessarily easy to obtain a large number of cells using LCM if the cell population is low in the tissue. Results of the study depend on various factors, such as quality of tissue specimens, efficiency of RNA extraction method, and sensitivity of the RT-PCR method. Therefore, pilot studies are required to determine the optimal condition of LCM/RT-PCR analysis. As few as 20 cells may be sufficient for generating reproducible results for RT-PCR in our laboratory. However, employment of such small amounts as materials for RT-PCR may result in technical artifacts through sectioning. Therefore, we usually collect approximately 500 cells from carcinoma tissues or 100 cells from normal tissues for LCM/RT-PCR analysis for STS (Miki *et al.*, 2002; Suzuki *et al.*, 2003). A combination of LCM with the microarray-based gene expression technique is also an attractive method for analyzing the genes related with STS expression. In this method, T7-based linear RNA amplification is generally employed to obtain sufficient amounts of RNA from microdissected samples (Fuller *et al.*, 2003). Accurate results may be obtained in a reproducible manner when using as few as 200 cells, and results are usually obtained from 2000 cells (Sgroi *et al.*, 1999).

Total RNA was extracted from laser-transferred cells according to the RNA microisolation protocol reported by Emmert-Buck *et al.* (1999) and Niino *et al.* (2001). In general, all amounts of the RNA were reverse transcribed using a reverse transcription kit (SuperScript II preamplification system, InVitrogen Life Technologies, Gaithersburg, MD) without the measurement of RNA volume of a specimen. Several kits are available commercially to support the procedures of LCM/RT-PCR. In the Super-Script III CellsDirect cDNA synthesis system (InVitrogen Life Technologies), cDNA can be synthesized efficiently from cells collected by LCM without RNA extraction, which is one of the most cumbersome steps.

The PCR procedure is the same as conventional methods using homogenized specimens. In our laboratory, PCR for STS was carried out with the Light Cycler System (Roche Diagnostics) using the DNA-binding dye SYBER Green I (Roche Diagnostics) for the detection of PCR products. The primer sequences for STS used in our study were as follows: FWD

5'-ACTGCAACGCCTACTTAAATG-3' and REV 5'-AGGGTCTGGGT GTGTCTGTC-3' [cDNA position (M16505); 1554–1842] (Utsumi *et al.*, 1999). As a positive control of LCM/RT-PCR for STS, we used syncytiotrophoblasts of human placental tissues, and the PCR products were confirmed by direct sequencing. Negative control experiments lacked the cDNA substrate to check for the possibility of exogenous contaminant DNA, and no amplified products were detected under these conditions. We also use the Light Cycler System (Roche Diagnostics) in order to semiquantify the level of STS mRNA in tissue. The mRNA level for STS in each tissue has been summarized as a ratio (%) of internal control such as glyceraldehyde-3-phosphate dehydrogenase (GAPDH) or ribosomal protein L 13a (RPL13A).

Conclusions

As described earlier, several methods are available to analyze the localization of STS in human tissues. Among these, immunohistochemistry may be the most preferable method because it is not cumbersome, is reliable, and reproducible results can be generally obtained from various specimens. However, antibodies for human STS, which are essential in immunohistochemistry, are not currently available commercially. Therefore, mRNA *in situ* hybridization for STS may be the most suitable method for examining the localization of STS in human tissues in a great majority of laboratories at this juncture. LCM/RT-PCR analysis, which is a newly developed method, can also be a useful tool if frozen specimens are available. LCM/RT-PCR is a highly sensitive method, and quantitative results can be obtained reliably according to the cell populations in tissues. Analysis for STS localization can provide important information and contributes to obtaining a better understanding of the biological significance of STS in human tissues.

Acknowledgments

We appreciate the skillful technical assistance of Ms. Chika Kaneko and Mr. Katsuhiko Ono (Department of Pathology, Tohoku University School of Medicine, respectively).

References

Burns, G. R. (1983). Purification and partial characterization of arylsulphatase C from human placental microsomes. *Biochim. Biophys. Acta* **759,** 199–204.
Dibbelt, L., and Kuss, E. (1986). Human placental steryl-sulfatase: Enzyme purification, production of antisera, and immunoblotting reactions with normal and sulfatase-deficient placentas. *Biol. Chem. Hoppe. Seyler* **367,** 1223–1229.

Dumoulin, F. L., Nischalke, H. D., Leifeld, L., von dem Bussche, A., Rockstroh, J. K., Sauerbruch, T., and Spengler, U. (2000). Semi-quantification of human C-C chemokine mRNAs with reverse transcription/real-time PCR using multi-specific standards. *J. Immunol. Methods* **241,** 109–119.

Emmert-Buck, M. R., Bonner, R. F., Smith, P. D., Chuaqui, R. F., Zhuang, Z., Goldstein, S. R., Weiss, R. A., and Liotta, L. A. (1996). Laser capture microdissection. *Science* **274,** 998–1001.

Evans, T. R., Rowlands, M. G., Law, M., and Coombes, R. C. (1994). Intratumoral oestrone sulphatase activity as a prognostic marker in human breast carcinoma. *Br. J. Cancer* **69,** 555–561.

Fend, F., Emmert-Buck, M. R., Chuaqui, R., Cole, K., Lee, J., Liotta, L. A., and Raffeld, M. (1999). Immuno-LCM: Laser capture microdissection of immunostained frozen sections for mRNA analysis. *Am. J. Pathol.* **154,** 61–66.

Fend, F., Kremer, M., and Quintanilla-Martinez, L. (2000). Laser capture microdissection: Methodical aspects and applications with emphasis on immuno-laser capture microdissection. *Pathobiology* **68,** 209–214.

Fuller, A. P., Palmer-Toy, D., Erlander, M. G., and Sgroi, D. C. (2003). Laser capture microdissection and advanced molecular analysis of human breast cancer. *J. Mammary Gland Biol. Neoplasia* **8,** 335–345.

Going, J. J., and Lamb, R. F. (1996). Practical histological microdissection for PCR analysis. *J. Pathol.* **179,** 121–124.

Kolble, K. (2000). The LEICA microdissection system: Design and applications. *J. Mol. Med.* **78,** B24–B25.

Miki, Y., Nakata, T., Suzuki, T., Darnel, A. D., Moriya, T., Kaneko, C., Hidaka, K., Shiotsu, Y., Kusaka, H., and Sasano, H. (2002). Systemic distribution of steroid sulfatase and estrogen sulfotransferase in human adult and fetal tissues. *J. Clin. Endocrinol. Metab.* **87,** 5760–5768.

Murakami, H., Liotta, L., and Star, R. A. (2000). IF-LCM: Laser capture microdissection of immunofluorescently defined cells for mRNA analysis rapid communication. *Kidney Int.* **58,** 1346–1353.

Niino, Y., Irie, T., Takaishi, M., Hosono, T., Huh, N., Tachikawa, T., and Kuroki, T. (2001). PKCtheta II, a new isoform of protein kinase C specifically expressed in the seminiferous tubules of mouse testis. *J. Biol. Chem.* **276,** 36711–36717.

Nitta, H., Kishimoto, J., and Grogan, T. M. (2003). Application of automated mRNA *in situ* hybridization for formalin-fixed, paraffin-embedded mouse skin sections: Effects of heat and enzyme pretreatment on mRNA signal detection. *Appl. Immunohistochem. Mol. Morphol.* **11,** 183–187.

Nussbaumer, P., and Billich, A. (2004). Steroid sulfatase inhibitors. *Med. Res. Rev.* **24,** 529–576.

Pasqualini, J. R., and Chetrite, G. S. (1999). Estrone sulfatase versus estrone sulfotransferase in human breast cancer: Potential clinical applications. *J. Steroid Biochem. Mol. Biol.* **69,** 287–292.

Reed, M. J., Purohit, A., Woo, L. W., Newman, S. P., and Potter, B. V. (2005). Steroid sulfatase: Molecular biology, regulation and inhibition. *Endocr. Rev.* **26,** 171–202.

Saeki, T., Takashima, S., Sasaki, H., Hanai, N., and Salomon, D. S. (1999). Localization of estrone sulfatase in human breast carcinomas. *Breast Cancer* **6,** 331–337.

Santen, R. J., and Harvey, H. A. (1999). Use of aromatase inhibitors in breast carcinoma. *Endocr. Relat. Cancer* **6,** 75–92.

Santner, S. J., Ohlsson-Wilhelm, B., and Santen, R. J. (1993). Estrone sulfate promotes human breast cancer cell replication and nuclear uptake of estradiol in MCF-7 cell cultures. *Int. J. Cancer* **54,** 119–124.

Saruta, M., Takahashi, K., Suzuki, T., Torii, A., Kawakami, M., and Sasano, H. (2004). Urocortin 1 in colonic mucosa in patients with ulcerative colitis. *J. Clin. Endocrinol. Metab.* **89,** 5352–5361.

Sasano, H., Matsuzaki, S., and Suzuki, T. (2001). Estrogen receptor mRNA *in situ* hybridization using microprobe system. *Methods Mol. Biol.* **176,** 317–325.

Schutze, K., and Lahr, G. (1998). Identification of expressed genes by laser-mediated manipulation of single cells. *Nature Biotechnol.* **16,** 737–742.

Sgroi, D. C., Teng, S., Robinson, G., Le Vangie, R., Hudson, J. R., Jr., and Elkahloun, A. G. (1999). *In vivo* gene expression profile analysis of human breast cancer progression. *Cancer Res.* **59,** 5656–5661.

Specht, K., Richter, T., Muller, U., Walch, A., and Hofler, M. W. (2000). Quantitative gene expression analysis in microdissected archival tissue by real-time RT-PCR. *J. Mol. Med.* **78,** B27.

Suzuki, T., Nakata, T., Miki, Y., Kaneko, C., Moriya, T., Ishida, T., Akinaga, S., Hirakawa, H., Kimura, M., and Sasano, H. (2003). Estrogen sulfotransferase and steroid sulfatase in human breast carcinoma. *Cancer Res.* **63,** 2762–2770.

Suzuki, T., Sasano, H., Sasaki, H., Fukaya, T., and Nagura, H. (1994). Quantitation of P450 aromatase immunoreactivity in human ovary during the menstrual cycle: Relationship between the enzyme activity and immunointensity. *J. Histochem. Cytochem.* **42,** 1565–1573.

Suzuki, T., Sasano, H., Sawai, T., Mason, J. I., and Nagura, H. (1992). Immunohistochemistry and *in situ* hybridization of P-45017 alpha (17 alpha-hydroxylase/17,20-lyase). *J. Histochem. Cytochem.* **40,** 903–908.

Tsuda, H., Sasano, H., Akiyama, F., Kurosumi, M., Hasegawa, T., Osamura, R. Y., and Sakamoto, G. (2002). Evaluation of interobserver agreement in scoring immunohisto-chemical results of HER-2/neu (c-erbB-2) expression detected by HercepTest, Nichirei polyclonal antibody, CB11 and TAB250 in breast carcinoma. *Pathol. Int.* **52,** 126–134.

Utsumi, T., Yoshimura, N., Takeuchi, S., Ando, J., Maruta, M., Maeda, K., and Harada, N. (1999). Steroid sulfatase expression is an independent predictor of recurrence in human breast cancer. *Cancer Res.* **59,** 377–381.

Whetsell, L., Maw, G., Nadon, N., Ringer, D. P., and Schaefer, F. V. (1992). Polymerase chain reaction microanalysis of tumors from stained histological slides. *Oncogene* **17,** 2355–2361.

[19] Metabolism of Phytoestrogen Conjugates

By Tracy L. D'Alessandro, Brenda J. Boersma-Maland,
T. Greg Peterson, Jeff Sfakianos, Jeevan K. Prasain, Rakesh P. Patel,
Victor M. Darley-Usmar, Nigel P. Botting, and Stephen Barnes

Abstract

Phytoestrogens are plant-derived compounds with physiologic estrogenic effects. They are present in the plant as glycosidic conjugates, some of which contain further chemical modifications (acetate, malonate, and 3-hydroxy-3-methylglutarate esters and 2,3-dihydroxysuccinate ether).

METHODS IN ENZYMOLOGY, VOL. 400
Copyright 2005, Elsevier Inc. All rights reserved.

0076-6879/05 $35.00
DOI: 10.1016/S0076-6879(05)00019-4

In the gastrointestinal tract, the conjugates undergo hydrolysis catalyzed by enzymes in the intestinal wall and by gut bacteria. On entering the systemic circulation, the phytoestrogens may undergo extensive metabolism to other compounds through reactions involving demethylation, methylation, hydroxylation, chlorination, iodination, and nitration. In addition, all these compounds can undergo conjugation to form β-glucuronides and sulfate esters. This chapter describes the methods of analysis of all these compounds, the sources of or methods to manufacture suitable standards, and the procedures for examining the enzymes that catalyze these reactions.

Introduction

Phytoestrogens are naturally occurring plant compounds, many of which have weak estrogenic or antiestrogenic activity in mammals. This class includes isoflavonoids, coumestanes, stilbenes, zearalones, and lignans. The soy isoflavones, genistein, daidzein, and glycitein, have been the most commonly studied of this class. Phytoestrogens have been proposed to have beneficial effects in chronic diseases such as atherosclerosis and cardiovascular diseases, hormone-dependent cancers, arthritis, neurodegeneration, and osteoporosis (Middleton, 2000). The mechanisms of action attributed to these disease-prevention effects range from their ability to elicit an estrogen-like response (Jacobs and Lewis, 2002) to tyrosine kinase inhibition (Akiyama et al., 1987) to antioxidant activity (Pietta, 2000). Many of the experiments demonstrating these effects have used phytoestrogens in their aglycone forms; however, phytoestrogens are present in foods and dietary supplements in the form of several types of conjugates as well as aglycones. In addition, in animals, the circulating forms of the phytoestrogens are mostly β-glucuronides and sulfates, with only small amounts (2–5%) as the aglycones. This discrepancy between the chemical forms used in laboratory experiments and those that are ingested and that circulate in the body requires a full appreciation of the metabolism of phytoestrogens during absorption, distribution, and excretion. This involves enzyme systems in the digestive tract, liver, and even at target tissue sites.

The reader is encouraged to read other reviews on the analysis of phytoestrogens (Wang et al., 2002; Wilkinson et al., 2002), particularly those involving mass spectrometry (Prasain et al., 2002, 2004). In many methods used for the analysis of phytoestrogens, for analytical convenience investigators convert the phytoestrogen conjugates to their aglycones. However, this provides an incomplete appreciation of phytoestrogen composition. Recent methodologies provide more accurate profiles of plasma, urine, feces, and other physiological samples. This chapter describes methods used for (1) the evaluation of phytoestrogen conjugation in foods, (2) the synthesis and biosynthesis of specific phytoestrogen metabolites found *in vivo*, (3) halogenated and nitrated metabolites formed during inflammation, and (4) quantification of these compounds *in vivo*.

Phytoestrogen Conjugates in Soybeans, Other Foods, and Dietary Supplements

The two major phytoestrogen classes contained in foods are isoflavones and lignans. The predominant isoflavones in soybeans are conjugates of genistein (5, 7, 4'-trihydroxyisoflavone) and daidzein (7, 4'-dihydroxyisoflavone). They are genistein-7-*O*-β-D-glycoside (genistin), 6''-*O*-malonylgenistin, daidzein-7-*O*-β-D-glycoside (daidzin), and 6''-*O*-malonyldaidzin (Fig. 1) (Barnes *et al.*, 1994; Wang and Murphy, 1994a). Glycitein (7, 4'-dihydroxy-6-methoxyisoflavone) and its conjugates are minor isoflavones in soybean cotyledons, but are major components in dietary supplements and foods made from the soybean hypocotyls (Kudou *et al.*, 1991). Red clover, also used to make dietary supplements, contains 4'-methylated

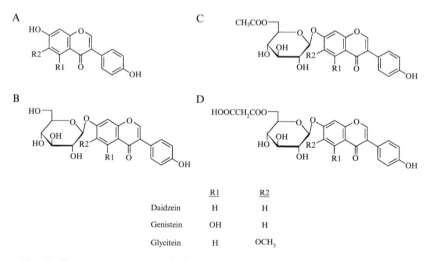

	R1	R2
Daidzein	H	H
Genistein	OH	H
Glycitein	H	OCH₃

Fig. 1. Chemical structures of isoflavones in soy foods and soy food products. (A) Aglycones, (B) β-glycoside conjugates, (C) acetyl-β-glycoside conjugates, and (D) malonyl-β-glycoside conjugates.

forms of daidzein (formononetin, Fig. 2A) and genistein (biochanin A, Fig. 2B) (He *et al.*, 1996). In dietary supplements made from kudzu root, C-glycosides of daidzein and genistein are the predominant conjugate forms (Fig. 3) (Prasain *et al.*, 2003b). Another source of isoflavone conjugates exists in soy sauce as ethers of 2,3-dihydroxysuccinic acid with genistein and daidzein (Kinoshita *et al.*, 1997).

Several other phytoestrogens are found in foods, including coumestrol (Fig. 2C), a coumestane in alfalfa and many legumes, zearalenone (Fig. 2D), a fungal estrogen that is a contaminant of stored feed, resveratrol (Fig. 2E), a stilbene of red grapes and red wines, and lignans. The latter are mainly found in the outer layers of cereals and grains, with flaxseed and rye among the most important. The predominant lignans in these foods are matairesinol (Fig. 2F) and secoisolariciresinol (Fig. 2G) with lesser amounts of pinoresinol, lariciresinol, and syringaresinol. Like the isoflavones, lignans exist as glycosides in foods and are converted to enterodiol and enterolactone by gut micoflora.

The isoflavone content in soybeans varies according to genetics, crop years, and growth location (Wang and Murphy, 1994b). As a consequence, the isoflavone content of soy foods derived from soybeans varies from year

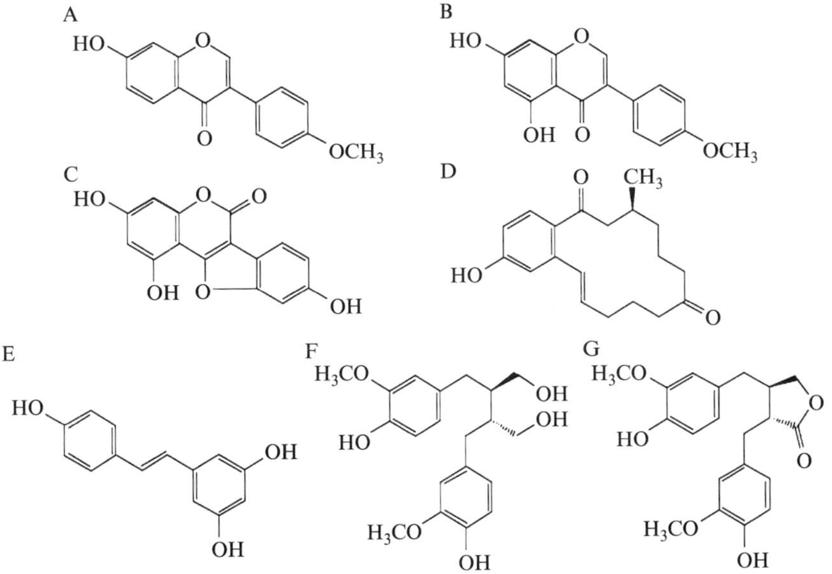

FIG. 2. Chemical structures of other phytoestrogens. (A) Formononetin, (B) biochanin A, (C) coumestrol, (D) zearalenone, (E) resveratrol, (F) matairesinol, and (G) secoisolariciresinol.

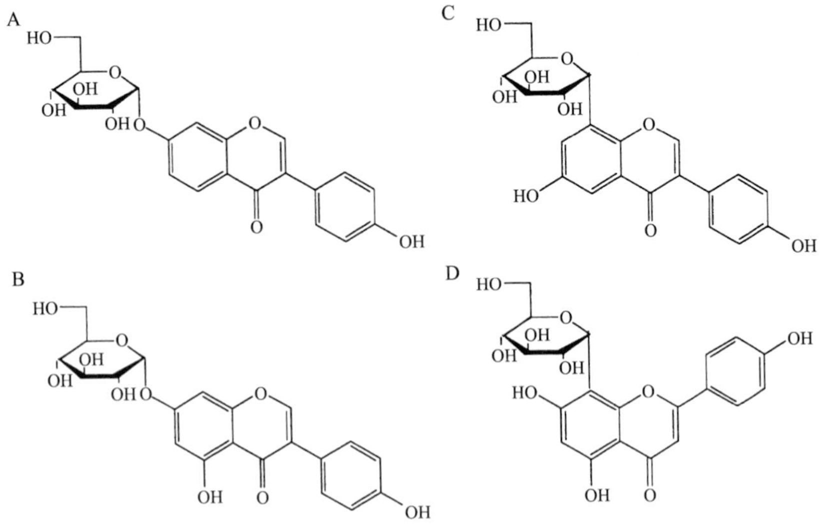

FIG. 3. Different chemistry of O- and C-glycosides of isoflavones. The kudzu root contains daidzin (A), genistin (B), and puerarin (daidzein-8-C-glycoside) (C). Millet contains vitexin (apigenin-8-C-glycoside) (D).

to year. This may also be a factor for other phytoestrogen-containing foods. Those foods that are used in preclinical experiments and clinical trials should therefore be examined carefully for their qualitative and quantitative content of phytoestrogens.

Phytoestrogen Standards

Many unlabeled phytoestrogens (Figs. 1–3) and their metabolites (Fig. 4) are available from Aldrich-Sigma Chemical Co. (St. Louis, MO), Indofine Chemical Co. (Hillsborough, NJ), LC Laboratories (Woburn, MA), and Plantech (Reading, UK).The β-glycosides of genistein and daidzein are available from Aldrich-Sigma Chemical Co. and LC Laboratories. It should be noted that the 6''-O-malonyl-β-glycosides are unstable as soon as they are dissolved in aqueous methanol. Radiolabeled phytoestrogens are not available commercially and may require a custom synthesis (Peterson *et al.*, 1996). However, the British Food Standards Agency funded a project for the preparation of multiply ^{13}C-labeled phytoestrogens. Investigators should contact Dr. Nigel Botting, Department of Chemistry, St. Andrew's University, Fife, Scotland, regarding the availability of these materials. U[14]C-labeled genistein is available from the National

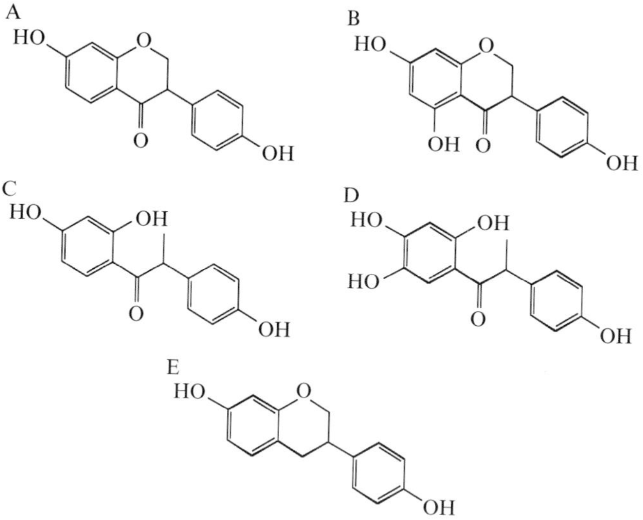

FIG. 4. Chemical structures of bacterial metabolites of isoflavones. (A) dihydrodaidzein, (B) dihydrogenistein, (C) O-desmethylangolensin, (D) 6-hydroxy-O-desmethylangolensin, and (E) equol.

Cancer Institute Chemical Carcinogen Reference Standard Repository. Another source of radiolabeled phytoestrogens is Dr. Mary Ann Lila at the University of Illinois at Urbana-Champaign, who utilizes plant cell cultures for the biosynthesis of isotopically labeled bioflavonoids (Lila *et al.*, 2005).

All standards should be verified by mass spectrometry (MS) and, when possible, by nuclear magnetic resonance spectrometry (NMR). The expected [M-H]⁻ molecular ions of many phytoestrogens are given in Tables I and II. A limited number of phytoestrogens have been studied by ^{1}H-NMR (Table III).

Extraction of Phytoestrogens from Foods

Several methods for the extraction of phytoestrogens, especially isoflavones, have been proposed. Some methods are used to determine the various conjugated forms of phytoestrogens, whereas others are used to determine total phytoestrogens within foods. To determine the individual phytoestrogen composition in foods, extractions are done without a hydrolysis step. To determine total phytoestrogen content, enzymatic hydrolytic with *Aspergillus niger* cellulase is carried out.

TABLE I

m/z Values of Monoisotopic [M-H]- Molecular Ions of Isoflavones and
Their Metabolites

Isoflavone	m/z
Daidzein	253.06
[$^{13}C_3$]Daidzein	256.06
Daidzin	415.11
6″-O-acetyldaidzin	457.12
6″-O-malonyldaidzin	501.11
Daidzein 4′-sulfate	333.01
Daidzein 7-sulfate	333.01
Daidzein 4′,7-disulfate	412.97
Daidzein 4′-glucuronide	429.09
Daidzein 7-glucuronide	429.09
Daidzein 4′,7-glucuronide	605.12
Daidzein 4′-sulfo-7-glucuronide	509.04
Daidzein 7-sulfo-4′-glucuronide	509.04
3′-Chlorodaidzein	287.02, 289.02
6-Chlorodaidzein	287.02, 289.02
8-Chlorodaidzein	287.02, 289.02
3′,6-Dichlorodaidzein	321.98, 323.98, 325.97
3′,8-Dichlorodaidzein	321.98, 323.98, 325.97
3′-Nitrodaidzein	298.04
Genistein	269.05
Genistin	431.11
6″-O-Acetylgenistin	473.12
6″-O-Malonylgenistin	517.11
Genistein 4′-sulfate	349.01
Genistein 7-sulfate	349.01
Genistein 4′,7-disulfate	429.97
Genistein 4′-glucuronide	445.09
Genistein 5-glucuronide	445.09
Genistein 7-glucuronide	445.09
Genistein 4′,7-diglucuronide	621.12
Genistein 4′-sulfo-7-glucuronide	525.04
Genistein 7-sulfo-4′-glucuronide	525.04
[$^{13}C_3$]Genistein	272.05
3′-Chlorogenistein	303.01, 305.01
6-Chlorogenistein	303.01, 305.01
8-Chlorogenistein	303.01, 305.01
3′,6-Dichlorogenistein	336.97, 338.97, 340.96
3′,8-Dichlorogenistein	336.97, 338.97, 340.96
3′-Nitrogenistein	314.04
Mono-chloro-nitrogenistein	348.00, 350.00
Dichloro-nitrogenistein	381.96, 383.96, 385.97
Glycitein	283.07
Glycitin	445.12
Glycitein 4′-glucuronide	459.10

(continued)

TABLE I *(continued)*

Isoflavone	m/z
Glycitein 7-glucuronide	459.10
Dihydrodaidzein	255.07
Dihydrodaidzein 7-glucuronide	431.11
Equol	241.09
Equol glucuronide	417.13
O-Desmethylangolensin	257.09
O-Desmethylangolensin glucuronide	433.12

TABLE II

m/z VALUES OF MONOISOTOPIC [M-H]⁻ MOLECULAR IONS
OF PHYTOESTROGENS

Phytoestrogen	m/z
Coumestrol	267.04
Coumestrol glucuronide	443.07
Resveratrol	227.08
Resveratrol glucuronide	403.11
Resveratrol sulfate	307.04
Zearalenone	317.15
Matairesinol	357.14
Matairesinol glycoside	519.19
Matairesinol glucuronide	533.17
Matairesinol diglucuronide	709.21
Secoisolariciresinol	361.17
Secoisolariciresinol glycoside	523.23
Secoisolariciresinol glucuronide	537.21
Secoisolariciresinol diglucuronide	713.24
Enterodiol	301.15
Enterodiol glucuronide	477.18
Enterolactone	297.12
Enterolactone glucuronide	473.15

Acidified Solvent Extraction of Phytoestrogens

A general procedure that is suitable for the extraction of phytoestrogens and their direct analysis is based on the method described by Murphy (1981) for the extraction of isoflavones from soybeans (Wang and Murphy, 1994b) and commercial soy foods (Wang and Murphy, 1994a). Raw soybean seeds (2 g) with their seed coats are ground, mixed with 2 ml 0.1 N HCl and 10 ml of acetonitrile, stirred for 2 h at room temperature, and

TABLE III
PROTON CHEMICAL SHIFTS OF ISOFLAVONES DETERMINED BY NMR

Compound	2-H	5-H	6-H	8-H	3',5'-H	2',6'-H	C_4OH	C_5OH	C_7OH
Daidzein[a]	8.30	7.98	6.95	6.84	6.84	7.40	n.m.	–	n.m.
Genistein[a]	8.40	N/A	6.41	6.25	6.81	7.40	n.m.	n.m.	n.m.
Formononetin[a]	8.40	8.00	6.90	6.90	7.00	7.50	–	–	n.m.
Daidzein-4'-sulfate[a]	8.35	7.98	6.95	6.90	7.21	7.46	–	–	n.m.
Daidzein-7-sulfate[a]	8.40	8.03	7.25	7.43	6.80	7.40	n.m.	–	-
Daidzein-4',7-disulfate[a]	8.30	8.02	7.23	7.40	7.16	7.45	–	–	-
Formononetin-7-sulfate[a]	8.40	8.02	7.20	7.39	7.00	7.50	–	–	-
Formononetin-7-glucuronide[a]	8.33	8.04	7.12	7.17	7.06	7.59	–	–	-
Genistein[b,c]	8.30	N/A	6.226	6.38	6.82	7.37	9.57	12.93	10.86
Monochlorogenistein[b,d]	8.416	N/A	6.220	6.380	6.815	7.380	9.648	12.916	n.m.
Dichlorogenistein[b,d]	8.360	N/A	N/A	N/A	6.816	7.379	9.600	13.720	n.m.

[a] Performed at 200 MHz (Clarke et al., 2002).
[b] Performed at 300 MHz (Barnes et al., 1994; Boersma et al., 1999).
[c] Relative to tetramethylsilane (0.00 ppm).
[d] Relative to the DMSO proton resonance at 2.49 ppm.

filtered through Whatman No. 42 filter paper. The filtrate is taken to dryness under vacuum at a temperature below 30°. The dried material is redissolved in 10 ml of 80% aqueous methanol and then filtered through a 0.45-μm filter unit. An aliquot of the filtrate (20 μl) is analyzed by reversed-phase HPLC. Although this extraction method preserves the isoflavone conjugates, degradation of the 6″-O-malonylglycosides in aqueous methanol occurs even at room temperature. Coward et al. (1998) have shown that extraction of isoflavones at 4°, where degradation is minimized, is possible if an internal standard such as fluorescein is used. In this latter method, the extract is not concentrated prior to HLPC analysis, but instead analyzed directly. Flavone and p-hydroxybenzoic acid have also been used as internal standards (Bednarek et al., 2001; Franke and Custer, 1994).

Solvent Extraction of Phytoestrogens

Isoflavones in solid foods are extracted into 80% aqueous methanol (10 ml/g) by stirring for 1 h at 60°. Other soy products (miso, soy milk, soy paste, and tofu) are freeze-dried and extracted whole. The mixture is centrifuged at 2500g for 10 min and the supernatant is transferred to a round-bottom flask. The pellet is extracted twice more in 5 ml each and centrifuged. Supernatants are combined in the round-bottom flask and evaporated to dryness in a rotary evaporator. The dried extracts are resolubilized in 5 ml of 50% aqueous methanol and then defatted by partitioning the neutral lipids into hexane (4 × 20 ml). The aqueous methanol is taken to dryness in a rotary evaporator, and the resulting dried residue is resolubilized in 10 ml 80% aqueous methanol. An aliquot is centrifuged at 14,000g for 2 min and analyzed by HPLC as described by Coward et al. (1993).

While Coward et al. (1993) used aqueous methanol, Griffith and Collison (2001) used acetonitrile for isoflavone extraction: soy foods (1 g) are dispersed into a 10-ml volume of acetonitrile followed by the addition of 6 ml double deionized water and 0.5 ml apigenin [2000 μg/ml in dimethyl sulfoxide (DMSO)]. Samples are shaken to mix and extracted on a rotary mixer for 2 h, after which the sample is recovered and the acetonitrile concentration is adjusted to 50% (v/v) with the addition of deionized water. Samples are centrifuged at 2000g for 10 min to pellet insoluble matter and eliminate foam. An aliquot is filtered through a 0.45-μm PVDF filter and analyzed by HPLC as described (Griffith and Collison, 2001).

Extraction Followed by Enzymatic Hydrolysis (Liggins et al., 1998)

Phytoestrogen-containing materials (2.5 g) are mixed with a minimum of 5 ml of 80% aqueous methanol. If the material under study is particularly enriched in phytoestrogens, then the amount taken for analysis should be reduced, as phytoestrogens have a limited solubility in aqueous methanol, which would thereby lead to an erroneously low result. Sonication for 10 min solubilizes the phytoestrogen glycosides in the methanol by breaking up cellular material. It is followed by a further hour of soaking in the solvent. Insoluble material is removed by filtration through a double layer of filter paper (Whatman No. 4 and then No. 1), and any adsorbed phytoestrogens are washed through with fresh 80% aqueous methanol (>5 ml). The methanol in the filtrate is evaporated, and 100 Fishman units of cellulase from *A. niger* are added to the sample in 5 ml of 0.1 *M* sodium acetate buffer, pH 5. Samples are sonicated and subsequently incubated overnight in a shaking water bath at 37°. The hydrolyzed phytoestrogen aglycones usually precipitate because of their reduced aqueous solubility. They are extracted from the aqueous hydrolyzate by the addition of 100% ethyl acetate. Three 2-ml extracts with ethyl acetate are combined; a 1-ml aliquot of the combined extract is pipetted into a separate vial and taken to dryness at 60° under nitrogen prior to chromatographic analysis of the phytoestrogen aglycones.

Extraction of Lignans

Milder *et al.* (2004) developed an assay to extract lignans from a variety of foods, including flaxseed, broccoli, whole wheat bread, and tea. Previous extraction methods allowed for the quantitation of the major lignans secoisolariciresinol and matairesinol but were lacking for pinoresinol and lariciresinol. To allow for quantification, the lignans need to be extracted from their food matrices. This can be difficult due to their ability to oligomerize with 3-hydroxy-3-methylglutaric acid through ester bonding. Therefore, lignan extractions require the use of alkaline hydrolysis to release the lignan glycosides from the 3-hydroxy-3-methylglutaric acid. There are many unknown lignan glycosides; therefore, lignan extraction methods typically have a hydrolysis step to allow for the release of the glycosides. This results in an estimate of the total lignan content in foods. The following describes the current extraction and analysis methods for lignans from foods.

Alkaline extraction of 1.0 g of dry food is performed with 24.0 ml of methanol/water (70/30, v/v) containing 0.3 *M* sodium hydroxide in a shaking water bath for 1 h at 150 rpm and 60°. After extraction, the pH is adjusted to 5–6 with 750 μl of 100% glacial acetic acid and the extract is centrifuged at 4,500*g* for 10 min at 10°. An aliquot (1 ml) is transferred to a preweighed test tube. Methanol is evaporated from this aliquot at 60°, under nitrogen, until the residual weight is ≤0.30 g. The volume is adjusted to approximately 1.2 ml with 0.05 *M* sodium acetate buffer, pH 5, and the extract is weighed again to calculate the dilution compared to the original aliquot of 1 ml.

A 1-ml aliquot of the weighed extract or 1 ml of beverage is hydrolyzed by the addition of *Helix pomatia* β-glucuronidase/sulfatase (0.83 mg in 1 ml 0.05 M sodium acetate buffer, pH 5). The samples are incubated overnight at 37°. Samples are extracted twice with 2 ml diethyl ether, and the two organic phases are combined. The diethyl ether is evaporated, and the dried samples are redissolved in 0.3 ml of methanol, mixed, and 0.7 ml water is added. A 240-μl aliquot of sample is added to 10 μl of internal standard solution containing 50 ng of secoisolariciresinol-d_8 and 50 ng of matairesinol-d_6 in 30% aqueous methanol. Samples are mixed and transferred to HPLC vials for LC-MRM-MS analysis.

Reversed-Phase HPLC Analysis of Phytoestrogens in Foods

Phytoestrogens are readily separated by reversed-phase HPLC on C_8 or C_{18} columns. The most commonly used methods involve a gradient of acetonitrile in a background of either acetic acid, formic acid, or trifluoroacetic acid (0.1–1.0%) (Fig. 5B). However, because of the complexity of phytoestrogen forms in the plant material (aglycones, β–glycosides, 6″-O-acetylglycosides, and 6″-O-malonylglycosides), there is considerable advantage in using a neutral buffer. This brings out differences in hydrophobicity based on changes in the charge state. For example, the malonate group at pH 7 is negatively charged, whereas it is unprotonated at an acidic pH. The 6″-O-malonylglycosides elute earlier than the β–glycosides (Fig. 5A). As a result, the major classes are better separated (Wang *et al.*, 2002).

By virtue of their phenolic groups and some heterocyclic ring systems, intact phytoestrogens have a strong molar absorbance. However, reduction and oxidation of the heterocyclic ring that typically occur during metabolism substantially reduce molar absorbance. Therefore, while phytoestrogens can be detected readily by their UV absorbance in foods containing them in moderate concentrations, the HPLC-UV analysis approach is inadequate for most studies of physiological fluids (serum, plasma, and urine). In the latter case, LC-MS provides both the sensitivity and the specificity for measurement of both phytoestrogens and their metabolites.

For example, lignan samples (50 μl) are separated on a 150 × 3.0-mm i.d. C_{18} column at a flow rate of 0.4 ml/min at 40°. The mobile phases consist of water (A) and methanol (B). The gradient is 0–0.5 min, 30% B; 0.5–12 min, linear gradient from 30 to 50% B; 12–15 min, isocratic at 50% B; 15–15.2 min, linear return to 30% B; and 15.2–19 min, isocratic at 30% B to equilibrate. The divert valve allows flow into the mass spectrometer from 8 to 19 min. A Micromass Quatro mass spectrometer equipped with an atmospheric pressure chemical ionization (APCI) source is operated in

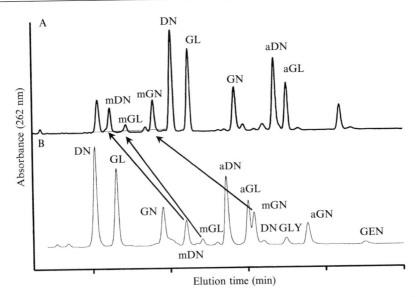

Fig. 5. Reversed-phase, gradient elution LC analysis of isoflavones and their conjugates in soy foods. Isoflavones were eluted with a 0–50% linear gradient of acetonitrile. In A, the background solvent is 10 m*M* ammonium acetate, whereas in B, it is 0.1% trifluoroacetic acid. The peaks are DN, daidzin; GL, glycitin, GN, genistin; mDN, malonyldaidzin; mGL, malonylglycitin; mGEN, malonylgenistin; and, aDN, acetyldaidzin; aGL, acetylglycitin; mGN, malonylgenistin; GEN, genistein. The mobility of the malonyl glycosides is much higher in A than in B.

the negative mode. For quantification, parent ion/product ion pairs are monitored using LC-MRM-MS analysis.

Gastrointestinal Tract Hydrolysis

Deconjugation by Intestinal Lactase Phlorizin Hydrolase (LPH)

Lactase phlorizin hydrolase is a membrane-spanning enzyme on the luminal side of the brush border of the small intestine. LPH is responsible for the hydrolysis of lactose, the main carbohydrate in mammalian milk. Structure studies have shown a second active site that is capable of hydrolyzing β-glycosylceramide (another component of milk) and phlorizin, a dihydrochalcone glycoside (Day *et al.*, 2000). Structural similarities among dihydrochalcones, flavonoids, and isoflavonoids led to the discovery that LPH has a role in the hydrolysis of the β-glycosides of these compounds (Day *et al.*, 2000; Wilkinson *et al.*, 2003).

LPH Enzyme Assay. Phytoestrogen glycosides (100 μM) are incubated in a volume of 0.1 ml with LPH (final concentration 1 $\mu g/ml$) in phosphate buffer, pH 6, for 20 min at 37°. The reaction is terminated by the addition of 0.15 ml acetonitrile–water (30/70, v/v) followed by centrifugation at 13,600g for 10 min at 4°. The supernatant is filtered through 0.22-μm PTFE filter units and analyzed by reversed-phase HPLC. For kinetic studies, various concentrations of phytoestrogen are added to LPH (final concentration 1 $\mu g/ml$) and phosphate buffer in a volume of 0.2 ml. Samples are incubated at 37° for 5–30 min during the linear phase of the reaction (depending on the substrate and its concentration). Reactions are terminated by the addition of 0.3 ml acetonitrile–water and analyzed by reversed-phase HPLC (Day *et al.*, 2000). The amount of substrate converted is calculated from standard curves of peak areas produced by injecting known amounts of various phytoestrogens directly onto the column.

HPLC analysis is carried out using a 250 × 4.6-mm i.d. column packed with Prodigy 5 μm ODS-3 reversed-phase silica (Phenomenex, Torrance, CA). A gradient sequence using solvents A (water–tetrahydrofuran–trifluoroacetic acid, 98:2:0.1, v/v) and B (acetonitrile) is run at a flow rate of 1 ml/min with 17% B for 2 min, increasing to 25% over 5 min, 35% B for 8 min, 50% B for 5 min, and then to 100% B over 5 min. A column clean-up step maintains B at 100% for 5 min followed by a reequilibration at 17% B (15 min). A diode array method is used to monitor the eluant (at 270 and 370 nm). The appropriate phytoestrogen glycoside for each enzyme kinetic analysis is used as an external standard at concentrations ranging from 0 to 250 μM (400 μM for daidzin). The product in each of the reactions is confirmed by coelution of peak with standard compounds and by matching UV spectra.

Metabolism in the Cannulated Everted Sac Model (Wilkinson et al., *2003).* Cannulated everted sacs of rat proximal jejunum are prepared from segments (5 cm) of male Wistar rat small intestine. The everted jejunal sacs, ligated at one end, are tied onto a 1-ml tapered disposable syringe precharged with 0.5 ml Krebs bicarbonate buffer (pH 7.2–7.4). The sacs are filled with Krebs buffer by depressing the plunger and can be emptied with minimal contamination of the serosal solutions by the reverse operation at the end of the experiment (Gee *et al.*, 1998). Daidzein and daidzin are dissolved in ethanol containing 2–3% (v/v) DMSO and then diluted to 1, 10, and 100 μM final concentrations using Krebs bicarbonate buffer (pH 7.3). Sacs are suspended in organ baths containing 8 ml Krebs buffer with or without isoflavone and incubated for 15 min at 37° with continuous aeration using oxygen:carbon dioxide (95:5). After incubation, the sacs are rinsed thoroughly using physiological saline (0.9% NaCl). The serosal solutions are then withdrawn from the sacs, transferred to Eppendorf tubes, and immediately stored at $-20°$. The mucosal solutions are also collected at the end of the incubation period and stored as described earlier. Samples are analyzed for aglycones by ELISA and HPLC.

To assess the role of the glucose transporters, incubations are carried out with daidzin or daidzein (100 μM) in the absence or presence of 250 μM N-(n-butyl)-deoxygalactonojirimycin (NB-DGJ), an inhibitor of LPH. A control incubation is carried out in Krebs buffer with inhibitor present but no isoflavonoid. Mucosal and serosal fluids are collected, frozen, and analyzed by ELISA and HPLC. Analysis of samples by ELISA is performed according to Creeke et al. (1998) using a fresh polyclonal antidaidzein serum.

Mucosal samples from the control (no isoflavones) and test daidzein and daidzin (10 and 100 μM) incubations are prepared for HPLC analysis by purification using a polyamide column. Briefly, a 0.3-ml sample is added to a preconditioned column and the column is eluted with 100% methanol followed by methanol/ammonium hydroxide (99.5:0.5%, v/v) to isolate neutral and acidic fractions, respectively. Both fractions are evaporated to dryness and reconstituted in methanol (1.0 and 0.2 ml).

In addition, serosal solutions from incubations in the absence and presence of the isoflavonoids (100 μM) are prepared with or without β-glucuronidase (25 U) treatment as follows. Duplicate aliquots of serosal fluid (50 or 100 μl) are spiked with genistein (250 ng, 5 μl) as an internal standard. An equivalent volume of Krebs buffer is added to one subsample followed by an equal volume of 0.8 mM ascorbic acid in methanol. β-Glucuronidase (25 U/25 μl) is added to the second aliquot, and the sample is incubated at $37°$ for 2 h. An equal volume of the methanol/ascorbic acid solution is again added. Both subsamples are evaporated and reconstituted in methanol (0.2 ml) before HPLC analysis.

Mucosal and serosal samples are analyzed by HPLC as described (Day et al., 2000). Peak identification is confirmed by coelution of peaks with authentic standard compounds by matching UV spectra and by positive ion electrospray LC/MS, such as on a Micromass Quattro II mass spectrometer equipped with a Z-spray ion source. Samples are introduced using a Hewlett Packard 1050 HPLC equipped with a diode array detector. Eluent flow (1 ml/min) is split between the diode array detector and the mass spectrometer ion source. Selected ion-monitoring measurements are performed for m/z 255.07 (daidzein $[M+H]^+$), 271.06 (genistein $[M+H]^+$), 417.12 and 439.10 (daidzin $[M+H]^+$, $[M+Na]^+$), 431.10 and 453.08 (daidzein-7-O-β-glucuronide $[M+H]^+$, $[M+Na]^+$). Diode array spectra are scanned from 190 and 450 nm, with an interval of 2 nm. Instrument control, data acquisition, and processing are performed using a Microsmass MassLynx version 3.4 system.

Metabolism of Phytoestrogens in the Liver

Microsomal Hydroxylation of Isoflavones (Fig. 6)

Rat liver microsomes (2 mg protein) are incubated with 50 nmol of isoflavone dissolved in 40 μl DMSO and a NADPH-generating system (3 mM MgCl$_2$, 1 mM NADP+, 8 mM D,L-isocitrate, and 0.5 U of isocitrate dehydrogenase) in 2 ml of 0.05 M potassium phosphate buffer, pH 7.4 (Kulling *et al.*, 2000, 2001). Microsomal reactions are preincubated for 2 min at 37° in a shaking water bath. Reactions are initiated with addition of the NADPH-generating system at 37° and terminated after 60 min by extracting with 4 × 2 ml of ice-cold ethyl acetate. The organic solvent layers are pooled and dried under reduced pressure at room temperature. Dried residues are resolubilized in 0.2 ml of methanol and analyzed by HPLC. Controls consist of either heat-inactivated microsomes or reaction mixtures without the NADPH-generating system.

Hu *et al.* (2003) used a similar protocol with the addition of monoclonal antibodies to different cytochrome P450s (CYP450s) to elucidate the CYP450 responsible for the hydroxylation and demethylation of isoflavones.

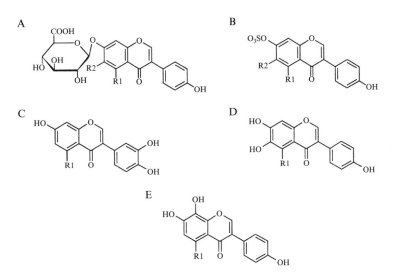

FIG. 6. Chemical structures of isoflavone metabolites in physiological fluids. (A) β-Glucuronides, (B) sulfate esters, and (C–E) hydroxylated metabolites. Individual isoflavones are daidzein (R1, R2 = H), glycitein (R1 = H, R2 = OCH3), and genistein (R1 = OH, R2 = H).

Glucuronidation of Phytoestrogens (Fig. 6)

In Vitro *Glucuronidation of Phytoestrogens.* The *in vitro* method de-scribed by Doerge *et al.* (2000) is well suited for phytoestrogens. Enzyme studies using purified UDP-glucuronosyltransferase (UGT) consist of reaction mixtures containing the following: 0.08 to 0.32 mg/ml UGT, 0 to 400 μM (final concentration) daidzein or genistein, and 5 to 10 mM MgCl$_2$. An inhibitor of β-glucuronidase, γ-saccharolactone (10 mM), is added to the incubation to prevent hydrolysis of β-glucuronides. The reactions are initiated by the addition of 0.08 to 3 mM uridine-5'-diphospho-β-D-glucuro-nic acid ester (UDPGA) in 0.05 M Tris-HCl, pH 7.4–7.5 (for UGT iso-zymes), or 0.1 M phosphate buffer, pH 8.0 [for bovine hepatic UGT or microsomes prepared from liver, kidney, and colon samples (0.1–0.25 mg/ml)], in a final volume of 125 μl for 2 h at 37°. The reactions are terminated with an equal volume of methanol followed by brief vortexing and centri-fugation at 10,000g for 10 min. Products are analyzed by reversed-phase LC with UV detection (260 nm for genistein) on a 5-μm Prodigy ODS-3 4.6 × 230-mm column (Phenomenex Co., Torrance, CA). The solvent system consists of solvents A (0.1% formic acid in water) and B (0.1% formic acid in acetonitrile). The gradient starts at 95% A and 5% B for 3 min followed by a linear gradient to 50% A and 50% B over 15 min and isocratic elution at 50% A and 50% B for 5 min. The flow rate is 1.0 ml/min.

In Vivo *Glucuronide Conjugation.* Conjugation *in vivo* can be carried out using the everted intestinal sac preparation (Sfakianos *et al.*, 1997; Wilkinson *et al.*, 2003) or the bile duct cannulation model (Sfakianos *et al.*, 1997). In the latter, the common bile duct of anesthetized rats is exposed and occluded distally with surgical silk. A polyethylene (PE-10) cannula is introduced into the common bile duct through a small opening made with eye scissors and tied in place with surgical silk. Rats are partic-ularly useful because they do not have a gallbladder and therefore bile can be collected quantitatively. The phytoestrogen aglycone is dissolved in a small volume of DMSO and diluted with physiological saline. It is intro-duced into either the femoral vein with a bevel-edged PE-10 cannula or the portal vein with a 27-gauge needle inserted into a PE-10 cannula. The veins are exposed, temporarily occluded with surgical silk, and the tubing in-serted into the vein. The tubing is held in place with surgical cement. The phytoestogen is slowly infused into the vein; bile is collected simultaneous-ly. During the experiment the body temperature of the rats is maintained at 38° with a heating pad.

Analysis of Biliary Phytoestrogen Glucuronides. Verification of the identity of phytoestrogen β-glucuronides in bile can be carried out by treating the bile sample with β-glucuronidase (in 100 mM ammonium acetate buffer, pH 5.0) and carrying out reversed-phase HPLC analysis of the products. The putative β-glucuronide peak will disappear and there will be a new peak eluting with the same retention time as the expected isoflavone aglycone.

An alternative method is to carry out LC-MS analysis. Bile extracts are separated by reversed-phase HPLC on a 15 × 0.21-cm i.d. Brownlee Aquapore C_8 column using a linear 0–50% gradient (5%/min) of acetonitrile in 10 mM ammonium acetate at a flow rate of 0.2 ml/min. The column eluate is split 1:1, and one stream is passed into the IonSpray interface of a PE-Sciex (Concord, ON, Canada) API III triple quadrupole mass spectrometer operating in the negative ion mode, with an orifice potential of −60 V. In the MS-MS mode, daughter ion spectra are obtained by selecting parent ions in the first quadrupole, which are then collided with argon/10% nitrogen gas in the second quadrupole and analyzed in the third quadrupole.

When the expected molecular [M-H]⁻ ion m/z 445 for genistein β-glucuronide is searched for, a complex chromatographic peak is observed (Fig. 7A) (Sfakianos $et\ al.$, 1997). However, by repeating the analysis and carrying out tandem mass spectrometry on the m/z 445 ion, the correct β-glucuronide peak can be discerned with the expected m/z 269 aglycone ion (Fig. 7B). The other m/z 445 peak produces m/z 80 and 97 daughter ions, suggesting that it is a phosphate or sulfate conjugate.

Isolation of Biliary Phytoestrogen β-*Glucuronides.* Purification of phytoestrogen conjugates from bile is carried out by chromatography using Sephadex LH-20 (Coward $et\ al.$, 1996; Sfakianos $et\ al.$, 1997). Pooled bile obtained after infusions with genistein is diluted with 50% aqueous ethanol and chromatographed on a 15 × 3-cm i.d. Sephadex LH-20 column pre-equilibrated with 50% aqueous methanol. Fractions containing β-glucuronides are passed over a 5 × 1.5-cm i.d. diethylaminoethyl-Sephadex (DEAE-Sephadex) column equilibrated with 70% aqueous methanol. Excess reagents are removed by washing with 70% aqueous methanol–0.2 M acetic acid. The β-glucuronide is eluted with 0.3 M LiCl in 70% aqueous methanol. The eluate is dried to remove the methanol; ammonium acetate (150 mM), pH 5, and triethylammonium sulfate (0.5 M) are added and the mixture is passed over several Sep-Pak C_{18} cartridges to adsorb the β-glucuronide. LiCl is removed by washing the cartridges with water. The β-glucuronide is eluted from the cartridges with several column volumes of methanol. Because the phytoestrogen glucuronides derived this way are heavily contaminated with the bile salt taurocholate (as determined by negative ion electrospray mass spectrometry), the partially purified material is treated with cholylglycine hydrolase to convert taurocholate to cholic acid and taurine. These can be removed by rechromatography of

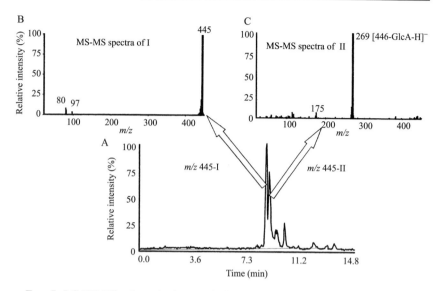

FIG. 7. LC-ESI-MS of genistein metabolites in rat bile. In A, the selected ion chromatogram for m/z 445 (the molecular [M-H]$^-$ ion for genistein 7-O-β-glucuronide) contains several peaks. Daughter ion spectra (B, C) of the two major peaks are quite distinct. The mass spectrum of the later eluting peak corresponded to genistein 7-O-β-glucuronide.

the partially purified material on the Sephadex LH-20 column. Fractions containing the phytoestrogen glucuronide are finally purified by preparative reversed-phase HPLC. Depending on the gender of the animal, sulfate esters of the phytoestrogen may also be present. Bile from female rats contains only β-glucuronides of the phytoestrogens (Coward et al., 1996).

Sulfation of Phytoestrogens (Fig. 6)

The method described by Doerge et al. (2000) is the preferred method. Individual sulfotransferases, SULT 1A1*2, 1A2*1, 1A3, and 1E, have been studied. The enzyme is diluted to 220 (1A1*2), 180 (1A2*1), 400 (1A3), or 100 (1E) ng/ml in a prechilled solution containing 5 mM phosphate buffer, pH 6.5, 1.5 mg/ml bovine serum albumin, and 10 mM dithiothreitol (DTT). To start the reaction, 100 μl of diluted SULT is mixed with 50 μl of 25 mM phosphate buffer, pH 6.5, containing 25 mM DTT, 1.28 μM adenosine-3'-phosphate 5'-phosphosulfate, and various concentrations of phytoestrogens (0–400 μM) in a final volume of 200 μl. The mixtures are incubated at 37° for 2 h and analyzed by LC-UV (as described for the UGT enzyme). Reactions containing SULT 2A1 and 1E also contain 0.25 mM MgCl$_2$.

In a similar manner for the phytoestrogen β-glucuronides, identification of phytoestrogen sulfate esters can be carried out by reversed-phase HPLC and/or LC-ESI-MS. The expected molecular ion for the sulfate ester is 80 Da above that of the aglycone. For genistein-7-sulfate it is m/z 349 in the negative ion mode. Electrospray ionization is preferred over heated nebulizer chemical ionization, which circumvents the effects of heating in this ionization process.

Metabolism of Isoflavones in Mammary Cell Lines

While Akiyama *et al.* (1987) demonstrated that biochanin A inhibits PTK activity 30-fold less effectively than genistein in cell-free systems, Peterson and Barnes (1991) reported that biochanin A and genistein inhibited breast cancer cell growth approximately equally in cell culture. Peterson *et al.* (1996) later showed that the metabolism of these compounds by human breast cancer MCF-7 cells was the reason for the discrepancy between cell-free and cell culture experiments. MCF-7 cells are able to rapidly demethylate biochanin A to genistein and to slowly sulfate genistein to genistein-7-sulfate (Peterson *et al.*, 1996, 1998). The following studies detail the methodologies employed to evaluate this type of metabolism.

Metabolism Studies of Isoflavones in Cells from the Mammary

Mammary cells (transformed human breast cancer MCF-7 cell line or cultured normal human mammary epithelial cells) are plated in 6-well plates and grown to 70% confluence. [4-^{14}C]Biochanin A or [4-^{14}C]genistein is added to the cells at a final concentration of 1 μg/ml in 100% DMSO [0.5% (v/v) final (DMSO) concentration] (0.083 and 0.087 μCi/ml for biochanin A and genistein, respectively). Blank wells containing media and DMSO but no cells are used as controls. After incubation for the indicated times, media are collected and the cells are scraped and collected in 2 ml modified proteinase K buffer (25 mM EDTA, 137 mM NaCl, and 10 mM Tris-HCl, pH 7.6). Cell lysates are sonicated and incubated with proteinase K (0.1 mg/ml) for 4 h at 37°. Aliquots (100 μl) of media and lysates are sampled for radioactivity analysis by scintillation counting. The remainder of the samples are passed through Sep-Pak C$_{18}$ cartridges to collect the [4-^{14}C]isoflavones and their metabolites. The cartridges are washed with 15 ml of distilled water to remove hydrophilic materials. Hydrophobic compounds are eluted with 10 ml of 80% aqueous methanol and dried at room temperature with air. Samples are resuspended in 100 μl

80% aqueous methanol and analyzed by reversed-phase HPLC. Samples are injected onto a 30 cm × 4.6 mm i.d. Aquapore C_8 column, which is eluted at a flow rate of 1 ml/min with a mobile phase consisting of a linear gradient of 0–45% acetonitrile (at 4.5%/min) in 0.1% (v/v) aqueous trifluoroacetic acid. Eluted substances are detected by their absorbance at 262 nm. The column eluate is collected in 30-s fractions and counted to determine the location of radioactive peaks using liquid scintillation counting.

Media extracts or individual radioactive metabolite peaks are analyzed by reversed-phase HPLC-MS as described earlier for the analysis of biliary isoflavone metabolites. In the case of mammary tumor cells, the principal metabolites are sulfate esters of phytoestrogens. Identification of the site of sulfation is possible by ^{1}H-NMR. The sulfate esters are isolated by preparative HPLC and dried thoroughly. When they are dissolved in d_6-DMSO, protons on the ring hydroxyl groups remain. These appear as singlets with marked downfield chemical shifts in the ^{1}H-NMR spectrum. The one that disappears from the spectrum readily allows identification of the site of sulfation. The amount of sulfotransferase activity varies considerably among the tumor cells, being the highest in ZR-75-1 cells (Peterson et al., 1998). The cells also exhibit marked demethylation of biochanin A.

Metabolism of Phytoestrogens by Products of Inflammatory Cells

Phytoestrogens described earlier all contain phenyl ring systems. This makes them capable of undergoing chemical reaction with highly reactive products of activated inflammatory cells. The latter generate hypohalogenic acids (HOCl, HOBr, and HOI) and peroxynitrite ($OONO^-$) once stimulated. These proinflammatory oxidants react with phenyl groups containing hydroxyl substituents. Proteins from sites of inflammation contain 3-chlorotyrosine and 3-nitrotyrosine in patients with atherosclerotic lesions. Genistein and daidzein both undergo chlorination and nitration when added to the medium of stimulated inflammatory cells (Boersma et al., 1999, 2003; D'Alessandro et al., 2003). Several isoflavones are converted to similar iodo derivatives when incubated with thyroid peroxidase, H_2O_2, and sodium iodide (Divi et al., 1997) as are many other polyphenols (Divi and Doerge, 1996) (Fig. 8).

Synthesis of Chlorinated and Nitrated Phytoestrogens

Solutions of phosphate-buffered saline with 1 mM diethylenetriamine pentaacetic acid (DTPA) with 50 μM of each isoflavone are reacted with HOCl (0 to 500 μM). Aliquots of a stock solution of 10 mM HOCl are

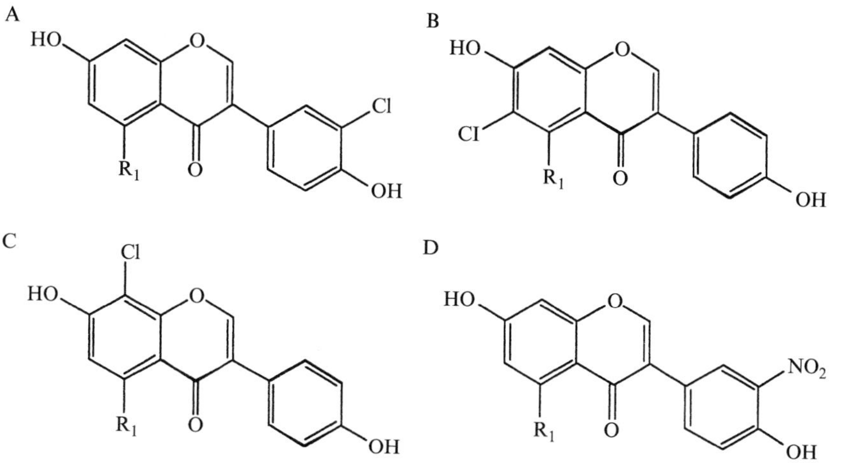

Fig. 8. Sites of chlorination and nitration in isoflavones. For daidzein $R_1 = H$, whereas for genistein $R_1 = OH$. In A, chlorination is in the 3' position; in B, it is in the 6 position, and in C it is in the 8 position. Nitration occurs in the 3' position (D).

added to the reaction mixture with continuous mixing to achieve the nominal concentration. Decomposed HOCl is made by reacting the acid with an equimolar solution of glutathione for 20 min (Winterbourn and Brennan, 1997). This mixture is added, while vortexing, to each isoflavone reaction mixture. Samples are extracted by adding 200 μl of the reaction mixture to 800 μl of water followed by 2 ml of diethyl ether. The samples are vortexed and then centrifuged at 3000g, whereupon the ethereal, upper layer is removed and the ether is evaporated under N_2. Prior to injection, 150 μl 80% methanol is added to redissolve the dried residues. HPLC analysis is carried out on a 4.6 mm $\times$ 25 cm C_8 Aquapore reversed-phase column with a linear 0–50% gradient of acetonitrile in 0.1% trifluoro-acetic acid at a flow rate of 1.0 ml/min. Isoflavones and their chlorinated derivatives are detected by their absorbance at 262 nm.

Analysis of chlorinated isoflavones by LC-MS reveals that addition of one chlorine atom increases the observed m/z by 34 or 36. Indeed, the relative amount of the chlorine isotopomers is a recognizable characteristic of a chlorinated compound. Also, several chlorinated isomers of each isoflavone are formed (Prasain et al., 2003a). The MS-MS spectrum of 3'-chlorogenistein is quite distinct from that of 6- or 8-chlorodaidzein. Dichloro derivatives of isoflavones (Boersma et al., 1999) and quercetin (Binsack et al., 2001) have been reported; these have an increase in m/z of

68, 70, and 72. Specific isomers of chlorodaidzein (3′-, 6-, and 8-monochloro-daidzein, and 3′,6- and 3′,8-dichlorodaidzein) have been synthesized by Dr. Nigel Botting, University of St. Andrew's, Fife, Scotland.

To prepare nitrated isoflavones, buffered solutions of 200 mM sodium phosphate (pH 7.4) with 1 mM DTPA and 50 μM of each isoflavone are reacted with OONO⁻ (0 to 500 μM). Aliquots of a 10 mM OONO⁻ stock solution are added to the reaction mixture with continuous mixing to obtain the final concentration. Decomposed OONO⁻ is prepared by adding the OONO⁻ to the buffer solution (pH 7.4) and allowing this to react for 10 min. The isoflavones are then added to the decomposed reaction mixture with continuous mixing. Genistein can also be nitrated by adding 500 μM tetranitromethane to 50 μM genistein in a buffer solution composed of a 1:1 mixture of sodium bicarbonate (50 mM, pH 9)/50% ethanol with 1 mM DTPA. To carry out reduction of nitrated groups, $Na_2S_2O_4$ (60 μM) is added to the nitrated isoflavones. Following each reaction, samples are extracted as described previously and subjected to HPLC and MS analysis. Changes in the UV absorption of the isoflavones are monitored at 262 nm. Mixing both HOCl and sodium nitrite with genistein resulted in a monochloronitrogenistein and a dichloronitrogenistein (Boersma et al., 1999). Nitration causes an increase in m/z of 45, whereas for chloronitration it is 79 and 81. For dichloronitration, the increase is 113, 115, and 117.

Formation of Chlorinated and Nitrated Isoflavones by Inflammatory Cells

Either the human leukemia cell line (HL-60) or isolated polymorpho-nuclear cells (PMNs) can be used for this experiment. When using HL-60 cells, the cells are cultured in the presence of 1.3% DMSO for 1 week in order for them to differentiate into neutrophil-like cells (Boersma et al., 2003), whereas PMNs are isolated using a Histopaque dual-phase gradient (D'Alessandro et al., 2003). Cells are suspended in Krebs-Henseleit buffer (K-H: 118 mM NaCl, 27.3 mM NaHCO₃, 4.8 mM KCl, 1.75 mM CaCl₂, 1.0 mM KH₂PO₄, 1.2 mM MgSO₄, 11.1 mM glucose, pH 7.4) at 1×10^6 cells/ml and activated with 10 μM phorbol 12-myristate-13-acetate (PMA) either in the presence or in the absence of 10 μM genistein/daidzein and 50 μM NaNO₂. Reaction occurs for 60 min at 37° followed by termination using 5 U/ml catalase and a 10-min incubation on ice. The cells are centrifuged at 800g at 4° and the supernatant (~950 μl) is extracted with diethyl ether (2 ml). The samples are vortexed and centrifuged at 2000g, whereupon the ethereal top layer is removed to a 13 × 100-mm glass tube. Ether extraction is repeated until a total volume of 6 ml of ether has been added. Ether extracts are combined and dried to completion under air. The dried extracts are redissolved in 100 μl 80% aqueous methanol and analyzed by liquid chromatography–multiple reaction monitoring–mass spectrometry (LC-MRM-MS) (Boersma *et al.*, 2003; D'Alessandro *et al.*, 2003).

Analysis of Chlorinated and Nitrated Isoflavones

Reaction mixtures are analyzed by LC–MS as described earlier except that the mobile phase is isocratic, composed of 40% acetonitrile in 10 mM NH$_4$OAc at a flow rate of 1.0 ml/min. To obtain quantitative data, specific parent ion product ion combinations are used in LC-MRM-MS analysis. After termination of the reaction, another phytoestrogen is used as the internal standard. For example, in the case of experiments examining the reaction of genistein, daidzein is added as the internal standard. A series of samples prepared with several known concentrations of the varying phytoestrogen and a single concentration of the internal standard are analyzed to generate an area response–concentration curve. These typically give correlation coefficients of 0.98 or better for a five-point curve (D'Alessandro *et al.*, 2003). Typically, concentrations as low as 5 nM can be measured, although newer triple quadrupole mass spectrometers are capable of quantitatively detecting as little as 10–50 fmol.

Acknowledgments

Support for research on the metabolism of isoflavones has been provided in part by a grant from the National Center for Complementary and Alternative Medicine to the Purdue-UAB Botanicals Center for Age-Related Disease (1 P50 AT-00477; Connie Weaver, PI), by the National Cancer Institute (1 R01 CA61668; SB, PI), and by the United Soybean Board (SB, PI). The purchase of the mass spectrometer used in this work was enabled by a NIH/NCRR Shared Instrumentation grant (1 S10 RR 06487).

References

Akiyama, T., Ishida, J., Nakagawa, S., Ogawara, H., Watanabe, S., Itoh, N., Shibuya, M., and Fukami, Y. (1987). Genistein, a specific inhibitor of tyrosine-specific protein kinases. *J. Biol. Chem.* **262,** 5592–5595.

Barnes, S., Kirk, M., and Coward, L. (1994). Isoflavones and their conjugates in soy foods: Extraction conditions and analysis by HPLC-mass spectrometry. *J. Agric. Food Chem.* **42,** 2466–2474.

Bednarek, P., Franski, R., Kerhoas, L., Einhorn, J., Wojtaszek, P., and Stobiecki, M. (2001). Profiling changes in metabolism of isoflavonoids and their conjugates in *Lupinus albus* treated with biotic elicitor. *Phytochemistry* **56,** 77–85.

Binsack, R., Boersma, B. J., Patel, R. P., Kirk, M. C., White, C. R., Darley-Usmar, V. M., Barnes, S., Zhou, F., and Parks, D. A. (2001). Enhanced antioxidant activity following chlorination of quercetin by hypochlorous acid. *Alcohol. Clin. Exp. Res.* **25,** 434–443.

Boersma, B. J., Patel, R. P., Kirk, M., Jackson, P. L., Muccio, D., Darley-Usmar, V. M., and Barnes, S. (1999). Chlorination and nitration of soy isoflavones. *Arch. Biochem. Biophys.* **368,** 265–275.

Boersma, B. J., D' Alessandro, T., Benton, M. R., Kirk, M., Wilson, L. S., Prasain, J., Botting, N. P., Barnes, S., Darley-Usmar, V. M., and Patel, R. P. (2003). Neutrophil myeloperoxidase chlorinates and nitrates soy isoflavones and enhances their antioxidant properties. *Free Radic. Biol. Chem.* **35,** 1317–1330.

Clarke, D. B., Lloyd, A. S., Botting, N. P., Oldfield, M. F., Needs, P. W., and Wiseman, H. (2002). Measurement of intact sulfate and glucuronide phytoestrogen conjugates in human urine using isotope dilution liquid chromatography-tandem mass spectrometry with [$^{13}C_3$] isoflavone internal standards. *Anal. Biochem.* **309,** 158–172.

Coward, L., Barnes, N. C., Setchell, K. D. R., and Barnes, S. (1993). Genistein, daidzein, and their β-glycoside conjugates: Antitumor isoflavones in soybean foods from American and Asian diets. *J. Agric. Food Chem.* **41,** 1961–1967.

Coward, L., Kirk, M., Albin, N., and Barnes, S. (1996). Analysis of plasma isoflavones by reversed-phase HPLC-multiple reaction ion monitoring-mass spectrometry. *Clin. Chem. Acta* **247,** 121–142.

Coward, L., Smith, M., Kirk, M., and Barnes, S. (1998). Chemical modification of isoflavones in soyfoods during cooking and processing. *Am. J. Clin. Nutr.* **68,** 1486S–1491S.

Creeke, P. I., Wilkinson, A. P., Lee, H. A., Morgan, M. R. A., Price, K. R., and Rhodes, M. J. C. (1998). Development of ELISAs for the measurement of the dietary phytoestrogens daidzein and equol in human plasma. *Food Agric. Immunol.* **10,** 325–333.

D'Alessandro, T., Prasain, J., Benton, M. R., Botting, N., Moore, R., Darley-Usmar, V., Patel, R., and Barnes, S. (2003). Polyphenols, inflammatory response, and cancer prevention: Chlorination of isoflavones by human neutrophils. *J. Nutr.* **133,** 3773S–3777S.

Day, A. J., Cañada, F. J., Díaz, J. C., Kroon, P. A., Mclauchlan, R., Faulds, C. B., Plumb, G. W., Morgan, M. R. A., and Williamson, G. (2000). Dietary flavonoid and isoflavone glycosides are hydrolysed by the lactase site of lactase phlorizin hydrolase. *FEBS Lett.* **468,** 166–170.

Divi, R. L., Chang, H. C., and Doerge, D. R. (1997). Anti-thyroid isoflavones from soybean: Isolation, characterization, and mechanisms of action. *Biochem. Pharmacol.* **54,** 1087–1096.

Divi, R. L., and Doerge, D. R. (1996). Inhibition of thyroid peroxidase by dietary flavonoids. *Chem. Res. Toxicol.* **9,** 16–23.

Doerge, D. R., Chang, H. C., Churchwell, M. I., and Holder, C. L. (2000). Analysis of soy isoflavone conjugation *in vitro* and in human blood using liquid chromatography-mass spectrometry. *Drug Metab. Dispos.* **28,** 298–307.

Franke, A. A., and Custer, L. J. (1994). High-performance liquid-chromatographic assay of isoflavonoids and coumestrol from human urine *J. Chromatogr. B.* **662,** 47–60.

Gee, J. M, DuPont, M. S., Rhodes, M. J. C., and Johnson, I. T. (1998). Quercetin glycosides interact with the intestinal glucose transport pathway. *Free Radic. Biol. Chem.* **25,** 19–25.

Griffith, A. P., and Collison, M. W. (2001). Improved methods for the extraction and analysis of isoflavones from soy-containing foods and nutritional supplements by reversed-phase high-performance liquid chromatography and liquid chromatography-mass spectrometry. *J. Chromatogr. A.* **913,** 397–413.

He, X.-G., Lin, L.-Z., and Lian, L.-Z. (1996). Analysis of flavonoids from red clover by liquid chromatography-electrospray mass spectrometry. *J. Chromatogr. A* **755**, 127–132.

Hu, M., Krausz, K., Chen, J., Ge, X., Li, J., Gelboin, H. L., and Gonzalez, F. J. (2003). Identification of CYP1A2 as the main isoform for the phase I hydroxylated metabolism of genistein and a pro-drug converting enzyme of methylated isoflavones. *Drug Metab. Dispos.* **31**, 924–931.

Jacobs, M. N., and Lewis, D. F. (2002). Steroid hormone receptors and dietary ligands: A selected review. *Proc. Nutr. Soc.* **61**, 105–122.

Kinoshita, E., Ozawa, Y., and Aishima, T. (1997). Novel tartaric acid isoflavone derivatives that play a key role in differentiating Japanese soy sauces. *J. Agric. Food Chem.* **45**, 3753–3759.

Kudou, S., Fleury, Y., Welti, D., Magnolato, D., Uchida, T., Kitamura, K., and Okubo, K. (1991). Malonyl isoflavone glycosides in soybean seeds (Glycine max MERRILL). *Agric. Biol. Chem.* **55**, 2227–2233.

Kulling, S. E., Honig, D. M., and Metzler, M. (2001). Oxidative metabolism of the soy isoflavones daidzein and genistein in humans *in vitro* and *in vivo*. *J. Agric. Food Chem.* **49**, 3024–3033.

Kulling, S. E., Honig, D. M., Simat, T. J., and Metlzer, M. (2000). Oxidative *in vitro* metabolism of the soy phytoestrogens daidzein and genistein. *J. Agric. Food Chem.* **48**, 4963–4972.

Liggins, J., Bluck, L. J. C., Coward, W. A., and Bingham, S. A. (1998). Extraction and quantification of daidzein and genistein in food. *Anal. Biochem.* **264**, 1–7.

Lila, M. A., Yousef, G. G., Jiang, Y., and Weaver, C. M. (2005). Sorting out bioactivity in flavonoid mixtures. *J. Nutr.* **135**, 1231–1235.

Middleton, E., Jr., Kandaswami, C., and Theoharides, T. C. (2000). The effects of plant flavonoids on mammalian cells: Implications for inflammation, heart disease, and cancer. *Pharmacol. Rev.* **52**, 673–751.

Milder, I. E. J., Arts, I. C. W., Venema, D. P., Lasaroms, J. J. P., Wahala, K., and Hollman, P. C. H. (2004). Optimization of liquid chromatography-tandem mass spectrometry method for quantification of the plant lignans secoisolariciresinol, matairesinol, lariciresinol, and pinoresinol in foods. *J. Agric. Food Chem.* **52**, 4643–4651.

Murphy, P. A. (1981). Separation of genistin, daidzin, and their aglycones and coumesterol by gradient high-performance liquid chromatography. *J. Chromatogr.* **211**, 166–169.

Peterson, G., and Barnes, S. (1991). Genistein inhibition of the growth of human breast cancer cells: Independence from estrogen receptors and multi-drug resistance gene product. *Biochem. Biophys. Res. Commun.* **179**, 661–667.

Peterson, T. G., Coward, L., Kirk, M., Falany, C. N., and Barnes, S. (1996). The role of metabolism in mammary epithelial cell growth inhibition by the isoflavones genistein and biochanin A. *Carcinogenesis* **17**, 1861–1869.

Peterson, T. G., Ji, G.-P., Kirk, M., Coward, L., Falany, C. N., and Barnes, S. (1998). Metabolism of the isoflavones genistein and biochanin A in human breast cancer cell lines. *Am. J. Clin. Nutr.* **68**, 1505–1511.

Pietta, P. G. (2000). Flavonoids as antioxidants. *J. Nat. Prod.* **63**, 1035–1042.

Prasain, J. K., Boersma, B. J., Kirk, M., Wilson, L., Patel, R., Darley-Usmar, V. M., Botting, N., and Barnes, S. (2003a). ESI-tandem mass spectrometric analysis of chlorinated and nitrated isoflavones. *J. Mass Spectrom.* **38**, 764–771.

Prasain, J. K., Jones, K., Kirk, M., Wilson, L., Smith-Johnson, M., Weaver, C., and Barnes, S. (2003b). Profiling and quantification of isoflavonoids in kudzu dietary supplements by high-performance liquid chromatography and electrospray ionization tandem mass spectrometry. *J. Agric. Food Chem.* **51**, 4213–4218.

Prasain, J. K., Wang, C.-C., and Barnes, S. (2002). Mass spectrometry in the analysis of phytoestrogens in biological samples. *In* "Phytoestrogens and Health" (G. S. Gilani and J. J. Anderson, eds.), pp. 147–177. AOCS Press, Champaign, IL.

Prasain, J. K., Wang, C. C., and Barnes, S. (2004). Mass spectrometric methods for determination of flavonoids in biological samples. *Free Radic. Biol. Med.* **37**, 1324–1350.

Sfakianos, J., Coward, L., Kirk, M., and Barnes, S. (1997). Intestinal uptake and biliary excretion of the isoflavone genistein in rats. *J. Nutr.* **127**, 1260–1268.

Wang, C-C., Prasain, J. K., and Barnes, S. (2002). Review of the methods used in the determination of phytoestrogens. *J. Chromatogr. B.* **777**, 3–28.

Wang, H.-J., and Murphy, P. A. (1994a). Isoflavone content in commercial soybean foods. *J. Agric. Food Chem.* **42**, 1666–1673.

Wang, H.-J., and Murphy, P. A. (1994b). Isoflavone composition of American and Japanese soybeans in Iowa: Effects of variety, crop year, and location. *J. Agric. Food Chem.* **42**, 1674–1677.

Wilkinson, A. P., Gee, J. M., DuPont, M. S., Needs, P. W., Mellon, F. A., Williamson, G., and Johnson, I. T. (2003). Hydrolysis by lactase phlorizin hydrolase is the first step in the uptake of daidzein glycosides by rat small intestine *in vitro*. *Xenobiotica* **33**, 255–264.

Wilkinson, A. P., Wahala, K., and Williamson, G. (2002). Identification and quantification of polyphenol phytoestrogens in foods and human biological fluids. *J. Chromatogr. B.* **777**, 93–109.

Winterbourn, C. C., and Brennan, S. O. (1997). Characterization of the oxidation products of the reaction between reduced glutathione and hypochlorous acid. *Biochem. J.* **326**, 87–92.

[20] Sulfation and Glucuronidation of Phenols: Implications in Coenyzme Q Metabolism

By Nandita Shangari, Tom S. Chan, and Peter J. O'Brien

Abstract

Phase II conjugation of phenolic compounds constitutes an important mechanism through which exogenous or endogenous toxins are detoxified and excreted. Species differences in the rates of glucuronidation or sulfation can lead to significant variation in the metabolism of this class of compounds. Conjugation of the hydroxyl groups of phenols can occur with glucuronate or sulfate. Quinone metabolism, deactivation, and detoxification are also affected by the same conjugatory systems as phenols; however, reduction of quinones to hydroquinols seems to be a prerequisite. This work reviews current knowledge on phenol conjugation and its implications on hydroquinone metabolism with special consideration for coenzyme Q metabolism.

METHODS IN ENZYMOLOGY, VOL. 400   0076-6879/05 $35.00

Phase II Biotransformation of Phenolic Compounds

The ubiquitous appearance of phenols in nature and in industry means that animal exposure from exogenous sources is inevitable. The diverse role of these agents, however, is by no means limited to toxicological importance, as vital and beneficial phenolic compounds (e.g., tyrosine and vitamin E) are also produced endogenously by all living things. Levels of many biogenic and industry-derived phenolic agents in the body are tightly controlled by bioavailability, amount of exposure, and rate of excretion. The fate of the majority of these phenolic agents is rapid biotransformation by oxidation and/or conjugation. It is well known that the former activity (most notably carried out by the cytochrome P450 superfamily of oxygenases) is often associated with an activation of phenolic compounds to toxic substances. Conjugation of phenols may occur through several enzymic systems located throughout the body. Glucuronidation and sulfation are among the most well-known phenolic conjugatory systems. A chemically activated species is required for such reactions to occur. This is accomplished through the prior conjugation of cofactors with UDP serving as stable leaving groups.

Glucuronidation of Phenols

The glucuronidation of phenolic compounds is carried out by glycosyltransferases specifically referred to as UDP-glucuronosyl transferases (UDPGTs), which are located predominantly in the endoplasmic reticulum. In glucuronidation, glucuronic acid is conjugated to a suitable nucleophilic functional group such as a hydroxyl, amine, or carboxylic acid. UDP-glucuronic acid is used as the cofactor (Fig. 1).

There are currently two different families of UGTs, which are denoted 1 and 2. In humans, family 1 contains nine different functional isoforms and four pseudogenes (Gong *et al.*, 2001). The isoform UGT1A6 has been denoted the simple, planar phenol glucuronosyl transferase largely responsible for glucuronidating model substrates such as 1-naphthol, *p*-nitrophenol, and acetaminophen. However, UGT1A9 is known to glucuronidate bulky phenolic compounds whose available hydroxyl functional group is sterically hindered by large substitutions around it (e.g., 2,6-diisopropyl phenol). UGT1A9 has also been demonstrated to accept more complex polyphenolic structures, such as flavonoids and flavopiridol (Ethell *et al.*, 2001; Oliveira *et al.*, 2002; Ramirez *et al.*, 2002). Levels of UGT1A6 and UGT1A9 are cell type specific and have been hypothesized to be controlled by variations in duplication of the first exon (Zhang *et al.*, 2004).

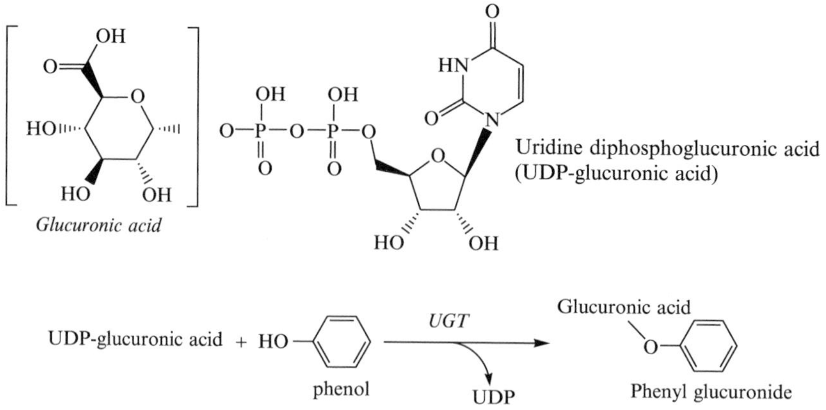

FIG. 1. Glucuronidation of phenol. Glucuronidation of phenolic compounds depends on the presence of UDP-glucuronic acid as a cofactor supplied from glycogen or glucose. UDP-glucuronic acid is transported into the endoplasmic reticulum where membrane-bound UGTs catalyze the conjugation of glucuronic acid to phenol via an ester linkage. The products of this reaction are the conjugated phenyl glucuronide and uridine diphosphate (UDP).

The UGT1A locus, which codes for all of the family 1 isoforms, is located on chromosome 2 at locus 2q.37 (Clarke *et al.*, 1997; Harding *et al.*, 1990). Expression of levels of these enzymes can be augmented by activation of the aryl hydrocarbon receptor (using TCDD or 3-methylcholanthrene as a substrate) and also by the steroid X receptor family [using pregnenolone-16α-carbonitrile (PCN)] (Chen *et al.*, 2003; Saarikoski *et al.*, 1998). Homologues of these two isoforms have been identified in other species, including the rat, cat, and dog; however, in the case of the cat, UGT1A6 codes for a pseudogene instead (Court, 2001). This finding is exemplified by the elevated susceptibility of these animal species to phenol-mediated toxicity (e.g., acetaminophen) (de Morais *et al.*, 1989). Although the UGT isoform principally responsible for simple, planar phenol glucuronidation is UGT1A6, overlapping phenolic specificity of various UGTs suggests that considerable redundancy exists. For example, acetaminophen glucuronidation is carried out primarily by UGT1A6 in rats. In Gunn rats, however, the UGT1A6 gene is nonfunctional due to a frameshift mutation in the UGT1A gene locus that inactivates the entire set of 1A enzymes. However, acetaminophen glucuronidation is still detectable, albeit at much lower levels, probably due to the contribution of the UGT2 isoforms. In cats, glucuronidation of acetaminophen can be compensated by isozymes 1A7 and possibly 1A9, which were shown to display lower, yet significant rates of acetaminophen conjugation, as seen in enzyme fractions prepared from cDNAs encoding multiple rat UGTs expressed in human embryonic kidney cells (Kessler *et al.*, 2002). In addition to this, there is considerable variation in the activity of UGT1A6 among different species. It is also a difficult task to compare the activity of the various isoforms of UGTs due to the limited selection of commercially available recombinant forms. Furthermore, more recent reports utilizing recombinant forms of UGTs from custom cDNA libraries vary in their use of substrates and in their reaction conditions or in their transfected cell type (Table I).

Assuming that acetaminophen represents a model phenol, the majority of phenolic glucuronidation can be considered to occur in the liver, as was shown from acetaminophen glucuronidation studies on rat and human microsomes (Bock *et al.*, 1993).

Sulfation of Phenols

Sulfotransferases (SULTs) are distributed ubiquitously in all animal cells but are particularly concentrated in the liver (Dunn *et al.*, 1998). They catalyze the conjugation of xenobiotics and endobiotics via the formation of sulfate esters on hydrophobic molecules (Fig. 2) or onto protein amino acids in the golgi apparatus. For the purposes of drug and endobiotic sulfation, SULTs localized in the soluble fraction of the cell are categorized as phenol and steroid SULTs. These types are considered major participants in drug and xenobiotic metabolism in animals (Yamazoe *et al.*, 1994). A standardized system of SULT nomenclature has been introduced for soluble SULTs that categorizes SULTs via their similarity in amino acid constitution (Blanchard *et al.*, 2004). The SULT1 family of sulfotransferases is responsible for catalyzing the sulfation of many of the simple endogenous and exogenous phenols such as acetaminophen, catecholamines, estrogen, and its related metabolites (Yamazoe *et al.*, 1994).

Unlike the UGT1A locus, the genetic localization of genes coding for the SULT1 family of enzymes is spread throughout the genome of humans. The SULT1A subfamily responsible for the sulfation of the majority of simple phenols and biogenic amines has been localized on adjacent hybrid intervals on chromosome 16p12.1-p11.2. This configuration likely arose from gene duplication leading to a divergence in sequence evolution (Dooley *et al.*, 1993, 1994). Further analysis of this loci revealed that polymorphisms do exist in all three subfamily isoforms, which likely play a prominent role in some individual variations in drug metabolism. SULT1A1 has been considered the major isoform

TABLE I
RATES OF PHENOL GLUCURONIDATION REPORTED FOR PHENOL GLUCURONOSYL TRANSFERASE (UGT 1A6 AND 1A9)

Animal	Substrate	Glucuronidating system	K_m (μM)	V_{max} (nmol/min/mg protein)	V_{max}/K_m	Condition	Ref.
Human	1-Naphthol	Recombinant UGT1A6	1.8 ± 0.1	43 ± 2	23.89	Lysate from HEK cells	Uchaipichat et al. (2004)
		Recombinant UGT1A9	1.3 ± 0.1	39 ± 0.8	30	Lysate from HEK cells	Uchaipichat et al. (2004)
	1-Hydroxypyrene	Recombinant UGT1A6	45 ± 14	0.15 ± 0.02	0.003	Lysate from V79 cells	Luukkanen et al. (2001)
		Recombinant UGT 1A9	1 ± 0.09	0.57 ± 0.03	0.57	Lysate from V79 cells	Luukkanen et al. (2001)
	2,6-Diisopropyl phenol	Microsomes	213.4	1.06	0.005	0.05% Brij 58	Shimizu et al. (2003)
	4-Methylumbelliferone	Recombinant UGT 1A6	8.8 ± 0.4	7.33 ± 0.116	0.83	Lysate from HEK cells	Uchaipichat et al. (2004)
		Recombinant UGT1A9	8.0 ± 0.6	9.016 ± 0.364	1.13	Lysate from HEK cells	Uchaipichat et al. (2004)

Rat	1-Naphthol	Recombinant UGT1A6	2.46	2.46	1	Yeast microsomes	Iwano et al. (1999)
	2,6-Diisopropyl phenol	Microsomes	32.5 ± 7.1	0.39 ± 0.2	0.012	0.05% Brij 58	Shimizu et al. (2003)
	Acetaminophen	Recombinant UGT1A6	3400 ± 400	0.60 ± 0.04	1.7×10^{-4}	Lysate from HEK cells	Kessler et al. (2002)
Ferret	Acetaminophen	Microsomes	4200 ± 2200	5.4 ± 2.6	1.3×10^{-3}	Ferret liver microsomes	Court (2001)
Dog	1-Naphthol	Recombinant UGT1A6	41 ± 5.9	0.07 ± 0.002	1.7×10^{-3}	Recombinantly expressed in V79 cells	Soars et al. (2001)
	4-Methylumbelliferone	Recombinant UGT1A6	300 ± 37	0.09 ± 0.004	3×10^{-4}	Recombinantly expressed in V79 cells	Soars et al. (2001)
Cat		UGT1A6[a]	ND	ND	—	ND	Court and Greenblatt (2000)

[a] Pseudogene.

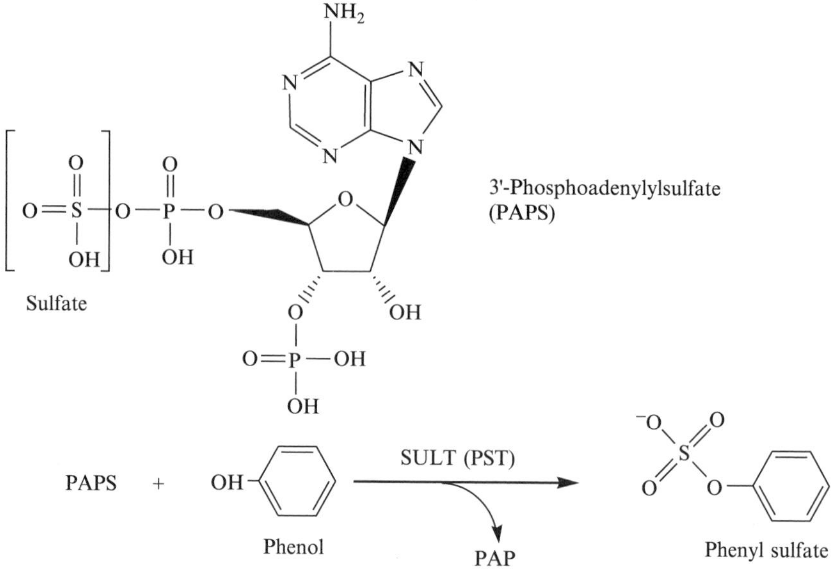

FIG. 2. Sulfation of phenol. Phenol sulfotransferases (SULT) require 3'-phosphoadeny-lylsulfate as a cofactor for the sulfation of substrates. PAPS is synthesized through a two-step mechanism involving ATP sulfurylase, which combines ATP and sulfate to form 5'-adenylylsulfate. The second reaction involves the enzyme APS kinase, which catalyzes the phosphorylation of APS to form PAPS. PST are located in the cytosol and catalyze the conjugation of sulfate from PAPS to phenol via an ether linkage. The products of this reaction are phenyl sulfate and 3'phosphoadenosine-5'-phosphate (PAP).

responsible for the sulfation of most simple planar phenols such as 4-nitrophenol, 1-naphthol, and acetaminophen in all species studied thus far (Table II), whereas SULT1A2 appears to perform the same task, albeit with considerably less efficiency. SULT1A3, however, preferentially sulfates catecholamines.

SULT1A1 polymorphisms have gained considerable attention, as variations in its activity have been implicated in individual differences in the ability to metabolize xenobiotics such as the dopamine agonist apomorphine (for treatment of Parkinson's disease) or to alter the susceptibility of individuals to polycyclic aromatic hydrocarbon-mediated genotoxicity (Engelke *et al.*, 2000; Hildebrandt *et al.*, 2004; Hung *et al.*, 2004; Thomas *et al.*, 2003; Tsukino *et al.*, 2004). A "gain of function" allele coding for SULT1A increases the age of onset for breast cancer while the chance of developing other types of cancer increases (Seth *et al.*, 2000). In addition,

TABLE II

COMPARISON OF SULTs THAT CONJUGATE PHENOLIC SUBSTANCES FROM VARIOUS SPECIES

Animal	Substrate[a]	Sulfonating system	$K_m(\mu M)$	V_{max}(nmol/min/mg protein)	V_{max}/K_m	Reference
Human	PNP	Purified SULT1A1	1.1 ± 0.2	581 ± 67	528	Brix et al. (1999)
	Dopamine		9.7 ± 1.4	22 ± 1.0	2.3	
	PNP	Purified SULT1A3	1024 ± 127	437 ± 19	0.4	
	Dopamine		3.7 ± 0.3	799 ± 30	216	
	PNP	Cytosol from bacteria expressing recombinant SULT1C2	13350 ± 4116	0.005 ± 0.001	3.7×10^{-7}	Hehonah et al. (1999)
	Dopamine		ND	ND	—	
Rat	T3	Cytosol from bacteria expressing recombinant SULT1B1	142 ± 9	1.156 ± 1.33	0.0081	Kester et al. (2003)
	3,3'-T2		100 ± 6	0.0508 ± 0.0063	0.000508	
Dog	PNP	Cytosol from bacteria expressing recombinant SULT1A1	0.069 ± 0.013	0.39 ± 0.014	5.7	Tsoi et al. (2002)
	Dopamine		180 ± 16	0.15 ± 0.0054	0.0086	
	Estrogen		14 ± 1.2	0.27 ± 0.0093	0.0043	
	PNP	Purified SULT1D1	200 ± 36	4.8 ± 0.35	0.02	Tsoi et al. (2001a)
	Dopamine		5.2 ± 0.75	5.7 ± 0.27	1.1	
	DHEA		ND	ND	—	
	PNP	Cytosol from bacteria expressing recombinant SULT1B1	65 ± 4.3	4.1 ± 0.078	0.063	Tsoi et al. (2001b)
	Dopamine		ND	ND	—	
	DHEA		ND	ND	—	
	T3		37 ± 17	5.1 ± 1.3	0.14	
	3,3'T2		33 ± 5.4	8.0 ± 0.42	0.24	
Rabbit	PNP	Cytosol from bacteria expressing recombinant SULT1A1	150 ± 40	897.5 ± 66.4	5.98	Riley et al. (2002)
	Dopamine		175.3±0.7	151 ± 5.7	0.86	
	PNP	Cytosol from bacteria expressing recombinant SULT1C2	2200 ± 1175	0.39 ± 0.1	1.8×10^{-4}	Hehonah et al. (1999)
	Dopamine		ND	ND	—	

[a]DHEA, dehydroepiandrosterone; PNP, 4-nitrophenol; T3, 3,5,3'-triiodothyronine ; 3,3-T2, 3,3'-diiodothyronine.

the functional activity of SULTs has been shown to be affected by a variety of stimuli, such as variations in expression due to exposure to xenobiotics, changes in cellular redox status, induced physical stress, and, in some species, through activation of the glucocorticoid or androgen receptor (Fang *et al.*, 2003; Maiti *et al.*, 2003, 2004; Marshall *et al.*, 2000).

The characterization of SULT isoforms in other species has benefited immensely from the expressed sequence tag (EST) database (http://www.ncbi.nlm.nih.gov/dbEST/). By matching known, expressed sequences from known SULTs, these linkage studies were able to identify a large number of orthologous gene products from different species. Currently 69 different SULT gene products have been identified (Blanchard *et al.*, 2004; Freimuth *et al.*, 2004; Lin *et al.*, 2004; Venkatachalam *et al.*, 2004). Remarkably, the functional similarities between species-specific SULTs mirror their nucleotide sequence similarity. For instance, rabbit SULT1A1 and SULT1C2 share 87.5 and 83.8%, respectively, of their nucleotide sequence with their human orthologues and both rabbit and human SULTs share a preference for the same substrates (Hehonah *et al.*, 1999; Riley *et al.*, 2002).

The low K_m of most substrates for SULTs allows for the conjugatory metabolism of xenobiotics whose levels may fall below the threshold for efficient glucuronidation. Thus, the combined activity of SULTs and UGTs has been classically thought to be highly complementary. Such activity has been demonstrated using various substrates of SULTs, including *p*-nitrophenol and acetaminophen (Kane *et al.*, 1990; Ring *et al.*, 1996). Evidence suggests that although SULT activity is widely distributed throughout the body, the primary means by which many different exogenous phenolic substrates are conjugated appear to be via glucuronidation. In humans, *in vivo* studies indicate that acetaminophen is excreted rapidly in its glucuronidated and sulfated forms in the proportion of approximately 7:3, which was mirrored by comparative *in vitro* studies using human hepatocytes cultures (Kane *et al.*, 1995). A slightly lower ratio of glucuronidation to sulfation conjugation was observed in rat and human liver slices conjugating *p*-nitrophenol (Kuhn *et al.*, 2001). Rates of SULT activity, however, may be largely limited by the available supply of PAPS.

Quinones in Biology

The ubiquitous appearance of quinones in all organisms is a testament to their fundamental importance in biology. Quinones catalyze important electron transfer reactions, such as those that occur during oxidative phosphorylation and photosynthesis (catalyzed by ubiquinones and plastoquinones, respectively), pyrroloquinoline quinone (PQQ), which is an essential cofactor for bacterial methanol dehydrogenase and glucose dehydrogenase (Duine *et al.*, 1980; Salisbury *et al.*, 1979). Quinones are also formed from the oxidation of plant-derived polyphenols (e.g., the ortho-quinone of caffeic acid by polyphenol oxidase).

The redox activity of quinones, however, also contributes to their well-known toxic effects in living systems. These toxic effects are due to their ability to receive and donate electrons to various compounds. As such, they possess a paradoxical nature that can be described as either antioxidant (electron donating) or prooxidant (electron abstracting) based on whether the quinone is in its reduced or oxidized form. For example, many plant-derived antioxidants exist as reduced forms of quinones (hydroquinones). Caffeic acid (3,4-dihydroxycinnamic acid) and its ester-conjugate, chlorogenic acid (3-[[3-(3,4-dihydroxyphenyl)-1-oxo-2-propenyl]oxy]-1,4,5-trihydroxycyclohexanecarboxylic acid), are well-known ortho-hydroquinone antioxidants found in many plant foods such as coffee beans (Nakayama, 1994; Yamada *et al.*, 1996). Coffee drinkers consume an average of 0.5–1 g of chlorogenic acid daily (Olthof *et al.*, 2001). Despite the beneficial antioxidant activity of these compounds, significant safety concerns still exist over their consumption at high doses. Both chlorogenic acid and caffeic acid consumption have been associated with cytotoxicity in cultured cell lines (Etzenhouser *et al.*, 2001), goitrogenic effects (Khelifi-Touhami *et al.*, 2003), and carcinogenicity (Hagiwara *et al.*, 1991). The formation of electrophilic orthoquinones as metabolites of ingested aromatic compounds has been implicated in the toxic mechanisms underlying polyphenol toxicity (Moridani *et al.*, 2002; Pierpoint, 1966). Another ortho-hydroquinol polyphenolic, quercetin (3,5,7,3′,4′-pentahydroxyflavone), has been shown to bind to DNA and protein in human intestinal and hepatic cell lines, Caco-2 and HepG2. The mechanism of binding involved the conversion of quercetin to a quinone or quinone methide intermediate(Walle *et al.*, 2003). The most popular mechanistic theory underlying the toxicity of quinones involves their propensity to bind to nucleophilic functional groups commonly found on many cellular components.

Abhorrent conjugation of quinones to proteins or DNA can result in mutation and/or protein dysfunction. Furthermore, the attachment of quinones to proteins may also lead to immunological damage through the recognition of quinone-bound epitopes from degraded protein. Quinone binding to proteins has been implicated in playing a causative role in the incidence of certain allergic or idiosyncratic drug reactions (Lepoittevin *et al.*, 1991; Parrish *et al.*, 1997; Petersen, 2002). For example, contact allergic reactions have been associated with 2-hydroxy-2,4-naphthoquinone (henna), which is a principal ingredient in many types of body dyes (Bolhaar et al., 2001).

Metabolic detoxification of quinones involves dissipating their electrophilic nature via the donation of electrons. This may occur via a simple quinone acceptor to donor chemical interactions or through more complicated processes involving enzymic systems. Conjugatory processes on the reduced hydroquinone forms prevent their reoxidation.

In preparation for conjugation, detoxification of quinones can occur through the donation of reducing equivalents from readily abundant and replenishable NAD(P)H in the cell. This transfer of electrons is catalyzed by a series of NAD(P)H-dependent quinone oxidoreductases located in the cytosol of the cell, or associated with the plasma membrane, or located in the mitochondrial membranes.

Sulfate and Glucuronic acid Conjugatory Activity Toward Coenyzme Q (CoQ) Hydroquinone Metabolites

In animals, coenzyme Q plays an important role in the transfer of electrons for various functions, most notably involving the mitochondrial electron transport chain and also for antioxidant protection of membrane components in the cell. Levels of CoQ in the body are most likely maintained by the rate of its synthesis and degradation considering that exogenous CoQ10 in rats has limited bioavailability preferentially partitioning in the spleen and in the liver (Zhang et al., 1996). This is not normally a problem, as all cells in the body are capable of synthesizing their own CoQ.

Pathways of coenzyme Q degradation, however, have not been studied as extensively. The half-life of CoQ9 (the native analogue in rats) in rats is between 49 and 125 days (Thelin et al., 1992). Radiolabeling studies have indicated that the degradation products of CoQ are likely oxidized products consisting of the intact 2,3-dimethoxy-5-methyl benzoquinone nucleus with shortened isoprenyl side chains and conjugated CoQ products, which are excreted into the urine or feces (Bentinger et al., 2003). The oxidation of CoQ may be carried out by cytochrome P450 and, based on the multitude of studies investigating the efficiency of short-chain analogues of CoQ, this oxidation reduces the ability of CoQ to participate in the mitochondrial electron transport chain (Lenaz, 1998). Furthermore, short chain derivatives of CoQ may endanger cells by imposing futile redox-cycling/oxygen activating activity (Degli et al., 1996; Esposti et al., 1996). However, several studies have identified potential antioxidant and therapeutic potential in the introduction of short chain analogues of CoQ to various models of oxidative stress. For instance, short chain CoQ

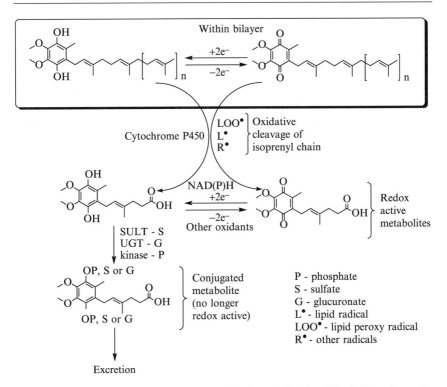

FIG. 3. Proposed order of metabolite formation from CoQ. Lipophilic CoQ carries out its physiological functions when situated within the lipid bilayer. Oxidative metabolism of the side chain via oxidative stress or cytochrome P450s releases CoQ metabolites into the soluble fraction of the cell where conjugatory reactions neutralize their redox activity. Conjugated metabolites are then excreted into the urine or feces.

analogues have been shown to be more effective at lowering lipid peroxidation caused by various models of oxidative stress (Kagan *et al.*, 1990; Littarru *et al.*, 1985). In our own studies, we found that CoQ1 hydroquinone was very effective at preventing rotenone-induced and cumene hydroperoxide-induced oxidative stress in isolated hepatocytes (Chan *et al.*, 2002, 2003). This antioxidant activity of CoQ and its short chain analogs, however, depends on an efficient reducing system in the cell such as mitochondrial complex I, the putative CoQ reductase, or DT-diaphorase (Kishi *et al.*, 2002; Landi *et al.*, 1997; Walker, 1992). The latter has been shown to participate (although to a limited extent) in the reduction of CoQ to its hydroquinone most likely because of the inaccessibility of CoQ to the soluble fractions of the cell where DT-diaphorase is located. Reduced forms of short chain analogues can be considered model substances that behave similarly to endogenous CoQ metabolites. They can penetrate membrane bilayers and are therefore more amenable to metabolism through conjugation with cytosolic mechanisms such as sulfation. CoQ1 was found to be sulfated to a large extent in the cytosolic fraction of isolated hepatocytes. Furthermore, this conjugatory process diminished the antioxidant activity of CoQ1 during oxidative insult (Chan et al., 2003). Sulfated forms of CoQ7 metabolites were also excreted from rats (Imada et al., 1970).

Conjugation of coenzyme Q with longer isoprenyl chains is a subject of ongoing debate. Analysis of membrane localization of CoQ has indicated that CoQ10 resides in between the two leaflets of the bilayer, with limited access to the soluble phase of the cell (Cornell et al., 1987). This localization, however, might be advantageous not only for mediated intramembrane electron transport, but also for preserving CoQ content in cells, as the hydroquinone form would not have access to cytosolic SULTs or even endoplasmic reticulum-anchored UGTs. Only through oxidation of the isoprenoid side chain of CoQ (through interactions with either free radicals or hydrocylation catalyzed by monooxygenases) can the hydrophilicity of CoQ increase sufficiently to expose the hydroquinone moiety to soluble conjugatory enzymes. Both sulfation and glucuronidation might therefore play a large role in quickly neutralizing water-soluble CoQ metabolites in the cell. However, because unmetabolized, labeled CoQ given ip to rats could be found in copious amounts in the feces, it is unlikely that conjugatory mechanisms are absolutely necessary for the clearance of exogenously administered CoQ. CoQ that has undergone metabolism in cells before being excreted via the urine or the feces would likely undergo oxidation prior to conjugation, as no conjugated CoQ10 in unmetabolized form has been documented (Fig. 3).

Because CoQ is synthesized ubiquitously in all cells, it is plausible that a similarly ubiquitous mechanism responsible for the metabolism of CoQ exists in each cell. Such a mechanism was alluded to by Bentinger et al. (2003) where a phosphorylated metabolite of CoQ was identified as the principal metabolite found in urine following CoQ10 intraperitoneal administration in rats. To our knowledge, outside of tyrosine, phosphorylation of hydroquinones or phenols has not been described previously. It would be interesting to surmise that the same tyrosine kinases are responsible for the phosphorylation of CoQ10 metabolites, as both targets are likely closely associated with the lipid bilayer. Because the presence of conjugated CoQ10 has not been observed, a scission of the side chain as observed with CoQ7 likely precedes any conjugatory metabolism. The presence of these metabolites in all cells of the body suggests that monooxygenases are not primarily responsible for isoprenyl chain scission. For example, CoQ metabolites were found to be distributed equally in the spleen and liver, even though the cytochrome P450 content is higher in the liver than in the spleen (Bentinger *et al.*, 2003). It is, however, possible that metabolites that form in the liver also accumulate in the spleen. Rather, oxidative stress may be the rate-limiting mechanism responsible for CoQ oxidation and ultimately conjugation.

References

Bentinger, M., Dallner, G., Chojnacki, T., and Swiezewska, E. (2003). Distribution and breakdown of labeled coenzyme Q(10) in rat. *Free Radic. Biol. Med.* **34,** 563–575.

Blanchard, R. L., Freimuth, R. R., Buck, J., Weinshilboum, R. M., and Coughtrie, M. W. (2004). A proposed nomenclature system for the cytosolic sulfotransferase (SULT) superfamily. *Pharmacogenetics* **14,** 199–211.

Bock, K. W., Forster, A., Gschaidmeier, H., Bruck, M., Munzel, P., Schareck, W., Fournel-Gigleux, S., and Burchell, B. (1993). Paracetamol glucuronidation by recombinant rat and human phenol UDP-glucuronosyltransferases. *Biochem. Pharmacol.* **45,** 1809–1814.

Bolhaar, S. T., Mulder, M., and van Ginkel, C. J. (2001). IgE-mediated allergy to henna. *Allergy* **56,** 248.

Brix, L. A., Barnett, A. C., Duggleby, R. G., Leggett, B., and McManus, M. E. (1999). Analysis of the substrate specificity of human sulfotransferases SULT1A1 and SULT1A3: Site-directed mutagenesis and kinetic studies. *Biochemistry* **38,** 10474–10479.

Chan, T. S., and O'Brien, P. J. (2003). Hepatocyte metabolism of coenzyme Q1 (ubiquinone-5) to its sulfate conjugate decreases its antioxidant activity. *Biofactors* **18,** 207–218.

Chan, T. S., Teng, S., Wilson, J. X., Galati, G., Khan, S., and O'Brien, P. J. (2002). Coenzyme Q cytoprotective mechanisms for mitochondrial complex I cytopathies involves NAD(P) H:quinone oxidoreductase 1(NQO1). *Free Radic. Res.* **36,** 421–427.

Chen, C., Staudinger, J. L., and Klaassen, C. D. (2003). Nuclear receptor, pregnane X receptor, is required for induction of UDP-glucuronosyltranferases in mouse liver by pregnenolone-16 alpha-carbonitrile. *Drug Metab. Dispos.* **31,** 908–915.

Clarke, D. J., Cassidy, A. J., See, C. G., Povey, S., and Burchell, B. (1997). Cloning of the human UGT1 gene complex in yeast artificial chromosomes: Novel aspects of gene structure and subchromosomal mapping to 2q37. *Biochem. Soc. Trans.* **25,** S562.

Cornell, B. A., Keniry, M. A., Post, A., Robertson, R. N., Weir, L. E., and Westerman, P. W. (1987). Location and activity of ubiquinone 10 and ubiquinone analogues in model and biological membranes. *Biochemistry* **26,** 7702–7707.

Court, M. H. (2001). Acetaminophen UDP-glucuronosyltransferase in ferrets: Species and gender differences, and sequence analysis of ferret UGT1A6. *J. Vet. Pharmacol. Ther.* **24,** 415–422.

Court, M. H., and Greenblatt, D. J. (2000). Molecular genetic basis for deficient acetaminophen glucuronidation by cats: UGT1A6 is a pseudogene, and evidence for reduced diversity of expressed hepatic UGT1A isoforms. *Pharmacogenetics* **10,** 355–369.

Degli, E. M., Ngo, A., McMullen, G. L., Ghelli, A., Sparla, F., Benelli, B., Ratta, M., and Linnane, A. W. (1996). The specificity of mitochondrial complex I for ubiquinones. *Biochem. J.* **313**(Pt 1), 327–334.

de Morais, S. M., and Wells, P. G. (1989). Enhanced acetaminophen toxicity in rats with bilirubin glucuronyl transferase deficiency1. *Hepatology* **10**, 163–167.

Dooley, T. P., Mitchison, H. M., Munroe, P. B., Probst, P., Neal, M., Siciliano, M. J., Deng, Z., Doggett, N. A., Callen, D. F., Gardiner, R. M., and Mole, S. E. (1994). Mapping of two phenol sulphotransferase genes, STP and STM, to 16p: Candidate genes for Batten disease. *Biochem. Biophys. Res. Commun.* **205**, 482–489.

Dooley, T. P., Obermoeller, R. D., Leiter, E. H., Chapman, H. D., Falany, C. N., Deng, Z., and Siciliano, M. J. (1993). Mapping of the phenol sulfotransferase gene (STP) to human chromosome 16p12.1-p11.2 and to mouse chromosome 7. *Genomics* **18**, 440–443.

Duine, J. A., Frank, J., and Verwiel, P. E. (1980). Structure and activity of the prosthetic group of methanol dehydrogenase. *Eur. J. Biochem.* **108**, 187–192.

Dunn, R. T., and Klaassen, C. D. (1998). Tissue-specific expression of rat sulfotransferase messenger RNAs. *Drug Metab. Dispos.* **26**, 598–604.

Engelke, C. E., Meinl, W., Boeing, H., and Glatt, H. (2000). Association between functional genetic polymorphisms of human sulfotransferases 1A1 and 1A2. *Pharmacogenetics* **10**, 163–169.

Esposti, M. D., Ngo, A., Ghelli, A., Benelli, B., Carelli, V., McLennan, H., and Linnane, A. W. (1996). The interaction of Q analogs, particularly hydroxydecyl benzoquinone (idebenone), with the respiratory complexes of heart mitochondria. *Arch. Biochem. Biophys.* **330**, 395–400.

Ethell, B. T., Beaumont, K., Rance, D. J., and Burchell, B. (2001). Use of cloned and expressed human UDP-glucuronosyltransferases for the assessment of human drug conjugation and identification of potential drug interactions. *Drug Metab. Dispos.* **29**, 48–53.

Etzenhouser, B., Hansch, C., Kapur, S., and Selassie, C. D. (2001). Mechanism of toxicity of esters of caffeic and dihydrocaffeic acids. *Bioorg. Med. Chem* **9**, 199–209.

Fang, H. L., Shenoy, S., Duanmu, Z., Kocarek, T. A., and Runge-Morris, M. (2003). Transactivation of glucocorticoid-inducible rat aryl sulfotransferase (SULT1A1) gene transcription. *Drug Metab. Dispos.* **31**, 1378–1381.

Freimuth, R. R., Wiepert, M., Chute, C. G., Wieben, E. D., and Weinshilboum, R. M. (2004). Human cytosolic sulfotransferase database mining: Identification of seven novel genes and pseudogenes. *Pharmacogenom. J.* **4**, 54–65.

Gong, Q. H., Cho, J. W., Huang, T., Potter, C., Gholami, N., Basu, N. K., Kubota, S., Carvalho, S., Pennington, M. W., Owens, I. S., and Popescu, N. C. (2001). Thirteen UDPglucuronosyltransferase genes are encoded at the human UGT1 gene complex locus. *Pharmacogenetics* **11**, 357–368.

Hagiwara, A., Hirose, M., Takahashi, S., Ogawa, K., Shirai, T., and Ito, N. (1991). Forestomach and kidney carcinogenicity of caffeic acid in F344 rats and C57BL/6N x C3H/HeN F1 mice. *Cancer Res.* **51**, 5655–5660.

Harding, D., Jeremiah, S. J., Povey, S., and Burchell, B. (1990). Chromosomal mapping of a human phenol UDP-glucuronosyltransferase, GNT1. *Ann. Hum. Genet.* **54**(Pt 1), 17–21.

Hehonah, N., Zhu, X., Brix, L., Bolton-Grob, R., Barnett, A., Windmill, K., and McManus, M. (1999). Molecular cloning, expression, localisation and functional characterisation of a rabbit SULT1C2 sulfotransferase. *Int. J. Biochem. Cell Biol.* **31**, 869–882.

Hildebrandt, M. A., Salavaggione, O. E., Martin, Y. N., Flynn, H. C., Jalal, S., Wieben, E. D., and Weinshilboum, R. M. (2004). Human SULT1A3 pharmacogenetics: Gene duplication and functional genomic studies. *Biochem. Biophys. Res. Commun.* **321**, 870–878.

Hung, R. J., Boffetta, P., Brennan, P., Malaveille, C., Hautefeuille, A., Donato, F., Gelatti, U., Spaliviero, M., Placidi, D., Carta, A., Scotto di Carlo, A., and Porru, S. (2004). GST, NAT,

SULT1A1, CYP1B1 genetic polymorphisms, interactions with environmental exposures and bladder cancer risk in a high-risk population. *Int. J. Cancer* **110**, 598–604.

Imada, I., Watanabe, M., Matsumoto, N., and Morimoto, H. (1970). Metabolism of ubiquinone-7. *Biochemistry* **9**, 2870–2878.

Iwano, H., Yokota, H., Ohgiya, S., and Yuasa, A. (1999). The significance of amino acid residue Asp446 for enzymatic stability of rat UDP-glucuronosyltransferase UGT1A6. *Arch. Biochem. Biophys.* **363**, 116–120.

Kagan, V. E., Serbinova, E. A., Koynova, G. M., Kitanova, S. A., Tyurin, V. A., Stoytchev, T. S., Quinn, P. J., and Packer, L. (1990). Antioxidant action of ubiquinol homologues with different isoprenoid chain length in biomembranes. *Free Radic. Biol. Med.* **9**, 117–126.

Kane, R. E., Li, A. P., and Kaminski, D. R. (1995). Sulfation and glucuronidation of acetaminophen by human hepatocytes cultured on Matrigel and type 1 collagen reproduces conjugation *in vivo*. *Drug Metab. Dispos.* **23**, 303–307.

Kane, R. E., Tector, J., Brems, J. J., Li, A. P., and Kaminski, D. L. (1990). Sulfation and glucuronidation of acetaminophen by cultured hepatocytes replicating *in vivo* metabolism. *ASAIO Trans.* **36**, M607–M610.

Kessler, F. K., Kessler, M. R., Auyeung, D. J., and Ritter, J. K. (2002). Glucuronidation of acetaminophen catalyzed by multiple rat phenol UDP-glucuronosyltransferases. *Drug Metab. Dispos.* **30**, 324–330.

Kester, M. H., Kaptein, E., Roest, T. J., van Dijk, C. H., Tibboel, D., Meinl, W., Glatt, H., Coughtrie, M. W., and Visser, T. J. (2003). Characterization of rat iodothyronine sulfotransferases. *Am. J. Physiol Endocrinol. Metab.* **285**, E592–E598.

Khelifi-Touhami, F., Taha, R. A., Badary, O. A., Lezzar, A., and Hamada, F. M. (2003). Goitrogenic activity of p-coumaric acid in rats. *J. Biochem. Mol. Toxicol.* **17**, 324–328.

Kishi, T., Takahashi, T., Mizobuchi, S., Mori, K., and Okamoto, T. (2002). Effect of dicumarol, a NAD(P)H:quinone acceptor oxidoreductase 1 (DT-diaphorase) inhibitor on ubiquinone redox cycling in cultured rat hepatocytes. *Free Radic. Res.* **36**, 413–419.

Kuhn, U. D., Rost, M., and Muller, D. (2001). Para-nitrophenol glucuronidation and sulfation in rat and human liver slices. *Exp. Toxicol. Pathol.* **53**, 81–87.

Landi, L., Fiorentini, D., Galli, M. C., Segura-Aguilar, J., and Beyer, R. E. (1997). DT-Diaphorase maintains the reduced state of ubiquinones in lipid vesicles thereby promoting their antioxidant function. *Free Radic. Biol. Med.* **22**, 329–335.

Lenaz, G. (1998). Quinone specificity of complex I. *Biochim. Biophys. Acta* **1364**, 207–221.

Lepoittevin, J. P., and Benezra, C. (1991). Allergic contact dermatitis caused by naturally occurring quinones. *Pharm. Weekbl. Sci.* **13**, 119–122.

Lin, Z., Lou, Y., and Squires, J. E. (2004). Molecular cloning and functional analysis of porcine SULT1A1 gene and its variant: A single mutation SULT1A1 causes a significant decrease in sulfation activity. *Mamm. Genome* **15**, 218–226.

Littarru, G. P., Lippa, S., De Sole, P., and Oradei, A. (1985). *In vitro* effect of different ubiquinones on the scavenging of biologically generated O2. *Drugs Exp. Clin. Res.* **11**, 529–532.

Luukkanen, L., Mikkola, J., Forsman, T., Taavitsainen, P., Taskinen, J., and Elovaara, E. (2001). Glucuronidation of 1-hydroxypyrene by human liver microsomes and human UDP-glucuronosyltransferases UGT1A6, UGT1A7, and UGT1A9: Development of a high-sensitivity glucuronidation assay for human tissue. *Drug Metab. Dispos.* **29**, 1096–1101.

Maiti, S., and Chen, G. (2003). Tamoxifen induction of aryl sulfotransferase and hydroxysteroid sulfotransferase in male and female rat liver and intestine. *Drug Metab. Dispos.* **31**, 637–644.

Maiti, S., Grant, S., Baker, S. M., Karanth, S., Pope, C. N., and Chen, G. (2004). Stress regulation of sulfotransferases in male rat liver. *Biochem. Biophys. Res. Commun.* **323**, 235–241.

Marshall, A. D., McPhie, P., and Jakoby, W. B. (2000). Redox control of aryl sulfotransferase specificity. *Arch. Biochem. Biophys.* **382**, 95–104.

Moridani, M. Y., Scobie, H., and O'Brien, P. J. (2002). Metabolism of caffeic acid by isolated rat hepatocytes and subcellular fractions. *Toxicol. Lett.* **133**, 141–151.

Nakayama, T. (1994). Suppression of hydroperoxide-induced cytotoxicity by polyphenols. *Cancer Res.* **54**, 1991s–1993s.

Oliveira, E. J., Watson, D. G., and Grant, M. H. (2002). Metabolism of quercetin and kaempferol by rat hepatocytes and the identification of flavonoid glycosides in human plasma. *Xenobiotica* **32**, 279–287.

Olthof, M. R., Hollman, P. C., and Katan, M. B. (2001). Chlorogenic acid and caffeic acid are absorbed in humans. *J Nutr.* **131**, 66–71.

Parrish, D. D., Schlosser, M. J., Kapeghian, J. C., and Traina, V. M. (1997). Activation of CGS 12094 (prinomide metabolite) to 1,4-benzoquinone by myeloperoxidase: Implications for human idiosyncratic agranulocytosis. *Fundam. Appl. Toxicol.* **35**, 197–204.

Petersen, K. U. (2002). From toxic precursors to safe drugs: Mechanisms and relevance of idiosyncratic drug reactions. *Arzneimittelforschung.* **52**, 423–429.

Pierpoint, W. S. (1966). The enzymic oxidation of chlorogenic acid and some reactions of the quinone produced. *Biochem. J.* **98**, 567–580.

Ramirez, J., Iyer, L., Journault, K., Belanger, P., Innocenti, F., Ratain, M. J., and Guillemette, C. (2002). *In vitro* characterization of hepatic flavopiridol metabolism using human liver microsomes and recombinant UGT enzymes. *Pharm. Res.* **19**, 588–594.

Riley, E., Bolton-Grob, R., Liyou, N., Wong, C., Tresillian, M., and McManus, M. E. (2002). Isolation and characterisation of a novel rabbit sulfotransferase isoform belonging to the SULT1A subfamily. *Int. J. Biochem. Cell Biol.* **34**, 958–969.

Ring, J. A., Ghabrial, H., Ching, M. S., Shulkes, A., Smallwood, R. A., and Morgan, D. J. (1996). Conjugation of para-nitrophenol by isolated perfused fetal sheep liver. *Drug Metab. Dispos.* **24**, 1378–1384.

Saarikoski, S. T., Ikonen, T. S., Oinonen, T., Lindros, K. O., Ulmanen, I., and Husgafvel-Pursiainen, K. (1998). Induction of UDP-glycosyltransferase family 1 genes in rat liver: Different patterns of mRNA expression with two inducers, 3-methylcholanthrene and beta-naphthoflavone. *Biochem. Pharmacol.* **56**, 569–575.

Salisbury, S. A., Forrest, H. S., Cruse, W. B., and Kennard, O. (1979). A novel coenzyme from bacterial primary alcohol dehydrogenases. *Nature* **280**, 843–844.

Seth, P., Lunetta, K. L., Bell, D. W., Gray, H., Nasser, S. M., Rhei, E., Kaelin, C. M., Iglehart, D. J., Marks, J. R., Garber, J. E., Haber, D. A., and Polyak, K. (2000). Phenol sulfotransferases: Hormonal regulation, polymorphism, and age of onset of breast cancer. *Cancer Res.* **60**, 6859–6863.

Shimizu, M., Matsumoto, Y., Tatsuno, M., and Fukuoka, M. (2003). Glucuronidation of propofol and its analogs by human and rat liver microsomes. *Biol. Pharm. Bull.* **26**, 216–219.

Soars, M. G., Smith, D. J., Riley, R. J., and Burchell, B. (2001). Cloning and characterization of a canine UDP-glucuronosyltransferase. *Arch. Biochem. Biophys.* **391**, 218–224.

Thelin, A., Schedin, S., and Dallner, G. (1992). Half-life of ubiquinone-9 in rat tissues. *FEBS Lett.* **313**, 118–120.

Thomas, N. L., and Coughtrie, M. W. (2003). Sulfation of apomorphine by human sulfotransferases: Evidence of a major role for the polymorphic phenol sulfotransferase, SULT1A1. *Xenobiotica* **33**, 1139–1148.

Tsoi, C., Falany, C. N., Morgenstern, R., and Swedmark, S. (2001a). Identification of a new subfamily of sulphotransferases: Cloning and characterization of canine SULT1D1. *Biochem. J.* **356,** 891–897.

Tsoi, C., Falany, C. N., Morgenstern, R., and Swedmark, S. (2001b). Molecular cloning, expression, and characterization of a canine sulfotransferase that is a human ST1B2 ortholog. *Arch. Biochem. Biophys.* **390,** 87–92.

Tsoi, C., Morgenstern, R., and Swedmark, S. (2002). Canine sulfotransferase SULT1A1: Molecular cloning, expression, and characterization. *Arch. Biochem. Biophys.* **401,** 125–133.

Tsukino, H., Kuroda, Y., Nakao, H., Imai, H., Inatomi, H., Osada, Y., and Katoh, T. (2004). Cytochrome P450 (CYP) 1A2, sulfotransferase (SULT) 1A1, and N-acetyltransferase (NAT) 2 polymorphisms and susceptibility to urothelial cancer. *J. Cancer Res. Clin. Oncol.* **130,** 99–106.

Uchaipichat, V., Mackenzie, P. I., Guo, X. H., Gardner-Stephen, D., Galetin, A., Houston, J. B., and Miners, J. O. (2004). Human UDP-glucuronosyltransferases: Isoform selectivity and kinetics of 4-methylumbelliferone and 1-naphthol glucuronidation, effects of organic solvents, and inhibition by diclofenac and probenecid. *Drug Metab. Dispos.* **32,** 413–423.

Venkatachalam, K. V., Llanos, D. E., Karami, K. J., and Malinovskii, V. A. (2004). Isolation, partial purification, and characterization of a novel petromyzonol sulfotransferase from *Petromyzon marinus* (lamprey) larval liver. *J. Lipid Res.* **45,** 486–495.

Walker, J. E. (1992). The NADH:ubiquinone oxidoreductase (complex I) of respiratory chains. *Q. Rev. Biophys.* **25,** 253–324.

Walle, T., Vincent, T. S., and Walle, U. K. (2003). Evidence of covalent binding of the dietary flavonoid quercetin to DNA and protein in human intestinal and hepatic cells. *Biochem. Pharmacol.* **65,** 1603–1610.

Yamada, J., and Tomita, Y. (1996). Antimutagenic activity of caffeic acid and related compounds. *Biosci. Biotechnol. Biochem.* **60,** 328–329.

Yamazoe, Y., Nagata, K., Ozawa, S., and Kato, R. (1994). Structural similarity and diversity of sulfotransferases. *Chem. Biol. Interact.* **92,** 107–117.

Zhang, T., Haws, P., and Wu, Q. (2004). Multiple variable first exons: A mechanism for cell- and tissue-specific gene regulation. *Genome Res.* **14,** 79–89.

Zhang, Y., Turunen, M., and Appelkvist, E. L. (1996). Restricted uptake of dietary coenzyme Q is in contrast to the unrestricted uptake of alpha-tocopherol into rat organs and cells. *J. Nutr.* **126,** 2089–2097.

[21] Synthesis of Bile Acid Coenzyme A Thioesters in the Amino Acid Conjugation of Bile Acids

By Erin M. Shonsey, James Wheeler, Michelle Johnson, Dongning He, Charles N. Falany, Josie Falany, and Stephen Barnes

Abstract

Bile acids are converted to their glycine and taurine *N*-acyl amidates by enzymes in the liver in a two-step process. This conjugation reaction increases the aqueous solubility of bile acids, particularly in the acidic environment of the initial portion of the small intestine. In the first step, bile acid coenzyme A (CoA) thioesters are formed by a bile acid CoA ligase (BAL). This chapter describes the methods used to purify BAL from rat liver microsomes and to isolate and clone the cDNAs encoding BAL from a rat liver cDNA library, the expression of BAL, the assays used to measure its activities, and the chemical synthesis of bile acid CoA thioesters.

Introduction

Physiological Importance and History

The physiological functions of bile acids have an ancient history (Haslewood, 1967). Once animals with gastrointestinal tracts evolved, the xenobiotics that were transported to the interior of the animal in turn had to be removed by excretion. This largely occurred via the kidneys (by filtration or active secretion into the urine) or by uptake from the blood by the liver and excretion into bile. Coincident with biliary xenobiotic excretion is excretion of the physiological bile acids. Bile acids are extracted from the blood with remarkable efficiency and actively excreted into bile at concentrations 100 times higher than those present in the blood. The amphipathic nature of bile acids and their property of forming micelles substantially increase the solubility in bile of otherwise hydrophobic compounds.

The earliest "bile acids" were actually conjugates of bile alcohols with sulfate (Hagey, 1992; Haslewood, 1967). These conjugates are common in fish and amphibians such as frogs. Crocodilians display an important chemical evolutionary step. Instead of conjugation with sulfate, the terminal carbon atom of the bile alcohol side chain is oxidized to a carboxylic acid. Although such a C_{27} bile acid is water soluble as its sodium salt, the pH of

METHODS IN ENZYMOLOGY, VOL. 400
0076-6879/05 $35.00
DOI: 10.1016/S0076-6879(05)00021-2

the upper intestinal tract causes its conversion to the free acid and as a result this bile acid precipitates (Hofmann, 1989). Taurine (aminoethane-sulfonic acid) evolved as an alternative to conjugation with sulfate to increase the aqueous solubility. The taurine-conjugated bile acid conjugates remain ionized at pH 1–2. Amino acid–bile acid conjugation became and remains the primary mechanism for bile acid conjugation in all but two mammals, the coypu and manatee—they still form sulfate conjugates of bile alcohols (Kuroki et al., 1988).

Enzymology of Bile Acid Amino Acid Conjugation

Conjugation of bile acids with amino acids is a two-step reaction (Fig. 1). Bile acids are first converted to their coenzyme A (CoA) thioesters by bile acid CoA ligase (BAL). These intermediates then become substrates for a second enzyme bile acid, CoA:amino acid N-acyltransferase (BAT) (see Shonsey et al., 2005). Bile acid CoA esters are also substrates for other enzymes that oxidize the C_{22}-C_{27} side chain. This leads to the production of the familiar C_{24} bile acids in most mammals (Hagey, 1992).

Substrate Specificity in Bile Acid Amino Acid Conjugation

BAL exhibits some substrate specificity. C_{24} bile acids such as chenodeoxycholic acid, deoxycholic acid, and lithocholic acid are the principal substrates. Nor-bile acids with 23 carbon atoms are not substrates (Kirkpatrick et al., 1988). This is observed in vivo where although C_{23} bile acids appear in bile, they are present as glucuronides and not amino acid conjugates (Yeh et al., 1997). Purified rat liver BAL does not use nor-bile acids as substrates (Wheeler et al., 1997) nor does the expressed rBAL cDNA (Falany et al., 2002). There is debate as to whether the BAL activity that forms thioesters of C_{24} and C_{27} bile acids is the same protein (Mihalik et al., 2002).

Standards for Analysis of Bile Acid CoA Thioesters

Many unconjugated bile acids are available from suppliers such as Aldrich-Sigma Chemical Co. (St. Louis, MO), Steraloids Inc. (Newport, RI), and Indofine Chemical Co. (Hillsborough, NJ). However, there are no sources of bile acid CoA standards and they must be synthesized (see later). [24-^{14}C]- and [11,12-^{3}H]-labeled bile acids used to be available from Amersham Biosciences (Piscataway, NJ) and New England Nuclear Corp. (Boston, MA). An alternative approach is to carry out reductive deuteration or tritiation of 11,12-unsaturated or 22,23-unsaturated bile acids (Duane et al., 1996; Ng et al., 1977). Labeling in these positions is stable and isotope exchange during sample processing is minimal.

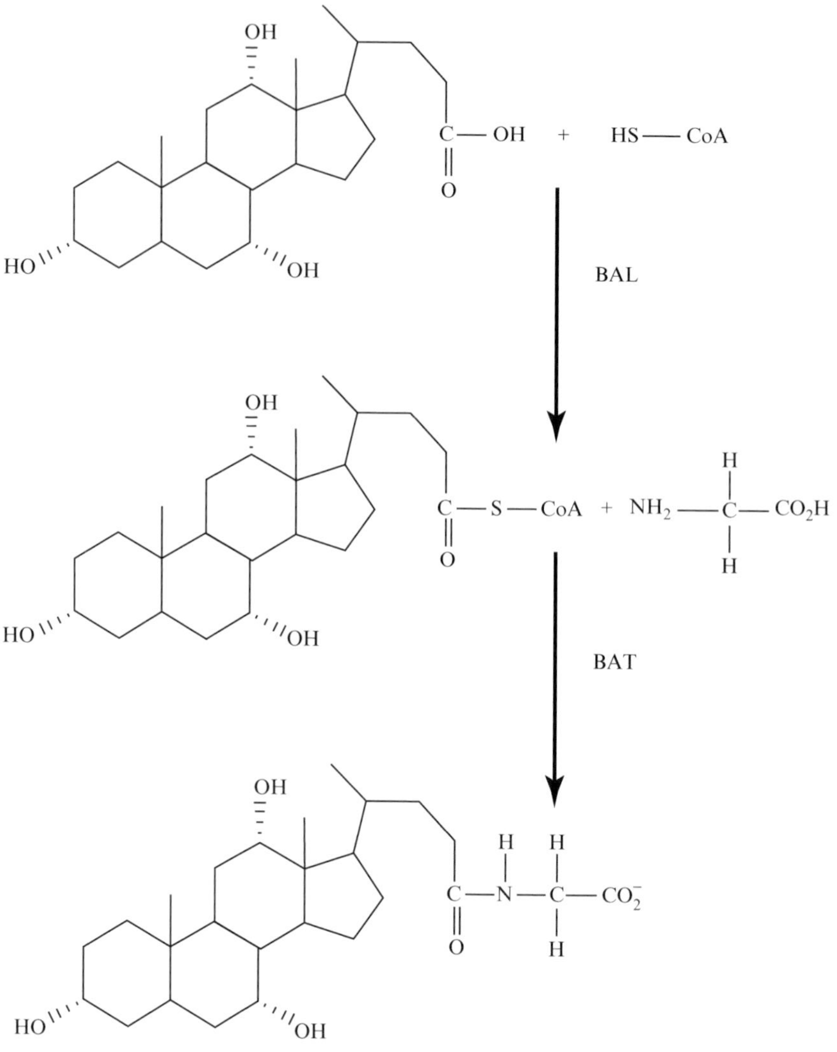

FIG. 1. The two-step reaction leading to the formation of bile acid N-acyl amidates.

Bile Acid CoA Thioester Synthesis (Johnson et al., 1989; Shah and Staple, 1968)

The synthesis of cholyl CoA is modified from an earlier procedure for the synthesis of palmitoyl CoA (Goldman and Vagelous, 1961). To synthesize cholyl CoA, α-collidine and 75 μmol (31 mg) cholic acid are added to 6 ml of dry methylene chloride. The reaction is left at room temperature for 10 min and 2 ml dry methylene chloride containing 0.012 mol ethyl chloroformate is added. The reaction is incubated at room temperature for 1 h, with occasional stirring. The solvent is then distilled in a rotary evaporator and the mixed anhydrides are dissolved in 5 ml of dry tetrahydrofuran. These compounds are then added slowly to a 12.5 mM solution of CoASH dissolved in water—the pH of 8 is maintained by the addition of 1 N NaOH. The solution sits for 15 min and is then acidified by the addition of 1% perchloric acid to a pH of 5. Tetrahydrofuran is removed by distillation in a rotary evaporator. The aqueous solution is acidified further by the addition of 1 ml of 10% perchloric acid and the precipitated cholyl CoA is collected by centrifugation at 15000g for 15 min. The precipitates are washed with 1 ml of diethyl ether. The precipitated cholyl CoA is dissolved in double distilled water and applied to an activated Sep-Pak C_{18} cartridge. The cartridge is washed with 3 ml of water, and the cholyl CoA is then eluted with 3 ml of 45% ethanol at pH 10. The Sep-Pak cartridge is washed again with 3 ml 100% ethanol containing a few drops of 0.880 ammonium hydroxide (to an apparent pH of 10). The cholyl CoA-containing eluent is dried under nitrogen and reconsitituted in water, and the extraction process is repeated (Johnson *et al.*, 1989).

It is crucial for the use of bile acid CoA in reactions with BAT that they do not have the unconjugated bile acid as a contaminant. The purity of bile acid CoAs is established by reversed-phase HPLC on a C_{18} column. The elution solvent consists of mixtures of buffer A–5% (v/v) isopropanol–10 mM ammonium acetate, pH 7, and buffer B–40% isopropanol (v/v)–6 mM ammonium acetate, pH 7. The gradient conditions are 0–8 min, 0–85% buffer B; 8–15 min, 85–100% buffer B; 15–30 min, 100% buffer B, and the flow rate is 1 ml/min. A peak at 11.8 min (trihydroxy-5β-cholestanoyl CoA) or 12 min (chenodeoxycholyl CoA) appears when purified active rBAL is added to the incubation mixture. However, the presence of unreacted, unconjugated bile acid is hard to detect because of its very low molar absorbance in the UV range. The preferred method to establish purity is the use of mass spectrometry. The bile acid CoA is dissolved in 50% aqueous acetronitrile and is infused into the electrospray ionization interface of a triple quadrupole mass spectrometer. In the positive ion mode, singly charged $[M + H]^+$ ions are observed. For cholyl CoA, the molecular ion is m/z 1159 (Fig. 2A). In the negative ion mode, singly $[M-H]^-$ and doubly charged $[M-2H]^-$ molecular ions are observed. For cholyl CoA, these are m/z 1157 and 578, respectively (Fig. 2B).

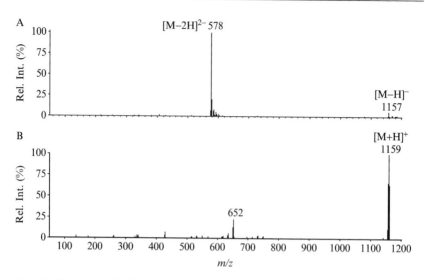

Fig. 2. Electrospray ionization mass spectra of cholyl CoA. This bile acid thioester was synthesized as described in the text. It was infused into the IonSpray interface of a Sciex API III triple quadrupole mass spectrometer. Spectra were collected in positive (A) and negative (B) modes.

Enzymatic Synthesis of Radiolabeled Bile Acid CoAs (Zhang et al., 1992)

BAL can be used to make [3]H- and ^{14}C-labeled bile acid CoA thioesters. They are reacted with BAL and unlabeled CoA to make radiolabeled bile acid CoA. It is important to have an excess of CoA and to allow the reaction to go to completion. Excess CoA must be removed using the Sep-Pak C_{18} cartridge as described earlier before the labeled bile acid CoA is incubated with BAL in kinetic experiments.

Bile Acid Coenzyme A Ligase

BAL activity was first identified and characterized in rat liver in the mid-1950s (Bremer, 1955). Since that time it has been characterized and isolated in a number of different species, including guinea pig (Vessey et al., 1987), pig (Vessey et al., 1987), and rat (Killenberg, 1978), and activity has even been observed in the rat kidney (Kwayke et al., 1993).

Microsome Preparations

In order to isolate BAL activity, microsomes are prepared from whole liver. Fresh livers are homogenized in ice-cold 10 mM Tris-HCl buffer, pH 8.4, containing 250 mM sucrose, 1 mM EGTA, and 1 mM dithiothreitol (DTT)(TDE-sucrose buffer). The homogenized livers are then centrifuged at 20,000g for 20 min at 4°. The supernatants are removed, combined, and centrifuged at 100,000g for 60 min at 4°. The microsomal pellet is resuspended in TDE-sucrose buffer, recentrifuged as a wash step, resolubilized in the TDE-sucrose buffer, and then stored in small aliquots at −80° until used (Wheeler *et al.*, 1997).

Activity Assay for BAL

Measurement of the activity of bile acid CoA ligase is based on an assay in which radiolabeled bile acid is converted to its more polar CoA thioester. The substrates and products are separated by partitioning them between ether and water under acidic conditions. The original method has been developed by Killenberg (1978) and makes use of a [^{14}C]bile acid to measure the amount of bile acid CoA produced.

A more recent assay has been developed as a modification to the original procedure. The reaction mixture of this assay consists of 5 mM ATP, 5 mM MgCl$_2$, 20 μM chenodeoxycholic acid, 50 μM CoA, 50 mM NaF, 100 mM Tris-HCl, pH 8.5, [11, 12-^{3}H] chenodeoxycholic acid (2 μCi; 25 Ci/mmol), and enzyme protein in a total volume of 0.1 ml. The reaction mixture is incubated at 37° for 2 min before adding 15 μl of the BAL enzyme preparation. This is then incubated for 20 min at 37°. In order to stop the reaction, 0.4 ml of 45% methanol-1.5% percholoric acid is added. Then extractions are performed twice with 3 ml of water-saturated diethyl ether, after which liquid scintillation counting is performed to determine the amount of [^{3}H]CDCA-CoA in the aqueous phase.

Isolation and Purification of BAL Activity

Three different procedures have been reported for the isolation and purification of BAL activity. Vessey *et al.* (1987) first described purification procedures for BAL activity from guinea pig and pig livers.

Wheeler *et al.* (1997) described the purification of bile acid CoA ligase to homogeneity from rat liver microsomes. First, rat liver microsomes are solubilized in Brij-58 to a final concentration of 2.5% (v/v). These are loaded onto a Q-Sepharose column equilibrated with TDE buffer. The column is washed with 2 column volumes of TDE buffer containing 0.1% Brij-58 (v/v). Fractions (8 ml) are collected and their absorbance monitored at 280 nm. Fractions with the greatest rBAL activity are pooled and loaded onto a hydroxyapatite column equilibrated with 20 mM potassium phosphate buffer, pH 6.8, containing 1 mM DTT and 0.02% (v/v) Brij-58

TABLE I
PURIFICATION OF RAT LIVER BAL

Fraction	Protein (mg)	Total act. (nmol min^{-1})	Specific act. (nmol min^{-1}/mg^{-1})	Yield (%)	Purification (fold)
Microsomes	176	341.4	1.94	100	1
Sol. microsomes	176	1010.2	5.74	295	2.96
Q-Sepharose pool	44	545.6	12.4	150	6.39
Hydroxyapatite	0.74	72.67	98.2	21	50.6
CM-Sepharose	0.055	21.18	385.0	6.2	198.4

(PD buffer). The column is washed with 10× volumes PD buffer, and rBAL activity is eluted with 80 mM NaCl in PD buffer. The eluent is diluted with 4 volumes of water and loaded onto a CM-Sepharose column equilibrated with 20 mM PD buffer containing 15 mM NaCl. (PD buffer contains Brij already) The column is washed with the equilibration buffer and eluted with 20 mM PD buffer containing 80 mM NaCl. This procedure showed a 200-fold purification (Wheeler et al., 1997) of rBAL (Table I).

The enzyme can be purified further from rat liver microsomes using the hydroxyapatite column procedure as the initial column after solubilization, followed by the use of an anti-rBAL affinity column prepared using a monoclonal antibody (anti-rBAL-3C1). Only one 65-kDa band detected by SDS–PAGE analysis (Fig. 3).

Cloning and Expression of rBAL

The rBAL cDNA was isolated in a two-step process using a combination of rBAL peptide-derived degenerate oligonucleotides and polymerase chain reaction (PCR) followed by screening of a λZAP II cDNA Sprague-Dawley rat liver library with the PCR product (Falany et al., 2002). In order to clone the rBAL cDNA, pure rBAL protein was digested proteolytically and the peptides were resolved by reversed-phase HPLC (Falany et al., 2002; Wheeler et al., 1997). This led to sequencing of two peptides and the synthesis of corresponding degenerate oligonucleotides. The oligonucleotides were used with oligo-dT(18) in RT-PCR reactions using rat liver mRNA as a template. These reactions resulted in the generation of a 423-bp PCR product that encoded the putative rBAL peptide sequences. In order to isolate the full-length rBAL cDNA, a λZAP II cDNA Sprague-Dawley rat liver library (Stratagene) was screened under high stringency conditions using the 423-bp DNA fragment of rBAL labeled with [^{32}P] dCTP as a probe (Wahl et al., 1987). A total of 300,000 plaques on ten

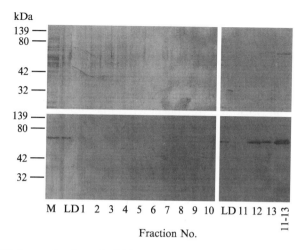

FIG. 3. Affinity purification of rBAL using an anti-rBAL-3C1-Sepharose-6B affinity column. Following the addition of solubilized microsomes (M) to the affinity column, no rBAL activity was eluted from the column at any point even with 1 M NaCl or acidic glycine buffer. Concentrating the protein in fractions 11–13 with a 10-kDa cutoff filter revealed a single band stained by Coomassie blue (A). Western blot analysis with the anti-rBAL-3C1 antibody demonstrates that the immunoreactive 65-kDa band elutes from the column only in a 50 mM glycine-HCl buffer, pH 2.5, wash (B).

150-mm-diameter plates were initially screened. Each plate was screened in duplicate and only plaques that were positive on both plates were selected for rescreening. Rescreening is carried out until several single pfu were isolated. The recombinant DNA sequences are recovered in pBluescript plasmids and sequenced. For rBAL, three cDNAs were initially sequenced and one of the clones (rBAL-1) consisted of a 2360-bp cDNA encoding full-length rBAL (Falany *et al.*, 2002). The other cDNA represented truncated cDNAs with identical nucleotide sequences to rBAL.

Recombinant Expression of rBAL

Characterization of rBAL activity has confirmed that it is a membrane-associated and lipid-dependent protein (Wheeler *et al.*, 1997). For efficient expression of rBAL, a baculoviral-dependent expression is used and active rBAL is expressed in Sf9 insect cells to allow for insertion of rBAL into the endoplasmic reticulum. Attempts to express full-length rBAL in *Escherichia coli* resulted in insertion of the protein into inclusion bodies. For expression in Sf9 cells, the full-length rBAL-1 cDNA is isolated from the pBluescript vector by *Eco*RI and *Spe*I digestion. The cDNA is ligated into the *Eco*RI and *Spe*I sites of the pFastBac1 vector to generate a native enzyme or the pFastBacHTb vector to generate the enzyme with a 6× His tag on the amino end to facilitate purification. Addition of the His tag to the amino-terminal end of the protein interferes with membrane insertion, and a substantial fraction of the protein will be present in the cytosolic fraction of the insect cell (although this is of some advantage for biochemical enzyme assays). Both of the baculoviral plasmids are transformed into competent *E. coli* DH5α cells (Falany *et al.*, 2002). The plasmids are purified from the DH5α cells and are then transformed into DH10Bac cells for transposition of the rBAL sequence into the bacmid. Recombinant bacmid DNA is isolated and transfected into Sf-900 cells. The cells are incubated at 27° for 72 h, and the baculovirus-containing supernatant fraction is removed and centrifuged at 500*g* for 5 min to remove cells and debris. To prepare rBAL-containing microsomes, Sf9 cells in midlog phase in liquid culture are infected with the recombinant virus and incubated at 27° for 72 h. The cells are then harvested by centrifugation at 500*g* for 5 min. The cells are resuspended in phosphate-buffered saline and a microsome preparation is performed. Microsomes prepared from Sf9 cells expressing native rBAL consistently show higher activity than microsomes from Sf9 cells expressing rBAL with the 6× His tag due to the amino-terminal His tag interfering with membrane insertion. However, the His-tagged rBAL in the microsome preparation can be purified further by Ni-NTA affinity chromatography. In contrast, control Sf9 cell microsomes have no rBAL activity (Falany *et al.*, 2002).

rBAL Amino Acid Sequence and Relationship to Other Enzymes

To date, only one BAL enzyme has had its amino acid sequence published (Table II) (Falany *et al.*, 2002). Interestingly, because the rBAL reading frame is translated into a protein of 690 amino acids and a molecular mass of 76.26 kDa, it is probable that rBAL isolated from rat liver microsomes (molecular mass of 66 kDa) undergoes cleavage. His tag expression data suggest that altering the N terminus prevents insertion into the endoplasmic reticulum.

Comparison of the rBAL sequence to other proteins shows a considerable similarity with sequences of six very long-chain acyl-CoA synthetases (VLACS) and two very long-chain acyl-CoA synthetase-related proteins. A mouse VLACS-related protein shares the highest homology with rBAL with 87.2% identity and 93.3% similarity. Other VLACS cloned from rat, mouse, and human share a 65.2–65.5% similarity and a 42.3–43.9% identity. Human VLACS-H2 or fatty acid transport protein 5 (FATP5) has been reported to have some cholyl CoA synthetase activity and to form CoA

TABLE II

AMINO ACID SEQUENCE OF RAT LIVER BAL

MGVWKKLTFL	LLSLLLLVGL	GQPLWPAATA	LALRWFLGDP	TCFVLLGLAF	LGRPWISSWI	60
PHWLSLAAAA	LTLSLLPPRP	PPELRWLHKD	VAFAFKLLFY	GLNLRRRLNR	HPPELFVDAL	120
EQQAQARPDQ	VALVCTGSEG	CSITNRELNA	KACQAAWALK	AKLKEATIQE	DKGATAILVL	180
PSKSISALSV	FLGLAKLGCP	VAWINPHSRG	MPLLHSVQSS	GASVLIVDPD	LQENLEEVLP	240
KLLAENIRCF	YLGHSSPTPG	VEALGAALDA	APSDPVPAKL	RANIKWKSPA	IFIYTSGTTG	300
LPKPAILSHE	RVIQMSNVLS	FCGRTADDVV	YNVLPLYHSM	GLVLGVLGCL	QLGATCVLAP	360
KFSASRYWAE	CRQYSVTVVL	YVGEVLRYLC	NVPGQPEDKK	HTVRFALGNG	LRADVWENFQ	420
QRFGPIQIWE	LYGSTEGNVG	LMNYVGHCGA	VGKTSCFIRM	LTPLELVQFD	IETAEPVRDK	480
QGFCIPVETG	KPGLLLTKIR	KNQPFLGYRG	SQDETKRKLV	ANVRQVGDLY	YNTGDVLALD	540
QEGFFYFRDR	LGDTFRWKGE	NVSTREVEGV	LSILDFLEEV	NVYGVTVPGC	EGKVGMAAVK	600
LAPGKTFDGQ	KLYQHVRSWL	PAYATPHFIR	IQDSLEITNT	YKLVKSQLAR	EGFDVGVIAD	660
PLYILDNKAE	TFRSLMPDVY	QAVCEGTWKL	690			

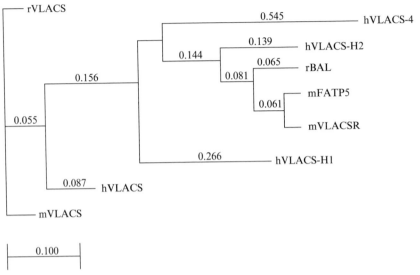

Fig. 4. Phylogenetic tree of rBAL and related proteins. The sequences of rBAL and several very long-chain acyl-CoA synthetase (VLACS) proteins were compared using the ClustalW alignment program and a phylogenetic tree generated using the McVector programs. The sequences and accession numbers of the proteins compared were rat BAL (AF242189), human VLACS-4 (AF030555), mouse VLACS-related protein (mVLACSR) (AJ223959), human VLACS-H2 (AF033031), human VLACS (D88308), human VLACS-H1 (AF064254), mouse VLACS (AJ223958), and rat VLACS (D85100).

esters of several bile acids (Mihalik *et al.*, 2002). This protein shares a high homology (71.6% identity, 85.5% similarity) with rBAL. These data support the inclusion of rBAL in the CoA synthetase/FATP gene family (Falany *et al.*, 2002) (Fig. 4).

Generation of Rabbit Anti-rBAL Polyclonal Antibody

A rabbit polyclonal antibody has been raised against a 463 amino acid fragment of rBAL expressed in *E. coli* to avoid problems associated with the insertion of full-length rBAL into inclusion bodies. The rBAL fragment is generated by subcloning the *Bam*HI–*Kpn*I restriction fragment of rBAL into pQE31 (Qiagen) to produce the peptide with an amino-terminal His tag. The rBAL463 fragment is expressed in *E. coli* M15 cells and purified by Ni-NTA agarose affinity chromatography. The protein is purified further by SDS–polyacrylamide gel electrophoresis and electroelution, and the identity of the protein is verified by peptide mass fingerprinting on a DE-Pro MALDI/TOF mass spectrometer (Applied Biosystems, Framingham, MA).

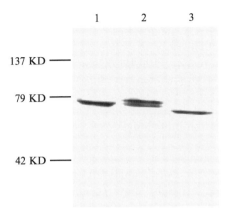

FIG. 5. Immunoblot analysis of rat, mouse, and human liver microsomes with a rabbit antirat BAL polyclonal antibody. Microsomes prepared from mouse, rat, and human livers were resolved in a 12.5% polyacrylamide gel and electrotransferred to a nitrocellulose membrane. Immunoblotting was carried out with a 1:1000 dilution of rabbit anti-rBAL antisera and 1:60,000 dilution of goat antirabbit IgG HRP conjugate (Falany *et al.*, 2002). Each lane contained 120 μg microsomal protein. Lane 1, rat microsomes; lane 2, mouse microsomes; and lane 3, human microsomes.

Pure rBAL463 protein (130 μg) is mixed with Freund's complete adjuvant and injected subcutaneously at several sites along the back of a female New Zealand white rabbit. The rabbit is subsequently boosted twice at 2-week intervals with 130 μg of rBAL463 in Freund's incomplete adjuvant. Two weeks later the animal is bled and serum is tested for antibody specificity by immunoblot analysis (Falany *et al.*, 2002). Figure 5 shows the reactivity of the rabbit anti-rBAL antibody with rat, mouse, and human liver microsomes. A single immunoreactive band is detected in rat and human microsomes, whereas two proteins are detected in mouse liver microsomes. The two bands in the mouse may represent two closely related proteins or a mixture of mBAL before and after cleavage of the signal sequence.

Identification of Products Generated by rBAL (Wheeler *et al.*, 1997)

A 30-μl aliquot of the unquenched rBAL assay reaction is applied directly to a 20 cm × 4.6 mm i.d. C_{18} reversed-phase C_{18} HPLC column with a mobile-phase flow rate of 1 ml/min. The elution solution is composed of buffer A [5% (v/v) isopropanol/10 mM ammonium acetate, pH 7] and buffer B [40% (v/v) isopropanol/6 mM ammonium acetate, pH 7]. A gradient is run under the conditions of 0–8 min, 0–85% buffer B; 8–15 min, 85–100% buffer B; 15–30 min, 100% buffer B, and the absorbance is monitored at 262 nm. A peak at 12.7 min appears after injection of either chemically synthesized CDC-CoA or the reaction mixture. The peaks are collected and analyzed by negative electrospray ionization mass spectrometry (ESI-MS) on a triple-quadrupole mass spectrometer, and spectra are collected over the range m/z 100 to 2000 and are recorded in both positive and negative ion modes (Falany et al., 2002). Chenodeoxycholyl CoA appears as singly positively charged m/z 1143 molecular ion $[M + H]^+$ and doubly negatively charged m/z 570 $[M-2H]^{2-}$ molecular ion. Corresponding ions for lithocholyl CoA and cholyl CoA would be m/z 1127 and m/z 562 and m/z 1159 and m/z 578, respectively. The ions for trihydroxy-5β-cholestanoyl CoA would be m/z 1201 and m/z 599. The bile acid CoA thioesters can also be analyzed by matrix-assisted laser desorption ionization–time of flight mass spectrometry. The bile acid CoA (1 pmol) is mixed with a saturated solution of α-cyano-4-hydroxycinnamic acid in 50% aqueous acetonitrile (1 μl) and spotted onto the MALDI target plate. Molecular ions are analyzed in positive and negative ion mode using a Voyager Elite mass spectrometer (Applied Biosystems, Inc., Foster City, CA). Using an acceleration voltage of 25 kV and a laser intensity of 2500 V, each spot is analyzed a minimum of three times, accumulating spectra composed of approximately 200 laser shots in total. The resulting spectra are analyzed by DataExplorer (Applied Biosystems, Inc). The instrument is calibrated using an external standard, and resulting spectra are Gaussian smoothed and baseline corrected.

Acknowledgments

Federal support for research on amino acid conjugation of bile acids has been in the form of grants-in-aid from the National Institute for Diabetes, Digestive and Kidney Disease (R01 DK46390 to SB). Funds for the mass spectrometers used in research at UAB were provided by a Shared Instrumentation Award from the National Center for Research Resources S10 RR06487 and RR11329).

References

Bremer, J. (1955). A method for the estimation of the taurine content and its conjugation with cholic acid in rat liver. Acta Chem. Scand. **9**, 683–688.

Duane, W. C., Schteingart, C. D., Ton-Nu, H. T., and Hofmann, A. F. (1996). Validation of [22,23-^{3}H]cholic acid as a stable tracer through conversion to deoxycholic acid in human subjects. J. Lipid Res. **37**, 431–436.

Falany, C. N., Xie, X., Wheeler, J. B., Wang, J., Smith, M., He, D., and Barnes, S. (2002). Molecular cloning and expression of rat liver bile acid CoA ligase. J. Lipid Res. **43**, 2062–2071.

Goldman, P., and Vagelous, P. R. (1961). The specificity of triglyceride synthesis from diglycerides in chicken adipose tissue. J. Biol. Chem. **236**, 2620–2623.

Hagey, L. R. (1992). "Bile Acid Biodiversity in Vertebrates and Its Chemistry and Evolutionary Implications." Ph.D. Thesis, University of California, San Diego.

Haslewood, G. A. D. (1967). Evolution of bile salts. *J. Lipid Res.* **8,** 535–550.

Hofmann, A. F. (1989). *In* "Handbook of Physiology, Section on Gastrointestinal System" (S. G. Schultz, ed.), pp. 549–566. American Physiological Society, Bethesda, MD.

Johnson, M. R., Barnes, S., and Diasio, R. B. (1989). Radioassay of bile acid coenzyme A:glycine/taurine:N-acyltransferase using an n-butanol solvent extraction procedure. *Anal. Biochem.* **182,** 360–365.

Killenberg, P. G. (1978). Measurement and subcellular distribution of choloyl-CoA synthetase and bile acid-CoA:amino acid N-acyltransferase activities in rat liver. *J. Lipid Res.* **19,** 24–31.

Kirkpatrick, R. B., Green, M. D., Hagey, L. R., Hofmann, A. F., and Tephly, T. R. (1988). Effect of side chain length on bile acid conjugation: Glucuronidation, sulfation and coenzyme A formation of nor-bile acids and their natural C_{24} homologs by human and rat liver fractions. *Hepatology* **8,** 353–357.

Kuroki, S., Schteingart, C. D., Hagey, L. R., Cohen, B. I., Mosbach, E. H., Rossi, S. S., Hofmann, A. F., Matoba, N., Une, M., and Hoshita, T. (1988). Bile salts of the West Indian manatee, *Trichechus manatus latirostris*: Novel bile alcohol sulfates and absence of bile acids. *J. Lipid Res.* **29,** 509–522.

Kwayke, J. B., Barnes, S., and Diasio, R. B. (1993). Identification of bile acid coenzyme A synthetase in rat kidney. *J. Lipid Res.* **34,** 95–99.

Mihalik, S. J., Steinberg, S. J., Pei, Z., Park, J., Kim do, G., Heinzer, A. K., Dacremont, G., Wanders, R. J., Cuebas, D. A., Smith, K. D., and Watkins, P. A. (2002). Participation of two members of the very long-chain acyl-CoA synthetase family in bile acid synthesis and recycling. *J. Biol. Chem.* **277,** 24771–24779.

Ng, P. Y., Allan, R. N., and Hofmann, A. F. (1977). Suitability of $[11,12\text{-}^3H_2]$chenodeoxy-cholic acid and $[11, 12\text{-}^3H_2]$lithocholic acid for isotope dilution studies of bile acid metabolism in man. *J. Lipid Res.* **18,** 753–758.

Shah, P. P., and Staple, E. (1968). Synthesis of coenzyme A esters of some bile acids. *Steroids* **12,** 571–576.

Shonsey, E. M., Sfakianos, M., Johnson, M., He, D., Falany, C. N., Falany, J., Merkler, D., and Barnes, S. (2005). Bile acid coenzyme A:amino acid *N*-acyltransferase in the amino acid conjugation of bile acids. *Methods Enzymol* **400**(22), this volume.

Vessey, D. A., Benfatto, A. M., and Kempner, E. S. (1987). Bile acid:CoASH ligases from guinea pig and porcine liver microsomes: Purification and characterization. *J. Biol. Chem.* **262,** 5360–5365.

Wahl, G. M., Berger, S. L., and Kimmel, A. R. (1987). Molecular hybridization of immobilized nucleic acids: Theoretical concepts and practical considerations. *Methods Enzymol.* **152,** 399–407.

Wheeler, J. B., Shaw, D. R., and Barnes, S. (1997). Purification and characterization of a rat liver bile acid coenzyme A ligase from rat liver microsomes. *Arch. Biochem. Biophys.* **348,** 15–24.

Yeh, H. Z., Schteingart, C. D., Hagey, L. R., Ton-Nu, H. T., Bolder, U., Gavrilkina, M. A., Steinbach, J. H., and Hofmann, A. F. (1997). Effect of side chain length on biotransformation, hepatic transport, and choleretic properties of chenodeoxycholyl homologues in the rodent: Studies with dinorchenodeoxycholic acid, norchenodeoxycholic acid, and chenodeoxycholic acid. *Hepatology* **26,** 374–385.

Zhang, R., Barnes, S., and Diasio, R. B. (1992). Differential intestinal deconjugation of taurine and glycine bile acid N-acyl amidates in rats. *Am. J. Physiol.* **262,** G351–G358.

[22] Bile Acid Coenzyme A: Amino Acid N-Acyltransferase in the Amino Acid Conjugation of Bile Acids

By Erin M. Shonsey, Mindan Sfakianos, Michelle Johnson, Dongning He, Charles N. Falany, Josie Falany, David J. Merkler, and Stephen Barnes

Abstract

Bile acids are converted to their glycine and taurine *N*-acyl amidates by enzymes in the liver in a two-step process. This increases their aqueous solubility, particularly in the acidic environment of the upper part of the small intestine. Bile acid coenzyme A (CoA) thioesters synthesized by bile acid CoA ligase (see Shonsey *et al.*, 2005) are substrates of bile acid CoA: amino acid *N*-acyltransferases (BAT) in the formation of bile acid *N*-acyl amidates. This chapter describes the methods used to purify BAT from human liver, to isolate and clone cDNAs encoding BAT from human, mouse, and rat liver cDNA libraries, the expression of BAT, the assays used to measure BAT activity, and the chemical syntheses of bile acid *N*-acylamidates. In addition, an enzyme that catalyzes further metabolism of glycine-conjugated bile acids is described.

Introduction

Conjugation of bile acids with amino acids is a two-step reaction (see Shonsey *et al.*, 2005). Bile acids are first converted to their coenzyme A (CoA) thioesters by bile acid CoA ligase (BAL) (see Shonsey *et al.*, 2005). These activated intermediates then become substrates for a second enzyme, bile acid CoA:amino acid *N*-acyltransferase (BAT). Bile acids are conjugated with taurine in many species. However, in recent evolutionary history (last 25 million years), conjugation with glycine has appeared in several parts of the mammalian kingdom (Hagey, 1992). Interestingly, most monkeys in the New World are exclusively taurine conjugators, whereas Old World primates, including human, conjugate with both amino acids. Some mammals (pigs and rabbits) are principally glycine conjugators. In human, glycine-conjugated bile acids in gallbladder and duodenal bile are usually in a three to four times greater abundance than taurine-conjugated bile acids (Vonk *et al.*, 1986). When bile acid turnover is accelerated by disease, glycine conjugation becomes even more dominant.

METHODS IN ENZYMOLOGY, VOL. 400
0076-6879/05 $35.00
DOI: 10.1016/S0076-6879(05)00022-4

In contrast, in cholestatic liver disease, taurine conjugate turnover is spared and taurine conjugates are found in greater proportions than normal (Williams, 1976).

Substrate Specificity in Bile Acid Amino Acid Conjugation

Bile acid amino acid conjugation *in vivo* is unusual in that it is restricted to the physiological amino acids glycine and taurine (Table I). This is a reflection of the substrate specificity of the BAT enzymes. In human and rat, glycine, but not L-alanine or β-alanine, is a substrate of BAT. However, a metabolite of the anticancer drug 5-fluorouracil, 2-fluoro-β-alanine (FBAL), is an excellent substrate for hBAT and rBAT (Falany *et al.*, 1994; He *et al.*, 2003; Johnson *et al.*, 1989, 1991). At first sight it would appear that this is due to the increased acidity of the carboxylic acid caused by the fluoro group on the 2 position, and thereby FBAL is structurally similar to taurine. However, this apparently does not apply to mouse liver BAT (Falany *et al.*, 1997). Although mBAT exclusively uses taurine instead of glycine to conjugate bile acid CoA esters, FBAL is not a substrate, nor is it an inhibitor of taurine conjugation. Two other members of the aminoalkylsulfonate family, in addition to taurine, are substrates of mBAT: aminomethanesulfonate (good substrate) and aminopropanesulfonate (poor substrate) (Table I) (Fig. 1). In summary, there are distinct structural restrictions in the amino acids that are substrates for BAT.

Relationships to Amino Acid Conjugation of Fatty Acids and Other Carboxylic Acids

Other organic acids that form N-acyl amidates predominantly undergo glycine conjugation (Caldwell *et al.*, 1980). The best known is hippuric acid, the glycine conjugate of benzoic acid. Conjugation with glutamine, ornithine, and taurine also occurs frequently. More rarely, conjugation with

TABLE I
AMINO ACID SUBSTRATES OF BAT

Formula	Chemical name	Trivial name	K_m (mM)
NH_2CH_2COOH	2-Aminoacetic acid	Glycine	5
$NH_2CH_2SO_3^-$	Aminomethanesulfonic acid	—	2
$NH_2CH_2CH_2SO_3^-$	2-Aminoethanesulfonic acid	Taurine	1
$NH_2CH_2CH_2CH_2SO_3^-$	3-Aminopropanesulfonic acid	—	>20
$NH_2CH_2CH_2COOH$	3-Aminopropanoic acid	β-Alanine	>200
$NH_2CH_2CHFCOOH$	3-Amino-2-fluoropropanoic acid	2-Fluoro-β-alanine	2

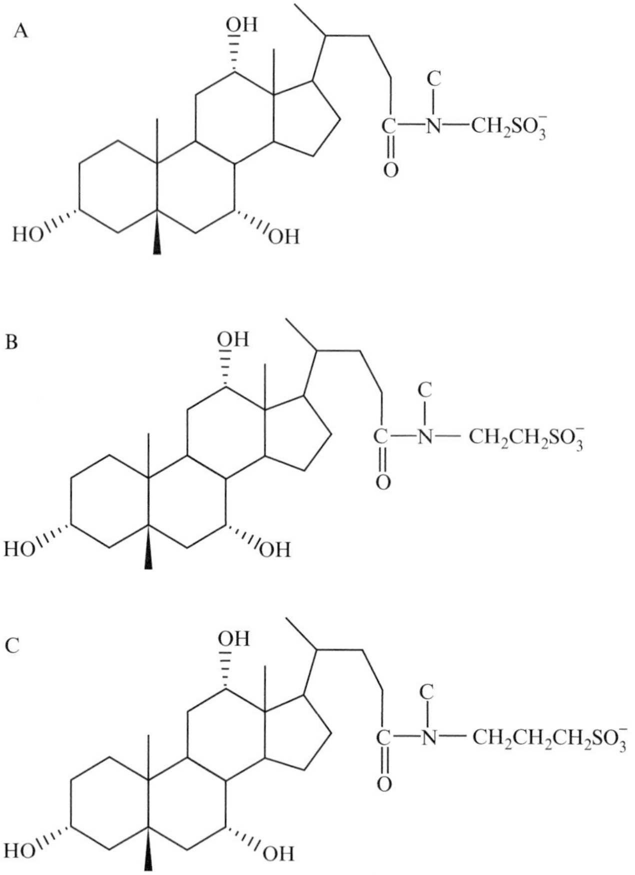

FIG. 1. Bile acid N-acyl amidates in the aminoalkane sulfonate series. A, aminomethane sulfonate; B, aminoethane sulfonate (taurine); C, aminopropane sulfonate.

alanine, serine, glutamic acid, or aspartic acid has been reported (van der Westhuizen et al., 2001). However, despite the apparent biochemical similarity, the N-acyltransferases that catalyze the formation of nonbile acid carboxylic acids are unrelated at the amino acid sequence level to bile acid CoA:amino acid N-acyltransferases.

Standards for Analysis of Bile Acid Amino Acid Conjugating Enzymes

Many glycine- and taurine-conjugated bile acids, as well as unconjugated bile acids, are available from suppliers such as Aldrich-Sigma Chemical Co. (St. Louis, MO), Steraloids Inc. (Newport, RI), and Indofine

Chemical Co. (Hillsborough, NJ). These may also be synthesized via both chemical and enzymatic approaches. The latter method is particularly useful for the preparation of radiolabeled bile acid amidates. 24-^{14}C-labeled and 11,12-^{3}H-labeled bile acids used to be available from Amersham Biosciences (Piscataway, NJ) and New England Nuclear Corp. (Boston, MA). An alternative approach is to carry out reductive deuteration or tritiation of 11,12-unsaturated or 22,23-unsaturated bile acids (Duane et al., 1996; Ng et al., 1977). [2-^{3}H] amino acids are available from Amersham Biosciences (Piscataway, NJ) and New England Nuclear Corp. (Boston, MA).

Chemical Synthesis of Bile Acid Amino Acid Conjugates

Taurine-conjugated bile acids are synthesized using a carbodiimide procedure (Tserng et al., 1977). Taurine (5.5 mmol) and 0.9 ml of triethylamine are added to a mixed solution of 5 mmol bile acid and 7 mmol N-ethoxycarbonyl-2-ethoxy-1,2-dihydroquinoline (EEDQ) in 10 ml of dimethylformamide. This solution is heated at 90° until the solution turns clear. It is stirred at 90° for 15 min and allowed to cool to room temperature, where it remains with stirring for 30 min. It is then poured slowly into 100 ml of stirred, cold diethyl ether on ice. The precipitated taurine conjugated bile acid is recovered by filtration.

Preparation of glycine-conjugated bile acids is similar to that of taurine-conjugated bile acids. The carboxylic acid of glycine is protected as its ethyl ester to prevent polymerization. Ethyl glycinate hydrochloride (7 mmol) is added to ethyl acetate (70 ml) containing triethylamine (1 ml) and is stirred at 25° for 30 min. EEDQ (7 mmol) and bile acid (5 mmol) are added, and the mixture is stirred at 25° for 10 min. The suspension is refluxed overnight on a steam bath and then cooled to room temperature. It is extracted with ethyl acetate and the ethyl acetate phase is washed sequentially with 50 ml of water, 0.5 M NaOH, water, 0.5 M HCl (twice), and water (twice). The ethyl acetate phase is dried under nitrogen and resuspended in 10 ml of absolute ethanol. The ethanol mixture is placed in a boiling water bath, and 10 ml of 10% potassium carbonate is added slowly. After a 15-min incubation, the mixture is removed and evaporated to dryness. The residue is resuspended in 2 ml of water and subsequently extracted twice with 2 volumes of butanol. The butanol phases are pooled and evaporated to dryness.

Bile acids have also been conjugated to novel amino acids formed by the metabolism of drugs. Sweeny et al. (1987) described a procedure in which bile acids were conjugated with 2-fluoro-β-alanine (FBAL), a metabolite of the anticancer agent 5-fluorouracil. These conjugates are found in the bile of patients treated with this drug (Sweeny *et al.*, 1987). In this synthesis procedure, FBAL is first converted to its methyl ester by incubating 100 mg in 3 ml of 20:1 methanol/acetyl chloride overnight. The clear solution is evaporated to dryness under a stream of nitrogen, and the residue is dried further over phosphorous pentoxide. The 2-fluoro-β-alanine methyl ester hydrochloride is used in the synthesis of the cholic acid conjugate following the procedure described for the synthesis of glycine conjugates of bile acids (see earlier discussion).

Enzymatic Synthesis of Radiolabeled Bile Acid Amino Acid Conjugates

The BAL and BAT enzymes can be used to make ^{3}H- and ^{14}C-labeled bile acid amino acid conjugates (Zhang *et al.*, 1992). Most of the bile acids are available in 24-^{14}C-labeled forms. They are reacted with BAL and unlabeled CoA to make 24-^{14}C-bile acid CoA. It is important to have an excess of CoA and to allow the reaction to go to completion. The excess CoA must be removed before the 24-^{14}C-bile acid CoA is incubated with BAT and either [1-^{3}H]glycine or [1,2-^{3}H]taurine to make doubly labeled bile acid conjugates or with unlabeled glycine or taurine. Again, the reaction is allowed to go to completion. The radiolabeled bile acid amino acid conjugate is purified by reversed-phase HPLC.

Bile Acid CoA: Amino Acid N-Acyltransferases

BAT activities have been identified in numerous species, including fish (Vessey *et al.*, 1990), seabream (Goto *et al.*, 1993), dog (Czuba and Vessey, 1981a), bovine (Czuba and Vessey, 1980), domestic fowl (Czuba and Vessey, 1981b), rat (Killenberg and Jordan, 1978), mouse (Falany *et al.*, 1997), and human (Kimura *et al.*, 1983). BAT activity has been characterized not only in the livers of these animals, but activity has also been observed in the rat kidney (Kwayke *et al.*, 1991), although neither rBAT message nor protein has been detected in kidney. This suggests that kidney BAT activity may be due to another uncharacterized enzyme. To date, only three of the BAT enzymes have been sequenced.

Assay for BAT Activity

A radioassay developed by Johnson *et al.* (1989) is recommended. The reaction mixture consists of potassium phosphate (10 mM), bile acid CoA (115 nmol), and unlabeled amino acid (25 nmol) with the corresponding ^{3}H-labeled amino acid (0.025 μCi) to a total volume of 100 μl at pH 8.4. The reaction mixture is then incubated at 37° for various time points over

90 min with differing amounts of cytosolic protein (140–725 μg). The reaction is initiated by the addition of bile acid CoA and is terminated by the addition of 1 ml of 100 mM potassium phosphate containing 1% (w/v) SDS, pH 2.0 (saturated with butanol). Butanol extractions are then performed to separate the labeled bile acid amino acid conjugates from the labeled amino acids, which are left in the aqueous phase. The upper n-butanol layer containing the ^{3}H-labeled bile acid amino acid conjugates is transferred to a clean 10 × 75-mm test tube and backwashed with 1 ml of the same potassium phosphate buffer. The mixture is vortexed again and centrifuged to separate the phases. A 850-μl aliquot of the n-butanol layer is added to 5.4 ml of scintillation cocktail and counted in a liquid scintillation counter.

Killenberg (1978) used an assay for BAT activity based on the release of CoASH from bile acid CoA. This leads to a change in the absorbance curve of the CoASH moiety that can be followed at 230 nm. Another assay uses the reaction of 5,5′-dithio-bis-2-nitrobenzoic acid with the sulfhydryl group in CoASH. However, both methods will also measure thioesterase activity, which may be a property of BAT or other thioesterases. To study bile acid amidation and thioesterase activity simultaneously, it is necessary to use LC-MS assay methods (see later section).

BAT Isolation and Purification

Several procedures have been described for the purification of BAT from rat liver (Killenberg and Jordan, 1978), human liver (Johnson et al., 1991; Kimura et al., 1983), domestic fowl liver (Czuba and Vessey, 1981b), and bovine liver (Vessey, 1979).

The preferred procedure is the one used by Johnson et al. (1991) to purify human liver BAT. In this procedure, human liver is first minced into small cubes and homogenized in 4 volumes of sucrose (0.25 M) and is then centrifuged at 16,000g for 30 min. The supernatant is filtered through gauze and centrifuged at 100,000g for 60 min. The clear supernatant is dialyzed at 4° against 10 liter of 50 mM Tris-HCl buffer, pH 8.25, overnight. The dialyzed cytosol is loaded onto a DEAE-cellulose column (5 × 30 cm) preequilibrated with 50 mM Tris-HCl, pH 8.25. The column is then washed with 1000 ml of 50 mM Tris-HCl, pH 8.25, and BAT activity is eluted with a linear gradient of 0–200 mM NaCl in 50 mM Tris-HCl buffer, pH 8.25.

Active fractions are pooled, diluted with an equal volume of water, and loaded onto a chromatofocusing column (1.6 × 70 cm) packed with PB-94 preequilibrated with imidazole-HCl (25 mM), 2-mercaptoethanol (10 mM), and 5% (v/v) glycerol, pH 6.8. The column is washed with 5 column volumes of the equilibration buffer, and bound proteins are eluted by a

TABLE II
PURIFICATION OF BAT FROM HUMAN LIVER[a]

Fraction	Total protein (mg)	Total activity[b] (nmol/min)	Specific activity[b] (nmol/min/mg)	Recovery (%)
Supernatant	18,000.0	1200.0	0.17	100
DEAE-cellulose	1764.0	987.0	1.91	82
Chromatofocusing	52.0	271.0	20.24	22
GC-Sepharose + gel filtration	7.7	246.7	63.50	20

[a] Adapted from data in Johnson et al. (1991).
[b] Activity measured using taurine as the substrate.

pH gradient using polybuffer 74 diluted 1:8 (v/v) with distilled water (final pH adjusted to 5.0 with HCl). BAT elutes at pH 6.0.

Fractions with BAT activity from the chromatofocusing column are pooled and loaded onto a glycocholate AH-Sepharose affinity column preequilibrated with 50 mM potassium phosphate buffer, pH 8.4. This affinity column is washed with 10 column volumes of 50 mM potassium phosphate buffer containing NaCl (100 mM), pH 8.4. BAT activity elutes with 50 mM potassium phosphate buffer containing 250 mM NaCl and 5 mM glycocholate, pH 8.4. This method results in a 480-fold enrichment in enzyme activity (Johnson et al., 1991) (Table II). The hBAT preparation is a single 46-kDa band by SDS-PAGE analysis (Fig. 2). After concentration, it is subjected to gel filtration on a TSK-250 column (2.5 × 60 cm) equilibrated with 35 mM potassium phosphate buffer containing 10 mM NaCl, pH 8.1, to separate BAT from glycocholate. This is essential to properly assess BAT enzymatic activity.

Cloning and Expression of hBAT

The initial BAT cDNA isolated encoded human liver BAT (Falany et al., 1994). Sequencing and expression of the hBAT cDNA validated the identity of the purified BAT and the cDNA. The hBAT cDNA is isolated from a human liver λZap XR cDNA library by screening with a polyclonal rabbit antihuman BAT antibody to detect hBAT-β-galactosidase fusion proteins (Falany et al., 1994). The rabbit anti-hBAT polyclonal antibody is raised against purified hBAT and immunoaffinity purified with hBAT bound to Sepharose Cl-4B (Johnson et al., 1991). The human liver cDNA library is grown in Escherichia coli XL-1 Blue cells on 150-mm petri dishes at 42° to allow lysis of the bacteria until the plaques are barely visible. Nitrocellulose membranes saturated with 10 mM isopropyl-β-D-thiogalac-

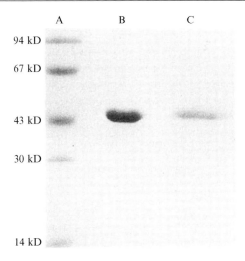

F<small>IG</small>. 2. SDS–polyacrylamide gel electrophoresis (12%) of purified bile acid-CoA: amino acid *N*-acyltransferase from human liver. Lane A contains molecular mass markers. Lane B contains 10 pg of the purified enzyme stained using Coomassie blue R-250. Lane C also contains 10 pg of purified enzyme; however, this lane was cut from the gel and visualized using a silver stain technique. kD, kilodaltons.

toside (IPTG) are placed onto each plate and incubated overnight at 37° for the lysogenic generation of β-galactosidase fusion proteins. The filters are removed from the plates, blocked in 3% bovine serum albumin for 1 h, and then incubated with the polyclonal rabbit anti-h BAT antibody followed by the goat antirabbit alkaline phosphatase-labeled second antibody. A total of 4×10^5 phage are screened using the immunodetection procedure, and phage plaques generating a positive response are purified by multiple cycles of dilution and immunodetection of hBAT reactive fusion protein. Following isolation of pure single plaque-forming units, cDNAs are recovered in Bluescript phagemids by coinfection with helper phage (R408). Two separate phage are ultimately isolated by immunoscreening, and sequence analysis indicates that one of the cDNAs possesses a full-length open-reading frame encoding hBAT.

The cloned hBAT sequence encodes the peptide sequences generated by partial sequencing of the purified hBAT protein. Expression of the hBAT cDNA in bacteria generates a protein that migrates with the same mass during SDS-PAGE as purified hBAT, reacts with the anti-hBAT polyclonal antibody, and is capable of amidating bile acid CoA esters.

cDNAs encoding mBAT and rBAT have been isolated from a mouse liver λZap XR cDNA library and a rat liver λZAP cDNA library using hBAT cDNA as a low stringency probe (Falany *et al.*, 1997; He *et al.*, 2003). The cDNA libraries are plated under lytic conditions until plaques are formed. The phages are lifted onto nitrocellulose membranes, denatured, and linked to membranes with UV light. The membranes are probed with the ^{32}P-labeled hBAT cDNA in 6× SSC buffer overnight at 55° (Wahl *et al.*, 1987). The membranes are washed under low stringency conditions, and positive clones are isolated by dilution and rescreening. BAT cDNAs are recovered in Bluescript phagemids using R408 helper phage, and the cDNAs are sequenced for comparison to peptides derived from the purified liver BAT proteins (Falany *et al.*, 1994). A comparison of the three BAT sequences is shown in Table III.

TABLE III
AMINO ACID SEQUENCES OF HUMAN, MOUSE, AND RAT BATs

```
rBAT  MAKLTAVPLS  ALVDEPVHIR  VTGLTPFQVV  CLQASLKDDK  GNLFNSQAFY   50
mBAT  MAKLTAVPLS  ALVDEPVHIQ  VTGLAPFQVV  CLQASLKDER  -KPVSSQAFY   49
hBAT  MIQLTATPVS  ALVDEPVHIR  ATGLIPFQMV  SFQASLEDEN  GDMFYSQAHY   50

rBAT  RASEVGEVDL  ERDSSLGGDY  MGVHPMGLFW  SMKPEKLLTR  LVKRDVMNRP  100
mBAT  RASEVGEVDL  EHDPSLGGDY  MGVHPMGLFW  SLKPEKLLGR  LIKRDVINSP   99
hBAT  RANEFGEVDL  NHASSLGGDY  MGVHPMGLFW  SLKPEKLLTR  LLKRDVMNRP  100

rBAT  HKVHIKLCHP  YFPVEGKVIS  SSLDSLILER  WYMAPGVTRI  HVKEGRIRGA  150
mBAT  YQIHIKACHP  YFPLQDLVVS  PPLDSLTLER  WYVAPGVKRI  QVKESRIRGA  149
hBAT  FQVQVKLYDL  ELIVNNKVAS  APKASLTLER  WYVAPGVTRI  KVREGRLRGA  150

rBAT  LFLPPGEGPF  PGVIDLFGGA  GGLFEFRASL  LASHGFATLA  LAYWGYDDLP  200
mBAT  LFLPPGEGPF  PGVIDLFGGA  GGLMEFRASL  LASRGFATLA  LAYWNYDDLP  199
hBAT  LFLPPGEGLF  PGVIDLFGGL  GGLLEFRASL  LASRGFASLA  LAYHNYEDLP  200

rBAT  SRLEKVDLEY  FEEGVEFLLR  HPKVLGPGVG  ILSVCIGAEI  GLSMAINLKQ  250
mBAT  SRLEKVDLEY  FEEGVEFLLR  HPKVLGPGVG  ILSVCIGAEI  GLSMAINLKQ  249
hBAT  RKPEVTDLEY  FEEAANFLLR  HPKVFGSGVG  VVSVCQGVQI  GLSMAIYLKQ  250

rBAT  ITATVLINGP  NFVSSNPHVY  RGKVFQPTPC  SEEFVTTNAL  GLVEFYRTFE  300
mBAT  IRATVLINGP  NFVSQSPHVY  HGQVYPPVPS  NEEFVVTNAL  GLVEFYRTFQ  299
hBAT  VTATVLINGT  NFPFGIPQVY  HGQIHQPLPH  SAQLISTNAL  GLLELYRTFE  300

rBAT  ETADKDSKYC  FPIEKAHGHF  LFVVGEDDKN  LNSKVHAKQA  IAQLMKSGKK  350
mBAT  ETADKDSKYC  FPIEKAHGHF  LFVVGEDDKN  LNSKVHANQA  IAQLMKNGKK  349
hBAT  TTQVGASQYL  FPIEEAQGQF  LFIVGEGDKT  INSKAHAEQA  IGQLKRHGKN  350

rBAT  NWTLLSYPGA  GHLIEPPYSP  LCSASRMPFV  IPSINWGGEV  IPH-AAAQEH  399
mBAT  NWTLLSYPGA  GHLIEPPYTP  LCQASRMPIL  IPSLSWGGEV  IPHSQAAQEH  399
hBAT  NWTLLSYPGA  GHLIEPPYSP  LCCASTTHDL  R--LHWGGEV  IPH-AAAQEH  397

rBAT  SWKEIQKFLK  QHLNPGFNSQL  420
mBAT  SWKEIQKFLK  QHLLPDLSSQL  420
hBAT  AWKEIQRFLR  KHLIPDVTSQL  418
```

Expression of BAT

Characterization of BAT proteins is facilitated greatly by the expression of active enzymes (Falany *et al.*, 1994). Analysis of the native form of hBAT utilizes expression of the enzyme in *E. coli* with the pKK233-2 bacterial expression vector. The hBAT cDNA is inserted into the *Nco*I–*Hind*III sites of pKK233-2 in two steps. A 3′-1261-bp *Nco*I–*Hind*III fragment is isolated from pBluescript-hBAT-8 (Falany *et al.*, 1994) and inserted directly into the pKK233-2 plasmid. A *Nco*I site incorporating the initial methionine of the hBAT cDNA is synthesized using polymerase chain reaction (PCR) and inserted into the pKK233-2 *Nco*I site. The correct orientation of the fragment is verified by restriction mapping and sequence analysis. hBAT is expressed in *E. coli* XL-1 Blue cells and induced with IPTG. Analysis of BAT activity in bacterial cytosol is complicated by the presence of esterase and hydrolase activity. Therefore, BAT activity is purified by DEAE anion-exchange chromatography to generate an active fraction with excellent kinetic characteristics (Falany *et al.*, 1994). Rat BAT has been expressed and purified from *E. coli* utilizing the incorporation of a histidine tag. However, expression of BAT protein in bacteria does exhibit difficulty in expressing the full-length enzyme with enzymatic activity due to proteolytic degradation. This has been observed with several prokaryotic expression vectors. The levels of recovery can be low if expression and purification are not carried out expeditiously.

In order to increase the reliability and levels of recovery of active BAT, expression in Sf9 insect cells has been utilized. BAT can be expressed in the insect cells with or without tag sequences to aid in purification. Rat BAT has been expressed efficiently in Sf9 cells with a His tag and purified by Ni-NTA affinity chromatography (He *et al.*, 2003). The pFastBacHTb baculovirus expression system (Life Technologies) is utilized to generate enzymatically active rBAT in Sf9 insect cells. Primers are used to amplify the open reading frame of rBAT by PCR incorporating a 5′- *Eco*R I site and a 3′-*Hind*III site to facilitate subcloning into the *Eco*R I–*Hind*III sites of the pFastBacHTb vector. This results in the generation of rBAT with an amino-terminal 6-His tag that can be cleaved by TEV protease. DH10Bac cells are then transformed with the pFastBacHTb-rBAT plasmid for transposition of the rBAT DNA into the bacmid to form recombinant bacmids. Cells with recombinant rBAT bacmids are selected to isolate the large (>23 kb) bacmid DNA with a modified miniprep technique (Life Technologies).

To generate recombinant baculovirus particles containing rBAT, Sf9 cells in midlog phase are seeded in six-well plates in 2 ml/well Sf-900 II serum-free medium (SFM) containing penicillin (50 units/ml) and streptomycin (50 mg/ml) and are then allowed to attach at 28° for 1.5 h.

For transfection, bacmid DNA and CellFECTIN reagent are each diluted in 100 ml Sf-900 II SFM without antibiotics, mixed gently, and incubated for 45 min at room temperature. The Sf9 cells are washed once with 2 ml Sf-900 II SFM without antibiotics and overlaid with the bacmid/CellFEC-TIN solution diluted with 200 ml Sf-900 II SFM. The cells are incubated at 28° for 5 h. The diluted bacmid/CellFECTIN complexes are removed and replaced with fresh Sf-900 II SFM containing antibiotics. After incubation at 28° for 72 h, the baculoviruses remain in the supernate after removal of cells by centrifugation at 500g for 5 min.

To obtain cytosolic rBAT, Sf9 cells in midlog phase in suspension culture are infected with the recombinant baculovirus and incubated in a 135-rpm orbital shaker at 28° for 48 h. Infected cells are harvested by centrifugation at 500g for 5 min. Cell pellets are resuspended in 50 mM Tris-HCl, pH 8.5, containing 5 mM 2-mercaptoethanol, 100 mM KCl, 1 mM phenylmethylsulfonyl fluoride, and 1% Nonidet P-40 and are lysed by brief sonication. The cytosolic fraction is obtained by centrifugation at 100,000g for 45 min. Ni-NTA affinity chromatography is performed to purify the His-tagged rBAT protein from the cytosol. The 6-His tags are subsequently cleaved by incubation with TEV protease at 4° for 6 h. Purified rBAT is recovered from the cleavage reaction in the flow-through fraction of a second Ni-NTA affinity column.

Expression and Recovery of Biotinylatable BAT

hBAT has also been expressed in a modified pET21a+ vector (pETGag-biotinHis plasmid) (Sfakianos et al., 2002). The procedure has been optimized for a better yield of purified hBAT. A biotinylation tag has been added so that the expressed protein could be biotinylated in vivo in an expression system that also expresses biotin ligase. The transformed cells are grown in 400 ml Luria broth containing both ampicillin (100 μg/ml) and chloramphenicol (100 μg/ml) at 37° with shaking. When the cells have grown to an A_{600} of ~0.6, 0.4 mM IPTG is added as well as 250 μM biotin in dimethyl sulfoxide. The temperature is decreased to 30° and incubation is continued for 4 h. These cells are then pelleted by centrifugation at 3000g at room temperature for 20 min. The pellets are resuspended in Bugbuster with the addition of Benzonase. The resuspension is incubated on a shaking platform for 20 min at room temperature and is then centrifuged at 16,000g at 4° for 40 min. The supernatant is then dialyzed against 5 L of double distilled water for 2 h at room temperature. The water is replaced, and dialysis continued at room temperature for a further 2 h. The water is replaced one more time, and the lysate is dialyzed overnight at 4°. The lysate is then assayed for BAT activity.

In order to purify expressed BAT, it is necessary to dissociate the 56-kDa bacterial GroEL protein that is responsible for the folding of this mammalian protein in a prokaryotic cell. As determined by electrospray ionization of the partially purified protein and deconvolution of the multiply charged ions, it copurifies (Fig. 3) with biotinylated hBAT when using the streptavidin affinity column (see later). To do this an ATP regeneration system composed of 80 mM creatine phosphate, creatine phosphokinase (500 μg/ml), and 5 mM ATP is mixed with 25 ml of cell lysate and is then loaded onto a Ni-NTA affinity column. The Ni-NTA column is equilibrated with 5 bed volumes of 50 mM Tris-HCl-5 mM imidazole, pH 8.5. The flowthrough from the Ni-NTA column is then loaded onto a Softlink avidin column, which is preequilibrated with 8 bed volumes of Bugbuster protein extraction reagent. The column is washed with 20 bed volumes of 50 mM Tris-HCl, pH 8.5, and the protein is eluted with the Tris-HCl solution, pH 8.5, containing 10 mM biotin. This procedure yields a single band of hBAT (Fig. 4).

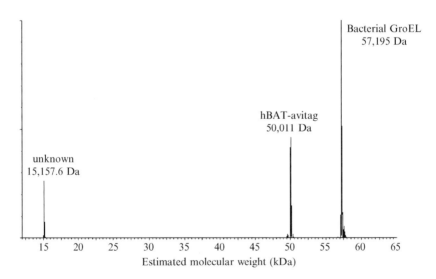

Fig. 3. Deconvoluted mass spectrum of biotinylated hBAT expressed in *E. coli.* Biotinylated hBAT was affinity absorbed to a soft avidin column and eluted with 10 mM biotin. The eluate was desalted and infused into the electrospray ionization interface of a Waters-Micromass Qtof2 mass spectrometer. The electrospray ionization mass spectrum was deconvoluted with MaxEnt software provided by the manufacturer. The molecular mass of hBAT (50,011 Da) includes the biotinylatable C-terminal tail. The peak at 57,195 Da is bacterial GroEL protein. The peak at 15,158 Da was not identified.

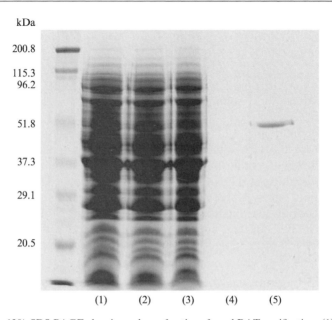

FIG. 4. 12% SDS-PAGE showing column fractions from hBAT purification. (1) Cytosol; (2) Nickel column flow through; (3) Avidin column flow through; (4) Avidin column wash; (5) Avidin column elution showing single hBAT band.

Relationship to Other Enzymes

To date, only three of the BAT enzymes have had their amino acid sequences published. The amino acid sequence of mBAT is 86% identical to that of rBAT. The amino acid sequence of hBAT is 69 and 70% identical to mBAT and rBAT, respectively. It is curious to note that both hBAT and rBAT conjugate glycine and taurine, but mBAT can only use taurine as a substrate. BAT enzymes also share high homology (>40%) with peroxisomal, mitochondrial, and cytosolic long chain fatty acid acylCoA thioesterases by amino acid sequence alignment (Fig. 5). They also share a homology with dienelactone hydrolases and other α/β hydrolases. The hydrolases have a conserved catalytic triad, which maps onto mammalian BAT at Cys-238, Asp-328, and His-362, a putative catalytic triad of hBAT (Sfakianos et al., 2002). The catalytic triad of the thioesterases contains a serine as opposed to a cysteine, and site-directed mutagensis approaches have shown that mutating the serine to a cysteine will confer acyltransferase activity (Witkowski et al., 1994). Mutating the Cys-238 in hBAT to a serine converted the enzyme from an N-acyltransferase to a thioesterase

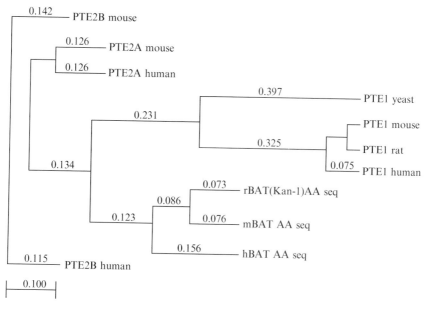

FIG. 5. Phylogenetic tree of BATs and related proteins. The sequences of hBAT, mBAT, and rBAT and several cytosolic and peroxisomal long-chain acyl-CoA thioesterase proteins were compared using the ClustalW alignment program and a phylogenetic tree generated using the McVector programs. The sequences and accession numbers of the proteins compared were human BAT (L34081), mouse BAT (U95215), rat BAT (D43964), PTE1 rat (Q8VHK0), PTE1 mouse (P58137), PTE1 human (O14734), PTE1 yeast (P41903), PTE2 human (P49753), PTE2B human (Q8N9L9), PTE2A mouse (Q9QYR7), and PTE2B mouse (Q8BWN8).

for glycine conjugation, but allowed both *N*-acyltransferase and thioesterase for reactions with taurine (Sfakianos *et al.*, 2002). These data suggest a close mechanistic relationship between thioesterase and *N*-acyltransferases.

Metabolism of Glycine-Conjugated Bile Acids

The peptidylglycine α-amidating monooxygenase (PAM) converts peptides with a C-terminal glycine group into their amides. Although well described for activation of neuropeptides, the activity of PAM is not restricted to peptides. Both glycine conjugates of fatty acids and bile acids are excellent substrates *in vitro*, being converted to fatty acid and bile acid amides. Fatty acid amides such as oleamide are endogenous sleep-inducing agents in sleep-deprived mammals (Cravatt *et al.*, 1995). This class of compounds has other important properties: linoleamide increases Ca^{2+}

flux in renal tubular cells (Huang and Jan, 2001), erucamide is angiogenic (Wakamatsu et al., 1990), and elaidamide is a potent inhibitor of epoxide hydrolase (Morisseau et al., 2001).

Assay of PAM Activity

Reactions at $37.0 \pm 0.1°$ are initiated by the addition of PAM (10–50 μg) into 2.4 ml of 100 mM MES/NaOH, pH 6.0, 30 mM NaCl, 1% (v/v) ethanol, 0.001% (v/v) Triton X-100, 1.0 μM Cu(NO$_3$)$_2$, 5.0 mM sodium ascorbate, and the bile acid glycine conjugate (generally 0.2 $K_{m,app}$ to 3.0 $K_{m,app}$). Initial rates are measured by following the PAM-dependent consumption of O$_2$ using a Yellow Springs Instrument Model 53 oxygen monitor. $V_{max,app}$ values are normalized to controls performed at 11.0 mM N-acetylglycine. Ethanol is added to protect the catalase against ascorbate-mediated inactivation, and Triton X-100 is included to prevent nonspecific absorption of the enzyme to the sides of the oxygen monitor chambers (King et al., 2000).

Use of Mass Spectrometry to Study Formation and Metabolism of Bile Acid N-Acyl Amidates

Mass spectrometry has been used in the study of bile acid-amino acid conjugating enzymes activities as well as the products of their catalysis. Procedures for this are described later. Bile acids are remarkably stable when analyzed using electrospray ionization—in fact, taurocholate has been used by mass spectrometry (MS) manufacturers as a calibrant for negative electrospray ionization (ESI)-MS.

Analysis of Bile Acid Amino Acid Conjugates in Rat and Mouse Bile

Biles are recovered under anesthesia from the gallbladder of mice (using a syringe with a 21-gauge needle) and the common bile duct of a rat (by inserting PE10 tubing). The biles are diluted with 1 ml of water and passed over an activated C$_{18}$ Sep-Pak cartridge. Bile acids are retained on the column and, after washing with 2 × 1 ml water, are eluted with 2 ml of methanol. Aliquots (5–20 μl) of the eluate are injected onto a 10 cm × 2.1 mm i.d. C$_8$ reversed-phase column preequilibrated with 30% acetonitrile in 0.1% acetic acid. The bile acid amino acid conjugates are separated using a linear gradient of 30–100% acetonitrile in 0.1% acetic acid over a 10-min period at a flow rate of 0.2 ml/min; the column eluate is split 1:1, with 0.1 ml/min being passed into the ESI of a triple quadrupole mass spectrometer operating in the negative ion mode, with an orifice potential of −60V. Spectra are recorded over 1.5-s intervals from *m/z* 300 to 800. Analysis of the mass spectra reveals that only taurine-conjugated bile acids are present in mouse bile (Falany *et al.*, 1997), whereas both taurine and glycine conjugates are present in rat bile (He *et al.*, 2003). The *m/z* values for [M-H]⁻ molecular ions of mono-, di-, and trihydroxy bile acids (lithocholate, chenodeoxycholate and deoxycholate, and cholate) are 375, 391, and 407, respectively. The molecular ions of the corresponding glycine conjugates are at *m/z* 432, 448, and 464, respectively, whereas those for the taurine conjugates are at *m/z* 482, 498, and 514, respectively.

Quantitative Assay for *N*-Acyltransferase and Thioesterase Activities

A liquid chromatography–electrospray ionization–mass spectrometry–multiple reaction monitoring (LC-ESI-MS-MRM) assay method may be used to simultaneously detect *N*-acyltransferase and thioesterase activities (Sfakianos *et al.*, 2002). Cholyl-CoA (50 μM) and glycine or taurine (2.5 mM) are incubated with 1 μg of wild-type or mutant BAT in a 10 mM potassium phosphate buffer, pH 8.25, at 37° for 5 min in a total reaction volume of 100 μl. The products of the reaction are extracted with 100 μl of *n*-butanol by vortexing for 1 min. The *n*-butanol extract is applied to a 20 cm × 4.6 mm i.d. C_{18} reversed-phase C_{18} HPLC column. Buffer A (10 mM ammonium acetate) and buffer B (10 mM ammonium acetate in 90% acetonitrile) are used in the elution with a linear gradient of buffer B (0–100%) applied to the column. The product is introduced into the electrospray ionization interface in the negative ion mode. Parent ion/daughter ion combinations for cholate (*m/z* 407/343), glycocholate (*m/z* 464/74), and taurocholate (*m/z* 514/124) (Fig. 6) are monitored to detect the formation of cholate, glycocholate, and taurocholate. A linear peak area–response curve is generated with standards containing known concentrations of cholate, glycocholate, and taurocholate. The estimated amounts of reaction products, cholate, glycocholate, and taurocholate, are determined by comparison to the standard response curve (Sfakianos *et al.*, 2002). The appearance of glycocholate and taurocholate is a measure of BAT activity, whereas the formation of cholate is a result of bile acid CoA thioesterase activity (Fig. 7).

LC-MS Analysis of the Metabolism of Glycine-Conjugated Bile Acids by PAM

The reaction of glycine-conjugated bile acids with PAM can be analyzed using LC-ESI-MS (King *et al.*, 2000). The products of this reaction are cholanamides. In the case of cholylglycine, it is converted to cholamide

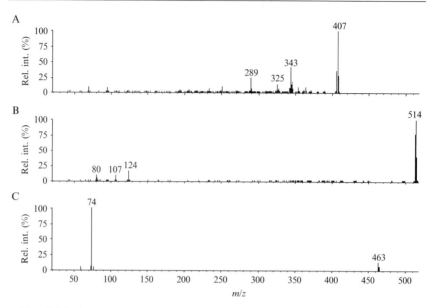

FIG. 6. Tandem mass spectra of bile acids. Cholic acid (A), taurocholic acid (B), and glycocholic acid (C) were infused into the electrospray ionization interface of a triple quadrupole mass spectrometer operating in the negative mode. Their $[M\text{-}H]^-$ molecular ions (m/z 407, 514, and 464, respectively) were subjected to collision-induced dissociation with argon gas. The resulting daughter ion spectra were recorded. Cholic acid underwent fragmentation by dehydration, whereas for the amino acid conjugates, the negative charge lay in the amino acid moiety. Therefore, the principal daughter ions were either glycine (m/z 74) or taurine (m/z 124).

(Fig. 8). Aliquots of reaction mixtures with PAM are analyzed by reversed-phase HPLC using a 2.1 mm × 15 cm i.d. Brownlee Aquapore C_8 5-μm column. The mobile phase is a linear 30–100% gradient of acetonitrile in 0.1% (v/v) aqueous formic acid over 10 min at a flow rate of 0.2 ml/min. The eluate is split 1:1 with 0.1 ml/min passed into the electrospray ionization interface of a triple quadrupole mass spectrometer. For negative ion spectra, the electrospray needle voltage and the orifice potential are set at −4900 and −60 to −70 V, respectively. Mass spectra are collected over a m/z range of 300–600. The reaction of glycocholate with PAM produces an ion of m/z 466, the $[M + 60]^-$ acetate adduct of the molecular ion of cholanamide, as well as m/z 480, the intermediate N-cholyl-α-hydroxygly-cine (King et al., 2000). Each of the glycine-conjugated bile acids reacts

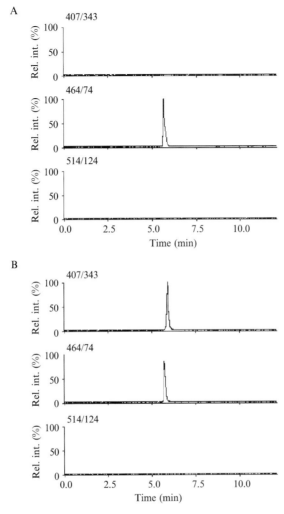

Fig. 7. LC-ESI-MRM-MS analyses of BAT and bile acid thioesterase activity. Reaction mixtures of wt-hBAT (A) and C235S mutant hBAT (B) incubated with cholyl CoA and glycine were analyzed by reversed-phase HPLC. The column eluate was split and a portion was introduced into the ESI interface of a triple quadrupole mass spectrometer. Cholic acid, glycocholic acid, and taurocholic acid were specifically measured using the parent ion/daughter ion combinations (*m/z* 407/343, 464/75, and 514/124, respectively). In A, glycocholic acid is essentially the only product, whereas in B, both glycocholic acid and cholic acid are formed, the latter due to thioesterase activity. Each channel is normalized to a single value; for A it was 40,000, whereas for B it was 5500.

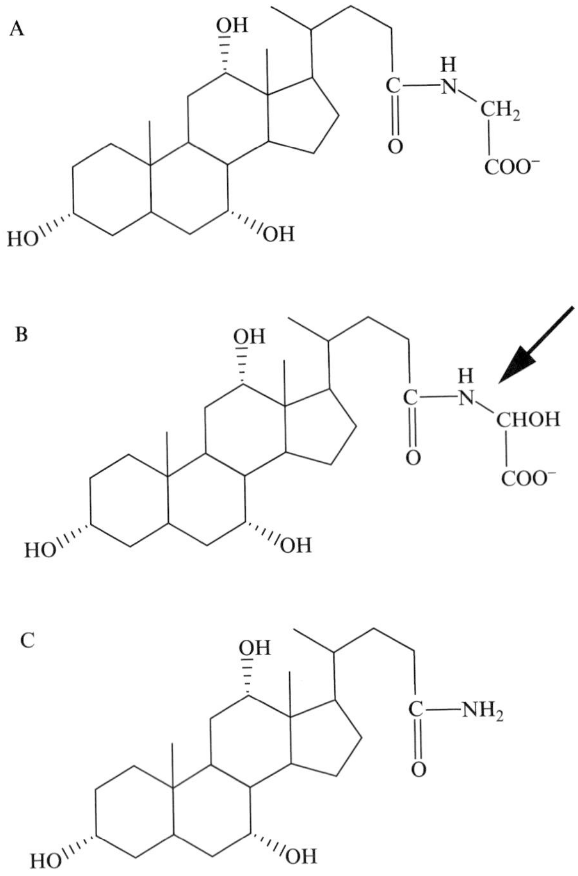

FIG. 8. The reaction of PAM with glycocholate (A). There is an intermediate (B) where the α-carbon atom of glycine is oxidized. This is further cleaved by PAM to produce cholanamide (C).

with PAM to produce expected products; in contrast, there is no reaction with taurine-conjugated bile acids.

Acknowledgments

Federal support for research on amino acid conjugation of bile acids has been in the form of grants-in-aid from the National Institute for Diabetes, Digestive, and Kidney Disease (R01 DK46390 to SB), the Shirley W. and William L. Griffin Foundation, and the NIH

(R15-GM067257) to DJM. Funds for the mass spectrometers used in research at UAB were provided by a Shared Instrumentation Award from the National Center for Research Resources (S10, RR06487, RR11329, and RR13795).

References

Caldwell, J., Idle, J. R., and Smith, R. L. (1980). The amino acid conjugations. *In* "Extrahepatic Metabolism of Drugs and Other Foreign Compounds" (T. E. Gram, ed.), pp. 453–493. SP Medical & Scientific Books, New York.

Cravatt, B. F., Prospero-Garcia, O., Siuzdak, G., Gilula, N. B., Henriksen, S. J., Boger, D. L., and Lerner, R. A. (1995). Chemical characterization of a family of brain lipids that induce sleep. *Science* **268**, 1506–1509.

Czuba, B., and Vessey, D. A. (1980). Kinetic characterization of cholyl-CoA glycine-taurine N-acyltransferase from bovine liver. *J. Biol. Chem.* **255**, 5296–5299.

Czuba, B., and Vessey, D. A. (1981a). Identification of a unique mammalian species of cholyl-CoA:amino acid N-acyltransferase. *Biochim. Biophys. Acta* **665**, 612–614.

Czuba, B., and Vessey, D. A. (1981b). Purification and characterization of cholyl-CoA:taurine N-acyltransferase from the liver of domestic fowl (*Gallus gallus*). *Biochem. J.* **195**, 263–266.

Duane, W. C., Schteingart, C. D., Ton-Nu, H. T., and Hofmann, A. F. (1996). Validation of [22,23-³H]cholic acid as a stable tracer through conversion to deoxycholic acid in human subjects. *J. Lipid Res.* **37**, 431–436.

Falany, C. N., Fortinberry, H., Leiter, E. H., and Barnes, S. (1997). Cloning, expression, and chromosomal localization of mouse liver bile acid CoA:amino acid N-acyltransferase. *J. Lipid Res.* **38**, 1139–1148.

Falany, C. N., Johnson, M. R., Barnes, S., and Diasio, R. B. (1994). Glycine and taurine conjugation of bile acids by a single enzyme: Molecular cloning and expression of human liver bile acid CoA: amino acid N-acyltransferase. *J. Biol. Chem.* **269**, 19375–19379.

Goto, T., Une, M., Kihira, K., Kuramoto, T., and Hoshita, T. (1993). Enzymatic formation of D-cysteinolic acid conjugated chenodeoxycholic acid in liver preparation from red seabream, *Pagrosomus major*. *Biol. Pharm. Bull.* **16**, 1216–1219.

Hagey, L. R. (1992). "Bile Acid Biodiversity in Vertebrates and Its Chemistry and Evolutionary Implications." Ph.D. Thesis, University of California, San Diego.

He, D., Barnes, S., and Falany, C. N. (2003). Rat liver bile acid CoA:amino acid N-acyltransferase: Expression, characterization, and peroxisomal localization. *J. Lipid Res.* **44**, 2242–2249.

Huang, J.-K., and Jan, C.-R. (2001). Linoleamide, a brain lipid that induces sleep, increases cytosolic Ca²⁺ levels in MDCK renal tubular cells. *Life Sci.* **68**, 997–1004.

Johnson, M. R., Barnes, S., and Diasio, R. B. (1989). Radioassay of bile acid coenzyme A: glycine/taurine:N-acyltransferase using an n-butanol solvent extraction procedure. *Anal. Biochem.* **182**, 360–365.

Johnson, M. R., Barnes, S., Kwakye, J. B., and Diasio, R. B. (1991). Purification of human liver cholyl CoA:amino acid N-acyltransferase. *J. Biol. Chem.* **262**, 10227–10233.

Killenberg, P. G. (1978). Measurement and subcellular distribution of choloyl-CoA synthetase and bile acid-CoA:amino acid N-acyltransferase activities in rat liver. *J. Lipid Res.* **19**, 24–31.

Killenberg, P. G., and Jordan, J. T. (1978). Purification and characterization of bile acid-CoA: amino acid N-acyltransferase from rat liver. *J. Biol. Chem.* **253**, 1005–1010.

Kimura, M., Okuno, E., Inada, J., Ohyama, H., and Kido, R. (1983). Purification and characterization of amino-acid N-choloyltransferase from human liver. *Hoppe Seylers Z. Physiol. Chem.* **364,** 637–645.

King, L., 3rd., Barnes, S., Glufke, U., Henz, M. E., Kirk, M., Merkler, K. A., Vederas, J. C., Wilcox, B. J., and Merkler, D. J. (2000). The enzymatic formation of novel bile acid primary amides. *Arch. Biochem. Biophys.* **374,** 107–117.

Kwayke, J. B., Johnson, M. R., Barnes, S., Grizzle, W. D., and Diasio, R. B. (1991). Identification of bile acid-CoA: amino acid N-acyltransferase in rat kidney. *Biochem. J.* **280,** 821–824.

Ng, P. Y., Allan, R. N., and Hofmann, A. F. (1977). Suitability of [11,12-^{3}H$_2$]chenodeoxycholic acid and [11,12-^{3}H$_2$]lithocholic acid for isotope dilution studies of bile acid metabolism in man. *J. Lipid Res.* **18,** 753–758.

Sfakianos, M. K., Wilson, L., Sakalian, M., Falany, C. N., and Barnes, S. (2002). Conserved residues in the putative catalytic triad of human bile acid coenzyme A:amino acid N-acyltransferase. *J. Biol. Chem.* **227,** 47270–47275.

Shonsey, E. M., Wheeler, J., Johnson, M., He, D., Falany, C. N., Falany, J., and Barnes, S. (2005). Synthesis of bile acid coenzyme A thioesters in the amino acid conjugation of bile acids. *Methods Enzymol.* **400,** 360–373.

Sweeny, D. J., Barnes, S., Heggie, G. D., and Diasio, R. B. (1987). Metabolies of 5-fluorouracil to an N-cholyl-2-fluror-β-alanine conjugate: Previously unrecognized role for bile acids in drug conjugation. *Proc. Natl. Acad. Sci. USA* **84,** 5439–5443.

Tserng, K., Hachey, D. L., and Klein, P. D. (1977). An improved procedure for the synthesis of glycine and taurine conjugates of bile acids. *J. Lipid Res.* **18,** 404–407.

van der Westhuizen, F. H., Pretorius, P. J., and Erasmus, E. (2001). The utilization of alanine, glutamic acid, and serine as amino acid substrates for glycine N-acyltransferase. *J. Biochem. Mol. Toxicol.* **14,** 102–109.

Vessey, D. A. (1979). The co-purification and common identity of cholyl CoA:glycine- and cholyl CoA:taurine-N-acyltransferase activities from bovine liver. *J. Biol. Chem.* **254,** 2059–2063.

Vessey, D. A., Benfatto, A. M., Zerweck, E., and Vestweber, C. (1990). Purification and characterization of the enzymes of bile acid conjugation from fish liver. *Comp. Biochem. Physiol. B* **95,** 647–652.

Vonk, R. J., Kneepkens, C. M., Havinga, R., Kuipers, F., and Bijleveld, C. M. (1986). Enterohepatic circulation in man: A simple method for the determination of duodenal bile acids. *J. Lipid Res.* **27,** 901–904.

Wahl, G. M., Berger, S. L., and Kimmel, A. R. (1987). Molecular hybridization of immobilized nucleic acids: Theoretical concepts and practical considerations. *Methods Enzymol.* **152,** 399–407.

Wakamatsu, K., Masaki, T., Itoh, F., Knodo, K., and Sudo, K. (1990). Isolation of fatty acid amide as an angiogenic principle from bovine necessary. *Biochem. Biophy. Res. Commun.* **168,** 423–429.

Williams, C. N. (1976). Bile-acid metabolism and the liver. *Clin. Biochem.* **9,** 149–152.

Witkowski, A., Witkowska, H. E., and Smith, S. (1994). Reengineering the specificity of a serine active-site enzyme: Two active-site mutations convert a hydrolase to a transferase. *J. Biol. Chem.* **269,** 379–383.

Zhang, R., Barnes, S., and Diasio, R. B. (1992). Differential intestinal deconjugation of taurine and glycine bile acid N-acyl amidates in rats. *Am. J. Physiol.* **262,** G351–G358.

[23] Multidrug Resistance Protein 1-Mediated Export of Glutathione and Glutathione Disulfide from Brain Astrocytes

By JOHANNES HIRRLINGER and RALF DRINGEN

Abstract

Many cell types are known to release glutathione (GSH) and glutathione disulfide (GSSG). Multidrug resistance proteins (Mrps) have been identified to be involved in these export processes. In the brain, astrocytes have key functions in GSH metabolism and in antioxidative defense. These cells release large amounts of GSH under unstressed conditions as well as GSSG during oxidative stress. This chapter describes experimental paradigms to analyze the release of the physiological Mrp substrates GSH and GSSG from cultured astrocytes. These assay systems can be used to screen for compounds that affect Mrp1-mediated export from astrocytes and therefore could interfere with the antioxidative defense system of the brain. In addition, our methods could be useful in investigating mechanisms of export of GSH and GSSG from other cell types.

Introduction

The tripeptide glutathione (GSH) is an important intracellular antioxidant that is oxidized to glutathione disulfide (GSSG) during the detoxification of peroxides and radicals. Within cells GSH is regenerated from GSSG in the reaction that is catalyzed by glutathione reductase. The cellular GSH content can be lowered by the formation of glutathione-S-conjugates as well as by release of GSH or GSSG (Dringen and Hirrlinger, 2003). Export of GSH and GSSG has been reported for various cell types from several organs (Akerboom and Sies, 1990; Dringen and Hirrlinger, 2003; Kaplowitz et al., 1996; Sies et al., 1972). Cellular release of GSH occurs under unstressed conditions and has important functions in providing extracellular GSH as an antioxidant, a substrate for the ectoenzyme γ-glutamyl transpeptidase (γGT), and as a cysteine source (Dringen and Hirrlinger, 2003; Ishikawa and Sies, 1989). In contrast, GSSG export from cells indicates the presence of oxidative stress and has been proposed as a mechanism of cellular self-defense (Akerboom and Sies, 1990; Akerboom et al., 1982; Keppler, 1999; Sies and Akerboom, 1984; Sies et al., 1972).

METHODS IN ENZYMOLOGY, VOL. 400
0076-6879/05 $35.00
DOI: 10.1016/S0076-6879(05)00023-6

The multidrug resistance proteins (Mrps) 1 and 2 have been identified to participate in the cosubstrate-independent cellular export of GSSG and GSH (Ballatori and Rebbeor, 1998; Fernandez-Checa et al., 1992; Hirrlinger et al., 2001, 2002b; Leier et al., 1996; Paulusma et al., 1999; Rebbeor et al., 2002). Mrps are members of the subgroup ABCC of the superfamily of ATP-binding cassette (ABC) transporters (Borst and Oude Elferink, 2002; Schinkel and Jonker, 2003). The genes of nine Mrp family members have been identified in the human genome; however, only 8 of them exist as orthologs in the mouse genome (Kruh and Belinsky, 2003). Mrps are large proteins that contain 12 to 17 transmembrane-spanning helices, which are organized in two or three membrane domains. Mrps are ATP-driven export pumps that mediate the cellular export of organic anions (Kruh and Belinsky, 2003). Their function was first described in the resistance of tumor cells against chemotherapeutic drugs (Cole et al., 1992). In vivo Mrps fulfill several essential transport functions, depending on the expressing cell type and tissue. Classical Mrp substrates are glutathione-S-conjugates, GSSG, conjugates of glucuronate, and cyclic nucleotides, as well as nucleotide analogs (Homolya et al., 2003; König et al., 1999; Kruh and Belinsky, 2003).

Within the central nervous system astrocytes have important functions in antioxidative metabolism and detoxification (Cooper, 1997; Dringen, 2000). The release of GSH from astrocytes is important for GSH homeostasis in brain, as GSH export from astrocytes is the first step in the supply of precursors for the GSH synthesis in neurons (Dringen and Hirrlinger, 2003; Dringen et al., 1999b). Under unstressed conditions, cultured astrocytes release GSH (Fig. 1) (Dringen et al., 1997; Hirrlinger et al., 2002b; Sagara et al., 1996; Stewart et al., 2002; Stone et al., 1999; Yudkoff et al., 1990) and even protect the GSH exported against oxidation by a factor released into the medium (Stewart et al., 2002; Stone et al., 1999). In contrast to astrocytes, only marginal amounts of GSH are released from other types of brain cells (Hirrlinger et al., 2002b). GSSG is only released from astrocytes during oxidative stress (Fig. 1). After application of a sustained H_2O_2-induced oxidative stress, astrocytes show a rapid and prolonged increase in intracellular GSSG that allows one to investigate the export of GSSG from these cells (Hirrlinger et al., 2001).

The presence of Mrp1 protein has been demonstrated for cultured astrocytes from rat (Decleves et al., 2000; Hirrlinger et al., 2001; Mercier et al., 2003), human (Marroni et al., 2003), and mouse (Gennuso et al., 2004) brain. In addition, Mrp1-protein has been localized in astrocytes of rat brain (Mercier et al., 2004) and in reactive astrocytes in human brain (Aronica et al., 2004; Lazarowski et al., 2004; Marroni et al., 2003; Sisodiya et al., 2002), but not in nonreactive astrocytes of the human brain (Nies et al., 2004).

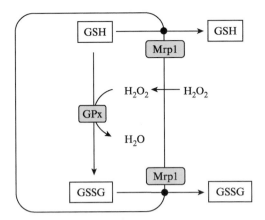

FIG. 1. Mrp1-mediated export of GSH and GSSG from astrocytes. Cultured rat astrocytes release about 10% of their cellular GSH per hour under unstressed conditions (Dringen and Hamprecht, 1997; Hirrlinger et al., 2002b). After application of H_2O_2 to cultured astrocytes, GSH is oxidized by glutathione peroxidases (GPx) to GSSG. During chronic H_2O_2 stress, 35% of the initial GSH content is exported from the cells as GSSG within 30 min (Hirrlinger et al., 2001).

Astrocytic export of GSH under unstressed conditions (Hirrlinger et al., 2002b) and of GSSG during oxidative stress (Hirrlinger et al., 2001) is strongly inhibited in the presence of cyclosporin A and MK571, known inhibitors of Mrp1 and Mrp2 (Büchler et al., 1996; Leier et al., 1994, 1996; Nies et al., 1998). Because mRNA for Mrp2 was not detected in rat brain, cultured rat astrocytes (Hirrlinger et al., 2001, 2002a), or individual astrocytes in mouse brain (unpublished data) nor was Mrp2-protein found in human brain (Nies et al., 2004), expression of Mrp2 and a contribution of this transporter in the export of GSH and GSSG from astrocytes can be excluded. Therefore, Mrp1 is the transporter that is most likely responsible for the MK571-sensitive export of GSSG and GSH from astrocytes.

Reagents, Solutions, and Cell Cultures

Reagents and Materials

Cyclosporin A and MK571 are from Alexis (Grünberg, Germany). NADH and NADPH are from Applichem (Darmstadt, Germany). Dulbecco's modified Eagle's medium is from Life Technologies (Eggenstein,

Germany). Catalase (EC 1.11.1.6; 260,000 U/ml), fetal calf serum, glutathione reductase from yeast (EC 1.6.4.2; 120 U/mg), GSH, GSSG, superoxide
dismutase (SOD; EC 1.15.1.1; 5000 U/mg), and xanthine oxidase (XO; EC
1.1.3.22; 1 U/mg) are from Roche Diagnostics (Mannheim, Germany).
Penicillin G, streptomycin sulfate, and Triton X-100 are from Serva
(Heidelberg, Germany). Acivicin, allopurinol, bovine serum albumin,
5,5'-dithio-bis(2-nitrobenzoic acid), hypoxanthine, sulfosalicylic acid, and
xylenol orange are from Sigma (Deisenhofen, Germany), and HEPES is
from Roth (Karlsruhe, Germany). All other chemicals are obtained at
analytical grade from E. Merck (Darmstadt, Germany). Sterile plastic
material, 24-well culture dishes for cell culture, and unsterile 96-well
microtiter plates are from Nunc (Wiesbaden, Germany) and Greiner
(Frickenhausen, Germany).

Incubation Buffers and Solutions

Solutions for Cell Incubations

Minimal medium (MM; 44 mM NaHCO$_3$, 110 mM NaCl, 1.8 mM
CaCl$_2$, 5.4 mM KCl, 0.8 mM MgSO$_4$, 0.92 mM NaH$_2$PO$_4$, 5 mM
glucose, adjusted with CO$_2$ to pH 7.4)
Incubation buffer (IB; 20 mM HEPES, 145 mM NaCl, 1.8 mM CaCl$_2$,
5.4 mM KCl, 1 mM MgCl$_2$, 0.8 mM Na$_2$HPO$_4$, 5 mM glucose, pH 7.4)
Phosphate-buffered saline (PBS; 10 mM potassium phosphate buffer,
150 mM NaCl, pH 7.4)

Solutions for GSx Quantification

1% (w/v) sulfosalicylic acid (SSA) in water
100 mM sodium phosphate buffer, 1 mM EDTA, pH 7.5

Cell Cultures

Astrocyte-rich primary cultures derived from the brains of neonatal
Wistar rats are prepared and maintained as described previously (Dringen
et al., 1999b; Hamprecht and Löffler, 1985). Three hundred thousand viable
cells are seeded per well of a 24-well dish in 2 ml medium consisting of 90%
Dulbecco's modified Eagle's medium, 10% fetal calf serum, 20 units/ml of
penicillin G, and 20 μg/ml of streptomycin sulfate and are cultivated in a
cell incubator (Hereaus, Hanau, Germany) containing a humidified atmosphere of 10% CO$_2$/90% air. The medium is renewed every seventh day.
Confluent cultures are used for experiments at a culture age between
14 and 23 days.

Determination Methods

Quantification of Total Glutathione (GSx) and GSSG in Cells and Media

GSx (amount of GSH plus twice the amount of GSSG) is measured in a microtiter plate assay as described previously (Dringen and Hamprecht, 1996; Dringen *et al.*, 1997, 1999b). The standards are carried through exactly the same procedure as the cell extracts. Ten microliters of the cell lysates or of GSSG standards (0–100 pmol GSx/10 μl in 1% SSA) is transferred into 90 μl of water prepared in wells of microtiter plates to determine the GSx content of cells, whereas 20 μl of mixtures of 1 volume of medium sample with 1 volume of 1% SSA or of mixtures of 1 volume of GSSG standards (in 1% SSA) with 1 volume of medium is transferred into 80 μl of water for determination of the GSx content of media. After addition to each well of 100 μl reaction mixture [100 mM sodium phosphate buffer pH 7.5, containing 1 mM EDTA, 0.3 mM 5,5'-dithio-bis(2-nitrobenzoic acid), 0.4 mM NADPH, and 1 U/ml glutathione reductase], the increase in extinction at 405 nm is recorded in 30-s intervals over 5 min using a microtiter plate reader (MRX TC Revelation, Dynex Technologies, Denkendorf, Germany). Glutathione contents are evaluated using a calibration curve established with standard samples using the software delivered with the plate reader.

Glutathione disulfide is measured after derivatization of GSH with 2-vinylpyridine (2VP; Griffith, 1980) as described previously (Dringen and Hamprecht, 1996). Briefly, 130 μl of the protein-free supernatant of cell lysates in 1% SSA or 130 μl of a mixture of 1 volume of medium sample with 1 volume of 1% SSA is mixed with 5 μl 2VP and adjusted with 0.2 M Tris to a pH between 5 and 7. Standard amounts of GSSG are treated identically. After 1 h incubation at room temperature, 10 μl of the 2VP-treated samples or standards is assayed for their GSx content as described earlier.

Quantification of Extracellular H_2O_2

Extracellular concentrations of H_2O_2 are determined as described previously (Dringen *et al.*, 1998). Briefly, 10 μl of the peroxide-containing incubation buffer is added directly to 170 μl 25 mM H_2SO_4 in a well of a microtiter plate. After collection of all samples, a 180-μl reaction mixture containing 0.5 mM $(NH_4)_2Fe(SO_4)_2$, 200 μM xylenol orange, and 200 mM sorbitol in 25 mM H_2SO_4 is added to each well of the microtiter plate. After a further 45 min of incubation at room temperature the extinction at 550 nm is determined using the microtiter plate reader and compared with the extinction read for standard H_2O_2 concentrations.

Test for Cell Viability and Cellular Protein

Cell viability is analyzed by determining the activity of lactate dehydrogenase (LDH) in the incubation medium and in the cells using the microtiter plate assay described previously in detail (Dringen *et al.*, 1998; Hirrlinger *et al.*, 2001). Cell lysates in the appropriate 1% Triton X-100-containing medium are used as control for maximal LDH release. The protein content per well of a 24-well dish is quantified after solubilization of the cells in NaOH (Hirrlinger *et al.*, 1999) according to the method described by Lowry *et al.* (1951) using bovine serum albumin as a standard.

Incubation of Cells to Test for GSH Export

Cultured astrocytes are washed with 2 ml of prewarmed (37°) MM and incubated in 1 ml MM containing 100 μM of the γGT inhibitor acivicin to prevent degradation of released GSH by γGT (Dringen *et al.*, 1997; Ruedig and Dringen, 2004). To quantify extracellular GSH it is essential to use a cystine-free medium to prevent the reaction of GSH with cystine to a mixed disulfide that is not detectable by the GSx assay used. For each time point and condition of incubation at least three wells with cells should be used. After the desired time of incubation, 100 μl of medium is collected and mixed with 100 μl 1% SSA on ice. Of this mixture, 130 μl is subjected to derivatization with 2VP as described earlier. For analysis of cellular GSx content, the remaining medium is aspirated, and the cells are washed with 2 ml ice-cold PBS and lysed with 500 μl ice-cold 1% SSA for 5 min on ice. Lysates are collected and centrifuged for 2 min at 14,000g in a microcentrifuge to remove protein precipitates. The cleared lysate is used for quantification of GSx or is subjected to derivatization with 2VP before determination of the GSSG content.

The experimental paradigm described earlier was used to analyze the export of GSH from cultured rat astrocytes (Fig. 2). Astrocytes release a substantial amount of GSH, which can be detected in the medium (Fig. 2, compare filled squares with filled circles). While almost no GSSG was detectable within the cells, up to 20% of the extracellular GSx was identified as GSSG after 6 h of incubation (Fig. 2). This GSSG was most likely generated by oxidation of released GSH. During an incubation for 6 h, the total amount of GSx in the cultures (sum of cells + medium) increased marginally (Fig. 2) probably due to synthesis of GSH by the cells.

If data on cellular GSx and GSSG contents, as well as extracellular GSSG-content, are not required, the experimental incubation can be simplified for studying only extracellular GSx accumulation by collecting 10-μl medium samples that are transferred directly into 90 μl of 0.11% (w/v) SSA

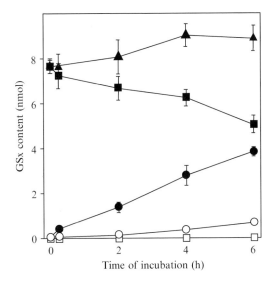

FIG. 2. Time course of intracellular (squares) and extracellular (circles) contents of GSx (filled symbols) and GSSG (open symbols) in cultured astrocytes. Filled triangles represent the sum of intracellular and extracellular GSx. Data of a representative experiment performed in triplicate on a 21-day-old astrocyte culture that contained $207 \pm 16\ \mu g$ protein per well are shown.

in wells of a microtiter plate. After collecting all the media samples, the microtiter plate can directly be used in the GSx assay described earlier. This modified approach has the advantages that (i) for each condition of incubation, only three wells of cells are required for triplicate measurement of all time points, (ii) extracellular GSx accumulation can be monitored for each well of cells separately, and (iii) the number of pipetting steps is reduced by directly collecting samples in the wells of a microtiter plate.

Linear regression of data obtained by the modified approach for incubation periods of up to 6 h allows calculation of GSH release rates for each individual well with cells. The specific release rate for GSH for cultured astrocytes from rat brain using the optimized GSH release test described earlier is 3.1 ± 0.6 nmol GSH/(h $\times$ mg protein) (Hirrlinger et al., 2002b). This value is almost identical to specific release rates of 2.1 and 3.2 nmol/(h $\times$ mg) that have been reported previously for cultured astrocytes using other experimental paradigms (Dringen et al., 1997; Sagara et al., 1996).

The GSH release test was used to screen for a number of compounds that might influence the GSH export rate from cultured astroglial cells (Hirrlinger et al., 2002b). The Mrp inhibitor MK571 (Hirrlinger et al., 2001; Leier et al., 1994) was identified as a potent modulator of GSH release from

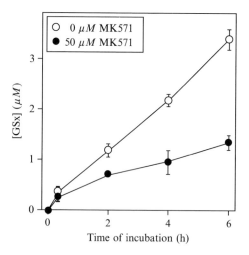

Fig. 3. Time course of accumulation of GSx in the incubation medium of cultured astrocytes during incubation in the absence (○) or the presence (●) of 50 μM MK571. Data of a representative experiment performed in triplicate on a 15-day-old astrocyte culture that contained 128 ± 6 μg protein per well are shown.

astrocytes (Hirrlinger *et al.*, 2002b). At a concentration of 50 μM, this compound slowed down the accumulation of extracellular GSx substantially (Fig. 3) and reduced the rate of GSH release from astrocytes by about 60% (Hirrlinger *et al.*, 2002b). Because astrocytes do not express Mrp2, these results suggest that the MK571-sensitive transporter Mrp1 is predominantly responsible for the observed GSH export from astrocytes. The observation that even in the presence of high concentrations of MK571, GSH export from astrocytes was only inhibited by 60% (Hirrlinger *et al.*, 2002b) indicates that other mechanisms are involved in the remaining 40% of GSH export from astrocytes. Candidate proteins that may contribute in the remaining MK571-insensitive GSH export from astrocytes could be other Mrps (Borst *et al.*, 1999; Lai and Tan, 2002; Rius *et al.*, 2003), organic anion transport proteins (Ballatori and Rebbeor, 1998; Li *et al.*, 1998, 2000), or the cystic fibrosis transmembrane conductance regulator (CFTR) (Kogan *et al.*, 2003).

Incubation of Cells with the H_2O_2-Generating System to Test for GSSG Export

Because untreated astrocytes contain only minute amounts of GSSG (Dringen and Hamprecht, 1996; Dringen *et al.*, 1999a), GSSG has to be generated in sufficient concentration within the cells in order to study

GSSG export from astrocytes. Due to the high capacity of astrocytes to clear extracellular H_2O_2 (Dringen et al., 1999a, 2005), bolus application of the peroxide to cultured astrocytes causes only a quick and transient increase in intracellular GSSG (Dringen and Hamprecht, 1997). Therefore, application of a H_2O_2-generating system such as the xanthine oxidase/hypoxanthine/SOD system (Hirrlinger et al., 1999) is necessary to produce a sufficiently high steady-state concentration of GSSG in cultured astrocytes to study export of the disulfide.

   Cultured astrocytes are washed with 2 ml of prewarmed (37°) incubation buffer and incubated at 37° in 0.5 ml IB containing XO (in the activity twice that needed during H_2O_2 production) and SOD (100 U). To initiate the production of H_2O_2, 0.5 ml of prewarmed (37°) 2 mM hypoxanthine (prepared by dilution of 1 volume of a 10 mM stock solution of hypoxanthine in 7.5 mM NaOH with 1 volume of double concentrated IB and 3 volumes of IB) is added per well. Superoxide radical anions generated by the XO-catalyzed oxidation of hypoxanthine are disproportionated by SOD to H_2O_2 and O_2. During incubation of the cells with the H_2O_2-generating system at 37° the extracellular concentration of H_2O_2 is monitored by analysis of 10-μl medium samples by the colorimetric method described earlier. After the desired time of incubation, 500 μl incubation medium is collected from the cells and mixed with 50 μl IB containing the xanthine oxidase inhibitor allopurinol (55 μM) and catalase (1100 U/ml) (Hirrlinger et al., 2001) to dispose of the H_2O_2 in the incubation medium before analysis of GSx content and LDH activity. Of this mixture, 200 μl is mixed with 200 μl 1% SSA and subjected to analysis of GSx and GSSG contents as described for GSH release. For analysis of cellular GSx and GSSG contents, the cells are washed with 2 ml PBS and lysed with 500 μl 1% (w/v) SSA, and 10-μl samples of the lysates are used in the GSx assay to quantify GSx and GSSG.

   Incubation of the cells with hypoxanthine and SOD in the absence of XO (open circles) did not generate substantial amounts of H_2O_2 in the medium (Fig. 4A) nor was GSSG detectable in the cells (Fig. 4B and C) or in the medium (Fig. 4F). In contrast, during incubation of cultured rat astrocytes with the H_2O_2-producing system (Hirrlinger et al., 1999, 2001), a steady-state concentration of H_2O_2 was established in the medium within 15 min (Fig. 4A). Immediately after application of the intact H_2O_2-generating system, the amount of intracellular GSSG increased strongly (Fig. 4B and C), while the level of GSx in the cells declined (Fig. 4D). In addition, in the presence of the H_2O_2-generating system, the extracellular content of GSx was elevated quickly (Fig. 4E) by accumulation of GSSG (Fig. 4F). Only small amounts of GSH (difference between the amounts of GSx and GSSG) were found in the medium under these conditions (compare

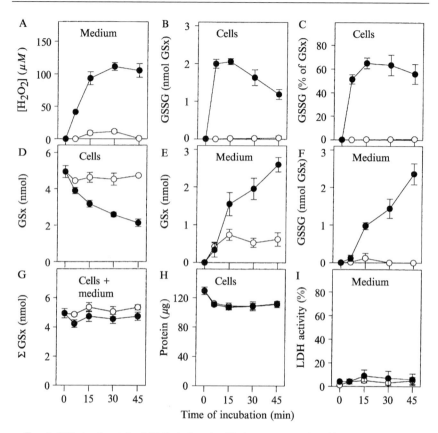

FIG. 4. Effects of sustained H_2O_2-induced oxidative stress on glutathione content in a 14-day-old astrocyte culture. Cells were incubated with hypoxanthine (1 mM) and SOD (100 U) in the absence (○) or the presence (●) of 20 mU of xanthine oxidase. Parameters measured in this representative experiment were calculated for cells per well ($n = 3$) or for medium (1 ml) per well ($n = 6$). (A) Concentration of H_2O_2; (B) GSSG content in cells; (C) GSSG content in cells expressed as a percentage of GSx; (D) GSx content in cells; (E) GSx content in the medium; (F) GSSG content in the medium; (G) sum of GSx in cells and medium; (H) protein content per well; and (I) LDH activity in the medium expressed as a percentage of total LDH activity per well. Modified from Hirrlinger *et al.* (2001).

Fig. 4E and F). All the GSx lost from the cells was recovered in the medium (Fig. 4E) as GSSG (Fig. 4F), leading to almost constant values of the sum of the GSx contents of cells plus medium (Fig. 4G). During incubation with the H_2O_2-generating system, the cellular protein content remained almost constant (Fig. 4H) and the viablity of the cells was not compromised (Fig. 4I), indicating that GSSG was released from viable cells.

MK571 and cyclosporine A inhibit GSSG release from astrocytes during oxidative stress by about 80% (Hirrlinger *et al.*, 2001), indicating that predominantly Mrp1 is responsible for the GSSG export from astrocytes. Thus, this system can be used to study the effect of modulators of Mrp1 transport function by monitoring export of the endogenously produced physiological substrate GSSG.

Concluding Remarks

The assay systems described here allow investigation of the export of GSH under unstressed conditions and of GSSG during oxidative stress from cultured astrocytes. The inhibition of GSH and GSSG export by MK571 and the absence of MK571-sensitive Mrp2 in astrocytes suggest that Mrp1 is responsible for about 60 and 80% of the export of GSH and GSSG from astrocytes, respectively (Hirrlinger *et al.*, 2001, 2002b). Thus, cultured astrocytes are a useful model system to test compounds for their ability to interfere with Mrp1-mediated transport processes. The test systems described here are robust and reliable and allow one to test substantial numbers of potential modulators in one experiment. There is no obvious reason why these assay systems should not be applicable for other types of adherend cells in wells of 24-well dishes. Because the assay conditions described here were optimized for astrocytes, they may have to be adapted to the properties of another cell type to be studied. For example, depending on the rate of GSH export from a given cell type, the volume of the incubation medium, as well as the duration of the incubation, may have to be adjusted. In addition, depending on the capacity of a cell type to detoxify H_2O_2, the amount of xanthine oxidase to maintain a substantial extracellular peroxide concentration, to induce intracellular oxidative stress, and to generate a substantial intracellular GSSG concentration may have to be adapted for other cell types.

Acknowledgment

RD is a recipient of a NeuroSciences Victoria Senior Research Fellowship. The authors' work was supported by the Deutsche Forschungsgemeinschaft (Grant DR262/7-1).

References

Akerboom, T. P., Bilzer, M., and Sies, H. (1982). The relationship of biliary glutathione disulfide efflux and intracellular glutathione disulfide content in perfused rat liver. *J. Biol. Chem.* **257,** 4248–4252.
Akerboom, T. P., and Sies, H. (1990). Glutathione transport and its significance in oxidative stress. *In* "Glutathione: Metabolism and Physiological Functions" (J. Vina, ed.), pp. 45–55. CRC Press, Boca Raton, FL.

Aronica, E., Gorter, J. A., Ramkema, M., Redeker, S., Ozbas-Gercerer, F., van Vliet, E. A., Scheffer, G. L., Scheper, R. J., van der Valk, P., Baayen, J. C., and Troost, D. (2004). Expression and cellular distribution of multidrug resistance-related proteins in the hippocampus of patients with mesial temporal lobe epilepsy. *Epilepsia* **45**, 441–451.

Ballatori, N., and Rebbeor, J. F. (1998). Roles of MRP2 and oatp1 in hepatocellular export of reduced glutathione. *Semin. Liver Dis.* **18**, 377–387.

Borst, P., Evers, R., Kool, M., and Wijnholds, J. (1999). The multidrug resistance protein family. *Biochim. Biophys. Acta* **1461**, 347–357.

Borst, P., and Oude Elferink, R. (2002). Mammalian ABC transporters in health and disease. *Annu. Rev. Biochem.* **71**, 537–592.

Büchler, M., König, J., Brom, M., Kartenbeck, J., Spring, H., Horie, T., and Keppler, D. (1996). cDNA cloning of the hepatocyte canalicular isoform of the multidrug resistance protein, cMrp, reveals a novel conjugate export pump deficient in hyperbilirubinemic mutant rats. *J. Biol. Chem.* **271**, 15091–15098.

Cole, S. P., Bhardwaj, G., Gerlach, J. H., Mackie, J. E., Grant, C. E., Almquist, K. C., Stewart, A. J., Kurz, E. U., Duncan, A. M., and Deeley, R. G. (1992). Overexpression of a transporter gene in a multidrug-resistant human lung cancer cell line. *Science* **258**, 1650–1654.

Cooper, A. J. L. (1997). Glutathione in the brain: Disorders of glutathione metabolism. *In* "The Molecular and Genetic Basis of Neurological Disease" (R. N. Rosenberg, S. B. Prusiner, S. DiMauro, R. L. Barchi, and L. M. Kunk, eds.), pp. 1195–1230. Butterworth-Heinemann, Boston, MA.

Decleves, X., Regina, A., Laplanche, J. L., Roux, F., Boval, B., Launay, J. M., and Scherrmann, J. M. (2000). Functional expression of P-glycoprotein and multidrug resistance-associated protein (Mrp1) in primary cultures of rat astrocytes. *J. Neurosci. Res.* **60**, 594–601.

Dringen, R. (2000). Metabolism and functions of glutathione in brain. *Prog. Neurobiol.* **62**, 649–671.

Dringen, R., and Hamprecht, B. (1996). Glutathione content as an indicator for the presence of metabolic pathways of amino acids in astroglial cultures. *J. Neurochem.* **67**, 1375–1382.

Dringen, R., and Hamprecht, B. (1997). Involvement of glutathione peroxidase and catalase in the disposal of exogenous hydrogen peroxide by cultured astroglial cells. *Brain Res.* **759**, 67–75.

Dringen, R., and Hirrlinger, J. (2003). Glutathione pathways in the brain. *Biol. Chem.* **384**, 505–516.

Dringen, R., Kranich, O., and Hamprecht, B. (1997). The γ-glutamyl transpeptidase inhibitor acivicin preserves glutathione released by astroglial cells in culture. *Neurochem. Res.* **22**, 727–733.

Dringen, R., Kussmaul, L., Gutterer, J. M., Hirrlinger, J., and Hamprecht, B. (1999a). The glutathione system of peroxide detoxification is less efficient in neurons than in astroglial cells. *J. Neurochem.* **72**, 2523–2530.

Dringen, R., Kussmaul, L., and Hamprecht, B. (1998). Detoxification of exogenous hydrogen peroxide and organic hydroperoxides by cultured astroglial cells assessed by microtiter plate assay. *Brain Res. Protoc.* **2**, 223–228.

Dringen, R., Pawlowski, P. G., and Hirrlinger, J. (2005). Peroxide detoxification by brain cells. *J. Neurosci. Res.* **79**(1–2), 157–165.

Dringen, R., Pfeiffer, B., and Hamprecht, B. (1999b). Synthesis of the antioxidant glutathione in neurons: Supply by astrocytes of CysGly as precursor for neuronal glutathione. *J. Neurosci.* **19**, 562–569.

Fernandez-Checa, J. C., Takikawa, H., Horie, T., Ookhtens, M., and Kaplowitz, N. (1992). Canalicular transport of reduced glutathione in normal and mutant Eisai hyperbilirubinemic rats. *J. Biol. Chem.* **267,** 1667–1673.

Gennuso, F., Fernetti, C., Tirolo, C., Testa, N., L'Episcopo, F., Caniglia, S., Morale, M. C., Ostrow, J. D., Pascolo, L., Tiribelli, C., and Marchetti, B. (2004). Bilirubin protects astrocytes from its own toxicity by inducing up-regulation and translocation of multidrug resistance-associated protein 1 (Mrp1). *Proc. Natl. Acad. Sci. USA* **101,** 2470–2475.

Griffith, O. W. (1980). Determination of glutathione and glutathione disulfide using glutathione reductase and 2-vinylpyridine. *Anal. Biochem.* **106,** 207–212.

Hamprecht, B., and Löffler, F. (1985). Primary glial cultures as a model for studying hormone action. *Methods Enzymol.* **109,** 341–345.

Hirrlinger, J., Hamprecht, B., and Dringen, R. (1999). Application and modulation of a permanent hydrogen peroxide-induced oxidative stress to cultured astroglial cells. *Brain Res. Protoc.* **4,** 223–229.

Hirrlinger, J., König, J., and Dringen, R. (2002a). Expression of mRNAs of multidrug resistance proteins (Mrps) in cultured rat astrocytes, oligodendrocytes, microglial cells and neurones. *J. Neurochem.* **82,** 716–719.

Hirrlinger, J., König, J., Keppler, D., Lindenau, J., Schulz, J. B., and Dringen, R. (2001). The multidrug resistance protein MRP1 mediates the release of glutathione disulfide from rat astrocytes during oxidative stress. *J. Neurochem.* **76,** 627–636.

Hirrlinger, J., Schulz, J. B., and Dringen, R. (2002b). Glutathione release from cultured brain cells: Multidrug resistance protein 1 mediates the release of GSH from rat astroglial cells. *J. Neurosci. Res.* **69,** 318–326.

Homolya, L., Varadi, A., and Sarkadi, B. (2003). Multidrug resistance-associated proteins: Export pumps for conjugates with glutathione, glucuronate or sulfate. *Biofactors* **17,** 103–114.

Ishikawa, T., and Sies, H. (1989). Glutathione as an antioxidant: Toxicological aspects. *In* "Glutathione: Chemical, Biochemical, and Medical Aspects" (B. Dolphin, O. Avramovic, and R. Poulson, eds.), pp. 85–109. Wiley, New York.

Kaplowitz, N., Fernandez-Checa, J. C., Kannan, R., Garcia-Ruiz, C., Ookhtens, M., and Yi, J. R. (1996). GSH transporters: Molecular characterization and role in GSH homeostasis. *Biol. Chem. Hoppe Seyler* **377,** 267–273.

Keppler, D. (1999). Export pumps for glutathione S-conjugates. *Free Radic. Biol. Med.* **27,** 985–991.

Kogan, I., Ramjeesingh, M., Li, C., Kidd, J. F., Wang, Y., Leslie, E. M., Cole, S. P., and Bear, C. E. (2003). CFTR directly mediates nucleotide-regulated glutathione flux. *EMBO J.* **22,** 1981–1989.

König, J., Nies, A. T., Cui, Y., Leier, I., and Keppler, D. (1999). Conjugate export pumps of the multidrug resistance protein (MRP) family: Localization, substrate specificity, and MRP2-mediated drug resistance. *Biochim. Biophys. Acta* **1461,** 377–394.

Kruh, G. D., and Belinsky, M. G. (2003). The MRP family of drug efflux pumps. *Oncogene* **22,** 7537–7552.

Lai, L., and Tan, T. M. (2002). Role of glutathione in the multidrug resistance protein 4 (MRP4/ABCC4)-mediated efflux of cAMP and resistance to purine analogues. *Biochem. J.* **361,** 497–503.

Lazarowski, A., Lubieniecki, F., Camarero, S., Pomata, H., Bartuluchi, M., Sevlever, G., and Taratuto, A. L. (2004). Multidrug resistance proteins in tuberous sclerosis and refractory epilepsy. *Pediatr. Neurol.* **30,** 102–106.

Leier, I., Jedlitschky, G., Buchholz, U., Center, M., Cole, S. P., Deeley, R. G., and Keppler, D. (1996). ATP-dependent glutathione disulphide transport mediated by the MRP gene-encoded conjugate export pump. *Biochem. J.* **314**, 433–437.

Leier, I., Jedlitschky, G., Buchholz, U., Cole, S. P., Deeley, R. G., and Keppler, D. (1994). The MRP gene encodes an ATP-dependent export pump for leukotriene C4 and structurally related conjugates. *J. Biol. Chem.* **269**, 27807–27810.

Li, L., Lee, T. K., Meier, P. J., and Ballatori, N. (1998). Identification of glutathione as a driving force and leukotriene C4 as a substrate for oatp1, the hepatic sinusoidal organic solute transporter. *J. Biol. Chem.* **273**, 16184–16191.

Li, L., Meier, P. J., and Ballatori, N. (2000). Oatp2 mediates bidirectional organic solute transport: A role for intracellular glutathione. *Mol. Pharmacol.* **58**, 335–340.

Lowry, O. H., Rosebrough, N. J., Farr, A. L., and Randall, R. J. (1951). Protein measurement with the Folin phenol reagent. *J. Biol. Chem.* **193**, 265–275.

Marroni, M., Agrawal, M. L., Kight, K., Hallene, K. L., Hossain, M., Cucullo, L., Signorelli, K., Namura, S., Bingaman, W., and Janigro, D. (2003). Relationship between expression of multiple drug resistance proteins and p53 tumor suppressor gene proteins in human brain astrocytes. *Neuroscience* **121**, 605–617.

Mercier, C., Decleves, X., Masseguin, C., Fragner, P., Tardy, M., Roux, F., Gabrion, J., and Scherrmann, J. M. (2003). P-glycoprotein (ABCB1) but not multidrug resistance-associated protein 1 (ABCC1) is induced by doxorubicin in primary cultures of rat astrocytes. *J. Neurochem.* **87**, 820–830.

Mercier, C., Masseguin, C., Roux, F., Gabrion, J., and Scherrmann, J. M. (2004). Expression of P-glycoprotein (ABCB1) and Mrp1 (ABCC1) in adult rat brain: Focus on astrocytes. *Brain Res.* **1021**, 32–40.

Nies, A. T., Cantz, T., Brom, M., Leier, I., and Keppler, D. (1998). Expression of the apical conjugate export pump, Mrp2, in the polarized hepatoma cell line, WIF-B. *Hepatology* **28**, 1332–1340.

Nies, A. T., Jedlitschky, G., König, J., Herold-Mende, C., Steiner, H. H., Schmitt, H. P., and Keppler, D. (2004). Expression and immunolocalization of the multidrug resistance proteins, MRP1-MRP6 (ABCC1-ABCC6), in human brain. *Neuroscience* **129**, 349–360.

Paulusma, C. C., van Geer, M. A., Evers, R., Heijn, M., Ottenhoff, R., Borst, P., and Oude Elferink, R. P. (1999). Canalicular multispecific organic anion transporter/multidrug resistance protein 2 mediates low-affinity transport of reduced glutathione. *Biochem. J.* **338**(Pt. 2), 393–401.

Rebbeor, J. F., Connolly, G. C., and Ballatori, N. (2002). Inhibition of Mrp2- and Ycf1p-mediated transport by reducing agents: Evidence for GSH transport on rat Mrp2. *Biochim. Biophys. Acta* **1559**, 171–178.

Rius, M., Nies, A. T., Hummel-Eisenbeiss, J., Jedlitschky, G., and Keppler, D. (2003). Cotransport of reduced glutathione with bile salts by MRP4 (ABCC4) localized to the basolateral hepatocyte membrane. *Hepatology* **38**, 374–384.

Ruedig, C., and Dringen, R. (2004). TNFα increases activity of γ-glutamyl transpeptidase in cultured rat astroglial cells. *J. Neurosci. Res.* **75**, 536–543.

Sagara, J., Makino, N., and Bannai, S. (1996). Glutathione efflux from cultured astrocytes. *J. Neurochem.* **66**, 1876–1881.

Schinkel, A. H., and Jonker, J. W. (2003). Mammalian drug efflux transporters of the ATP binding cassette (ABC) family: An overview. *Adv. Drug Deliv. Rev.* **55**, 3–29.

Sies, H., and Akerboom, T. P. (1984). Glutathione disulfide (GSSG) efflux from cells and tissues. *Methods Enzymol.* **105**, 445–451.

Sies, H., Gerstenecker, C., Menzel, H., and Flohe, L. (1972). Oxidation in the NADP system and release of GSSG from hemoglobin-free perfused rat liver during peroxidatic oxidation of glutathione by hydroperoxides. *FEBS Lett.* **27,** 171–175.

Sisodiya, S. M., Lin, W. R., Harding, B. N., Squier, M. V., and Thom, M. (2002). Drug resistance in epilepsy: Expression of drug resistance proteins in common causes of refractory epilepsy. *Brain* **125,** 22–31.

Stewart, V. C., Stone, R., Gegg, M. E., Sharpe, M. A., Hurst, R. D., Clark, J. B., and Heales, S. J. (2002). Preservation of extracellular glutathione by an astrocyte derived factor with properties comparable to extracellular superoxide dismutase. *J. Neurochem.* **83,** 984–991.

Stone, R., Stewart, V. C., Hurst, R. D., Clark, J. B., and Heales, S. J. (1999). Astrocytes nitric oxide causes neuronal mitochondrial damage, but antioxidants release limits neuronal cell death. *Ann. NY Acad. Sci.* **893,** 400–403.

Yudkoff, M., Pleasure, D., Cregar, L., Lin, Z. P., Nissim, I., Stern, J., and Nissim, I. (1990). Glutathione turnover in cultured astrocytes: Studies with [^{15}N]glutamate. *J. Neurochem.* **55,** 137–145.

[24] The Genetics of ATP-Binding Cassette Transporters

By MICHAEL DEAN

Abstract

The ATP-binding cassette (ABC) superfamily consists of membrane proteins that transport a wide variety of substrates across membranes. Mutations in ABC transporters cause or contribute to a number of different Mendelian disorders, including adrenoleukodystrophy, cystic fibrosis, retinal degeneration, cholesterol, and bile transport defects. In addition, the genes are involved in an increasing number of complex disorders. The proteins play essential roles in the protection of organisms from toxic metabolites and compounds in the diet and are involved in the transport of compounds across the intestine, blood–brain barrier, and the placenta. There are 48 ABC genes in the human genome divided into seven subfamilies based in gene structure, amino acid alignment, and phylogenetic analysis. These seven subfamilies are found in all other sequenced eukaryotic genomes and are of ancient origin. Further characterization of all ABC genes from humans and model organisms will lead to additional insights into normal physiology and human disease.

ATP-Binding Cassette (ABC) Gene Organization and Protein Structure

ABC transporters bind and hydrolyze ATP and use the energy from ATP hydrolysis to pump compounds across the membrane or to flip molecules from the inner to the outer leaflet of the membrane (Childs and Ling,

METHODS IN ENZYMOLOGY, VOL. 400
0076-6879/05 $35.00
DOI: 10.1016/S0076-6879(05)00024-8

1994; Dean and Allikmets, 1995; Higgins, 1992). Genes are classified into the ABC superfamily based on the sequence identity of the ATP-binding domain(s), also known as nucleotide-binding folds (NBFs)(Dean and Allikmets, 1995; Dean et al., 2001b). NBFs contain residues that are found in other ATP-binding proteins (the Walker A and B motifs), separated by 90–120 amino acids and an additional element, the signature or C motif, located just in front of the Walker B site (Hyde et al., 1990). The proteins also possess two transmembrane (TM) domains composed of 6–11 membrane-spanning α helices. The functional transporter can either be a single protein with two NBFs and two TM domains (a full transporter) or be a dimer consisting of two half transporters. In vertebrates the ABCA and ABCC subfamilies are composed exclusively of full transporters, the ABCD, ABCG, and ABCH subfamily of half transporters, and the ABCB subfamily contains both half and full transporters. The ABCE and ABCF subfamilies consist of proteins with two NBFs and no TM domains. These proteins are not transporters but they are clearly evolutionarily related based on analysis of their NBFs.

As the ABC genes encode large membrane proteins, they are very difficult to express and purify in quantities sufficient for crystal structure determination. A small number of bacterial NBDs have been crystallized and provide some structural information of the whole protein (Locher, 2004). In addition, the structure of a few bacterial ABC transporters has been solved (Chang, 2003; Chang and Roth, 2001; Locher et al., 2002). These studies have shown that ATP residues are bound by residues from both NBFs. This provides a mechanism where the binding of ATP can induce a substantial structural change in the molecule, sufficient to force the transported compound across the membrane. The other surprising finding is that the transmembrane helices are often at considerable angles to the bilayer or even parallel to the membrane. A structure from two-dimensional crystals of the ABCB1/MDR1 protein has been presented (Rosenberg, 2005).

Human ABC Genes

Table I shows all of the 48 known ABC genes present in the human genome with their cytogenetic location, protein size, genomic size, and conservation to dog, human, and mouse orthologs. The genes are mostly dispersed in the genome with a few clusters of two to five genes that are believed to be the result of recent gene duplications. Most of the human genes are conserved in other vertebrates (Dean et al., 2001b), and within mammals there is little variation in gene content (Dean and Annilo, 2005). Lineage-specific genes present only in one group of animals include the ABCA10 gene found only in primates and the Abcb1b and Abcg3 genes

TABLE I
LIST OF HUMAN ABC GENES, CHROMOSOMAL LOCATION, AND FEATURES[a]

Symbol	Location	MIM[b]	AA	Gene size	dN/dS Dog	dN/dS Mouse	dN/dS Rat	Transcript	Protein
ABCA1	9q31.1	600046	2262	147154	0.078	0.048	0.072	1	1
ABCA2	9q34.3	600047	2467	21689		0.043	0.03	2	2
ABCA3	16p13.3	601615	1705	53974	0.072	0.069	0.072	1	1
ABCA4	1p21.3	601691	2274	128287	0.151	0.128	0.126	1	1
ABCA5	17q24.3		1643	80499	0.132	0.091	0.087	2	2
ABCA6	17q24.3		1618	63169		0.356	0.282	1	1
ABCA7	19p13.3	605414	2147	10347		0.118	0.13	2	2
ABCA8	17q24.3		1582	88101	0.256	0.256	0.298	1	1
ABCA9	17q24.3		1625	86155	0.239	0.245	0.271	2	2
ABCA10	17q24.3		1544	96808	NA	NA	NA	1	1
ABCA12	2q34	607800	2595	206885	0.113	0.075	0.101	2	2
ABCA13	7p12.3	607807	5004	449249	0.397	0.221	0.218	1	1
ABCB1	7q21.12	171050	1281	209617	0.18	0.16	0.145	1	1
ABCB2	6p21	170260	809	8765	0.016	0.031	0.026	1	1
ABCB3	6p21	170261	704	16912	0.225	0.204	0.207	2	2
ABCB4	7q21.12	171060	1287	75506	0.152	0.097	0.091	3	3
ABCB5	7p21.1		813	108203	0.283	0.212	0.197	1	1
ABCB6	2q35	605452	843	9179	0.201	0.134	0.156	1	1
ABCB7	Xq21–22	300135	754	103026	0.215	0.11	0.059	1	1
ABCB8	7q36.1	605464	719	17116	0.135	0.137	0.15	1	1
ABCB9	12q24.31	605453	767	46214	0.045	0.036	0.037	4	3
ABCB10	1q42.13	605454	739	42113	0.1	0.12	0.161	1	1
ABCB11	2q24.3	603201	1322	108385	0.173	0.212		1	1
ABCC1	16p13.12	158343	1532	192840	0.064	0.06	0.062	7	7
ABCC2	10q24.2	601107	1546	69011	0.313	0.243	0.18	1	1
ABCC3	17q21.33	604323	1528	56836	0.169	0.164	0.204	1	1
ABCC4	13q32.1	605250	1326	281594	0.137	0.099	0.06	1	1
ABCC5	3q27.1	605251	1438	97956	0.069	0.073	0.053	1	1
ABCC6	16p13.12	603234	1504	73325	0.098	0.149	0.16	1	1
ABCC7	7q31.31	602421	1481	188699	0.172	0.23	0.238	1	1
ABCC8	11p15.1	600509	1582	84017	0.093	0.045	0.055	1	1
ABCC9	12p12.1	601439	1550	135631	0.021	0.039		3	3
ABCC10	6p21.1		1465	18675	0.17	0.195	0.171	1	1
ABCC11	16q12.1	607040	1383	68267	0.304	NA	NA	2	3
ABCC12	16q12.1	607041	1360	63798	0.2	0.184	0.166	1	1
ABCD1	Xq28	300371	746	19846	0.045	0.053	0.058	1	1
ABCD2	12q11	601081	741	67424	0.085	0.066	0.073	1	1
ABCD3	1p22.1	170995	660	100072	0.05	0.051	0.058	1	1
ABCD4	14q24.3	603214	607	17540	0.188	0.112	0.11	1	1
ABCE1	4q31.31	601213	600	30851	0.003	0.061	0.002	1	1
ABCF1	6p21.1	603429	807	19920	0.022	0.057	0.055	1	1
ABCF2	7q36.1		635	19395	0.036	0.011	0.01	2	2
ABCF3	3q27.1		710	7908	0.027	0.03	0.035	1	1

(continued)

TABLE I (continued)

Symbol	Location	MIM[b]	AA	Gene size	dN/dS Dog	Mouse	Rat	Transcript	Protein
ABCG1	21q22.3	603076	645	97556	0.012	0.012	0.01	7	7
ABCG2	4q22	603756	656	66883	0.269	0.206	0.202	1	1
ABCG4	11q23	607784	647	13626	0.026	0.064	0.027	1	1
ABCG5	2p21	605459	652	26348	0.302	0.175	0.157	1	1
ABCG8	2p21	605460	674	39503	0.152	0.126	0.206	1	1

[a] AA, amino acid; dN/dS, ratio of nonsynonymous to synonymous substitutions of human compared to dog, mouse, or rat from www.ensembl.org; number of transcripts and protein are from NCBI (http://www.ncbi.nlm.nih.gov).

[b] MIM, Mendelian inheritance in Man number (http://www.ncbi.nlm.nih.gov/entrez/query.fcgi?db=OMIM).

found only in rodents. The *ABCC13* gene is present in fish and dog and is also present in all primates except the great apes. An *ABCH1* gene is present in fish but not in mammals. Zebrafish has duplicates of many ABC genes (*ABCA1*, *ABCA4*, *ABCB1*, *ABCB3*, *ABCB6*, *ABCB11*, *ABCG2*, and *ABCG4*) that appear to be vestiges of the whole genome duplication.

Human ABC proteins range in size from 600 to 5004 amino acids and their genes are from 7.9 to 449 kb. All of the human ABC genes have introns, and range from 10 to 60 exons. While differential splicing has not been studied systematically, several of the genes are known to express from two to seven different transcripts and protein products. The ratio of nonsynonymous (amino acid altering) to synonymous nucleotide changes is one measure of the degree of conservation of a gene. Table I shows this ratio (dN/dS) for each gene compared to the predicted dog, mouse, and rat orthologs. Data indicate that *ABCA1*, *ABCA2*, *ABCB2*, *ABCB9*, *ABCC1*, *ABCC8*, *ABCD1*, *ABCD3*, *ABCE1*, *ABCF1*, *ABCF2*, *ABCF3*, *ABCG1*, and *ABCG4* genes are highly conserved and that *ABCA8*, and *ABCA9* genes have diverged. These data are consistent with data on amino acid identities (not shown).

ABC Gene Diseases

To date, mutations in 18 of the human ABC genes have been found to cause human diseases or phenotypes (Table II).

Cystic Fibrosis/CFTR

Cystic fibrosis (CF, OMIM 602421) is caused by mutations in the *CFTR* gene. CFTR is the only known ABC protein that functions as a chloride channel. The CFTR protein is found in all exocrine tissues of the body and, in response to cAMP stimulation, the channel opens and chloride ions

TABLE II
DISEASES AND PHENOTYPES CAUSED BY ABC GENES[a]

Gene	Mendelian disorder	Complex disease	Animal model
ABCA1	Tangier disease, FHDLD		Mouse, chicken
ABCA3	Surfactant deficiency		
ABCA4	Stargardt/FFM, RP, CRD	AMD	Mouse
ABCA12	Lamellar and harlequin ichthyosis		
ABCB1	Ivermectin sensitivity	Digoxin uptake	Mouse, dog
ABCB2	Immune deficiency		Mouse
ABCB3	Immune deficiency		Mouse
ABCB4	PFIC-3	ICP	
ABCB7	XLSA/A		
ABCB11	PFIC-2		
ABCC2	Dubin–Johnson syndrome		Rat, sheep, monkey
ABCC6	Pseudoxanthoma elasticum		Mouse
ABCC7	Cystic fibrosis, CBAVD	Pancreatitis, bronchiectasis	Mouse
ABCC8	FPHHI		Mouse
ABCC9	DCVT		
ABCD1	ALD		Mouse
ABCG5	Sitosterolemia		Mouse
ABCG8	Sitosterolemia		Mouse

[a] FHDLD, familial hypoapoproteinemia; FFM, fundus flavimaculatis; RP, retinitis pigmentosum 19; CRD, cone–rod dystrophy; AMD, age-related macular degeneration; PFIC, progressive familial intrahepatic cholestasis; ICP, intrahepatic cholestasis of pregnancy; XLSA/A, X-linked sideroblastosis and anemia; FPHHI, familial persistent hyperinsulinemic hypoglycemia of infancy; ALD, adrenoleukodystrophy; IDDM, insulin-dependent diabetes mellitus; DCVT, dilated cardiomyopathy with ventricular tachycardia.

flow through the channel. Cystic fibrosis is the most common fatal genetic disease in European populations. CF is characterized by intestinal obstructions, deficiency in digestive enzymes produced in the pancreas, and lung infections. CF patients also have an abnormally high level of chloride and sodium ions in sweat, and males with CF are typically infertile, due to a defect in the development of the vas deferens.

The most common mutation in the CF gene is a 3-bp deletion that results in the deletion of the phenylalanine at position 508 (ΔF508). This allele has a frequency of 50–88% of CF alleles in European populations, being most frequent in Northern and Western nations. Over 1000 additional mutations in the *CFTR* gene have been described. As many as 1 in 25 Europeans are carriers for a *CFTR* mutation. Because of the high frequency of the disease, it has been suggested that there might have been a selective advantage to *CFTR* mutations. CFTR channel activity is activated by several bacterial toxins, such as cholera toxin, and *CFTR* heterozygotes are speculated to be less susceptible to diarrhea due to bacterial infections. Data from the analysis of *Cftr*-deficient mice are equivocal on the topic. CFTR has also been shown to be a receptor for *Salmonella typhus* (Pier *et al.*, 1998). In addition to the frequent ΔF508 mutation in Europeans, the W1282X mutation is frequent in Askenazi Jewish populations and the 1677delTA allele is very common in Georgian and other Black Sea peoples (Angelicheva *et al.*, 1994; Estivill *et al.*, 1997). These data are consistent with a selection for multiple alleles in different populations.

Mutations in CFTR are also associated with congenital bilateral absence of the vas deferens, a genetic cause of male infertility(Anguiano *et al.*, 1992). CBAVD patients have none of the other symptoms of CF, suggesting that the vas deferens is highly sensitive to mutations in the *CFTR* gene. *CFTR* mutations have also been associated with bronchiectasis and pancreatitis (Cohn *et al.*, 1998; Pignatti *et al.*, 1995).

*Adrenoleukodystrophy/*ABCD1

Adrenoleukodystrophy is an X-linked disorder caused by mutations in the *ABCD1* gene and is characterized by neurodegenerative phenotypes with onset typically in late childhood (Aubourg *et al.*, 1993). Adrenal deficiency commonly occurs, and the presentation of ALD is highly variable. Childhood ALD, adrenomyeloneuropathy, and adult onset forms are recognized, but there is no apparent correlation to *ABCD1* alleles (Kemp *et al.*, 2001). Female heterozygotes display symptoms such as spastic paraparesis and peripheral neuropathy (Holmberg *et al.*, 1991).

More than 406 mutations have been documented in the *ABCD1* gene. Although most mutations are point mutations, several large intragenic deletions have also been described (Kutsche *et al.*, 2002). A contiguous gene syndrome, contiguous *ABCD1 DXS1357E* deletion syndrome (CADDS), has been described that includes *ABCD1* and the adjacent *DXS1357E* gene. These patients present with symptoms at birth, as opposed to X-ALD, which present after 3 years of age (Corzo *et al.*, 2002).

ALD patients have an accumulation of unbranched saturated fatty acids, with a chain length of 24–30 carbons (VLCFA), in the cholesterol esters of the brain and in the adrenal cortex. The ALD protein is located in the peroxisome. A treatment consisting of erucic acid, a C22 monounsaturated fat, and oleic acid, a C18 monounsaturated fat (Lorenzo's oil), was developed that results in normalization of VLCFA levels in the blood of patients but does not appear to dramatically slow the progression of the disease (Aubourg *et al.*, 1993). This is probably because the treatment fails to lower fatty acid levels in the brain (Poulos *et al.*, 1994). An *Abcd1* –/– mouse has been generated, and the animals display accumulation of VLCFAs in kidney and brain; however, they do not show the severe neurological abnormalities of the childhood cerebral form of X-ALD (Forss-Petter *et al.*, 1997; Yamada *et al.*, 2000). The mice do show evidence of a late-onset neurological disorder characterized by slower nerve conduction and myelin and axonal anomalies detectable in the spinal cord and sciatic nerve (Pujol *et al.*, 2002).

Macular Degeneration/ABCA4

The *ABCA4* (ABCR) gene maps to chromosome 1p21.3 and is expressed exclusively in photoreceptors. Mutations in the *ABCA4* gene have been associated with multiple eye disorders (Allikmets, 2000). A complete loss of ABCA4 function leads to retinitis pigmentosa, whereas patients with at least one missense allele have Startgardt disease (Allikmets *et al.*, 1997a,b; Martinez-Mir *et al.*, 1998). Startgardt disease is characterized by juvenile to early adult-onset macular dystrophy with loss of central vision (OMIM:248200). Most patients with recessive cone rod dystrophy also have mutations in *ABCA4* (Cremers *et al.*, 1998). Therefore, three different recessive retinal degeneration disorders are caused by *ABCA4* mutations and there is a correlation between the predicted functional activity of the protein and severity of the disease.

ABCA4 is believed to transport retinol (vitamin A)/phospholipid derivatives from the photoreceptor outer segment disks into the cytoplasm (Allikmets *et al.*, 1997b; Azarian and Travis, 1997). These compounds stimulate the ATP hydrolysis activity of the purified protein (Sun *et al.*, 1999). Mice lacking *Abca4* show increased all-*trans*-retinaldehyde after light exposure, elevated phosphatidylethanolamine (PE) in outer segments, accumulation of the protonated Schiff base complex of all-*trans*-retinaldehyde and PE (*N*-retinylidene-PE), and striking deposition of a major lipofuscin fluorophore (A2-E) in retinal pigment epithelium (Weng *et al.*, 1999). These data suggest that ABCA4 is an outwardly directed flippase for *N*-retinylidene-PE.

ABCA4 mutation carriers are also increased in frequency in age-related macular degeneration (AMD) patients (Allikmets *et al.*, 1997a). AMD patients display a variety of phenotypic features, including the loss of central vision, after 60 years of age. The causes of this complex trait are poorly understood, but a combination of genetic and environmental factors plays a role. The abnormal accumulation of retinoids attributable to ABCA4 deficiency may be one mechanism by which this process could be initiated. Defects in ABCA4 lead to an accumulation of retinal derivatives in the retinal pigment epithelium behind the retina. Consistent with this idea is the demonstration in _ABCA4_ +/– mice of light-dependent accumulation of pigmented deposits in the retinal pigment epithelium, very reminiscent of AMD (Mata _et al._, 2001). The drug isotretinoin is an inhibitor of the synthesis of 11-cis-retinaldehyde and partially prevents lipofuscin accumulation and visual loss in _Abca4_ -/- mice (Radu _et al._, 2004).

Cholesterol and Lipid Transport

Several ABC genes are involved in the transport of phospholipids and/ or cholesterol across the membrane. This includes all or nearly all members of the ABCA subfamily and most of the ABCG genes. Mutations in several of these genes are associated with lipid and sterol deficiencies.

Tangier Disease/ABCA1. Mutations in _ABCA1_ cause the recessive disease Tangier disease, a disorder of cholesterol transport between tissues and the liver, mediated by binding of the cholesterol onto high-density lipoprotein (HDL) particles (Bodzioch _et al._, 1999; Brooks-Wilson _et al._, 1999; Marcil _et al._, 1999; Remaley _et al._, 1999; Rust _et al._, 1999). Patients with familial hypoalphalipoproteinemia have also been described that have mutations in the _ABCA1_ gene, demonstrating that these disorders are allelic (Marcil _et al._, 1999). Other patients with reduced levels of HDLs without the classical symptoms of Tangier disease have also been described with _ABCA1_ mutations (Ishii _et al._, 2002). _ABCA1_ controls the export of membrane phospholipid and cholesterol toward specific extracellular acceptors; however, the exact role of the protein in this process is not known. It has been proposed that _ABCA1_ flips membrane phospholipid, principally phosphatidylcholine, toward the lipid-poor apolipoprotein particles, which can then accept cholesterol (Dean _et al._, 2001a). ABCA1 also plays a role in the engulfment of apoptotic bodies. Furthermore, the _ced-7_ gene, an _ABCA1_ ortholog in _Caenorhabditis elegans_, plays a role in phagocytosis by precluding the redistribution of phagocyte receptors around the apoptotic particle (Moynault _et al._, 1998; Wu and Horvitz, 1998).

Disruption of the mouse _Abca1_ gene results in similarly low levels of HDLs and accumulation of cholesterol in tissues (McNeish _et al._, 2000; Orso _et al._, 2000). Analyses of _Abca1_ –/– mice indicate that the transport of lipids from the golgi to the plasma membrane is defective (Orso _et al._, 2000). However, these mice have normal secretion of cholesterol into bile, indicating that Abca1 does not play a role in this process (Groen _et al._, 2001). In contrast, the constitutive overexpression of Abca1 results in a protection of animals against an atherosclerotic diet (Joyce _et al._, 2002; Singaraja _et al._, 2002). The Wisconsin hypoalpha mutant (WHAM) chicken has a mutant ABCA1 gene and is a model for Tangier disease (Schreyer *et al.*, 1994).

Because of the important role of ABCA1 in cholesterol transport, several groups have examined the *ABCA1* gene for polymorphisms that might be associated with plasma lipid levels and cardiovascular disease. A common variation in noncoding regions of *ABCA1* may significantly alter the severity of atherosclerosis, without necessarily influencing plasma lipid levels (Zwarts *et al.*, 2002).

Sitosterolemia/ABCG5 *and* ABCG8. Sitosterolemia is a disorder characterized by the defective transport of plant and fish sterols and cholesterol (Bhattacharyya and Connor, 1974). The *ABCG5* gene maps to chromosome 2p21 and is adjacent to and is arranged head to head with the *ABCG8* gene (Berge *et al.*, 2000; Lee *et al.*, 2001; Shulenin *et al.*, 2001). Mutations in either gene cause sitosterolemia, and the two half transporters form a functional heterodimer and are regulated by the same promoter (Remaley *et al.*, 2002). Patients mutated in either *ABCG5* or *ABCG8* have similarly elevated levels of sitosterol, suggesting that the heterodimer is the principal transporter of sitosterol (Lu *et al.*, 2001). However, Asian sitosterolemia patients almost exclusively have mutations in *ABCG5*, whereas Caucasian patients have mutations in *ABCG8* (Heimer *et al.*, 2002; Lu *et al.*, 2001). This suggests that there are some independent functions of the two genes and that they may also form heterodimers to transport some of the wide variety of noncholesterol sterols found in plants and shellfish.

The levels of sterols in blood were demonstrated to be highly heritable, and at least two variants in *ABCG8* (D19H and T400K) were shown to be associated with lower concentrations of sterols in parents and their offspring (Berge *et al.*, 2002). Several additional frequent missense variants in the *ABCG8* gene were also described (Hubacek *et al.*, 2001; Lu *et al.*, 2001), suggesting that both *ABCG5* and *ABCG8* are functionally polymorphic, perhaps in response to selection based on dietary sterol exposure.

Lamellar Ichthyosis Type 2 (LI2)/ABCA12

Lamellar ichthyosis type 2 is a genetically heterogeneous skin disorder characterized by large, dark, pigmented scales. The disease was mapped to chromosome 2q33–35, the region where ABCA12 is located. Mutations in *ABCA12* were identified in several LI2 families (Lefevre *et al.*, 2003). It is likely that ABCA12 plays a role in lipid secretion or membrane organization in the developing skin. The *ABCA12* gene is also expressed in the stomach, indicating that it may play a role in mucous secretion (Annilo *et al.*, 2002).

*Surfactant Deficiency/*ABCA3

Respiratory distress syndrome is an important cause of neonatal mortality and morbidity and is often caused by a deficiency in lung surfactants (Nogee, 2004). Surfactant forms a lipid-rich monolayer that coats the pulmonary airways and is essential for the inflation of the lung. Surfactant is produced and secreted by alveolar type II cells and consists of lipids, cholesterol, and specialized proteins. The *ABCA3* gene is expressed in alveolar type II cells, and the protein is localized to lamellar bodies (Mulugeta *et al.*, 2002; Yamano *et al.*, 2001) Mutations in the *ABCA3* gene are an important cause of this disease, and patients display abnormal surfactant, elevated surface tension, and abnormal lamellar bodies (Shulenin *et al.*, 2004). Although typically fatal, mild cases have been identified and associated with missense mutations in the *ABCA3* gene (Shulenin *et al.*, 2004).

*Cholestasis/*ABCB4/ABCB11/ABCC2

Several ABC transporters play an important role in the secretion of substances from the liver into the bile (Elferink and Groen, 2002). In addition to the role of ABCG5 and ABCG8 in secreting cholesterol into bile, the ABCB4, ABCB11, and ABCC2 proteins also transport bile components. The ABCB4 protein is involved in the secretion of phosphotidylcholine into the bile. Mutations in this gene lead to progressive familial intrahepatic cholestasis type 3 (PFIC-3), a disorder characterized by early onset of cholestasis that can progress to cirrhosis and liver failure before adulthood (Deleuze *et al.*, 1996). The *ABCB11/BSEP* gene transports bile salts such as taurochenodeoxycholate, tauroursodeoxycholate, taurocholate, glycocholate, and cholate into the bile. Mutations in *ABCB11* lead to PFIC-2 (Strautnieks *et al.*, 1998) and a mouse model supports this result (Wang *et al.*, 2001). Heterozygotes for *ABCB4* or *ABCB11* mutations have also been found to suffer more frequently from cholestasis of pregnancy (Dixon *et al.*, 2000; Pauli-Magnus *et al.*, 2004).

The *ABCC2/MRP2* gene is a transporter of organic anions out of the liver (Borst *et al.*, 2000b). Mutations in *ABCC2* lead to Dubin–Johnson syndrome, a recessive disease characterized by hyperbilirubinemia, an increase in the urinary excretion of coproporphyrin isomer I, deposition of melanin-like pigment in hepatocytes, and prolonged retention of sulfobromophthalein (Toh *et al.*, 1999; Wada *et al.*, 1998). A rat model of the disease, the TR- rat, has also been characterized and is homozygous for an *Abcc2* mutation (Paulusma *et al.*, 1996).

Ivermectin Sensitivity/ABCB1

Ivermectin is a highly useful drug that is effective against a variety of invertebrates, including helminthes, *Onchocerca volvulus* (the worm causing river blindness), and mites (including those causing scabies) (Lawrence *et al.*, 2005). Ivermectin is in wide use in both veterinary and human medicine. Collie dogs frequently display sensitivity to the drug, as do mice that are *Abcb1a −/−*. It was found that collies have a 4-bp deletion in the gene and that at least nine other breeds carry this mutation (Neff *et al.*, 2004). ABCB1 is a drug transporter that is expressed in many tissues, but plays a particularly important role in the blood–brain barrier. ABCB1 mutant collies are also sensitive to a number of other drugs, including doramectin, loperamide, and several anticancer drugs (Geyer *et al.*, 2005).

*Immune Deficiency/*TAP1 *and* TAP2

The *TAP1/ABCB2* and *TAP2/ABCB3* genes are half transporters that form the pump in the endoplasmic reticulum that complexes peptides with HLA class I molecules for antigen presentation on the cell surface. Rare mutations in each of these genes have been identified in patients with immune deficiency (de la Salle *et al.*, 1999; Teisserenc *et al.*, 1997). An allele of the *TAP2/ABCB3* gene (M577V) has been identified that is present at 5% in the general population and is associated with the presence of autoantibodies in patients with Sjogren syndrome (Kumagai *et al.*, 1997). This allele is also associated with the altered presentation of peptides on HLA class I molecules (Kageyama *et al.*, 2004). Tumor cells can potentially evade the immune system by failing to present class I antigens, and mutations in TAP genes have been found in cancer cell lines that are class I negative (Chen *et al.*, 1996; Seliger *et al.*, 2001; Yang *et al.*, 2003).

Sideroblastic Anemia/ABCB7

Sideroblastic anemia and ataxia (301310) is an X-linked disorder characterized by anemia and incoordination (ataxia). Mutations in these patients have been identified in the *ABCB7* gene, a half transporter in the mitochondria, implicated in the transport of iron/sulfur complexes into the cytoplasm (Allikmets *et al.*, 1999; Bekri *et al.*, 2000; Maguire *et al.*, 2001). This pathway is potentially relevant to Freidrich's ataxia, as the frataxin protein may act as a chaperone to deliver iron to the Fe–S assembly complex (Napier *et al.*, 2005).

Pseudoxanthoma Elasticum (PXE)/ABCC6

Pseudoxanthoma elasticum is an autosomal-recessive disease characterized by skin laxity and vision impairment characterized by angioid streaks and occlusion of blood vessels. Calcification of elastic fibers is a diagnostic feature. PXE is caused by mutations in the *ABCC6* gene (Bergen *et al.*, 2000; Le Saux *et al.*, 2000; Ringpfeil *et al.*, 2000). Interestingly, there is considerable variation in the presentation of PXE even within affected individuals in the same family, suggesting that the clinical manifestations of PXE are biochemically removed from the function. *ABCC6* is expressed predominantly in the liver and kidney and is proposed to transport a critical metabolite into or out of the blood. Mutation carriers of *ABCC6* variants have been associated with an increased risk of cardiovascular disease (Trip *et al.*, 2002).

Familial Persistent Hyperinsulinemic Hypoglycemia of Infancy/ABCC8

Familial persistent hyperinsulinemic hypoglycemia of infancy (MIM 256450) is an autosomal-recessive disorder characterized by unregulated insulin secretion. The *ABCC8* gene codes for a high-affinity receptor for the drug sulfonylurea, and sulfonylureas result in increased insulin secretion in patients with noninsulin-dependent diabetes. ABCC8 is known to bind to and regulate the inward rectifying potassium channel KIR6.2 (KCNJ11). Mutations in the *ABCC8* gene are found in the majority of PHHI families (Thomas *et al.*, 1995), and *KCNJ11* mutations account for most of the rest of the families (Tornovsky *et al.*, 2004). Multiple studies have reported association of the E23K variant of Kir6.2 with risk of type 2 diabetes. However, this variant has a very strong allelic association with the A1369S variant in the *ABCC8* gene (Florez *et al.*, 2004). Thus the association cannot be ascribed to either gene and may be a compound effect of both variants.

Dilated Cardiomyopathy/ABCC9

Individuals with dilated cardiomyopathy present with heart failure and rhythm disturbances. *ABCC9* displays low-affinity binding to sulfonylurea and is a major regulator of ATP-dependent potassium channels in muscle. *ABCC9* mutations have been identified in two patients with dilated cardiomyopathy and these variants have been shown to disrupt catalytic K(ATP) channel gating (Bienengraeber *et al.*, 2004).

Perspectives

ABC genes are increasingly found to be the cause of diverse genetic disorders. Virtually all of these are recessive diseases, consistent with the role of these genes as nonimprinted functional proteins for which a single copy of the gene is sufficient for most activities. A dominant-negative mutation in the yeast ortholog of ABCE1 (RLI1) suggests that other such mutations may exist (Dong *et al.*, 2004). Few ABC genes appear to be essential from human and models systems such as *Saccharomyces, Drosophila,* and *C. elegans.* Only one gene (*Abce1*) has emerged as essential so far from lethal locus screens in zebrafish (Amsterdam *et al.*, 2004). Continued functional characterization of the ABC genes is likely to lead to further insight into human biology, and the completion of drafts of zebrafish, chicken, and dog genomes will aid the development of model organisms greatly.

ABC genes are increasingly being shown to play a role in complex disease. In most cases these associations follow logically from the phenotype of the Mendelian disease. Thus *ABCA4* mutations are involved in macular degeneration and are associated with late-onset, disorder age-related macular degeneration. Similarly, two of the genes that cause inherited cholestasis (*ABCB4, ABCB11*) are associated with cholestasis in pregnancy. It is likely that variation in structure or expression of ABC genes is involved in other liver, skin, eye, brain, lipid transport, and mitochondrial phenotypes.

A few trends can be reasonably predicted to emerge in the coming years. The ABCA gene family appears to encode genes that transport phospholipids in specialized tissues from the insight into ABCA1, ABCA3, ABCA4, and ABCA12. It is likely that the remaining genes have similar functions in other tissues, such as the predicted role of ABCA2 in myelination of the brain (Zhou *et al.*, 2001). The major drug transporters ABCB1, ABCC1, and ABCG2 (Gottesman *et al.*, 2002) are involved in the access of metabolites and drugs in such tissues as the intestine, liver, placenta, and brain. Variants in these genes, or rare mutations, are likely to be found in patients with abnormal drug metabolism or severe drug reactions. ABCC1/MRP1-related proteins are likely to also play important roles in specific drug transport and metabolism (Borst *et al.*, 2000a). Several ABC genes, mostly ABCB family members, are present in mitochondria and play roles in iron–sulfur complex generation and transport as well as heme transport. In addition, at least one ABC protein, ABCE1, contains Fe–S clusters, and actions of the ABCD family have been implicated in protecting mitochondrial function. The most conserved ABC genes are the soluble or nontransporters of the ABCE and ABCF family. Roles for these genes in the processing and formation of ribosomal components and the initiation of translation may explain the deep conservation of these genes in all eukaryotes and archae.

The emerging view that all cancers contain a population of cells with stem cell properties and that these cancer stem cells are the critical cell that must be eliminated in chemotherapy focuses attention on the transporters in stem cells. While ABCB1 is known to be expressed in these cells, ABCG2 appears to be the most widely expressed drug pump in early stem cells. However, high levels of expression of other ABC genes, such as *ABCA3*, have also been observed in cancer stem cells. Strategies for inhibiting these pumps in cancer stem cells without incurring excessive toxicity to normal stem cells may be one of the most important quests of the field in the coming years.

Acknowledgments

I thank Tarmo Annilo, Stefan Stefanof, Z.Q. Chen, Sergey Shulenin, Lauren Thomas, Julie Costantino, and Jeff Dean for unpublished data and compiling results. I apologize to all whose primary papers could not be cited due to lack of space. This research was supported by the Intramural Research Program of the NIH, National Cancer Institute, Center for Cancer Research.

References

Allikmets, R. (2000). Simple and complex ABCR: Genetic predisposition to retinal disease. *Am. J. Hum. Genet.* **67**, 793–799.

Allikmets, R., Raskind, W. H., Hutchinson, A., Schueck, N. D., Dean, M., and Koeller, D. M. (1999). Mutation of a putative mitochondrial iron transporter gene (ABC7) in X-linked sideroblastic anemia and ataxia (XLSA/A). *Hum. Mol. Genet.* **8**, 743–749.

Allikmets, R., Shroyer, N. F., Singh, N., Seddon, J. M., Lewis, R. A., Bernstein, P. S., Peiffer, A., Zabriskie, N. A., Li, Y., Hutchinson, A., *et al.* (1997a). Mutation of the Stargardt disease gene *(ABCR)* in age-related macular degeneration. *Science* **277**, 1805–1807.

Allikmets, R., Singh, N., Sun, H., Shroyer, N. F., Hutchinson, A., Chidambaram, A., Gerrard, B., Baird, L., Stauffer, D., Peiffer, A., *et al.* (1997b). A photoreceptor cell-specific ATP-binding transporter gene *(ABCR)* is mutated in recessive Stargardt macular dystrophy. *Nat. Genet.* **15**, 236–246.

Amsterdam, A., Nissen, R. M., Sun, Z., Swindell, E. C., Farrington, S., and Hopkins, N. (2004). Identification of 315 genes essential for early zebrafish development. *Proc. Natl. Acad. Sci. USA* **101**, 12792–12797.

Angelicheva, D., Boteva, K., Jordanova, A., Savov, A., Kufardjieva, A., Tolun, A., Telatar, M., Akarsubasi, A., Koprubasi, F., Aydogdu, S., *et al.* (1994). Cystic fibrosis patients from the Black Sea region: The 1677delTA mutation. *Hum. Mutat.* **3**, 353–357.

Anguiano, A., Oates, R. D., Amos, J. A., Dean, M., Gerrard, B., Stewart, C., Maher, T. A., White, M. B., and Milunsky, A. (1992). Congenital bilateral absence of the vas deferens: A primarily genital form of cystic fibrosis. *JAMA* **267**, 1794–1797.

Annilo, T., Shulenin, S., Chen, Z. Q., Arnould, I., Prades, C., Lemoine, C., Maintoux-Larois, C., Devaud, C., Dean, M., Denefle, P., and Rosier, M. (2002). Identification and characterization of a novel ABCA subfamily member, ABCA12, located in the lamellar ichthyosis region on 2q34. *Cytogenet. Genome Res.* **98**, 169–176.

Aubourg, P., Mosser, J., Douar, A. M., Sarde, C. O., Lopez, J., and Mandel, J. L. (1993). Adrenoleukodystrophy gene: Unexpected homology to a protein involved in peroxisome biogenesis. *Biochimie* **75**, 293–302.

Azarian, S. M., and Travis, G. H. (1997). The photoreceptor rim protein is an ABC transporter encoded by the gene for recessive Stargardt's disease (ABCR). *FEBS Lett.* **409**, 247–252.

Bekri, S., Kispal, G., Lange, H., Fitzsimons, E., Tolmie, J., Lill, R., and Bishop, D. F. (2000). Human ABC7 transporter: Gene structure and mutation causing X-linked sideroblastic anemia with ataxia with disruption of cytosolic iron-sulfur protein maturation. *Blood* **96**, 3256–3264.

Berge, K. E., Tian, H., Graf, G. A., Yu, L., Grishin, N. V., Schultz, J., Kwiterovich, P., Shan, B., Barnes, R., and Hobbs, H. H. (2000). Accumulation of dietary cholesterol in sitosterolemia caused by mutations in adjacent ABC transporters. *Science* **290**, 1771–1775.

Berge, K. E., von Bergmann, K., Lutjohann, D., Guerra, R., Grundy, S. M., Hobbs, H. H., and Cohen, J. C. (2002). Heritability of plasma noncholesterol sterols and relationship to DNA sequence polymorphism in ABCG5 and ABCG8. *J. Lipid Res.* **43**, 486–494.

Bergen, A. A., Plomp, A. S., Schuurman, E. J., Terry, S., Breuning, M., Dauwerse, H., Swart, J., Kool, M., van Soest, S., Baas, F., ten Brink, J. B., and de Jong, P. T. (2000). Mutations in ABCC6 cause pseudoxanthoma elasticum. *Nat. Genet.* **25**, 228–231.

Bhattacharyya, A. K., and Connor, W. E. (1974). Beta-sitosterolemia and xanthomatosis: A newly described lipid storage disease in two sisters. *J. Clin. Invest.* **53**, 1033–1043.

Bienengraeber, M., Olson, T. M., Selivanov, V. A., Kathmann, E. C., O'Cochlain, F., Gao, F., Karger, A. B., Ballew, J. D., Hodgson, D. M., Zingman, L. V., Pang, Y. P., Alekseev, A. E., and Terzic, A. (2004). ABCC9 mutations identified in human dilated cardiomyopathy disrupt catalytic KATP channel gating. *Nat. Genet.* **36**, 382–387.

Bodzioch, M., Orso, E., Klucken, J., Langmann, T., Bottcher, A., Diederich, W., Drobnik, W., Barlage, S., Buchler, C., Porsch-Ozcurumez, M., Kaminski, W. E., Hahmann, H. W., Oette, K., Rothe, G., Aslanidis, C., Lackner, K. J., and Schmitz, G. (1999). The gene encoding ATP-binding cassette transporter 1 is mutated in Tangier disease. *Nat. Genet.* **22**, 347–351.

Borst, P., Evers, R., Kool, M., and Wijnholds, J. (2000a). A family of drug transporters: The multidrug resistance-associated proteins. *J. Natl. Cancer Inst.* **92**, 1295–1302.

Borst, P., Zelcer, N., and van Helvoort, A. (2000b). ABC transporters in lipid transport. *Biochim. Biophys. Acta* **1486**, 128–144.

Brooks-Wilson, A., Marcil, M., Clee, S. M., Zhang, L. H., Roomp, K., van Dam, M., Yu, L., Brewer, C., Collins, J. A., Molhuizen, H. O., Loubser, O., Ouelette, B. F., Fichter, K., Ashbourne-Excoffon, K. J., Sensen, C. W., Scherer, S., Mott, S., Denis, M., Martindale, D., Frohlich, J., Morgan, K., Koop, B., Pimstone, S., Kastelein, J. J., Hayden, M. R., *et al.* (1999). Mutations in ABC1 in Tangier disease and familial high-density lipoprotein deficiency. *Nat. Genet.* **22**, 336–345.

Chang, G. (2003). Structure of MsbA from Vibrio cholera: A multidrug resistance ABC transporter homolog in a closed conformation. *J. Mol. Biol.* **330**, 419–430.

Chang, G., and Roth, C. B. (2001). Structure of MsbA from *E. coli*: A homolog of the multidrug resistance ATP binding cassette (ABC) transporters. *Science* **293**, 1793–1800.

Chen, H. L., Gabrilovich, D., Tampe, R., Girgis, K. R., Nadaf, S., and Carbone, D. P. (1996). A functionally defective allele of TAP1 results in loss of MHC class I antigen presentation in a human lung cancer. *Nat. Genet.* **13**, 210–213.

Childs, S., and Ling, V. (1994). The MDR superfamily of genes and its biological implications. *Important Adv. Oncol.* **2**, 1–36.

Cohn, J. A., Friedman, K. J., Noone, P. G., Knowles, M. R., Silverman, L. M., and Jowell, P. S. (1998). Relation between mutations of the cystic fibrosis gene and idiopathic pancreatitis. *N. Engl. J. Med.* **339**, 653–658.

Corzo, D., Gibson, W., Johnson, K., Mitchell, G., LePage, G., Cox, G. F., Casey, R., Zeiss, C., Tyson, H., Cutting, G. R., Raymond, G. V., Smith, K. D., Watkins, P. A., Moser, A. B., Moser, H. W., and Steinberg, S. J. (2002). Contiguous deletion of the X-linked adrenoleukodystrophy gene (ABCD1) and DXS1357E: A novel neonatal phenotype similar to peroxisomal biogenesis disorders. *Am. J. Hum. Genet.* **70**, 1520–1531.

Cremers, F. P., van de Pol, D. J., van Driel, M., den hollander, A. I., van Haren, F. J., Knoers, N. V., Tijmes, N., Bergen, A. A., Rohrschneider, K., Blankenagel, A., Pinckers, A. J., Deutman, A. F., and Hoyng, C. B. (1998). Autosomal recessive retinitis pigmentosa and cone-rod dystrophy caused by splice site mutations in the Stargardt's disease gene *ABCR*. *Hum. Mol. Genet.* **7**, 355–362.

Dean, M., and Allikmets, R. (1995). Evolution of ATP-binding cassette transporter genes. *Curr. Opin. Genet. Dev.* **5**, 779–785.

Dean, M., and Annilo, T. (2005). Evolution of the ATP-binding cassette (ABC) transporter superfamily in vertebrates. *Annu. Rev. Hum. Genet. Genom.* **6**, 123–142.

Dean, M., Hamon, Y., and Chimini, G. (2001a). The human ATP-binding cassette (ABC) transporter superfamily. *J. Lipid Res.* **42**, 1007–1017.

Dean, M., Rzhetsky, A., and Allikmets, R. (2001b). The human ATP-binding cassette (ABC) transporter superfamily. *Genome Res.* **11**, 1156–1166.

de la Salle, H., Zimmer, J., Fricker, D., Angenieux, C., Cazenave, J. P., Okubo, M., Maeda, H., Plebani, A., Tongio, M. M., Dormoy, A., and Hanau, D. (1999). HLA class I deficiencies due to mutations in subunit 1 of the peptide transporter TAP1. *J. Clin. Invest.* **103**, R9–R13.

Deleuze, J. F., Jacquemin, E., Dubuisson, C., Cresteil, D., Dumont, M., Erlinger, S., Bernard, O., and Hadchouel, M. (1996). Defect of multidrug-resistance 3 gene expression in a subtype of progressive familial intrahepatic cholestasis. *Hepatology* **23**, 904–908.

Dixon, P. H., Weerasekera, N., Linton, K. J., Donaldson, O., Chambers, J., Egginton, E., Weaver, J., Nelson-Piercy, C., de Swiet, M., Warnes, G., Elias, E., Higgins, C. F., Johnston, D. G., McCarthy, M. I., and Williamson, C. (2000). Heterozygous MDR3 missense mutation associated with intrahepatic cholestasis of pregnancy: Evidence for a defect in protein trafficking. *Hum. Mol. Genet.* **9**, 1209–1217.

Dong, J., Lai, R., Nielsen, K., Fekete, C. A., Qiu, H., and Hinnebusch, A. G. (2004). The essential ATP-binding cassette protein RLI1 functions in translation by promoting preinitiation complex assembly. *J. Biol. Chem.* **279**, 42157–42168.

Elferink, R. O., and Groen, A. K. (2002). Genetic defects in hepatobiliary transport. *Biochim. Biophys. Acta* **1586**, 129–145.

Estivill, X., Bancells, C., and Ramos, C. (1997). Geographic distribution and regional origin of 272 cystic fibrosis mutations in European populations: The Biomed CF Mutation Analysis Consortium. *Hum. Mutat.* **10**, 135–154.

Florez, J. C., Burtt, N., de Bakker, P. I., Almgren, P., Tuomi, T., Holmkvist, J., Gaudet, D., Hudson, T. J., Schaffner, S. F., Daly, M. J., Hirschhorn, J. N., Groop, L., and Altshuler, D. (2004). Haplotype structure and genotype-phenotype correlations of the sulfonylurea receptor and the islet ATP-sensitive potassium channel gene region. *Diabetes* **53**, 1360–1368.

Forss-Petter, S., Werner, H., Berger, J., Lassmann, H., Molzer, B., Schwab, M. H., Bernheimer, H., Zimmermann, F., and Nave, K. A. (1997). Targeted inactivation of the X-linked adrenoleukodystrophy gene in mice. *J. Neurosci. Res.* **50**, 829–843.

Geyer, J., Doring, B., Godoy, J. R., Moritz, A., and Petzinger, E. (2005). Development of a PCR-based diagnostic test detecting a nt230(del4) MDR1 mutation in dogs: Verification in a moxidectin-sensitive Australian shepherd. *J. Vet. Pharmacol. Ther.* **28**, 95–99.

Gottesman, M. M., Fojo, T., and Bates, S. E. (2002). Multidrug resistance in cancer: Role of ATP-dependent transporters. *Nature Rev. Cancer* **2**, 48–58.

Groen, A. K., Bloks, V. W., Bandsma, R. H., Ottenhoff, R., Chimini, G., and Kuipers, F. (2001). Hepatobiliary cholesterol transport is not impaired in Abca1-null mice lacking HDL. *J. Clin. Invest.* **108**, 843–850.

Heimer, S., Langmann, T., Moehle, C., Mauerer, R., Dean, M., Beil, F. U., Von Bergmann, K., and Schmitz, G. (2002). Mutations in the human ATP-binding cassette transporters ABCG5 and ABCG8 in sitosterolemia. *Hum. Mutat.* **20**, 151.

Higgins, C. F. (1992). ABC transporters: From micro-organisms to man. *Annu. Rev. Cell Biol.* **8**, 67–113.

Holmberg, B. H., Hagg, E., and Hagenfeldt, L. (1991). Adrenomyeloneuropathy: Report on a family. *J. Intern. Med.* **230**, 535–538.

Hubacek, J. A., Berge, K. E., Cohen, J. C., and Hobbs, H. H. (2001). Mutations in ATP-cassette binding proteins G5 (ABCG5) and G8 (ABCG8) causing sitosterolemia. *Hum. Mutat.* **18**, 359–360.

Hyde, S. C., Emsley, P., Hartshorn, M. J., Mimmack, M. M., Gileadi, U., Pearce, S. R., Gallagher, M. P., Gill, D. R., Hubbard, R. E., and Higgins, C. F. (1990). Structural model of ATP-binding proteins associated with cystic fibrosis, multidrug resistance and bacterial transport. *Nature* **346**, 362–365.

Ishii, J., Nagano, M., Kujiraoka, T., Ishihara, M., Egashira, T., Takada, D., Tsuji, M., Hattori, H., and Emi, M. (2002). Clinical variant of Tangier disease in Japan: Mutation of the ABCA1 gene in hypoalphalipoproteinemia with corneal lipidosis. *J. Hum. Genet.* **47**, 366–369.

Joyce, C. W., Amar, M. J., Lambert, G., Vaisman, B. L., Paigen, B., Najib-Fruchart, J., Hoyt, R. F., Jr., Neufeld, E. D., Remaley, A. T., Fredrickson, D. S., Brewer, H. B., Jr., and Santamarina-Fojo, S. (2002). The ATP binding cassette transporter A1 (ABCA1) modulates the development of aortic atherosclerosis in C57BL/6 and apoE-knockout mice. *Proc. Natl. Acad. Sci. USA* **99**, 407–412.

Kageyama, G., Kawano, S., Kanagawa, S., Kondo, S., Sugita, M., Nakanishi, T., Shimizu, A., and Kumagai, S. (2004). Effect of mutated transporters associated with antigen-processing 2 on characteristic major histocompatibility complex binding peptides: Analysis using electrospray ionization tandem mass spectrometry. *Rapid Commun. Mass. Spectrom.* **18**, 995–1000.

Kemp, S., Pujol, A., Waterham, H. R., van Geel, B. M., Boehm, C. D., Raymond, G. V., Cutting, G. R., Wanders, R. J., and Moser, H. W. (2001). ABCD1 mutations and the X-linked adrenoleukodystrophy mutation database: Role in diagnosis and clinical correlations. *Hum. Mutat.* **18**, 499–515.

Kumagai, S., Kanagawa, S., Morinobu, A., Takada, M., Nakamura, K., Sugai, S., Maruya, E., and Saji, H. (1997). Association of a new allele of the TAP2 gene, TAP2*Bky2 (Val577), with susceptibility to Sjogren's syndrome. *Arthritis Rheum.* **40**, 1685–1692.

Kutsche, K., Ressler, B., Katzera, H. G., Orth, U., Gillessen-Kaesbach, G., Morlot, S., Schwinger, E., and Gal, A. (2002). Characterization of breakpoint sequences of five rearrangements in L1CAM and ABCD1 (ALD) genes. *Hum. Mutat.* **19**, 526–535.

Lawrence, G., Leafasia, J., Sheridan, J., Hills, S., Wate, J., Wate, C., Montgomery, J., Pandeya, N., and Purdie, D. (2005). Control of scabies, skin sores and haematuria in children in the Solomon Islands: Another role for ivermectin. *Bull. World Health Org.* **83**, 34–42.

Lee, M. H., Lu, K., Hazard, S., Yu, H., Shulenin, S., Hidaka, H., Kojima, H., Allikmets, R., Sakuma, N., Pegoraro, R., Srivastava, A. K., Salen, G., Dean, M., and Patel, S. B. (2001). Identification of a gene, ABCG5, important in the regulation of dietary cholesterol absorption. *Nat. Genet.* **27**, 79–83.

Lefevre, C., Audebert, S., Jobard, F., Bouadjar, B., Lakhdar, H., Boughdene-Stambouli, O., Blanchet-Bardon, C., Heilig, R., Foglio, M., Weissenbach, J., Lathrop, M., Prud'homme, J. F., and Fischer, J. (2003). Mutations in the transporter ABCA12 are associated with lamellar ichthyosis type 2. *Hum. Mol. Genet.* **12**, 2369–2378.

Le Saux, O., Urban, Z., Tschuch, C., Csiszar, K., Bacchelli, B., Quaglino, D., Pasquali-Ronchetti, I., Pope, F. M., Richards, A., Terry, S., et al. (2000). Mutations in a gene encoding an ABC transporter cause pseudoxanthoma elasticum. Nat. Genet. 25, 223–227.

Locher, K. P. (2004). Structure and mechanism of ABC transporters. Curr. Opin. Struct. Biol. 14, 31.

Locher, K. P., Lee, A. T., and Rees, D. C. (2002). The E. coli BtuCD structure: A framework for ABC transporter architecture and mechanism. Science 296, 1091–1098.

Lu, K., Lee, M. H., Hazard, S., Brooks-Wilson, A., Hidaka, H., Kojima, H., Ose, L., Stalenhoef, A. F., Mietinnen, T., Bjorkhem, I., Bruckert, E., Pandya, A., Brewer, H. B., Jr., Salen, G., Dean, M., Srivastava, A., and Patel, S. B. (2001). Two genes that map to the STSL locus cause sitosterolemia: Genomic structure and spectrum of mutations involving sterolin-1 and sterolin-2, encoded by ABCG5 and ABCG8, respectively. Am. J. Hum. Genet. 69, 278–290.

Maguire, A., Hellier, K., Hammans, S., and May, A. (2001). X-linked cerebellar ataxia and sideroblastic anaemia associated with a missense mutation in the ABC7 gene predicting V411L. Br. J. Haematol. 115, 910–917.

Marcil, M., Brooks-Wilson, A., Clee, S. M., Roomp, K., Zhang, L. H., Yu, L., Collins, J. A., van Dam, M., Molhuizen, H. O., Loubster, O., Ouellette, B. F., Sensen, C. W., Fichter, K., Mott, S., Denis, M., Boucher, B., Pimstone, S., Genest, J., Jr., Kastelein, J. J., and Hayden, M. R. (1999). Mutations in the ABC1 gene in familial HDL deficiency with defective cholesterol efflux. Lancet 354, 1341–1346.

Martinez-Mir, A., Paloma, E., Allikmets, R., Ayuso, C., del Rio, T., Dean, M., Vilageliu, L., Gonzalez-Duarte, R., and Balcells, S. (1998). Retinitis pigmentosa caused by a homozygous mutation in the Stargardt disease gene ABCR. Nat. Genet. 18, 11–12.

Mata, N. L., Tzekov, R. T., Liu, X., Weng, J., Birch, D. G., and Travis, G. H. (2001). Delayed dark-adaptation and lipofuscin accumulation in abcr$^{+/-}$ mice: Implications for involvement of ABCR in age-related macular degeneration. Invest. Ophthalmol. Vis. Sci. 42, 1685–1690.

McNeish, J., Aiello, R. J., Guyot, D., Turi, T., Gabel, C., Aldinger, C., Hoppe, K. L., Roach, M. L., Royer, L. J., de Wet, J., Broccardo, C., Chimini, G., and Francone, O. L. (2000). High density lipoprotein deficiency and foam cell accumulation in mice with targeted disruption of ATP-binding cassette transporter-1. Proc. Natl. Acad. Sci. USA 97, 4245–4250.

Moynault, A., Luciani, M. F., and Chimini, G. (1998). ABC1, the mammalian homologue of the engulfment gene ced-7, is required during phagocytosis of both necrotic and apoptotic cells. Biochem. Soc. Trans. 26, 629–635.

Mulugeta, S., Gray, J. M., Notarfrancesco, K. L., Gonzales, L. W., Koval, M., Feinstein, S. I., Ballard, P. L., Fisher, A. B., and Shuman, H. (2002). Identification of LBM180, a lamellar body limiting membrane protein of alveolar type ii cells, as the ABC transporter protein ABCA3. J. Biol. Chem. 277, 22147–22155.

Napier, I., Ponka, P., and Richardson, D. R. (2005). Iron trafficking in the mitochondrion: Novel pathways revealed by disease. Blood 105, 1867–1874.

Neff, M. W., Robertson, K. R., Wong, A. K., Safra, N., Broman, K. W., Slatkin, M., Mealey, K. L., and Pedersen, N. C. (2004). Breed distribution and history of canine mdr1–1Delta, a pharmacogenetic mutation that marks the emergence of breeds from the collie lineage. Proc. Natl. Acad. Sci. USA 101, 11725–11730.

Nogee, L. M. (2004). Genetic mechanisms of surfactant deficiency. Biol. Neonate 85, 314–318.

Orso, E., Broccardo, C., Kaminski, W. E., Bottcher, A., Liebisch, G., Drobnik, W., Gotz, A., Chambenoit, O., Diederich, W., Langmann, T., Spruss, T., Luciani, M. F., Rothe, G., Lackner, K. J., Chimini, G., and Schmitz, G. (2000). Transport of lipids from golgi to plasma membrane is defective in tangier disease patients and Abc1-deficient mice. *Nat. Genet.* **24,** 192–196.

Pauli-Magnus, C., Lang, T., Meier, Y., Zodan-Marin, T., Jung, D., Breymann, C., Zimmermann, R., Kenngott, S., Beuers, U., Reichel, C., Kerb, R., Penger, A., Meier, P. J., and Kullak-Ublick, G. A. (2004). Sequence analysis of bile salt export pump (ABCB11) and multidrug resistance p-glycoprotein 3 (ABCB4, MDR3) in patients with intrahepatic cholestasis of pregnancy. *Pharmacogenetics* **14,** 91–102.

Paulusma, C. C., Bosma, P. J., Zaman, G. J. R., Bakker, C. T. M., Otter, M., Scheffer, G. L., Scheper, R. J., Borst, P., and Oude Elferink, R. P. J. (1996). Congenital jaundice in rats with a mutation in a multidrug resistance-associated protein gene. *Science* **271,** 1126–1128.

Pier, G. B., Grout, M., Zaidi, T., Meluleni, G., Mueschenborn, S. S., Banting, G., Ratcliff, R., Evans, M. J., and Colledge, W. H. (1998). *Salmonella typhi* uses CFTR to enter intestinal epithelial cells. *Nature* **393,** 79–82.

Pignatti, P. F., Bombieri, C., Marigo, C., Benetazzo, M., and Luisetti, M. (1995). Increased incidence of cystic fibrosis gene mutations in adults with disseminated bronchiectasis. *Hum. Mol. Genet.* **4,** 635–639.

Poulos, A., Gibson, R., Sharp, P., Beckman, K., and Grattan-Smith, P. (1994). Very long chain fatty acids in X-linked adrenoleukodystrophy brain after treatment with Lorenzo's oil. *Ann. Neurol.* **36,** 741–746.

Pujol, A., Hindelang, C., Callizot, N., Bartsch, U., Schachner, M., and Mandel, J. L. (2002). Late onset neurological phenotype of the X-ALD gene inactivation in mice: A mouse model for adrenomyeloneuropathy. *Hum. Mol. Genet.* **11,** 499–505.

Radu, R. A., Mata, N. L., Bagla, A., and Travis, G. H. (2004). Light exposure stimulates formation of A2E oxiranes in a mouse model of Stargardt's macular degeneration. *Proc. Natl. Acad. Sci. USA* **101,** 5928–5933.

Remaley, A. T., Bark, S., Walts, A. D., Freeman, L., Shulenin, S., Annilo, T., Elgin, E., Rhodes, H. E., Joyce, C., Dean, M., Santamarina-Fojo, S., and Brewer, H. B. (2002). Comparative genome analysis of potential regulatory elements in the ABCG5-ABCG8 gene cluster. *Biochem. Biophys. Res. Commun.* **295,** 276–282.

Remaley, A. T., Rust, S., Rosier, M., Knapper, C., Naudin, L., Broccardo, C., Peterson, K. M., Koch, C., Arnould, I., Prades, C., Duverger, N., Funke, H., Assman, G., Dinger, M., Dean, M., Chimini, G., Santamarina-Fojo, S., Fredrickson, D. S., Denefle, P., and Brewer, H. B., Jr. (1999). Human ATP-binding cassette transporter 1 (ABC1): Genomic organization and identification of the genetic defect in the original Tangier disease kindred. *Proc. Natl. Acad. Sci. USA* **96,** 12685–12690.

Ringpfeil, F., Lebwohl, M. G., Christiano, A. M., and Uitto, J. (2000). Pseudoxanthoma elasticum: Mutations in the MRP6 gene encoding a transmembrane ATP-binding cassette (ABC) transporter. *Proc. Natl. Acad. Sci. USA* **97,** 6001–6006.

Rust, S., Rosier, M., Funke, H., Real, J., Amoura, Z., Piette, J. C., Deleuze, J. F., Brewer, H. B., Duverger, N., Denefle, P., and Assmann, G. (1999). Tangier disease is caused by mutations in the gene encoding ATP-binding cassette transporter 1. *Nat. Genet.* **22,** 352–355.

Schreyer, S. A., Hart, L. K., and Attie, A. D. (1994). Hypercatabolism of lipoprotein-free apolipoprotein A-I in HDL-deficient mutant chickens. *Arterioscler. Thromb.* **14,** 2053–2059.

Seliger, B., Ritz, U., Abele, R., Bock, M., Tampe, R., Sutter, G., Drexler, I., Huber, C., and Ferrone, S. (2001). Immune escape of melanoma: First evidence of structural alterations in two distinct components of the MHC class I antigen processing pathway. *Cancer Res.* **61,** 8647–8650.

Shulenin, S., Nogee, L. M., Annilo, T., Wert, S. E., Whitsett, J. A., and Dean, M. (2004). ABCA3 gene mutations in newborns with fatal surfactant deficiency. *N. Engl. J. Med.* **350,** 1296–1303.

Shulenin, S., Schriml, L. M., Remaley, A. T., Fojo, S., Brewer, B., Allikmets, R., and Dean, M. (2001). An ATP-binding cassette gene (ABCG5) from the ABCG (White) gene subfamily maps to human chromosome 2p21 in the region of the Sitosterolemia locus. *Cytogenet. Cell Genet.* **92,** 204–208.

Singaraja, R. R., Fievet, C., Castro, G., James, E. R., Hennuyer, N., Clee, S. M., Bissada, N., Choy, J. C., Fruchart, J. C., McManus, B. M., Staels, B., and Hayden, M. R. (2002). Increased ABCA1 activity protects against atherosclerosis. *J. Clin. Invest.* **110,** 35–42.

Strautnieks, S., Bull, L. N., Knisely, A. S., Kocoshis, S. A., Dahl, N., Arnell, H., Sokal, E., Dahan, K., Childs, S., Ling, V., Tanner, M. S., Kagalwalla, A. F., Nemeth, A., Pawlowska, J., Baker, A., Mieli-Vergani, G., Freimer, N. B., Gardiner, R. M., and Thompson, R. J. (1998). A gene encoding a liver-specific ABC transporter is mutated in progressive familial intrahepatic cholestasis. *Nat. Genet.* **20,** 233–238.

Sun, H., Molday, R. S., and Nathans, J. (1999). Retinal stimulates ATP hydrolysis by purified and reconstituted ABCR, the photoreceptor-specific ATP-binding cassette transporter responsible for Stargardt disease. *J. Biol. Chem.* **274,** 8269–8281.

Teisserenc, H., Schmitt, W., Blake, N., Dunbar, R., Gadola, S., Gross, W. L., Exley, A., and Cerundolo, V. (1997). A case of primary immunodeficiency due to a defect of the major histocompatibility gene complex class I processing and presentation pathway. *Immunol. Lett.* **57,** 183–187.

Thomas, P. M., Cote, G. J., Wohllk, N., Haddad, B., Mathew, P. M., Rabl, W., Aguilar-Bryan, L., Gagel, R. F., and Bryan, J. (1995). Mutations in the sulfonylurea receptor gene in familial persistent hyperinsulinemic hypoglycemia of infancy. *Science* **268,** 426–429.

Toh, S., Wada, M., Uchiumi, T., Inokuchi, A., Makino, Y., Horie, Y., Adachi, Y., Sakisaka, S., and Kuwano, M. (1999). Genomic structure of the canalicular multispecific organic anion-transporter gene (MRP2/cMOAT) and mutations in the ATP-binding-cassette region in Dubin-Johnson syndrome. *Am. J. Hum. Genet.* **64,** 739–746.

Tornovsky, S., Crane, A., Cosgrove, K. E., Hussain, K., Lavie, J., Heyman, M., Nesher, Y., Kuchinski, N., Ben-Shushan, E., Shatz, O., Nahari, E., Potikha, T., Zangen, D., Tenenbaum-Rakover, Y., de Vries, L., Argente, J., Gracia, R., Landau, H., Eliakim, A., Lindley, K., Dunne, M. J., Aguilar-Bryan, L., and Glaser, B. (2004). Hyperinsulinism of infancy: Novel ABCC8 and KCNJ11 mutations and evidence for additional locus heterogeneity. *J. Clin. Endocrinol. Metab.* **89,** 6224–6234.

Trip, M. D., Smulders, Y. M., Wegman, J. J., Hu, X., Boer, J. M., ten Brink, J. B., Zwinderman, A. H., Kastelein, J. J., Feskens, E. J., and Bergen, A. A. (2002). Frequent mutation in the ABCC6 gene (R1141X) is associated with a strong increase in the prevalence of coronary artery disease. *Circulation* **106,** 773–775.

Wada, M., Toh, S., Taniguchi, K., Nakamura, T., Uchiumi, T., Kohno, K., Yoshida, I., Kimura, A., Sakisaka, S., Adachi, Y., and Kuwano, M. (1998). Mutations in the canilicular multispecific organic anion transporter (cMOAT) gene, a novel ABC transporter, in patients with hyperbilirubinemia II/Dubin-Johnson syndrome. *Hum. Mol. Genet.* **7,** 203–207.

Wang, R., Salem, M., Yousef, I. M., Tuchweber, B., Lam, P., Childs, S. J., Helgason, C. D., Ackerley, C., Phillips, M. J., and Ling, V. (2001). Targeted inactivation of sister of P-glycoprotein gene (spgp) in mice results in nonprogressive but persistent intrahepatic cholestasis. *Proc. Natl. Acad. Sci. USA* **98,** 2011–2016.

Weng, J., Mata, N. L., Azarian, S. M., Tzekov, R. T., Birch, D. G., and Travis, G. H. (1999). Insights into the function of Rim protein in photoreceptors and etiology of Stargardt's disease from the phenotype in abcr knockout mice. *Cell* **98**, 13–23.

Wu, Y. C., and Horvitz, H. R. (1998). The *C. elegans* cell corpse engulfment gene ced-7 encodes a protein similar to ABC transporters. *Cell* **93**, 951–960.

Yamada, T., Shinnoh, N., Kondo, A., Uchiyama, A., Shimozawa, N., Kira, J., and Kobayashi, T. (2000). Very-long-chain fatty acid metabolism in adrenoleukodystrophy protein-deficient mice. *Cell. Biochem. Biophys.* **32**, 239–246.

Yamano, G., Funahashi, H., Kawanami, O., Zhao, L. X., Ban, N., Uchida, Y., Morohoshi, T., Ogawa, J., Shioda, S., and Inagaki, N. (2001). ABCA3 is a lamellar body membrane protein in human lung alveolar type II cells. *FEBS Lett.* **508**, 221–225.

Yang, T., McNally, B. A., Ferrone, S., Liu, Y., and Zheng, P. (2003). A single-nucleotide deletion leads to rapid degradation of TAP-1 mRNA in a melanoma cell line. *J. Biol. Chem.* **278**, 15291–15296.

Zhou, C., Zhao, L., Inagaki, N., Guan, J., Nakajo, S., Hirabayashi, T., Kikuyama, S., and Shioda, S. (2001). Atp-binding cassette transporter ABC2/ABCA2 in the rat brain: A novel mammalian lysosome-associated membrane protein and a specific marker for oligodendrocytes but not for myelin sheaths. *J. Neurosci.* **21**, 849–857.

Zwarts, K. Y., Clee, S. M., Zwinderman, A. H., Engert, J. C., Singaraja, R., Loubser, O., James, E., Roomp, K., Hudson, T. J., Jukema, J. W., *et al.* (2002). ABCA1 regulatory variants influence coronary artery disease independent of effects on plasma lipid levels. *Clin. Genet.* **61**, 115–125.

[25]   Functional Analysis of Detergent-Solubilized and Membrane-Reconstituted ATP-Binding Cassette Transporters

By BERT POOLMAN, MARK K. DOEVEN, ERIC R. GEERTSMA, ESTHER BIEMANS-OLDEHINKEL, WIL N. KONINGS, and DOUGLAS C. REES

Abstract

ATP-binding cassette (ABC) transporters are vital to any living system and are involved in the translocation of a wide variety of substances, from ions and nutrients to high molecular weight proteins. This chapter describes methods used to purify and membrane reconstitute ABC transporters in a fully functional state. The procedures are largely based on our experience with substrate-binding protein-dependent ABC uptake systems from bacteria, but the approaches should be applicable to multisubunit membrane complexes in general. Also, we present simple methods, based on substrate binding or translocation, to follow the activity of the protein complexes in detergent-solubilized and/or membrane-reconstituted state(s).

METHODS IN ENZYMOLOGY, VOL. 400   0076-6879/05 $35.00
Copyright 2005, Elsevier Inc. All rights reserved.   DOI: 10.1016/S0076-6879(05)00025-X

Introduction

ATP-binding cassette (ABC) proteins comprise one of the largest superfamilies of proteins known to date. The majority of ABC proteins are involved in the translocation of solutes across the membrane, for example, nutrient uptake, drug and antibiotic excretion, cell volume regulation, lipid trafficking, and biogenesis. Additionally, a subset of ABC proteins is involved in DNA maintenance and protein synthesis, for example, recombination, DNA repair, chromosome condensation and segregation, and translation elongation. These latter proteins exert their functions in the cytoplasm and/or nucleus and are not considered here; for a comprehensive overview on the different types of ABC proteins and their functions, we refer to Holland *et al.* (2003). ABC proteins facilitating solute translocation, referred to as ABC transporters, reside in the cytoplasmic membrane of bacteria, archaea, and eukaryotes and can also be found in the organellar membranes of the higher organisms, that is, the inner mitochondrial membrane, endoplasmic reticulum, and peroxisomal and vacuolar membranes.

ABC transporters use the hydrolysis of ATP to translocate solutes across cellular membranes. The translocator component of the ABC transporters is composed of two transmembrane and two intracellular ATP-binding subunits (Fig. 1A), with the individual subunits expressed as separate polypeptides or fused to each other in any possible combination (Holland *et al.*, 2003). In addition to these ubiquitous components, prokaryotic ABC transporters involved in solute uptake employ a specific extracellular ligand-binding protein to capture the substrate (Fig. 1B–D). These substrate-binding proteins (SBPs), which are the main determinants

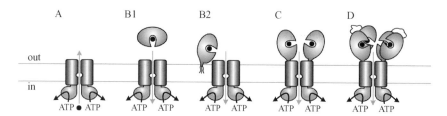

Fig. 1. Schematic representation of the domain organization of ABC transporters. (A) Efflux system. (B) Conventional SBP-dependent uptake system with periplasmic (B1) or lipid-anchored SBP (B2). The chimeric substrate-binding/translocator systems with two and four substrate-binding sites per functional complex are shown in C and D, respectively. "Out" and "in" indicate the extra- and intracellular side of the membrane, respectively; the translocator and NBD subunits are in gray and orange, respectively. (See color insert.)

of the specificity of SBP-dependent ABC transporters, were first identified in gram-negative bacteria, where they reside in the periplasmic space (Neu *et al.*, 1965) (Fig. 1B1). In gram-positive bacteria and Archaea, which lack a periplasm, SBPs are anchored to the outer surface of the cell membrane via a N-terminal lipid moiety (Sutcliffe *et al.*, 1995) (Fig. 1B2), a N-terminal transmembrane segment (observed for Archaea only) (Albers *et al.*, 1999), or fused to either N or C terminus of the translocator protein (van der Heide *et al.*, 2002). With systems that have one or two substrate-binding domains fused to the translocator protein, there may be two (Fig. 1C) or even four (Fig. 1D) substrate-binding sites functioning in the translocation process (Biemans-Oldehinkel *et al.*, 2003).

Crystal structures of ABC transporters are available for the lipid A exporter MsbA from *Escherichia coli* (Chang *et al.*, 2001) and *Vibrio cholera* (Chang, 2003) and the vitamin B_{12} uptake system BtuCD from

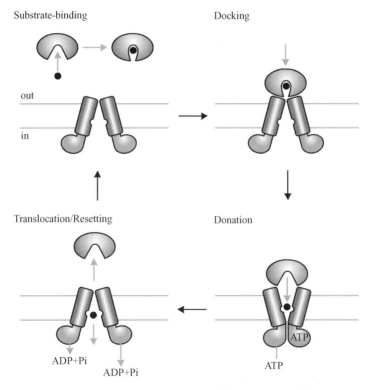

Fig. 2. Model for the transport cycle of a SBP-dependent ABC transporter. For explanation, see text. (See color insert.)

E. coli (Locher *et al.*, 2002). In addition to these structures of complete complexes, crystal structures are available for many of the SBPs and ATP-binding cassette subunits (Holland *et al.*, 2003). From these structures and mutational analyses and kinetic studies on efflux and uptake systems, an understanding of the translocation mechanism of ABC transporters is beginning to emerge (indicated schematically for a SBP-dependent uptake system in Fig. 2). The model is based on data obtained from both SBP-dependent uptake and SBP-independent efflux systems (Chen *et al.*, 2001; Davidson, 2002; Higgins *et al.*, 2004; Liu *et al.*, 1999; Locher *et al.*, 2002). Although details of the model may not be the same for all ABC transporters, the main mechanistic steps seem to be well conserved. Translocation via SBP-dependent ABC transporters starts with binding of the substrate to the SBP. Upon docking of the liganded SBP onto the transmembrane domains (TMDs), a signal is transmitted to the nucleotide-binding domains (NBDs). This enhances the cooperative binding of two ATP molecules, which in turn facilitates closed dimer formation. The closing of the NBD dimer is coupled mechanistically to critical rearrangements in the TMDs, and the affinity for substrate is reduced by opening of the SBP, facilitating the donation of the substrate to a binding site in the TMDs or to the cytoplasm directly. After the substrate has crossed the membrane, the SBP dissociates from the translocator. Hydrolysis of two ATPs initiates resetting of the system for another translocation cycle, that is, after the sequential release of inorganic phosphate and ADP. In an extension of this model (van der Does *et al.*, 2004), it has been postulated that in some ABC transporters, the dimer may be stabilized by one ATP and subsequent hydrolysis of a single ATP may be sufficient for translocation. The issue of the number of ATP molecules hydrolysed per substrate translocated has not been completely settled, and an experimental strategy for ATP/substrate stoichiometry determination is outlined later. Finally, in the model presented in Fig. 2, a single SBP is involved in substrate delivery to the translocator; in case of ABC transporters with multiple substrate-binding domains (SBDs), cooperative interactions between SBDs and the translocator domain may occur (Biemans-Oldehinkel *et al.*, 2003).

This chapter deals with the analysis of purified ABC transporters in detergent-solubilized and membrane-reconstituted states. The focus is on the strategies to incorporate ABC transport systems into lipid vesicles, so-called large unilamellar vesicles (LUVs; diameter of 100–300 nm) and giant unilamellar vesicles (GUVs; diameter of 5–100 μm). In addition, we provide protocols based on radiolabel distribution and fluorescence measurements, yielding kinetic information on substrate binding and substrate translocation by ABC transporters.

ABC Transporters in the Detergent-Solubilized State

Protein Purification and Stability of Oligomeric Complexes

Most membrane proteins, including ABC transporters, are purified at present by affinity chromatography, taking advantage of, in most cases, an amino- or carboxyl-terminal 6- or 10-his tag, typically fused to one of the subunits of the transporter. The affinity tag is often flanked by a specific proteolytic cleavage site to remove the tag after purification. Protocols used to purify proteins by affinity chromatography can be found in numerous papers and are not detailed here. However, because many translocator complexes of ABC transporters are composed of multiple subunits and tend to dissociate upon solubilization, giving rise to specific complications and opportunities, we provide pointers to obtain complexes of stoichiometric amounts of polypeptides after purification.

Chromatography. To facilitate purification, the expression of ABC genes is often increased by using a strong promoter, which can lead to nonstoichiometric production of the individual subunits. In such cases, it is desirable to have the his tag fused to the subunit that is produced in the lowest amount, which allows removal of an excess nontagged subunit(s) by washing of the affinity resin. Alternatively, size exclusion

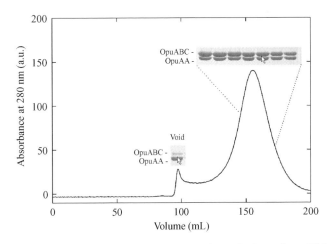

Fig. 3. Size-exclusion chromatography of OpuA from *L. lactis* (unpublished experiment). OpuA was purified by Ni-NTA chromatography, essentially as described (van der Heide *et al.*, 2000), concentrated to 20 mg/ml on Amicon Ultra (100,000 MWCO,) and 40 mg of protein was loaded onto Sephadex S300 in 10 m*M* Na-HEPES, pH 8.0, 100 m*M* NaCl, 15% (v/v) glycerol, and 0.15 % (w/v) CyMal5. (Inset) A Coomassie brilliant blue-stained SDS–polyacrylamide gel of the individual protein fractions. (See color insert.)

chromatography (SEC) can be used to remove the free, noncomplexed subunit, which has the additional advantage that aggregated protein can also be separated from the native complex. An example of purification of the osmoregulatory transporter OpuA from *Lactococcus lactis* (Fig. 1C) is presented in Fig. 3. The overexpression of *opuA* genes yielded a small but significant excess of the ATPase subunit (OpuAA), which in SEC experiments appeared as aggregated material in the void volume of the column.

Cosolvents. To obtain stoichiometric amounts of polypeptides after purification, composition of the chromatography media is critical. Relevant parameters are the choice of detergent, the presence of cosolvents, ionic strength, and the presence or absence of specific additives. In the case of OpuA from *L. lactis* (Fig. 1C), minimally 15% (v/v) glycerol is needed as a cosolvent to prevent the complex from dissociating during the chromatography steps (metal affinity chromatography followed by SEC) (Biemans-Oldehinkel *et al.*, 2003). Because dissociation of the OpuA complex (two substrate-binding/translocator subunits, OpuABC, plus two ATPase subunits, OpuAA) proved to be reversible, we took advantage of this property to form unique heterodimeric complexes (e.g., one wild type and one mutated "nonfunctional" OpuABC subunit plus two OpuAA subunits). This allowed us to dissect the roles of the individual subunits in the oligomeric complex (Biemans-Oldehinkel *et al.*, 2003).

PROCEDURE 1. Purified wild-type and mutant OpuA complexes in buffer A [50 mM KPi, pH 8.0, 200 mM KCl, 20% (v/v) glycerol, 0.05% (w/v) dodecyl-maltoside (DDM)] are mixed at different ratios (final protein concentration of 0.4 mg/ml) and, subsequently, dissociated by decreasing the glycerol concentration to 5% (v/v). For this purpose, the protein mixture is diluted fourfold with buffer A without glycerol. After 30 min of incubation at 4°, the glycerol concentration is increased again to 20% by addition of buffer A containing 60% (v/v) glycerol. Reassembly of the complexes is allowed to continue for 30 min at 4°, after which the proteins are incorporated into liposomes (see later).

Imidazole. The oligopeptide transporter Opp from *L. lactis* is composed of a lipid-anchored substrate-binding protein (OppA), two integral membrane proteins (OppB and C), and two ATP-binding cassettes (OppD and F) (Fig. 1B2). Although metal affinity-based purification protocols generally employ low concentrations (5–30 mM) of imidazole in the protein binding to the resin and washing steps, imidazole severely compromises the stability of the OppBCDF translocator (Doeven *et al.*, 2004). Initial conditions for maintaining an intact OppBCDF complex during purification were screened for ionic strength (0–500 mM KCl) and pH (6.0–8.0) using 20 mM imidazole, 0.05% (w/v) DDM, and 20% (v/v) glycerol as basal medium. This did not lead to purification of the complete translocator, but rather resulted in purification of the His$_6$-tagged component OppC only. However, when imidazole was omitted from the buffer during solubilization and binding of the complex to the metal affinity resin, the OppBCDF proteins could be obtained in an approximate 1:1:1:1 ratio. In contrast to OpuA, varying the glycerol concentration from 0 to 40% (v/v) did not have any effect on the polypeptide stoichiometry of Opp obtained after purification.

Detergents. ABC transporters OpuA (Fig. 1C) and GlnPQ (Fig. 1D) have been solubilized and purified successfully in alkyl-maltosides [decyl to tridecyl; critical micelle concentration (CMC) values ranging from 1.8 to 0.03 mM], cyclohexyl-alkyl-maltosides [CyMal-5, -6, and -7; CMC values ranging from 2.4 to 0.19 mM], alkyl-phosphocholines (FOS-choline-10, -12, and -14; CMC values ranging from 11 to 0.14 mM), and Triton X-100 (CMC of 0.23 mM); detergents are Anagrade and obtained from Anatrace Inc. (Maumee, USA). In general, the longer the alkyl chains (lower CMC), the more stable the protein complexes, but in the case of OpuA, this parameter was less critical than the glycerol concentration. The membrane complexes MalFGK$_2$ and HisQMP$_2$ of the well-characterized maltose transport system from *Escherichia coli* and the histidine transport system from *Salmonella typhimurium*, respectively, were initially purified in octyl β-D-glucopyranoside (OG) (Ames *et al.*, 2001; Davidson *et al.*, 1991). Although OG with its high CMC ($\sim$30 mM) has the advantage that it is removed readily by detergent dilution or dialysis in membrane reconstitution experiments, membrane protein complexes are generally not very stable in this detergent (Knol *et al.*, 1996). Indeed, MalFGK$_2$ was inactivated by OG when used in purification; activity was observed when the protein complex was isolated and purified in DDM and reconstitution was mediated by OG (Davidson *et al.*, 1991).

Binding of Substrates to SBPs and Translocator Subunits

Here, we discriminate between SBP-dependent uptake systems that employ a specific extracellular receptor to deliver the substrate to the translocator (as in Fig. 1B–D) and SBP-independent efflux systems where substrate binding exclusively takes place in the transmembrane domains (Fig. 1A). The general mechanism of substrate association to SBPs is described by the Venus flytrap mechanism (Quiocho *et al.*, 1996), that is, the ligand binds in the cleft between two globular domains and, upon binding of the ligand, the protein closes. The binding of substrate to SBPs is often of relatively high affinity with dissociation constants (K_D) in the submicromolar to low micromolar range and can be monitored by equilibrium dialysis (Silhavy *et al.*, 1975), rapid filtration of protein trapped with ligand (Detmers *et al.*, 2000; Richarme *et al.*, 1983), or spectroscopic (Miller *et al.*, 1983) methods.

For glycine betaine binding to OpuA, we have successfully employed the ammonium sulfate precipitation method described by Richarme *et al.* (1983).

PROCEDURE 2. Detergent-solubilized protein (0.5–2 μM) and [^{3}H] glycine betaine (0.1–20 μM) are mixed in a total volume of 0.1 ml and binding is allowed to proceed for 2 min at 30° (Biemans-Oldehinkel *et al.*, 2003). The reaction is stopped by adding 2 ml ice-cold 70% (v/v) saturated ammonium sulfate and rapid filtration of the mixture through 0.45-μm pore-size cellulose nitrate filters (Schleicher and Schuell GmbH, Dassel, Germany). The filters are washed twice with 2 ml ice-cold 70% (v/v) saturated ammonium sulfate and radioactivity is counted by liquid scintillation spectrometry.

This method is very simple but is not suitable for every protein; the ammonium sulfate-precipitated SBP may not sufficiently trap the substrate and dissociation may occur during the washing steps or the ammonium sulfate may also precipitate free ligand [oligopeptides such as bradykinin start precipitating above 50% (v/v) saturated ammonium sulfate]. Therefore, for the peptide-binding protein OppA from *L. lactis*, an alternative strategy was devised (Detmers *et al.*, 2000). Because OppA is synthesized with an N-terminal lipid moiety, the protein was tethered to the surface of liposomes (for procedure, see later), which increased the retention of the protein on the filters.

PROCEDURE 3. Peptide binding to OppA is measured by making use of the high-affinity ligand [3,4(n)-^{3}H]bradykinin, which is a cationic peptide with the sequence RPPGFSPFR. Bradykinin (0.02–20 μM) is incubated with liposome-tethered OppA at 25° for 4 min in assay buffer [final volume of 0.1 ml; final OppA concentration is 20 μg/ml ($\sim$0.3 μM)], followed by a 1-min incubation with antibodies raised against OppA using a titer of 1:10. Subsequently, the assay mix is diluted with 2 ml ice-cold 8% (w/v) PEG 6000 and filtered over 0.2-μm pore-size cellulose acetate (OE66) filters (Schleicher and Schuell GmbH, Dassel, Germany), after which the filters are washed again with 2 ml ice-cold 8% (w/v) PEG 6000 and radioactivity is counted.

Some aspects of this assay require explanation. First, cellulose acetate instead of cellulose nitrate filters were used to minimize nonspecific binding of bradykinin to the filters; cationic compounds such as bradykinin bind strongly to nitrocellulose filters. Second, antibodies raised against OppA, together with PEG 6000, were used to collect the proteoliposomes on the filters more effectively. Without these treatments, more than 60% of the material passed through the filters. Although similar K_D values were observed in the absence of the antibodies, the amount of peptide binding, reflecting the maximal number of binding sites (B_{max}), was lower due to the loss of liposomes with OppA. The smaller the proteoliposomes, the less well retained they are by the cellulose acetate filters. Third, with neutral or anionic substrates, cellulose nitrate filters can be used, and SBP-specific antibodies plus PEG 6000 are not needed to quantitatively recover the proteoliposomes because these filters trap (proteo)liposomes more efficiently than those made of cellulose acetate due to charge interactions between the lipids and the filter.

For most SBPs that we and others have analyzed, it appears that substrate binding to SBPs elicits conformational changes that can be probed by fluorescence spectroscopy (Lanfermeijer et al., 1999, 2000; Miller et al., 1983). With this method, presteady-state kinetics of binding and dissociation can also be monitored accurately. The spectroscopic method is often preferred over equilibrium dialysis (slow and time-consuming) and rapid filtration (possible loss of ligand) methods because measurements can be performed on-line and estimates of binding constants can be obtained for both high- and low-affinity ligands (Lanfermeijer et al., 1999). One needs to be aware of bleaching of the protein fluorophores by the excitation light, however, and perform the appropriate mock controls. Moreover, with some SBPs, the change in conformation upon substrate binding does not elicit a significant change in protein fluorescence.

PROCEDURE 4. Fluorescence spectra of OppA are obtained with 1 ml of protein solution (0.5–2 μM) in filtered and thoroughly degassed buffers in a quartz cuvette (stirred continuously and kept at 15° with a circulating water bath). Effects of peptides on fluorescence are measured by exciting at 280 $\pm$ 2 nm and measuring the emission at 315 $\pm$ 8 nm. The effect of a saturating concentration of peptide on the intrinsic protein fluorescence of OppA is shown in Fig. 4A; the concentration dependence of the fluorescence increase induced by peptide is shown in Fig. 4B. For a full analysis of steady-state and presteady-state data of ligand binding to SBP, refer to Miller et al. (1983) and Lanfermeijer et al. (1999).

With respect to substrate binding and transport in ABC transporters, it is worth emphasizing that for the initiation of translocation the kinetically relevant species is the liganded substrate-binding protein rather than the free substrate. In case of the histidine transporter from S. typhimurium and maltose transporter from E. coli (Dean et al., 1992; Prossnitz et al., 1989), it has been shown that unliganded SBP competes with liganded SBP for binding to the translocator and that too high a SBP concentration may thus inhibit transport. This finding could not be confirmed for the Opp oligopeptide transporter from L. lactis (Doeven et al., 2004), where the

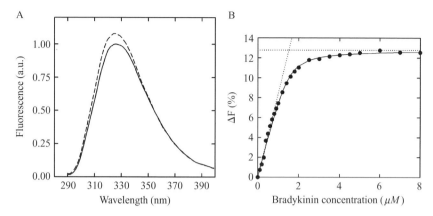

FIG. 4. Peptide binding to OppA. (A) Effect of saturating concentrations of peptide on the intrinsic protein fluorescence of OppA. Emission spectra were recorded in the absence (solid line) and presence (dashed line) of 30 μM of SLSQSKVLP. (B) Concentration dependence of the fluorescence increase (ΔF) induced by bradykinin. The intercept of the two dotted lines is at a peptide concentration of 1.53 μM. Because the concentration of OppA in this experiment was 1.49 μM, approximately 1 mol of bradykinin is bound per mole of OppA. The solid line through the data points represents the best fit; data were taken from Lanfermeijer *et al.* (1999).

affinity of the OppBCDF translocator for liganded OppA appears to be much higher than for unliganded OppA. The affinity of the OppBCDF translocator for liganded OppA was determined by measuring in proteoliposomes (see later) uptake rates at increasing OppA concentrations and saturating amounts of substrate (essentially all the OppA molecules had a peptide bound). The affinity for unliganded OppA was estimated at limiting concentration of substrate, thereby keeping the concentration of liganded OppA essentially constant and low compared to the concentration of unliganded OppA. Although these transport-based assays do not yield direct estimates of K_D values for SBP binding (transport instead of binding activity is measured), the experiments can provide important information on the initial steps in translocation by ABC transporters.

In an elegant series of experiments, Davidson and co-workers (Chen *et al.*, 2001; Davidson, 2002) characterized the binding of maltose to maltose-binding protein (MalE) in the presence of DDM-solubilized MalFGK$_2$ and under conditions where Mg-ATP (or Mg-ADP) was present, either with or without added *ortho*-vanadate. Vanadate inhibits ABC transporters and, in case of the maltose system, trapped ADP in one of the two nucleotide-binding sites immediately after ATP hydrolysis. [γ-^{32}P] ATP-Mg and [α-^{32}P]ATP-Mg were used to monitor binding of ATP and

ADP, respectively, to establish the nucleotide-bound state(s) of the transporter. In these experiments, glycerol was used as a cosolvent to stabilize the oligomeric complex. MalE, MalFGK$_2$, and radiolabeled substrates were separated by ion-exchange chromatography, and the radioactivity associated with each of the protein fractions was determined. These experiments demonstrated that in the vanadate-trapped state, MalE was bound tightly to MalFGK$_2$, whereas both molecules eluted separately in the absence of vanadate. Moreover, they showed that [^{14}C]maltose, tightly bound to free MalE, did not coelute with MalE-MalFGK$_2$. Apparently, upon docking of liganded MalE onto MalFGK$_2$, a "vanadate-stabilized" transition state is formed, provided ATP was present. ATP binding to NBDs opens MalE and allows maltose to enter the translocator and subsequently be released to the cytoplasm. In Fig. 2, these steps are indicated schematically as "docking" and "donation." Thus, more complex ligand–protein and protein–protein-binding assays are possible with ABC transporter complexes in the detergent-solubilized state, and these studies have been instrumental in dissecting the individual steps of the translocation process.

For SBP-independent efflux systems, the substrate binds only in the transmembrane domains and one needs the detergent-solubilized or membrane-reconstituted translocator complex for binding studies. The best studied example is P-glycoprotein[1] (Lugo *et al.*, 2005), but data on drug binding to the multidrug transporter LmrA (Alqawi *et al.*, 2003) from *L. lactis* have also been reported. These systems bind and export relatively large and hydrophobic substrates, and a wide variety of fluorescent and photoaffinity probes are available as reporters of substrate binding and/or transport. Other substrates, such as daunomycin, verapamil, steroid and bile acid conjugates, glucuronide, and glutathione conjugates, are available in radiolabeled form (see Holland *et al.*, 2003). Despite complications in data analysis, there is compelling evidence that P-glycoprotein and members of the MRP family have multiple, partially overlapping, substrate-binding sites in the hydrophobic domain. The broad specificity of these systems for hydrophobic substrates offers an advantage in the choice of ligands, but it also has the disadvantage that these molecules readily partition into detergent micelles and lipid bilayers so that binding and transport data are often complex. Moreover, certain detergents and/or lipids can represent substrates of the multiple drug resistance (MDR)-type

---

[1] There is an enormous amount of data on the binding and transport of substrates to P-glycoprotein and other MDR types of ABC transporters for which Holland *et al.* (2003) is a good starting point; the majority of these studies involve crude membrane preparations with amplified levels of transporter rather than purified and reconstituted proteins.

transporters, which complicates the analysis even further (Borst *et al.*, 2003; Putman *et al.*, 1999). It is beyond the scope of this chapter to discuss these issues further.

Binding of Nucleotides and Fluorescent Nucleotide Analogues to NBDs

ATP, ADP, nonhydrolyzable ATP analogues such as AMP-PNP and ATP-γS, and azido derivatives of ATP and ADP, either with or without vanadate-induced trapping of nucleotides, have been used in numerous studies to delineate the mechanism by which ATP binding and hydrolysis are coupled to substrate translocation. It is beyond the scope of this chapter to evaluate each of these methods. This section describes a simple protocol to evaluate the functional integrity of ABC transporters in the detergent-solubilized state by monitoring binding of the fluorescent nucleotide deriv-ative 2'(3')-*O*-(2,4,6-trinitrophenyl)adenosine 5'-triphosphate (TNP-ATP) or the ADP derivative TNP-ADP. These compounds bind to ATP-binding cassettes with much higher affinity than ATP or ADP and are hydrolyzed slowly (Qu *et al.*, 2003). TNP-labeled nucleotides are weakly fluorescent in aqueous solution, but their quantum yield is enhanced greatly upon trans-fer to a hydrophobic environment, such as the nucleotide-binding site of a protein. The high affinity of TNP-ATP, compared to ATP, is most likely caused by hydrophobic interactions of the TNP moiety with the protein, whereas the specificity of binding is fully conserved by the ATP moiety.

In ABC transporters such as P-glycoprotein, ATP hydrolysis and drug transport can be poorly coupled and, in the absence of substrate, significant ATPase activity can be observed even in the detergent-solubilized state.[2] In transporters such as OpuA, ATP hydrolysis and substrate uptake are tightly coupled and, irrespective of the presence of substrate, ATPase activity is negligible in the detergent-solubilized state. However, upon membrane reconstitution, the OpuA system is fully functional and hydrolyzes ATP in the presence of substrate. Instead of reconstituting individual protein frac-tions, it is much more convenient to obtain first an indication of the func-tional and structural integrity of a system by measuring TNP-ATP binding, for example, for crystallization trials and detergent or cosolvent screens. TNP-ATP binding to a number of ABC transporters has been reported, including P-glycoprotein (Qu *et al.*, 2003), the nucleotide-binding domains of human CFTR (Kidd *et al.*, 2004) and OpuA from *B. subtilis* (Horn *et al.*, 2003), the MDR transporter BmrA from *B. subitilis* (Steinfels *et al.*, 2004),

---

[2] In some cases, it is not clear if the lipid or detergent served as substrate and thereby enhanced ATPase activity.

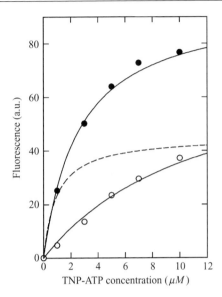

Fig. 5. Binding of TNP-ATP to OpuA (unpublished experiment). TNP-ATP was added stepwise to 50 mM potassium phosphate, pH 7.0, 200 mM KCl, 20% (v/v) glycerol, and 0.05% (w/v) DDM in the absence (○; A) and presence of 0.5 μM OpuA (●; B). The dashed line represents B–A, from which a K_D for TNP-ATP binding to OpuA of 0.9 μM was estimated. Intrinsic protein fluorescence was measured at excitation wavelength settings of 408 ± 2 nm, and fluorescence emission was collected from 520 to 560 nm; measurements were carried out at 20° in a total volume of 1.0 ml.

and OpuA from *L. lactis* (unpublished), but the method is not yet widely used.

PROCEDURE 5. Binding of TNP-ATP (or TNP-ADP) is conveniently observed at protein concentrations of 0.5 μM and fluorescence excitation at 408 ± 2 nm and emission settings at 520–560 nm. To obtain the K_D of TNP nucleotide binding, titration data in the absence of protein need to be subtracted from data in the presence of protein; a typical example of TNP-ATP binding to DDM-solubilized OpuA is shown in Fig. 5. Moreover, by titrating protein-bound TNP-ATP with ATP or ADP, one can obtain an estimate of the affinity constant for binding of these genuine nucleotides. An analysis of the stoichiometry of TNP nucleotide binding to P-glycoprotein is presented in Qu *et al.* (2003). In case of OpuA, negligible hydrolysis of ATP or TNP-ATP is observed at 20–30°. For systems that hydrolyze TNP-nucleotides, measurements may have to be carried out at 4° and/or in the absence of Mg^{2+} to minimize hydrolysis.

Membrane Reconstitution

Strategy and Procedure

The most commonly employed methods for the reconstitution of membrane proteins are based on mixing detergent-solubilized protein together with lipid, either dispersed in detergent or in the form of detergent-destabilized preformed liposomes, followed by subsequent removal of detergent by dilution, dialysis, gel filtration, or adsorption to polystyrene beads (Rigaud et al., 2003). Although each of these methods has been used in combination with a variety of detergents, the dilution, dialysis, and gel filtration techniques are not very efficient in removing low CMC detergents. Because proteins are generally more stable in low CMC detergents such as DDM (and equivalents of the Fos-choline or cyclohexyl-alkyl series) and Triton X-100, these detergents are often preferred for membrane reconstitution and are removed most efficiently by adsorption to polystyrene beads (BioBeads SM2, Bio-Rad Inc.). These beads have an adsorption capacity for detergents of 0.2–0.45 mmol/g of wet beads, whereas the binding of lipid is low, provided the bead-to-detergent ratio is low in the initial stages of reconstitution. In the following experiments, the ABC transporters are generally purified in 0.05% (w/v) DDM or 0.05% Triton X-100, corresponding to $\sim$1 mM of detergent. Due to binding to the proteins, however, the actual concentration of detergent in a protein solution will be significantly higher (see later).

To find optimal conditions for membrane reconstitution, the strategy based on the stepwise solubilization of preformed liposomes and protein incorporation at the different stages of liposome solubilization is used (Knol et al., 1998; Lichtenberg, 1985; Rigaud et al., 1988). The physical state of the liposomes during the titration with detergent is followed by measuring the optical density at 540 nm. The equilibration of detergent and lipid is temperature dependent, and the mixture is generally kept at 20° when the liposomes are titrated with detergent. The solubilization of preformed liposomes can be divided into three stages (Lichtenberg, 1985). During stage I, detergent molecules partition between the aqueous buffer and the bilayer. Stage II starts when liposomes are saturated with detergent, which defines the onset of solubilization ($=R_{sat}$) and continues upon a further increase of the detergent concentration, thereby inducing liposome solubilization and the formation of micelles. At stage III, when the optical density has reached its minimal value (R_{sol}), the mixture consists of micelles at varying detergent/lipid ratios. The parameters R_{sat} and R_{sol} for various detergents describing the solubilization of liposomes can be found in Rigaud et al. (2003).

A good starting point is to mix purified detergent-solubilized protein with preformed liposomes that have been treated with an amount of detergent corresponding to R_{sat} plus 10–20%. For liposomes at a lipid concentration of 4 mg/ml and suspended in 100 mM potassium phosphate, pH 7.0, this corresponds to 4–5 mM of DDM or 1.6–2 mM of Triton X-100. Little or no insertion of protein occurs at detergent concentrations below R_{sat}. The efficiency of membrane reconstitution beyond R_{sat} depends heavily on the detergents used (Knol et al., 1998; Rigaud et al., 1988). Triton X-100 mediates an efficient reconstitution in the range of R_{sat} to R_{sol}, whereas other detergents such as DDM are most effective at R_{sat}. These variations reflect in part the types of structures formed when preformed liposomes are treated with different detergents (Knol et al., 1998). In addition, the optimal detergent concentration to use also depends on the protein. In our hands, Triton X-100-mediated incorporation consistently leads to a better reconstitution efficiency in terms of "translocation activity" than when DDM or other detergents are used. However, we prefer DDM (and other detergents) over Triton X-100 for solubilization and purification because of the high UV absorbance associated with Triton X-100, which precludes accurate protein determination and turbidity measurements by UV spectroscopy. Therefore, many of our reconstitutions involve two detergents: DDM (or other amphiphile) to purify the protein and Triton X-100 to destabilize the preformed liposomes and facilitate reconstitution.

The equilibration of liposomes with detergent requires seconds to minutes for Triton X-100, whereas DDM equilibrates very slowly when the detergent is added stepwise to the liposomes. The low equilibration rate is probably caused by a slow flip-flop of detergent molecules in the bilayer and accompanying rearrangement of lipid and detergent molecules. It is therefore important to follow the titration carefully by measuring the A_{540} to ensure that equilibrium is reached prior to adding the protein. The reconstitution efficiency may be low if the system has not come to equilibrium. Following equilibration of the liposomes with the appropriate concentration of detergent, which has to be determined empirically, the mixture is equilibrated with purified protein in detergent. We routinely use protein stock solutions of ~1 mg/ml for these preparations. Following addition of protein, the detergent is subsequently removed by adsorption onto polystyrene beads (BioBeads SM2).[3] Detergent-solubilized ABC transporter is mixed with the detergent-destabilized liposomes to yield a

---

[3] Routinely, the purified protein and titrated liposomes are mixed and incubated for 30 min at room temperature under gentle agitation before BioBeads are added at a wet weight of 40 mg per milliliter of sample. After 15 min of incubation, fresh BioBeads (40 mg/ml) are added. The sample is incubated at 4° and subsequent additions of BioBeads (40 mg/ml) take place after intervals of 15 and 30 min, overnight, and 2 h before the BioBeads are removed by filtration. Before use, the BioBeads are washed extensively with methanol and water and stored in water at 4°.

final lipid-to-protein of 5000–50,000 to 1 (mol/mol); for a transporter complex with a molecular mass of 200 kDa (OpuA), this corresponds to 17.5–175 to 1 (g/g). At these lipid-to-protein ratios, the amount of detergent brought into the system via the protein is small compared to that needed for destabilization of the preformed liposomes (R_{sat} plus 10–20%). One can minimize the amount carried over by concentrating the protein, for example, by ultrafiltration, which may be needed when reconstitution experiments are carried out at lipid-to-protein ratios below 5000 to 1 (mol/mol). It is our experience that the membrane cutoff of Amicon Ultra (100,000 MWCO) allows passage of detergent micelles (generally <50,000 Da), while concentrating the protein–detergent complex (generally >100,000 Da). In general, we keep the volume of purified protein solution below 10% of the final volume of the liposome mixture. To calculate the contribution of detergent brought into the system via the protein, not only does the "free" detergent concentration need to be considered but also the amount bound to the protein. The amount of bound detergent is evaluated most conveniently by equilibrating the protein with isotopically labeled detergent, with subsequent determination of the protein concentration and radioactivity in fractions following elution from affinity or ion-exchange resins (Friesen et al., 2000; LeMaire et al., 2000; Moller et al., 1993).

In case of the major facilitator superfamily transporters LacS and XylP (12 predicted transmembrane segments per subunit), a binding stoichiometry of ~200 mol of DDM/mol of polypeptide was determined (Friesen et al., 2000; Heuberger et al., 2002); for a protein solution at 1 mg/ml (~14 μM LacS) purified in 1 mM DDM, this corresponds to a total detergent concentration of $1 + 2.8 = 3.8$ mM. For the ABC transporter BmrA, the number of DDM molecules bound per dimeric BmrA (12 transmembrane α helices) was found to be 380 (Ravaud et al., 2005), which is significantly higher than observed for LacS and XylP but consistent with data for DDM-purified OpuA. By SEC, OpuA in DDM runs at an apparent molecular mass of 460 kDa (unpublished). With a protein mass of ~200 kDa, this corresponds to binding of approximately 500 molecules of DDM per dimeric OpuA (16 predicted transmembrane segments). Thus, from these examples of different transporters, there seems to be no simple rule of thumb to assess the amount of detergent binding to the hydrophobic domain of the proteins.

The membrane reconstitution strategy outlined earlier (i.e., insertion of purified protein complexes into detergent-destabilized preformed liposomes) has been successful for several ABC transporters, that is, the MDR efflux systems LmrA (Margolles et al., 1999) from L. lactis and BmrA (Steinfels et al., 2004) from Bacillus subtilis (Fig. 1A), the oligopeptide transporter Opp (Doeven et al., 2004) (Fig. 1B2), the osmoregulatory glycine betaine uptake system OpuA (van der Heide et al., 2000) (Fig. 1C), and the glutamine–glutamic acid uptake system GlnPQ (Schuurman-Wolters *et al.*, 2005) (Fig. 1D) from *L. lactis*. The histidine uptake system HisJQMP$_2$ (Fig. 1B1) from *Salmonella typhimurium* (Ames *et al.*, 2001) and the maltose transport system MalEFGK$_2$ (Fig. 1B1) from *E. coli* (Davidson *et al.*, 1991) have been reconstituted via detergent dilution. In case of HisJQMP$_2$ and MalEFGK$_2$, the complexes were purified in *n*-decanoyl-sucrose and DDM, respectively, and membrane reconstitution was achieved after detergent exchange to OG, followed by detergent dilution (Ames *et al.*, 2001; Chen *et al.*, 2001; Davidson *et al.*, 1991). Membrane reconstitution of P-gp has been achieved by dilution of OG-purified protein in the presence of lipids (Ramachandra *et al.*, 1998). The cystic fibrosis transmembrane conductance regulator (CFTR), an ABC protein that acts as a chloride channel, has been reconstituted by extensive dialysis of lithium dodecyl sulfate- or sodium pentadecafluorooctanoic acid-solublized protein in the presence of sonicated lipids (Ramjeesingh *et al.*, 1999). Given our experience with the membrane reconstitution of several ABC and non-ABC type of transporters, we would not generally recommend the use of these ionic detergents.

The proteoliposomes obtained after detergent removal by polystyrene beads can either be used directly for functional assays or frozen and stored in liquid nitrogen. ABC efflux systems that are reconstituted inside out can be studied by adding Mg-ATP to the external medium and monitoring the uptake of substrate. However, ABC uptake systems reconstituted right side out have the ATP-binding cassettes on the inside, and the ATP or ATP-regenerating system needs to be included in the vesicle lumen to determine substrate uptake. The inclusion of components in the vesicle lumen is often accomplished by multiple cycles of freezing and thawing, followed by extrusion of the proteoliposomes through polycarbonate filters. By freezing and slow thawing of (proteo)liposomes, the membranes fuse, with water-soluble components trapped in the lumen, but the vesicles also become multilamellar in this process. The multilamellar (proteo)liposomes can be made homogeneous and largely unilamellar by extrusion through polycarbonate filters (average pore diameter of 400, 200, or 100 nm; Avestin Inc., Ottawa, Canada). The smaller the pore diameter, the more homogeneous the vesicles are, but the more the specific internal volume decreases, which is disadvantageous for assaying transport reactions. In general, we use filters of 400 or 200 nm through which the (proteo)liposomes are extruded 11 times. A typical protocol for the reconstitution of ABC transport proteins is given in procedure 6.

PROCEDURE 6. A stock solution of liposomes at 20 mg/ml lipid concentration in 50 m*M* KPi, pH 7.0, is extruded through 400-nm pore-size polycarbonate filters, diluted to 4 mg/ml [final buffer composition 50 m*M*

KPi, pH 7.0, 20% (w/v) glycerol], and titrated using Triton X-100. The Ni^{2+}-NTA-purified ABC transporter at a protein concentration of 0.1–1 mg/ml in elution buffer [50 mM KPi, pH 7.0, 200 mM KCl, 20% (w/v) glycerol, 0.05% (w/v) DDM plus 200 mM imidazole] is mixed with the detergent-destabilized liposomes to give a protein:lipid ratio of 1:100 (w/w). The 20% (w/v) glycerol in the buffers is only used for ABC transporter complexes that disassemble in the absence of the cosolvent. In case imidazole has a destabilizing effect, it can be replaced by histidine. The protein and the liposomes are incubated for 30 min at room temperature while shaking gently. To remove the detergent, 40 mg/ml wet weight BioBeads SM2 are added, followed by a 15-min incubation at room temperature. BioBeads SM2 are added four more times, and the incubation times are 15 min, 30 min, overnight, and 1 h at 4°. After five times dilution with 50 mM KPi, pH 7.0 (to lower the glycerol concentration), the ABC transporter-containing proteoliposomes are collected by centrifugation for 1.5 h at 150,000g and 4°, resuspended to 20 mg/ml of lipid in 50 mM KPi, pH 7.0, flash frozen, and stored in liquid nitrogen.

Membrane reconstitution of LmrA, BmrA, OpuA, and GlnPQ has been performed in one or two steps, that is, the insertion of the translocator complex into the membrane either with or without freeze–thaw extrusion to incorporate ATP or an ATP-regenerating system into the (proteo) liposome lumen as appropriate (see later). For the reconstitution of Opp or equivalent systems, depicted schematically in Fig. 1B2, the procedure is somewhat more complex because of the requirement for lipid-anchored substrate-binding protein. Membrane reconstitution of the oligopeptide transporter was achieved via a three-step procedure (Doeven et al., 2004). First, the purified translocator complex OppBCDF in DDM was incorporated into Triton X-100-destabilized liposomes. Thereafter, purified OppA was anchored to the outside of OppBCDF containing liposomes via its N-terminal lipid modification by absorbing the purified protein in 0.05% (w/v) DDM to the (proteo)liposomes, followed by removal of residual detergent with polystyrene beads. This resulted in proteoliposomes containing all five component proteins of the Opp system. Finally, ATP or an ATP-regenerating system was incorporated into the vesicle lumen by freeze–thaw extrusion.

ATP versus ATP-Regenerating System

ABC transporters are driven by ATP but are strongly inhibited by the hydrolysis product ADP. In experiments where ATP is added to the external medium of a proteoliposome suspension, the amount of ADP formed (micromolar) is often low compared to the amount of ATP present (millimolar), and ADP inhibition may be low. However, when ATP is included in the vesicle lumen, the decrease in ATP and the accompanying increase in ADP are substantial and transport may halt after a few minutes, even when residual ATP is still present (Patzlaff *et al.*, 2003). Higher levels of uptake can be attained by incorporating an ATP-regenerating system in the vesicle lumen, but even then uptake will eventually level off due to the accumulation of ADP.

PROCEDURE 7. In a typical experiment (final concentrations are indicated), ATP[-Mg] at 3–10 mM and adjusted to pH 7.0 together with creatine kinase (2 mg/ml; Roche Diagnostics, Mannheim, Germany), creatine-monophosphate[-Na] (20–30 mM) plus 50 mM phosphate[-K or -Na salt] are mixed with the proteoliposomes (20 mg of lipid/ml; see procedure 6), and the mixture is frozen in liquid nitrogen and thawed slowly (tubes with 0.5 ml of proteoliposomes are placed in contact with the wall of a Styrofoam block) at room temperature. The freeze–thaw cycle is repeated two to five times (see later), after which the proteoliposomes are made homogeneous by extrusion through polycarbonate filters. It is critical to flush the extruder with buffer plus Mg-ATP or ATP-regenerating system before and after the final extrusion of the proteoliposomes through polycarbonate filters. Subsequently, external components are removed by ultracentrifugation (300,000g for 15 min), and the proteoliposomes are washed and resuspended in isotonic buffers, usually 50–100 phosphate[-K or -Na salt] at pH 7.0.

In many cases, two to three cycles of freezing and thawing are sufficient to trap ATP or ATP-regenerating system and other components in the vesicle lumen. It is our experience that many membrane proteins insert into preformed liposomes in a particular orientation (Knol *et al.*, 1996), predominantly inside out or right side out, and multiple cycles (five or more) of freezing and thawing are needed to obtain a random orientation of the protein (unpublished observations); an example is given in the Section "ABC Transporters in Large Unicellular Vesicles," illustrates the advantages of a random protein orientation.

Choice of Lipids

Although mixtures of synthetic lipids can be used as sources of exogenous lipids, most bacterial transport systems tested to date show good activity when reconstituted into liposomes composed of a 3 to 1 mixture of *E. coli* total lipids and egg PC. The *E. coli* total lipid mixture contains ~75% phosphatidylethanolamine (PE), ~20% phosphatidylglycerol (PG) and ~5% cardiolipin (CL) or ~75% PE, ~5% PG plus ~20% CL, depending on the growth phase at which the cells were harvested for the isolation of the lipids. Although purified *E. coli* lipids can be purchased from Avanti Polar Lipids Inc., it is our experience that one obtains a much better preparation by self-purifying the crude total lipid extract of Avanti with an acetone/diethylether wash following published procedures (Newman and Wilson, 1980). For systems that depend critically on a particular lipid composition, it is desirable to use synthetic lipids, as the quality of commercial total lipid extracts can vary from batch to batch. A good starting point for a mixture of synthetic lipids is 1,2-dioleoyl-*sn*-glycero-3-phosphatidylcholine (DOPC, bilayer-forming zwitterionic lipid), 1,2-dioleoyl-*sn*-glycero-3-phosphatidylglycerol (DOPG, bilayer-forming anionic lipid), and 1,2-dioleoyl-*sn*-glycero-3-phosphatidylethanolamine (DOPE, nonbilayer-forming zwitterionic lipid) in a ratio of 1:1:2 (van der Heide *et al.*, 2001). In place of dioleoyl (18:1), one can also use dipalmitoleoyl (16:1) lipids.

ABC Transporters in Large Unilamellar Vesicles (LUVs)

Substrate Import versus Export

Proteoliposomes obtained after extrusion through 200-nm polycarbonate filters have an average diameter of 170 ± 50 nm and are often described as large unilamellar vesicles. In order to monitor uptake into these vesicles, one could have a preference for ABC efflux systems reconstituted "inside out" and ABC uptake systems "right side out," but it is not possible to predict beforehand whether a system will be incorporated in a particular orientation. A random orientation has the disadvantage that half of the molecules do not participate in the transport reaction. However, one can also take advantage of a random orientation, as illustrated in the experiment presented in Fig. 6. After five cycles of freeze–thawing and subsequent extrusion through 200-nm polycarbonate filters, the OpuA molecules are oriented randomly. This results in half of the molecules having their substrate-binding domains on the outside and the ATP-hydrolyzing subunits on the inside ("*in vivo* or right-side-out orientation"), whereas the other half has the "inside-out orientation." Because transport of glycine betaine by OpuA is unidirectional and dependent on access of the ABC subunits to the membrane-impermeant cosubstrate, ATP, molecules with the right-side-out and inside-out orientation can be studied separately. Figure 6A shows the uptake of glycine betaine via right-side-out reconstituted OpuA. After approximately 10 min, the uptake halted because of depletion of ATP and build up of ADP. At that point (Fig. 6B), Mg-ATP was added to the assay medium to effect the exit of glycine betaine from the proteoliposomes via inside-out reconstituted OpuA. Figure 6B shows that inside-out reconstituted OpuA was only activated when a high

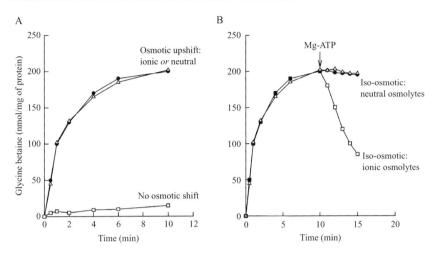

FIG. 6. Glycine betaine uptake and efflux in proteoliposomes containing OpuA. (A) Uptake of glycine betaine in proteoliposomes preloaded with ATP-regenerating system and resuspended in 50 m*M* KPi, 100 m*M* KCl, pH 7.0. Uptake assays were performed under isoosmotic (50 m*M* KPi, 100 m*M* KCl, pH 7.0, □) or hyperosmotic conditions (50 m*M* KPi plus 400 m*M* KCl, pH 7.0, ● 50 m*M* KPi plus 430 m*M* sucrose, pH 7.0, △). (B) After 10 min of uptake, the efflux of glycine betaine was stimulated by the addition of 50 m*M* KPi, pH 7.0, containing 9 m*M* ATP/Mg^{2+} in the presence of 430 m*M* sucrose (△) or 400 m*M* KCl (□). Modified after Patzlaff *et al.* (2003).

concentration of electrolyte was present externally and not with an equivalent amount of sugar, indicating that OpuA is in fact sensing ionic strength rather than osmotic stress (for details, see van der Heide *et al.*, 2001; Poolman *et al.*, 2004). The unidirectionality of transport and the two populations of OpuA molecules allowed us to distinguish between effects exerted by effector molecules at the external and cytoplasmic face of the transporter. In case of OpuA, studies of right-side-out and inside-out OpuA were important for elucidating the osmosensing mechanism of the transporter (Poolman *et al.*, 2004; van der Heide *et al.*, 2001) and determining the ATP/substrate stoichiometry (Patzlaff *et al.*, 2003) (next section).

ATP/Substrate Stoichiometry

In addition to monitoring transport in proteoliposomes, one can also follow the hydrolysis of ATP. With ATP inside the vesicles, the decrease in ATP (and increase in ADP), associated with substrate transport via right-side-out-oriented ABC uptake, can be determined by the luciferin-luciferase assay (ATPlite-M Packard Inc., Groningen, The Netherlands)

(McElroy and DeLuca, 1983). For inside-out-oriented ABC uptake systems or right-side-out-oriented efflux systems, the decrease in ATP concentration is too low to be determined by luciferin-luciferase due to the large external volume. An alternative method, not dependent on a decrease in ATP concentration, involves measurement of the appearance of inorganic phosphate (malachite green-based phosphate assay) (Hess *et al.*, 1975), associated with ATP-driven transport. In this experiment, the concentration of external ATP is kept constant (e.g., around 10 mM). This more sensitive detection method requires phosphate-free buffers. In the case of OpuA and some other transporters, we have noted that the transport activity in proteoliposomes after reconstitution from phosphate buffers is consistently higher than when phosphate-free media are used. In these cases, proteoliposomes were prepared in phosphate-based buffers, after which external phosphate was removed by extensive washing (gel filtration and/or ultracentrifugation) with an alternative isotonic buffer.

To date, there is no consensus on the ATP/substrate stoichiometry for members of the ABC transport family. Indeed, quite the opposite is true, with variations in stoichiometries ranging from 1 to 50 reported even when measured on the same ABC transporter (see Patzlaff *et al.*, 2003). In initial stoichiometry measurements performed with right-side-out-oriented OpuA, we observed variations in stoichiometries in different batches of proteoliposomes that were beyond our control. By sizing the proteoliposomes through 400-, 200-, and 100-nm polycarbonate filters, we could demonstrate a systematic increase in stoichiometry of right-side-out-reconstituted OpuA with decreasing size of the proteoliposomes. Measurement of ATP/substrate stoichiometries on the population of inside-out-oriented OpuA proved that the size of the proteoliposome did not affect the stoichiometry. In each sample, we observed an ATP/substrate stoichiometry of approximately two with no deviation with respect to proteoliposome size. With right-side-out OpuA, ATP/substrate stoichiometry was determined from the decrease in luminal ATP levels, and the ATP concentration (and ATP/ADP ratio) decreased significantly in the course of the experiment. With inside-out-oriented OpuA, the ATP was present on the outside and could be kept constant at approximately 10 mM, which apparently ensures a better coupling than when the ATP and the ADP concentration vary. As stated earlier, it is possible that some transporters hydrolyze one ATP per substrate translocated, whereas other systems such as OpuA use two molecules of ATP, but we feel that many of the stoichiometry values reported in the literature suffer from experimental artifacts, for example, poor coupling between ATP hydrolysis and transport, and substrate leakage (see next paragraph) (Patzlaff *et al.*, 2003), so that most of the earlier experiments may need to be reevaluated.

Fluorescent Substrates to Monitor Transport

Most transport assays are based on determining the distribution of radiolabeled substrates between the inner and the outer compartment of a cell or vesicle. External label is separated from accumulated label by rapid filtration, followed by the measurement of filter-bound radioactivity; the procedures and materials are essentially the same as described for the substrate-binding assays (procedures 2 and 3). However, the nature of the substrate can pose problems, particularly when hydrophobic substrates are used. Hydrophobic substrates, for example, those used by ABC efflux systems such as P-glycoprotein and homologues, leak out relatively easily (during the washing of the filters), resulting in underestimation of the transport activities (and overestimation of ATP/substrate stoichiometries). The leakage of hydrophobic substrates is very prominent in proteoliposomal systems, where the membrane–lipid surface area is relatively large and passive leakage is even more prominent than in native membranes. Fluorescent substrates can offer an alternative as measurements can be performed on-line and separation of internalized and external substrate is not necessary. Particularly powerful are substrates such as ethidium that become highly fluorescent when intercalated to DNA. These measurements, however, require inclusion of DNA in the proteoliposome lumen and removal of external DNA. Because ethidium is relatively hydrophobic, it is important to perform the appropriate controls and compare liposomes with and without ABC transporter (or use inactivated transporter). If DNA intercalation is not the rate-limiting process, one observes a difference in the rate of fluorescence increase when transporter-mediated ethidium uptake is faster than passive diffusion of ethidium.

Hoechst 33342, however, not only becomes highly fluorescent in the presence of DNA, but also when trapped inside the hydrophobic part of the membrane, and this compound is used frequently to assay the activity of MDR type transporters, not only in proteoliposomes, but also in intact cells (Lubelski *et al.*, 2004; Lugo *et al.*, 2005; Putman *et al.*, 2000). It has been noted that the interpretation of transport data obtained with Hoechst 33342 is not as straightforward as often assumed because of the complex pH dependence of the fluorescence spectrum of the compound (Mazurkiewicz, 2004). When Hoechst 33342 is dissolved in aqueous media, the fluorescence quantum yield is much higher at pH 5.0 than at 8.0, but the opposite is true in the presence of liposomes. The increase of Hoechst 33342 fluorescence with increasing pH in the presence of membranes is most likely due to a more efficient partitioning of deprotonated probe into the lipid bilayer. As Hoechst 33342 can exist in four different ionisation states, that is, from 0 to 3+, each with its own lipid-partitioning coefficient, the interpretation of fluorescence traces can thus be highly complex.

ABC Transporters in Giant Unilamellar Vesicles (GUVs)

Giant Unilamellar Vesicles

Giant unilamellar vesicles have been widely used for studies on lipid mobility, membrane dynamics, and lipid domain (raft) formation using single molecule techniques such as fluorescence correlation spectroscopy (FCS) (Kahya et al., 2003). Reports on membrane protein dynamics in these types of model membranes are by far less advanced due to the difficulty of incorporating proteins into GUVs in a functional state.

Protein-containing GUVs can be prepared by drying proteoliposomes (LUVs) followed by the addition of aqueous medium. Water penetrates the dried lamellar structures and GUVs are formed spontaneously due to membrane fusion processes, particularly when 10–25% (w/w) of anionic lipid is present or 10 mM of Mg^{2+} is added to neutral lipids after prewetting (Akashi et al., 1998). In addition, AC electrical fields have been reported to facilitate or impede GUV formation. The major bottleneck for direct incorporation of membrane proteins into GUVs is the dehydration step preceding the formation process. When (proteo)liposomes prepared from unsaturated lipids are dried, the transition temperature (T$_M$) increases by 70–80° (Ricker et al., 2003). This causes the lipids to go from a liquid crystalline to a gel phase, which induces lateral phase separation. These events may cause the protein to aggregate and lose activity. Kahya et al. (2001) circumvented this problem by using peptide-induced fusion of LUVs, containing the membrane protein of interest, with preformed GUVs. Although this method has been applied successfully to study the dynamics and aggregation state of bacteriorhodopsin in GUVs (Kahya et al., 2002), the method is laborious and requires the presence of an unusual lipid (not commercially available) and a fusogenic peptide in the model membranes. Alternative methods (Folgering et al., 2004; Girard et al., 2004) for incorporating polytopic membrane proteins into GUVs involve (partial) dehydration of LUVs containing (purified) membrane proteins, followed by rehydration in the presence of an AC electrical field (Angelova et al., 1992). This method is suitable for highly stable membrane proteins, but not for more labile complexes such as ABC transporters. For single molecule techniques such as FCS, fluorescence resonance energy transfer (FRET), and atomic force microscopy (AFM), it is essential that heterogeneities due to nonproductive protein conformations can be ruled

out and that 100% protein activity can be recovered. The following procedure takes advantage of the stabilizing properties of disaccharides on
membranes and proteins to incorporate membrane protein(s) (complexes)
into GUVs (Doeven *et al.*, 2005), including, among others, the lipid-anchored peptide-binding protein OppA and the translocator complex
OppBCDF of the oligopeptide ABC transporter from *L. lactis*. Sucrose
(and trehalose) prevents lateral phase separation and presumably protein
aggregation during drying by maintaining the membrane in the liquid
crystalline phase; the sugars keep the phase transition temperature, T_M,
of the membrane low (Ricker *et al.*, 2003). In addition, sucrose may have a
stabilizing effect during drying by direct interaction (hydrogen bonding) of
the sugar with polar groups of the protein.

Conversion of LUVs into GUVs

In the following method, protein activity can be preserved by adding
sucrose during drying of protein-containing LUVs; depending on the system, as little as 20–100 mg sucrose/g of lipid is sufficient for the recovery of
full activity.

PROCEDURE 8. LUVs (10 μl of 20 mg/ml lipids), containing Alexa Fluor
488-labeled membrane protein at a given protein-to-lipid ratio in 50 mM
NH$_4$HCO$_3$, pH 8.0, are dried overnight under vacuum at 4° on UV-ozone-
cleaned glass or ITO-coated coverslips (custom coated by GeSim, Dresden,
Germany) in a custom-built sample chamber. UV-ozone cleaning is not
essential for GUV formation but increases the wetting properties of the
coverslip surface, making it easier to dry liposomes from aqueous solution.
Sucrose is added to stabilize the proteins during dehydration. A side effect
of sugars during drying is that they inhibit membrane fusion (Hincha *et al.*,
2003). However, there appears to be an optimal sucrose concentration at
which membrane protein activity is retained (20–100 mg sucrose/g lipid)
and membrane fusion is still possible (<0.86 g sucrose/g lipid) (Doeven
et al., 2005). Rehydration is performed by adding 0.5 ml of 10 mM potassium phosphate, pH 7.0, at room temperature. For electroformation of
GUVs (Angelova *et al.*, 1992), lipids are dried onto ITO-coated coverslips,
and a Pt wire is assembled 1 mm above the ITO-coated slide in the sample
chamber, after which an AC electric field is applied (10 Hz, 1.2 V). GUV
formation is monitored by fluorescence microscopy.

Fluorescence Correlation Spectroscopy

The lateral mobility of OppA and OppBCDF in GUVs was determined
by FCS measurements as described (Doeven *et al.*, 2005). Confocal images
were made of GUVs containing fluorescent-labeled protein, and the focal

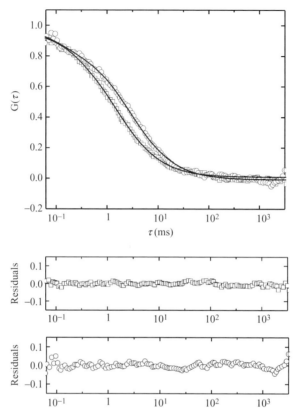

FIG. 7. Diffusion of Opp ABC transporter components in GUVs (unpublished result). Autocorrelation curves for OppA I602C (□) and OppBC(I296C)DF (○) in GUVs. Proteins were labeled with Alexa Fluor 488 as described (Doeven *et al.*, 2005). Curves were fitted with a one-component, two-dimensional diffusion model (solid lines) using Origin software (OriginLab Corporation, Northampton, MA); the residuals of the fits are shown below the figure.

volume was focused on the pole of the GUVs. Representative auto-correlation curves for OppA and OppBCDF are shown in Fig. 7. The lipid-anchored oligopeptide-binding protein OppA diffused with the same speed as fluorescent lipid probes, indicating that OppA does not have interactions with the membrane other than through its lipid anchor. The integral membrane protein complex OppBCDF diffused two times slower compared to OppA, which is in accordance with the Saffman and Delbruck model for diffusion in biological membranes (Saffman *et al.*, 1975). We are currently analyzing, at the single molecule level in GUVs by fluorescence auto- and cross-correlation spectroscopy, the effects of substrate (peptides) and activation state (ATP or nonhydrolyzable nucleotides) on the interactions of OppA with OppBCDF.

Concluding Remarks

At present more than 10 different ABC transporters have been purified and reconstituted in artificial membranes in functional form, and the methods presented here should with minor, if any, modifications be applicable to many more systems. A concern with multisubunit membrane protein complexes is their instability in the detergent-solubilized state. As the functional state of ABC transporters is often characterized by measurements of nucleotide binding and/or hydrolysis, loss of functional and structural integrity of the membrane-embedded domain may go unnoticed. In fact, the relatively high ATPase activity of many ABC transporters in the absence of substrate may actually be caused by dissociation of the complexes in the presence of detergent. The high membrane protein concentrations employed in crystallization trials will necessarily involve significantly higher detergent concentrations bound to the protein (see earlier discussion), which could have profound consequences for the subunit–subunit associations and conformation of the membrane-spanning region. For example, the subunit arrangement reported for the MsbA lipid A exporter from *E. coli* may reflect a partial dissociation of the subunits, mediated by the detergent, during the crystallization process. By screening a range of detergent and (co-)solvent conditions, it has been possible, for the vast majority of membrane proteins studied since the mid-1990s, to obtain intact transporter complexes. The full functionality of the systems can only be assessed after insertion of the proteins into lipid vesicles, for which lipid to protein ratios of 5000–50,000 to 1 on mole basis are generally sufficient. In our hands, it has proven difficult to incorporate proteins into artificial lipid vesicles at lipid-to-protein ratios below 5000, that is as required for various types of biophysical analyses (ESR, FTIR, NMR). Although protein may seem associated with the lipid vesicles as assessed from density centrifugation experiments (unpublished observations), a significant fraction of the protein may not be inserted correctly into the membrane (and may not be translocation competent). In fact, the specific activity of the systems studied to date drops below a lipid-to-protein ratio of 5000. The recent achievement of incorporating ABC transporters and other membrane proteins into giant unilamellar vesicles offers opportunities to study the proteins not only in ensemble, but also at the single molecule level.

Acknowledgments

This work was supported by "Top-subsidie" Grant 700-50-302 from the Netherlands Organization for Scientific Research-Chemische Wetenschappen, a Fulbright grant from NACEE, and EU-FP6 (E-Mep programme). We thank Nushin Aghajari, Arnold Driessen, Marc LeMaire, and Gea Schuurman-Wolters for sharing information prior to publication.

References

Akashi, K., Miyata, H., Itoh, H., and Kinosita, K., Jr. (1998). Formation of giant liposomes promoted by divalent cations: Critical role of electrostatic repulsion. *Biophys. J.* **74**, 2973–2982.

Albers, S. V., Elferink, M. G., Charlebois, R. L., Sensen, C. W., Driessen, A. J., and Konings, W. N. (1999). Glucose transport in the extremely thermoacidophilic *Sulfolobus solfataricus* involves a high-affinity membrane-integrated binding protein. *J. Bacteriol.* **181**, 4285–4291.

Alqawi, O., Poelarends, G., Konings, W. N., and Georges, E. (2003). Photoaffinity labeling under non-energized conditions of a specific drug-binding site of the ABC multidrug transporter LmrA from *Lactococcus lactis*. *Biochem. Biophys. Res. Commun.* **311**, 696–701.

Ames, G. F., Nikaido, K., Wang, I. X., Liu, P. Q., Liu, C. E., and Hu, C. (2001). Purification and characterization of the membrane-bound complex of an ABC transporter, the histidine permease. *J. Bioenerg. Biomem.* **33**, 79–92.

Angelova, M. I., Soléau, S., Ph., Méléard, Faucon, J. F., and Bothorel, P. (1992). Preparation of giant vesicles by external AC electric fields: Kinetics and applications. *Progr. Colloid Polym. Sci.* **89**, 127–131.

Biemans-Oldehinkel, E., and Poolman, B. (2003). On the role of the two extracytoplasmic substrate-binding domains in the ABC transporter OpuA. *EMBO J.* **22**, 5983–5993.

Borst, P., *et al.* (2003). *In* "ABC Proteins: From Bacteria to Man" (I. B. Holland, S. P. C. Cole, K. Kuchler, and C. F. Higgins, eds.). Academic Press, San Diego.

Chang, G. (2003). Structure of MsbA from Vibrio cholera: A multidrug resistance ABC transporter homolog in a closed conformation. *J. Mol. Biol.* **330**, 419–430.

Chang, G., and Roth, C. B. (2001). Structure of MsbA from *E. coli*: A homolog of the multidrug resistance ATP binding cassette (ABC) transporters. *Science* **293**, 1793–1800.

Chen, J., Sharma, S., Quiocho, F. A., and Davidson, A. L. (2001). Trapping the transition state of an ATP-binding cassette transporter: Evidence for a concerted mechanism of maltose transport. *Proc. Natl. Acad. Sci. USA* **98**, 1525–1530.

Davidson, A. L. (2002). Mechanism of coupling of transport to hydrolysis in bacterial ATP-binding cassette transporters. *J. Bacteriol.* **184**, 1225–1233.

Davidson, A. L., and Nikaido, H. (1991). Purification and characterization of the membrane-associated components of the maltose transport system from *Escherichia coli. J. Biol. Chem.* **266**, 8946–8951.

Dean, D. A., Hor, L. I., Shuman, H. A., and Nikaido, H. (1992). Interaction between maltose-binding protein and the membrane-associated maltose transporter complex in *Escherichia coli. Mol. Microbiol.* **6**, 2033–2040.

Detmers, F. J., Lanfermeijer, F. C., Abele, R., Jack, R. W., Tampe, R., Konings, W. N., and Poolman, B. (2000). Combinatorial peptide libraries reveal the ligand-binding mechanism of the oligopeptide receptor OppA of *Lactococcus lactis. Proc. Natl. Acad. Sci. USA* **97**, 12487–12492.

Doeven, M. K., Abele, R., Tampe, R., and Poolman, B. (2004). The binding specificity of OppA determines the selectivity of the oligopeptide ATP-binding cassette transporter. *J. Biol. Chem.* **279**, 32301–32307.

Doeven, M. K., Folgering, J. H., Krasnikov, V., Geertsma, E. R., van den Bogaart, G., and Poolman, B. (2005). Distribution, lateral mobility and function of membrane proteins incorporated into giant unilamellar vesicles. *Biophys. J.* **88**, 1134–1142.

Folgering, J. H., Kuiper, J. M., de Vries, A. H., Engberts, J. B., and Poolman, B. (2004). Lipid-mediated light activation of a mechanosensitive channel of large conductance. *Langmuir* **20**, 6985–6987.

Friesen, R. H., Knol, J., and Poolman, B. (2000). Quaternary structure of the lactose transport protein of *Streptococcus thermophilus* in the detergent-solubilized and membrane-reconstituted state. *J. Biol. Chem.* **275**, 33527–33535.

Girard, P., Pecreaux, J., Lenoir, G., Falson, P., Rigaud, J. L., and Bassereau, P. (2004). A new method for the reconstitution of membrane proteins into giant unilamellar vesicles. *Biophys. J.* **87**, 419–429.

Hess, H. H., and Derr, J. E. (1975). Assay of inorganic and organic phosphorus in the 0.1–5 nanomole range. *Anal. Biochem.* **63**, 607–613.

Heuberger, E. H. M. L., Veenhoff, L. M., Duurkens, H. H., Friesen, R. H. E., and Poolman, B. (2002). Oligomeric state of membrane transport proteins analyzed with blue native electrophoresis and analytical ultracentrifugation. *J. Mol. Biol.* **317**, 617–626.

Higgins, C. F., and Linton, K. J. (2004). The ATP switch model for ABC transporters. *Nature Struct. Mol. Biol.* **11**, 918–926.

Hincha, D. K., Zuther, E., and Heyer, A. G. (2003). The preservation of liposomes by raffinose family oligosaccharides during drying is mediated by effects on fusion and lipid phase transitions. *Biochim. Biophys. Acta* **1612**, 172–177.

Holland, I. B., Cole, S. P. C., Kuchler, K., and Higgins, C. F. (eds.) (2003). "ABC Proteins: From Bacteria to Man." Academic Press, San Diego.

Horn, C., Bremer, E., and Schmitt, L. (2003). Nucleotide dependent monomer/dimer equilibrium of OpuAA, the nucleotide-binding protein of the osmotically regulated ABC transporter OpuA from *Bacillus subtilis*. *J. Mol. Biol.* **334**, 403–419.

Kahya, N., Pecheur, E. I., de Boeij, W. P., Wiersma, D. A., and Hoekstra, D. (2001). Reconstitution of membrane proteins into giant unilamellar vesicles via peptide-induced fusion. *Biophys. J.* **81**, 1464–1474.

Kahya, N., Scherfeld, D., Bacia, K., Poolman, B., and Schwille, P. (2003). Probing lipid mobility of raft-exhibiting model membranes by fluorescence correlation spectroscopy. *J. Biol. Chem.* **278**, 28109–28115.

Kahya, N., Wiersma, D. A., Poolman, B., and Hoekstra, D. (2002). Spatial organization of bacteriorhodopsin in model membranes: Light-induced mobility changes. *J. Biol. Chem.* **277**, 39304–39311.

Kidd, J. F., Ramjeesingh, M., Stratford, F., Huan, L. J., and Bear, C. E. (2004). A heteromeric complex of the two nucleotide binding domains of cystic fibrosis transmembrane conductance regulator (CFTR) mediates ATPase activity. *J. Biol. Chem.* **279**, 41664–41669.

Knol, J., Sjollema, K., and Poolman, B. (1998). Detergent-mediated reconstitution of membrane proteins. *Biochemistry* **37**, 16410–16415.

Knol, J., Veenhoff, L., Liang, W. J., Henderson, P. J., Leblanc, G., and Poolman, B. (1996). Unidirectional reconstitution into detergent-destabilized liposomes of the purified lactose transport system of *Streptococcus thermophilus*. *J. Biol. Chem.* **271**, 15358–15366.

Lanfermeijer, F. C., Detmers, F. J., Konings, W. N., and Poolman, B. (2000). On the binding mechanism of the peptide receptor of the oligopeptide transport system of *Lactococcus lactis*. *EMBO J.* **19**, 3649–3656.

Lanfermeijer, F. C., Picon, A., Konings, W. N., and Poolman, B. (1999). Kinetics and consequences of binding of nona- and dodecapeptides to the oligopeptide binding protein (OppA) of *Lactococcus lactis*. *Biochemistry* **38,** 14440–14450.

LeMaire, M., Champeil, P., and Moller, J. V. (2000). Interaction of membrane proteins and lipids with solubilizing detergents. *Biochim. Biophys. Acta* **1508,** 86–111.

Lichtenberg, D. (1985). Characterization of the solubilization of lipid bilayers by surfactants. *Biochim. Biophys. Acta* **821,** 470–478.

Liu, C. E., Liu, P. Q., Wolf, A., Lin, E., and Ames, G. F. (1999). Both lobes of the soluble receptor of the periplasmic histidine permease, an ABC transporter (traffic ATPase), interact with the membrane-bound complex: Effect of different ligands and consequences for the mechanism of action. *J. Biol. Chem.* **274,** 739–747.

Locher, K. P., Lee, A. T., and Rees, D. C. (2002). The *E. coli* BtuCD structure: A framework for ABC transporter architecture and mechanism. *Science* **296,** 1091–1098.

Lubelski, J., Mazurkiewicz, P., van Merkerk, R., Konings, W. N., and Driessen, A. J. (2004). *ydaG* and *ydbA* of *Lactococcus lactis* encode a heterodimeric ATP-binding cassette-type multidrug transporter. *J. Biol. Chem.* **279,** 34449–34455.

Lugo, M. R., and Sharom, F. J. (2005). Interaction of LDS-751 with P-glycoprotein and mapping of the location of the R drug binding site. *Biochemistry* **44,** 643–655.

Margolles, A., Putman, M., van Veen, H. W., and Konings, W. N. (1999). The purified and functionally reconstituted multidrug transporter LmrA of *Lactococcus lactis* mediates the transbilayer movement of specific fluorescent phospholipids. *Biochemistry* **38,** 16298–16306.

Mazurkiewicz, P. (2004). University of Groningen, The Netherlands.

McElroy, W. D., and DeLuca, M. A. (1983). Firefly and bacterial luminescence: Basic science and applications. *J. Appl. Biochem.* **3,** 197–209.

Miller, D. M., Olson, J. S., Pflugrath, J. W., and Quiocho, F. A. (1983). Rates of ligand binding to periplasmic proteins involved in bacterial transport and chemotaxis. *J. Biol. Chem.* **258,** 13665–13672.

Moller, J. V., and LeMaire, M. (1993). Detergent binding as a measure of hydrophobic surface area of integral membrane proteins. *J. Biol. Chem.* **268,** 18659–18672.

Neu, H. C., and Heppel, L. A. (1965). The release of enzymes from *Escherichia coli* by osmotic shock and during the formation of spheroplasts. *J. Biol. Chem.* **240,** 3685–3692.

Newman, N. J., and Wilson, T. H. (1980). Solubilization and reconstitution of the lactose transport system from *Escherichia coli*. *J. Biol. Chem.* **255,** 10583–10586.

Patzlaff, J. S., van der Heide, T., and Poolman, B. (2003). The ATP/substrate stoichiometry of the ATP-binding cassette (ABC) transporter OpuA. *J. Biol. Chem.* **278,** 29546–29551.

Poolman, B., Spitzer, J., and Wood, J. M. (2004). Bacterial osmosensing: Roles of membrane structure and electrostatics in lipid-protein and protein-protein interactions. *Biochim. Biophys. Acta* **1666,** 88–104.

Prossnitz, E., Gee, A., and Ames, G. F. (1989). Reconstitution of the histidine periplasmic transport system in membrane vesicles: Energy coupling and interaction between the binding protein and the membrane complex. *J. Biol. Chem.* **264,** 5006–5014.

Putman, M., van Veen, H. W., and Konings, W. N. (2000). Molecular properties of bacterial multidrug transporters. *Microbiol. Mol. Biol. Rev.* **64,** 672–693.

Putman, M., van Veen, H. W., Poolman, B., and Konings, W. N. (1999). Restrictive use of detergents in the functional reconstitution of the secondary multidrug transporter LmrP. *Biochemistry* **38,** 1002–1008.

Qu, Q., Russell, P. L., and Sharom, F. J. (2003). Stoichiometry and affinity of nucleotide binding to P-glycoprotein during the catalytic cycle. *Biochemistry* **42,** 1170–1177.

Quiocho, F. A., and Ledvina, P. S. (1996). Atomic structure and specificity of bacterial periplasmic receptors for active transport and chemotaxis: Variation of common themes. *Mol. Microbiol.* **20,** 17–25.

Ramachandra, M., Ambudkar, S. V., Chen, D., Hrycyna, C. A., Dey, S., Gottesman, M. M., and Pastan, I. (1998). Human P-glycoprotein exhibits reduced affinity for substrates during a catalytic transition state. *Biochemistry* **37,** 5010–5019.

Ramjeesingh, M., Garami, E., Galley, K., Li, C., Wang, Y., and Bear, C. E. (1999). Purification and reconstitution of epithelial chloride channel cystic fibrosis transmembrane conductance regulator. *Methods Enzymol.* **294,** 227–246.

Ravaud, S., Do-Cao, M. A., Jidenko, M., Ebel, C., le Maire, M., Jault, J. M., di Pietro, A., Haser, R., and Aghajari, N. (2005). Manuscript in preparation.

Richarme, G., and Kepes, A. (1983). Study of binding protein-ligand interaction by ammonium sulfate-assisted adsorption on cellulose esters filters. *Biochim. Biophys. Acta* **742,** 16–24.

Ricker, J. V., Tsvetkova, N. M., Wolkers, W. F., Leidy, C., Tablin, F., Longo, M., and Crowe, J. H. (2003). Trehalose maintains phase separation in an air-dried binary lipid mixture. *Biophys. J.* **84,** 3045–3051.

Rigaud, J. L., and Levy, D. (2003). Reconstitution of membrane proteins into liposomes. *Methods Enzymol.* **372,** 65–86.

Rigaud, J. L., Paternostre, M. T., and Bluzat, A. (1988). Mechanisms of membrane protein insertion into liposomes during reconstitution procedures involving the use of detergents. 2. Incorporation of the light-driven proton pump bacteriorhodopsin. *Biochemistry* **27,** 2677–2688.

Saffman, P. G., and Delbruck, M. (1975). Brownian motion in biological membranes. *Proc. Natl. Acad. Sci. USA* **72,** 3111–3113.

Schuurman-Wolters, G. K., and Poolman, B. (2005). GlnPQ from *Lactococcus lactis*: An ABC transporter with four extracytoplasmic substrate-binding domains. *J. Biol. Chem* **280,** 23785–23790.

Silhavy, T. J., Szmelcman, S., Boos, W., and Schwartz, M. (1975). On the significance of the retention of ligand by protein. *Proc. Natl. Acad. Sci. USA* **72,** 2120–2124.

Steinfels, E., Orelle, C., Fantino, J. R., Dalmas, O., Rigaud, J. L., Denizot, F., Di Pietro, A., and Jault, J. M. (2004). Characterization of YvcC (BmrA), a multidrug ABC transporter constitutively expressed in *Bacillus subtilis*. *Biochem.* **43,** 7491–7502.

Sutcliffe, I. C., and Russell, R. R. (1995). Lipoproteins of gram-positive bacteria. *J.Bacteriol.* **177,** 1123–1128.

van der Does, C., and Tampé, R. (2004). How do ABC transporters drive transport? *Biol. Chem.* **385,** 927–933.

van der Heide, T., and Poolman, B. (2000). Osmoregulated ABC-transport system of *Lactococcus lactis* senses water stress via changes in the physical state of the membrane. *Proc. Natl. Acad. Sci. USA* **97,** 7102–7106.

van der Heide, T., and Poolman, B. (2002). ABC transporters: One, two or four extracytoplasmic substrate-binding sites? *EMBO Rep.* **3,** 938–943.

van der Heide, T., Stuart, M. C., and Poolman, B. (2001). On the osmotic signal and osmosensing mechanism of an ABC transport system for glycine betaine. *EMBO J.* **20,** 7022–7032.

[26] Yeast ATP-Binding Cassette Transporters: Cellular Cleaning Pumps

By ROBERT ERNST, ROBIN KLEMM, LUTZ SCHMITT, and
KARL KUCHLER

Abstract

Numerous ATP-binding cassette (ABC) proteins have been implicated in multidrug resistance, and some are also intimately connected to genetic diseases. For example, mammalian ABC proteins such as P-glycoproteins or multidrug resistance-associated proteins are associated with multidrug resistance phenomena (MDR), thus hampering anticancer therapy. Likewise, homologues in bacteria, fungi, or parasites are tightly associated with multidrug and antibiotic resistance. Several orthologues of mammalian MDR genes operate in the unicellular eukaryote *Saccharomyces cerevisiae.* Their functions have been linked to stress response, cellular detoxification, and drug resistance. This chapter discusses those yeast ABC transporters implicated in pleiotropic drug resistance and celluar detoxification. We describe strategies for their overexpression, biochemical purification, functional analysis, and a reconstitution in phospholipid vesicles, all of which are instrumental to better understanding their mechanisms of action and perhaps their physiological function.

Introduction

The genome of baker's yeast *Saccharomyces cerevisiae* harbors some 30 distinct genes encoding ATP-binding cassette (ABC) proteins (Bauer *et al.*, 1999; Decottignies and Goffeau, 1997; Taglicht and Michaelis, 1998). Many ABC proteins have transmembrane domains (TMDs), with several predicted membrane-spanning α helices, constituting a subfamily of transporters with overwhelming substrate diversity and a remarkable variety of cellular functions. In *S. cerevisiae,* they are implicated in maintenance of mitochondrial function, maturation of cytosolic Fe/S proteins, pheromone transport, peroxisome biogenesis, stress response, and lipid bilayer homeostasis (Bauer *et al.*, 2003; Schüller *et al.*, 2003). Moreover, a subset of yeast ABC transporters mediate pleiotropic drug resistance (PDR) (Bauer *et al.*, 2003), a phenomenon rather similar to the clinically relevant multidrug resistance (MDR) occuring in mammalian cells, parasites, fungal pathogens, and even in bacteria (Borst and Elferink, 2002; Li and Nikaido, 2004; Litman *et al.*, 2001; Ouellette *et al.*, 2001).

METHODS IN ENZYMOLOGY, VOL. 400   0076-6879/05 $35.00
          DOI: 10.1016/S0076-6879(05)00026-1

ATP-binding cassette transporters can execute membrane translocation of ions, nutrients such as sugars and amino acids, drugs, bile acids, steroids, phospholipids, peptides, and even whole proteins (Benabdelhak et al., 2003; Borst and Elferink, 2002; Litman et al., 2001; Schüller et al., 2003). Nevertheless, the substrate specificities of different ABC transporters range from an extreme narrow substrate specificity to being rather promiscuous and unspecific. ABC transporters of broad substrate specificity display a fascinating yet enigmatic interplay of substrate affinity, diversity, and efficiency in binding and transporting structurally unrelated xenobiotics. In addition, it is questionable whether standard textbook knowledge on substrate binding such as the "key-lock" or the "induced-fit" principle are applicable at all to these proteins.

Despite their different functions and substrate specificities, most, if not all, ABC transporters share a basic blueprint in their domain organization. A minimal ABC transporter consists of two nucleotide-binding domains (NBDs) and two TMDs (Borst and Elferink, 2002; Schmitt and Tampé, 2002; Schüller et al., 2003). TMDs are believed to be part of the substrate binding site(s) and they are certainly instrumental for the architecture of the elusive translocation pore. The highly conserved NBDs energize the transport step by binding and hydrolysis of nucleotides such as ATP, but they also contribute to specific functions of individual ABC transporters. Whereas most bacterial ABC transporters are encoded by separate genes, TMDs and NBDs in yeast, as well as in other eukaryotes, are normally fused and duplicated, yielding a forward (TMD-NBD)$_2$ or reverse (NBD-TMD)$_2$ topology. Furthermore, half-size transporters are also known as having only one NBD fused to a single TMD. Many half-size ABC transporters act as homo- or heterodimers. For example, a functional transporter, the *Escherichia coli* hemolysin A transporter *HlyB* (Benabdelhak et al., 2005) or the human transporter associated with antigen processing (TAP) (Chen et al., 2004; Schmitt and Tampé, 2000), requires a complex of two half transporters.

The hallmark domain of all ABC proteins is the *C-loop* or *signature motif* (LSGGQ) (Higgins, 1992). This consensus sequence and its spacing between conserved *Walker A* and *Walker B* motifs (Walker et al., 1982), which are also found in other nucleotide-binding proteins, define a diagnostic feature of the entire family of ABC proteins. This chapter summarizes methods and approaches to study properties, structure, and function of fungal ABC transporters. We restrict our discussion to characterized *S. cerevisiae* ABC pumps whose expression or overexpression mediates PDR. We have chosen these transporters because they have been studied extensively and because many methods are applicable to studies on other fungal ABC transporters.

Pleiotropic Drug Resistance in Yeast

The inventory of 30 genes encoding ABC proteins in *S. cerevisiae* has been classified into five distinct subfamilies (Schüller *et al.*, 2003; Taglicht and Michaelis, 1998). Based on sequence and structural similarities to mammalian ABC proteins, three major subfamilies were named: the MDR, MRP/CFTR, and ALDP family (Table I). The PDR family constitutes the fourth group, containing several ABC transporters with reverse domain organization. The fifth family is named YEF3/RLi and comprises non-transporter forming yeast ABC proteins.

Several genes of the PDR and MRP/CFTR subfamilies have been implicated in PDR, which can be described as cross-resistance to many structurally and functionally unrelated drugs and xenobiotics. In principle, development of PDR can be the consequence of distinct and very often simultaneously operating molecular mechanisms. Mutations in drug target genes or their transcriptional activators/deactivators can lower the sensitivity to certain drugs. Moreover, reduction of the effective intracellular drug concentration can also be achieved by intracellular drug inactivation, reduced uptake, changes in plasma membrane permeability, or vacuolar sequestration (Fig. 1). Importantly, increased drug efflux due to the overexpression of

TABLE I

INVENTORY OF *SACCHAROMYCES CEREVISIAE* ABC TRANSPORTERS MEDIATING PDR[a]

ABC pump	Substrates	AA	Topology	Location
PDR family				
Pdr5p	Drugs, steroids, antifungal, PL	1511	$(NBD-TMD)_2$	PM
Pdr10p	Detergents	1564	$(NBD-TMD)_2$	PM
Pdr15p	Detergents, phenol herbicides	1529	$(NBD-TMD)_2$	PM
Snq2p	Mutagenic agents, drugs	1501	$(NBD-TMD)_2$	PM
Pdr12p	Weak organic acids	1511	$(NBD-TMD)_2$	PM
Pdr11p	Cholesterol uptake?	1411	$(NBD-TMD)_2$	PM??
Aus1p	Cholesterol uptake?	1394	$(NBD-TMD)_2$	PM??
MRP/CFTR family				
Yor1p	Oligomycin, reveromycin	1477	$NTE-NBD-TMS_6-NBD$	PM
Ycf1p	GS-conj. Cd^{2+}, BA, UCB	1515	$NTE-NBD-TMS_6-NBD$	Vacuole
Ybt1p	BA	1661	$NTE-NBD-TMS_6-NBD$	Vacuole
Bpt1p	UCB	1559	$NTE-NBD-TMS_6-NBD$	Vacuole
Vmr1p	?	1592	$NTE-NBD-TMS_6-NBD$	Vacuole
Nft1p	?	1524	$NTE-NBD-TMS_6-NBD$	?

[a] AA, amino Acids; NBD, nucleotide-binding domain; TMD, transmembrane domain; NTE, N-terminal extension; GS, glutathione S; UCB, unconjugated bilirubin; BA, bile acids; PL, phospholipids; PM, plasma membrane; ER, endoplasmic reticulum.

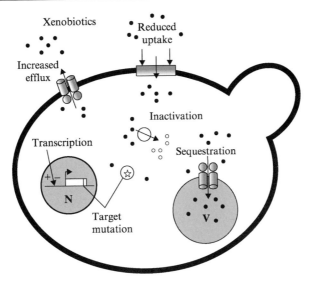

FIG. 1. Principal mechanisms of drug resistance. Drug resistance phenotypes can arise from several distinct molecular principles. Pleiotropic drug resistance often results from overexpression of cell surface ABC efflux pumps, leading to an increased efflux of xenobiotics. Each mechanism on its own or in combination with another one can cause a compound drug resistance phenotype in yeast cells. N, nucleus; V, vacuole.

plasma membrane ABC transporters represents a major cause for acquired drug resistance phenotypes in fungi (Schüller *et al.*, 2003; Taglicht and Michaelis, 1998), as well as in higher cells (Borst and Elferink, 2002).

Numerous studies unraveled a complex regulatory network, which controls the development of PDR in yeast. The PDR network comprises—as also extensively previously reviewed (Moye-Rowley, 2003; Wolfger *et al.*, 2001)—dedicated transcriptional regulators and stress response factors controlling ABC gene dosage positively and negatively (Bauer *et al.*, 2003). Plasma membrane ABC transporters are prime targets of regulation, but also membrane permeases of the major facilitator family (Bauer *et al.*, 1999). Notably, global microarray profiling identified new candidate genes of the PDR network, many of which appear implicated in stress response, lipid and cell wall metabolism, as well as a number of genes of unknown function (DeRisi *et al.*, 2000).

The First Defense Line: Plasma Membrane Pumps Mediating PDR

All cytotoxic compounds must enter cells crossing the plasma membrane. Thus, a number of drug efflux pumps guard the cell surface to protect cells from unwanted or damaging xenobiotics. However, the same efflux pumps may also be involved in extruding endogenous metabolites that may be detrimental to cells (Wolfger *et al.*, 2004), hence playing a dual role in detoxification. The major abundant *S. cerevisiae* drug pumps are Pdr5p, Snq2p, and Yor1p, whereas Pdr10p and Pdr15p are much lower expressed. Pdr5p, Snq2p, and Yor1p recognize hundreds of structurally and functionally unrelated hydrophobic drugs, many of which are cationic (Egner *et al.*, 1998; Golin *et al.*, 2003; Katzmann *et al.*, 1995; Kolaczkowski *et al.*, 1996, 1998; Mahé *et al.*, 1996a; Servos *et al.*, 1993). Notably, the closest homologue of Snq2p, Pdr12p, is highly stress regulated and extrudes weak organic acid anions as well as carboxylate anions (Piper *et al.*, 1998). Many pumps may recognize and extrude at least some of their substrates from the inner leaflet of the lipid bilayer, but this remains a somewhat controversial issue. A lack of transporters leads to viable but drug-hypersensitive cells. Conversely, expression or induced overexpression of ABC pumps causes PDR due to increased drug efflux. The spectrum of identified Snq2p and Pdr5p substrates covers mutagens, agricultural fungicides, antifungal azoles, mycotoxins, herbicides, steroids, nigericin, anticancer drugs, and a great many others (Bissinger and Kuchler, 1994; Kolaczkowski *et al.*, 1998; Mahé *et al.*, 1996a; Wolfger *et al.*, 2004).

Pdr15p is the closest homologue of Pdr5p, sharing 74% primary sequence identity. In contrast to Pdr5p, Pdr15p is markedly induced by general stress conditions such as low pH, starvation, high osmolarity, or heat shock (Wolfger *et al.*, 1997). Interestingly, cells entering diauxic growth have strongly elevated Pdr15p levels, whereas expression of *PDR5* is almost completely abolished (Mamnun *et al.*, 2004; Wolfger *et al.*, 2004). Pdr15p has been implicated in tolerance against membrane-perturbing agents and detergents (Mamnun and Kuchler, in preparation). Pdr10p, another close homologue of Pdr5p, is also regulated within the PDR network (Wolfger *et al.*, 1997) and is strongly induced by stress conditions such as osmostress. However, no transport function or substrate has been identified for Pdr10p. Likewise, a distinct drug efflux activity has not been established for the PDR transporters Aus1p and Pdr11p, which may also localize to the plasma membrane (Wilcox *et al.*, 2002).

Yor1p is also found at the plasma membrane and is probably among the best-studied members of the MRP family. The gene product was initially identified as a transporter conferring oligomycin resistance (Katzmann *et al.*, 1995). Yor1p shares overlapping substrate specificity with PDR pumps such as Pdr5p and Snq2p. It mediates efflux of a large spectrum of structurally unrelated drugs. Moreover, *yor1*Δ cells display marked cadmium hypersensitivity. Yor1p is the only MRP member found in the plasma membrane. All other yeast MRP-like transporters characterized so far appear to home in the vacuolar membrane.

Looking Out My Back Door: Vacuolar Sequestration

Another important mechanism of PDR, the second or intracellular defense line, involves the sequestration of toxic compounds, metabolites, and xenobiotics into the vacuole. Hence, vacuolar ABC pumps of the MRP/CFTR subfamily, such as Ycf1p, Bpt1p, and Ybt1p, constitute a second defense line in the vacuolar membrane. The yeast cadmium factor (Ycf1p) mediates vacuolar detoxification of heavy metals and glutathione S conjugates (GS-conjugates) (Li *et al.*, 1996; Szczypka *et al.*, 1994), as well as the red pigment accumulating in *ade2* mutant cells (Sharma *et al.*, 2003). Loss of Ycf1p leads to hypersensitivity of cells against cadmium and defects in vacuolar uptake of As-GS$_3$ (Ghosh *et al.*, 1999). Bpt1p, a homologue of Ycf1p, accounts for one-third of the total vacuolar uptake of glutathione conjugates. Moreover, it mediates transport of cadmium and unconjugated bile pigments (Klein *et al.*, 2002; Sharma *et al.*, 2002). Finally, the yeast bile transporter Ybt1p mediates vacuolar uptake of bile acids such as taurocholate (Ortiz *et al.*, 1997). Two other vacuolar homologues of Ybt1p, Vmr1p and Nft1p, which appears in the *Saccharomyces* Genome Database (YGD; *http://www.yeastgenome.org/*) as two adjacent open reading frames, *YKR103w* and *YKR104w*, have not been studied in detail (Mason *et al.*, 2003). Their physiological cargo and their cellular localization remain to be elucidated.

Altogether, a battery of yeast ABC transporters build up the external and internal defense lines against a multitude of exogenous xenobiotics but also endogenous toxic metabolites. Notably, more and more players of the PDR network emerge and numerous data about the role of PDR-mediating ABC transporters has accumulated in the last years. However, tantalizing questions about the physiological roles and substrates of Pdr5p, Pdr10p, Pdr15p, and Snq2p, apart from their protective function, remain as yet unanswered. Interestingly, several eukoroytic ABC transporters are present in lipid rafts (Pohl *et al.*, 2005), highly specialized lipid domains within biological membranes with a number of proposed biological functions (Simons and Vaz, 2004). Hence, a hypothesis was put forward claiming that certain ABC transporters function in membrane lipid homeostasis, regulation of membrane permeability, and phospholipid bilayer distribution (Pohl *et al.*, 2005; Pomorski *et al.*, 2003). For instance, yeast Pdr10p is also associated with raft-like domains and its absence causes abnormal trafficking of some other cell surface components (J. Thorner and N. Rockwell, personal communication). Moreover, many Pdr5p, Snq2p, Pdr15p, and Yor1p substrates include lipid-like molecules (Decottignies *et al.*, 1998; Mahé *et al.*, 1996a). ABC pumps might also play a role in the asymmetric distribution of phospholipids in membranes or in the membrane removal of oxidized lipids. Indeed, membrane-damaging agents such as detergents and lyso-phospholipids, all of which impose strong membrane stress, also strongly induce Pdr15p levels (Mamnun and Kuchler, in preparation). Interestingly, the less well-characterized PDR transporters Aus1p and Pdr11p may be involved in sterol uptake under conditions of impaired ergosterol biogenesis (Wilcox et al., 2002). Furthermore, several studies link the master transcriptional regulators of the PDR network, Pdr1p and Pdr3p, to asymmetric lipid distribution and phospholipd metabolism (Grant et al., 2001; Hallström et al., 2001; Hanson and Nichols, 2001; Hanson et al., 2002, 2003; Kean et al., 1997; Nichols, 2002). Certain Pdr5p homologues from fungal pathogens, Cdr1p and Cdr2p, have been implicated in phospholipid flipping (Smriti et al., 2002). Likewise, mammalian ABC transporters may mediate phospholipid flipping (Kalin et al., 2004; Ruetz and Gros, 1994). However, most of these studies used fluorescent NBD lipids as transport or "flippase"substrates, which may be recognized as "drugs"rather than natural membrane lipids. While new "magic" lipid substrates emerge (van Meer and Liskamp, 2005), natural lipid substrates will certainly be necessary for studies on membrane lipid asymmetry.

Despite accumulating evidence, a role of yeast ABC pumps in membrane lipid homeostasis remains controversial. The literature certainly indicates a connection between ABC transporter expression regulation and stress response. For instance, the Pdr12p efflux pump is essential for adaptation to weak organic acid stress exerted by benzoate or sorbate or other lipophilic carboxylate anions (Piper et al., 1998; Schüller et al., 2004). The yeast PDR system and its regulation might have evolved to allow cells to cope with rapidly changing growth and stress conditions, many of which strongly influence membrane lipid composition as well as membrane function. The PDR system may be necessary to maintain the most appropriate transporter composition at the right cellular membrane. The functional dissection of the PDR network is thus extremely challenging because of its complexity, dynamics, and intrinsic redundancy. Nevertheless, the clinical relevance of PDR/MDR phenomena and the fundamental importance of the proposed roles in membrane lipid homeostasis put eukaryotic ABC pumps in the focus of major research efforts.

Genetic Analysis and Phenotypic Characterization

To study ABC proteins in their cellular context, it is indispensable to analyze their overexpression and deletion phenotypes. Chromosomal deletions or disruptions of all yeast ABC pumps are available to aid these experiments (Winzeler *et al.*, 1999). It is remarkable that none of the membrane-embedded yeast ABC pumps is essential for viability. However, in the presence of xenobiotics, anticancer, or antifungal drugs, cells lacking Pdr5p, Yor1p, or Snq2p display marked drug hypersensitivities (Bauer *et al.*, 2003; Wolfger *et al.*, 2001). Double deletions of two plasma membrane transporters showed additive effects on drug sensitivity, implying functional overlaps. For example, a *snq2Δ* deletion strain shows increased sensitivity to Na^+, Li^+, and Mn^{2+}, whereas a *snq2Δ pdr5Δ* double deletion strain exhibits an even more pronounced effect regarding metal sensitivity and intracellular metal ion accumulation (Miyahara *et al.*, 1996). However, the formal possibility remains that Pdr5p and/or Snq2p somehow functions in membrane lipid distribution. Hence, plasma membrane permeability changes due to the absence of these pumps may, at least in part, be responsible for the observed drug susceptibility phenotypes. Surprisingly, for certain drugs, single deletion strains lacking *PDR5* and *YOR1* display no hypersensitivity, but a strong increase in resistance. The mechanism by which the absence of ABC pumps creates hyperresistance—regardless if it is due to altered drug uptake, altered membrane permeability, a different expression profile of other transport systems (see later), or altered vacuolar sequestration—remains an enigma.

Functional overlap also exists between plasma membrane and vacuolar transporters. Whereas Ycf1p transports cadmium into the vacuolar lumen, loss of the plasma membrane transporter Yor1p causes hypersensitivity to oligomycin, various organic anions, and strikingly also cadmium (Cui *et al.*, 1996; Szczypka *et al.*, 1994). This indicates that cadmium detoxification is accomplished by transport across both the vacuolar and the plasma membranes. Single and double deletions of the vacuolar ABC pumps Ycf1p and Bpt1p revealed their implication in vacuolar sequestration of free glutathione, GS-conjugates, and cadmium detoxification (Klein *et al.*, 2002; Sharma *et al.*, 2002). A 60% reduction of glucuronide conjugate (E217βG) transport in *bpt1Δ* cells predicted that *ycf1Δ* cells would show a reduction of about 40%. However, deletion of *YCF1* caused a reduction of transport activity of more than 95%. Such nonadditive phenotypes may be due to an intimately functional connection or cross talk between Ycf1p and the Bpt1p transporter.

Two reports demonstrated the pivotal importance for careful deletion strain analysis (Kihara and Igarashi, 2004; Zhang and Moye-Rowley, 2001). Strains lacking Pdr5p, Yor1p, or both gene products show strongly elevated expression of Rsb1p, an ER, and plasma membrane-associated long chain base (LCB) transporter (Kihara and Igarashi, 2004). These observations imply that these strains not only lack one gene product, but have an altered expression profile of certain other genes in the background, which may compensate or counteract loss of gene function within the same or parallel pathways. Because a deletion of other putative lipid translocases, such as Dnf1p, Dnf2p, and Ros3p, also causes increased Rsb1p expression, the authors suggest a sophisticated cross-talk between sphingolipid and glycerol-phospholipid homeostasis in the establishment of membrane asymmetry (Kihara and Igarashi, 2004; Kato et al., 2002; Pomorski et al., 2003).

In the other example, a single gene deletion strongly affects the expression of two other gene products. Mutations that influenced the functional status of the F_O subunit of the mitochondrial F_1F_O-ATPase led to activation of PDR3 and PDR5 expression, and thus cycloheximide hyperresistance (Zhang and Moye-Rowley, 2001; Zhang et al., 2005). These examples show how little is known about the complex interplays within the PDR network, its dedicated sensors (e.g., for plasma membrane stress), and the downstream signal transduction events. Hence, it is imperative to isolate individual components and transporters to study their biochemical and biophysical properties in vitro, to avoid ambiguities, and to obtain answers about their molecular functions.

Substrate Specificity and Structure–Function Analysis of PDR Transporters

The PDR network comprises several individual ABC pumps, building up a broad drug transport profile due to distinct or partially overlapping substrate specificities (Kolaczkowski et al., 1998). Among those, Pdr5p is perhaps one of the best characterized with respect to the structure–function determinants of substrate binding and transport. The simultaneous identification of Pdr5p by several groups already implied broad drug substrate specificity, including cycloheximide, mycotoxins, glycocorticoids, or cross-resistance to cerulenenin and staurosporine (Bissinger and Kuchler, 1994; Balzi et al., 1994; Hirata et al., 1994; Kralli et al., 1995). Meanwhile, hundreds of substrates were identified by standard resistance assays, in vivo efflux and accumulation assays, their ability to inhibit Pdr5p-mediated rhodamine 6G transport (Conseil et al., 2000, 2001; Egner et al., 1998; Golin et al., 2003; Kolaczkowski et al., 1996; Mahé et al., 1996a). Notably, phospholipids are also suspected transport substrates of Pdr5p (Kean et al., 1997). This was also demonstrated by the accumulation of fluorescent phosphatidylethanolamine (PE) in vivo, and by the observation that cells lacking Pdr5p and Yor1p have an increased PE level in the outer leaflet of the plasma membrane (Decottignies et al., 1998; Pomorski et al., 2003). The molecular basis for this remarkably broad substrate specificity is a major objective for in many studies, including attempts to determine the crystal structure of ABC pumps. The first structural insight into transporter–substrate interactions was obtained from determination of the X-ray structure of the proton-motive, force-driven *Escherichia coli* multidrug efflux pump *AcrB* (Murakami *et al.*, 2002). Diffracting *AcrB* crystals were soaked with four different substrates, revealing the importance of diverse residues for individual substrates and even stabilizing interactions of bound ligand molecules with each other (Yu *et al.*, 2003).

The first studies on Pdr5p along these lines entailed a mutational analysis to pinpoint residues involved in protein folding, inhibitor susceptibility, substrate recognition, and/or drug transport (Egner *et al.*, 1998). Consequently, various mutations in Pdr5p were identified, affecting substrate specificity, inhibitor susceptibility, protein folding, and cell surface targeting (Egner *et al.*, 1998; Gnann *et al.*, 2004; Plemper *et al.*, 1998). Remarkably, the S1360F mutation in the predicted transmembrane helix 10 of Pdr5p led to altered substrate specificity as well as eliminated inhibitor susceptibility for the immunosuppressive drug FK506 (Egner *et al.*, 2000). Strikingly, the substitution S1360A resulted in a mutant Pdr5p variant suddenly hypersensitive to FK506 inhibition. The drug resistance profile of this mutant transporter was highly restricted to the antifungal drug ketoconazole. These studies emphasize the importance of single residues in substrate recognition and inhibitor susceptibility. The genetic approach also implies the existence of more than one drug-binding site in fungal efflux pumps of broad substrate specificity (Egner *et al.*, 1998).

Similar to Pdr5p, detailed mutational structure–function analysis of the MRP/CFTR family members Ycf1p and Yor1p has been performed. Mutations analogous to the most prominent F508Δ mutation in human CFTR, namely the deletion of F713 in Ycf1p and F670 in Yor1p, were analyzed (Katzmann *et al.*, 1999; Wemmie and Moye-Rowley, 1997). Loss of the conserved phenylalanine led to ER retention and degradation. As in the case of Pdr5p, the generated Ycf1p mutations fell in two groups: (1) those that caused defective transporter biogenesis and (2) those that led to a transporter with impaired cadmium tolerance or GS-conjugated leukotriene C_4 (LTC_4) transport. Notably, genetic separation of cadmium resistance and LTC_4 transport is possible by mutations in the regulatory (R) domain and in the cytoplasmic loop 4 (Falcon-Perez *et al.*, 1999). In addition, phosphorylation of two residues within the R domain, S908 and T911, is required for Ycf1p-mediated cadmium detoxification (Eraso *et al.*, 2004; Szczypka *et al.*, 1994).

With the available structural and biochemical information from diverse bacterial and eukaryotic NBDs (Gaudet and Wiley, 2001; Hopfner *et al.*, 2000; Hung *et al.*, 1998; Lewis *et al.*, 2004; Schmitt *et al.*, 2003; Smith *et al.*, 2002), the specific functions of some residues for ATP-binding and hydrolysis were elucidated to an extent, which allows for prediction of their influence on transport activity. Such detailed knowledge of structure–function relationships in the TMDs may aid in the rational design to identify PDR pump inhibitors for improved treatment of fungal infections (Schuetzer-Muehlbauer *et al.*, 2003). However, due to inherent difficulties associated with attempts to solve three-dimensional structures of membrane proteins, it appears extremely challenging to obtain crystal structures of fully assembled and functional ABC drug pumps. Thus, only three ABC pump structures have been solved so far: the lipid-A flippase *MsbA* from *E. coli* (Chang and Roth, 2001) and *Vibrio cholera* (Chang, 2003), as well as the *E. coli* vitamin B_{12} transporter *BtuCD* (Locher *et al.*, 2002), all of which are bacterial ABC transporters. As for Pdr5p, information from a low-resolution reconstructed structure has not led to new insights in the mechanism of Pdr5p function (Ferreira-Pereira *et al.*, 2003).

Life of PDR Pumps: Where Do They Come From— Where Are They Going?

It is reasonably clear that cellular defense against environmental xenobiotics, toxic metabolites, breakdown products, and heavy metals involves increased drug efflux through plasma membrane-residing pumps and vacuolar sequestration. Thus, it is not surprising that most PDR-mediating transporters, such as Pdr5p, Snq2p, Yor1p, and Pdr15p, localize to the plasma membrane (Decottignies *et al.*, 1995; Egner *et al.*, 1995; Katzmann *et al.*, 1999; Mahé *et al.*, 1996b). In contrast, transporters Ycf1p, Bpt1p, and Ybt1p reside in the vacuolar membrane (Fig. 2) where they are responsible for heavy metal detoxification and sequestration of stress-derived breakdown products or metabolites (Li *et al.*, 1996; Ortiz *et al.*, 1997; Sharma *et al.*, 2002).

Some methods of choice to study protein localization include fluorescence microscopy and subcellular fractionation employing sucrose gradient centrifugation. Protein degradation can be analyzed by pulse-chase experiments and subcellular fractionation in different strain backgrounds, carrying mutations in either one of the major proteolytic systems represented by the vacuole or the cytoplasmic proteasome (Plemper *et al.*, 1998). Localization of the vacuolar MRP family member Ycf1p requires its hydrophobic N-terminal extension (NTE), which is comprised by a transmembrane domain (MSD0) and a cytosolic linker region (Sharma *et al.*, 2002). In Ycf1p variants lacking MSD0, vacuolar trafficking is impaired. Notably, the linker region is dispensable for Ycf1p localization, but is required for transport function. These studies strikingly parallel the findings for human

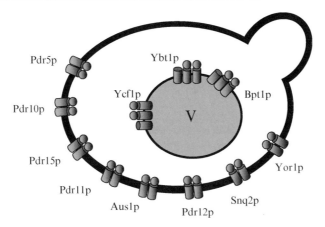

FIG. 2. Subcellular localization of known yeast ABC proteins mediating PDR. Subcellular localization of major yeast ABC proteins in various cellular membranes or compartments is shown. For a list of ABC proteins implicated in PDR and further details, see text and Table I.

MRP2, where MSD0 plays a pivotal role in apical membrane localization (Fernandez *et al.*, 2002).

The life cycle of yeast plasma membrane ABC pumps appears to follow a general scheme of trafficking. After Sec61p-dependent insertion into the ER membrane and quality control of proper protein folding, yeast ABC transporters are delivered to the plasma membrane via the classical exocytic machinery (Egner *et al.*, 1995). Interestingly, a detailed truncation and alanine replacement study of Yor1p identified two DXE sequence motifs that are essential for efficient ER exit and forward traffic to the plasma membrane (Epping and Moye-Rowley, 2002). Once at the plasma membrane, Pdr5p and Yor1p, as well as others, are rather short-lived proteins with a half-life ranging from 45 to 90 min (Egner *et al.*, 1995; Katzmann *et al.*, 1999). Their proteolytic turnover proceeds in the vacuole, to which they are delivered by constitutive endocytosis. Yeast mutants defective in vacuolar protein degradation confirmed that Pdr5p and Yor1p undergo constitutive endocytosis and terminal degradation in the vacuole (Egner *et al.*, 1995). The molecular signal for their endocytosis is a mono-ubiquitin modification at the plasma membrane, which triggers vacuolar delivery of Pdr5p, Snq2, and Yor1p, as well as the related Ste6p **a**-factor mating pheromone transporter (Egner and Kuchler, 1996; Katzmann *et al.*, 1999; Kölling and Hollenberg, 1994; Loayza and Michaelis, 1998). Notably, this modification is limited to a single ubiquitin, which is not sufficient to target the proteins for proteasomal degradation. Instead, ubiquitination functions as an internalization signal for several yeast plasma membrane proteins and may even represent the rate-limiting step in internalization and turnover (Egner and Kuchler, 1996; Kölling and Losko, 1997; Shih *et al.*, 2000). In addition, phosphorylation of Pdr5p by yeast casein kinase I isoforms (Yck1p; Yck2p) seems to stabilize the protein and may play a role in Pdr5p trafficking and turnover (Decottignies *et al.*, 1999).

Extensive mutational studies on Pdr5p and Yor1p identified mutant proteins that do not localize to the plasma membrane. The C1427Y mutation in Pdr5p, which disrupts a disulfide bond (Egner *et al.*, 1998), and the F670 deletion in Yor1p (Katzmann *et al.*, 1999), analogous to the most frequent CFTR mutation, ΔF508, lead to misfolded proteins. They fail to exit the ER membrane and are thus subject to removal by the ER-associated degradation (ERAD) system devoted to the rapid removal of misfolded proteins immediately after or even during their synthesis (Fewell *et al.*, 2001). Whereas wild-type ABC pumps are degraded in the vacuole, misfolded mutants are extracted directly from the ER membrane through a Sec61p-dependent retrograde pathway, become polyubiquitin-modified, and subsequently degraded by the proteasome (Plemper *et al.*, 1998). Similar data have been obtained for Ycf1p (Wemmie and Moye-Rowley, 1997), the mating pheromone transporter Ste6p (Loayza *et al.*, 1998), as well as the human CFTR expressed in yeast (Gnann *et al.*, 2004). Interestingly, complementation studies of ERAD-deficient yeast cells by mammalian orthologues underscore the similarity of membrane protein folding and ERAD in mammals and yeast (Gnann *et al.*, 2004). Even more important, they emphasize the value of yeast as a model system to study the principles of sorting, localization, and degradation of multispanning membrane proteins.

Functional Assays for Yeast ABC Pumps Mediating Drug Resistance

For most ABC transporters of the PDR network, numerous functional assays have been established. Due to their simplicity and low cost, the most widely used tests for drug resistance are growth inhibition assays on agar plates (Bissinger and Kuchler, 1994). However, more complicated assays, such as cross-linking studies and *in vivo* accumulation studies, as well as real-time *in vitro* transport studies on isolated plasma membrane vesicles using fluorescent dyes, were also performed (Decottignies *et al.*, 1998; Egner *et al.*, 1998; Kolaczkowski *et al.*, 1996). Drug susceptibilities of various yeast strains can be analyzed qualitatively and even semiquantitatively using drug gradient plates, allowing for estimation of minimal inhibitory concentrations (Katzmann *et al.*, 1999). Serial dilutions of yeast cultures are spotted on agar plates containing a putative substrate at different concentrations. Alternatively, the halo assay can be employed. Filter disks soaked with drug solutions are placed onto lawns of tester cells. The resulting zone of inhibition surrounding the filter disk is a quantitative measure of toxicity (Nakamura *et al.*, 2001). However, the hydrophobic nature of many drug substrates and their limited diffusibility in agar plates may produce artifacts in the halo assay and may lead to underestimation and misinterpretation of toxicity.

Certain PDR transporters were overexpressed and partially purified for further analysis. The affinity constants of diverse substrates and inhibitors were calculated from tryptophane-quenching studies using partially purified and solubilized Pdr5p (Conseil *et al.*, 2000). However, in the case of Pdr5p, it seems questionable whether all 21 tryptophane residues participate directly and specifically in substrate interaction, as judged from a 100% quenching efficiency reported in this study. Furthermore, it is hard to envision that the concentration of a certain compound necessary to inhibit 50% of transport is smaller than the concentration necessary to saturate 50% of the binding sites in the absence of transport. In addition, the effective affinity of a drug depends strongly on the partition coefficient between the membrane (or the detergent micelle) and the aqueous phase, making it nearly impossible to obtain solid functional data without reconstitution of purified protein into proteoliposomes.

An excellent tool used to study the function of PDR transporters in real time is the fluorescent-based, energy-dependent rhodamine 6G quenching assay (Kolaczkowski *et al.*, 1996). Here, enriched plasma membranes from Pdr5p-overexpressing or *pdr5Δ* cells are incubated with the fluorescent dye rhodamine 6G. Upon addition of Mg^{2+}/ATP, rhodamine 6G fluorescence decays in a time-dependent manner for membranes isolated from Pdr5p-overexpressing cells, but not for membranes derived from *pdr5Δ* cells. As suggested by the authors, rhodamine 6G is either transported to the lumen of the inside-out plasma membrane vesicles, followed by rapid rebinding to the inner leaflet, or simply flipped from the outer to the inner membrane leaflet (Kolaczkowski *et al.*, 1996). In both scenarios, rhodamine 6G accumulates in the inner leaflet, causing concentration-dependent formation of nonfluorescent dye dimers and trimers. Therefore, the rate of fluorescence decay and the total change of fluorescence correlate directly with the rate of rhodamine 6G transport. The observed quenching could be competitively inhibited by micromolar concentrations of anticancer drugs, ionophoric peptides, and steroids. Upon addition of oligomycin, an inhibitor of Pdr5p ATPase activity, the fluorescence signal recovered completely, confirming the tight coupling between drug transport and ATP hydrolysis (Kolaczkowski *et al.*, 1996).

Alternatively, intracellular accumulation of fluorescent dyes within yeast cells can be quantified with a fluorescence-activated cell sorter (FACS). This can be used to determine the activity of mutant Pdr5p variants and to screen for inhibitors of Pdr5p-mediated transport (Egner et al., 1998). Intracellular accumulation of radiolabeled substrates was also used to identify steroids as substrates of Pdr5p and Snq2p, as overexpression of these pumps lowered the intracellular estradiol concentration (Mahé et al., 1996a). Efflux assays measuring the extrusion of diverse radiolabeled compounds from yeast cells were used to decipher the size dependence of Pdr5p substrates for xenobiotic efflux, suggesting at least three partially overlapping but independent drug-binding sites (Golin et al., 2003; Yu et al., 2003), which is in excellent agreement with earlier genetic studies (Egner et al., 1998, 2000).

A main problem in analyzing the molecular events of drug transport is the hydrophobic nature of substrates, resulting in high membrane partitioning. Thus, although there is a partition of substrates between the lipid bilayer and the aqueous phase, the substrates are not transported from one aqueous environment to another (e.g., into a vesicular lumen), but rather between two membrane leaflets. To understand the mechanism of transport, to identify substeps of translocation, and to learn more about the complex interactions among membrane lipids, protein, and drugs, new functional assays of high time resolution and sensitivity have to be established.

In contrast to Pdr5p, the vacuolar Ycf1p transports organic compounds after their conjugation to cellular gluthatione. Conjugation of substrates increases their solubility and prevents rebinding to membranes. This circumstance allowed—after isolation of vacuolar membrane vesicles—for a detailed in vitro uptake analysis using radiolabeled substrate complexes. For example, vacuolar transport of Cd-GS$_2$, but not Cd-GS, requires Ycf1p (Li et al., 1997; Rebbeor et al., 1998). Similar to the rhodamine 6G quenching assay, this type of assay allows for the investigation of transport inhibition by ATPase inhibitors or by competition with nonlabeled substrates.

Because ABC transporters act as ATP-dependent pumps, the mechanism of ATP hydrolysis and how the release of chemical energy is coupled to substrate transport is of central interest. So far, the ATPase activities of partially purified Pdr5p, Snq2p, and Yor1p have been analyzed (Decottignies et al., 1994, 1995, 1998). Pdr5p hydrolyzes CTP, GTP, UTP, or ATP to a comparable degree (NTPase activity) and is inhibited by vanadate and oligomycin. ATP-binding by Pdr5p and Yor1p was confirmed by photolabeling of these proteins with TNP-8-azido-ATP. However, all these data were obtained from yeast plasma membranes enriched in ABC transporters or from pump-enriched, detergent-solubilized preparations. Obviously, it is highly challenging, if not impossible, to use such assays to paint a detailed qualitative and quantitative picture of how substrates and nucleotides interact with these pumps, not even to mention thermodynamic or kinetic parameters governing this process. Hence, the best way to obtain such essential information about ABC pump function is by the use of highly purified, monodisperse, and, most importantly, functional protein—either in detergent solution or reconstituted in liposomes. Such a strategy should allow one to dissect and answer some major open questions regarding the molecular mode of action of yeast PDR pumps.

Overexpression and Purification of Yeast ABC Pumps

The first bottleneck in biochemical and biophysical investigations of eukaryotic membrane proteins is their functional overexpression, particularly since many ABC transporters are of rather low abundance. Yeast cells harboring a *PDR1-3* allele encoding a hyperactive master regulator of the PDR network (Balzi *et al.*, 1994; Meyers *et al.*, 1992) have become a highly useful tool for biochemical studies on efflux pumps. The hyperactive Pdr1-3p transcription factor leads to overexpression of plasma membrane ABC pumps such as Pdr5p and Snq2p (Mahé *et al.*, 1996b; Mamnun *et al.*, 2002), as well as numerous other proteins involved in lipid and cell wall metabolism (DeRisi *et al.*, 2000). *PDR1-3* cells were employed for overexpression of *PDR5*, *SNQ2*, and *YOR1* (Decottignies *et al.*, 1994, 1995, 1998) and partially purified membranes were used for biochemical studies. Later, Pdr5p was purified by virtue of a C-terminal hexa-histidine affinity tag and analyzed by electron microscopy and single particle techniques (Ferreira-Pereira *et al.*, 2003). The three-dimensional reconstruction suggested a dimeric organization, which is in line with the previous suggested domain–domain interaction of the N-terminal NBD with itself (Subba Rao *et al.*, 2002). However, this is in contrast to earlier genetic studies suggesting that Pdr5p is functional as a monomer (Egner *et al.*, 1998). Because Snq2p and Pdr5p cofractionate on glycerol gradients, the partial purification of Snq2p was improved by "genetic purification" through the deletion of *PDR5* and other ABC pumps (Decottignies *et al.*, 1998). As for Yor1p, its expression level, which even in the *PDR1-3* background is rather low, was enhanced drastically by the substitution of the *YOR1* promotor with the *PDR5* promotor, allowing for partial purification and characterization of ATP-binding and ATP hydrolysis.

These approaches raised the question whether the *PDR5* promotor in combination with specifically improved affinity tags can also be used for functional overexpression and purification of other yeast membrane proteins in general. Based on deletion cassettes currently used for gene deletion approaches (Wach *et al.*, 1994, 1997), we constructed a versatile tool for a one-step polymerase chain reaction (PCR)-mediated chromosomal tagging of any yeast gene, such that its expression is driven from the *PDR5* promoter. This new tagging cassette is titratable through Pdr1-3p, perhaps enabling overexpression and purification of most genes in yeast (Ernst *et al.*, in preparation). Using the heterologous *Schizosaccharomyces pombe his5+* module as a selectable marker, a removable N-terminal deca-histidine affinity tag can be fused to any yeast gene (Fig. 3).

A

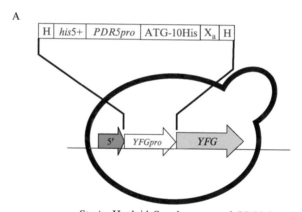

Strain: Haploid *S.c.* chromosomal *PDR1-3*

B

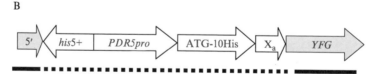

FIG. 3. Tagging cassette for genomic tagging and overexpression of yeast genes. (A) A specialized, haploid *S. cerevisiae* strain with a *PDR1-3* background is transformed with a PCR product carrying the cassette for overexpression and affinity tagging of your favorite gene (YFG) using a short homology flanking region of about 40 bp (H) on both sides of the cassette. The cassette contains the *Schizosaccharomyces pombe his5+* (*his5+*) module as the selection marker, the sequence of the *PDR5* promotor (*PDR5pro*), a start codon (ATG), and a deca-histidine affinity tag (10His), as well as a factor X_a cleavage site (X_a). (B) The correctly targeted cassette substitutes the promotor of the gene of interest *(YFG)*, driving overexpression from the *PDR5* promoter. The expression protein of interest is fused to a removable N-terminal deca-histidine tag, which allows for rapid and efficient purification by chelating metal affinity chromatography. (See color insert.)

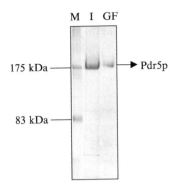

FIG. 4. Affinity purification of Pdr5p via a deca-histidine affinity tag. Cells overexpressing Pdr5p with an N-terminal deca-histidine tag were harvested from a 10-liter culture and disrupted using a cell rupture machine. Pdr5p was extracted and solubilized in detergent from a plasma membrane-enriched fraction. Pdr5p was purified with immobilized metal affinity chromatography (I) followed by gel filtration chromatography (GF). A Coomassie-stained 7.5% SDS–PAGE, containing highly purified Pdr5p after immobilized metal affinity chromatography (I) and gel filtration (GF), is shown. M indicates protein molecular mass marker in kilodaltons. Further details of the purification scheme will be reported elsewhere (Ernst *et al.*, in preparation). (See color insert.)

Furthermore, if gene tagging is done in a *PDR1–3* genomic background, expression levels are dramatically higher than in wild type cells. Depending on the protein in question, this system can yield very high protein levels, constituting up to 10% of the total plasma membrane protein. We have exploited and optimized this system for the overexpression and purification of Pdr5p, yielding up to 1 mg of close to homogeneity-purified Pdr5p ABC transporter per liter of yeast culture (Fig. 4).

The purified Pdr5p transporter has been reconstituted into vesicles and subsequently used for studies on the Pdr5p transport mechanism (Ernst *et al.*, in preparation). Hence, we expect and predict that this system is well suited for the functional overexpression and purification of other ABC transporters, as well as many other plasma membrane proteins.

Conclusions and Perspectives

Although hundreds of different compounds have been identified as substrates for yeast ABC transporters, their normal cellular functions and physiological roles remain ill-defined. A hypothetical function of these pumps may be membrane lipid homeostasis, and at least some fungal ABC pumps have indeed been shown to be involved in membrane lipid transport. However, the experimental evidence is vague and still controversial at best. What else could be the physiological function of PDR pumps? Considering the environmental cues yeast cells encounter in nature, requiring constant and rapid adaptation to new environments and the need to defend in a continuous ongoing microbiological warfare on ecological niches, a normal function of PDR mediation transporters in detoxification and stress adaptation remains as the most highly plausible hypothesis. The possibility of obtaining highly purified and functional yeast ABC pumps allows for mechanistic and structural investigations, which will provide important insights in the molecular mode of action, and perhaps contribute to a better understanding of other medically relevant ABC proteins.

Acknowledgments

Research was supported by grants from the Austrian Science Foundation (FWF-15934-B08) and the Vienna Science and Technology Fund from the City of Vienna (WWTF-Project LS113), and in part by an FPG Marie-Curie Training Network "Flippases" (MRTN-CT-2004-005330). Robert Ernst received a short-term fellowship of the European Molecular Biology Organization EMBO (ASTF 297-2003 and ASTF 193-2004). Lutz Schmitt was supported by the Emmy Noether Programme of the DFG.

References

Balzi, E., Wang, M., Leterme, S., Van Dyck, L., and Goffeau, A. (1994). *PDR5*, a novel yeast multidrug resistance conferring transporter controlled by the transcription regulator *PDR1. J. Biol. Chem.* **269**, 2206–2214.

Bauer, B. E., Schüller, C., and Kuchler, K. (2003). Fungal ABC proteins in clinical drug resistance and cellular detoxification. *In* "ABC Proteins: From Bacteria to Man" (I. B. Holland, K. Kuchler, and C. F. Higgins, eds.), pp. 295–316. Academic Press/Elsevier Science, Amsterdam.

Bauer, B. E., Wolfger, H., and Kuchler, K. (1999). Inventory and function of yeast ABC proteins: About sex, stress, pleiotropic drug and heavy metal resistance. *Biochim. Biophys. Acta* **1461**, 217–236.

Benabdelhak, H., Kiontke, S., Horn, C., Ernst, R., Blight, M. A., Holland, I. B., and Schmitt, L. (2003). A specific interaction between the NBD of the ABC-transporter HlyB and a C-terminal fragment of its transport substrate haemolysin A. *J. Mol. Biol.* **327**, 1169–1179.

Benabdelhak, H., Schmitt, L., Horn, C., Jumel, K., Blight, M. A., and Holland, I. B. (2005). Positive co-operative activity and dimerization of the isolated ABC ATPase domain of *HlyB* from *Escherichia coli. Biochem. J.* **386**, 489–495.

Bissinger, P. H., and Kuchler, K. (1994). Molecular cloning and expression of the *Saccharomyces cerevisiae STS1* gene product: A yeast ABC transporter conferring mycotoxin resistance. *J. Biol. Chem.* **269**, 4180–4186.

Borst, P., and Elferink, R. O. (2002). Mammalian ABC transporters in health and disease. *Annu. Rev. Biochem.* **71**, 537–592.

Chang, G., and Roth, C. B. (2001). Structure of *MsbA* from *E. coli*: A homolog of the multidrug resistance ATP-binding cassette (ABC) transporters. *Science* **293**, 1793–1800.

Chang, G. (2003). Structure of *MsbA* from *Vibrio cholera*: A multidrug resistance ABC transporter homolog in a closed conformation. *J. Mol. Biol.* **330**, 419–430.

Chen, M., Abele, R., and Tampé, R. (2004). Functional non-equivalence of ATP-binding cassette signature motifs in the transporter associated with antigen processing (TAP). *J. Biol. Chem.* **279,** 46073–46081.

Conseil, G., Decottignies, A., Jault, J. M., Comte, G., Barron, D., Goffeau, A., and Di Pietro, A. (2000). Prenyl-flavonoids as potent inhibitors of the Pdr5p multidrug ABC transporter from *Saccharomyces cerevisiae. Biochemistry* **39,** 6910–6917.

Conseil, G., Perez-Victoria, J. M., Jault, J. M., Gamarro, F., Goffeau, A., Hofmann, J., and Di Pietro, A. (2001). Protein kinase C effectors bind to multidrug ABC transporters and inhibit their activity. *Biochemistry* **40,** 2564–2571.

Cui, Z., Hirata, D., Tsuchiya, E., Osada, H., and Miyakawa, T. (1996). The multidrug resistance-associated protein (MRP) subfamily (Yrs1/Yor1) of *Saccharomyces cerevisiae* is important for the tolerance to a broad range of organic anions. *J. Biol. Chem.* **271,** 14712–14716.

Decottignies, A., and Goffeau, A. (1997). Complete inventory of the yeast ABC proteins. *Nature Genet.* **15,** 137–145.

Decottignies, A., Grant, A. M., Nichols, J. W., de Wet, H., McIntosh, D. B., and Goffeau, A. (1998). ATPase and multidrug transport activities of the overexpressed yeast ABC protein Yor1p. *J. Biol. Chem.* **273,** 12612–12622.

Decottignies, A., Kolaczkowski, M., Balzi, E., and Goffeau, A. (1994). Solubilization and characterization of the overexpressed *PDR5* multidrug resistance nucleotide triphosphatase of yeast. *J. Biol. Chem.* **269,** 12797–12803.

Decottignies, A., Lambert, L., Catty, P., Degand, H., Epping, E. A., Moye-Rowley, W. S., Balzi, E., and Goffeau, A. (1995). Identification and characterization of *SNQ2*, a new multidrug ATP binding cassette transporter of the yeast plasma membrane. *J. Biol. Chem.* **270,** 18150–18157.

Decottignies, A., Owsianik, G., and Ghislain, M. (1999). Casein kinase I-dependent phosphorylation and stability of the yeast multidrug transporter Pdr5p. *J. Biol. Chem.* **274,** 37139–37146.

DeRisi, J., van den Hazel, B., Marc, P., Balzi, E., Brown, P., Jacq, C., and Goffeau, A. (2000). Genome microarray analysis of transcriptional activation in multidrug resistance yeast mutants. *FEBS Lett.* **470,** 156–160.

Egner, R., Bauer, B. E., and Kuchler, K. (2000). The transmembrane domain 10 of the yeast Pdr5p ABC antifungal efflux pump determines both substrate specificity and inhibitor susceptibility. *Mol. Microbiol.* **35,** 1255–1263.

Egner, R., and Kuchler, K. (1996). The yeast multidrug transporter Pdr5 of the plasma membrane is ubiquitinated prior to endocytosis and degradation in the vacuole. *FEBS Lett.* **378,** 177–181.

Egner, R., Mahé, Y., Pandjaitan, R., and Kuchler, K. (1995). Endocytosis and vacuolar degradation of the plasma membrane-localized Pdr5 ATP-binding cassette multidrug transporter in *Saccharomyces cerevisiae. Mol. Cell. Biol.* **15,** 5879–5887.

Egner, R., Rosenthal, F. E., Kralli, A., Sanglard, D., and Kuchler, K. (1998). Genetic separation of FK506 susceptibility and drug transport in the yeast Pdr5 ATP-binding cassette multidrug resistance transporter. *Mol. Biol. Cell* **9,** 523–543.

Epping, E. A., and Moye-Rowley, W. S. (2002). Identification of interdependent signals required for anterograde traffic of the ATP-binding cassette transporter protein Yor1p. *J. Biol. Chem.* **277,** 34860–34869.

Eraso, P., Martinez-Burgos, M., Falcon-Perez, J. M., Portillo, F., and Mazon, M. J. (2004). Ycf1-dependent cadmium detoxification by yeast requires phosphorylation of residues Ser908 and Thr911. *FEBS Lett.* **577,** 322–326.

Falcon-Perez, J. M., Mazon, M. J., Molano, J., and Eraso, P. (1999). Functional domain analysis of the yeast ABC transporter Ycf1p by site-directed mutagenesis. *J. Biol. Chem.* **274**, 23584–23590.

Fernandez, S. B., Hollo, Z., Kern, A., Bakos, E., Fischer, P. A., Borst, P., and Evers, R. (2002). Role of the N-terminal transmembrane region of the multidrug resistance protein MRP2 in routing to the apical membrane in MDCKII cells. *J. Biol. Chem.* **277**, 31048–31055.

Ferreira-Pereira, A., Marco, S., Decottignies, A., Nader, J., Goffeau, A., and Rigaud, J. L. (2003). Three-dimensional reconstruction of the *Saccharomyces cerevisiae* multidrug resistance protein Pdr5p. *J. Biol. Chem.* **278**, 11995–11999.

Fewell, S. W., Travers, K. J., Weissman, J. S., and Brodsky, J. L. (2001). The action of molecular chaperones in the early secretory pathway. *Annu. Rev. Genet.* **35**, 149–191.

Gaudet, R., and Wiley, D. C. (2001). Structure of the ABC ATPase domain of human TAP1, the transporter associated with antigen processing. *EMBO J.* **20**, 4964–4972.

Ghosh, M., Shen, J., and Rosen, B. P. (1999). Pathways of As(III) detoxification in *Saccharomyces cerevisiae. Proc. Natl. Acad. Sci. USA* **96**, 5001–5006.

Gnann, A., Riordan, J. R., and Wolf, D. H. (2004). Cystic fibrosis transmembrane conductance regulator degradation depends on the lectins Htm1p/EDEM and the Cdc48 protein complex in yeast. *Mol. Biol. Cell* **15**, 4125–4135.

Golin, J., Ambudkar, S. V., Gottesman, M. M., Habib, A. D., Sczepanski, J., Ziccardi, W., and May, L. (2003). Studies with novel Pdr5p substrates demonstrate a strong size dependence for xenobiotic efflux. *J. Biol. Chem.* **278**, 5963–5969.

Grant, A. M., Hanson, P. K., Malone, L., and Nichols, J. W. (2001). NBD-labeled phosphatidylcholine and phosphatidylethanolamine are internalized by transbilayer transport across the yeast plasma membrane. *Traffic* **2**, 37–50.

Hallström, T. C., Lambert, L., Schorling, S., Balzi, E., Goffeau, A., and Moye-Rowley, W. S. (2001). Coordinate control of sphingolipid biosynthesis and multidrug resistance in *Saccharomyces cerevisiae. J. Biol. Chem.* **276**, 23674–23680.

Hanson, P. K., Grant, A. M., and Nichols, J. W. (2002). NBD-labeled phosphatidylcholine enters the yeast vacuole via the pre-vacuolar compartment. *J. Cell Sci.* **115**, 2725–2733.

Hanson, P. K., Malone, L., Birchmore, J. L., and Nichols, J. W. (2003). Lem3p is essential for the uptake and potency of alkylphosphocholine drugs, edelfosine and miltefosine. *J. Biol. Chem.* **278**, 36041–36050.

Hanson, P. K., and Nichols, J. W. (2001). Energy-dependent flip of fluorescence-labeled phospholipids is regulated by nutrient starvation and transcription factors, PDR1 and PDR3. *J. Biol. Chem.* **276**, 9861–9867.

Higgins, C. F. (1992). ABC transporters: From microorganisms to man. *Annu. Rev. Cell Biol.* **8**, 67–113.

Hirata, D., Yano, K., Miyahara, K., and Miyakawa, T. (1994). *Saccharomyces cerevisiae YDR1*, which encodes a member of the ATP-binding cassette (ABC) superfamily, is required for multidrug resistance. *Curr. Genet.* **26**, 285–294.

Hopfner, K. P., Karcher, A., Shin, D. S., Craig, L., Arthur, L. M., Carney, J. P., and Tainer, J. A. (2000). Structural biology of Rad50 ATPase: ATP-driven conformational control in DNA double-strand break repair and the ABC-ATPase superfamily. *Cell* **101**, 789–800.

Hung, L. W., Wang, I. X., Nikaido, K., Liu, P. Q., Ames, G. F., and Kim, S. H. (1998). Crystal structure of the ATP-binding subunit of an ABC transporter. *Nature* **396**, 703–707.

Kalin, N., Fernandes, J., Hrafnsdottir, S., and van Meer, G. (2004). Natural phosphatidylcholine is actively translocated across the plasma membrane to the surface of mammalian cells. *J. Biol. Chem.* **279**, 33228–33236.

Kato, U., Emoto, K., Fredriksson, C., Nakamura, H., Ohta, A., Kobayashi, T., Murakami-Murofushi, K., and Umeda, M. (2002). A novel membrane protein, Ros3p, is required for phospholipid translocation across the plasma membrane in *Saccharomyces cerevisiae*. *J. Biol. Chem.* **277**, 37855–37862.

Katzmann, D. J., Epping, E. A., and Moye-Rowley, W. S. (1999). Mutational disruption of plasma membrane trafficking of *Saccharomyces cerevisiae* Yor1p, a homologue of mammalian multidrug resistance protein. *Mol. Cell. Biol.* **19**, 2998–3009.

Katzmann, D. J., Hallström, T. C., Voet, M., Wysock, W., Golin, J., Volckaert, G., and Moye-Rowley, W. S. (1995). Expression of an ATP-binding cassette transporter-encoding gene (*YOR1*) is required for oligomycin resistance in *Saccharomyces cerevisiae*. *Mol. Cell. Biol.* **15**, 6875–6883.

Kean, L. S., Grant, A. M., Angeletti, C., Mahé, Y., Kuchler, K., Fuller, R. S., and Nichols, J. W. (1997). Plasma membrane translocation of fluorescent-labeled phosphatidylethanolamine is controlled by transcription regulators, *PDR1* and *PDR3*. *J. Cell Biol.* **138**, 255–270.

Kihara, A., and Igarashi, Y. (2004). Cross talk between sphingolipids and glycerophospholipids in the establishment of plasma membrane asymmetry. *Mol. Biol. Cell* **15**, 4949–4959.

Klein, M., Mamnun, Y. M., Eggmann, T., Schüller, C., Wolfger, H., Martinoia, E., and Kuchler, K. (2002). The ATP-binding cassette (ABC) transporter Bpt1p mediates vacuolar sequestration of glutathione conjugates in yeast. *FEBS Lett.* **520**, 63–67.

Kolaczkowski, M., Kolaczowska, A., Luczynski, J., Witek, S., and Goffeau, A. (1998). *In vivo* characterization of the drug resistance profile of the major ABC transporters and other components of the yeast pleiotropic drug resistance network. *Microb. Drug Resist.* **4**, 143–158.

Kolaczkowski, M., van der Rest, M., Cybularz-Kolaczkowska, A., Soumillion, J. P., Konings, W. N., and Goffeau, A. (1996). Anticancer drugs, ionophoric peptides, and steroids as substrates of the yeast multidrug transporter Pdr5p. *J. Biol. Chem.* **271**, 31543–31548.

Kölling, R., and Hollenberg, C. P. (1994). The ABC-transporter Ste6 accumulates in the plasma membrane in a ubiquitinated form in endocytosis mutants. *EMBO J.* **13**, 3261–3271.

Kölling, R., and Losko, S. (1997). The linker region of the ABC-transporter Ste6 mediates ubiquitination and fast turnover of the protein. *EMBO J.* **16**, 2251–2261.

Kralli, A., Bohen, S. P., and Yamamoto, K. R. (1995). *LEM1*, an ATP-binding-cassette transporter, selectively modulates the biological potency of steroid hormones. *Proc. Natl. Acad. Sci. USA* **92**, 4701–4705.

Lewis, H. A., Buchanan, S. G., Burley, S. K., Conners, K., Dickey, M., Dorwart, M., Fowler, R., Gao, X., Guggino, W. B., Hendrickson, W. A., Hunt, J. F., Kearins, M. C., Lorimer, D., Maloney, P. C., Post, K. W., Rajashankar, K. R., Rutter, M. E., Sauder, J. M., Shriver, S., Thibodeau, P. H., Thomas, P. J., Zhang, M., Zhao, X., and Emtage, S. (2004). Structure of nucleotide-binding domain 1 of the cystic fibrosis transmembrane conductance regulator. *EMBO J.* **23**, 282–293.

Li, X. Z., and Nikaido, H. (2004). Efflux-mediated drug resistance in bacteria. *Drugs* **64**, 159–204.

Li, Z. S., Lu, Y. P., Zhen, R. G., Szczypka, M., Thiele, D. J., and Rea, P. A. (1997). A new pathway for vacuolar cadmium sequestration in *Saccharomyces cerevisiae: YCF1*-catalyzed transport of bis(glutathionato)cadmium. *Proc. Natl. Acad. Sci. USA* **94**, 42–47.

Li, Z. S., Szczypka, M., Lu, Y. P., Thiele, D. J., and Rea, P. A. (1996). The yeast cadmium factor protein (YCF1) is a vacuolar glutathione S-conjugate pump. *J. Biol. Chem.* **271**, 6509–6517.

Litman, T., Druley, T. E., Stein, W. D., and Bates, S. E. (2001). From MDR to MXR: New understanding of multidrug resistance systems, their properties and clinical significance. *Cell. Mol. Life Sci.* **58**, 931–959.

Loayza, D., and Michaelis, S. (1998). Role for the ubiquitin-proteasome system in the vacuolar degradation of Ste6p, the a-factor transporter in *Saccharomyces cerevisiae*. *Mol. Cell. Biol.* **18**, 779–789.

Loayza, D., Tam, A., Schmidt, W. K., and Michaelis, S. (1998). Ste6p mutants defective in exit from the endoplasmic reticulum (ER) reveal aspects of an ER quality control pathway in *Saccharomyces cerevisiae*. *Mol. Biol. Cell* **9**, 2767–2784.

Locher, K. P., Lee, A. T., and Rees, D. C. (2002). The *E. coli BtuCD* structure: A framework for ABC transporter architecture and mechanism. *Science* **296**, 1091–1098.

Mahé, Y., Lemoine, Y., and Kuchler, K. (1996a). The ATP-binding cassette transporters Pdr5 and Snq2 of *Saccharomyces cerevisiae* can mediate transport of steroids *in vivo*. *J. Biol. Chem.* **271**, 25167–25172.

Mahé, Y., Parle-McDermott, A., Nourani, A., Delahodde, A., Lamprecht, A., and Kuchler, K. (1996b). The ATP-binding cassette multidrug transporter Snq2 of *Saccharomyces cerevisiae*: A novel target for the transcription factors Pdr1 and Pdr3. *Mol. Microbiol.* **20**, 109–117.

Mamnun, Y. M., Pandjaitan, R., Mahé, Y., Delahodde, A., and Kuchler, K. (2002). The yeast zinc finger regulators Pdr1p and Pdr3p control pleiotropic drug resistance (PDR) as homo- and heterodimers *in vivo*. *Mol. Microbiol.* **46**, 1429–1440.

Mamnun, Y. M., Schüller, C., and Kuchler, K. (2004). Expression regulation of the yeast *PDR5* ATP-binding cassette (ABC) transporter suggests a role in cellular detoxification during the exponential growth phase. *FEBS Lett.* **559**, 111–117.

Mason, D. L., Mallampalli, M. P., Huyer, G., and Michaelis, S. (2003). A region within a lumenal loop of *Saccharomyces cerevisiae* Ycf1p directs proteolytic processing and substrate specificity. *Eukaryot. Cell* **2**, 588–598.

Meyers, S., Schauer, W., Balzi, E., Wagner, M., Goffeau, A., and Golin, J. (1992). Interaction of the yeast pleiotropic drug resistance genes *PDR1* and *PDR5*. *Curr. Genet.* **21**, 431–436.

Miyahara, K., Mizunuma, M., Hirata, D., Tsuchiya, E., and Miyakawa, T. (1996). The involvement of the *Saccharomyces cerevisiae* multidrug resistance transporters Pdr5p and Snq2p in cation resistance. *FEBS Lett.* **399**, 317–320.

Moye-Rowley, W. S. (2003). Transcriptional control of multidrug resistance in the yeast *Saccharomyces*. *Prog. Nucleic Acid Res. Mol. Biol.* **73**, 251–279.

Murakami, S., Nakashima, R., Yamashita, E., and Yamaguchi, A. (2002). Crystal structure of bacterial multidrug efflux transporter AcrB. *Nature* **419**, 587–593.

Nakamura, K., Niimi, M., Niimi, K., Holmes, A. R., Yates, J. E., Decottignies, A., Monk, B. C., Goffeau, A., and Cannon, R. D. (2001). Functional expression of *Candida albicans* drug efflux pump Cdr1p in a *Saccharomyces cerevisiae* strain deficient in membrane transporters. *Antimicrob. Agents Chemother.* **45**, 3366–3374.

Nichols, J. W. (2002). Internalization and trafficking of fluorescent-labeled phospholipids in yeast. *Semin. Cell Dev. Biol.* **13**, 179–184.

Ortiz, D. F., St Pierre, M. V., Abdulmessih, A., and Arias, I. M. (1997). A yeast ATP-binding cassette-type protein mediating ATP-dependent bile acid transport. *J. Biol. Chem.* **272**, 15358–15365.

Ouellette, M., Legare, D., and Papadopoulou, B. (2001). Multidrug resistance and ABC transporters in parasitic protozoa. *J. Mol. Microbiol. Biotechnol.* **3**, 201–206.

Piper, P., Mahé, Y., Thompson, S., Pandjaitan, R., Holyoak, C., Egner, R., Mühlbauer, M., Coote, P., and Kuchler, K. (1998). The Pdr12 ABC transporter is required for the development of weak organic acid resistance in yeast. *EMBO J.* **17**, 4257–4265.

Plemper, R. K., Egner, R., Kuchler, K., and Wolf, D. H. (1998). Endoplasmic reticulum degradation of a mutated ATP-binding cassette transporter Pdr5 proceeds in a concerted action of Sec61 and the proteasome. *J. Biol. Chem.* **273**, 32848–32856.

Pohl, A., Devaux, P. F., and Herrmann, A. (2005). Function of prokaryotic and eukaryotic ABC proteins in lipid transport. *Biochim. Biophys. Acta* **1733**, 29–52.

Pomorski, T., Lombardi, R., Riezman, H., Devaux, P. F., van Meer, G., and Holthuis, J. C. (2003). Drs2p-related P-type ATPases Dnf1p and Dnf2p are required for phospholipid translocation across the yeast plasma membrane and serve a role in endocytosis. *Mol. Biol. Cell* **14**, 1240–1254.

Rebbeor, J. F., Connolly, G. C., Dumont, M. E., and Ballatori, N. (1998). ATP-dependent transport of reduced glutathione on *YCF1*, the yeast orthologue of mammalian multidrug resistance associated proteins. *J. Biol. Chem.* **273**, 33449–33454.

Ruetz, S., and Gros, P. (1994). Phosphatidylcholine translocase: A physiological role for the *mdr2* gene. *Cell* **77**, 1071–1081.

Schmitt, L., Benabdelhak, H., Blight, M. A., Holland, I. B., and Stubbs, M. T. (2003). Crystal structure of the nucleotide-binding domain of the ABC-transporter haemolysin B: Identification of a variable region within ABC helical domains. *J. Mol. Biol.* **330**, 333–342.

Schmitt, L., and Tampé, R. (2000). Affinity, specificity, diversity: A challenge for the ABC transporter TAP in cellular immunity. *Chembiochemistry* **1**, 16–35.

Schmitt, L., and Tampe, R. (2002). Structure and mechanism of ABC transporters. *Curr. Opin. Struct. Biol.* **12**, 754–760.

Schützer-Muehlbauer, M., Willinger, B., Egner, R., Ecker, G., and Kuchler, K. (2003). Reversal of antifungal resistance mediated by ABC efflux pumps from *Candida albicans* functionally expressed in yeast. *Int. J. Antimicrob. Agents* **22**, 291–300.

Schüller, C., Bauer, B. E., and Kuchler, K. (2003). Inventory and evolution of fungal ABC protein genes. *In* "ABC Proteins: From Bacteria to Man" (I. B. Holland, K. Kuchler, and C. F. Higgins, eds.), pp. 279–293. Academic Press/Elsevier Science, Amsterdam.

Schüller, C., Mamnun, Y. M., Mollapour, M., Krapf, G., Schuster, M., Bauer, B. E., Piper, P. W., and Kuchler, K. (2004). Global phenotypic analysis and transcriptional profiling defines the weak acid stress response regulon in *Saccharomyces cerevisiae*. *Mol. Biol. Cell* **15**, 706–720.

Servos, J., Haase, E., and Brendel, M. (1993). Gene SNQ2 of *Saccharomyces cerevisiae*, which confers resistance to 4-nitroquinoline-N-oxide and other chemicals, encodes a 169 kDa protein homologous to ATP-dependent permeases. *Mol. Gen. Genet.* **236**, 214–218.

Sharma, K. G., Kaur, R., and Bachhawat, A. K. (2003). The glutathione-mediated detoxification pathway in yeast: An analysis using the red pigment that accumulates in certain adenine biosynthetic mutants of yeasts reveals the involvement of novel genes. *Arch. Microbiol.* **180**, 108–117.

Sharma, K. G., Mason, D. L., Liu, G., Rea, P. A., Bachhawat, A. K., and Michaelis, S. (2002). Localization, regulation, and substrate transport properties of Bpt1p, a *Saccharomyces cerevisiae* MRP-type ABC transporter. *Eukaryot. Cell* **1**, 391–400.

Shih, S. C., Sloper-Mould, K. E., and Hicke, L. (2000). Monoubiquitin carries a novel internalization signal that is appended to activated receptors. *EMBO J.* **19**, 187–198.

Simons, K., and Vaz, W. L. (2004). Model systems, lipid rafts, and cell membranes. *Annu. Rev. Biophys. Biomol. Struct.* **33**, 269–295.

Smith, P. C., Karpowich, N., Millen, L., Moody, J. E., Rosen, J., Thomas, P. J., and Hunt, J. F. (2002). ATP binding to the motor domain from an ABC transporter drives formation of a nucleotide sandwich dimer. *Mol. Cell* **10**, 139–149.

Smriti, Krishnamurthy, S., Dixit, B. L., Gupta, C. M., Milewski, S., and Prasad, R. (2002). ABC transporters Cdr1p, Cdr2p and Cdr3p of a human pathogen *Candida albicans* are general phospholipid translocators. *Yeast* **19**, 303–318.

Subba Rao, G., Bachhawat, A. K., and Gupta, C. M. (2002). Two-hybrid-based analysis of protein-protein interactions of the yeast multidrug resistance protein, Pdr5p. *Funct. Integr. Genom.* **1**, 357–366.

Szczypka, M. S., Wemmie, J. A., Moye-Rowley, W. S., and Thiele, D. J. (1994). A yeast metal resistance protein similar to human cystic fibrosis transmembrane conductance regulator (CFTR) and multidrug resistance-associated protein. *J. Biol. Chem.* **269**, 22853–22857.

Taglicht, D., and Michaelis, S. (1998). *Saccharomyces cerevisiae* ABC proteins and their relevance to human health and disease. *Methods Enzymol.* **292**, 130–162.

van Meer, G., and Liskamp, R. M. (2005). Brilliant lipids. *Nature Methods* **2**, 14–15.

Wach, A., Brachat, A., Alberti-Segui, C., Rebischung, C., and Philippsen, P. (1997). Heterologous *HIS3* marker and GFP reporter modules for PCR-targeting in *Saccharomyces cerevisiae*. *Yeast* **13**, 1065–1075.

Wach, A., Brachat, A., Pohlmann, R., and Philippsen, P. (1994). New heterologous modules for classical or PCR-based gene disruptions in *Saccharomyces cerevisiae*. *Yeast* **10**, 1793–1808.

Walker, J. E., Saraste, M., Runswick, M. J., and Gay, N. J. (1982). Distantly related sequences in the alpha- and beta-subunits of ATP synthase, myosin, kinases and other ATP-requiring enzymes and a common nucleotide binding fold. *EMBO J.* **1**, 945–951.

Wemmie, J. A., and Moye-Rowley, W. S. (1997). Mutational analysis of the *Saccharomyces cerevisiae* ATP-binding cassette transporter protein Ycf1p. *Mol. Microbiol.* **25**, 683–694.

Wilcox, L. J., Balderes, D. A., Wharton, B., Tinkelenberg, A. H., Rao, G., and Sturley, S. L. (2002). Transcriptional profiling identifies two members of the ATP-binding cassette transporter superfamily required for sterol uptake in yeast. *J. Biol. Chem.* **277**, 32466–32472.

Winzeler, E. A., Shoemaker, D. D., Astromoff, A., Liang, H., Anderson, K., Andre, B., Bangham, R., Benito, R., Boeke, J. D., Bussey, H., Chu, A. M., Connelly, C., Davis, K., Dietrich, F., Dow, S. W., El Bakkoury, M., Foury, F., Friend, S. H., Gentalen, E., Giaever, G., Hegemann, J. H., Jones, T., Laub, M., Liao, H., Davis, R. W., Liebundhuth, N., Lockhart, D. J., Lucau-Danila, A., Lussier, M., M'Rabet, N., Menard, P., and Mittmann, M.Pai, C., Rebischung, C., Revuelta, J. L., Riles, L., Roberts, C. J., Ross-MacDonald, P., Scherens, B., Snyder, M., Sookhai-Mahadeo, S., Storms, R. K., Veronneau, S., Voet, M., Volckaert, G., Ward, T. R., Wysocki, R., Yen, G. S., Yu, K., Zimmermann, K., Phillippsen, P., and Johnston, M. (1999). Functional characterization of the *S. cerevisiae* genome by gene deletion and parallel analysis. *Science* **285**, 901–906.

Wolfger, H., Mahé, Y., Parle-McDermott, A., Delahodde, A., and Kuchler, K. (1997). The yeast ATP binding cassette (ABC) protein genes *PDR10* and *PDR15* are novel targets for the Pdr1 and Pdr3 transcriptional regulators. *FEBS Lett.* **418**, 269–274.

Wolfger, H., Mamnun, Y. M., and Kuchler, K. (2001). Fungal ABC proteins: Pleiotropic drug resistance, stress response and cellular detoxification. *Res. Microbiol.* **152**, 375–389.

Wolfger, H., Mamnun, Y. M., and Kuchler, K. (2004). The yeast Pdr15p ATP-binding cassette (ABC) protein is a general stress response factor implicated in cellular detoxification. *J. Biol. Chem.* **279**, 11593–11599.

Yu, E. W., McDermott, G., Zgurskaya, H. I., Nikaido, H., and Koshland, D. E., Jr. (2003). Structural basis of multiple drug-binding capacity of the AcrB multidrug efflux pump. *Science* **300**, 976–980.

Zhang, X., Kolaczkowska, A., Devaux, F., Panwar, S. L., Hallstrom, T. C., Jacq, C., and Moye-Rowley, W. S. (2005). Transcriptional regulation by Lge1p requires a function independent of its role in histone H2B ubiquitination. *J. Biol. Chem.* **280**, 2759–2770.

Zhang, X., and Moye-Rowley, W. S. (2001). *Saccharomyces cerevisiae* multidrug resistance gene expression inversely correlates with the status of the F(0) component of the mitochondrial ATPase. *J. Biol. Chem.* **276**, 47844–47852.

[27] High-Speed Screening of Human ATP-Binding Cassette Transporter Function and Genetic Polymorphisms: New Strategies in Pharmacogenomics

By Toshihisa Ishikawa, Aki Sakurai, Yoichi Kanamori,
Makoto Nagakura, Hiroyuki Hirano, Yutaka Takarada,
Kazunari Yamada, Kazuhisa Fukushima, and Masato Kitajima

Abstract

Drug transporters represent an important mechanism in cellular uptake and efflux of drugs and their metabolites. Hitherto a variety of drug transporter genes have been cloned and classified into either solute carriers or ATP-binding cassette (ABC) transporters. Such drug transporters are expressed in various tissues such as the intestine, brain, liver, kidney, and, importantly, cancer cells, where they play critical roles in the absorption, distribution, and excretion of drugs. We developed high-speed functional screening and quantitative structure–activity relationship analysis methods to study the substrate specificity of ABC transporters and to evaluate the effect of genetic polymorphisms on their function. These methods would provide powerful and practical tools for screening synthetic and natural compounds, and the deduced data can be applied to the molecular design of new drugs. Furthermore, we demonstrate a new "SNP array" method to detect genetic polymorphisms of ABC transporters in human samples.

Introduction

ATP-binding cassette (ABC) proteins form one of the largest protein families encoded in the human genome (Dean *et al.*, 2001; Holland *et al.*, 2003). Hitherto more than 48 human ABC protein genes have been identified and sequenced (Klein *et al.*, 1999). It has been reported that mutations of ABC protein genes are causative of several genetic disorders in humans (Dean *et al.*, 2001). Many of the human ABC proteins are involved in membrane transport of drugs, xenobiotics, endogenous substances, or ions, thereby exhibiting a wide spectrum of biological functions (Schinkel and Jonker, 2003). Based on the arrangement of molecular structure components, that is, nucleotide-binding domains and topologies of transmembrane domains, the hitherto reported human ABC proteins have been classified into seven different subfamilies (A to G) (Borst and Oude Elferink, 2002;

METHODS IN ENZYMOLOGY, VOL. 400
0076-6879/05 $35.00
DOI: 10.1016/S0076-6879(05)00027-3

Ishikawa, 2003; Klein *et al.*, 1999). The HUGO Human Gene Nomenclature Committee developed a new system of nomenclature for the human ABC transporter family. The new nomenclature scheme was implemented in 1999, and detailed information is available on the website at *http://www.gene.ucl.ac.uk/nomenclature/genefamily/abc.html.*

Metabolic systems for xenobiotics including drugs are widely referred to as phase I and II systems, where phase I includes oxidation of xenobiotics and phase II deals with the conjugation of phase I products (Fig. 1). Oxidative metabolism in the phase I system is mediated by cytochrome P450 (CYP) or flavin mixed function oxidases. Some of the activated xenobiotics can interact with DNA and/or proteins in cells to cause toxic effects. In the phase II system, however, activated hydrophobic xenobiotics are converted into hydrophilic forms via conjugation reactions with glutathione, sulfate, or glucuronide. This phase II metabolism is regarded as the detoxification process for xenobiotics. In some cases, however, the phase II system is a critical step in the formation of genotoxic electrophiles. Furthermore, accumulation of the resulting metabolites in cells can lead to a decrease in the detoxification activity of the phase II system. Therefore, a phase III system must take on the task of eliminating phase II metabolites from cells (Ishikawa, 1992). Several ABC transporters, including ABCB1,

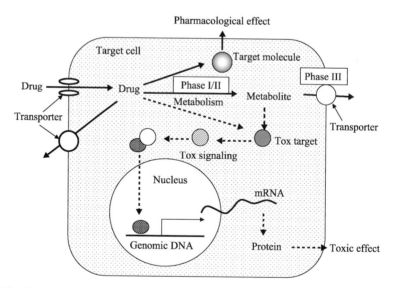

FIG. 1. Schematic illustration of drug actions in the target cell. Drug transporters play a significant role in the absorption and excretion of drugs, thereby modulating pharmacological and toxic effects. Phase I, II, and III systems of drug metabolism and excretion are indicated.

ABCB11, ABCC1, ABCC2, ABCC3, ABCC4, ABCC5, ABCC6, and ABCG2, are considered to be major players in the phase III detoxification system (Borst and Oude Elferink, 2002; Schinkel and Jonker, 2003).

Expression of Human ABC Transporter ABCG2 in Insect Cells

Human ABC Transporter ABCG2

A breast cancer-resistant protein (BCRP) has been discovered in doxorubicin-resistant breast cancer cells (Doyle *et al.*, 2001). Because the same transporter has also been found in the human placenta (Allikmets *et al.*, 1998), as well as in drug-resistant cancer cells selected in mitoxantrone (Miyake *et al.*, 1999), it was also called ABCP or MXR1. This ABC transporter protein is now named ABCG2 and has been classified in the G subfamily of human ABC transporter genes, according to the new nomenclature.

The *ABCG2* gene is located on chromosome 4q22 and spans over 66 kb, consisting of 16 exons and 15 introns (Bailey-Dell *et al.*, 2001). ABCG2 is expressed endogenously in placental trophoblast cells, in the epithelium of the small intestine, and in the liver canalicular membrane, as well as in ducts and lobules of the breast. In addition, expression of ABCG2 is detected in venous and capillary endothelium. In particular, the high levels of ABCG2 expression in trophoblast cells suggest that this pump is responsible either for transporting compounds into the fetal blood supply or for removing toxic metabolites. Apical localization in the epithelium of the small intestine and colon indicates a possible role of ABCG2 in regulation of the uptake of *po* administered drugs as well as protection against toxic xenobiotics (Jonker *et al.*, 2002; Maliepaard *et al.*, 2001; Zhou *et al.*, 2001). In addition, we have provided direct evidence that ABCG2 transports SN-38, an active metabolite of irinotecan (CPT-11), and its glucuronide conjugate (Nakatomi *et al.*, 2001; Yoshikawa *et al.*, 2004).

Construction of Expression Vector

Human ABCG2 cDNA was cloned from cDNA of the MCF7/BCRP clone-8 cell line by polymerase chain reaction (PCR), as described previously (Mitomo *et al.*, 2003). The PCR product is first inserted into the pCR2.1 TOPO vector, and its sequence is analyzed by automated DNA sequencing. The cDNA is then inserted into the pNNK vector (Fig. 2) that was modified from pPSC8 (Protein Sciences Co.) by Nosan Corporation (Yokohama, Japan). Briefly, the ABCG2 cDNA-containing pCR2.1 TOPO vector is digested by *Eco*RI, and ABCG2 cDNA is removed. After

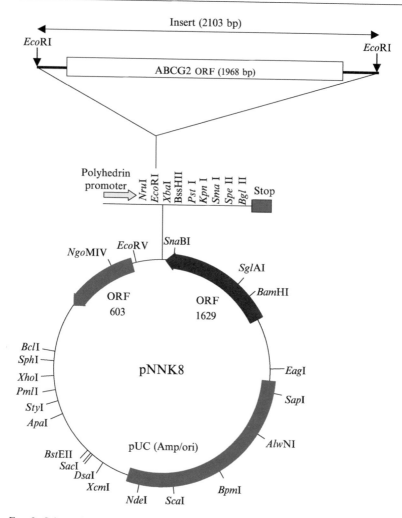

FIG. 2. Schematic diagram of the pNNK8 transfer vector for the expression of human ABCG2 in insect cells. (See color insert.)

treatment with alkaline phosphatase, ABCG2 cDNA is ligated to the *Eco*RI site of the pNNK vector using the rapid DNA ligation kit (Roche Diagnostics Co., Indianapolis, IN). The recombinant virus is constructed according to the methods described by Summers and Smith (1987) and O'Reilly *et al.* (1994).

Transfection to Insect Cells

Sf9 insect cells are diluted with Sf-900II serum-free medium at a cell density of 0.2×10^6 cells/ml. Five milliliters (total of 1.0×10^6 cells) of the cell suspension is kept in a 25-cm^2 culture dish. Cells are allowed to attach to the dish for 1 hr at room temperature, and the following reagents are added: 5 μl (358 ng/μl) of the ABCG2/pNNK vector, 5 μl (17 ng/μl) of linear AcNPV DNA (Protein Sciences Co.), 20 μl of Cellfectin reagent (Invitrogen Co.), and 180 μl of Sf-900II serum-free medium (Invitrogen Co.). Following incubation at 28° for 7 days, the cell suspension is centrifuged at 3000g for 15 min at 4°. The resulting supernatant is collected and defined as the primary viral solution.

Purification of Baculovirus by Plaque Assay

The primary viral solution is diluted 10^4-, 10^5-, 10^6-, and 10^7-fold with Sf-900II serum-free medium. For the plaque assay, 4 ml of the Sf9 cell suspension (0.5×10^6 cells/ml) is put into each 60-mm dish. After the Sf9 cells form a monolayer, the culture medium is removed from the dish. One milliliter each of the diluted viral solutions is added to each dish containing the cell monolayer. Sf9 cells are then incubated with gentle shaking at room temperature for 1 h. After the viral solution is removed, Sf-900II serum-free medium with 0.5% SeaKem GTG agarose (FMC Co.) is added to the dish, and Sf9 cells are cultured further at 27° for 7 days. Recombinant plaques (without polyhedrin) are collected with a Pasteur pipette. Each plaque, together with agarose, is then suspended in 1 ml of serum-free Sf-900II medium to obtain the primary viral solution.

Amplification of Recombinant Virus and Titer Check

Dilute Sf9 cells in Sf-900II serum-free medium at a cell density of 1.0×10^6 cells/ml and add 2 ml (total of 2.0×10^6 cells) of the cell suspension to each 25-cm^2 culture dish. Allow the cells to attach to the dish for 1 h at room temperature. Thereafter, 0.5 ml of the primary viral solution and 3 ml of serum-free Sf-900II medium are added to the dish. The cell suspension, thus obtained, is incubated at 27° for 72 h. After centrifugation at 3000g for 15 min at 4°, the resulting supernatant is recovered and defined as the first-generation viral solution. Five milliliters of this viral solution and 10 ml of fresh serum-free Sf-900II medium are added into a 75-cm^2 culture dish containing Sf9 cells (6×10^6 cells). After 72 h postinfection, the cell suspension is centrifuged at 3000 g for 15 min at 4°, and the supernatant (15 ml) is recovered and stored at 4° as the second-generation viral solution.

A culture of *expres*SF+ cells (Protein Sciences Co.) is maintained in serum-free Sf-900II medium. Cells in the exponential multiplication stage are withdrawn and diluted to 1.5×10^6 cells/ml with the same medium. Then, 1 ml of the second-generation viral solution is added to the cell suspension (100 ml) in a 250-ml Erlenmeyer flask. The cell suspension is maintained with gentle rotation (130 rpm) at 27°, and cells are harvested at 72 h postinfection. After centrifugation at 3000g for 15 min at 4°, the supernatant is recovered and defined as the third-generation viral solution.

To determine the viral titer, the third generation of viral solution is diluted to 10^5-, 10^6-, 10^7-, and 10^8-fold with Sf-900II serum-free medium. Four milliliters of the Sf9 cell suspension (0.5×10^6 cells/ml) is put into each 60-mm dish. After the Sf9 cells have formed a monolayer, the culture medium is removed from the dish. One milliliter each of the diluted viral solutions is added to each dish containing the cell monolayer. Sf9 cells are then incubated with gentle shaking at room temperature for 1 h. After the viral solution is removed, Sf-900II serum-free medium with 0.5% SeaKem GTG agarose (FMC Co.) is added to the dish. Sf9 cells are cultured further at 27° for 7 days. Resultant plaques are counted, and the viral titer is determined.

Expression of ABCG2 in expres*SF+* Cells

*expres*SF+ cells in the exponential multiplication stage are diluted to a cell density of 1.5×10^6 cells/ml with Sf-900II serum-free medium, and 100 ml of the cell suspension is mixed with the third-generation viral solution in a 250-ml Erlenmeyer flask. Multiplicity of infection (MOI) at this time should be set at 1.0. Cells are maintained with a gentle rotation (130 rpm) at 27°. Three days after infection, *expres*SF+ cells are harvested by centrifugation. Cells are subsequently washed with phosphate-buffered saline (PBS) at 4°, collected by centrifugation, and stored at –30° until used.

Preparation of Plasma Membrane Vesicles and Detection of ABCG2 Protein

Plasma Membrane Vesicles from Insect Cells

Plasma membrane vesicles can be prepared from either Sf9 cells or *expres*SF+ cells according to the flowchart in Fig. 3. The harvested and frozen cells are thawed quickly, diluted with 30 ml of an ice-cold hypotonic buffer (0.5 m*M* Tris/HEPES, pH 7.4, and 0.1 m*M* EGTA), and then homogenized with a Potter–Elvehjem homogenizer. After centrifugation

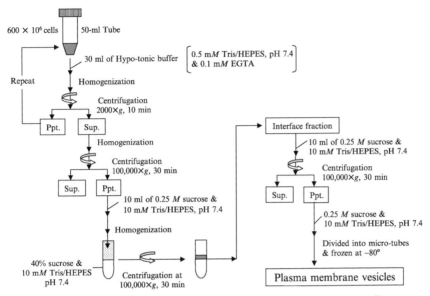

FIG. 3. Procedure of plasma membrane vesicle preparation from insect cells.

at 2000g for 10 min at 4°, the supernatant is collected, whereas the precipitate is homogenized further with a Potter–Elvehjem homogenizer in another 30 ml of the hypotonic buffer. After centrifugation at 2000g, the resulting supernatant is collected and combined with the first supernatant fraction. The supernatant (total of 60 ml) is centrifuged at 100,000g for 30 min. The pellet, thus obtained, is suspended in 10 ml of 250 mM sucrose containing 10 mM Tris/HEPES, pH 7.4, and is homogenized with a Potter–Elvehjem homogenizer. The crude membrane fraction is carefully layered over a 40% (w/v) sucrose solution and centrifuged at 100,000g for 30 min. The turbid layer at the interface is then collected with a Pasteur pipette, suspended in 250 mM sucrose containing 10 mM Tris/HEPES, pH 7.4, and centrifuged at 100,000g for 30 min. The membrane fraction is collected and resuspended in a small volume (250 to 500 μl) of 250 mM sucrose containing 10 mM Tris/HEPES, pH 7.4. After the measurement of protein concentration by the BCA protein assay kit (Pierce, Rockford, IL), the membrane preparations should be frozen and stored either at −80° or in liquid N_2 until used.

Immunological Detection of ABCG2 in Plasma Membrane Vesicles

The amount of ABCG2 expressed in cell membrane vesicles is determined by immunoblotting with BXP-21 (Signet, Dedham, MA), a specific antibody to human ABCG2. Briefly, proteins of the plasma membrane are

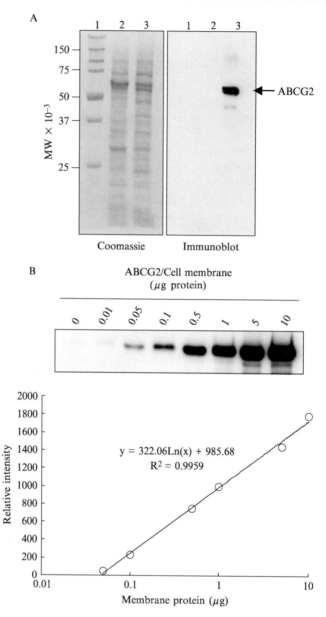

FIG. 4. Detection and quantitative analysis of human ABCG2 expressed in the plasma membrane of insect cells. (A) Coomassie staining and immunoblot analyses of membrane proteins. Plasma membrane proteins (2 μg) were separated by SDS–PAGE (10% acrylamide gel), and ABCG2 was detected by immunoblotting. (B) Quantitative analysis of ABCG2

separated by electrophoresis on 10% sodium dodecyl sulfate (SDS)–poly-acrylamide slab gels and subsequently electroblotted onto Hy-bond ECL nitrocellulose membranes (Amersham, Buckinghamshire, UK). Immuno-blotting is performed using BXP-21 (1:250 dilution) as the first antibody and an antimouse IgG– horseradish peroxidase (HRP) conjugate (Cell Signaling Technology, Beverly, MA) (1:3000 dilution) as the secondary antibody. HRP-dependent luminescence is developed using Western Lighting Chemiluminescent Reagent Plus (PerkinElmer Life Sciences, Boston, MA) and detected by Lumino Imaging Analyzer FAS-1000 (Toyobo, Osaka, Japan). According to this immunoblotting procedure, the ABCG2 monomer expressed in insect cell membranes is detected at a molecular weight of 68,000; the sample is treated with mercaptoethanol (Fig. 4A).

To analyze the transport activity of ABCG2 variants quantitatively, it is critically important to normalize the expression level of each variant protein. Figure 4B clearly demonstrates a linear relationship between the signal intensity of immunoblotting and the logarithmic value of the amount of protein applied to the electrophoresis. Based on the linear relation-ship, the expression levels of ABCG2 and its variants in different plasma membrane preparations can be estimated and normalized quantitatively.

High-Speed Screening to Measure the Transport Activity of ABCG2 and Its Variants

Development of a High-Speed Screening System

The original assay method used to measure ATP-dependent transport of organic anions into inside-out plasma membrane vesicles was developed by Ishikawa (1989) and subsequently used by other researchers (Ishikawa et al., 1990; Keppler et al., 1998). We have improved the method to enhance the speed of screening and to profile the substrate specificity of ABC transporters. To detect the transport activity of ABCG2, we use metho-trexate (MTX) as a substrate (Ishikawa et al., 2003; Mitomo et al., 2003).

The frozen stocked membrane is thawed quickly, and membrane vesi-cles are formed by passing the membrane suspension through a 27-gauge needle. To measure the ABCG2-mediated MTX transport, the standard

protein in the plasma membrane preparation. Different amounts of plasma membrane proteins (0, 0.01, 0.05, 0.1, 0.5, 1, 5, and 10 μg) were subjected to SDS–PAGE, and ABCG2 protein was detected by immunoblotting. The signal intensity of immunoblots is expressed as a function of the logarithmic value of the amount of membrane proteins applied to SDS–PAGE. (See color insert.)

incubation medium should contain plasma membrane vesicles (10 or 50 μg of protein), 200 μM [3',5',7'-^{3}H]MTX (Amersham, Buckinghamshire, UK), 0.25 M sucrose, 10 mM Tris/HEPES, pH 7.4, 10 mM MgCl2, 1 mM ATP, 10 mM creatine phosphate, and 100 μg/ml of creatine kinase in a final volume of 100 μl. The incubation is carried out at 37°. After a specified time (20 min for the standard condition), the reaction medium is mixed with 1 ml of the ice-cold stop solution (0.25 M sucrose, 10 mM Tris/HEPES, pH 7.4, and 2 mM EDTA) to terminate the transport reaction. Subsequently, aliquots (280 μl per well) of the resulting mixture are transferred to MultiScreen plates (Nihon Millipore KK, Tokyo, Japan) (Fig. 5A). Under aspiration, each well of the plate is rinsed with the 0.25 M sucrose solution containing 10 mM Tris/HEPES, pH 7.4, four times (4 × 200 μl for each well) in an EDR384S system (BioTec, Tokyo, Japan) (Fig. 5B). [^{3}H]MTX thus incorporated into the vesicles is measured by counting the radioactivity remaining on the filter of MultiScreen plates, where each filter is placed in 2 ml of liquid scintillation fluid (Ultima Gold, Packard BioScience).

Figure 6 depicts the time course of MTX transport into plasma membrane vesicles. MTX transport into ABCG2-expressing plasma membrane vesicles was enhanced greatly in the presence of ATP. In plasma membrane vesicles prepared from mock-infected cells, ATP-dependent transport of MTX was not observed.

Functional Screening of Genetic Polymorphisms of ABCG2

There is accumulating information on the genetic polymorphism of ABCG2 (for reviews, see Cervenak et al., 2005; Ishikawa et al., 2004b). Figure 7A shows the hitherto reported alterations of amino acid residues that are due to nonsynonymous SNPs and acquired mutations. Quantitative studies should be carried out to precisely evaluate functional changes associated with such genetic polymorphisms. For this purpose, variant forms (V12M, G51C, Q126stop, Q141K, T153M, Q166E, I206L, E334stop, N590Y, D620N, R482G, and R482T) have been created by site-directed mutagenesis with the QuikChange site-directed mutagensis kit (Stratagene, La Jolla, CA). Recombinant baculoviruses to express the aforementioned variant forms of ABCG2 in insect cells are generated with BAC-TO-BAC baculovirus expression systems (Invitrogen), as described previously (Ishikawa et al., 2004b). Insect Sf9 cells (1 × 10^6 cell/ml) are infected with the recombinant baculoviruses and cultured in EX-CELL 420 insect serum-free medium (JRH Bioscience, Levexa, KS) at 27° with gentle shaking. Fourty-eight hours after the infection, cells are harvested by centrifugation. Cell membranes are prepared as described earlier (Fig. 3). Figure 7B shows data on the MTX transport activity of those ABCG2

A

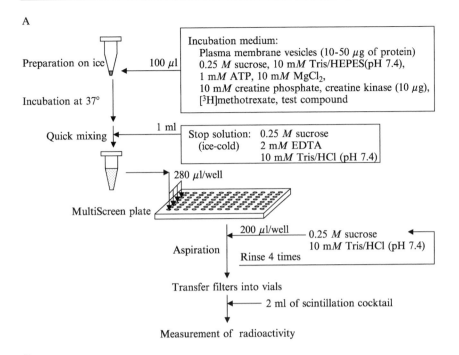

Preparation on ice   100 μl

Incubation medium:
    Plasma membrane vesicles (10-50 μg of protein)
    0.25 M sucrose, 10 mM Tris/HEPES(pH 7.4),
    1 mM ATP, 10 mM MgCl$_2$,
    10 mM creatine phosphate, creatine kinase (10 μg),
    [^{3}H]methotrexate, test compound

Incubation at 37°

Quick mixing   1 ml

Stop solution: 0.25 M sucrose
(ice-cold)   2 mM EDTA
               10 mM Tris/HCl (pH 7.4)

280 μl/well

MultiScreen plate

Aspiration   200 μl/well   0.25 M sucrose
            Rinse 4 times   10 mM Tris/HCl (pH 7.4)

Transfer filters into vials

2 ml of scintillation cocktail

Measurement of radioactivity

B

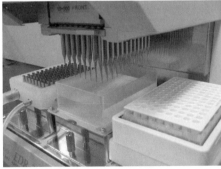

FIG. 5. High-speed assay of ATP-dependent MTX transport mediated by ABCG2 in plasma membrane vesicles. (A) Procedure for the vesicle transport assay using 96-well MultiScreen plates. (B) High-speed screening system (BioTec EDR384S) used for the vesicle transport assay. (See color insert.)

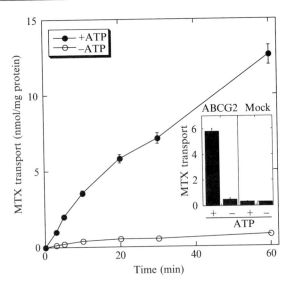

Fig. 6. Time courses of MTX transport in plasma membrane vesicles prepared from insect cells infected with ABCG2 virus. (Insert) MTX transport in plasma membrane vesicles prepared from insect cells infected with ABCG2 virus and mock virus. Transport was measured for 20 min in the presence and absence of ATP.

variants after the normalization of expression levels. No MTX transport activity was observed in the variants Q126stop, E334stop, R482G, and R482T.

High-Speed Screening of Human ABCB1 and Its Variants

Human ABC Transporter ABCB1

Human ABCB1 (P-glycoprotein or MDR1) was identified because of its overexpression in cultured cancer cells associated with an acquired cross-resistance to multiple anticancer drugs, such as colchicine, doxorubicin, daunorubicin, vincristine, and VP16 (Ambudkar *et al.*, 1999; Ling, 1997). In addition to cancer cells, ABCB1 is expressed in many normal tissues, for example, the apical domain of enterocytes of the gastrointestinal tract (jejunum and duodenum), endothelial cells lining the small vessels of the human cortex, the lumenal membrane of capillary endothelial cells of the brain, the canalicular domain of hepatocytes, the brush border membrane of proximal renal tubules, the placenta, and the testis.

A

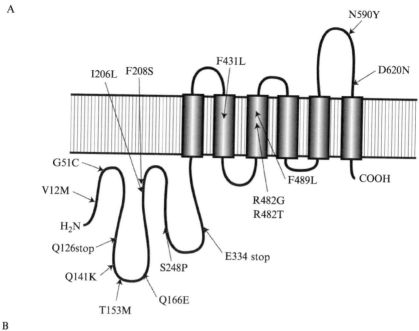

B

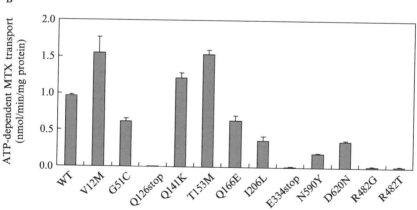

FIG. 7. Nonsynonymous polymorphisms of human ABCG2. (A) Schematic illustration of the structure of ABCG2 protein and the locations of amino acid changes. (B) The effect of nonsynonymous polymorphisms on MTX transport activity of human ABCG2. R482G and R482T are acquired mutations. MTX transport was measured as described in Fig. 6.

The human *ABCB1* gene is located on chromosome 7q21, and the gene product is an integral membrane protein consisting of 1280 amino acid residues (Fig. 8). ABCB1 is composed of two homologous halves, each of which consists of an N-terminal, hydrophobic, membrane-associated domain (approximately 250 amino acid residues) and a C-terminal, hydrophilic, nucleotide-binding fold (approximately 300 amino acid residues). Plasma membrane-associated domains in the two halves of ABCB1 each consist of six transmembrane domains, which are followed by an intracellular ATP-binding cassette. There are two ATP-binding cassettes in one molecule of the ABCB1 protein. Those ATP-binding cassettes are functionally nonidentical, but essential for the transport function of ABCB1 (Sauna and Ambudkar, 2000; Senior *et al.*, 1995). Because ABCB1 is an ATP-dependent active transporter, drug transport is coupled with ATP hydrolysis. ATPase activity can be determined by measuring inorganic phosphate liberation (Carter and Karl, 1982; Sarkadi *et al.*, 1992). This property is advantageous to screen for the function and substrate specificity of ABCB1 (Garrigues *et al.*, 2002; Onishi *et al.*, 2003).

High-Speed Screening System to Measure ABCB1 ATPase Activity

We have developed a high-speed screening system using 96-well plates to analyze the substrate specificity of ABCB1 and its genetic variants (Ishikawa *et al.*, 2004a; Onishi *et al.*, 2003). ABCB1 has been expressed in

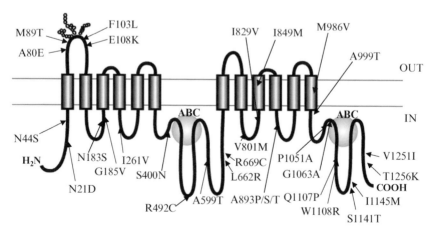

FIG. 8. Nonsynonymous polymorphisms of human ABCB1. (A) Schematic illustration of the structure of ABCB1 protein and the locations of amino acid changes. G185V is an acquired mutation. ABC, ATP-binding cassette.

insect cells. Figure 9 demonstrates schematically the protocol for the ATPase activity assay. Briefly, cell membranes expressing ABCB1 (2 μg of membrane protein per well) are suspended in 10 μl of the incubation medium containing 50 mM Tris-Mes (pH 6.8), 2 mM EGTA, 2 mM dithiothreitol, 50 mM potassium chloride, 5 mM sodium azide, and 2 mM ouabain. This medium is mixed with 10 μl of a test compound solution and is then preincubated at 37° for 3 min. The ATPase reaction is started by adding 20 μl of 4 mM ATP/Mg solution to the reaction mixture, and the incubation is maintained at 37° for 30 min. The reaction is stopped by the addition of 20 μl of 5% trichloroacetic acid followed by 42 μl of "solution A" and 18 μl of "solution B" (see Fig. 9 for details). Thereafter, 120 μl of "solution C" is added to the mixture. These mixing processes are automatically carried out in the HALCS-I system (BioTec Co. Ltd., Tokyo, Japan). The absorbance of each reaction mixture in the 96-well plates is measured photometrically at a wavelength of 630 nm in a Multiskan JX system (Dainippon Pharmaceuticals Co., Osaka, Japan). The amount of liberated phosphate is quantified based on the calibration line established with inorganic phosphate standards. In addition, each 96-well plate contains a positive control in which Sf9 cell membranes are incubated with 10 μM

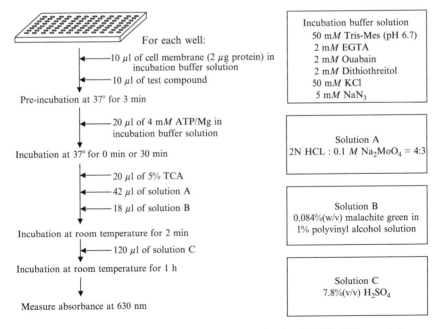

For each well:
—10 μl of cell membrane (2 μg protein) in incubation buffer solution
—10 μl of test compound

Pre-incubation at 37° for 3 min

—20 μl of 4 mM ATP/Mg in incubation buffer solution

Incubation at 37° for 0 min or 30 min

—20 μl of 5% TCA
—42 μl of solution A
—18 μl of solution B

Incubation at room temperature for 2 min

—120 μl of solution C

Incubation at room temperature for 1 h

Measure absorbance at 630 nm

Incubation buffer solution
50 mM Tris-Mes (pH 6.7)
2 mM EGTA
2 mM Ouabain
2 mM Dithiothreitol
50 mM KCl
5 mM NaN$_3$

Solution A
2N HCL : 0.1 M Na$_2$MoO$_4$ = 4:3

Solution B
0.084%(w/v) malachite green in 1% polyvinyl alcohol solution

Solution C
7.8%(v/v) H$_2$SO$_4$

Fig. 9. Schematic diagram of high-speed screening for ABCB1 ATPase activity.

verapamil. Figure 10A shows a Michaelis–Menten-type relationship between ABCB1 ATPase activity and verapamil concentration.

Genetic Polymorphisms of ABCB1 vs ATPase Activity

To date, extensive studies have been carried out on genetic variations of the human *ABCB1* gene (for reviews, see Kerb *et al.*, 2001; Sparreboom *et al.*, 2003). Figure 8 depicts hitherto identified nonsynonymous polymorphisms in the ABCB1 protein. Several preclinical and clinical studies have provided evidence for naturally occurring polymorphisms in ABCB1 and their effects on drug absorption, distribution, and elimination. Quantitative studies are required, however, to precisely evaluate functional changes associated with such genetic polymorphisms. For this purpose, we have prepared several variant forms (i.e., N183S, S400N, R492C, R669C, I849M, A893T, M986V, A999T, P1051A, and G1063A) by site-directed mutagenesis. These variants of ABCB1 were then expressed in Sf9 cells by using the pFASTBAC1 vector and recombinant baculoviruses, as described previously (Ishikawa *et al.* 2004a; A. Sakurai *et al.*, manuscript in preparation). ABCB1 variant proteins expressed in Sf9 cell membranes are detected by the Western blot method with the C219 monoclonal antibody. With membranes prepared from insect cells expressing those ABCB1 variants, ATPase activity has been measured in the presence of verapamil at various concentrations. Figure 10B summarizes the kinetic parameters (K_m and V_{max} values) observed with verapamil for those variant forms as well as for the wild type of ABCB1. K_m and V_{max} values for verapamil vary to some extent among the variants. The V_{max} values of those variants were normalized to that of the wild type by referring to the intensity of each variant protein on the immunoblotting as shown in Fig. 4B. Kinetic parameters of nonsynonymous polymorphisms of ABCB1 observed with different substrates are reported elsewhere.

Quantitative Structure–Activity Relationship (SAR) Analysis to Evaluate the Substrate Specificity of ABCB1

To understand the impact of nonsynonymous polymorphisms on the function of ABCB1, it is critically important to quantitatively analyze the functional difference among such variants. We have developed a method of quantitative SAR analysis. Using the high-speed screening system, we first measure ABCB1 ATPase activity toward a total of 41 different therapeutic drugs and compounds. The tested compounds are classified into seven groups, that is, A, neurotransmitters; B, Ca^{2+} channel blockers; C, steroids; D, potassium channel modulators; E, nonsteroidal anti-inflammatory drugs

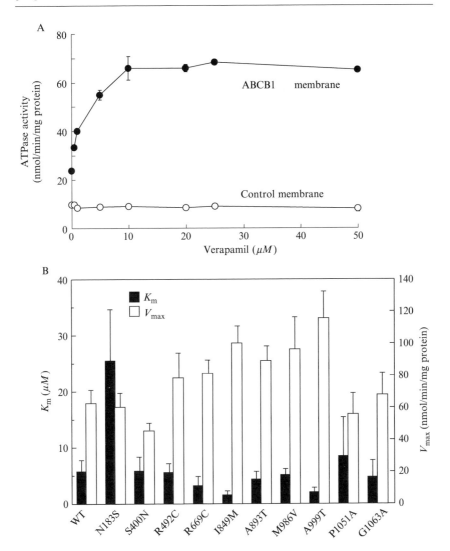

FIG. 10. ATPase activity of ABCB1 and its variants. (A) Verapamil-enhanced ATPase activity of ABCB1 measured in plasma membranes prepared from Sf9 cells. Plasma membranes were incubated with verapamil at different concentrations (0, 1, 2, 5, 10, 20, 25, and 50 μM). Closed circles, ABCB1-expressing Sf9 cells; open circles, control Sf9 cells. (B) the effect of nonsynonymous polymorphisms on ATPase activity. Verapamil-enhanced ATPase activity was measured under the same conditions described in Fig. 9. For each variant form of ABCB1, K_m and V_{max} values were calculated from Lineweaver–Burk plots by referring to the immunoblot intensity of plasma membrane preparations.

(NSAIDs); F, anticancer drugs; and G, miscellaneous. Figure 11 shows the effects of those test compounds on ABCB1 ATPase activity. Test compounds were evaluated at a concentration of 10 μM, and data are expressed relative to the ATPase activity measured with 10 μM verapamil. Among 41 different therapeutic drugs and compounds tested in this study, Ca^{2+} channel blockers, such as verapamil (B-1), bepridil (B-4), fendiline (B-5), prenylamine (B-6), nicardipine (B-7), and FK506 (G-4), stimulated ATPase activity. At the concentration of 100 μM, paclitaxel (F-5), doxorubicin (F-7), and quinidine (G-1) stimulated ABCB1 ATPase activity more significantly, even though the extent of ATPase stimulation was relatively smaller than that of Ca^{2+} channel blockers (data not shown).

Chemical fragmentation codes are practical and useful for describing the chemical structures of a variety of substrates and nonsubstrates for

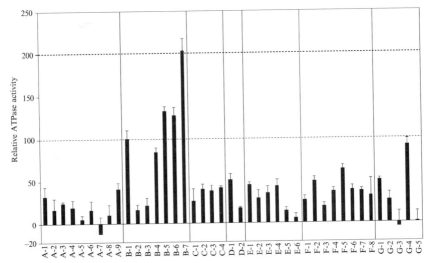

Fig. 11. The effect of therapeutic drugs and compounds on ABCB1 ATPase activity. ATPase activity was measured in the presence of 10 μM of a test compound. All activities are expressed as relative values+SD as compared with the activity measured with 10 μM verapamil (100%). The tested drugs and compounds are glycine (A-1), glutamic acid (A-2), dopamine (A-3), norepinephrine (A-4), epinephrine (A-5), γ-aminobutyric acid (A-6), histamine (A-7), serotonin (A-8), melatonin (A-9), verapamil (B-1), nifedipine (B-2), diltiazem (B-3), bepridil (B-4), fendiline (B-5), prenylamine (B-6), nicardipine (B-7), dexamethasone (C-1), betamethasone (C-2), prednisolone (C-3), cortisone (C-4), nicorandil (D-1), pinacidil (D-2), acetylsalicylic acid (E-1), indomethacin (E-2), acemetacin (E-3), ibuprofen (E-4), naproxen (E-5), mepirizole (E-6), vinblastine (F-1), etoposide (F-2), actinomycin D (F-3), daunorubicin (F-4), paclitaxel (F-5), methotrexate (F-6), doxorubicin (F-7), 5-fluorouracil (F-8), quinidine (G-1), p-aminohippuric acid (G-2), penicillin G (G-3), FK506 (G-4), and novobiocin (G-5).

TABLE I

MULTIPLE LINEAR REGRESSION ANALYSIS MODELS TO PREDICT ABCB1 ATPase ACTIVITY TOWARD TESTED COMPOUNDS[a]

Chemical fragmentation code	Model 1	Model 2	Model 3	Model 4	Model 5	Model 6
J581	96.87	90.10	90.67	82.86	84.69	97.04
G100	54.47	59.33	51.62	55.01	54.80	49.17
M331	38.42	46.44	42.30	42.91	43.67	37.71
M270	0	0	0	0	11.61	0
M272	0	0	0	11.40	0	0
M531	−61.63	−64.51	−61.04	−62.64	−63.38	−59.43
F014	−28.31	−22.44	−29.09	−19.80	−20.92	−22.41
H100	0	0	0	0	0	−15.64
M321	0	0	−14.60	0	0	0
M370	0	−12.56	0	0	0	0
M391	−14.29	0	0	0	0	0
Constant	43.44	36.61	45.01	26.99	26.40	38.79
R =	0.953	0.952	0.953	0.952	0.952	0.954

[a] ABCB1 ATPase activity is formulated as a linear combination of chemical fragmentation codes weighted by the corresponding coefficient, where the symbol of "i" in parentheses designates a specific chemical fragmentation code. ABCB1 ATPase activity (predicted) = $\Sigma C(i) \times$ chem. frag. code (i) + constant. R: Correlation coefficient.

ABCB1. Derwent Information, Ltd., developed this structure-indexing language suitable for describing chemical patents. Markush TOPFRAG is the software that generates the chemical fragment codes from chemical structure information.

We used Markush TOPFRAG to generate chemical fragmentation codes for each compound tested in Fig. 11 and then carried out a multiple linear regression analysis to obtain a relationship between ABCB1 ATPase activity and chemical fragmentation codes thus generated. In this way, we identified several sets of chemical fragmentation codes related to the substrate specificity of ABCB1. For example, based on data shown in Fig. 11, a total of six best-fitting models were created. Table I summarizes the contents of those multiple linear regression analysis models, and Table II provides explanations for the chemical fragmentation codes generated in the analysis. These results demonstrate that the moieties represented by the chemical fragmentation codes of J581, G100, and M331 contribute positively to ATPase activity, whereas those of M531 and F014 have negative contributions. Among those codes, J581 has the greatest contribution (Table I), suggesting that an oxo group bonded to an

TABLE II

EXPLANATION OF CHEMICAL FRAGMENTATION CODES USED FOR PREDICTION OF
ABCB1 ATPASE ACTIVITY

Chemical fragmentation code		Ext. code	Explanation
J58	Oxo group bonded to aliphatic C	J581	One oxo group bonded to aliphatic C
G1	Unfused aromatic rings	G100	Unfused aromatic ring(s) present, no other carbocyclic ring systems are present
M33	Straight or branched carbon chains	M331	Straight carbon chain with -CH3, -C = CH2, and/or -C ≡ CH
M27	Chain bonded to U	M270	Chain bonded to U
M27	Chain bonded to U	M272	Chain bonded to O
M53	Carbocyclic systems with at least one aromatic ring	M531	One M53 code
F01	Positions substituted	F014	Position 4 substituted
H10	Type of amine	H100	One primary amine
M32	Multipliers for subset M31 M31: number of C atoms in polyvalent chain	G321	One or more M31 code used once
M37	Carbon chain bonded to ring C and (U and/or C = U and/or C ≡ CH) but not V, C = V, C ≡ V	M370	Carbon chain bonded to ring C and (U and/or C = U and/or C ≡ CH) but not V, C = V, C ≡ V
M39	Multipliers for codes M350 to M383 (polyvalent carbon chain attachments)	M391	One or more of codes used once

(see *http://thomsonderwent.com/products/patentresearch/markushtopfrag* and *http://thom-sonderwent.com/derwenthome/media/support/userguides/chemindguide* for more information.)
U = C, H, O, S, Se,Te, or N.
V = atom other than U.

aliphatic carbon (Table II) is an important moiety for the recognition and/ or transport by the ABCB1 protein. In addition, it is suggested that an unfused aromatic ring (G100) and a straight carbon chain (M331) are important chemical moieties for the substrate specificity of ABCB1.

The uniqueness of this approach derives from the facts that ABCB1 ATPase activity is described as a linear combination of chemical fragmentation codes and that the coefficient for each chemical fragment code reflects the extent of the contribution of a specific chemical moiety to the ATPase activity. The point in the catalytic cycle at which substrate binding takes place and details of how ATP hydrolysis drives transport may be critical for understanding the mechanism of substrate specificity. This quantitative SAR analysis can be applied for each variant form of ABCB1 to gain more insight into the effect of nonsynonymous polymorphisms on the substrate specificity of ABCB1.

SNP Array to Detect Genetic Polymorphisms of ABC Transporters

It is estimated that about 3 million SNPs can be derived by comparing genomic sequences from individuals among several populations. These variants may be used as an important tool in association studies or linkage disequilibrium mapping to elucidate the genetic foundation of multifactorial disorders and individual differences in drug response. It is not realistic, however, to analyze all SNPs for each patient, as the number of SNPs is so large and each SNP must ultimately be evaluated in terms of functionality. Therefore, at first we validate SNPs of drug transporters as well as drug-metabolizing enzymes from a functional point of view. Through functional validation, the number of SNPs can be reduced significantly. A limited set of SNPs closely related pharmacokinetically will be analyzed by a cost-effective and automated analytical technology.

We have developed a "SNP array" method to meet such practical and clinical needs. Based on our functional assay data as well as currently available SNP databases on drug transporter genes, SNP-specific probes are carefully designed and spotted on the epoxy-activated surface of glass plates by means of the GENESHOT system (NGK Insulators, Ltd., Japan). Each probe spot is 50 μm in diameter. Each SNP is detected by one set of probes (usually five oligonucleotide probes) that have different stringencies of hybridization with DNA samples.

Figure 12 shows a flowchart of the SNP detection process, that is, DNA sample preparation, multiplex PCR, hybridization on the SNP array, and fluorescence signal measurement. DNA samples for SNP array analysis are amplified by multiplex PCR with biotin-linked primers and genomic DNA extracted from human white blood cells. The multiplex PCR conditions are as follows: 94° for 5 min, 35 cycles of 95° for 30 s, 62.5° for 30 s, and 72° for 30 s, and finally 72° for 2 min. The resulting PCR product (2.5 μl) is subsequently mixed with 2.5 μl of 0.6 M NaOH. Hybridization with SNP probes is performed in 20 μl of a hybridization buffer solution comprising 200 mM citrate/phosphate, pH 6.0, 2% SDS, 750 mM NaCl, and 0.1% NaN$_3$. The hybridization is maintained at 55° overnight. After the hybridization buffer solution and excess PCR product are removed by rinsing as outlined in Fig. 12, the SNP array is treated with Cy5-linked streptavidin in TNB solution (0.1 M Tris/HCl, pH 7.6, 0.15 M NaCl, and 0.5% NEN blocking reagent) at room temperature for 30 min. Because biotin-labeled

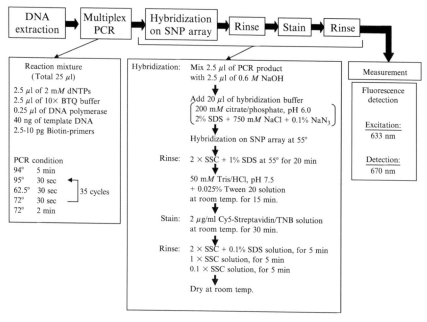

FIG. 12. SNP detection by the SNP array method. The detection process includes DNA sample preparation, multiplex PCR, hybridization on the SNP array, and fluorescence signal measurement.

primers are used in the multiplex PCR, the hybridized PCR product can be conjugated with streptavidin. Thus, Cy5-linked streptavidin enables fluorescence detection of the hybridized PCR product with excitation light at 633 nm (He/Ne laser) or 647 nm (Ar/Kr laser). Figure 13 shows the result of SNP array detection where nonsynonymous polymorphisms of ABCG2, that is, Q126stop and Q141K, were analyzed. Homo- and heteroalleles of 141-Gln (Q) and 141-Lys (K) in human samples were detected and clearly distinguished with this SNP array.

Imai *et al.* (2002) identified three allelic variants in the ABCG2 gene, of which two were nonsynonymous SNPs (V12M and Q141K) and the third was a splice variant with deletion of nucleotides 944–949 that lacks Ala-315 and Thr-316 (Δ315-6). Compared with wild-type transfected cells, ABCG2 Q141K variant-transfected cells showed a low level of drug resistance that is associated with decreased protein expression. The SNP (Q141K) was postulated to cause increased sensitivity of normal cells to anticancer agents that are ABCG2 substrates such as SN-38. Moreover, a novel SNP in exon 4, 376C>T, substituting a stop codon for Gln-126, was found in the

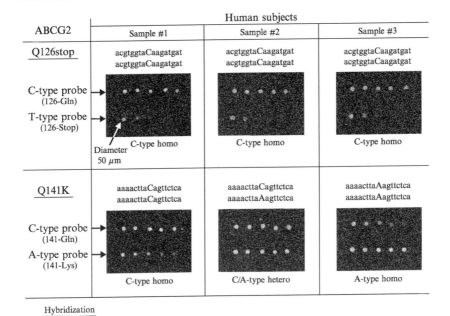

FIG. 13. Detection of nonsynonymous polymorphisms (Q126stop and Q141K) of ABCG2 with the SNP array. (See color insert.)

Japanese population with an allelic frequency of 2.4% (Honjo *et al.*, 2002). It has been postulated that the 376C>T SNP may have a higher impact than the 421C>A polymorphism causing Q141K because active ABCG2 protein will not be synthesized from the variant allele.

Concluding Remarks

Pharmacogenomics is recognized to be increasingly important for predicting pharmacokinetic profiles and/or adverse reactions of drugs (Kalow *et al.*, 2001). Drug transporters, as well as drug-metabolizing enzymes, play pivotal roles in determining the pharmacokinetic profiles of drugs (i.e., drug absorption, distribution, metabolism, and elimination, as well as drug concentration at the target site). The effects of drug transporters on the pharmacokinetic profile of a drug depend on their expression and functionality. There are an increasing number of reports that address genetic polymorphisms of drug transporters. It is also true, however, that there is still considerable discrepancy among hitherto reported results. Functional analysis of the polymorphism of drug transporters is one of such important approaches that provide clear insight into the biochemical significance of genetic polymorphisms (Ishikawa et al., 2004b).

This chapter conveys a new strategy for analyzing the relationship between the substrate specificity of ABC transporters (e.g., ABCG2 and ABCB1) and the chemical structure of substrates. This approach is applicable for the functional analysis of genetic polymorphisms of ABC transporters. The effect of SNPs on the transport activity may depend on the substrates tested, and therefore the functional analysis of SNPs using a wide variety of substrates is of great interest. One amino acid substitution can alter interactions between the active site of an ABC transporter and substrate molecules. Therefore, it is critically important to quantitatively analyze and evaluate such structure-related interactions. In this context, the new SAR analysis with chemical fragmentation codes will provide a powerful tool to quantify the impact of genetic polymorphisms on the function of ABC transporters. Furthermore, as demonstrated in this chapter, the SNP array method is a simple and practical tool for detecting SNPs that can affect the function and/or expression of drug transporters. The combination of SNP detection with functional evaluations will provide clear insights into the molecular mechanisms underlying individual differences in the drug response of patients.

Acknowledgments

This study was supported, in part, by a research grant entitled "Toxicoproteomics: Expression of ABC transporter genes and drug-drug interactions" (H14-Toxico-002) from the Japanese Ministry of Health and Welfare, a Grant-in-Aid for Creative Scientific Research (No. 13NP0401), and a research grant (No. 14370754) from the Japan Society for the Promotion of Science.

References

Allikmets, R., Schriml, L. M., Hutchinson, A., Romano-Spica, V., and Dean, M. (1998). A human placenta-specific ATP-binding cassette gene (ABCP) on chromosome 4q22 that is involved in multidrug resistance. Cancer Res. 58, 5337–5339.

Ambudkar, S. V., Dey, S., Hrycyna, C. A., Ramachandra, M., Pastan, I., and Gottesman, M. M. (1999). Biochemical, cellular, and pharmacological aspects of the multidrug transporter. Annu. Rev. Pharmacol. Toxicol. 39, 361–398.

Bailey-Dell, K. J., Hassel, B., Doyle, L. A., and Ross, D. D. (2001). Promoter characterization and genomic organization of the human breast cancer resistance protein (ATP-binding cassette transporter G2) gene. Biochim. Biophys. Acta 1520, 234–241.

Borst, P., and Elferink, R. O. (2002). Mammalian ABC transporters in health and disease. Annu. Rev. Biochem. 71, 537–592.

Carter, S. G., and Karl, D. W. (1982). Inorganic phosphate assay with malachite green: An improvement and evaluation. J. Biochem. Biophys. Methods 7, 7–13.

Cervenak, J., Andrikovics, H., Özvegy-Laczka, C., Tordai, A., Német, K., Vásadi, A., and Sarkadi, B. (2005). The role of the human ABCG2 multidrug transporter and its variants in cancer therapy and toxicology. *Cancer Lett.*

Dean, M., Rzhetsky, A., and Allikmets, R. (2001). The human ATP-binding cassette (ABC) transporter superfamily. *Genome Res.* **11,** 1156–1166.

Doyle, L. A., Yang, W., Abruzzo, L. V., Krogmann, T., Gao, Y., Rishi, A. K., and Ross, D. D. (1998). A multidrug resistance transporter from human MCF-7 breast cancer cells. *Proc. Natl. Acad. Sci. USA* **95,** 15665–15670.

Garrigues, A., Nugier, J., Orlowski, S., and Ezan, E. (2002). A high-throughput screening microplate test for the interaction of drugs with P-glycoprotein. *Anal. Biochem.* **305,** 106–114.

Gottesman, M. M., and Pastan, I. (1988). The multidrug transporter, a double-edged sword. *J. Biol. Chem.* **263,** 12153–12166.

Holland, I. B., Cole, S. P. C., Kuchler, K., and Higgins, C. F. (2003). "ABC Proteins: From Bacteria to Man." Academic Press, Amsterdam.

Honjo, Y., Morisaki, K., Huff, L. M., Robey, R. W., Hung, J., Dean, M., and Bates, S. E. (2002). Single-nucleotide polymorphism (SNP) analysis in the ABC half-transporter ABCG2 (MXR/BCRP/ABCP1). *Cancer Biol. Ther.* **1,** 696–702.

Imai, Y., Nakane, M., Kage, K., Tsukahara, S., Ishikawa, E., Tsuruo, T., Miki, Y., and Sugimoto, Y. (2002). C421A polymorphism in human breast cancer resistance protein gene is associated with low expression of Q141K protein and low-level drug resistance. *Mol. Cancer Ther.* **1,** 611–616.

Ishikawa, T. (1989). ATP/Mg^{2+}-dependent cardiac transport system for glutathione S-conjugates: A study using rat heart sarcolemma vesicles. *J. Biol. Chem.* **264,** 17343–17348.

Ishikawa, T. (1992). The ATP-dependent glutathione S-conjugate export pump. *Trends Biochem. Sci.* **17,** 433–468.

Ishikawa, T. (2003). Multidrug resistance: Genomics of ABC transporters. *In* "Nature Encyclopedia of the Human Genome" (D. N. Cooper, ed.), Vol. 4, pp. 154–160. Nature Publishing Group, London.

Ishikawa, T., Kasamtsu, S., Hagiwara, Y., Mitomo, H., Kato, R., and Sumino, Y. (2003). Expression and functional characterization of human ABC transporter ABCG2 variants in insect cells. *Drug Metab. Pharamcokin.* **18,** 194–202.

Ishikawa, T., Müller, M., Klünemann, C., Schaub, T., and Keppler, D. (1990). ATP-dependent primary active transport of cysteinyl leukotrienes across rat liver canalicular membrane: The role of glutathione S-conjugate carrier. *J. Biol. Chem.* **265,** 19279–19286.

Ishikawa, T., Onishi, Y., Hirano, H., Oosumi, K., Nagakura, M., and Tarui, S. (2004a). Pharmacogenomics of drug transporters: A new approach to functional analysis of the genetic polymorphisms of ABCB1 (P-glycoprotein/MDR1). *Bio. Pharm. Bull.* **27,** 939–948.

Ishikawa, T., Tsuji, A., Inui, K., Sai, Y., Anzai, N., Wada, M., Endou, H., and Sumino, Y. (2004b). The genetic polymorphism of drug transporters: Functional analysis approaches. *Pharmacogenomics* **5,** 67–99.

Jonker, J. W., Buitelaar, M., Wagenaar, E., Van der Valk, M. A., Scheffer, G. L., Scheper, R. J., Plosch, T., Kuipers, F., Elferink, R. P., Rosing, H., Beijinen, J. H., and Schinkel, A. H. (2002). The breast cancer resistance protein protects against a major chlorophyll-derived dietary phototoxin and protoporphyria. *Proc. Natl. Acad. Sci. USA* **99,** 15649–15654.

Kalow, W., Meyer, U. A., and Tyndale, R. F. (2001). "Pharmacogenomics." Dekker, New York.

Keppler, D., Jedlitschky, G.., and Leier, I. (1998). Transport function and substrate specificity of multidrug resistance protein. *Methods Enzymol.* **292,** 607–616.

Kerb, R., Hoffmeyer, S., and Brinkmann, U. (2001). ABC drug transporters: Hereditary polymorphisms and pharmacological impact in MDR1, MRP1 and MRP2. *Pharmacogenomics* **2,** 51–64.

Klein, I., Sarkadi, B., and Váradi, A. (1999). An inventory of the human ABC proteins. *Biochim. Biophys. Acta* **1461,** 237–262.

Ling, V. (1997). Multidrug resistance: Molecular mechanisms and clinical relevance. *Cancer Chemother. Pharmacol.* **40,** S3–S8.

Maliepaard, M., Scheffer, G. L., Faneyte, I. F., van Gastelen, M. A., Pijnenborg, A. C., Schinkel, A. H., van De Vijver, M. J., Scheper, R. J., and Schellens, J. H. (2001). Subcellular localization and distribution of the breast cancer resistance protein transporter in normal human tissues. *Cancer Res.* **61,** 3458–3464.

Mitomo, H., Kato, R., Ito, A., Kasamatsu, S., Ikegami, Y., Kii, I., Kudo, A., Kobatake, E., Sumino, Y., and Ishikawa, T. (2003). A functional study on polymorphism of the ATP-binding cassette transporter ABCG2: Critical role of Arg-482 in methotrexate transport. *Biochem. J.* **373,** 767–774.

Miyake, K., Mickley, L., Litman, T., Zhan, Z., Robey, R., Cristensen, B., Brangi, M., Greenberger, L., Dean, M., Fojo, T., and Bates, S. E. (1999). Molecular cloning of cDNAs which are highly overexpressed in mitoxantrone-resistant cells: Demonstration of homology to ABC transport genes. *Cancer Res.* **59,** 8–13.

Nakatomi, K., Yoshikawa, M., Oka, M., Ikegami, Y., Hayasaka, S., Sano, K., Shiozawa, K., Kawabata, S., Soda, H., Ishikawa, T., Tanabe, S., and Kohno, S. (2001). Transport of 7-ethyl-10-hydroxycamptothecin (SN-38) by breast cancer resistance protein ABCG2 in human lung cancer cells. *Biochem. Biophys. Res. Commun.* **288,** 827–832.

Onishi, Y., Hirano, H., Nakata, K., Oosumi, K., Nagakura, M., Tarui, S., and Ishikawa, T. (2003). High-speed screening and structure-activity relationship analysis for the substrate specificity of P-glycoprotein (ABCB1). *Chem.-Biol. Inform. J.* **3,** 175–193.

O'Reilly, D. R., Miller, L. K., and Luckow, V. A. (1994). *In* "Baculovirus Expression Vectors, a Laboratory Manual." Oxford Univ. Press, New York.

Sarkadi, B., Price, E. M., Boucher, R. C., Germann, U. A., and Scarborough, G. A. (1992). Expression of the human multidrug resistance cDNA in insect cells generates a high activity drug-stimulated membrane ATPase. *J. Biol. Chem.* **267,** 4854–4858.

Sauna, Z. E., and Ambudkar, S. V. (2000). Evidence for a requirement for ATP hydrolysis at two distinct steps during a single turnover of the catalytic cycle of human P-glycoprotein. *Proc. Natl. Acad. Sci. USA* **97,** 2515–2520.

Schinkel, A. H., and Jonker, J. W. (2003). Mammalian drug efflux transporters of the ATP binding cassette (ABC) family: An overview. *Adv. Drug Deliv. Rev.* **55,** 3–29.

Senior, A. E., Al-Shawi, M. K., and Urbatsch, I. L. (1995). The catalytic cycle of P-glycoprotein. *FEBS Lett.* **377,** 285–289.

Sparreboom, A., Danesi, R., and Ando, Y. (2003). Pharmacogenomics of ABC transporters and its role in cancer chemotherapy. *Drug Resistance Update* **6,** 71–84.

Summers, M. D., and Smith, G. E. (1987). A manual of methods for baculovirus vectors and insect cell culture procedures. *Texas Agric. Exp. Stat. Bull.* **1555,** 1–57.

Yoshikawa, M., Ikegami, Y., Sano, K., Yoshida, H., Mitomo, H., Sawada, S., and Ishikawa, T. (2004). Transport of SN-38 by the wild type of human ABC transporter ABCG2 and its inhibition by quercetin, a natural flavonoid. *J. Exp. Ther. Oncol.* **4,** 25–35.

Zhou, S., Schuetz, J. D., Bunting, K. D., Colapietro, A. M., Sampath, J., Morris, J. J., Lagutina, I., Grosveld, G. C., Osawa, M., Nakauchi, H., and Sorrentino, B. P. (2001). The ABC transporter Bcrp1/ABCG2 is expressed in a wide variety of stem cells and is a molecular determinant of the side-population phenotype. *Nat. Med.* **7,** 1028–1034.

[28] Coordinate Transcriptional Regulation of Transport and Metabolism

By Jyrki J. Eloranta, Peter J. Meier, and
Gerd A. Kullak-Ublick

Abstract

Intestinal absorption and hepatic clearance of drugs, xenobiotics, and bile acids are mediated by transporter proteins expressed at the plasma membranes of intestinal epithelial cells and liver parenchymal cells in a polarized manner. Within enterocytes and hepatocytes, these exogenous or endogenous, potentially toxic compounds may be metabolized by phase I cytochrome P450 (CYP) and phase II conjugating enzymes. Many transporter proteins and metabolizing enzymes are subject to direct translational modification, enabling very rapid changes in their activity. However, to achieve intermediate and longer term changes in transport and enzyme activities, the genes encoding drug and bile acid transporters, as well as the CYP and conjugating enzymes, are regulated by a complex network of transcriptional cascades. These are typically mediated by specific members of the nuclear receptor family of transcription factors, particularly FXR, SHP, PXR, CAR, and HNF-4α. Most nuclear receptors are activated by specific ligands, including numerous xenobiotics (PXR, CAR) and bile acids (FXR). The fine-tuning of transcriptional control of drug and bile acid homeostasis depends on regulated interactions of specific nuclear receptors with their target genes.

Overview of Transport and Metabolism of Bile Acids, Drugs, and Xenobiotics

The highly efficient enterohepatic circulation of bile acids, the products of cholesterol catabolism, is essential for hepatic excretory function (Kullak-Ublick *et al.*, 2004; Trauner and Boyer, 2003). Perturbations in the function and expression of plasma membrane-associated and intracellular bile acid transporters, as well as enzymes involved in synthesis and conjugation of bile acids, can result in impaired bile formation and cholestatic liver disease. Bile acids are physiological detergents that solubilize cholesterol in bile and emulsify lipids in the intestine, thus facilitating their absorption and excretion. Bile acids have been further identified as ligands for specific members of the nuclear receptor family of transcription factors

METHODS IN ENZYMOLOGY, VOL. 400   0076-6879/05 $35.00
          DOI: 10.1016/S0076-6879(05)00028-5

involved in bile acid homeostasis and as initiators of signaling cascades important for metabolic regulation and liver function. The ability of bile acids to modulate transcription and signaling allows regulatory positive and negative feedback and feedforward circuits, necessary to finely adjust their enterohepatic circulation.

Transport of exogenous compounds, such as drugs and environmental xenobiotics, is also mediated by specialized transport proteins and, indeed, certain bile acid transporters also mediate uptake or efflux of these exogenous agents. For certain transporters expressed in the liver and in the intestine, mainly drugs and xenobiotics have been identified as transport substrates. Similar to bile acid metabolism, both the transport and the oxidative (phase I) and conjugative (phase II) metabolism of xenobiotics is subject to extensive feedback and feedforward regulation at the level of transcription. This is achieved through the ability of numerous xenobiotics to serve as ligands for metabolic nuclear receptors, which regulate the genes encoding transporters and enzymes involved in drug metabolism. The range of nuclear receptors and their target genes responding to either bile acids or xenobiotics is overlapping, resulting in intricate regulatory cross talk between bile acid and drug metabolism. The aim of this chapter is to illustrate these intertwined transcriptional circuits using selected target genes encoding transporters and metabolizing enzymes as examples. The transcriptional events discussed here are summarized in Fig. 1.

Metabolic Nuclear Receptors as Transcription Factors

Most nuclear receptors have a modular structure, including a conserved DNA-binding domain and a C-terminal region mediating ligand-binding, dimerization, and ligand-dependent transactivation (Kumar and Thompson, 1999), and bind to DNA as either homo- or heterodimers. Consistent with dimerization, DNA-binding sites for nuclear receptors typically contain two hexameric half-sites (consensus AGGTCA) that may be arranged as direct (DR), inverted (IR), or everted (ER) repeats (Khoranizahed and Rastinejad, 2001). Most nuclear receptors are crucially enhanced in their ability to transactivate by small lipophilic ligands. In the absence of agonistic ligands or in the presence of antagonistic ligands, C-terminal domains of nuclear receptors associate with transcriptional corepressors. Upon binding to agonistic ligands, C termini undergo a conformational shift, leading to dissociation of corepressors and association with transcriptional coactivators (Collingwood et al., 1999). Many coactivators act by acetylating or methylating histones or nonhistone proteins assembled on the target promoters, thus leading to an enhanced rate of transcription (Sterner and Berger, 2000; Stallcup 2001). The dependence

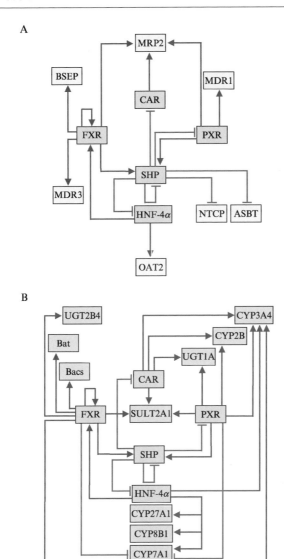

FIG. 1. Nuclear receptors FXR, SHP, PXR, CAR, and HNF-4α regulate a complex network of genes involved in (A) transport and (B) metabolism of drugs and bile acids. Nuclear receptors are shown within blue boxes, bile acid transporter genes in yellow boxes, and phase I and phase II metabolizing genes in light green boxes. Green and red lines indicate transcriptional activation and suppression, respectively. (See color insert.)

of most nuclear receptors on specific ligands, such as hormones, fatty acids, bile acids, or xenobiotics, allows them to act as sensors of, and consequently orchestrate transcriptional responses to, the metabolic information within a cell.

Specific nuclear receptors function as master regulators of the defense against xeno- and endo-biotic toxicity (PXR, pregnane X receptor and CAR, constitutive androstane receptor) and of bile acid metabolism (FXR, farnesoid X receptor). Ligands for the metabolic nuclear receptors are commonly present in cells in much higher (micromolar) concentrations than ligands for steroid receptors (nanomolar range), such as estrogen receptor (ER) and glucocorticoid receptor (GR), which are mediators of endocrine signaling. Common to most metabolic nuclear receptors is the dependence of their DNA binding on heterodimerization with another nuclear receptor, retinoid X receptor (RXR), the ligand for which is 9-*cis*-retinoic acid. Certain nuclear receptor heterodimers do not require, or respond to, the RXR ligand, whereas others, such as FXR-RXR, do respond to ligands that bind to both partners and are often synergistically stimulated in the presence of both (Blumberg and Evans, 1998). This dependence on the RXR ligand may vary for a particular nuclear receptor heterodimer depending on the exact promoter context (Tzameli *et al.*, 2003).

Role of FXR in Bile Acid Homeostasis

Due to their detergent nature, bile acids are intrinsically toxic to cells, hence their intracellular concentration must be tightly controlled. Many transcriptional regulatory events initiated by increased levels of bile acids are defense mechanisms against potential bile acid toxicity: by acting as ligands for nuclear receptors, bile acids can induce expression of hepatocellular bile acid efflux mechanisms, downregulate bile acid uptake mechanisms, repress the enzymes required for *de novo* synthesis of bile acids, and enhance detoxification of bile acids. Bile acids function as endogenous ligands for the nuclear receptor FXR (Makishima *et al.*, 1999; Parks *et al.*, 1999; Wang *et al.*, 1999). Naturally occurring bile acids are not equally efficient as FXR agonists, but differ in their potency: chenodeoxycholic acid (CDCA) is the most efficient FXR activator, followed by deoxycholic acid (DCA), cholic acid (CA), and ursodeoxycholic acid (UDCA) (Lew *et al.*, 2004; Makishima *et al.*, 1999; Parks *et al.*, 1999). While the secondary bile acid lithocholic acid (LCA) can by itself serve as a weak agonistic ligand for FXR, it can strongly antagonize CDCA-stimulated activation of FXR (Yu *et al.*, 2002), providing a possible molecular mechanism for LCA-induced cholestasis.

FXR is expressed abundantly and fulfills most of its physiologically important functions in the liver. It is also present in other tissues exposed to bile acids, the intestine and the kidney (Forman *et al.*, 1995). In order to activate its target promoters, FXR generally binds DNA as a hetero-dimer with RXR. The preferred DNA-binding motif for the FXR-RXR heterodimer is IR-1 (inverted repeat of AGGTCA hexamers, separated by 1 bp), present in most target promoters directly activated by FXR (Laffitte *et al.*, 2000). Increasing evidence shows that FXR can regulate its own expression. The human *FXR* promoter is induced by bile acids in transfection assays, an effect that can be blocked by cotransfection of a transactivation-deficient FXR mutant (Chen *et al.*, 2004). Also, treatment of cultured hepatoma cells with bile acids increases endogenous levels of the FXR protein (Lew *et al.*, 2004). Supporting *in vitro* findings, FXR mRNA and protein levels are reduced upon removal of the endogenous bile acid pool in rabbits and are restored when the depleted bile acid pool is replaced by individual bile acids (Xu *et al.*, 2002). The positive autoregulation may serve to amplify FXR-mediated transcriptional responses to bile acids.

Positive Transcriptional Regulation by FXR

The gene encoding BSEP, a major canalicular bile salt export pump of the ATP-binding cassette (ABC) transporter family (Meier and Stieger, 2002), is a well-known target for transactivation by FXR. FXR binds to an IR-1 element in the *BSEP* promoter (Ananthanarayanan *et al.*, 2001; Plass *et al.*, 2002; Schuetz *et al.*, 2001), and BSEP expression is reduced in mice lacking the FXR gene (Sinal *et al.*, 2000). Thus, during conditions of increased bile acid load in hepatocytes, bile acids enhance their own efflux into bile by activating FXR and consequently increasing BSEP expression. Another ABC transporter, the gene for which is transactivated by FXR, is MDR3 in humans (Huang *et al.*, 2003). MDR3 transports phospholipids across the canalicular membrane of hepatocytes. This function of MDR3 counteracts the toxicity of bile acids in the biliary tree by promoting the formation of mixed micelles that contain cholesterol, bile acids, and phospholipids.

In addition to transporter genes, FXR also positively regulates genes encoding phase II drug-metabolizing enzymes. The human *UGT2B4* gene, the product of which converts hydrophobic bile acids to more hydrophilic and less toxic glucuronide derivatives, is transactivated by FXR (Barbier *et al.*, 2003). *UGT2B4* is the only gene described to date that is positively regulated by FXR binding DNA as a monomer. Not only can FXR transactivate the *UGT2B4* gene without heterodimerizing with RXR, but activation of RXR by its ligands inhibits DNA binding of FXR to the *UGT2B4* promoter, both *in vitro* and *in vivo*. RXR-mediated inhibition of mono-meric DNA binding by FXR may involve a squelching mechanism: RXR ligands may promote formation of FXR-RXR dimers, thus reducing the pool of monomeric FXR capable of interacting with the *UGT2B4* promoter. Other FXR-regulated genes that encode bile acid-modifying enzymes include *Bacs* (bile acid-CoA synthetase) and *Bat* (bile acid-CoA: aminoacid *N*-acetyltransferase) in rats (Pircher *et al.*, 2003). Bacs and Bat enzymes mediate the conjugation of bile acids to taurine and glycine, and both *Bacs* and *Bat* promoters contain IR-1 elements, which bind FXR-RXR heterodimers. In human bile, concentrations of glycine- and taurine-conjugated bile acids greatly exceed those of the more toxic unconjugated forms. As BSEP only transports conjugated bile acids (Noe *et al.*, 2002), it seems logical to upregulate the conjugation machinery coordinately with BSEP.

Small Heterodimer Partner (SHP) as a Mediator of Negative Transcriptional Regulation by FXR

FXR can also negatively regulate transcription of specific target genes in a ligand-dependent manner. Although evidence shows that this may, in specific promoter contexts, be achieved through direct binding by FXR to negative bile acid response elements (Claudel *et al.*, 2002, 2003), more often FXR downregulates target genes via an indirect mechanism involving another nuclear receptor, SHP (Seol *et al.*, 1996). The *SHP* gene is broadly expressed, with the liver being a site of strong expression. The SHP protein is an atypical nuclear receptor, lacking the conserved DNA-binding domain. However, it does contain the di-merization domain and an apparent ligand-binding domain, although no SHP ligands regulating its activity have been identified. SHP can in-teract with, and negatively affect, the transcriptional activity of several other members of the nuclear receptor family, as well as with trans-cription factors belonging to other families (reviewed in Eloranta and Kullak-Ublick, 2005). The *SHP* promoter is transactivated by FXR in response to its ligands. Furthermore, the livers of FXR-null mice exhibit reduced SHP expression and a complete loss of SHP induction by cholic acid (Sinal *et al.*, 2000). At least in *in vitro* transfection assays, the transcriptional coactivator PGC-1 enhances FXR/RXR-dependent trans-cription of the *SHP* promoter (Kanaya *et al.*, 2004). Interestingly, exogenously expressed SHP can suppress the PGC-1-mediated coacti-vation of its own promoter, suggesting a negative feedback loop to adjust or attenuate bile acid-stimulated transcriptional responses executed by FXR-induced SHP.

FXR Negatively Regulates Bile Acid Uptake Systems

The Na$^+$-taurocholate-cotransporting polypeptide (NTCP) is chiefly responsible for conjugated bile acid uptake across the basolateral membrane of hepatocytes. In cholestatic rodent models, such as bile acid feeding and bile duct ligation (Fickert *et al.*, 2001; Gartung *et al.*, 1996; Wolters *et al.*, 2002), expression of the both Ntcp mRNA and protein are suppressed. Thus, in addition to enhancing bile acid efflux through upregulation of BSEP, bile acids suppress expression of the major bile acid uptake system when hepatocellular bile acid concentrations are already high. It has been proposed that SHP induction by bile acid-induced FXR is responsible for decreased expression of Ntcp in rats through the interference of SHP with the retinoic acid receptor (RAR)-RXR heterodimer that binds to its response element in the rat *Ntcp* promoter (Denson *et al.*, 2001). However, the downregulation of *Ntcp* in SHP-null mice in response to bile acid feeding is maintained (Wang *et al.*, 2002), implying that parallel or compensatory pathways exist. One suggested SHP-independent mechanism involves bile acid-induced activation of the c-Jun N-terminal kinase (JNK) signaling pathway (Gupta *et al.*, 2001), which leads to phosphorylation of RXR and consequently to a decrease in DNA binding of the RAR-RXR heterodimer (Li *et al.*, 2002). The sequence of the RAR-RXR response element of the rat *Ntcp* promoter is not conserved in the human *NTCP* gene, suggesting that the mechanism for transcriptional suppression by bile acids is not conserved between the two species. We have shown that bile acid treatment of cultured hepatoma cells, as well as overexpression of SHP, can strongly suppress glucocorticoid receptor-mediated activation of the human *NTCP* promoter (Eloranta *et al.*, 2005), implying that SHP negatively regulates the *NTCP/Ntcp* gene in different species, but via distinct mechanisms.

The apical sodium-dependent bile acid transporter (ASBT) belongs to the same family of transporter proteins as NTCP and is the major apical sodium-dependent bile salt uptake system in enterocytes. Supporting the crucial role of ASBT in ileal absorption of bile acids, mutations in the *ASBT* gene result in considerable bile acid malabsorption in humans (Oelkers *et al.*, 1997). *ASBT/Asbt* genes from different species exhibit differences in their response to bile acids (Arrese *et al.*, 1998; Chen *et al.*, 2003). The mouse, but not the rat, *Asbt* gene appears to be downregulated by bile acids in both *in vivo* and *in vitro* models. The human *ASBT* gene is also suppressed by treatment of cultured cells with bile acids (Neimark *et al.*, 2004). This has been proposed to be caused by interference of SHP with RAR-RXR-dependent transactivation of the human *ASBT* promoter. Similarly to the human *NTCP* gene, the *ASBT* gene is strongly activated by the glucocorticoid receptor in an agonist-dependent manner in humans (Jung *et al.*, 2004). Induction of the human *ASBT* gene by glucocorticoids can also be targeted negatively by exogenously expressed SHP in transient transfections (Eloranta *et al.*, 2005), further demonstrating coordinate regulation between the bile salt uptake systems in both human liver and intestine and emphasizing the divergent mechanisms of SHP-mediated negative regulation of these genes between species.

Further examples of transcriptional regulation via the FXR-SHP cascade are presented in the following sections.

Role of PXR and CAR in Drug Transport and Metabolism

Phase I cytochrome P450 (CYP) monooxygenase enzymes are the main catalysts of metabolic conversion of xenobiotics and foreign chemicals, as well as endogenous substrates such as steroid hormones, into less toxic derivatives (Nebert and Russell, 2002). CYP substrates include a wide range of environmental pollutants and carcinogens, as well as many prescription drugs and dietary xenobiotics. Substrates for the CYP enzymes often regulate the expression levels of the CYPs themselves, for example, dexamethasone is both a potent inducer of the *CYP3A4* gene and a substrate of the CYP3A4 enzyme. PXR is a crucial component of the xenobiotic response, which is aimed at protecting the body against harmful substances. Accordingly, PXR is highly expressed in the liver and moderately in the intestine, both of which are active sites for the detoxification of endo- and exogenous chemicals (Kliewer *et al.*, 1998). PXR binds DNA as a heterodimer with RXR and is flexible in its binding specificity. Known functional PXR response elements within natural target promoters include DR-3, DR-4, ER-6, and ER-8 configurations (Handschin and Meyer, 2003; Wang and LeCluyse, 2003). Furthermore, PXR has promiscuous and often low-affinity ligand specificity. PXR ligands include xenobiotics such as rifampicin and paclitaxel and natural and synthetic steroids such as pregnelenone, dexamethasone, and antiglucocorticoids. Accordingly, elucidation of the crystal structure of PXR has revealed considerable flexibility within the ligand-binding pocket (Watkins *et al.*, 2001).

PXR is a master regulator of the *CYP3A4* gene (Bertilsson *et al.*, 1998; Blumberg *et al.*, 1998; Kliewer *et al.*, 1998). Members of the CYP3A family of enzymes are pharmacologically important, as they metabolize a large proportion of all prescription drugs. Concurrent administration of drugs that serve both as agonistic ligands for PXR and as substrates for the drug-metabolizing enzyme CYP3A4 may cause adverse drug–drug interactions, in which metabolism of one drug is accelerated by the other.

Another PXR target gene commonly involved in untoward drug–drug interactions is *MDR1*. The *MDR1* gene encodes P-glycoprotein, an ATP-dependent efflux pump with a broad specificity and an important role in presystemic elimination of xenobiotics at the brush border membrane of enterocytes (Geick *et al.*, 2001). An illustrative example of drug–drug interactions is provided by the case of two heart transplant patients receiving cyclosporine as an immunosuppressant (Ruschitzka *et al.*, 2000). Following the patients' self-medication with the putative antidepressant Saint-John's-wort, plasma concentrations of cyclosporine were reduced, resulting in acute rejections. The molecular mechanism for the interaction between cyclosporine and Saint-John's-wort became apparent when the compound hyperforin present in Saint-John's-wort extract was found to be a potent PXR agonist. Via activation of PXR, hyperforin may induce the PXR target genes *MDR1* and *CYP3A4*, leading to decreased absorption and enhanced metabolism of cyclosporine, a known substrate for both P-glycoprotein and CYP3A4 (Moore *et al.*, 2000).

Another nuclear receptor expressed abundantly in the liver and the intestine, and activated by xenobiotics, is the constitutive androstane receptor (CAR). CAR also binds DNA as a heterodimer with RXR (Choi *et al.*, 1997) and is best known for its role as an inducer of the phenobarbital-responsive *CYP2B* gene family (Honkakoski *et al.*, 1998a,b). CAR-RXR and PXR-RXR heterodimers can bind to an overlapping range of hexameric repeat DNA configurations within target promoters (Handschin and Meyer, 2003; Wang and LeCluyse, 2003). It is thus not surprising that there is extensive cross talk between PXR and CAR: PXR can regulate *CYP2B* genes through recognition of the phenobarbital response element. Reciprocally, CAR can bind to and activate the PXR response element within the *CYP3A4* promoter (Xie *et al.*, 2000). This functional symmetry between the two metabolic sensors may provide a two-layered defense mechanism against the toxic compounds that the CYP enzymes metabolize.

In addition to *CYP* genes, PXR and CAR regulate genes encoding phase II drug-metabolizing enzymes, such as the UDP-glucuronosyl-transferase 1A (*UGT1A*) gene (Xie *et al.*, 2003). The regulatory region of the *UGT1A* locus is highly complex and contains multiple promoters governing expression of an overlapping, yet distinct, range of mRNAs (Gong *et al.*, 2001). These mRNA species encode multiple UGT1A enzyme isoforms with glucuronidation specificities for different substrate spectra. PXR and CAR both appear capable of inducing the expression of specific UGT1A isoforms, including those involved in bilirubin and estrogen metabolism, via direct binding to a DR-3 element.

Intertwined Transcriptional Regulation of Bile Acid and Drug
  Metabolism by FXR, PXR, and CAR

FXR is not the only nuclear receptor that can bind bile acids as ligands.
Certain bile acids, such as the toxic secondary bile acid LCA, can also serve
as agonistic ligands for PXR (Staudinger *et al.*, 2001; Xie *et al.*, 2001).
Activation of PXR by LCA suggests that xenobiotics and drugs are handled similarly to toxic bile acids (Handschin *et al.*, 2002). Thus, xenobiotics
elicit transcriptional responses in hepatocytes similar to those invoked by
elevated levels of endogenous bile acids. PXR activation can protect mouse
livers against injury caused by LCA (Staudinger *et al.*, 2001). Thus, under
cholestatic conditions, when bile acids accumulate in the liver, causing cell
damage, PXR provides additional mechanisms necessary to adjust bile acid
metabolism. This notion is supported by the phenotype of mice lacking
both FXR and PXR, as these mice exhibit more severe disturbances in bile
acid homeostasis than mice lacking only one of the two nuclear receptors
(Guo *et al.*, 2003). As discussed earlier, PXR is a master regulator of the
CYP3A4 gene, which, in addition to detoxifying xenobiotics, can also
metabolize bile acids to less toxic derivatives. Thus, bile acids, being
both activators of the *CYP3A4* gene and substrates of the CYP3A4 enzyme, can, by activating PXR, initiate a feedforward mechanism in order to
minimize hepatocellular damage caused by bile acids. It has been reported
that FXR activated by the primary bile acid CDCA or the synthetic FXR
agonist GW4064 (Willson *et al.*, 2001) can also induce the *CYP3A4* promoter in transient transfection assays via direct DNA binding (Gnerre
et al., 2004).

Regulation of the gene encoding MRP2, the multidrug resistance-
associated protein-2, provides another example of the complex interactions
between metabolic nuclear receptors. MRP2 is a member of the ABC
family of transporters and mediates efflux of several conjugated anions,
including glucuronidated and sulfated bile acids, across the canalicular
membrane of hepatocytes (Konig *et al.*, 1999). Mutations in the *MRP2*
gene are linked to the Dubin–Johnson syndrome, characterized by impaired transfer of conjugated bilirubin into bile. Human and rodent
MRP2/Mrp2 promoters are activated by ligands specific for FXR, PXR,
or CAR (Kast *et al.*, 2002). Within the MRP2 promoter, ligand-induced
stimulation has been mapped to an ER-8 DNA element, capable of binding
any of the three nuclear receptors as RXR heterodimers.

Another gene regulated by FXR, PXR, and CAR encodes dehydroepiandrosterone sulfotransferase (SULT2A1/Sult2A1), an enzyme that catalyzes phase II sulfoconjugation of a range of compounds in the liver and in
the intestine, including bile acids and drugs (Strott, 2002). Sulfation of

lithocholic acid and taurolithocholic acid renders them more soluble and thus more amenable for rapid excretion. Accordingly, the Sult2a1 enzyme provides protection against LCA-induced toxicity in the mouse liver (Kitada *et al.*, 2003). PXR, CAR, and FXR can all bind to, and transactivate through, the same IR-0 DNA response element within the rodent *Sult2a1* promoter (Assem *et al.*, 2004; Saini *et al.*, 2004; Song *et al.*, 2001; Sonoda *et al.*, 2002).

The relative contributions and the degree of redundancy of the three nuclear receptors, PXR, CAR, and FXR, in the regulation of *CYP3A4*, *MRP2/Mrp2*, and *SULT2A1/Sult2a1* genes remain to be clarified in further studies. It seems conceivable that the involvement of multiple bile acid- and drug-responsive transcription factors provides a fail-safe mechanism to upregulate crucial drug transporters and detoxifiers, when appropriate.

The transcriptional repressor SHP has been shown to interact with both PXR and CAR and to consequently inhibit their ability to transactivate their respective target promoters in transient transfection assays (Bae *et al.*, 2004; Ourlin *et al.*, 2003). Given the inducibility of SHP expression by bile acid-activated FXR, this shows another mechanism by which PXR and CAR may respond to changes in intracellular bile acid concentrations, without having to bind bile acids as ligands. Interestingly, in addition to its activity being regulated by SHP, PXR also appears to be a regulator of SHP expression: in human hepatoma cells the *SHP* gene is a direct target for transactivation not only by FXR but also by ligand-activated PXR (Frank *et al.*, 2005), adding further complexity to the close integration of bile acid and drug metabolism.

Role of HNF-4α in the Regulation of Bile Acid and Drug Metabolism

HNF-4α is a liver-enriched nuclear receptor with a crucial role in hepatocyte differentiation and maintenance of hepatic gene expression. It binds as a homodimer to its preferred DNA response elements of the DR-1 or DR-2 configuration, as shown for a range of natural target promoters and by polymerase chain reaction-based binding site selection (Fraser *et al.*, 1998). Elucidation of the crystal structure of the HNF-4α ligand-binding domain has suggested that it may be constitutively bound by endogenous fatty acids (Dhe-Paganon *et al.*, 2002; Wisely *et al.*, 2002) and that its activity may not thus be modulated readily by other exogenous or endogenous ligands. HNF-4α target genes encode proteins involved in a range of physiological processes, notably cholesterol and glucose metabolism (Cereghini, 1996; Hayhurst *et al.*, 2001). Emphasizing its important role in the control of glucose homeostasis, mutations in HNF-4α have been linked to the disease "maturity onset of diabetes of the young, type 1" (MODY1), characterized by impaired insulin secretion (Gupta and Kaestner, 2004).

HNF-4α has emerged as one of the key regulators of hepatic transport and metabolism of drugs and bile acids. We have identified the gene encoding the human drug transporter OAT2 (organic anion transporter 2, *SLC22A7*) as a target for transactivation by HNF-4α via its direct binding to a DR-1 element within the promoter region (Popowski *et al.*, 2005). Depletion of endogenous HNF-4α protein in cultured hepatoma cells using a pool of specific siRNAs completely abolishes OAT2 mRNA expression, indicating that *OAT2* promoter activity is critically dependent on HNF-4α. Treatment of human hepatoma cells with the bile acid CDCA suppresses the expression of the endogenous OAT2 mRNA through two potential mechanisms targeting HNF-4α. First, by activating FXR, bile acids induce expression of SHP, which in turn interferes with the activity of the HNF-4α homodimer bound to the *OAT2* promoter. Second, the expression of both the HNF-4α mRNA and the protein are down-regulated by bile acids through a mechanism not yet elucidated, thus decreasing the amount of HNF-4α available to bind to and transactivate the *OAT2* promoter. OAT2 is expressed at the basolateral membrane of hepatocytes and mediates the uptake of numerous drugs, such as salicylates, cephalosporins, and prostaglandins (Sekine *et al.*, 2000). The bile acid-mediated downregulation of OAT2 expression may lead to a decreased uptake of these and other OAT2 substrates into hepatocytes in cholestatic conditions.

An important group of genes transactivated by HNF-4α and downregulated by bile acids are those encoding the CYP enzymes CYP7A1 (Crestani *et al.*, 1998; De Fabiani *et al.*, 2001, 2003), CYP8B1 (Yang *et al.*, 2002; Zhang and Chiang, 2001), and CYP27A1 (Chen and Chiang, 2003) involved in the synthesis of bile acids from cholesterol. CYP7A1, a liver-specific microsomal enzyme, catalyzes the first rate-limiting step in the neutral pathway leading to the synthesis of the primary bile acids cholic acid and chenodeoxycholic acid. CYP8B1, in turn, controls the ratio of cholic to chenodeoxycholic acid, thus largely determining the overall hydrophobicity of the bile acid pool. CYP27A1 is a mitochondrial enzyme, which catalyzes the first step in the acidic pathway of bile acid biosynthesis. Transcriptional suppression of these enzymes in response to elevated levels of bile acids presents a negative feedback mechanism by which bile acids inhibit their own undesired *de novo* production. The promoter regions of these three *CYP* genes contain a negative bile acid response element (BARE) capable of binding the transcriptional transactivator HNF-4α. In the context of *CYP7A1* and *CYP8B1* genes, another nuclear receptor, liver receptor homologue-1 can also bind to the negative BARE as a monomer and may serve as an alternative or parallel target for bile acid-mediated suppression (del Castillo-Olivares and Gil, 2000, 2001; Goodwin *et al.*, 2000; Lu *et al.*, 2000).

Similarly to the *OAT2* gene, bile acids may negatively target the HNF-4α-mediated transactivation of *CYP7A1*, *CYP8B1*, and *CYP27A1* genes via the FXR-SHP cascade or through suppression of HNF-4α expression. While mice lacking SHP expression exhibit impaired feedback regulation of bile acid synthesis, it is not abolished completely, implying that additional SHP-independent repression pathways exist (Kerr *et al.*, 2002; Wang *et al.*, 2002). One such SHP-independent mechanism by which FXR may downregulate the human *CYP7A1* gene is through direct activation of the promoter of the *FGF-19* (fibroblast growth factor-19) gene (Holt *et al.*, 2003). FGF-19 is a growth factor that elicits a signaling cascade upon binding to the FGFR4 receptor tyrosine kinase located on the surface of responsive cells. Activated FGFR4 stimulates the intracellular JNK pathway, eventually leading to suppression of the *CYP7A1* promoter. Furthermore, bile acids may interfere negatively with HNF-4α-mediated activation of the *CYP7A1* promoter independently not only of SHP, but apparently also of FXR. In support of this notion, bile acids can block the recruitment of transcriptional coactivators to HNF-4α bound to the negative BARE of the *CYP7A1* promoter or disrupt the association between coactivators and HNF-4α, thereby abolishing transactivation by HNF-4α (De Fabiani *et al.*, 2003). The molecular mechanisms by which bile acids may interfere with the transcriptional activator–coactivator complexes assembled on the *CYP7A1* are unknown, but may involve bile acid-induced posttranslational modifications of the participating proteins.

Bile acid activation of PXR similarly leads to repression of the *CYP7A1* gene (Staudinger *et al.*, 2001). Thus, two distinct nuclear receptors, FXR and PXR, can both downregulate the bile acid biosynthetic gene *CYP7A1* in response to increased levels of intracellular bile acids. Similarly to FXR, suppression of *CYP7A1* by PXR may involve induction of SHP (Frank *et al.*, 2005) and its subsequent interference with the activity of HNF-4α and/or LRH-1 on the *CYP7A1* promoter. Alternative SHP-independent models have also been suggested, where ligand-bound PXR would interfere negatively with the interaction between HNF-4α and its coactivator PGC-1, thus leading to decreased transcription of the *CYP7A1* gene (Bhalla *et al.*, 2004; Li and Chiang, 2004). Contrary to mutual antagonism between PXR and HNF-4α in the context of the *CYP7A1* promoter, HNF-4α is capable of augmenting PXR-mediated activation of the *CYP3A4* promoter through direct binding to the promoter (Tirona *et al.*, 2003).

Finally, in addition to being subject to negative regulation by bile acids and the bile acid receptor FXR, HNF-4α has been identified as a transactivator of the *FXR* gene itself (Zhang *et al.*, 2004). Thus HNF-4α may be a central target for negative feedback regulation and attenuation of FXR-dependent transcriptional events.

Summary

The transcriptional regulation of genes involved in bile acid metabolism and drug detoxification is highly complex and contains numerous feedback and feedforward loops. The nuclear receptors chiefly responsible for these transcriptional events interact functionally with each other to adjust the expression levels of genes encoding transporter proteins and metabolic enzymes that mediate appropriate responses to endo- and xenobiotic challenges in the cellular environment. The overlapping range of ligands and target genes shared by the metabolic nuclear receptors serves as a redundant safety mechanism to induce protective responses against xenobiotic and cholestatic attacks so that when one pathway is compromised, a salvage pathway may take over.

Acknowledgment

This work was supported by Grant PPOOB-108511/1 from the Swiss National Science Foundation.

References

Ananthanarayanan, M., Balasubramanian, N., Makishima, M., Mangelsdorf, D. J., and Suchy, F. J. (2001). Human bile salt export pump promoter is transactivated by the farnesoid X receptor/bile acid receptor. *J. Biol. Chem.* **276**, 28857–28865.

Arrese, M., Trauner, M., Sacchiero, R. J., Crossman, M. W., and Shneider, B. L. (1998). Neither intestinal sequestration of bile acids nor common bile duct ligation modulate the expression and function of the rat ileal bile acid transporter. *Hepatology* **28**, 1081–1087.

Assem, M., Schuetz, E. G., Leggas, M., Sun, D., Yasuda, K., Reid, G., Zelcer, N., Adachi, M., Strom, S., Evans, R. M., Moore, D. D., Borst, P., and Schuetz, J. D. (2004). Interactions between hepatic Mrp4 and Sult2a as revealed by the constitutive androstane receptor and Mrp4 knockout mice. *J. Biol. Chem.* **279**, 22250–22257.

Bae, Y., Kemper, J. K., and Kemper, B. (2004). Repression of CAR-mediated transactivation of CYP2B genes by the orphan nuclear receptor, short heterodimer partner (SHP). *DNA Cell Biol.* **23**, 81–91.

Barbier, O., Torra, I. P., Sirvent, A., Claudel, T., Blanquart, C., Duran-Sandoval, D., Kuipers, F., Kosykh, V., Fruchart, J. C., and Staels, B. (2003). FXR induces the UGT2B4 enzyme in hepatocytes: A potential mechanism of negative feedback control of FXR activity. *Gastroenterology* **124**, 1926–1940.

Bertilsson, G., Heidrich, J., Svensson, K., Asman, M., Jendeberg, L., Sydow-Backman, M., Ohlsson, R., Postlind, H., Blomquist, P., and Berkenstam, A. (1998). Identification of a human nuclear receptor defines a new signaling pathway for CYP3A induction. *Proc. Natl. Acad. Sci. USA* **95**, 12208–12213.

Bhalla, S., Ozalp, C., Fang, S., Xiang, L., and Kemper, J. K. (2004). Ligand-activated pregnane X receptor interferes with HNF-4 signaling by targeting a common coactivator PGC-1alpha: Functional implications in hepatic cholesterol and glucose metabolism. *J. Biol. Chem.* **279**, 45139–45147.

Blumberg, B., and Evans, R. M. (1998). Orphan nuclear receptors: New ligands and new possibilities. *Genes Dev.* **12**, 3149–3155.

Blumberg, B., Sabbagh, W., Jr., Juguilon, H., Bolado, J., Jr., van Meter, C. M., Ong, E. S., and Evans, R. M. (1998). SXR, a novel steroid and xenobiotic-sensing nuclear receptor. *Genes Dev.* **12**, 3195–3205.

Cereghini, S. (1996). Liver-enriched transcription factors and hepatocyte differentiation. *FASEB J.* **10**, 267–282.

Chen, F., Ananthanarayanan, M., Emre, S., Neimark, E., Bull, L. N., Knisely, A. S., Strautnieks, S. S., Thompson, R. J., Magid, M. S., Gordon, R., Balasubramanian, N., Suchy, F. J., and Shneider, B. L. (2004). Progressive familial intrahepatic cholestasis, type 1, is associated with decreased farnesoid X receptor activity. *Gastroenterology* **126**, 756–764.

Chen, F., Ma, L., Dawson, P. A., Sinal, C. J., Sehayek, E., Gonzalez, F. J., Breslow, J., Ananthanarayanan, M., and Shneider, B. L. (2003). Liver receptor homologue-1 mediates species- and cell line-specific bile acid-dependent negative feedback regulation of the apical sodium-dependent bile acid transporter. *J. Biol. Chem.* **278**, 19909–19916.

Chen, W., and Chiang, J. Y. (2003). Regulation of human sterol 27-hydroxylase gene (CYP27A1) by bile acids and hepatocyte nuclear factor 4alpha (HNF4alpha). *Gene* **313**, 71–82.

Choi, H. S., Chung, M., Tzameli, I., Simha, D., Lee, Y. K., Seol, W., and Moore, D. D. (1997). Differential transactivation by two isoforms of the orphan nuclear hormone receptor CAR. *J. Biol. Chem.* **272**, 23565–23571.

Claudel, T., Inoue, Y., Barbier, O., Duran-Sandoval, D., Kosykh, V., Fruchart, J., Fruchart, J. C., Gonzalez, F. J., and Staels, B. (2003). Farnesoid X receptor agonists suppress hepatic apolipoprotein CIII expression. *Gastroenterology* **125**, 544–555.

Claudel, T., Sturm, E., Duez, H., Torra, I. P., Sirvent, A., Kosykh, V., Fruchart, J. C., Dallongeville, J., Hum, D. W., Kuipers, F., and Staels, B. (2002). Bile acid-activated nuclear receptor FXR suppresses apolipoprotein A-I transcription via a negative FXR response element. *J. Clin. Invest.* **109**, 961–971.

Collingwood, T. N., Urnov, F. D., and Wolffe, A. P. (1999). Nuclear receptors: Coactivators, corepressors and chromatin remodeling in the control of transcription. *J. Mol. Endocrinol.* **23**, 255–275.

Crestani, M., Sadeghpour, A., Stroup, D., Galli, G., and Chiang, J. Y. (1998). Transcriptional activation of the cholesterol 7alpha-hydroxylase gene (CYP7A) by nuclear hormone receptors. *J. Lipid Res.* **39**, 2192–2200.

De Fabiani, E., Mitro, N., Anzulovich, A. C., Pinelli, A., Galli, G., and Crestani, M. (2001). The negative effects of bile acids and tumor necrosis factor-alpha on the transcription of cholesterol 7alpha-hydroxylase gene (CYP7A1) converge to hepatic nuclear factor-4: A novel mechanism of feedback regulation of bile acid synthesis mediated by nuclear receptors. *J. Biol. Chem.* **276**, 30708–30716.

De Fabiani, E., Mitro, N., Gilardi, F., Caruso, D., Galli, G., and Crestani, M. (2003). Coordinated control of cholesterol catabolism to bile acids and of gluconeogenesis via a novel mechanism of transcription regulation linked to the fasted-to-fed cycle. *J. Biol. Chem.* **278**, 39124–39132.

del Castillo-Olivares, A., and Gil, G. (2000). Alpha 1-fetoprotein transcription factor is required for the expression of sterol 12alpha-hydroxylase, the specific enzyme for cholic acid synthesis: Potential role in the bile acid-mediated regulation of gene transcription. *J. Biol. Chem.* **275**, 17793–17799.

del Castillo-Olivares, A., and Gil, G. (2001). Suppression of sterol 12alpha-hydroxylase transcription by the short heterodimer partner: Insights into the repression mechanism. *Nucleic Acids Res.* **29**, 4035–4042.

Denson, L. A., Sturm, E., Echevarria, W., Zimmerman, T. L., Makishima, M., Mangelsdorf, D. J., and Karpen, S. J. (2001). The orphan nuclear receptor, shp, mediates bile acid-induced inhibition of the rat bile acid transporter, ntcp. *Gastroenterology* **121**, 140–147.

Dhe-Paganon, S., Duda, K., Iwamoto, M., Chi, Y. I., and Shoelson, S. E. (2002). Crystal structure of the HNF4 alpha ligand binding domain in complex with endogenous fatty acid ligand. *J. Biol. Chem.* **277**, 37973–37976.

Eloranta, J. J., Jung, D., and Kullak-Ublick, G. A. (2005). The human Na+ taurocholate cotransporting polypeptide gene (NTCP, SLC10A1) is activated by the glucocorticoid receptor and its coactivator PGC-1 and suppressed by the small heterodimer partner SHP. *Hepatology* **40**, 519a.

Eloranta, J. J., and Kullak-Ublick, G. A. (2005). Coordinate transcriptional regulation of bile acid homeostasis and drug metabolism. *Arch. Biochem. Biophys.* **433**, 397–412.

Fickert, P., Zollner, G., Fuchsbichler, A., Stumptner, C., Pojer, C., Zenz, R., Lammert, F., Stieger, B., Meier, P. J., Zatloukal, K., Denk, H., and Trauner, M. (2001). Effects of ursodeoxycholic and cholic acid feeding on hepatocellular transporter expression in mouse liver. *Gastroenterology* **121**, 170–183.

Forman, B. M., Goode, E., Chen, J., Oro, A. E., Bradley, D. J., Perlmann, T., Noonan, D. J., Burka, L. T., McMorris, T., Lamph, W. W., Evans, R. M., and Weinberger, C. (1995). Identification of a nuclear receptor that is activated by farnesol metabolites. *Cell* **81**, 687–693.

Frank, C., Makkonen, H., Dunlop, T. W., Vaisanen, M. M. S., and Carlberg, C. (2005). Identification of pregnane X receptor binding sites in the regulatory regions of genes involved in bile acid homeostasis. *J. Mol. Biol.* **346**, 505–519.

Fraser, J. D., Martinez, V., Straney, R., and Briggs, M. R. (1998). DNA binding and transcription activation specificity of hepatocyte nuclear factor 4. *Nucleic Acids Res.* **26**, 2702–2707.

Gartung, C., Ananthanarayanan, M., Rahman, M. A., Schuele, S., Nundy, S., Soroka, C. J., Stolz, A., Suchy, F. J., and Boyer, J. L. (1996). Down-regulation of expression and function of the rat liver Na+/bile acid cotransporter in extrahepatic cholestasis. *Gastroenterology* **110**, 199–209.

Geick, A., Eichelbaum, M., and Burk, O. (2001). Nuclear receptor response elements mediate induction of intestinal MDR1 by rifampin. *J. Biol. Chem.* **276**, 14581–14587.

Gnerre, C., Blattler, S., Kaufmann, M. R., Looser, R., and Meyer, U. A. (2004). Regulation of CYP3A4 by the bile acid receptor FXR: Evidence for functional binding sites in the CYP3A4 gene. *Pharmacogenetics* **14**, 635–645.

Gong, Q. H., Cho, J. W., Huang, T., Potter, C., Gholami, N., Basu, N. K., Kubota, S., Carvalho, S., Pennington, M. W., Owens, I. S., and Popescu, N. C. (2001). Thirteen UDP glucuronosyltransferase genes are encoded at the human UGT1 gene complex locus. *Pharmacogenetics* **11**, 357–368.

Goodwin, B., Jones, S. A., Price, R. R., Watson, M. A., McKee, D. D., Moore, L. B., Galardi, C., Wilson, J. G., Lewis, M. C., Roth, M. E., Maloney, P. R., Willson, T. M., and Kliewer, S. A. (2000). A regulatory cascade of the nuclear receptors FXR, SHP-1, and LRH-1 represses bile acid biosynthesis. *Mol. Cell* **6**, 517–526.

Guo, G. L., Lambert, G., Negishi, M., Ward, J. M., Brewer, H. B., Jr., Kliewer, S. A., Gonzalez, F. J., and Sinal, C. J. (2003). Complementary roles of farnesoid X receptor, pregnane X receptor, and constitutive androstane receptor in protection against bile acid toxicity. *J. Biol. Chem.* **278**, 45062–45071.

Gupta, R. K., and Kaestner, K. H. (2004). HNF-4alpha: From MODY to late-onset type 2 diabetes. *Trends Mol. Med.* **10**, 521–524.

Gupta, S., Stravitz, R. T., Dent, P., and Hylemon, P. B. (2001). Down-regulation of cholesterol 7alpha-hydroxylase (CYP7A1) gene expression by bile acids in primary rat hepatocytes is mediated by the c-Jun N-terminal kinase pathway. *J. Biol. Chem.* **276**, 15816–15822.

Handschin, C., and Meyer, U. A. (2003). Induction of drug metabolism: The role of nuclear receptors. *Pharmacol. Rev.* **55**, 649–673.

Handschin, C., Podvinec, M., Amherd, R., Looser, R., Ourlin, J. C., and Meyer, U. A. (2002). Cholesterol and bile acids regulate xenosensor signaling in drug-mediated induction of cytochromes P450. *J. Biol. Chem.* **277**, 29561–29567.

Hayhurst, G. P., Lee, Y. H., Lambert, G., Ward, J. M., and Gonzalez, F. J. (2001). Hepatocyte nuclear factor 4alpha (nuclear receptor 2A1) is essential for maintenance of hepatic gene expression and lipid homeostasis. *Mol. Cell. Biol.* **21**, 1393–1403.

Holt, J. A., Luo, G., Billin, A. N., Bisi, J., McNeill, Y. Y., Kozarsky, K. F., Donahee, M., Wang da, Y., Mansfield, T. A., Kliewer, S. A., Goodwin, B., and Jones, S. A. (2003). Definition of a novel growth factor-dependent signal cascade for the suppression of bile acid biosynthesis. *Genes Dev.* **17**, 1581–1591.

Honkakoski, P., Moore, R., Washburn, K. A., and Negishi, M. (1998a). Activation by diverse xenochemicals of the 51-base pair phenobarbital-responsive enhancer module in the CYP2B10 gene. *Mol. Pharmacol.* **53**, 597–601.

Honkakoski, P., Zelko, I., Sueyoshi, T., and Negishi, M. (1998b). The nuclear orphan receptor CAR-retinoid X receptor heterodimer activates the phenobarbital-responsive enhancer module of the CYP2B gene. *Mol. Cell. Biol.* **18**, 5652–5658.

Huang, L., Zhao, A., Lew, J. L., Zhang, T., Hrywna, Y., Thompson, J. R., de Pedro, N., Royo, I., Blevins, R. A., Pelaez, F., Wright, S. D., and Cui, J. (2003). Farnesoid X receptor activates transcription of the phospholipid pump MDR3. *J. Biol. Chem.* **278**, 51085–51090.

Jung, D., Fantin, A. C., Scheurer, U., Fried, M., and Kullak-Ublick, G. A. (2004). Human ileal bile acid transporter gene ASBT (SLC10A2) is transactivated by the glucocorticoid receptor. *Gut* **53**, 78–84.

Kanaya, E., Shiraki, T., and Jingami, H. (2004). The nuclear bile acid receptor FXR is activated by PGC-1alpha in a ligand-dependent manner. *Biochem. J.* **382**, 913–921.

Kast, H. R., Goodwin, B., Tarr, P. T., Jones, S. A., Anisfeld, A. M., Stoltz, C. M., Tontonoz, P., Kliewer, S., Willson, T. M., and Edwards, P. A. (2002). Regulation of multidrug resistance-associated protein 2 (ABCC2) by the nuclear receptors pregnane X receptor, farnesoid X-activated receptor, and constitutive androstane receptor. *J. Biol. Chem.* **277**, 2908–2915.

Kerr, T. A., Saeki, S., Schneider, M., Schaefer, K., Berdy, S., Redder, T., Shan, B., Russell, D. W., and Schwarz, M. (2002). Loss of nuclear receptor SHP impairs but does not eliminate negative feedback regulation of bile acid synthesis. *Dev. Cell.* **2**, 713–720.

Khorasanizadeh, S., and Rastinejad, F. (2001). Nuclear-receptor interactions on DNA-response elements. *Trends Biochem. Sci.* **26**, 384–390.

Kitada, H., Miyata, M., Nakamura, T., Tozawa, A., Honma, W., Shimada, M., Nagata, K., Sinal, C. J., Guo, G. L., Gonzalez, F. J., and Yamazoe, Y. (2003). Protective role of hydroxysteroid sulfotransferase in lithocholic acid-induced liver toxicity. *J. Biol. Chem.* **278**, 17838–17844.

Kliewer, S. A., Moore, J. T., Wade, L., Staudinger, J. L., Watson, M. A., Jones, S. A., McKee, D. D., Oliver, B. B., Willson, T. M., Zetterstrom, R. H., Perlmann, T., and Lehmann, J. M. (1998). An orphan nuclear receptor activated by pregnanes defines a novel steroid signaling pathway. *Cell* **92**, 73–82.

Konig, J., Nies, A. T., Cui, Y., Leier, I., and Keppler, D. (1999). Conjugate export pumps of the multidrug resistance protein (MRP) family: Localization, substrate specificity, and MRP2-mediated drug resistance. *Biochim. Biophys. Acta* **1461**, 377–394.

Kullak-Ublick, G. A., Stieger, B., and Meier, P. J. (2004). Enterohepatic bile salt transporters in normal physiology and liver disease. *Gastroenterology* **126**, 322–342.

Kumar, R., and Thompson, E. B. (1999). The structure of the nuclear hormone receptors. *Steroids* **64**, 310–319.

Laffitte, B. A., Kast, H. R., Nguyen, C. M., Zavacki, A. M., Moore, D. D., and Edwards, P. A. (2000). Identification of the DNA binding specificity and potential target genes for the farnesoid X-activated receptor. *J. Biol. Chem.* **275**, 10638–10647.

Lew, J. L., Zhao, A., Yu, J., Huang, L., De Pedro, N., Pelaez, F., Wright, S. D., and Cui, J. (2004). The farnesoid X receptor controls gene expression in a ligand- and promoter-selective fashion. *J. Biol. Chem.* **279**, 8856–8861.

Li, D., Zimmerman, T. L., Thevananther, S., Lee, H. Y., Kurie, J. M., and Karpen, S. J. (2002). Interleukin-1 beta-mediated suppression of RXR:RAR transactivation of the Ntcp promoter is JNK-dependent. *J. Biol. Chem.* **277**, 31416–31422.

Li, T., and Chiang, J. Y. (2004). Mechanism of rifampicin and pregnane X receptor inhibition of human cholesterol 7 alpha-hydroxylase gene transcription. *Am. J. Physiol. Gastrointest. Liver Physiol.* **288**, G74–G84.

Lu, T. T., Makishima, M., Repa, J. J., Schoonjans, K., Kerr, T. A., Auwerx, J., and Mangelsdorf, D. J. (2000). Molecular basis for feedback regulation of bile acid synthesis by nuclear receptors. *Mol. Cell* **6**, 507–515.

Makishima, M., Okamoto, A. Y., Repa, J. J., Tu, H., Learned, R. M., Luk, A., Hull, M. V., Lustig, K. D., Mangelsdorf, D. J., and Shan, B. (1999). Identification of a nuclear receptor for bile acids. *Science* **284**, 1362–1365.

Meier, P. J., and Stieger, B. (2002). Bile salt transporters. *Annu. Rev. Physiol.* **64**, 635–661.

Moore, L. B., Goodwin, B., Jones, S. A., Wisely, G. B., Serabjit-Singh, C. J., Willson, T. M., Collins, J. L., and Kliewer, S. A. (2000). St. John's wort induces hepatic drug metabolism through activation of the pregnane X receptor. *Proc. Natl. Acad. Sci. USA* **97**, 7500–7502.

Nebert, D. W., and Russell, D. W. (2002). Clinical importance of the cytochromes P450. *Lancet* **360**, 1155–1162.

Neimark, E., Chen, F., Li, X., and Shneider, B. L. (2004). Bile acid-induced negative feedback regulation of the human ileal bile acid transporter. *Hepatology* **40**, 149–156.

Noe, J., Stieger, B., and Meier, P. J. (2002). Functional expression of the canalicular bile salt export pump of human liver. *Gastroenterology* **123**, 1659–1666.

Oelkers, P., Kirby, L. C., Heubi, J. E., and Dawson, P. A. (1997). Primary bile acid malabsorption caused by mutations in the ileal sodium-dependent bile acid transporter gene (SLC10A2). *J. Clin. Invest.* **99**, 1880–1887.

Ourlin, J. C., Lasserre, F., Pineau, T., Fabre, J. M., Sa-Cunha, A., Maurel, P., Vilarem, M. J., and Pascussi, J. M. (2003). The small heterodimer partner interacts with the pregnane X receptor and represses its transcriptional activity. *Mol. Endocrinol.* **17**, 1693–1703.

Parks, D. J., Blanchard, S. G., Bledsoe, R. K., Chandra, G., Consler, T. G., Kliewer, S. A., Stimmel, J. B., Willson, T. M., Zavacki, A. M., Moore, D. D., and Lehmann, J. M. (1999). Bile acids: Natural ligands for an orphan nuclear receptor. *Science* **284**, 1365–1368.

Pircher, P. C., Kitto, J. L., Petrowski, M. L., Tangirala, R. K., Bischoff, E. D., Schulman, I. G., and Westin, S. K. (2003). Farnesoid X receptor regulates bile acid-amino acid conjugation. *J. Biol. Chem.* **278**, 27703–27711.

Plass, J. R., Mol, O., Heegsma, J., Geuken, M., Faber, K. N., Jansen, P. L., and Muller, M. (2002). Farnesoid X receptor and bile salts are involved in transcriptional regulation of the gene encoding the human bile salt export pump. *Hepatology* **35**, 589–596.

Popowski, K., Eloranta, J. J., Saborowski, M., Fried, M., Meier, P. J., and Kullak-Ublick, G. A. (2005). The human organic anion transporter 2 gene is transactivated by hepatocyte nuclear factor-4α and suppressed by bile acids. *Mol. Pharmacol.* **67,** 1629–1638.

Ruschitzka, F., Meier, P. J., Turina, M., Luscher, T. F., and Noll, G. (2000). Acute heart transplant rejection due to Saint John's wort. *Lancet* **355,** 548–549.

Saini, S. P., Sonoda, J., Xu, L., Toma, D., Uppal, H., Mu, Y., Ren, S., Moore, D. D., Evans, R. M., and Xie, W. (2004). A novel constitutive androstane receptor-mediated and CYP3A-independent pathway of bile acid detoxification. *Mol. Pharmacol.* **65,** 292–300.

Schuetz, E. G., Strom, S., Yasuda, K., Lecureur, V., Assem, M., Brimer, C., Lamba, J., Kim, R. B., Ramachandran, V., Komoroski, B. J., Venkataramanan, R., Cai, H., Sinal, C. J., Gonzalez, F. J., and Schuetz, J. D. (2001). Disrupted bile acid homeostasis reveals an unexpected interaction among nuclear hormone receptors, transporters, and cytochrome P450. *J. Biol. Chem.* **276,** 39411–39418.

Sekine, T., Cha, S. H., and Endou, H. (2000). The multispecific organic anion transporter (OAT) family. *Pflug. Arch.* **440,** 337–350.

Seol, W., Choi, H. S., and Moore, D. D. (1996). An orphan nuclear hormone receptor that lacks a DNA binding domain and heterodimerizes with other receptors. *Science* **272,** 1336–1339.

Sinal, C. J., Tohkin, M., Miyata, M., Ward, J. M., Lambert, G., and Gonzalez, F. J. (2000). Targeted disruption of the nuclear receptor FXR/BAR impairs bile acid and lipid homeostasis. *Cell* **102,** 731–744.

Song, C. S., Echchgadda, I., Baek, B. S., Ahn, S. C., Oh, T., Roy, A. K., and Chatterjee, B. (2001). Dehydroepiandrosterone sulfotransferase gene induction by bile acid activated farnesoid X receptor. *J. Biol. Chem.* **276,** 42549–42556.

Sonoda, J., Xie, W., Rosenfeld, J. M., Barwick, J. L., Guzelian, P. S., and Evans, R. M. (2002). Regulation of a xenobiotic sulfonation cascade by nuclear pregnane X receptor (PXR). *Proc. Natl. Acad. Sci. USA* **99,** 13801–13806.

Stallcup, M. R. (2001). Role of protein methylation in chromatin remodeling and transcriptional regulation. *Oncogene* **20,** 3014–3020.

Staudinger, J. L., Goodwin, B., Jones, S. A., Hawkins-Brown, D., MacKenzie, K. I., LaTour, A., Liu, Y., Klaassen, C. D., Brown, K. K., Reinhard, J., Willson, T. M., Koller, B. H., and Kliewer, S. A. (2001). The nuclear receptor PXR is a lithocholic acid sensor that protects against liver toxicity. *Proc. Natl. Acad. Sci. USA* **98,** 3369–3374.

Sterner, D. E., and Berger, S. L. (2000). Acetylation of histones and transcription-related factors. *Microbiol. Mol. Biol. Rev.* **64,** 435–459.

Strott, C. A. (2002). Sulfonation and molecular action. *Endocr. Rev.* **23,** 703–732.

Tirona, R. G., Lee, W., Leake, B. F., Lan, L. B., Cline, C. B., Lamba, V., Parviz, F., Duncan, S. A., Inoue, Y., Gonzalez, F. J., Schuetz, E. G., and Kim, R. B. (2003). The orphan nuclear receptor HNF4alpha determines PXR- and CAR-mediated xenobiotic induction of CYP3A4. *Nat. Med.* **9,** 220–224.

Trauner, M., and Boyer, J. L. (2003). Bile salt transporters: Molecular characterization, function, and regulation. *Physiol. Rev.* **83,** 633–671.

Tzameli, I., Chua, S. S., Cheskis, B., and Moore, D. D. (2003). Complex effects of rexinoids on ligand dependent activation or inhibition of the xenobiotic receptor, CAR. *Nucl. Recept.* **1,** 2.

Wang, H., Chen, J., Hollister, K., Sowers, L. C., and Forman, B. M. (1999). Endogenous bile acids are ligands for the nuclear receptor FXR/BAR. *Mol. Cell* **3,** 543–553.

Wang, H., and LeCluyse, E. L. (2003). Role of orphan nuclear receptors in the regulation of drug-metabolising enzymes. *Clin. Pharmacokinet.* **42,** 1331–1357.

Wang, L., Lee, Y. K., Bundman, D., Han, Y., Thevananther, S., Kim, C. S., Chua, S. S., Wei, P., Heyman, R. A., Karin, M., and Moore, D. D. (2002). Redundant pathways for negative feedback regulation of bile acid production. *Dev. Cell* **2**, 721–731.

Watkins, R. E., Wisely, G. B., Moore, L. B., Collins, J. L., Lambert, M. H., Williams, S. P., Willson, T. M., Kliewer, S. A., and Redinbo, M. R. (2001). The human nuclear xenobiotic receptor PXR: Structural determinants of directed promiscuity. *Science* **292**, 2329–2333.

Willson, T. M., Jones, S. A., Moore, J. T., and Kliewer, S. A. (2001). Chemical genomics: Functional analysis of orphan nuclear receptors in the regulation of bile acid metabolism. *Med. Res. Rev.* **21**, 513–522.

Wisely, G. B., Miller, A. B., Davis, R. G., Thornquest, A. D., Jr., Johnson, R., Spitzer, T., Sefler, A., Shearer, B., Moore, J. T., Willson, T. M., and Williams, S. P. (2002). Hepatocyte nuclear factor 4 is a transcription factor that constitutively binds fatty acids. *Structure (Camb.)* **10**, 1225–1234.

Wolters, H., Elzinga, B. M., Baller, J. F., Boverhof, R., Schwarz, M., Stieger, B., Verkade, H. J., and Kuipers, F. (2002). Effects of bile salt flux variations on the expression of hepatic bile salt transporters in vivo in mice. *J. Hepatol.* **37**, 556–563.

Xie, W., Barwick, J. L., Simon, C. M., Pierce, A. M., Safe, S., Blumberg, B., Guzelian, P. S., and Evans, R. M. (2000). Reciprocal activation of xenobiotic response genes by nuclear receptors SXR/PXR and CAR. *Genes Dev.* **14**, 3014–3023.

Xie, W., Radominska-Pandya, A., Shi, Y., Simon, C. M., Nelson, M. C., Ong, E. S., Waxman, D. J., and Evans, R. M. (2001). An essential role for nuclear receptors SXR/PXR in detoxification of cholestatic bile acids. *Proc. Natl. Acad. Sci. USA* **98**, 3375–3380.

Xie, W., Yeuh, M. F., Radominska-Pandya, A., Saini, S. P., Negishi, Y., Bottroff, B. S., Cabrera, G. Y., Tukey, R. H., and Evans, R. M. (2003). Control of steroid, heme, and carcinogen metabolism by nuclear pregnane X receptor and constitutive androstane receptor. *Proc. Natl. Acad. Sci. USA* **100**, 4150–4155.

Xu, G., Pan, L. X., Li, H., Forman, B. M., Erickson, S. K., Shefer, S., Bollineni, J., Batta, A. K., Christie, J., Wang, T. H., Michel, J., Yang, S., Tsai, R., Lai, L., Shimada, K., Tint, G. S., and Salen, G. (2002). Regulation of the farnesoid X receptor (FXR) by bile acid flux in rabbits. *J. Biol. Chem.* **277**, 50491–50496.

Yang, Y., Zhang, M., Eggertsen, G., and Chiang, J. Y. (2002). On the mechanism of bile acid inhibition of rat sterol 12alpha-hydroxylase gene (CYP8B1) transcription: Roles of alpha-fetoprotein transcription factor and hepatocyte nuclear factor 4alpha. *Biochim. Biophys. Acta* **1583**, 63–73.

Yu, J., Lo, J. L., Huang, L., Zhao, A., Metzger, E., Adams, A., Meinke, P. T., Wright, S. D., and Cui, J. (2002). Lithocholic acid decreases expression of bile salt export pump through farnesoid X receptor antagonist activity. *J. Biol. Chem.* **277**, 31441–31447.

Zhang, M., and Chiang, J. Y. (2001). Transcriptional regulation of the human sterol 12alpha-hydroxylase gene (CYP8B1): Roles of hepatocyte nuclear factor 4alpha in mediating bile acid repression. *J. Biol. Chem.* **276**, 41690–41699.

Zhang, Y., Castellani, L. W., Sinal, C. J., Gonzalez, F. J., and Edwards, P. A. (2004). Peroxisome proliferator-activated receptor-gamma coactivator 1alpha (PGC-1alpha) regulates triglyceride metabolism by activation of the nuclear receptor FXR. *Genes Dev.* **18**, 157–169.

[29] Uptake and Efflux Transporters for Conjugates in Human Hepatocytes

By Dietrich Keppler

Abstract

Conjugates of endogenous substances and of xenobiotics, formed extrahepatically or inside hepatocytes, undergo vectorial transport into bile. Substances conjugated with glucuronate, sulfate, or glutathione are substrates for organic anion uptake transporters in the basolateral (sinusoidal) membrane as well as substrates for the unidirectional ATP-driven conjugate efflux pump in the apical (canalicular) membrane, termed multidrug resistance protein 2 (MRP2; systematic name ABCC2). Localization of the efflux pumps ABCC3 and ABCC4 to the basolateral membrane of human hepatocytes has provided insight into the molecular mechanisms of conjugate efflux from hepatocytes into blood, as exemplified by the efflux of bilirubin glucuronosides mediated by ABCC3. The cloning and stable expression of the complementary DNAs encoding the organic anion transporters in the basolateral membrane of human hepatocytes and of members of the ABCC subfamily of efflux pumps in the apical as well as in the basolateral membrane have improved our understanding of hepatobiliary elimination and of the substrate specificity with respect to anionic conjugates. The stable expression of human hepatocyte uptake and efflux transporters in polarized cell lines, as described in this chapter, provides valuable tools for the *in vitro* analysis of human hepatobiliary transport in general and specifically for uptake and efflux of the anionic conjugates formed in various phase 2 reactions.

Introduction

One of the most important functions of hepatocytes is the removal, and possibly detoxification, of endogenous and xenobiotic substances from the blood circulation and their excretion into bile. Anionic conjugates may be formed extrahepatically or inside hepatocytes. Two processes play a decisive role in the vectorial transport of anionic conjugates: The basolateral (sinusoidal) uptake by members of the organic anion transporter family, specifically by OATP1B1, OATP1B3, and OATP2B1, and the apical (canalicular) efflux into bile mediated by the ATP-dependent conjugate export pump ABCC2, also known as multidrug resistance

METHODS IN ENZYMOLOGY, VOL. 400
0076-6879/05 $35.00
DOI: 10.1016/S0076-6879(05)00029-7

protein 2. Vectorial transport of substances at a sufficient rate does not occur in the absence of the respective transport proteins in the basolateral and in the apical membrane of polarized cells. This is evident from *in vivo* studies in mutant rats (Huber *et al.*, 1987; Jansen *et al.*, 1985, 1987) lacking Abcc2 in the apical membrane (Büchler *et al.*, 1996; Paulusma *et al.*, 1996) and from kinetic analyses in polarized cells stably expressing an uptake transporter and an efflux pump (Cui *et al.*, 2001a; Sasaki *et al.*, 2002). Many substances taken up by hepatocytes only become substrates for efflux mediated by ABCC2 after their oxidation in phase 1 reactions and/or after conjugation with glucuronate, sulfate, or glutathione in phase 2 reactions (Fig. 1).

A considerable number of uptake transporters, in part with overlapping substrate specificities, have been localized to the basolateral membrane of human hepatocytes. Members of the OATP family expressed predominantly in human hepatocytes differ markedly in sequence and substrate specificity from hepatic rodent Oatps (Abe *et al.*, 1999; Hagenbuch and Meier, 2004; König *et al.*, 2000a). This chapter, therefore, focuses on human hepatocyte transporters. Members of the OATP family expressed preferentially in human hepatocytes comprise OATP1B1 (Abe *et al.*, 1999; Hsiang *et al.*, 1999; König *et al.*, 2000a), OATP1B3 (König *et al.*, 2000b),

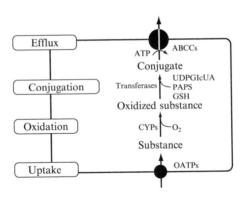

FIG. 1. Uptake, biotransformation, and ATP-dependent efflux of endogenous and xenobiotic substances by members of the ABCC subfamily. Many substances, after their transport into the cell, first undergo cytochrome P450-mediated oxidation, then conjugation with glucuronate, sulfate, or glutathione, and finally efflux mediated by ABCC2, ABCC3, or another member of the ABCC subfamily. Several conjugates may also be formed outside hepatocytes, as exemplified by part of the bilirubin glucuronosides, which are substrates for OATP1B1 and OATP1B3 (Cui *et al.*, 2001b; König *et al.*, 2000a), and are taken up into hepatocytes. They may then undergo, without further biotransformation, ATP-dependent efflux by ABCC2 in the apical membrane (Kamisako *et al.*, 1999) or by ABCC3 in the basolateral hepatocyte membrane (Lee *et al.*, 2004).

and OATP2B1 (Kullak-Ublick *et al.*, 2001; Tamai *et al.*, 2000). Additional uptake transporters in the hepatocyte basolateral membrane include the sodium-dependent bile salt transporter NTCP (Hagenbuch and Dawson, 2004), the organic cation transporter OCT1 (Koepsell *et al.*, 2003), and the organic anion transporter OAT2 (Sekine *et al.*, 2000; Sun *et al.*, 2001) (Fig. 2).

Anionic conjugates are generally not substrates for ATP-dependent efflux mediated by the long-known ABC transporter MDR1 P-glycoprotein

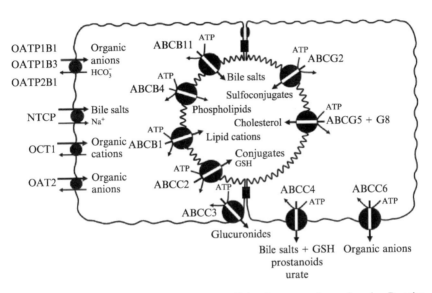

FIG. 2. Transporter proteins in human hepatocellular plasma membrane domains. Proteins mediating the uptake of organic anions and organic cations across the basolateral membrane are indicated on the left. OATP1B1 (previously termed OATP-C, OATP2, or LST-1), OATP1B3 (formerly termed OATP8), and OATP2B1 (formerly termed OATP-B) are thought to function as exchangers. NTCP, sodium-dependent taurocholate cotransporting polypeptide; OCT1, organic cation transporter 1; OAT2, organic anion transporter 2. Many ATP-dependent export pumps mediating the unidirectional transport of substances into bile have been localized to the apical (canalicular) membrane. ABCB11 designates the bile salt export pump, BSEP; ABCB4 is the phospholipid transporter also known as MDR3; ABCB1 is the MDR1 P-glycoprotein; ABCC2 is the apical multidrug resistance protein MRP2; the heterodimer ABCG5 + ABCG8 functions as a sterol transporter; and ABCG2, also termed breast cancer resistance protein, BCRP, or mitoxantrone resistance protein, MXR, transports sulfoconjugates (Suzuki *et al.*, 2003) in addition to porphyrins and a number of drugs and toxins (Jonker *et al.*, 2002). Basolateral ATP-binding cassette transporters include ABCC3, the basolateral multidrug resistance protein MRP3; ABCC4, the multidrug resistance protein MRP4; and ABCC6, the multidrug resistance protein MRP6. Only some of the substrates transported by these proteins are indicated.

(ABCB1). Rather, members of the subfamily of multidrug resistance proteins, particularly ABCC1, ABCC2, ABCC3, and ABCC4, accept conjugates with glutathione, glucuronate, or sulfate as substrates (Borst et al., 2000; Haimeur et al., 2004; Jedlitschky et al., 1994, 1996; Kruh and Belinsky, 2003; König et al., 1999). In the apical membrane of hepatocytes, only ABCC2 functions as a conjugate export pump of broad substrate specificity. Some conjugates, such as sulfoconjugates transported by ABCG2 (Suzuki et al., 2003), and the taurine and glycine conjugates of bile salts transported by the bile salt export pump ABCB11 (Noe et al., 2002) can be effluxed into the biliary space by ABC transporters that are not members of the ABCC subfamily. Several members of the ABCC subfamily have been localized to the basolateral membrane of human hepatocytes and may thus contribute to the efflux of conjugates from hepatocytes into blood. Accordingly, ABCC3 in the basolateral hepatocyte membrane (König et al., 1999) transports bilirubin glucuronosides (Lee et al., 2004), ABCC4, also localized to the basolateral membrane of hepatocytes (Rius et al., 2003) transports, among others, sulfoconjugates (Kruh and Belinsky, 2003; Zelcer et al., 2003), and the basolateral ABCC6 (Scheffer et al., 2002) was reported to transport glutathione conjugates (Ilias et al., 2002).

Analysis of hepatic transporter protein function has previously included studies on mutant animals deficient in transporter genes (Huber et al., 1987; Jansen et al., 1985; Smit et al., 1993; Takenaka et al., 1995), transporter studies, particularly on uptake transporters, in transiently or stably transfected cell lines (Jacquemin et al., 1994; König et al., 2000a,b; Kullak-Ublick et al., 2001), and inside-out vesicles from apical membranes of the liver, particularly for studies on ATP-dependent transport and ATPase activity (Akerboom et al., 1991; Ishikawa et al., 1990; Nicotera et al., 1985), and from stably transfected cell lines (Borst et al., 2000; Cui et al., 1999; Evers et al., 1998; Keppler et al., 1998). These previous studies have not allowed for an integrated analysis of uptake and efflux transporters in polarized cells such as hepatocytes. The recent development of polarized cell lines, stably expressing human genes encoding a hepatocellular uptake transporter in the basolateral membrane and an ATP-driven export pump in the apical membrane, enables studies on vectorial transport and reflects more closely the elimination of substances by the liver (Cui et al., 2001a; Letschert et al., 2005; Mita et al., 2005; Sasaki et al., 2002; Shitara et al., 2005). Such double-transfected cell lines may be, in addition, stably transfected with cDNAs encoding enzymes of phase 1 oxidation or phase 2 conjugation.

Experimental Procedures

Cell Culture and Transfection

MDCKII cells, a subline of Madin–Darby canine kidney cells, are cultured in minimal essential medium supplemented with 10% fetal bovine serum, 100 U/ml penicillin, and 100 μg/ml streptomycin at 37° and 5% CO_2 as described (Cui et al., 2001a; Letschert et al., 2005). MDCKII cells are transfected with the respective plasmid (Cui et al., 1999, 2001a; Letschert et al., 2004, 2005): pcDNA3.1(+)-ABCC2 for ABCC2 and pcDNA3.1/ Hygro(-)-OATP1B3 for OATP1B3. After geneticin and hygromycin selection, single colonies are screened for OATP1B3 and ABCC2 protein by immunoblot analysis and immunofluorescence microscopy (Cui et al., 2001a). The MDCKII cell lines are grown on cell culture inserts to confluence for 3 days and induced with 10 mM sodium butyrate for 24 h prior to analysis to obtain higher levels of the recombinant proteins (Cui et al., 1999). Polyethylene terephthalate cell culture inserts with a diameter of 24 mm, a pore size of 0.4 μM, and a pore density of 1×10^8 pores per cm^2 are preferably used (Thin Cert, Greiner Bio-One, Frickenhausen, Germany) (Letschert et al., 2005). OATP1B3 is detected by immunofluorescence microscopy in the lateral membrane of virtually all transfected cells, and ABCC2 is found in $82 \pm 8\%$ of the cells in the apical membrane domain when high-density pore cell culture inserts are used (Letschert et al., 2005).

Modifications and different transporter combinations in stably transfected MDCKII cells have been described. These include ABCC2/ OATP1B1 double-transfected cells (Fehrenbach et al., 2003; Sasaki et al., 2002), rat Abcc2/Oatp1b2 double-transfected MDCKII cells (Sasaki et al., 2004), rat Abcb11/Ntcp double-transfected MDCKII cells (Mita et al., 2005), and other combinations, such as ABCC2 together with several OATP uptake transporters (Fig. 3).

Transport Assays

MDCKII cells are grown on cell culture inserts to confluence and induced with 10 mM butyrate for 24 h as described earlier. For transport measurements, cells are first washed with buffer (142 mM NaCl, 5 mM KCl, 1 mM K_2HPO_4, 1.2 mM $MgSO_4$, 1.5 mM $CaCl_2$, 5 mM glucose, and 12.5 mM HEPES, pH 7.3). Subsequently, radioactively labeled substrates are added in transport buffer (1.5 ml) to the basolateral compartment, and 1 ml of transport buffer is added to the apical compartment (Cui et al., 2001a). After various time periods, the buffer from the apical compartment is collected and counted for radioactivity to determine vectorial transport

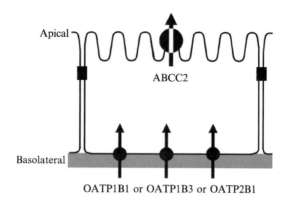

FIG. 3. Scheme of MDCKII cells stably expressing the human apical conjugate export pump ABCC2 together with one of several hepatocellular uptake transporters of the OATP family. Polarized cells are grown on a filter membrane support for studies on vectorial transport of substances that may be substrates for the respective OATP as well as for ABCC2 (Cui *et al.*, 2001a; Letschert *et al.*, 2005).

in the basolateral-to-apical direction. The intracellular accumulation of radioactivity is determined after washing of the cells three times with cold transport buffer and solubilization of cells with 2 ml of 0.2% sodium dodecyl sulfate in water (Cui *et al.*, 2001a). Rates of substrate uptake into cells and of vectorial transport are based on cellular protein determined in the cell lysate. The use of fluorescent substrates, such as Fluo-3, or the measurement of substrate concentrations by absorbance, as shown for rifampicin, in studies on vectorial transport has been described (Cui *et al.*, 2001a).

Vectorial Transport by Double-Transfected Cells

MDCKII cells are grown in a polarized fashion on cell culture inserts (Letschert *et al.*, 2005) and are shown by confocal laser-scanning microscopy to express ABCC2 in the apical membrane domain and an OATP uptake transporter in their basolateral or lateral membrane domain (Cui *et al.*, 2001a; Letschert *et al.*, 2005). ^{3}H-labeled bromosulfophthalein (BSP) is used as a standard substrate, which is transported by ABCC2 as well as by OATP1B1, OATP1B3, or OATP2B1 (Cui *et al.*, 2001a; Kullak-Ublick *et al.*, 2001). The transcellular transport of BSP by MDCK-ABCC2/OATP1B3 cells (Cui *et al.*, 2001a) is a vectorial process, as only basolateral-to-apical transport is observed, whereas apical-to-basolateral transport is negligible (Fig. 4). Corresponding results were obtained with other

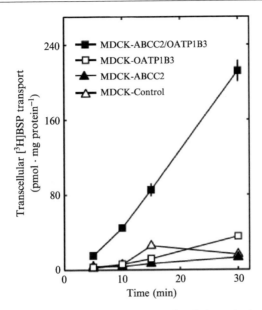

FIG. 4. Transport of labeled bromosulfophthalein (BSP) from the basolateral into the apical compartment. [³H]BSP was added to the basolateral compartment at a concentration of 1 μM and radioactivity was determined in the apical compartment at the times indicated. Note that a high rate of vectorial transport is only observed with double-transfected MDCKII cells expressing OATP1B3 in their basolateral and ABCC2 in their apical membrane as described earlier (Cui *et al.*, 2001a).

substrates, including pravastatin (Sasaki *et al.*, 2002) and sulfated cholecystokinin octapeptide, CCK-8 (Letschert *et al.*, 2005). These results are in line with the unidirectional action of ABCC2 as the efflux pump in the apical membrane.

The choice of substrates for vectorial transport by double-transfected cells is critical and compounds should be substrates for both the uptake transporter (e.g., OATP1B3) and the efflux pump (e.g., ABCC2). The rate-limiting step in this system depends not only on the relative affinity of the respective substrate for uptake or efflux, but also on the level of the recombinant protein expressed in the basolateral or the apical membrane domain of MDCKII cells. The kinetics of vectorial transport may be influenced by endogenous transporters, for example, by canine Abcc2 of MDCKII cells, and by endogenous enzymes of biotransformation. This is indicated, for instance, by a low rate of BSP transport into the apical compartment in MDCKII cells only expressing recombinant OATP1B3, but not human ABCC2 (Fig. 4). The experiment shown in Fig. 4 furthermore illustrates the decisive role of transport proteins for vectorial transport at a sufficient rate, whereas paracellular flow and passive diffusion are negligible in this system.

Substrates and Inhibitors for Vectorial Transport by Double-Transfected Cells

A number of anionic conjugates and other organic anions were shown to undergo vectorial transport by double-transfected cells expressing OATP1B3 together with ABCC2. For additional conjugates it was demonstrated in separate assays that they are substrates for recombinant OATP1B3 in transfected cells (Cui et al., 2001b; König et al., 2000b; Kullak-Ublick et al., 2001), as well as substrates for ATP-dependent transport by human ABCC2 in inside-out membrane vesicle transport assays (Cui et al., 1999; Kamisako et al., 1999; König et al., 1999). Prototypic glucuronosides, which are substrates for OATP1B3, as well as for ABCC2, include 17β–glucuronosyl estradiol and monoglucuronosyl bilirubin. The vectorial transport of 17β–glucuronosyl estradiol has been studied extensively in double-transfected cell lines (Cui et al., 2001a; Sasaki et al., 2002). The sulfoconjugate dehydroepiandrosterone 3-sulfate and the glutathione conjugate leukotriene C$_4$ are substrates for vectorial transport by MDCKII-ABCC2/OATP1B3 cells (Cui et al., 2001a). In the case of leukotriene C$_4$, rapid degradation of the glutathione residue by ectoenzymes, leading to leukotriene E$_4$, has to be taken into account, as this decreases transport efficiency. Synthetic organic anions that are not conjugates but useful substrates for studies on vectorial transport by MDCKII-ABCC2/OATP1B3 cells include BSP, Fluo-3, and rifampicin (Cui et al., 2001a), as well as the peptide CCK-8 (Letschert et al., 2005).

Inhibition of vectorial transport of a known labeled substrate, such as BSP, may serve to identify inhibitors, including drugs and drug candidates. A selective or preferential inhibition of the efflux pump results in an intracellular accumulation of labeled BSP and in a decrease of its efflux into the apical compartment. This has been observed with 1-chloro-2,4-dinitrobenzene (CDNB), which, after its intracellular conjugation yielding S-(2,4-dinitrophenyl) glutathione, inhibits ABCC2-mediated efflux of BSP (Cui et al., 2001a). However, several known inhibitors interfere with the uptake transporter OATP1B3 as well as with the efflux pump ABCC2. This is exemplified by cyclosporin A and by the quinoline derivative MK571, which both potently inhibit uptake mediated by OATP1B3 with K_i values of 1.2 and 0.6 μM, respectively; in addition, both compounds inhibit ABCC2 with K_i values of 24 and 8.5 μM, respectively (Letschert et al.,

2005). Thus, dual inhibition of uptake and efflux transporter may strongly interfere with vectorial transport across these polarized cells.

Acknowledgments

Studies in the author's laboratory were supported by the German Cancer Research Center, Heidelberg, the Fonds der Chemischen Industrie, Frankfurt, and the Deutsche Forschungsgemeinschaft, Bonn, Germany, and by a research collaboration with Pfizer Global Research and Development, Groton, Connecticut. Contributions to this work from past and present members of our laboratory, particularly from Yunhai Cui, Jörg König, and Katrin Letschert, are gratefully acknowledged.

References

Abe, T., Kakyo, M., Tokui, T., Nakagomi, R., Nishio, T., Nakai, D., Nomura, H., Unno, M., Suzuki, M., Naitoh, T., Matsuno, S., and Yawo, H. (1999). Identification of a novel gene family encoding human liver-specific organic anion transporter LST-1. *J. Biol. Chem.* **274,** 17159–17163.

Akerboom, T. P., Narayanaswami, V., Kunst, M., and Sies, H. (1991). ATP-dependent S-(2,4-dinitrophenyl)glutathione transport in canalicular plasma membrane vesicles from rat liver. *J. Biol. Chem.* **266,** 13147–13152.

Borst, P., Evers, R., Kool, M., and Wijnholds, J. (2000). A family of drug transporters: The multidrug resistance-associated proteins. *J. Natl. Cancer Inst.* **92,** 1295–1302.

Büchler, M., König, J., Brom, M., Kartenbeck, J., Spring, H., Horie, T., and Keppler, D. (1996). cDNA cloning of the hepatocyte canalicular isoform of the multidrug resistance protein, cMrp, reveals a novel conjugate export pump deficient in hyperbilirubinemic mutant rats. *J. Biol. Chem.* **271,** 15091–15098.

Cui, Y., König, J., Buchholz, J. K., Spring, H., Leier, I., and Keppler, D. (1999). Drug resistance and ATP-dependent conjugate transport mediated by the apical multidrug resistance protein, MRP2, permanently expressed in human and canine cells. *Mol. Pharmacol.* **55,** 929–937.

Cui, Y., König, J., and Keppler, D. (2001a). Vectorial transport by double-transfected cells expressing the human uptake transporter SLC21A8 and the apical export pump ABCC2. *Mol. Pharmacol.* **60,** 934–943.

Cui, Y., König, J., Leier, I., Buchholz, U., and Keppler, D. (2001b). Hepatic uptake of bilirubin and its conjugates by the human organic anion transporter SLC21A6. *J. Biol. Chem.* **276,** 9626–9630.

Evers, R., Kool, M., van Deemter, L., Janssen, H., Calafat, J., Oomen, L. C., Paulusma, C. C., Oude Elferink, R. P., Baas, F., Schinkel, A. H., and Borst, P. (1998). Drug export activity of the human canalicular multispecific organic anion transporter in polarized kidney MDCK cells expressing *cMOAT (MRP2)* cDNA. *J. Clin. Invest.* **101,** 1310–1319.

Fehrenbach, T., Cui, Y., Faulstich, H., and Keppler, D. (2003). Characterization of the transport of the bicyclic peptide phalloidin by human hepatic transport proteins. *Naunyn-Schmiedeberg's Arch. Pharmacol.* **368,** 415–420.

Hagenbuch, B., and Dawson, P. (2004). The sodium bile salt cotransport family SLC10. *Pflüg. Arch. Eur. J. Physiol.* **447,** 566–570.

Hagenbuch, B., and Meier, P. J. (2004). Organic anion transporting polypeptides of the OATP/ SLC21 family: Phylogenetic classification as OATP/ SLCO superfamily, new nomenclature and molecular/functional properties. *Pflüg. Arch. Eur. J. Physiol.* **447,** 653–665.

Haimeur, A., Conseil, G., Deeley, R. G., and Cole, S. P. (2004). The MRP-related and BCRP/ ABCG2 multidrug resistance proteins: Biology, substrate specificity and regulation. *Curr. Drug Metab.* **5,** 21–53.

Hsiang, B., Zhu, Y., Wang, Z., Wu, Y., Sasseville, V., Yang, W. P., and Kirchgessner, T. G. (1999). A novel human hepatic organic anion transporting polypeptide (OATP2): Identification of a liver-specific human organic anion transporting polypeptide and identification of rat and human hydroxymethylglutaryl-CoA reductase inhibitor transporters. *J. Biol. Chem.* **274,** 37161–37168.

Huber, M., Guhlmann, A., Jansen, P. L., and Keppler, D. (1987). Hereditary defect of hepatobiliary cysteinyl leukotriene elimination in mutant rats with defective hepatic anion excretion. *Hepatology* **7,** 224–228.

Ilias, A., Urban, Z., Seidl, T. L., Le Saux, O., Sinko, E., Boyd, C. D., Sarkadi, B., and Varadi, A. (2002). Loss of ATP-dependent transport activity in pseudoxanthoma elasticum-associated mutants of human ABCC6 (MRP6). *J. Biol. Chem.* **277,** 16860–16867.

Ishikawa, T., Müller, M., Klünemann, C., Schaub, T., and Keppler, D. (1990). ATP-dependent primary active transport of cysteinyl leukotrienes across liver canalicular membrane: Role of the ATP-dependent transport system for glutathione S-conjugates. *J. Biol. Chem.* **265,** 19279–19286.

Jacquemin, E., Hagenbuch, B., Stieger, B., Wolkoff, A. W., and Meier, P. J. (1994). Expression cloning of a rat liver Na$^+$-independent organic anion transporter. *Proc. Natl. Acad. Sci. USA* **91,** 133–137.

Jansen, P. L., Groothuis, G. M., Peters, W. H., and Meijer, D. F. (1987). Selective hepatobiliary transport defect for organic anions and neutral steroids in mutant rats with hereditary-conjugated hyperbilirubinemia. *Hepatology* **7,** 71–76.

Jansen, P. L., Peters, W. H., and Lamers, W. H. (1985). Hereditary chronic conjugated hyperbilirubinemia in mutant rats caused by defective hepatic anion transport. *Hepatology* **5,** 573–579.

Jedlitschky, G., Leier, I., Buchholz, U., Barnouin, K., Kurz, G., and Keppler, D. (1996). Transport of glutathione, glucuronate, and sulfate conjugates by the *MRP* gene-encoded conjugate export pump. *Cancer Res.* **56,** 988–994.

Jedlitschky, G., Leier, I., Buchholz, U., Center, M., and Keppler, D. (1994). ATP-dependent transport of glutathione S-conjugates by the multidrug resistance-associated protein. *Cancer Res.* **54,** 4833–4836.

Jonker, J. W., Buitelaar, M., Wagenaar, E., Van Der Valk, M. A., Scheffer, G. L., Scheper, R. J., Plosch, T., Kuipers, F., Elferink, R. P., Rosing, H., Beijnen, J. H., and Schinkel, A. H. (2002). The breast cancer resistance protein protects against a major chlorophyll-derived dietary phototoxin and protoporphyria. *Proc. Natl. Acad. Sci. USA* **99,** 15649–15654.

Kamisako, T., Leier, I., Cui, Y., König, J., Buchholz, U., Hummel-Eisenbeiss, J., and Keppler, D. (1999). Transport of monoglucuronosyl and bisglucuronosyl bilirubin by recombinant human and rat multidrug resistance protein 2. *Hepatology* **30,** 485–490.

Keppler, D., Jedlitschky, G., and Leier, I. (1998). Transport function and substrate specificity of multidrug resistance protein. *Methods Enzymol.* **292,** 607–616.

Koepsell, H., Schmitt, B. M., and Gorboulev, V. (2003). Organic cation transporters. *Rev. Physiol. Biochem. Pharmacol.* **150,** 36–90.

König, J., Cui, Y., Nies, A. T., and Keppler, D. (2000a). A novel human organic anion transporting polypeptide localized to the basolateral hepatocyte membrane. *Am. J. Physiol. Gastrointest. Liver Physiol.* **278,** G156–G164.

König, J., Cui, Y., Nies, A. T., and Keppler, D. (2000b). Localization and genomic organization of a new hepatocellular organic anion transporting polypeptide. *J. Biol. Chem.* **275**, 23161–23168.

König, J., Nies, A. T., Cui, Y., Leier, I., and Keppler, D. (1999). Conjugate export pumps of the multidrug resistance protein (MRP) family: Localization, substrate specificity, and MRP2-mediated drug resistance. *Biochim. Biophys. Acta* **1461**, 377–394.

Kruh, G. D., and Belinsky, M. G. (2003). The MRP family of drug efflux pumps. *Oncogene* **22**, 7537–7552.

Kullak-Ublick, G. A., Ismair, M. G., Stieger, B., Landmann, L., Huber, R., Pizzagalli, F., Fattinger, K., Meier, P. J., and Hagenbuch, B. (2001). Organic anion-transporting polypeptide B (OATP-B) and its functional comparison with three other OATPs of human liver. *Gastroenterology* **120**, 525–533.

Lee, Y. M., Cui, Y., König, J., Risch, A., Jager, B., Drings, P., Bartsch, H., Keppler, D., and Nies, A. T. (2004). Identification and functional characterization of the natural variant MRP3-Arg1297His of human multidrug resistance protein 3 (MRP3/ABCC3). *Pharmacogenetics* **14**, 213–223.

Letschert, K., Keppler, D., and König, J. (2004). Mutations in the *SLCO1B3* gene affecting the substrate specificity of the hepatocellular uptake transporter OATP1B3 (OATP8). *Pharmacogenetics* **14**, 441–452.

Letschert, K., Komatsu, M., Hummel-Eisenbeiss, J., and Keppler, D. (2005). Vectorial transport of the peptide CCK-8 by double-transfected MDCKII cells stably expressing the organic anion transporter OATP1B3 (OATP8) and the export pump ABCC2. *J. Pharmacol. Exp. Ther.* **313**, 549–556.

Mita, S., Suzuki, H., Akita, H., Stieger, B., Meier, P. J., Hofmann, A. F., and Sugiyama, Y. (2005). Vectorial transport of bile salts across MDCK cells expressing both rat Na$^+$-taurocholate cotransporting polypeptide and rat bile salt export pump. *Am. J. Physiol. Gastrointest. Liver Physiol.* **288**, G159–G167.

Nicotera, P., Moore, M., Bellomo, G., Mirabelli, F., and Orrenius, S. (1985). Demonstration and partial characterization of glutathione disulfide-stimulated ATPase activity in the plasma membrane fraction from rat hepatocytes. *J. Biol. Chem.* **260**, 1999–2002.

Noe, J., Stieger, B., and Meier, P. J. (2002). Functional expression of the canalicular bile salt export pump of human liver. *Gastroenterology* **123**, 1659–1666.

Paulusma, C. C., Bosma, P. J., Zaman, G. J., Bakker, C. T., Otter, M., Scheffer, G. L., Scheper, R. J., Borst, P., and Oude Elferink, R. P. (1996). Congenital jaundice in rats with a mutation in a multidrug resistance-associated protein gene. *Science* **271**, 1126–1128.

Rius, M., Nies, A. T., Hummel-Eisenbeiss, J., Jedlitschky, G., and Keppler, D. (2003). Cotransport of reduced glutathione with bile salts by MRP4 (ABCC4) localized to the basolateral hepatocyte membrane. *Hepatology* **38**, 374–384.

Sasaki, M., Suzuki, H., Aoki, J., Ito, K., Meier, P. J., and Sugiyama, Y. (2004). Prediction of in vivo biliary clearance from the in vitro transcellular transport of organic anions across a double-transfected Madin-Darby canine kidney II monolayer expressing both rat organic anion transporting polypeptide 4 and multidrug resistance associated protein 2. *Mol. Pharmacol.* **66**, 450–459.

Sasaki, M., Suzuki, H., Ito, K., Abe, T., and Sugiyama, Y. (2002). Transcellular transport of organic anions across a double-transfected Madin-Darby canine kidney II cell monolayer expressing both human organic anion-transporting polypeptide (OATP2/SLC21A6) and multidrug resistance-associated protein 2 (MRP2/ABCC2). *J. Biol. Chem.* **277**, 6497–6503.

Scheffer, G. L., Hu, X., Pijnenborg, A. C., Wijnholds, J., Bergen, A. A., and Scheper, R. J. (2002). MRP6 (ABCC6) detection in normal human tissues and tumors. *Lab. Invest.* **82**, 515–518.

Sekine, T., Cha, S. H., and Endou, H. (2000). The multispecific organic anion transporter (OAT) family. *Pflüg. Arch. Eur. J. Physiol.* **440**, 337–350.

Shitara, Y., Sato, H., and Sugiyama, Y. (2005). Evaluation of drug-drug interaction in the hepatobiliary and renal transport of drugs. *Annu. Rev. Pharmacol. Toxicol.* **45,** 689–723.

Smit, J. J., Schinkel, A. H., Oude Elferink, R. P., Groen, A. K., Wagenaar, E., van Deemter, L., Mol, C. A., Ottenhoff, R., van der Lugt, N. M., van Roon, M. A., Van der Valk, M. A., Offerhaus, G. J. A., Berns, A. J. M., and Borst, P. (1993). Homozygous disruption of the murine mdr2 P-glycoprotein gene leads to a complete absence of phospholipid from bile and to liver disease. *Cell* **75,** 451–462.

Sun, W., Wu, R. R., van Poelje, P. D., and Erion, M. D. (2001). Isolation of a family of organic anion transporters from human liver and kidney. *Biochem. Biophys. Res. Commun.* **283,** 417–422.

Suzuki, M., Suzuki, H., Sugimoto, Y., and Sugiyama, Y. (2003). ABCG2 transports sulfated conjugates of steroids and xenobiotics. *J. Biol. Chem.* **278,** 22644–22649.

Takenaka, O., Horie, T., Kobayashi, K., Suzuki, H., and Sugiyama, Y. (1995). Kinetic analysis of hepatobiliary transport for conjugated metabolites in the perfused liver of mutant rats (EHBR) with hereditary conjugated hyperbilirubinemia. *Pharm. Res.* **12,** 1746–1755.

Tamai, I., Nezu, J., Uchino, H., Sai, Y., Oku, A., Shimane, M., and Tsuji, A. (2000). Molecular identification and characterization of novel members of the human organic anion transporter (OATP) family. *Biochem. Biophys. Res. Commun.* **273,** 251–260.

Zelcer, N., Reid, G., Wielinga, P., Kuil, A., van der Heijden, I., Schuetz, J. D., and Borst, P. (2003). Steroid and bile acid conjugates are substrates of human multidrug-resistance protein (MRP) 4 (ATP-binding cassette C4). *Biochem. J.* **371,** 361–367.

[30] Biliary Transport Systems: Short-Term Regulation

By RALF KUBITZ, ANGELIKA HELMER, and DIETER HÄUSSINGER

Abstract

Bile secretion is conveyed by a large set of transporter proteins. Their activity is controlled on long- and short-term timescales. Short-term regulation of transcellular transport has to guarantee intra- and extracellular molecular homeostasis and has to meet the actual cellular metabolic needs. As transport activity depends not only on transporter expression, measurements of mRNA or protein levels will not fully predict functionality. Transporter activity is also determined by covalent modifications (e.g., phosphorylation), substrate competition, and subcellular transporter localization. The latter is a major target of short-term regulation of bile secretion and involves rapid endo- and exocytosis of transporter-bearing vesicles from and into the respective cell membrane. In liver parenchymal cells, several signaling pathways were identified that govern these processes; however, the underlying molecular mechanisms still need to be characterized. Different techniques have been employed in studies on transporter retrieval and insertion, which are discussed in this chapter.

METHODS IN ENZYMOLOGY, VOL. 400   0076-6879/05 $35.00
DOI: 10.1016/S0076-6879(05)00030-3

Regulated Hepatobiliary Transport

Bile formation is a complex process that involves sinusoidal (basolateral) and canalicular (apical) transporter proteins. Regulation of these transporters occurs at the level of gene expression (Trauner et al., 1998), transporter degradation (Xia et al., 2004), covalent modifications of transporters (Gottesman and Pastan, 1993; Noe et al., 2002), and their regulated exocytic insertion into or endocytic retrieval from the membrane (Kubitz et al., 1997, 1999; Schmitt et al., 2000, 2001). Short-term regulation also involves substrate availability or substrate competition (Häussinger et al., 2000). Short-term regulation of canalicular secretion is mandatory because the load of cholephilic compounds toward the liver may vary considerably within short time periods. For example, the bile salt load to the liver changes in response to food intake and gallbladder contraction [involving the enterohepatic circulation of bile salt (Kullak-Ublick et al., 2004)] and exhibits circadian rhythmicity. Intracellular bile salt homeostasis is required because bile salts per se influence many other liver activities, such as cholesterol biosynthesis (Oude Elferink and Groen, 2002), immune function (Kawamata et al., 2003), and apoptosis (Higuchi and Gores, 2003). Furthermore, secretion of GSH or glutathione conjugates into bile or back into the blood exerts control on the intracellular glutathione disulfide/glutathione (GSSG/GSH) ratio and thereby on the potential to counteract oxidative stress. Also, bilirubin, which is generally considered as a metabolic end product, may serve as an antioxidant (Stocker et al., 1987a,b).

Hepatic transport processes are finely tuned to meet many of these demands. While many of the transporter proteins have been characterized in the past years at the genomic and functional level, their regulation in physiological and pathophysiological conditions comes more into focus; however, many issues are still unsolved.

Canalicular and Sinusoidal Transport Systems

Various transport systems in the sinusoidal and the canalicular membrane of the hepatocyte participate in the vectorial transport of biliary constituents from the sinusoidal blood or the cellular interior into the bile canalicular lumen (for reviews, see Häussinger et al., 2000, 2004; Kullak-Ublick et al., 2004; Meier and Stieger, 2002). Na^+/K^+-ATPase is located in the sinusoidal membrane and creates an electrochemical Na^+ gradient, which provides the driving force for Na^+-coupled transport systems, such as the Na^+-taurocholate cotransporting polypeptide (Ntcp/ Slc10a1) (Hagenbuch and Meier, 1994; Stieger et al., 1994). Ntcp/NTCP (for the protein in rodents and humans, respectively) represents the major uptake system for conjugated bile salts and is functionally complemented by different members of the organic anion transporting polypeptide (Oatp/OATP) family (Hagenbuch and Meier, 2003) (Fig. 1). Some OATPs can transport bile salts in a Na^+-independent fashion apart from other anions such as conjugates of estrogens, leukotrienes, ajmalin, or ouabain (Hagenbuch and Meier, 2003).

The organic anion transporters OAT/Oat, including human OAT2 (Slc22A7) and OAT5 (Slc22A10) (Sekine et al., 1998; Sun et al., 2001), mediate Na^+-independent uptake of substrates such as methotrexate, p-aminohippurate, or salicylate (Sekine et al., 1998). Transport of organic cations is accomplished by the organic cation transporters OCT/Oct (Grundemann et al., 1994), including the liver-specific human OCT1 (Slc22A1) (Zhang et al., 1997). Apart from transporters mediating substrate uptake, primary active transporters for substrate export are also present in the lateral membrane of hepatocytes, albeit normally at a low expression level. They belong to the ATP-binding cassette (ABC) transporter family (Borst et al., 2000) with common structural and functional properties (Chang, 2003). These transporters include the multidrug resistance associated protein 3 (MRP3/ABCC3) (Kiuchi et al., 1998) with a substrate specificity that overlaps with that of MRP2 (Keppler and König, 2000). Basolateral Mrp4 (ABCC4) is an export pump for reduced glutathione (GSH) and bile salts (Rius et al., 2003) and Mrp5 (ABCC5) functions as an export pump for cyclic nucleotides (Jedlitschky et al., 2000). Furthermore, MRP6 (ABCC6) (Kool et al., 1999) is expressed at the basolateral membrane of hepatocytes; its physiological substrate is still unclear. In proliferating liver cells, Mrp1 (ABCC1), which is expressed in many tissues, also becomes detectable in the sinusoidal/lateral membrane (Roelofsen et al., 1997) of hepatocytes.

Canalicular secretion in liver is also brought about by specific transport ATPases, all of which belong to the ABC transporter superfamily. These primary active transport systems transport bile acids, organic anions, conjugates, and xenobiotics against steep concentration gradients across the canalicular membrane. They include the bile salt export pump (BSEP/SPGP, ABCB11), the canalicular multidrug resistance protein MRP2 (cMRP/cMOAT, ABCC2) for transport of anionic conjugates (e.g., glutathione and glucuronide conjugates), the human MDR3 P-glycoprotein (in mouse Mdr2) for phospholipid excretion, and the multidrug resistance P-glycoprotein (MDR1 and Mdr1a/b for human and mouse, respectively) for the excretion of amphiphilic organic cations and xenobiotic compounds. The two half-transporter ABCG5 and ABCG8, which form an obligate heterodimer, secrete cholesterol into bile (Graf et al., 2003; Yu

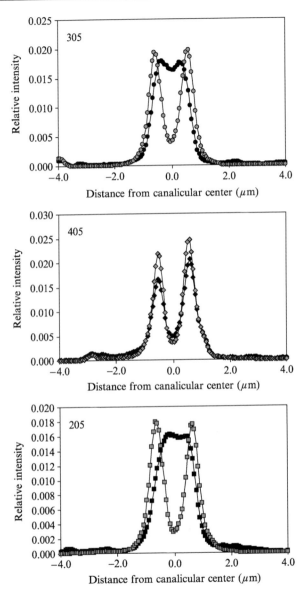

FIG. 1. Fluorescence distribution of apical proteins determined by confocal laser-scanning microscopy. Rat livers were perfused with normoosmotic (305 mosmol/liter) buffer followed by two 30-min perfusion periods with hyperosmotic (405 mosmol/liter) and hypoosmotic (205 mosmol/liter) buffer. Small tissue samples were excised after each perfusion period, snap frozen, kryosectioned, and stained for Bsep (black symbols) and ZO-1 (gray symbols). Fluorescence distribution was measured in pictures of confocal laser-scanning microscopy. After hyperosmotic perfusion, a clear lateralization of Bsep was detected approaching the distribution of ZO-1. Hypoosmotic exposure reversed this process.

et al., 2002)[Kubitz *et al.* (2005) summarizes the different transporter proteins and their localization.]

Short-Term Regulation of Canalicular Transport: Levels of Control

Environmental challenges and variations of substrate availability necessitate efficient short-term regulation of transporters, which occurs at the levels of substrate availability, allosteric and covalent modification of canalicular transporter proteins, and their regulated exocytic insertion into or endocytic retrieval from the membrane. These processes underlie the control by multiple intracellular signaling pathways comprising different second messenger and protein kinase/phosphatase systems. Transporter insertion and retrieval are discussed in more detail later.

In general, the transport capacity of sinusoidal transporters exceeds that of canalicular transport systems, which led to the widely accepted view that under physiological conditions the canalicular excretion step is rate limiting for overall transcellular transport of most cholephilic compounds. However, the control strength theory (Groen *et al.*, 1982) has not yet been applied to transcellular transport. Thus, the possibility is not ruled out that significant control on bile formation is exerted also at the step of uptake across the sinusoidal membrane. This may be relevant, especially at physiologically low bile acid concentrations or under pathophysiologic conditions, such as sepsis or estrogen treatment, in which the expression of Ntcp and other transporters is downregulated (Green *et al.*, 1996; Simon *et al.*, 1996). The view that sinusoidal bile acid uptake exerts significant control on canalicular excretion is further suggested by the finding that taurocholate transport capacity via Ntcp underlies short-term control by cAMP, $Ca^{2+}/$calmodulin, and okadaic acid-sensitive protein phosphatases (Grüne *et al.*, 1993; Mukhopadhyay *et al.*, 1997, 1998a,b). Here, cAMP enhances the V_{max} of the transporter within minutes by a microfilament-dependent translocation of intracellularly stored Ntcp molecules to the plasma membrane (Dranoff *et al.*, 1999; Mukhopadhyay *et al.*, 1997). Thus, modulation of Ntcp transport activity by transporter recruitment to the sinusoidal membrane or by changes in the driving force of the transporter (i.e., the electrochemical Na^+ potential of the plasma membrane) is expected to exert indirect control on canalicular bile acid secretion by the canalicular bile salt export pump (Bsep).

Methods Used to Study Subcellular Transporter Distribution

Various techniques were employed in order to study the subcellular localization of transporter proteins in hepatocytes. They include immunofluorescent studies on isolated cells or kryosectioned tissues using confocal laser-scanning microscopy (Kubitz *et al.*, 1997, 1999; Mühlfeld *et al.*, 2003; Schmitt *et al.*, 2000, 2001), studies at the level of electron microscopy (Beuers *et al.*, 2001; Dombrowski *et al.*, 2000), the use of differential centrifugation (Kipp and Arias, 2000) or sucrose gradient centrifugation (Mukhopadhyay *et al.*, 1998a), and measurement of transport capacity as a surrogate marker of transporter availability (Häussinger *et al.*, 1992, 1993; Kubitz *et al.*, 2004), as well as biotinylation of membrane proteins (Webster *et al.*, 2000). Furthermore, a method that uses a FLAG-tagged Ntcp in combination with flow cytometry in order to determine the distribution of Ntcp between plasma membrane and intracellular sites has been developed.

Determination of Subcellular Localisation of Transporters by Confocal Laser-Scanning Microscopy

Rapid and reversible transporter retrieval and insertion from and into the canalicular membrane of liver parenchymal cells in response to aniso-osmotic challenge was first demonstrated in our laboratory (Kubitz *et al.*, 1997). The process of transporter withdrawal from the canalicular membrane was shown for Mrp2 (Kubitz *et al.*, 1997) and Bsep (Schmitt *et al.*, 2001) in response to lipopolysaccharide (LPS) (Kubitz *et al.*, 1999), oxidative stress (Schmitt *et al.*, 2000), tauroursodeoxycholate (Kurz *et al.*, 2001) and hyperosmolarity. This technique was also employed for the investigation of estrogen-induced retrieval of Mrp2 (Mottino *et al.*, 2002) and Bsep (Crocenzi *et al.*, 2003) and phalloidin-induced bulk retrieval of canalicular proteins (Rost *et al.*, 1999).

An important step in the determination of canalicular transporter localization was covisualization of the transporter proteins together with the tight junction associated protein *zonula occludens* protein 1 (ZO-1). ZO-1 accurately delineates the canalicular domain from the (baso-)lateral membrane domain and the cellular interior. In liver tissue, ZO-1 is arranged along two sharp lines (comparable to train tracks). Canalicular transporter proteins within the canalicular membrane are located between these two lines, whereas retrieved transporter proteins are situated on either side of these lines (Kubitz *et al.*, 1997, 1999, 2004; Mühlfeld *et al.*, 2003; Schmitt *et al.*, 2000, 2001). Under conditions of transporter retrieval, more vesicle-like, transporter-bearing structures were found aside of the ZO-1 staining.

By staining of different canalicular membrane proteins (coimmuno-fluorescence), selectivity of the retrieval process could be established in accordance with functional data (Dombrowski *et al.*, 2000; Kubitz *et al.*, 1999). The canalicular marker protein dipeptidylpeptidase IV (Dpp IV) was not retrieved in endotoxemic rats in contrast to Mrp2 (Kubitz et al., 1999). However, hyperosmotic exposure induced retrieval of Bsep and Mrp2, but only 15% of endocytic vesicles were shared by both transporters, while they were otherwise distributed in distinct vesicle populations (Schmitt et al., 2001).

In order to quantify the amount of retrieved protein, a densitometric analytic procedure was developed (Kubitz et al., 1997). Here, distribution of the fluorescence attached to the transporter proteins (via antibodies) was measured in a perpendicular orientation to the canalicular axis with a digital camera. When densitometry was combined with conventional fluorescence microscopy, the fluorescence distributions approximated Gaussian distributions. The width of the Gaussian distribution can be expressed by the variance, which then can be used for statistical analysis. This method was applied for different conditions (Kubitz et al., 1997, 1999; Kurz et al., 2001; Mühlfeld et al., 2003; Schmitt et al., 2000, 2001). Interestingly, when fluorescence profiles were measured in very thin optical planes by confocal laser-scanning microscopy, fluorescence distribution within the canalicular area was even more precisely resolved. In livers perfused with hyperosmotic medium, that is, when certain transporters were retrieved, the distribution of Bsep resembled the distribution of tight junctions, whereas under normo- and hypoosmotic conditions, Bsep was enriched within the canalicular area (delineated by ZO-1) (Fig. 1).

Determination of Distribution of Mrp2 by Differential Centrifugation

Administration of endotoxin/LPS to rats causes a rapid retrieval of canalicular transporter proteins into intracellular vesicles, as demonstrated originally by confocal laser-scanning microscopy (Kubitz et al., 1999) and electron microscopy (Dombrowski et al., 2000). Retrieval of the Mrp2 was associated with decreased secretion of Mrp2 substrates into bile, a process that was reversible in early stages of LPS treatment only (Kubitz et al., 1999). Using differential centrifugation adapted to the particularities of the liver, retrieval of Mrp2 can also be quantified by Western blotting. Due to its dense architecture comprising many structural, filamentous proteins (Tsukada et al., 1995), bile canaliculi are characterized by a high mass density. Therefore, centrifugation with relatively low centrifugal forces allows the enrichment of canalicular membranes (Kipp and Arias, 2000). This section describes a simplified method to enrich canalicular transporters such as Mrp2 (Fig. 2). After treatment with or without LPS, rat livers are shortly perfused with ice-cold phosphate-buffered saline via the portal vein in order to remove blood. Small liver pieces are cut in 10 volumes of a homogenization buffer (20 mmol/liter Tris, pH 7.4, 250 mmol/liter sucrose,

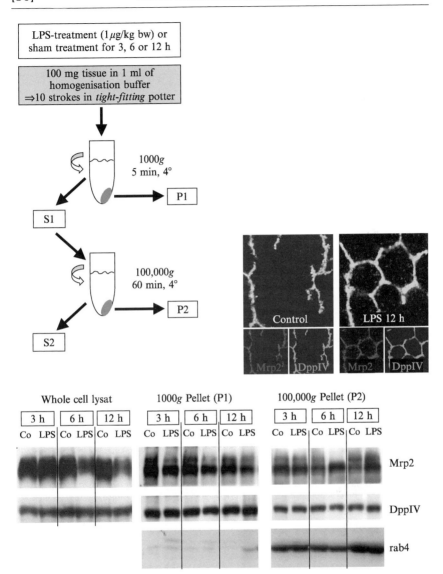

FIG. 2. Differential centrifugation in order to enrich canalicular proteins. Liver tissue was homogenized by tight-fitting pottering, and fractions enriched in canalicular proteins (P1) or endomembranes (P2) were collected as illustrated. LPS treatment induced a retrieval (see confocal pictures) and a reduction in Mrp2. Neither the localization nor the amount of the canalicular marker protein dipeptidylpeptidase IV (DppIV) was altered by LPS treatment. Despite the decrease in total Mrp2 by LPS treatment (whole cell lysate), Mrp2 in the endomembrane fraction increased slightly (100,000g pellet, P2). No changes were observed for DppIV. Note that the 1000g pellet was almost free of the endocytic marker protein rab4. S1/S2, supernatants. (See color insert.)

5 mmol/liter EGTA, 1 mmol/liter MgCl$_2$,, protease inhibitors) and homogenized by 10 strokes in a tight-fitting potter. Homogenates are centrifuged at low speed (1000g, 5 min, 4°). It is of note that no nuclear pellet is discarded. Pellet P1 of the low-speed centrifugation contains most of the Mrp2, while it is virtually free of marker proteins of endocytic vesicles such as rab4 or clathrin (Fig. 2). Supernatant S1 is centrifuged at high speed (100,000g, 60 min, 4°) in order to enrich endomembranes. Pellet P2 contains a smaller fraction of Mrp2, but significant amounts of rab4 or clathrin. Under cholestatic conditions such as LPS-induced cholestasis, an increase of Mrp2 in the endomembrane fraction (P2) can be seen, whereas the total amount of Mrp2 (as seen in the whole lysate or the "canalicular" fraction (P1) is diminished (Fig. 2).

When livers are perfused with thymeleatoxin or the phorbol ester phorbol-12-myristate-13-acetate (PMA), activators of protein kinase C, which have been shown to induce cholestasis (Kubitz et $al.$, 2004), retrieval of Mrp2 can be demonstrated by confocal laser-scanning microscopy, but also by differential centrifugation as described earlier (Fig. 3).

Determination of Distribution of Mrp2 or Ntcp by Sucrose
   Gradient Centrifugation

Another possibility to make use of the different densities of integral membrane proteins from the cell membrane compared to the same proteins from intracellular vesicles is the widely applied method of sucrose gradient centrifugation. In order to prepare a continuous sucrose gradient, three solutions with increasing sucrose content are employed [e.g., 20, 36,

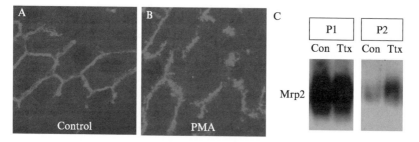

FIG. 3. Protein kinase C (PKC)-dependent retrieval of Mrp2. Livers were perfused with PMA or thymeleatoxin (Ttx), activators of PKC. PMA and Ttx induced a vesicular retrieval of Mrp2. As shown by laser-scanning microscopy, PKC activation resulted in fussy Mrp2 staining around the canaliculi in PMA-treated liver (B) compared to controls (A). In the pellet of high-speed centrifugation (P2, see text for details), an enrichment of Mrp2 was seen, suggesting a redistribution of Mrp2 in an intracellular compartment (C). Con, control.

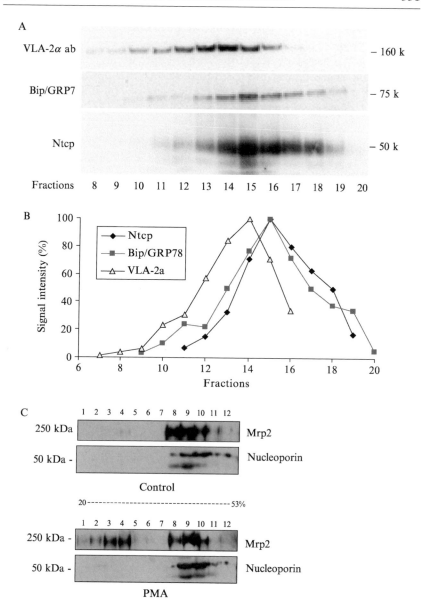

FIG. 4. Distribution of Ntcp and Mrp2. (A and B) HepG2 cells stably expressing rat Ntcp were treated with a proteasome inhibitor. Ntcp accumulated intracellularly. Intracellular Ntcp displayed the same distribution within the sucrose gradient as compared to the ER marker Bip/GRP, whereas the plasma membrane marker VLA-2α showed a distinct distribution. (C) Rat livers were perfused with or without the phorbolester PMA (10 nmol/liter) for 30 min

and 53% sucrose (w/v) with specific weights ranging from 1.08 to 1.25 g/liter]. The three solutions are carefully layered on top of each other in a centrifuge tube, starting with the dense solution at the bottom. The tube needs to be closed tightly and tilted to the horizontal position in order to increase the surface for the equilibration of sucrose concentration. Within 2 h (at 4°) a continuous gradient is established.

Cells or tissue must be homogenized, for example, by pottering in a 10% sucrose solution in order to obtain a suspension of membrane vesicles. The specific weights of these vesicles depend on the composition of the proteins within the membranes. This homogenate can be layered on top of the gradient (in the upright tube) and can be ultracentrifuged (e.g., at 100,000g for 60 min at 4°). Equal fractions (e.g., 10 to 25) of the gradient are collected carefully, starting at the bottom of the gradient using a Hamilton syringe. Equal amounts of these fractions are separated by SDS–PAGE (after solubilization of the proteins/vesicles with an appropriate lysis buffer) and analyzed by standard Western blot techniques. The distribution of proteins within this gradient can be quantified by densitometry. The proteins can then be assigned to a subcellular compartment through the comparison to the distribution of marker proteins for these compartments (Fig. 4).

Determination of Distribution of Ntcp by FLAG-Tag and Flow Cytometry

Ntcp is a key transporter of bile acid uptake. In order to measure the distribution of Ntcp between cell membrane and intracellular sites, Ntcp is FLAG tagged by cloning the FLAG pentapeptide to the extracellular N terminus of rat Ntcp. The intracellularly localized C terminus is tagged with the enhanced green fluorescent protein (EGFP). HepG2-cells stably expressing FLAG-Ntcp-EGFP are established. Ntcp is targeted correctly to the basolateral cell membrane in these cells (Fig. 5). Ntcp localized at the plasma membrane is detected by (fluorophore-labeled) anti-FLAG antibodies, whereas in intact cells, intracellularly stored Ntcp is not accessible to the anti-FLAG antibody.

After treatment of these cells with various stimuli, the distribution of membrane-bound Ntcp compared to total Ntcp can be determined in

---

to induce a cholestasis. Liver pieces were homogenized in a non-lysing buffer and vesicle suspensions were separated by continuous sucrose gradients. Twelve fractions were collected from the gradients. In control livers, almost the entire amount of Mrp2 was found as fractions of higher density (fractions 8 to 10), whereas in PMA-treated livers, significant amounts of Mrp2 appeared in fractions of lower density (fractions 2 to 4) suggestive for the redistribution in a low-density compartment.

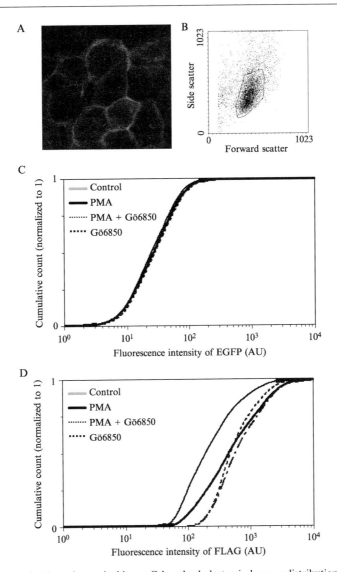

FIG. 5. Activation of protein kinase C by phorbolester induces redistribution of Ntcp. FLAG-Ntcp-EGFP expressing HepG2 cells (A) were treated with the phorbolester PMA (100 nmol/liter, 30 min) and/or the PKC-inhibitor Gö6850 (1 μmol/liter). After treatment, cells were trypsinized, stained with a fluorophore-labeled anti-FLAG antibody, and analyzed by flow cytometry. Single cells gave rise to a distinct population (B), which was analyzed further: The total amount of Ntcp, as measured by green fluorescence of the Ntcp-bound EGFP (C), was not altered by the treatment. In contrast, membrane-bound Ntcp, as detected the FLAG antibody (D) decreased in PMA-treated cells (black line) compared to control cells (gray line). Pretreatment with the PKC inhibitor Gö6850 prevented the decrease of membrane-bound Ntcp (dotted line). For easier interpretation, frequencies of fluorescence intensities are shown in cumulative histograms. AU, arbitrary units.

several thousand single cells within a few minutes by flow cytometry. Figure 5 gives an example in which treatment with the protein kinase C activator PMA induces within minutes a retrieval of Ntcp from the plasma membrane: while the total amount of Ntcp, as measured by fluorescence of the green fluorescent protein was unaltered (Fig. 5C), membrane-bound Ntcp decreased. The FLAG tag of intracellular Ntcp was not accessible for the anti-FLAG antibody; therefore, less FLAG-bound fluorescence was detectable in PMA-treated cells (Fig. 5D), resulting in a shift toward the left of the cumulative histogram.

Concluding Remarks

Several independent techniques can be used in order to demonstrate transporter insertion or retrieval in hepatocytes; all of them supporting the concept that bile formation is controlled at the short-term timescale by a regulated insertion/retrieval of transporter molecules. Future developments may aim toward the visualization of such transporter movements in the living cell and the disclosure of protein/protein interactions involved in this dynamic process; fluorescence resonance energy transfer techniques may have some potential here. In any case, however, all such studies need to be complemented by functional assessments such as measurement of transporter function in order to obtain information on structural–functional relationships.

Acknowledgments

Our own studies reported herein were supported by Deutsche Forschungsgemeinschaft through Sonderforschungsbereich SFB 575 "Experimental Hepatology" (Düsseldorf). Expert work in the preparation of sucrose gradients from cells by Thomas Kühlkamp is acknowledged.

References

Beuers, U., Bilzer, M., Chittattu, A., Kullak-Ublick, G. A., Keppler, D., Paumgartner, G., and Dombrowski, F. (2001). Tauroursodeoxycholic acid inserts the apical conjugate export pump, Mrp2, into canalicular membranes and stimulates organic anion secretion by protein kinase C-dependent mechanisms in cholestatic rat liver. *Hepatology* **33**, 1206–1216.

Borst, P., Evers, R., Kool, M., and Wijnholds, J. (2000). A family of drug transporters: The multidrug resistance-associated proteins. *J. Natl. Cancer Inst.* **92**, 1295–1302.

Chang, G. (2003). Multidrug resistance ABC transporters. *FEBS Lett.* **555**, 102–105.

Crocenzi, F. A., Mottino, A. D., Cao, J., Veggi, L. M., Pozzi, E. J., Vore, M., Coleman, R., and Roma, M. G. (2003). Estradiol-17beta-D-glucuronide induces endocytic internalization of Bsep in rats. *Am. J. Physiol. Gastrointest Liver Physiol.* **285**, G449–G459.

Dombrowski, F., Kubitz, R., Chittattu, A., Wettstein, M., Saha, N., and Häussinger, D. (2000). Electron-microscopic demonstration of multidrug resistance protein 2 (Mrp2) retrieval from the canalicular membrane in response to hyperosmolarity and lipopolysaccharide. *Biochem. J.* **348**(Pt. 1), 183–188.

Dranoff, J. A., McClure, M., Burgstahler, A. D., Denson, L. A., Crawford, A. R., Crawford, J. M., Karpen, S. J., and Nathanson, M. H. (1999). Short-term regulation of bile acid uptake by microfilament-dependent translocation of rat ntcp to the plasma membrane. *Hepatology* **30**, 223–229.

Gottesman, M. M., and Pastan, I. (1993). Biochemistry of multidrug resistance mediated by the multidrug transporter. *Annu. Rev. Biochem.* **62**, 385–427.

Graf, G. A., Yu, L., Li, W. P., Gerard, R., Tuma, P. L., Cohen, J. C., and Hobbs, H. H. (2003). ABCG5 and ABCG8 are obligate heterodimers for protein trafficking and biliary cholesterol excretion. *J. Biol. Chem.* **278**, 48275–48282.

Green, R. M., Beier, D., and Gollan, J. L. (1996). Regulation of hepatocyte bile salt transporters by endotoxin and inflammatory cytokines in rodents. *Gastroenterology* **111**, 193–198.

Groen, A. K., van der Meer, R., and Westerhoff, H. V. (1982). Control of metabolic fluxes. *In* "Metabolic Compartmentation" (H. Sies, ed.). Academic Press, London.

Grundemann, D., Gorboulev, V., Gambaryan, S., Veyhl, M., and Koepsell, H. (1994). Drug excretion mediated by a new prototype of polyspecific transporter. *Nature* **372**, 549–552.

Grüne, S., Engelking, L. R., and Anwer, M. S. (1993). Role of intracellular calcium and protein kinases in the activation of hepatic Na+/taurocholate cotransport by cyclic AMP. *J. Biol. Chem.* **268**, 17734–17741.

Hagenbuch, B., and Meier, P. J. (1994). Molecular cloning, chromosomal localization, and functional characterization of a human liver Na+/bile acid cotransporter. *J. Clin. Invest.* **93**, 1326–1331.

Hagenbuch, B., and Meier, P. J. (2003). The superfamily of organic anion transporting polypeptides. *Biochim. Biophys. Acta* **1609**, 1–18.

Häussinger, D., Hallbrucker, C., Saha, N., Lang, F., and Gerok, W. (1992). Cell volume and bile acid excretion. *Biochem. J.* **288**(Pt. 2), 681–689.

Häussinger, D., Kubitz, R., Reinehr, R., Bode, J. G., and Schliess, F. (2004). Molecular aspects of medicine: From experimental to clinical hepatology. *Mol. Aspects Med.* **25**, 221–360.

Häussinger, D., Saha, N., Hallbrucker, C., Lang, F., and Gerok, W. (1993). Involvement of microtubules in the swelling-induced stimulation of transcellular taurocholate transport in perfused rat liver. *Biochem. J.* **291**(Pt. 2), 355–360.

Häussinger, D., Schmitt, M., Weiergräber, O., and Kubitz, R. (2000). Short-term regulation of canalicular transport. *Semin. Liver Dis.* **20**, 307–321.

Higuchi, H., and Gores, G. J. (2003). Bile acid regulation of hepatic physiology. IV. Bile acids and death receptors. *Am. J. Physiol. Gastrointest. Liver Physiol.* **284**, G734–G738.

Jedlitschky, G., Burchell, B., and Keppler, D. (2000). The multidrug resistance protein 5 functions as an ATP-dependent export pump for cyclic nucleotides. *J. Biol. Chem.* **275**, 30069–30074.

Kawamata, Y., Fujii, R., Hosoya, M., Harada, M., Yoshida, H., Miwa, M., Fukusumi, S., Habata, Y., Itoh, T., Shintani, Y., Hinuma, S., Fujisawa, Y., and Fujino, M. (2003). A G protein-coupled receptor responsive to bile acids. *J. Biol. Chem.* **278**, 9435–9440.

Keppler, D., and König, J. (2000). Hepatic secretion of conjugated drugs and endogenous substances. *Semin. Liver Dis.* **20**, 265–272.

Kipp, H., and Arias, I. M. (2000). Newly synthesized canalicular ABC transporters are directly targeted from the Golgi to the hepatocyte apical domain in rat liver. *J. Biol. Chem.* **275**, 15917–15925.

Kiuchi, Y., Suzuki, H., Hirohashi, T., Tyson, C. A., and Sugiyama, Y. (1998). cDNA cloning and inducible expression of human multidrug resistance associated protein 3 (MRP3). *FEBS Lett.* **433,** 149–152.

Kool, M., van der, L. M., de Haas, M., Baas, F., and Borst, P. (1999). Expression of human MRP6, a homologue of the multidrug resistance protein gene MRP1, in tissues and cancer cells. *Cancer Res.* **59,** 175–182.

Kubitz, R., D'Urso, D., Keppler, D., and Häussinger, D. (1997). Osmodependent dynamic localization of the multidrug resistance protein 2 in the rat hepatocyte canalicular membrane. *Gastroenterology* **113,** 1438–1442.

Kubitz, R., Keitel, V., and Häussinger, D. (2005). Inborn errors of biliary canalicular transport systems. *Methods Enzymol.* **400**[31] this volume.

Kubitz, R., Saha, N., Kuhlkamp, T., Dutta, S., vom, D. S., Wettstein, M., and Haussinger, D. (2004). Ca^{2+}-dependent protein kinase C isoforms induce cholestasis in rat liver. *J. Biol. Chem.* **279,** 10323–10330.

Kubitz, R., Wettstein, M., Warskulat, U., and Häussinger, D. (1999). Regulation of the multidrug resistance protein 2 in the rat liver by lipopolysaccharide and dexamethasone. *Gastroenterology* **116,** 401–410.

Kullak-Ublick, G., Stieger, B., and Meier, P. J. (2004). Enterohepatic bile salt transporters in normal physiology and liver disease. *Gastroenterology* **126,** 322–342.

Kurz, A. K., Graf, D., Schmitt, M., vom Dahl, S., and Häussinger, D. (2001). Tauroursodesoxycholate-induced choleresis involves p38(MAPK) activation and translocation of the bile salt export pump in rats. *Gastroenterology* **121,** 407–419.

Meier, P. J., and Stieger, B. (2002). Bile salt transporters. *Annu. Rev. Physiol.* **64,** 635–661.

Mottino, A. D., Cao, J., Veggi, L. M., Crocenzi, F., Roma, M. G., and Vore, M. (2002). Altered localization and activity of canalicular Mrp2 in estradiol-17beta-D-glucuronide-induced cholestasis. *Hepatology* **35,** 1409–1419.

Mühlfeld, A., Kubitz, R., Dransfeld, O., Häussinger, D., and Wettstein, M. (2003). Taurine supplementation induces Mrp2 and Bsep expression in rats and prevents endotoxin-induced cholestasis. *Arch. Biochem. Biophys.* **413,** 32–40.

Mukhopadhyay, S., Ananthanarayanan, M., Stieger, B., Meier, P. J., Suchy, F. J., and Anwer, M. S. (1997). cAMP increases liver Na+-taurocholate cotransport by translocating transporter to plasma membranes. *Am. J. Physiol.* **273,** G842–G848.

Mukhopadhyay, S., Ananthanarayanan, M., Stieger, B., Meier, P. J., Suchy, F. J., and Anwer, M. S. (1998a). Sodium taurocholate cotransporting polypeptide is a serine, threonine phosphoprotein and is dephosphorylated by cyclic adenosine monophosphate. *Hepatology* **28,** 1629–1636.

Mukhopadhyay, S., Webster, C. R., and Anwer, M. S. (1998b). Role of protein phosphatases in cyclic AMP-mediated stimulation of hepatic Na+/taurocholate cotransport. *J. Biol. Chem.* **273,** 30039–30045.

Noe, J., Stieger, B., and Meier, P. J. (2002). Functional expression of the canalicular bile salt export pump of human liver. *Gastroenterology* **123,** 1659–1666.

Oude Elferink, R. P., and Groen, A. K. (2002). Genetic defects in hepatobiliary transport. *Biochim. Biophys. Acta* **1586,** 129–145.

Rius, M., Nies, A. T., Hummel-Eisenbeiss, J., Jedlitschky, G., and Keppler, D. (2003). Cotransport of reduced glutathione with bile salts by MRP4 (ABCC4) localized to the basolateral hepatocyte membrane. *Hepatology* **38,** 374–384.

Roelofsen, H., Vos, T. A., Schippers, I. J., Kuipers, F., Koning, H., Moshage, H., Jansen, P. L., and Müller, M. (1997). Increased levels of the multidrug resistance protein in lateral membranes of proliferating hepatocyte-derived cells. *Gastroenterology* **112,** 511–521.

Rost, D., Kartenbeck, J., and Keppler, D. (1999). Changes in the localization of the rat canalicular conjugate export pump Mrp2 in phalloidin-induced cholestasis. *Hepatology* **29**, 814–821.

Schmitt, M., Kubitz, R., Lizun, S., Wettstein, M., and Häussinger, D. (2001). Regulation of the dynamic localization of the rat Bsep gene-encoded bile salt export pump by anisoosmolarity. *Hepatology* **33**, 509–518.

Schmitt, M., Kubitz, R., Wettstein, M., vom Dahl, S., and Häussinger, D. (2000). Retrieval of the mrp2 gene encoded conjugate export pump from the canalicular membrane contributes to cholestasis induced by tert-butyl hydroperoxide and chloro-dinitrobenzene. *Biol. Chem.* **381**, 487–495.

Sekine, T., Cha, S. H., Tsuda, M., Apiwattanakul, N., Nakajima, N., Kanai, Y., and Endou, H. (1998). Identification of multispecific organic anion transporter 2 expressed predominantly in the liver. *FEBS Lett.* **429**, 179–182.

Simon, F. R., Fortune, J., Iwahashi, M., Gartung, C., Wolkoff, A., and Sutherland, E. (1996). Ethinyl estradiol cholestasis involves alterations in expression of liver sinusoidal transporters. *Am. J. Physiol.* **271**, G1043–G1052.

Stieger, B., Hagenbuch, B., Landmann, L., Hoechli, M., Schröder, A., and Meier, P. J. (1994). In situ localisation of the hepatocytic Na$^+$/taurocholate cotransporting polypeptide in rat liver. *Gastroenterology* **107**, 1781–1787.

Stocker, R., Glazer, A. N., and Ames, B. N. (1987a). Antioxidant activity of albumin-bound bilirubin. *Proc. Natl. Acad. Sci. USA* **84**, 5918–5922.

Stocker, R., Yamamoto, Y., McDonagh, A. F., Glazer, A. N., and Ames, B. N. (1987b). Bilirubin is an antioxidant of possible physiological importance. *Science* **235**, 1043–1046.

Sun, W., Wu, R. R., van Poelje, P. D., and Erion, M. D. (2001). Isolation of a family of organic anion transporters from human liver and kidney. *Biochem. Biophys. Res. Commun.* **283**, 417–422.

Trauner, M., Meier, P. J., and Boyer, J. L. (1998). Molecular pathogenesis of cholestasis. *N. Engl. J Med.* **339**, 1217–1227.

Tsukada, N., Ackerley, C. A., and Phillips, M. J. (1995). The structure and organization of the bile canalicular cytoskeleton with special reference to actin and actin-binding proteins. *Hepatology* **21**, 1106–1113.

Webster, C. R., Blanch, C. J., Phillips, J., and Anwer, M. S. (2000). Cell swelling-induced translocation of rat liver Na(+)/taurocholate cotransport polypeptide is mediated via the phosphoinositide 3-kinase signaling pathway. *J. Biol. Chem.* **275**, 29754–29760.

Xia, X., Roundtree, M., Merikhi, A., Lu, X., Shentu, S., and Le Sage, G. (2004). Degradation of the apical sodium-dependent bile acid transporter (ASBT) by the ubiquitin-proteasome pathway in cholangiocytes. *J. Biol. Chem.* **279**(43), 44931–44937.

Yu, L., Hammer, R. E., Li-Hawkins, J., Von Bergmann, K., Lutjohann, D., Cohen, J. C., and Hobbs, H. H. (2002). Disruption of Abcg5 and Abcg8 in mice reveals their crucial role in biliary cholesterol secretion. *Proc. Natl. Acad. Sci. USA* **99**, 16237–16242.

Zhang, L., Dresser, M. J., Gray, A. T., Yost, S. C., Terashita, S., and Giacomini, K. M. (1997). Cloning and functional expression of a human liver organic cation transporter. *Mol. Pharmacol.* **51**, 913–921.

[31] Inborn Errors of Biliary Canalicular Transport Systems

By Ralf Kubitz, Verena Keitel, and Dieter Häussinger

Abstract

Cholestatic syndromes are inborn or acquired disorders of bile formation. In recent years, several inherited cholestatic syndromes were characterized at the molecular level: progressive familial intrahepatic cholestasis (PFIC) and benign recurrent intrahepatic cholestasis (BRIC). Both PFIC and BRIC were divided phenotypically in distinct subtypes; however, at the genotype level, these clinical entities overlap. PFIC starts in early childhood and progresses toward liver cirrhosis, which often requires liver transplantation within the first decade of life. The diagnosis of PFIC is usually made on the basis of clinical and laboratory findings but needs to be confirmed by genetic and histological analysis. Only recently was it recognized that BRIC, which was estimated as a milder form of PFIC-1, may be caused by more than one gene.

Disorders of Hepatobiliary Transport: Cholestatic Syndromes

Cholestatic disorders are characterized by deficient bile formation, which resides in either a disturbance of hepatocellular secretion of cholephilic compounds (intrahepatic or hepatocellular cholestasis) or a mechanical obstruction of bile ducts (obstructive cholestasis). Cholestasis most often is acquired during life but some forms of cholestasis are inherited. The identification of hepatobiliary transporter proteins provided the basis for the molecular understanding of inborn hepatocellular cholestatic syndromes (Fig. 1).

Progressive Familial Intrahepatic Cholestasis

Clinically, three types of progressive familial intrahepatic cholestasis are distinguished and their underlying molecular defects identified in recent years. In 1969, Clayton and colleagues described a cholestatic syndrome affecting several siblings of an Amish family, which was named Byler's disease according to the family's name. This inherited form of cholestasis was later termed progressive familial intrahepatic cholestasis (PFIC). The underlying gene defect of Byler's disease, or PFIC type 1

METHODS IN ENZYMOLOGY, VOL. 400
0076-6879/05 $35.00
DOI: 10.1016/S0076-6879(05)00031-5

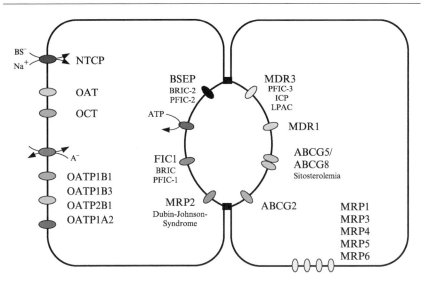

FIG. 1. Expression of hepatobiliary transporters in human liver parenchymal cells. Hepatobiliary transporters are expressed differentially at the sinusoidal (basolateral) or canalicular (apical) membrane. Mutations of certain transporters are associated with distinct hereditary diseases (common names beneath transporter abbreviations). NTCP, Na$^+$-taurocholate cotransporting protein; OAT, organic anion transporter; OCT, organic cation transporter; OATP, organic anion transporting polypeptide; BSEP, bile salt export pump; FIC, familial intrahepatic cholestasis; BRIC, benign recurrent intrahepatic cholestasis; PFIC, progressive familial intrahepatic cholestasis; MRP, multidrug resistance associated protein; MDR, multidrug resistance; ICP, intrahepatic cholestasis of pregnancy; LPAC, low phospholipid associated cholelithiasis; ABCG, ABC-transporter subfamily G.

(PFIC-1), involves a defect of the *familial intrahepatic cholestasis1* gene (FIC1/ATP8B1). FIC1, a P-type ATPase, is expressed in liver parenchymal cells and more strongly in enterocytes and may play a role in the entero-hepatic circulation of bile acids (Bull *et al.*, 1998). It probably functions as an aminophospholipid translocase (Ujhazy *et al.*, 2001). PFIC-1 is associated with elevated bile salt concentrations in serum, normal γ-GT serum activities, decreased bile acid secretion, pruritus, jaundice, and diarrhea. Impaired bile secretion causes fat malabsorption and a deficiency of lipophilic vitamins. Therapeutic options for affected patients include partial external biliary diversion (Kurbegov *et al.*, 2003), administration of rifampicin (van Ooteghem *et al.*, 2002), which only seems to be effective in some PFIC-1 patients, and the HMG-CoA reductase inhibitor simvastatin, which is applied in order to reduce bile acid synthesis from its precursor cholesterol (van Mil *et al.*, 2001). FIC1 mutations with milder functional defects are the molecular basis for "benign recurrent intrahepatic cholestasis" (BRIC), which presents with recurrent cholestatic/icteric attacks but without progression toward liver cirrhosis (Bull et al., 1997, 1998). Greenland familial cholestasis (GFC) is also caused by mutations of FIC1, but represents a more severe form of FIC1 disease (Klomp et al., 2000). The reason for elevated bile acids during attacks of BRIC or in PFIC-1 and GFC is not fully elucidated yet. However, a decrease in expression of the farnesoid-X-receptor (FXR) (Makishima et al., 1999; Parks et al., 1999), the principal regulator of bile acid homeostasis, which counteracts elevated bile acid levels, has been demonstrated in PFIC-1 patients (Alvarez et al., 2004; Chen et al., 2004). It has been shown that decreased FXR levels lead to increased bile acid reabsorption in the small intestine and reduced bile acid secretion into bile by inverse regulation of the apical sodium-dependent bile acid transporter (ASBT) in enterocytes and of the bile salt export pump (BSEP) in hepatocytes. An increase in hydrophobic, that is, more toxic bile acid, is associated with hepatocellular injury at least in part due to bile acid-dependent apoptosis (Higuchi and Gores, 2003).

There are at least two further phenotypes of PFIC. The defect in PFIC-2 was mapped to the gene of the bile salt export pump BSEP (SPGP, ABCB11) (Jansen et al., 1999; Strautnieks et al., 1998; Thompson and Strautnieks, 2001) located to chromosome 2q24-31 (Strautnieks et al., 1997). BSEP is the major canalicular bile salt transporter that transports conjugated bile salts from the hepatocytes into the bile (Gerloff et al., 1998). It has been documented that some patients with recurrent intrahepatic cholestasis of the BRIC phenotype do not have a FIC1 mutation but also carry a BSEP mutation (van Mil et al., 2004), which was consequently named BRIC-2. Interestingly, almost two-thirds of the patients investigated in that study suffered from gallstones (van Mil et al., 2004), which is unusual for "BRIC-1" patients with a FIC1 mutation.

Because bile acid excretion is diminished in PFIC-2 (and PFIC-1), bile canaliculi and cholangiocytes experience less contact with toxic bile constituents and are therefore less prone to damage, which may explain the normal serum γ-GT activities in both PFIC-1 and PFIC-2. Indeed, elevations of serum γ-GT activities are so far considered a hallmark of PFIC-3, but not of PFIC-1 and -2. PFIC-3 is due to a defect of the MDR3 gene product (Deleuze et al., 1996; De Vree et al., 1998; Jacquemin, 2001), which normally brings about phospholipid secretion into bile. Phospholipids are required in bile in order to protect the canaliculi and cholangiocytes from the toxic effects of high bile salt concentrations, which are achieved by mixed micelle formation (Oude Elferink et al., 1997). Liver damage starts later in PFIC-3 patients, but is more pronounced. Due to the damage of cholangiocytes, strong bile duct proliferation can be detected histopathologically. Together with jaundice and pruritus, PFIC-3 patients develop hepatosplenomegaly and portal hypertension. An important therapy for PFIC-3 is the administration of ursodeoxycholic acid (UDCA), but it has been recognized that only patients with some remaining MDR3 activity benefit from UDCA (Jacquemin *et al.*, 1997; Jansen *et al.*, 1999). The protective effect of UDCA is explained by its inhibitory effect on hepatocyte apoptosis, which is otherwise triggered by an intrahepatocellular accumulation of hydrophobic bile acids (Faubion and Gores, 1999; Graf *et al.*, 2002; Higuchi and Gores, 2003). All forms of PFIC are inherited in an autosomal recessive way.

Other Clinical Manifestations of MDR3 Mutations

Mutations of MDR3 not only cause PFIC-3, but are also found in about half of the cases of intrahepatic cholestasis of pregnancy (ICP) (Jacquemin, 2001; Jacquemin *et al.*, 1999). ICP is a reversible form of cholestasis that develops in late pregnancy, characterized by jaundice, elevated serum bile acids, and pruritus, while other causes of cholestasis are absent (such as obstruction of the biliary tree). Even though ICP resolves after delivery and is not detrimental to the mother, it is associated with an increased incidence of fetal complications (Lammert *et al.*, 2000).

Some mutations of MDR3 predispose for a rare condition of gallstone susceptibility called "low phospholipid associated cholelithiasis" (LPAC) syndrome (Rosmorduc *et al.*, 2001). LPAC characteristically includes two or three of the following features: (1) it occurs in patients below an age of 40 years, (2) the patients experience recurrence of gallstone disease after cholecystectomy, and (3) intrahepatic sludge or stones are found on abdominal ultrasonography (Rosmorduc *et al.*, 2003). Furthermore, a positive family history of gallstone disease is frequently observed.

Dubin–Johnson Syndrome

The Dubin–Johnson syndrome is an inherited form of conjugated hyperbilirubinemia (Dubin and Johnson, 1954). It is caused by mutations of canalicular multidrug resistance associated protein 2 (MRP2/cMRP/cMOAT/ABCC2) (Kartenbeck *et al.*, 1996; Keitel *et al.*, 2000, 2002; Shoda *et al.*, 2003; Suzuki and Sugiyama, 2002). Livers of Dubin–Johnson syndrome patients contain a black pigment within the lysosomes. Affected patients present with chronic jaundice, but have a normal life expectancy and the Dubin–Johnson syndrome has a benign clinical course. The minor impairment of overall liver function may be due to the compensatory upregulation of MRP3 (ABCC3) at the lateral membrane of liver cells of

Dubin–Johnson syndrome patients (König *et al.*, 1999), which may mediate reflux of conjugated bilirubin (and other MRP2 substrates) into the blood (Oude Elferink and Groen, 2002).

Sitosterolemia

Plant sterols are almost completely excluded from the body due to efficient enteric excretion. In the rare inherited disease called sitosterolemia, mutations of either one of the half-transporters ABCG5 (Lee *et al.*, 2001) or ABCG8 (Berge *et al.*, 2000) lead to accumulation of the plant sterol sitosterol (24-ethyl-cholesterol) due to the impaired secretion of sitosterol (and cholesterol) from the enterocytes back into the gut lumen and from the liver into the bile (Graf *et al.*, 2003). Clinically, sitosterolemia is characterized by atherosclerosis at a young age, tendon xanthomas, hemolytic episodes, and arthralgias (Salen *et al.*, 1992). Mutations of ABCG5/ABCG8, which represent the major hepatic transporter for cholesterol secretion, also affect cholesterol secretion: however, an increase of cholesterol is partly antagonized by a reduction of HMG-Co reductase activity (Salen *et al.*, 1992).

Genetic Abnormalities in Sinusoidal Transporter Proteins

So far no distinct inherited diseases are attributed to defects of sinusoidal transport systems in humans. However, some nonsynonymous single nucleotide polymorphisms (SNP), affecting single amino acids in transporter proteins from this membrane, have been identified. For example, Ho and co-workers (2004) identified five SNPs in the human sodium-dependent taurocholate cotransporting polypeptide (NTCP), the major uptake system for bile acids into hepatocytes. One of these SNPs was associated with a defect in maturation and/or targeting of NTCP (Ho *et al.*, 2004). In addition to NTCP, members of the organic anion transporting polypeptide (OATP) superfamily (Hagenbuch and Meier, 2003) mediate bile acid uptake in addition to many other organic anions. For OATP1B1 (OATP-C/OATP2/LST-1), SNPs were correlated with alterations in the transport of estrogen derivatives (Tirona *et al.*, 2001). Another SNP of OATP1B1 was identified, which was associated with a defect in maturation and intracellular retention of OATP1B1 (Michalski *et al.*, 2002). Similarly, several nonsynonymous mutations were identified within OATP1B3 (OATP8), resulting in alterations in targeting and/or transport of OATP1B3 (Letschert *et al.*, 2004). For human MRP3, a member of the ABCC subfamily of ABC transporters, a frequently nonsynonymous SNP, was identified close to the second nucleotide-binding domain, but without functional consequences (Lee *et al.*, 2004). Interestingly, another study reported about polymorphisms of the promoter region of human MRP3, which were associated with decreased MRP3 mRNA expression (Lang *et al.*, 2004). To what extent such SNPs predispose to diseases is currently unknown.

Diagnostic Procedures for Progressive Familial Intrahepatic Cholestasis

The diagnosis of PFIC can be suspected on the basis of clinical and laboratory findings (see earlier discussion). PFIC-1 and PFIC-2 may be clinically discriminated by the manifestation of extrahepatic symptoms, such as diarrhea or pancreatic insufficiency, which occur in PFIC-1 but not in PFIC-2. PFIC-3 is differentiated more easily from PFIC-1/2 due to high γGT activities; however, some patients are borderline in terms of their γGT activities, that is, PFIC-2 patients can exhibit slightly elevated γGT activities and vice versa, whereas some PFIC-3 patients have normal γGT values (Keitel *et al.*, 2005). Therefore, accurate diagnosis must rely on criteria distinct from the clinical phenotype, as prognosis and treatment differ between PFIC subtypes.

One approach to diagnose PFIC subtypes includes identification of mutations at the genome level. Mutations may be detected by cDNA sequencing when (liver) tissue is available for reverse transcription of the mRNA of the respective gene. Normally, a biopsy cylinder of 1 cm length derived from the standard liver puncture procedure yields enough material for cDNA sequencing. If mRNA is not available, mutation analysis can be performed using genomic DNA. The polymerase chain reaction (PCR)-based technique of single strand conformation polymorphism allows searching for single nucleotide exchanges within short stretches of DNA in the range of 150 to 250 bp. In order to identify new mutations of a gene, multiple PCR fragments of this size are produced from genomic DNA by the use of specific primers covering the entire coding regions and the adjacent exon/intron boundaries. The PCR fragments are separated by agarose gel electrophoresis at different temperatures. Single nucleotide exchanges within the PCR product result in differences of electrophoretic mobility compared to fragments of the wild-type gene and give rise to different bands in the agarose gels. These fragments can then be analyzed by sequencing. Direct sequencing of the entire gene is more time-consuming and expensive.

When more disease-causing mutations (DCM) are identified, a more rapid screen for such DCMs in individual patients will be possible using a plate reader or high-throughput technologies. Techniques that can be applied include hybridization-based methods (such as TaqMan technology or melting curve analysis) or pyro-sequencing.

Depending on the type of polymorphism/mutation detected, the relevance of the polymorphism/mutation for the disease has to be established. While homozygous deletion or frameshift mutations are correlated easily, it is more difficult with nonsense mutations. Here, direct proof of the

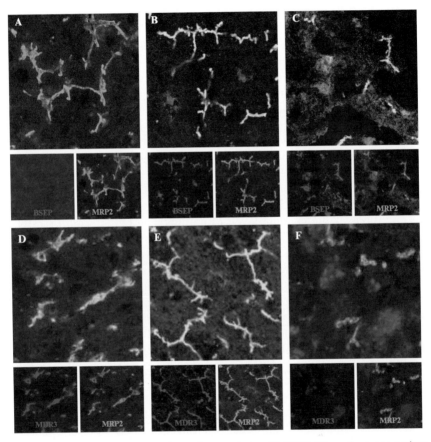

FIG. 2. Immunofluorescence of livers from patients with PFIC. Livers from one patient with PFIC-2 (A, D) and two patients with PFIC-3 (B and E; C and F) were examined by immunofluorescence and confocal laser-scanning microscopy. In the upper row, the bile salt export pump BSEP (red), which is mutated in PFIC-2, was counterstained by the multidrug resistance protein 2 (MRP2, green). In PFIC-2 patients, no canalicular BSEP staining was detectable. In contrast, some patients have MDR3 mutations (B, E; middle), which display a normal canalicular MDR3-staining (MDR3 in red; MRP2 in green). Other MDR3 mutations (such as patient of C and F) have absent canalicular MDR3 (right). (See color insert.)

functional consequence, for example, by insertion of the mutation into a cell system, is often necessary. Possible effects of a mutation are decreased mRNA stability due to nonsense-mediated decay (Thermann *et al.*, 1998) and defects in protein maturation and targeting (Dixon *et al.*, 2000; Keitel *et al.*, 2000, 2002; Plass *et al.*, 2004; Wang *et al.*, 2002). Other mutations may only affect transporter activity (Hashimoto *et al.*, 2002; Keitel *et al.*, 2002; Mor-Cohen *et al.*, 2001), but some may also be irrelevant.

cDNA and gene sequencing studies need to be complemented by either functional assays or immunohistochemistry in order to assess transporter localization and protein expression levels.

For example, PFIC-2 can be diagnosed by the absence of BSEP immunoreactivity from the canaliculi (Fig. 2) when BSEP-specific antibodies are used. Thus, BSEP staining in livers of PFIC-2 patients might be a reliable tool for diagnosis (Thompson and Strautnieks, 2001), as shown in a study on five PFIC-2 patients (Keitel *et al.*, 2005). Liver immunofluorescence studies may especially be important in cases with borderline γGT activities in order to differentiate PFIC-2 from PFIC-3. While all BSEP mutations identified so far were associated with a clear targeting defect in human liver, several MDR3 mutations were identified that display normal apical protein targeting (see examples in Fig. 2) (Jacquemin, 2001; Jacquemin *et al.*, 2001). So far, these mutations have not been assessed for their functional impact.

Perspectives

During the last decades many transporter proteins were cloned and were characterized at the molecular level. With the help of current technologies, many transporter polymorphisms will be discovered—polymorphisms that account for monogenetic diseases such as PFIC, but also polymorphisms that predispose for or protect from the development of illnesses. Because hepatobiliary transporters are major determinants of pharmocokinetics, individual drug tolerance and drug response may find their explanation in transporter SNPs. However, genetic as well as morphological and functional studies will be necessary to answer upcoming questions.

References

Alvarez, L., Jara, P., Sanchez-Sabate, E., Hierro, L., Larrauri, J., Diaz, M. C., Camarena, C., De, L. V., Frauca, E., Lopez-Collazo, E., and Lapunzina, P. (2004). Reduced hepatic expression of Farnesoid X Receptor in hereditary cholestasis associated to mutation in ATP8B1. *Hum. Mol. Genet.* **13**(20), 2451–2460.

Berge, K. E., Tian, H., Graf, G. A., Yu, L., Grishin, N. V., Schultz, J., Kwiterovich, P., Shan, B., Barnes, R., and Hobbs, H. H. (2000). Accumulation of dietary cholesterol in sitosterolemia caused by mutations in adjacent ABC transporters. *Science* **290,** 1771–1775.

Bull, L. N., Carlton, V. E., Stricker, N. L., Baharloo, S., DeYoung, J. A., Freimer, N. B., Magid, M. S., Kahn, E., Markowitz, J., DiCarlo, F. J., McLoughlin, L., Boyle, J. T., Dahms, B. B., Faught, P. R., Fitzgerald, J. F., Piccoli, D. A., Witzleben, C. L., O'Connell, N. C., Setchell, K. D., Agostini, R. M., Jr., Kocoshis, S. A., Reyes, J., and Knisely, A. S. (1997). Genetic and morphological findings in progressive familial intrahepatic cholestasis (Byler disease [PFIC-1] and Byler syndrome): Evidence for heterogeneity. *Hepatology* **26,** 155–164.

Bull, L. N., van Eijk, M. J., Pawlikowska, L., DeYoung, J. A., Juijn, J. A., Liao, M., Klomp, L. W., Lomri, N., Berger, R., Scharschmidt, B. F., Knisely, A. S., Houwen, R. H., and Freimer, N. B. (1998). A gene encoding a P-type ATPase mutated in two forms of hereditary cholestasis. *Nature Genet.* **18,** 219–224.

Chen, F., Ananthanarayanan, M., Emre, S., Neimark, E., Bull, L. N., Knisely, A. S., Strautnieks, S. S., Thompson, R. J., Magid, M. S., Gordon, R., Balasubramanian, N., Suchy, F. J., and Shneider, B. L. (2004). Progressive familial intrahepatic cholestasis, type 1, is associated with decreased farnesoid X receptor activity. *Gastroenterology* **126,** 756–764.

Clayton, R. J., Iber, F. L., Ruebner, B. H., and McKusick, V. A. (1969). Byler disease: Fatal familial intrahepatic cholestasis in an Amish kindred. *Am. J. Dis. Child.* **117,** 112–124.

Deleuze, J. F., Jacquemin, E., Dubuisson, C., Cresteil, D., Dumont, M., Erlinger, S., Bernard, O., and Hadchouel, M. (1996). Defect of multidrug-resistance 3 gene expression in a subtype of progressive familial intrahepatic cholestasis. *Hepatology* **23,** 904–908.

De Vree, J. M., Jacquemin, E., Sturm, E., Cresteil, D., Bosma, P. J., Aten, J., Deleuze, J. F., Desrochers, M., Burdelski, M., Bernard, O., Oude Elferink, R. P., and Hadchouel, M. (1998). Mutations in the MDR3 gene cause progressive familial intrahepatic cholestasis. *Proc. Natl. Acad. Sci. USA* **95,** 282–287.

Dixon, P. H., Weerasekera, N., Linton, K. J., Donaldson, O., Chambers, J., Egginton, E., Weaver, J., Nelson-Piercy, C., de Swiet, M., Warnes, G., Elias, E., Higgins, C. F., Johnston, D. G., McCarthy, M. I., and Williamson, C. (2000). Heterozygous MDR3 missense mutation associated with intrahepatic cholestasis of pregnancy: Evidence for a defect in protein trafficking. *Hum. Mol. Genet.* **9,** 1209–1217.

Dubin, I. N., and Johnson, F. B. (1954). Chronic idiopathic jaundice with unidentified pigment in liver cells; a new clinicopathologic entity with a report of 12 cases. *Medicine (Baltimore)* **33,** 155–197.

Faubion, W. A., and Gores, G. J. (1999). Death receptors in liver biology and pathobiology. *Hepatology* **29,** 1–4.

Gerloff, T., Stieger, B., Hagenbuch, B., Madon, J., Landmann, L., Roth, J., Hofmann, A. F., and Meier, P. J. (1998). The sister of P-glycoprotein represents the canalicular bile salt export pump of mammalian liver. *J. Biol. Chem.* **273,** 10046–10050.

Graf, D., Kurz, A. K., Fischer, R., Reinehr, R., and Häussinger, D. (2002). Taurolithocholic acid-3 sulfate induces CD95 trafficking and apoptosis in a c-Jun N-terminal kinase-dependent manner. *Gastroenterology* **122,** 1411–1427.

Graf, G. A., Yu, L., Li, W. P., Gerard, R., Tuma, P. L., Cohen, J. C., and Hobbs, H. H. (2003). ABCG5 and ABCG8 are obligate heterodimers for protein trafficking and biliary cholesterol excretion. *J. Biol. Chem.* **278,** 48275–48282.

Hagenbuch, B., and Meier, P. J. (2003). The superfamily of organic anion transporting polypeptides. *Biochim. Biophys. Acta* **1609,** 1–18.

Hashimoto, K., Uchiumi, T., Konno, T., Ebihara, T., Nakamura, T., Wada, M., Sakisaka, S., Maniwa, F., Amachi, T., Ueda, K., and Kuwano, M. (2002). Trafficking and functional defects by mutations of the ATP-binding domains in MRP2 in patients with Dubin-Johnson syndrome. *Hepatology* **36**, 1236–1245.

Higuchi, H., and Gores, G. J. (2003). Bile acid regulation of hepatic physiology. IV. Bile acids and death receptors. *Am. J. Physiol. Gastrointest. Liver Physiol.* **284**, G734–G738.

Ho, R. H., Leake, B. F., Roberts, R. L., Lee, W., and Kim, R. B. (2004). Ethnicity-dependent polymorphism in Na+-taurocholate cotransporting polypeptide (SLC10A1) reveals a domain critical for bile acid substrate recognition. *J. Biol. Chem.* **279**(8), 7213–7222.

Jacquemin, E. (2001). Role of multidrug resistance 3 deficiency in pediatric and adult liver disease: One gene for three diseases. *Semin. Liver Dis.* **21**, 551–562.

Jacquemin, E., Cresteil, D., Manouvrier, S., Boute, O., and Hadchouel, M. (1999). Heterozygous non-sense mutation of the MDR3 gene in familial intrahepatic cholestasis of pregnancy. *Lancet* **353**, 210–211.

Jacquemin, E., De Vree, J. M., Cresteil, D., Sokal, E. M., Sturm, E., Dumont, M., Scheffer, G. L., Paul, M., Burdelski, M., Bosma, P. J., Bernard, O., Hadchouel, M., and Elferink, R. P. (2001). The wide spectrum of multidrug resistance 3 deficiency: From neonatal cholestasis to cirrhosis of adulthood. *Gastroenterology* **120**, 1448–1458.

Jacquemin, E., Hermans, D., Myara, A., Habes, D., Debray, D., Hadchouel, M., Sokal, E. M., and Bernard, O. (1997). Ursodeoxycholic acid therapy in pediatric patients with progressive familial intrahepatic cholestasis. *Hepatology* **25**, 519–523.

Jansen, P. L., Strautnieks, S. S., Jacquemin, E., Hadchouel, M., Sokal, E. M., Hooiveld, G. J., Koning, J. H., Jager-Krikken, A., Kuipers, F., Stellaard, F., Bijleveld, C. M., Gouw, A., Van Goor, H., Thompson, R. J., and Muller, M. (1999). Hepatocanicular bile salt export pump deficiency in patients with progressive familial intrahepatic cholestasis. *Gastroenterology* **117**, 1370–1379.

Kartenbeck, J., Leuschner, U., Mayer, R., and Keppler, D. (1996). Absence of the canalicular isoform of the MRP gene-encoded conjugate export pump from the hepatocytes in Dubin-Johnson syndrome. *Hepatology* **23**, 1061–1066.

Keitel, V., Burdelski, M., Warskulat, U., Kühlkamp, T., Keppler, D., Häussinger, D., and Kubitz, R. (2005). Expression and localization of hepatobiliary transport proteins in progressive familial intrahepatic cholestasis. *Hepatology* **41**, 1160–1172.

Keitel, V., Kartenbeck, J., Nies, A. T., Spring, H., Brom, M., and Keppler, D. (2000). Impaired protein maturation of the conjugate export pump multidrug resistance protein 2 as a consequence of a deletion mutation in Dubin-Johnson syndrome. *Hepatology* **32**, 1317–1328.

Keitel, V., Nies, A. T., Brom, M., Hummel-Eisenbeiss, J., Spring, H., and Keppler, D. (2002). A common Dubin-Johnson syndrome mutation impairs protein maturation and transport activity of MRP2 (ABCC2). *Am. J. Physiol. Gastrointest. Liver Physiol.* **284**, G165–G174.

Klomp, L. W., Bull, L. N., Knisely, A. S., Der Doelen, M. A., Juijn, J. A., Berger, R., Forget, S., Nielsen, I. M., Eiberg, H., and Houwen, R. H. (2000). A missense mutation in FIC1 is associated with Greenland familial cholestasis. *Hepatology* **32**, 1337–1341.

König, J., Rost, D., Cui, Y., and Keppler, D. (1999). Characterization of the human multidrug resistance protein isoform MRP3 localized to the basolateral hepatocyte membrane. *Hepatology* **29**, 1156–1163.

Kurbegov, A. C., Setchell, K. D., Haas, J. E., Mierau, G. W., Narkewicz, M., Bancroft, J. D., Karrer, F., and Sokol, R. J. (2003). Biliary diversion for progressive familial intrahepatic cholestasis: Improved liver morphology and bile acid profile. *Gastroenterology* **125**, 1227–1234.

Lammert, F., Marschall, H. U., Glantz, A., and Matern, S. (2000). Intrahepatic cholestasis of pregnancy: Molecular pathogenesis, diagnosis and management. *J. Hepatol.* **33**, 1012–1021.

Lang, T., Hitzl, M., Burk, O., Mornhinweg, E., Keil, A., Kerb, R., Klein, K., Zanger, U. M., Eichelbaum, M., and Fromm, M. F. (2004). Genetic polymorphisms in the multidrug resistance-associated protein 3 (ABCC3, MRP3) gene and relationship to its mRNA and protein expression in human liver. *Pharmacogenetics* **14**, 155–164.

Lee, M. H., Lu, K., Hazard, S., Yu, H., Shulenin, S., Hidaka, H., Kojima, H., Allikmets, R., Sakuma, N., Pegoraro, R., Srivastava, A. K., Salen, G., Dean, M., and Patel, S. B. (2001). Identification of a gene, ABCG5, important in the regulation of dietary cholesterol absorption. *Nature Genet.* **27**, 79–83.

Lee, Y. M., Cui, Y., Konig, J., Risch, A., Jager, B., Drings, P., Bartsch, H., Keppler, D., and Nies, A. T. (2004). Identification and functional characterization of the natural variant MRP3-Arg1297His of human multidrug resistance protein 3 (MRP3/ABCC3). *Pharmacogenetics* **14**, 213–223.

Letschert, K., Keppler, D., and Konig, J. (2004). Mutations in the SLCO1B3 gene affecting the substrate specificity of the hepatocellular uptake transporter OATP1B3 (OATP8). *Pharmacogenetics* **14**, 441–452.

Makishima, M., Okamoto, A. Y., Repa, J. J., Tu, H., Learned, R. M., Luk, A., Hull, M. V., Lustig, K. D., Mangelsdorf, D. J., and Shan, B. (1999). Identification of a nuclear receptor for bile acids. *Science* **284**, 1362–1365.

Michalski, C., Cui, Y., Nies, A. T., Nuessler, A. K., Neuhaus, P., Zanger, U. M., Klein, K., Eichelbaum, M., Keppler, D., and Konig, J. (2002). A naturally occurring mutation in the SLC21A6 gene causing impaired membrane localization of the hepatocyte uptake transporter. *J. Biol. Chem.* **277**, 43058–43063.

Mor-Cohen, R., Zivelin, A., Rosenberg, N., Shani, M., Muallem, S., and Seligsohn, U. (2001). Identification and functional analysis of two novel mutations in the multidrug resistance protein 2 gene in Israeli patients with Dubin-Johnson syndrome. *J. Biol. Chem.* **276**, 36923–36930.

Oude Elferink, R. P., and Groen, A. K. (2002). Genetic defects in hepatobiliary transport. *Biochim. Biophys. Acta* **1586**, 129–145.

Oude Elferink, R. P., Tytgat, G. N., and Groen, A. K. (1997). Hepatic canalicular membrane 1: The role of mdr2 P-glycoprotein in hepatobiliary lipid transport. *FASEB J.* **11**, 19–28.

Parks, D. J., Blanchard, S. G., Bledsoe, R. K., Chandra, G., Consler, T. G., Kliewer, S. A., Stimmel, J. B., Willson, T. M., Zavacki, A. M., Moore, D. D., and Lehmann, J. M. (1999). Bile acids: Natural ligands for an orphan nuclear receptor. *Science* **284**, 1365–1368.

Plass, J. R., Mol, O., Heegsma, J., Geuken, M., de Bruin, J., Elling, G., Muller, M., Faber, K. N., and Jansen, P. L. (2004). A progressive familial intrahepatic cholestasis type 2 mutation causes an unstable, temperature-sensitive bile salt export pump. *J. Hepatol.* **40**, 24–30.

Rosmorduc, O., Hermelin, B., Boelle, P. Y., Parc, R., Taboury, J., and Poupon, R. (2003). ABCB4 gene mutation-associated cholelithiasis in adults. *Gastroenterology* **125**, 452–459.

Rosmorduc, O., Hermelin, B., and Poupon, R. (2001). MDR3 gene defect in adults with symptomatic intrahepatic and gallbladder cholesterol cholelithiasis. *Gastroenterology* **120**, 1459–1467.

Salen, G., Shefer, S., Nguyen, L., Ness, G. C., Tint, G. S., and Shore, V. (1992). Sitosterolemia. *J. Lipid Res.* **33**, 945–955.

Shoda, J., Suzuki, H., Suzuki, H., Sugiyama, Y., Hirouchi, M., Utsunomiya, H., Oda, K., Kawamoto, T., Matsuzaki, Y., and Tanaka, N. (2003). Novel mutations identified in the human multidrug resistance-associated protein 2 (MRP2/ABCC2) gene in a Japanese patient with Dubin-Johnson syndrome. *Hepatol. Res.* **27**, 323–326.

Strautnieks, S. S., Bull, L. N., Knisely, A. S., Kocoshis, S. A., Dahl, N., Arnell, H., Sokal, E., Dahan, K., Childs, S., Ling, V., Tanner, M. S., Kagalwalla, A. F., Nemeth, A., Pawlowska, J., Baker, A., Mieli-Vergani, G., Freimer, N. B., Gardiner, R. M., and Thompson, R. J. (1998). A gene encoding a liver-specific ABC transporter is mutated in progressive familial intrahepatic cholestasis. *Nature Genet.* **20**, 233–238.

Strautnieks, S. S., Kagalwalla, A. F., Tanner, M. S., Knisely, A. S., Bull, L., Freimer, N., Kocoshis, S. A., Gardiner, R. M., and Thompson, R. J. (1997). Identification of a locus for progressive familial intrahepatic cholestasis PFIC2 on chromosome 2q24. *Am. J. Hum. Genet.* **61**, 630–633.

Suzuki, H., and Sugiyama, Y. (2002). Single nucleotide polymorphisms in multidrug resistance associated protein 2 (MRP2/ABCC2): Its impact on drug disposition. *Adv. Drug Deliv. Rev.* **54**, 1311–1331.

Thermann, R., Neu-Yilik, G., Deters, A., Frede, U., Wehr, K., Hagemeier, C., Hentze, M. W., and Kulozik, A. E. (1998). Binary specification of nonsense codons by splicing and cytoplasmic translation. *EMBO J.* **17**, 3484–3494.

Thompson, R., and Strautnieks, S. (2001). BSEP: Function and role in progressive familial intrahepatic cholestasis. *Semin. Liver Dis.* **21**, 545–550.

Tirona, R. G., Leake, B. F., Merino, G., and Kim, R. B. (2001). Polymorphisms in OATP-C: Identification of multiple allelic variants associated with altered transport activity among European- and African-Americans. *J. Biol. Chem.* **276**, 35669–35675.

Ujhazy, P., Ortiz, D., Misra, S., Li, S., Moseley, J., Jones, H., and Arias, I. M. (2001). Familial intrahepatic cholestasis 1: Studies of localization and function. *Hepatology* **34**, 768–775.

van Mil, S. W., Klomp, L. W., Bull, L. N., and Houwen, R. H. (2001). FIC1 disease: A spectrum of intrahepatic cholestatic disorders. *Semin. Liver Dis.* **21**, 535–544.

van Mil, S. W., Van Der Woerd, W. L., Van Der, B. G., Sturm, E., Jansen, P. L., Bull, L. N., Van, D. B. I., Berger, R., Houwen, R. H., and Klomp, L. W. (2004). Benign recurrent intrahepatic cholestasis type 2 is caused by mutations in ABCB11. *Gastroenterology* **127**, 379–384.

van Ooteghem, N. A., Klomp, L. W., Berge-Henegouwen, G. P., and Houwen, R. H. (2002). Benign recurrent intrahepatic cholestasis progressing to progressive familial intrahepatic cholestasis: Low GGT cholestasis is a clinical continuum. *J. Hepatol.* **36**, 439–443.

Wang, L., Soroka, C. J., and Boyer, J. L. (2002). The role of bile salt export pump mutations in progressive familial intrahepatic cholestasis type II. *J. Clin. Invest.* **110**, 965–972.

[32] Epoxide Hydrolases: Structure, Function, Mechanism, and Assay

By Michael Arand, Annette Cronin, Magdalena Adamska, and Franz Oesch

Abstract

Epoxide hydrolases are a class of enzymes important in the detoxification of genotoxic compounds, as well as in the control of physiological signaling molecules. This chapter gives an overview on the function, structure, and enzymatic mechanism of structurally characterized epoxide

METHODS IN ENZYMOLOGY, VOL. 400   0076-6879/05 $35.00
               DOI: 10.1016/S0076-6879(05)00032-7

hydrolases and describes selected assays for the quantification of epoxide hydrolase activity.

Introduction

Epoxide hydrolases (EH) are a class of enzymes that cleave oxiran derivatives to yield the corresponding diols. If viewed from the perspective of the classical phase concept of xenobiotic metabolism, EH lead a life at the interface of phase I and phase II (Fig. 1), thus representing a kind of chimera between a functionalizing and a conjugating enzyme: typical hydrolases, such as the esterases, belong to the phase I (functionalization) of xenobiotic metabolism, yet epoxide hydrolysis may as well be regarded as a conjugation with water because water is added to the molecule without splitting it into fragments.

The first EH to be characterized was the membrane-bound mammalian microsomal epoxide hydrolase (Oesch, 1973), which plays a major role in the control of chemically reactive and thus potentially cytotoxic/genotoxic epoxides. These arise as intermediates in the metabolism of numerous xenobiotics, and their metabolic detoxification is therefore of primary

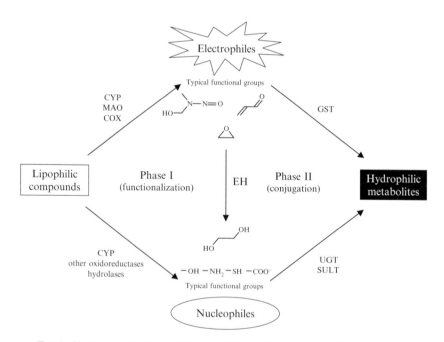

FIG. 1. Phase concept of xenobiotic metabolism: The role of epoxide hydrolase.

importance (Guengerich, 1982). A few years later, a soluble mammalian epoxide hydrolase (sEH) was identified that was able to convert the insect signaling molecule juvenile hormone III, an oxiran derivative, to the corresponding diol (Hammock *et al.*, 1976). In addition, this enzyme turned out to complement the substrate profile of mEH in that it hydrolyzed 1,2-*trans*-substituted epoxides efficiently, a class of compounds that is only poorly, if at all, processed by mEH (Ota and Hammock, 1980). mEH and sEH turned out to be distantly related by structure and phylogeny (Arand *et al.*, 1994; Lacourciere and Armstrong, 1994; Pries *et al.*, 1994), and most other EH identified to date belong to this family of enzymes. Three more families of epoxide hydrolyzing enzymes of separate evolutionary origin have been identified, as will be described in more detail later. Thus, EH has expanded to a large heterogeneous group of enzymes, and these are found in all phylae of life. It appears that all complex organisms that have been analyzed so far possess actually more than one EH gene (M. Arand, unpublished observation), indicating the essential function(s) of this class of enzymes.

Physiological Functions

Detoxification

The first function of EH to be well understood was its role in the detoxification of genotoxic epoxides (Jerina and Daly, 1974). Many epoxides are sufficiently reactive electrophiles that can chemically attack electron-rich structures in nucleic acids, thus leading to the formation of DNA adducts (Kim *et al.*, 1984). Depending on the miscoding potential of those adducts, gene mutations can occur during DNA replication and, if a protooncogene or a tumor suppressor gene was the target, may initiate tumor development.

The chemical reactivity of epoxides is heavily dependent on the substitution pattern at the oxiran ring (Frantz *et al.*, 1985). As a rule of thumb, asymmetric substitution, as well as substitution with electron withdrawing binding partners, enhances the reactivity of epoxides. The spontaneous hydrolysis of epoxides in aqueous solution may be taken as a reasonable indicator of the electrophilic reactivity of a given compound. A highly reactive epoxide, such as the aflatoxin B_1 8,9-*exo*-epoxide, displays a half life of <1 s in aqueous solution, whereas compounds such as 9,10-epoxystearic acid, a *quasi*-symmetric aliphatic oxiran derivative, show virtually no spontaneous hydrolysis under similar conditions.

In the mammalian organism, the mEH has shown to be particularly important for the control of genotoxic epoxides (Guengerich, 1982). The enzyme has broad substrate specificity and displays a surprisingly high apparent affinity to a wide variety of structurally divergent substrates (Arand et al., 2003). While the hydrolysis of these substrates usually results in termination of their chemical reactivity, the resulting metabolites, in few cases, are the precursors of third-generation reactive metabolites of some-times extraordinary genotoxic potential. Famous examples for such a case are the bay region dihydrodiol epoxides of polycyclic aromatic compounds, such as the benzo[a]pyrene-7,8-dihydrodiol-9,10-epoxide (Sims et al., 1974). In this particular context, the net result of mEH metabolism is not beneficial for the organism. This has been well documented in respective trials with mEH knockout mice (Miyata et al., 1999). However, it should be kept in mind that, in general, mEH plays a protective role, as has been demonstrated with the same animal strain in a different experimental setting (Wickliffe et al., 2003). As already mentioned, there is a group of epoxides that are poor or no substrates for mEH; these are trans-substituted oxiran derivatives. If not too bulky, these compounds usually are good substrates for the sEH (Hammock et al., 1997). Furthermore, the glutathione S-transferases described extensively elsewhere in this vol-ume are capable of conjugating many epoxides, thereby cleaving the oxiran ring. However, these enzymes often display a significantly higher K_m with epoxides than mEH, with the latter thus being more efficient in the detoxification of epoxides if these are present at low concentration.

Processing of Signal Molecules

The first substrate described for mammalian sEH was, by chance, the insect juvenile hormone III, definitely a signal molecule (Hammock et al., 1976). Thereafter, much emphasis was put on the fact that sEH has a substrate selectivity that can be described as complementary to the one of mEH, which suggested that detoxification of a particular class of reactive epoxides might be the physiological role of this enzyme (Oesch et al., 1984). In recent years, it has become evident that fatty acid epoxides are the major physiological substrates of sEH and that these and/or their diols obtained after enzymatic hydrolysis serve as signaling molecules in (patho)physio-logical processes. First, a role of sEH in the development of adult res-piratory distress syndrome and multiple organ failure has been established by the identification of leukotoxin diol, a sEH metabolite of the fatty acid epoxide leukotoxin, as the important trigger of those fatal effects (Moghaddam et al., 1997). Second, the rapid hydrolysis of arachidonic acid 14,15-epoxide, a potent endothelial-derived hyperpolarizing factor, affects blood pressure significantly, as evidenced by knockout and inhibitor experiments (Sinal et al., 2000; Yu et al., 2000). Meanwhile, it appears that in particular the metabolism of the different arachidonic acid epoxide regiomers by sEH mediates a plethora of different effects and represents a separate branch of the arachidonic acid-derived signaling pathway, as reviewed in Spector *et al.* (2004).

Exploitation of Energy Sources

About one-fifth of the microorganisms that have been characterized genomically appear to have epoxide hydrolase genes, the physiological function of which is mostly obscure. In some cases, however, it is well defined. In particular, the limonene epoxide hydrolase (LEH) from *Rhodococcus erythropolis* has been identified in the frame of the characterization of an operon of this particular bacterium that allows its growth on limonene as the single carbon source (van der Werf *et al.*, 1998). It has been found that LEH catalyzes an essential step in the pathway of limonene degradation that counteracts accumulation of the 1,2-epoxide of this compound, the first metabolite in this metabolic cascade (van der Werf *et al.*, 1999). Thus, in this case the epoxide hydrolase is integrated into a biochemical pathway that permits exploitation of a nonconventional energy source.

Structure and Enzymatic Mechanism

α/β Hydrolase Fold EHs

The vast majority of known EH belong to the family of α/β hydrolase fold enzymes (Barth *et al.*, 2004). The common three-dimensional structure of these enzymes is composed of the α/β hydrolase fold domain with a lid domain on top and an optional N-terminal domain (Argiriadi *et al.*, 1999; Nardini *et al.*, 1999; Zou *et al.*, 2000). The substrate-binding pocket is situated between the α/β hydrolase fold domain and the lid domain. Three residues of the α/β hydrolase fold domain at the interface to the lid domain constitute a catalytic triad. One of these, invariably an aspartate, acts as a catalytic nucleophile and forms an ester intermediate with the substrate in a first step of the catalytic reaction (Fig. 2), while the other two, a histidine and an acidic acid, build a charge relay system that activates a water molecule through proton abstraction (Arand *et al.*, 1996). This affords hydrolysis of the ester intermediate in the second step of the catalytic cycle. Two more residues essential for catalysis reach from the lid domain into the substrate-binding pocket. These two, always two tyrosines, form hydrogen bonds to the oxygen atom in the epoxide ring of the substrate, thereby increasing its electrophilic reactivity and optimizing its position for the nucleophilic attack by the aspartate.

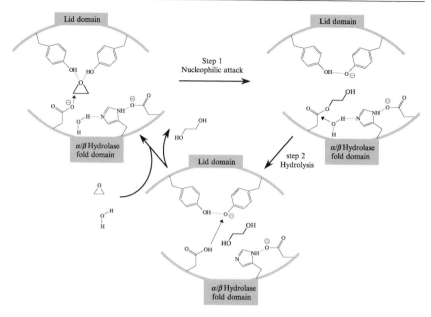

FIG. 2. Enzymatic mechanism of α/β hydrolase fold epoxide hydrolases. For further details, see text.

An analysis of the phylogenetic relationship within this group of EH (Barth *et al.*, 2004) has divided it into two superfamilies, namely the microsomal EH with four families and the cytosolic (better: soluble) EH with eight families. Microsomal EHs share an N-terminal extension that appears to have a stabilizing function (Zou *et al.*, 2000). It forms a large meander that extends from the bottom of the α/β hydrolase fold domain to the top of the lid and back, kind of tying the two domains together. In the soluble EH superfamily, a few members occur as fusion proteins with additional catalytic domains. The best-documented examples are mammalian sEH. They all have an N-terminal phosphatase domain preceding the α/β hydrolase fold (Cronin *et al.*, 2003; Newman *et al.*, 2003). One of the many mycobacterial EH, EphD, is fused to a C-terminal domain that, on the basis of sequence similarity, belongs to the family of short chain dehydrogenases (TubercuList: http://genolist.pasteur.fr/TubercuList/); however, no enzymatic activity has been documented with this domain so far. In all cases, the physiological significance of the fusion is not yet understood.

The human mEH and sEH are both subject to polymorphism. Several alleles have been described for both enzymes that apparently have moderate effects on the enzyme protein half-life (Hosagrahara *et al.*, 2004; Sandberg *et al.*, 2000) but do not appear to affect enzyme activity. Nevertheless, a number of pathologic conditions have been associated with the allelic status of EH (Raijmakers *et al.*, 2004; Sato *et al.*, 2004; Smith and Harrison, 1997), the significance of which have to be evaluated after further studies.

Limonene Epoxide Hydrolase

The limonene epoxide hydrolase from *R. erythropolis* has been isolated and cloned (Barbirato *et al.*, 1998; van der Werf *et al.*, 1998). Simply on the basis of its comparatively small size of 150 amino acids, it was predictably not a member of the α/β hydrolase fold enzymes. Three-dimensional structure analysis revealed a homodimeric enzyme, with the subunit having a cone shell-like single domain fold composed of six ß strands and three α helices (Arand *et al.*, 2003). The lumen of the cone shell represents the substrate-binding cavity. At its bottom, two aspartates, Asp[101] and Asp[132], are present that are major players in catalysis. Arginine Arg[99] situated between these two builds up a dense network of hydrogen bonds that supports catalysis. According to the present concept of the catalytic mechanism (Fig. 3), which is based on the X-ray structure of the enzyme with the competitive inhibitor valpromide bound in the active site, as well as kinetic data of active site mutants (Arand *et al.*, 2003), Asp[132] activates the catalytic water through proton abstraction. It is supported by two other

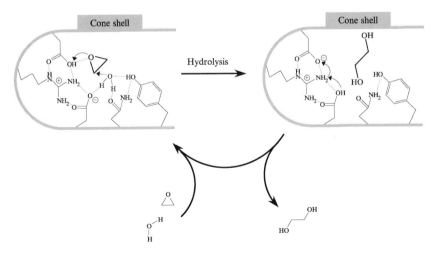

FIG. 3. Enzymatic mechanism of limonene epoxide hydrolases. For further details, see text.

residues lining the substrate-binding cavity, namely tyrosine Tyr[53] and asparagine Asn[55]. These also form hydrogen bonds to the water and thus hold it in the proper position for nucleophilic attack of the substrate. At the same time, Asp[101] donates its proton to the epoxide oxygen. Later, after release of the hydrolysis product, the proton that has shifted from Asp[101] to Asp[132] can be shuttled back rapidly via Arg[99]. Thus, epoxide hydrolysis proceeds in a single step, in contrast to the catalysis by α/β hydrolase fold EHs, and regeneration of the enzyme can proceed very fast.

Up until now, this class of EH has only been found in bacteria. However, the observation that a mycobacterial relative of LEH can hydrolyze the stereochemically demanding substrate cholesterol 5,6-epoxide (Johansson *et al.*, 2005) lends support to the speculation that the long-sought mammalian cholesterol epoxide hydrolase may belong to this family of EH.

FosX

A third class of epoxide hydrolases has evolved from the observation that a small enzyme structurally related to bacterial glutathione/cysteine transferases confers resistance to the epoxide antibiotic fosfomycin (Fillgrove *et al.*, 2003). In contrast to its aforementioned relatives, it inactivates fosfomycin by hydrolysis rather than by thiol conjugation. X-ray structure analysis revealed a homodimeric enzyme with a single domain subunit that harbors a Mn(II) in a metal-binding site that is conserved between FosX and its aforementioned relatives. The potential mechanism has not been completely worked out yet, but it has been shown that Glu[44], a residue in the vicinity of the Mn(II) that is not present in the aforementioned transferases, is essential for catalysis.

Like LEH and its relatives, FosX-related EH have been found exclusively in bacteria so far.

Leukotriene A_4 Hydrolase

Leukotriene A_4 hydrolase (LTA$_4$H) represents an untypical EH and is therefore discussed only briefly in this chapter. It is the only EH reported so far to form a nonvicinal diol (McGee and Fitzpatrick, 1985). The product of the reaction, leukotriene B_4, is a 5,12-diol, resulting from hydrolysis of the substrate leukotriene A_4, a 5,6-epoxide. The hydroxy group is introduced far from the site of ring opening because of a shift of the π-electron system of the three conjugated double bonds in the positions $\Delta 7$, $\Delta 9$, and $\Delta 11$. In addition to EH activity, LTA$_4$H has peptidase activity (Haeggstrom *et al.*, 1990; Minami *et al.*, 1990).

LTA$_4$H is a 69-kDa protein composed of three domains: an N-terminal domain, a catalytic domain, and a C-terminal domain (Thunnissen *et al.*,

2001). The catalytic domain has a thermolysin-like fold, including a zinc-binding site. Catalytic sites for the EH and peptidase activity overlap, but are not identical. In particular, the position of the water molecule used for hydrolysis is different in the two reactions. According to the present concept of the mechanisms, Asp[375] is important for water activation in the EH reaction, whereas Glu[296] activates the water in the peptidase reaction. For further details, see Haeggstrom (2004).

EH as Tools in Biocatalysis

Epoxides and, to some extent, vicinal diols represent interesting building blocks in the synthesis of a variety of drugs and fine chemicals (Archelas and Furstoss, 1998). Typically, the resulting molecules are optically active and exist in two enantiomeric forms. In the case of drugs, the desired therapeutic effect is usually associated with one of the two enantiomers, whereas the other one only contributes to unwanted side effects. Thus, enantioselective synthesis is highly desirable for optically active drugs. Because the desired enantioselectivity can be introduced at the level of the epoxide or diol, EH have a high potential as selective biocatalysts in this field of chemical synthesis. Over the last decade, there have been numerous approaches toward this direction, which are reviewed in de Vries and Janssen (2003), Genzel et al. (2002), and Steinreiber and Faber (2001). Meanwhile, a number of recombinant EH of microbial origin are available commercially (2005 Fluka catalog) that facilitate exploration of the suitability of these enzymes as enantioselective catalysts for the broad scientific community.

In recent times, the first approaches used to improve the catalytic properties of recombinant EH by directed evolution have been reported (Reetz et al., 2004; van Loo et al., 2004). It can be expected that in the near future, EH with desired substrate-, enantio-, and regioselectivities will be available through the construction of enzyme expression libraries that harbor millions of mutants, each with individual, potentially useful properties.

Recombinant Expression of EH

A variety of different expression systems have been employed for the recombinant production of EH. The first EH to be produced this way was the mammalian mEH. Although expression in *Escherichia coli* is possible (Bell and Kasper, 1993), *Saccharomyces cerevisiae* seems to be a somewhat superior expression host for this enzyme (Arand et al., 1999; Eugster et al., 1991; Gautier et al., 1993), probably because it has an endoplasmic reticulum. Another, yet more cost-intense alternative turned out to be baculovirus-mediated expression in insect cells (Bentley *et al.*, 1994). For analysis of the detoxification function, stable expression in mammalian cell lines was applied successfully (Friedberg *et al.*, 1994; Herrero *et al.*, 1997).

The sEH was expressed successfully in *E. coli* (Knehr *et al.*, 1993); however, production in insect cells turned out to be far superior because much higher yields of correctly folded expression product were obtained per volume of culture (Beetham *et al.*, 1993).

In the meantime, many different EH, mainly of bacterial, fungal, or plant origin, have been recombinantly expressed, usually with *E. coli* as the host (Arand *et al.*, 1999; Misawa *et al.*, 1998; Monterde *et al.*, 2004; Rink *et al.*, 1997; Visser *et al.*, 2000). However, overexpression of α/β hydrolase fold EH in this system often results in the formation of large amounts of inclusion bodies, representing misfolded recombinant protein. Sometimes, this can be overcome by tightly controlling the degree of expression using titratable expression systems or by reducing the expression temperature, but this does not guarantee success. The 11 α/β hydrolase EH-related enzymes from *Mycobacterium. tuberculosis*, for instance, display inclusion body formation in excess of 98% of the total recombinant protein in our hands, even when the aforementioned measures are employed (M. Arand *et al.*, unpublished observation). Under such circumstances, a suitable protocol for the successful refolding of active enzyme, starting from inclusion bodies, would be desirable. We have been successful for the first time in obtaining enzymatically active EH from inclusion bodies by such a refolding attempt—albeit with moderate efficiency—using the following protocol.

Refolding of Recombinant EH from Inclusion Bodies

The present protocol has been used to refold recombinant mycobacterial EphA and EphB (Cole *et al.*, 1998) from inclusion bodies obtained by expression in *E. coli*, with a yield in enzymatically active enzyme between 5 and 10% with respect to the amount of starting material.

Refolding Procedure

1. Wash 200 mg of inclusion body pellet (wet weight) twice in 2 ml of buffer 1 (Tris-HCl, 100 mM; NaCl, 200 mM; EDTA, 1 mM; Triton X-100, 0.5%; NaN$_3$, 0.02%; pH 7.8) by vortexing and subsequent centrifugation at 6000g for 5 min.
2. Resolve/resuspend the pellet in 7 ml of buffer 2 (urea, 8 M; Tris-HCl, 100 mM; NaN$_3$, 0.02%; pH 8,5) as fast as possible, aiming at a final protein concentration of about 15 mg/ml.

3. Incubate for 2 h at room temperature or, alternatively, overnight at 4° and then centrifuge at 4° with 10,000g for 15 min
4. Pour the supernatant into 15 volumes of ice-cold buffer 3 (sodium phosphate, 20 mM; glycerol, 20%; NaN$_3$, 0.02%; pH 7,4), mix rapidly, and incubate overnight
5. Centrifuge at 4° with 10,000g for 30 min, reduce the volume of the supernatant to 5 ml, and dialyze two times against 250 ml of buffer 3.

If successful, this procedure should yield an EH solution with a protein content in excess of 0.5 mg/ml, which should be suitable for further purification or storage at 4°.

Assay of EH

Numerous assays for the analysis of specific epoxide hydrolase activities are available, and the methodologies employed cover a broad range of analytical techniques. The following section describes three representative examples.

Hydrolysis of Styrene 7,8-oxide, a Generic Assay for Epoxide Hydrolases (Arand et al., 2003)

This assay has the particular advantage that almost every epoxide hydrolase tested so far is capable of catalyzing the hydrolysis of styrene 7,8-oxide. It is based on a simple partition procedure, taking advantage of the fact that the difference in hydrophilicity between the mother compound and the resulting metabolite is sufficient for an adequate separation of the two by a single extraction step. The drawback, however, is that the radiolabeled compound required for this procedure is not usually available from commercial sources and has to be prepared in advance or ordered as a custom synthesis. The preparation itself is not complicated because the compound can be obtained in a single step from radiolabeled styrene by turnover with m-chloroperoxybenzoic acid in methylene chloride and subsequent purification of the product (Oesch et al., 1971).

Assay Procedure

1. On ice, dilute your enzyme source in Tris-HCl, 50 mM, pH 7.4, to a final volume of 196 μl.
2. Incubate the mixture at 37° for 2 min.
3. Add 4 μl ^{3}H-labeled styrene 7,8-oxide from a 1 mM stock solution in acetonitrile, corresponding to about 10^5 cpm (dilute with unlabeled styrene 7,8-oxide as appropriate), and mix thoroughly.
4. Incubate for a suitable duration, usually 5–15 minutes, at 37°.

5. Terminate by the addition of 200 μl chloroform and immediate vortexing for 10–15 s.
6. Centrifuge for 3 min in a benchtop centrifuge.
7. Transfer 100 μl of the upper, aqueous phase to a scintillation vial containing 2 ml of a scintillation cocktail suitable for aqueous samples; place in a scintillation counter for analysis.

Important Considerations

1. Prepare a blank by substituting the enzyme source with an equal amount of bovine serum albumine, as protein concentration affects the extraction procedure.
2. Determine the total cpm employed in the reaction; this is best done by counting 100 μl of an extra reaction setup without prior chloroform extraction.
3. Run series of different incubation times and different enzyme source concentrations to determine the respective linear range for substrate turnover.
4. Avoid substrate turnover in excess of 20%; keep in mind that styrene 7,8-oxide usually is a racemic mixture of the *R* and *S* enantiomers and that many enzymes display pronounced enantio-selectivity; product inhibition, however, is rarely observed with EH.

Calculation

$$\text{turnover} = \frac{\left(\text{cpm}_{\text{sample}} - \text{cpm}_{\text{blank}}\right) * 1.5}{\text{cpm}_{\text{setup}}} * 4[\text{nmol}]$$

$$\text{specific activity} = \frac{\text{turnover}}{\text{amount of protein} * \text{incubation time}} \left[\frac{\text{nmol}}{\text{mg} * \text{min}}\right]$$

The use of chloroform as the organic solvent has two advantages that made it difficult to substitute it against a less harmful chemical in the present protocol. First, the partition coefficient of substrate and product in the water:chloroform system is close to ideal. Only negligible amounts of styrene 7,8-oxide partition into the aqueous phase, whereas about 65% of the diol remain in this fraction. This percentage is slightly influenced by buffer composition and protein content, but is very reproducible if these parameters are kept constant. Second, because chloroform has a higher density as compared to water, the upper aqueous phase is easily accessible for radiometric analysis without the danger of contamination with the organic phase, which usually contains the vast majority of radioactive material. The main caveat of the procedure is that not only the diol but also a respective glutathione conjugate would remain in the aqueous phase, and the formation of this type of metabolite would thus lead to erroneous results. Therefore, care must be taken to avoid the presence of glutathione, in particular together with a glutathione S-transferase, in the incubation mixture.

Hydrolysis of 9,10-cis-Epoxystearic Acid (Müller et al., 1997)

For a variety of EH substrates, the difference in hydrophilicity as compared to their hydrolysis product is not sufficient to allow separation in a single extraction step. In such case, separation by thin-layer chromatography (TLC), although more tedious than an extraction procedure, may be the method of choice. The higher effort that must be taken is rewarded by the fact that the R_f value of the product obtained by the procedure is a reasonably specific property that helps discriminate diol formation from other events, such as glutathione conjugation. Again, the method is based on a radiometric substrate that has to be acquired beforehand. A rather convenient synthesis is based on direct formation of the compound from radiolabeled oleic acid with excess dimethyl dioxiran in acetone (Müller et al., 1997). After completion of the reaction (typically after 1 h at room temperature), excess dimethyl dioxiran and acetone can be evaporated under a gentle stream of nitrogen, yielding the desired product in sufficient purity.

Assay Procedure

1. On ice, dilute your enzyme source in a final volume of 49 μl sodium phosphate buffer, 100 mM, pH 7.4, containing 50 mM sodium chloride.
2. Incubate the mixture at 37° for 2 min.
3. Add 1 μl ^{14}C-labelled 9,10-epoxystearic acid from a 1 mM stock solution in acetonitrile, corresponding to about 10^5 cpm, and mix thoroughly.
4. Incubate for a suitable duration, usually 15 min, at 37°.
5. Terminate by the addition of 100 μl ethyl acetate and immediate vortexing for 10–15 s.
6. Centrifuge for 3 min in a benchtop centrifuge.
7. Carefully apply 25 μl of the upper organic phase onto the start line of a silica gel 60 TLC sheet (Merk, Darmstadt, Germany) and, after drying of the spots, develop the chromatogram in a chromatography chamber using n-hexane/diethyl ether/formic acid (70:30:2) as the eluent; under

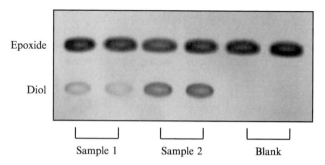

FIG. 4. 9,10-Epoxystearic acid hydrolysis assay. Autoradiography of the chromatographic separation of substrate and product after incubation of 9,10-epoxystaric acid with epoxide hydrolases. Samples and blank incubation (absence of epoxide hydrolase) were run in duplicates. The positions of epoxide and diol on the plate are indicated.

these conditions, epoxide and diol typically show R_f values of 0.5 and 0.2, respectively (Fig. 4)

8. After drying of the TLC sheet under a fume hood, quantify the radioactive spots using a phosphoimager or similar device suitable for ^{14}C quantification; if no such machine is available, signals may be visualized by autoradiography on an X-ray film; precise quantification is then possible by subsequent excision of the detected spots from the TLC plate followed by scintillation counting.

Important Considerations

1. The ethyl acetate used as the solvent for simultaneous extraction of substrate and product has a high vapor pressure; although this is convenient for evaporation of the solvent during application of the sample to the TLC plate, pipetting itself is somewhat difficult because the liquid tends to leak off the pipette tip, which also affects precision of the quantitation; this may be avoided by drawing ethyl acetate up and down the application pipette several times before pipetting the sample.

2. Saturation of the chromatography chamber atmosphere with the eluent vapor is important for an even development of the chromatogram, which itself is a prerequisite for proper quantification.

3. Like with styrene 7,8-oxide, avoid substrate turnover in excess of 20%; 9,10-epoxystearic acid also usually is a racemic mixture, and enantioselectivity has been observed with this substrate in several cases.

Calculation

$$\text{turnover} = \frac{\text{cpm}_{\text{product}}}{\left(\text{cpm}_{\text{product}} + \text{cpm}_{\text{substrate}}\right)} \ [\text{nmol}]$$

$$\text{specific activity} = \frac{\text{turnover}}{\text{amount of protein} * \text{incubation time}} \left[\frac{\text{nmol}}{\text{mg} * \text{min}}\right]$$

The Adrenaline Test (Wahler and Reymond, 2002)

This method is a universal procedure suitable for the semiquantitative activity determination of epoxide hydrolase and some other hydrolase reactions. It is based on the quantitative reaction of periodate with vicinal diols at neutral pH and subsequent back titration of remaining periodate with adrenaline, resulting in the formation of adrenochrome as a diagnostic chromophore. The advantages of this procedure are that (i) it is a general method applicable to almost any possible EH substrate, as long as it is sufficiently soluble in the assay mixture and a vicinal diol is the resulting metabolite, (ii) it does not require a labeled substrate, and (iii) it is suitable for high-throughput screening.

Assay Procedure

1. In a well of a microplate reader dish (96 well), place 75 μl of your enzyme source diluted in sodium phosphate buffer, 20 mM, pH 7.2.
2. Add 10 μl of sodium periodate from a 10 mM aqueous stock solution.
3. Add 5 μl of substrate from a respective stock solution (40–200 mM in acetonitrile or other suitable solvent).
4. Mix and incubate at 37° for 30 min.
5. Add 10 μl of adrenalin from a 15 mM aqueous stock solution.
6. Incubate for 5 min at 26°.
7. Determine the OD$_{490}$ in a microplate reader.

Important Considerations

1. Prepare the blank reaction as similar as possible to the sample setup, ideally with the presence of the epoxide hydrolase being the only difference.

2. The setup is designed for 96-well microplates and yields an OD$_{490}$ of approximately 0.7–0.8 in the absence of periodate-reducing reaction partners and would require modulation (reduction) of the periodate and adrenaline concentration if to be used with standard cuvettes.

3. If carried out as described here, this method is a semiquantitative procedure, thus no mathematical equation for the calculation of turnover is given to avoid the impression of a respective precision; however, it may be possible to turn it into a quantitative procedure when all potentially interfering factors are well under control, particularly when working with purified enzymes; in theory, reduction of the OD_{490} by 50% should equal a substrate turnover of 50 nmol.

The obvious restriction of this procedure is its sensitivity to redox-reactive compounds that might reduce the periodate or oxidize the adrenaline, the absence of which has to be proven by respective control experiments. However, in our hands it has even been possible to adapt the method to monitor the activity of recombinant EH in intact bacteria (M. Adamska, to be published), thus allowing the high-throughput screening of EH mutant libraries.

References

Arand, M., Cronin, A., Oesch, F., Mowbray, S. L., and Jones, T. A. (2003). The telltale structures of epoxide hydrolases. *Drug Metab. Rev.* **35,** 365–383.

Arand, M., Grant, D. F., Beetham, J. K., Friedberg, T., Oesch, F., and Hammock, B. D. (1994). Sequence similarity of mammalian epoxide hydrolases to the bacterial haloalkane dehalogenase and other related proteins: Implication for the potential catalytic mechanism of enzymatic epoxide hydrolysis. *FEBS Lett.* **338,** 251–256.

Arand, M., Hallberg, B. M., Zou, J. Y., Bergfors, T., Oesch, F., van der Werf, M. J., de Bont, J. A. M., Jones, T. A., and Mowbray, S. L. (2003). Structure of *Rhodococcus erythropolis* limonene-1,2-epoxide hydrolase reveals a novel active site. *EMBO J.* **22,** 2583–2592.

Arand, M., Hemmer, H., Dürk, H., Baratti, J., Archelas, A., Furstoss, R., and Oesch, F. (1999). Cloning and molecular characterization of a soluble epoxide hydrolase from *Aspergillus niger* that is related to mammalian microsomal epoxide hydrolase. *Biochem. J.* **344,** 273–280.

Arand, M., Müller, F., Mecky, A., Hinz, W., Urban, P., Pompon, D., Kellner, R., and Oesch, F. (1999). Catalytic triad of microsomal epoxide hydrolase: Replacement of Glu(404) with Asp leads to a strongly increased turnover rate. *Biochem. J.* **337,** 37–43.

Arand, M., Wagner, H., and Oesch, F. (1996). Asp(333), Asp(495), and His(523) form the catalytic triad of rat soluble epoxide hydrolase. *J. Biol. Chem.* **271,** 4223–4229.

Archelas, A., and Furstoss, R. (1998). Epoxide hydrolases: New tools for the synthesis of fine organic chemicals. *Trends Biotechnol.* **16,** 108–116.

Argiriadi, M. A., Morisseau, C., Hammock, B. D., and Christianson, D. W. (1999). Detoxification of environmental mutagens and carcinogens: Structure, mechanism, and evolution of liver epoxide hydrolase. *Proc. Natl. Acad. Sci. USA* **96,** 10637–10642.

Barbirato, F., Verdoes, J. C., de Bont, J. A. M., and van der Werf, M. J. (1998). The *Rhodococcus erythropolis* DCL14 limonene-1,2-epoxide hydrolase gene encodes an enzyme belonging to a novel class of epoxide hydrolases. *FEBS Lett.* **438,** 293–296.

Barth, S., Fischer, M., Schmid, R. D., and Pleiss, J. (2004). Sequence and structure of epoxide hydrolases: A systematic analysis. *Proteins* **55,** 846–855.

Beetham, J. K., Tian, T. G., and Hammock, B. D. (1993). cDNA cloning and expression of a soluble epoxide hydrolase from human liver. *Arch. Biochem. Biophys.* **305,** 197–201.

Bell, P. A., and Kasper, C. B. (1993). Expression of rat microsomal epoxide hydrolase in *Escherichia coli*: Identification of a histidyl residue essential for catalysis. *J. Biol. Chem.* **268,** 14011–14017.

Bentley, W. E., Kebede, B., Franey, T., and Wang, M. Y. (1994). Segregated characterization of recombinant epoxide hydrolase synthesis via the baculovirus-insect cell expression system. *Chem. Eng. Sci.* **49,** 4133–4141.

Cole, S. T., Brosch, R., Parkhill, J., Garnier, T., Churcher, C., Harris, D., Gordon, S. V., Eiglmeier, K., Gas, S., Barry, C. E., 3rd, Tekaia, F., Badcock, K., Basham, D., Brown, D., Chillingworth, T., Connor, R., Davies, R., Devlin, K., Feltwell, T., Gentles, S., Hamlin, N., Holroyd, S., Hornsby, T., Jagels, K., Barrell, B. G., *et al.* (1998). Deciphering the biology of *Mycobacterium tuberculosis* from the complete genome sequence. *Nature* **393,** 537–544.

Cronin, A., Mowbray, S., Dürk, H., Homburg, S., Fleming, I., Fisslthaler, B., Oesch, F., and Arand, M. (2003). The N-terminal domain of mammalian soluble epoxide hydrolase is a phosphatase. *Proc. Natl. Acad. Sci. USA* **100,** 1552–1557.

de Vries, E. J., and Janssen, D. B. (2003). Biocatalytic conversion of epoxides. *Curr. Opin. Biotechnol.* **14,** 414–420.

Eugster, H. P., Sengstag, C., Hinnen, A., Meyer, U. A., and Würgler, F. E. (1991). Heterologous expression of human microsomal epoxide hydrolase in *Saccharomyces cerevisiae*: Study of the valpromide carbamazepine epoxide interaction. *Biochem. Pharmacol.* **42,** 1367–1372.

Fillgrove, K. L., Pakhomova, S., Newcomer, M. E., and Armstrong, R. N. (2003). Mechanistic diversity of fosfomycin resistance in pathogenic microorganisms. *J. Am. Chem. Soc.* **125,** 15730–15731.

Frantz, S. W., van den Eeckhout, E., Sinsheimer, J. E., Yoshihara, M., and Koreeda, M. (1985). Mutagenicity in Salmonella assays of cyclohexane epoxide derivatives. *Toxicol. Lett.* **25,** 265–271.

Friedberg, T., Becker, R., Oesch, F., and Glatt, H. (1994). Studies on the importance of microsomal epoxide hydrolase in the detoxification of arene oxides using the heterologous expression of the enzyme in mammalian cells. *Carcinogenesis* **15,** 171–175.

Gautier, J. C., Urban, P., Beaune, P., and Pompon, D. (1993). Engineered yeast-cells as model to study coupling between human xenobiotic metabolizing enzymes: Simulation of the 2 1st steps of benzo[a]pyrene activation. *Eur. J. Biochem.* **211,** 63–72.

Genzel, Y., Archelas, A., Broxterman, Q. B., Schulze, B., and Furstoss, R. (2002). Microbial transformations 50: Selection of epoxide hydrolases for enzymatic resolution of 2-, 3- or 4-pyridyloxirane. *J. Mol. Cat. B Enz.* **16,** 217–222.

Guengerich, F. P. (1982). Epoxide hydrolase: Properties and metabolic roles. *Rev. Biochem. Toxicol.* **4,** 5–30.

Haeggstrom, J. Z. (2004). Leukotriene A(4) hydrolase/aminopeptidase, the gatekeeper of chemotactic leukotriene B(4) biosynthesis. *J. Biol. Chem.* **279,** 50639–50642.

Haeggstrom, J. Z., Wetterholm, A., Vallee, B. L., and Samuelsson, B. (1990). Leukotriene A4 hydrolase: An epoxide hydrolase with peptidase activity. *Biochem. Biophys. Res. Commun.* **173,** 431–437.

Hammock, B. D., Gill, S. S., Stamoudis, V., and Gilbert, L. I. (1976). Soluble mammalian epoxide hydratase: Action on juvenile hormone and other terpenoid epoxides. *Comp. Biochem. Physiol. B* **53,** 263–265.

Hammock, B. D., Storms, D., and Grant, D. (1997). Epoxide hydrolases. *In* "Comprehensive Toxicology" (F. P. Guengerich, ed.), pp. 283–305. Pergamon Press, Oxford.

Herrero, M. E., Arand, M., Hengstler, J. G., and Oesch, F. (1997). Recombinant expression of human microsomal epoxide hydrolase protects V79 Chinese hamster cells from styrene oxide, but not from ethylene oxide-induced DNA strand breaks. *Environ. Mol. Mutagen.* **30**, 429–439.

Hosagrahara, V. P., Rettie, A. E., Hassett, C., and Omiecinski, C. J. (2004). Functional analysis of human microsomal epoxide hydrolase genetic variants. *Chem.- Biol. Interact.* **150**, 149–159.

Jerina, D. M., and Daly, J. W. (1974). Arene oxides: A new aspect of drug metabolism. *Science* **185**, 573–582.

Johansson, P., Unge, T., Cronin, A., Arand, M., Bergfors, T., Jones, T. A., and Mowbray, S. L. (2005). Structure of an atypical epoxide hydrolase from *Mycobacterium tuberculosis* gives insights into its function. *J. Mol. Biol.* **351**, 1048–1056.

Kim, M. H., Geacintov, N. E., Pope, M., and Harvey, R. G. (1984). Structural effects in reactivity and adduct formation of polycyclic aromatic epoxide and diol epoxide derivatives with DNA: Comparison between 1-oxiranylpyrene and benzo[a]pyrenediol epoxide. *Biochemistry* **23**, 5433–5439.

Knehr, M., Thomas, H., Arand, M., Gebel, T., Zeller, H. D., and Oesch, F. (1993). Isolation and characterization of a cDNA encoding rat liver cytosolic epoxide hydrolase and its functional expression in *Escherichia coli*. *J. Biol. Chem.* **268**, 17623–17627.

Lacourciere, G. M., and Armstrong, R. N. (1994). Microsomal and soluble epoxide hydrolases are members of the same family of C-X bond hydrolase enzymes. *Chem. Res. Toxicol.* **7**, 121–124.

McGee, J., and Fitzpatrick, F. (1985). Enzymatic hydration of leukotriene A4: Purification and characterization of a novel epoxide hydrolase from human erythrocytes. *J. Biol. Chem.* **260**, 12832–12837.

Minami, M., Ohishi, N., Mutoh, H., Izumi, T., Bito, H., Wada, H., Seyama, Y., Toh, H., and Shimizu, T. (1990). Leukotriene A_4 hydrolase is a zinc-containing aminopeptidase. *Biochem. Biophys. Res. Commun.* **173**, 620–623.

Misawa, E., Chion, C., Archer, I. V., Woodland, M. P., Zhou, N. Y., Carter, S. F., Widdowson, D. A., and Leak, D. J. (1998). Characterisation of a catabolic epoxide hydrolase from a *Corynebacterium* sp. *Eur. J. Biochem.* **253**, 173–183.

Miyata, M., Kudo, G., Lee, Y. H., Yang, T. J., Gelboin, H. V., Fernandez-Salguero, P., Kimura, S., and Gonzalez, F. J. (1999). Targeted disruption of the microsomal epoxide hydrolase gene: Microsomal epoxide hydrolase is required for the carcinogenic activity of 7,12-dimethylbenz[a]anthracene. *J. Biol. Chem.* **274**, 23963–23968.

Moghaddam, M. F., Grant, D. F., Cheek, J. M., Greene, J. F., Williamson, K. C., and Hammock, B. D. (1997). Bioactivation of leukotoxins to their toxic diols by epoxide hydrolase. *Nature Med.* **3**, 562–566.

Monterde, M. I., Lombard, M., Archelas, A., Cronin, A., Arand, M., and Furstoss, R. (2004). Enzymatic transformations. 58: Enantioconvergent biohydrolysis of styrene oxide derivatives catalysed by the *Solanum tuberosum* epoxide hydrolase. *Tetrahedron Asymmetry* **15**, 2801–2805.

Müller, F., Arand, M., Frank, H., Seidel, A., Hinz, W., Winkler, L., Hänel, K., Blée, E., Beetham, J. K., Hammock, B. D., and Oesch, F. (1997). Visualization of a covalent intermediate between microsomal epoxide hydrolase, but not cholesterol epoxide hydrolase, and their substrates. *Eur. J. Biochem.* **245**, 490–496.

Nardini, M., Ridder, I. S., Rozeboom, H. J., Kalk, K. H., Rink, R., Janssen, D. B., and Dijkstra, B. W. (1999). The x-ray structure of epoxide hydrolase from *Agrobacterium* radiobacter AD1: An enzyme to detoxify harmful epoxides. *J. Biol. Chem.* **274**, 14579–14586.

Newman, J. W., Morisseau, C., Harris, T. R., and Hammock, B. D. (2003). The soluble epoxide hydrolase encoded by EPXH2 is a bifunctional enzyme with novel lipid phosphate phosphatase activity. *Proc. Natl. Acad. Sci. USA* **100**, 1558–1563.

Oesch, F. (1973). Mammalian epoxide hydrases: Inducible enzymes catalysing the inactivation of carcinogenic and cytotoxic metabolites derived from aromatic and olefinic compounds. *Xenobiotica* **3**, 305–340.

Oesch, F., Jerina, D. M., and Daly, J. (1971). A radiometric assay for hepatic epoxide hydrase activity with [7-3H]styrene oxide. *Biochim. Biophys. Acta* **227**, 685–691.

Oesch, F., Timms, C. W., Walker, C. H., Guenthner, T. M., Sparrow, A., Watabe, T., and Wolf, C. R. (1984). Existence of multiple forms of microsomal epoxide hydrolase with radically different substrate specifities. *Carcinogenesis* **5**, 7–9.

Ota, K., and Hammock, B. D. (1980). Cytosolic and microsomal epoxide hydrolases: Differential properties in mammalian liver. *Science* **207**, 1479–1481.

Pries, F., Kingma, J., Pentenga, M., Vanpouderoyen, G., Jeronimusstratingh, C. M., Bruins, A. P., and Janssen, D. B. (1994). Site-directed mutagenesis and oxygen isotope incorporation studies of the nucleophilic aspartate of haloalkane dehalogenase. *Biochemistry* **33**, 1242–1247.

Raijmakers, M. T. M., de Galan-Roosen, T. E. M., Schilders, G. W., Merkus, J., Steegers, E. A. P., and Peters, W. H. M. (2004). The Tyr113His polymorphism in exon 3 of the microsomal epoxide hydrolase gene is a risk factor for perinatal mortality. *Acta Obstet. Gynecol. Scand.* **83**, 1056–1060.

Reetz, M. T., Torre, C., Eipper, A., Lohmer, R., Hermes, M., Brunner, B., Maichele, A., Bocola, M., Arand, M., Cronin, A., Genzel, Y., Archelas, A., and Furstoss, R. (2004). Enhancing the enantioselectivity of an epoxide hydrolase by directed evolution. *Org. Lett.* **6**, 177–180.

Rink, R., Fennema, M., Smids, M., Dehmel, U., and Janssen, D. B. (1997). Primary structure and catalytic mechanism of the epoxide hydrolase from *Agrobacterium* radiobacter AD1. *J. Biol. Chem.* **272**, 14650–14657.

Sandberg, M., Hassett, C., Adman, E. T., Meijer, J., and Omiecinski, C. J. (2000). Identification and functional characterization of human soluble epoxide hydrolase genetic polymorphisms. *J. Biol. Chem.* **275**, 28873–28881.

Sato, K., Emi, M., Ezura, Y., Fujita, Y., Takada, D., Ishigami, T., Umemura, S., Xin, Y. P., Wu, L. L., Larrinaga-Shum, S., Stephenson, S. H., Hunt, S. C., and Hopkins, P. N. (2004). Soluble epoxide hydrolase variant (Glu287Arg) modifies plasma total cholesterol and triglyceride phenotype in familial hypercholesterolemia: Intrafamilial association study in an eight-generation hyperlipidemic kindred. *J. Hum. Genet.* **49**, 29–34.

Sims, P., Grover, P. L., Swaisland, A., Pal, K., and Hewer, A. (1974). Metabolic activation of benzo(a)pyrene proceeds by a diol-epoxide. *Nature* **252**, 326–328.

Sinal, C. J., Miyata, M., Tohkin, M., Nagata, K., Bend, J. R., and Gonzalez, F. J. (2000). Targeted disruption of soluble epoxide hydrolase reveals a role in blood pressure regulation. *J. Biol. Chem.* **275**, 40504–40510.

Smith, C. A. D., and Harrison, D. J. (1997). Association between polymorphism in gene for microsomal epoxide hydrolase and susceptibility to emphysema. *Lancet* **350**, 630–633.

Spector, A. A., Fang, X., Snyder, G. D., and Weintraub, N. L. (2004). Epoxyeicosatrienoic acids (EETs): Metabolism and biochemical functions. *Prog. Lipid Res.* **43**, 55–90.

Steinreiber, A., and Faber, K. (2001). Microbial epoxide hydrolases for preparative biotransformations. *Curr. Opin. Biotechnol.* **12**, 552–558.

Thunnissen, M., Nordlund, P., and Haeggstrom, J. Z. (2001). Crystal structure of human leukotriene A(4) hydrolase, a bifunctional enzyme in inflammation. *Nat. Struct. Biol.* **8**, 131–135.

van der Werf, M. J., Overkamp, K. M., and de Bont, J. A. M. (1998). Limonene-1,2-epoxide hydrolase from *Rhodococcus erythropolis* DCL14 belongs to a novel class of epoxide hydrolases. *J. Bacteriol.* **180**, 5052–5057.

van der Werf, M. J., Swarts, H. J., and de Bont, J. A. M. (1999). *Rhodococcus erythropolis* DCL14 contains a novel degradation pathway for limonene. *Appl. Environ. Microbiol.* **65,** 2092–2102.

van Loo, B., Spelberg, J. H. L., Kingma, J., Sonke, T., Wubbolts, M. G., and Janssen, D. B. (2004). Directed evolution of epoxide hydrolase from A-radiobacter toward higher enantioselectivity by error-prone PCR and DNA shuffling. *Chem. Biol.* **11,** 981–990.

Visser, H., Vreugdenhil, S., de Bont, J. A. M., and Verdoes, J. C. (2000). Cloning and characterization of an epoxide hydrolase-encoding gene from *Rhodotorula glutinis*. *Appl. Microbiol. Biotechnol.* **53,** 415–419.

Wahler, D., and Reymond, J.-L. (2002). The adrenaline test for enzymes. *Angew. Chem. Int. Ed.* **41,** 1229–1232.

Wickliffe, J. K., Ammenheuser, M. M., Salazar, J. J., Abdel-Rahman, S. Z., Hastings-Smith, D. A., Postlethwait, E. M., Lloyd, R. S., and Word, J. B. (2003). A model of sensitivity: 1,3-Butadiene increases mutant frequencies and genomic damage in mice lacking a functional microsomal epoxide hydrolase gene. *Environ. Mol. Mutagen.* **42,** 106–110.

Yu, Z., Xu, F., Huse, L. M., Morisseau, C., Draper, A. J., Newman, J. W., Parker, C., Graham, L., Engler, M. M., Hammock, B. D., Zeldin, D. C., and Kroetz, D. L. (2000). Soluble epoxide hydrolase regulates hydrolysis of vasoactive epoxyeicosatrienoic acids. *Circ. Res.* **87,** 992–998.

Zou, J. Y., Hallberg, B. M., Bergfors, T., Oesch, F., Arand, M., Mowbray, S. L., and Jones, T. A. (2000). Structure of *Aspergillus niger* epoxide hydrolase at 1.8 angstrom resolution: Implications for the structure and function of the mammalian microsomal class of epoxide hydrolases. *Structure* **8,** 111–122.

[33] Pregnane X Receptor-Mediated Transcription

By Thomas K. H. Chang and David J. Waxman

Abstract

The pregnane X receptor (PXR, receptor NR1I2) is a ligand-activated transcription factor that is activated by structurally diverse endogenous steroids and foreign chemicals and serves as an important steroid and xenobiotic sensor. This member of the nuclear receptor superfamily is highly expressed in liver and in the gastrointestinal tract, where it regulates transcription of a large set of genes that contribute to foreign compound metabolism and to the metabolism and transcellular transport of steroid hormones, bile acids, and other endogenous substances. This chapter summarizes studies of PXR and its biological functions and describes a cell culture-based luciferase reporter gene assay for determination of PXR transcriptional activity. This assay can be used to identify novel drugs and environmental chemicals that serve as PXR ligands and thereby modulate PXR activity and may aid in the prediction of drug–drug interactions and foreign chemical-induced toxicities associated with the activation of PXR transcriptional responses.

METHODS IN ENZYMOLOGY, VOL. 400   0076-6879/05 $35.00
Copyright 2005, Elsevier Inc. All rights reserved.   DOI: 10.1016/S0076-6879(05)00033-9

Introduction

The pregnane X receptor (PXR) is a ligand-activated transcription factor. It belongs to the nuclear receptor superfamily and is designated NR1I2 according to the unified nomenclature for nuclear receptors (Giguere, 1999). PXR (also designated SXR and PAR) and the closely related nuclear receptor constitutive androstane receptor (CAR; NR1I3) both act as cellular sensors for foreign chemicals ("xenosensors") (Handschin and Meyer, 2003; Waxman, 1999) and exhibit overlapping but distinct ligand specificities (Moore *et al.*, 2000b). Both mammalian xenosensors may have evolved from a common ancestor related to chicken X receptor (CXR), which carries out analogous functions in the chicken (Handschin *et al.*, 2004). Studies of PXR-null mice (Xie *et al.*, 2000) demonstrate that the activation of PXR target genes can impart beneficial effects, for example, protection from the toxicity of cholestatic bile acids (Staudinger *et al.*, 2001; Xie *et al.*, 2001), but may also contribute to deleterious responses, such as acetaminophen-induced hepatotoxicity (Guo *et al.*, 2004).

PXR Target Genes

PXR transcriptional activity is activated following the binding of ligand. Ligand-bound PXR forms a heterodimer with a second nuclear receptor, retinoid X receptor (RXR). The resulting PXR-RXR heterodimer binds to DNA response elements (PXREs) in the promoter regions of PXR target genes and recruits coactivator proteins, which in turn stimulate gene transcription (Watkins *et al.*, 2003). PXR target genes encode cytochrome P450 (CYP) monooxygenases (phase I enzymes), phase II conjugation enzymes, and phase III transporters, among others (Rosenfeld *et al.*, 2003). Thus, in addition to *CYP3A* genes (Goodwin *et al.*, 2002; Waxman, 1999), PXR activates transcription of *CYP2B6* (Goodwin *et al.*, 2001), CYP2C9 (Chen *et al.*, 2004), UDP-glucuronosyltransferases (UGTs) *1A1*, *1A3*, *1A4* (Gardner-Stephen *et al.*, 2004; Maglich *et al.*, 2002), and *1A6* (Xie *et al.*, 2003), glutathione *S*-transferase A2 (*GSTA2*) (Falkner *et al.*, 2001), and dehydroepiandrosterone sulfotransferase *SULT2A1* (Echchgadda *et al.*, 2004). Drug transporters whose expression is induced by ligand-activated PXR include P-glycoprotein, various multidrug resistance-associated proteins, and organic anion transporting polypeptides (Rosenfeld *et al.*, 2003).

Tissue Distribution, Ontogeny, and Interindividual Expression

PXR is expressed at high levels in liver and, to a lesser extent, in extrahepatic tissues, such as small intestines and colon (Bertilsson *et al.*, 1998; Blumberg *et al.*, 1998; Lehmann *et al.*, 1998; Zhang *et al.*, 1999). Other tissues with lower levels of PXR expression include kidney, lung, uterus, ovary, breast, and placenta. In rats, hepatic PXR mRNA is detectable in embryonic tissue and increases postnatally, reaching a maximum by day 28 of age (Balasubramaniyan *et al.*, 2005). No apparent changes in mouse hepatic PXR mRNA levels occur during senescence, evaluated up to 24 months of age (Echchgadda *et al.*, 2004).

Interindividual variability in PXR mRNA levels is seen in human liver (Chang *et al.*, 2003; Gardner-Stephen *et al.*, 2004). Consistent with this finding, PXR expression can be altered by various factors, including drug exposure (e.g., dexamethasone treatment) (Pascussi *et al.*, 2000) and disease state (e.g., inflammation and ulcerative colitis) (Beigneux *et al.*, 2002; Langmann *et al.*, 2004). Several naturally occurring variants of PXR have been identified and may contribute to interindividual differences in PXR activity. For example, a human PXR variant splice variant with an in-frame deletion of 37 amino acids has reduced ability to activate *UGT1A1, UGT1A3,* and *UGT1A4* (Gardner-Stephen *et al.*, 2004) and presumably other PXR target genes, whereas the human allelic variant PXR-D163G displays altered responsiveness to several drugs and environmental chemicals (Hurst and Waxman, 2004; Hustert *et al.*, 2001).

PXR Ligands

PXR was initially shown to be activated by certain steroid hormones, including naturally occurring pregnanes. Other endogenous PXR ligands subsequently identified include bilirubin, cholesterol, bile acids, and certain vitamins (Handschin and Meyer, 2005; Traber, 2004). Many CYP3A inducers were found to be activating ligands of PXR, highlighting the importance of PXR for CYP3A induction by drugs and steroids (Waxman, 1999). Later studies showed that PXR ligands also induce other CYP enzymes, including CYP2B6 and CYP2C9, as well as phase II conjugation enzymes and phase III transporters (Handschin and Meyer, 2005), as noted earlier. PXR ligands include hyperforin, an active constituent of Saint-John's-wort, a widely used natural product and a strong inducer of liver CYP3A expression (Moore *et al.*, 2000a). Environmental chemicals that activate PXR include phthalate monoesters (Hurst and Waxman, 2004), organochlorine pesticides (Coumoul *et al.*, 2002; Lemaire *et al.*, 2004), and polychlorinated biphenyls (PCBs) (Hurst and Waxman, 2005; Tabb *et al.*, 2004).

Species Specificity of PXR Activation

Species differences in the induction of drug-metabolizing enzymes, including conjugation enzymes, are well established. A fuller understanding of the underlying molecular basis for these species differences was provided by the discovery of substantial differences in the ligand-binding specificity of PXR cloned from different species. These species differences are dictated by differences in the PXR ligand-binding domain. Notably, the ligand-binding domain of human PXR is only ~76% identical in amino acid sequence to its mouse and rat counterparts, which corresponds to a rather low degree of identity for a nuclear receptor between rodent and human orthologs (compare ~96% identity of mouse and human PXR DNA-binding domains) (Jones et al., 2000). The species-specific effects that certain ligands have on PXR transcriptional activity are governed by a limited number of specific individual amino acid residues at the ligand-binding site of PXR. For example, residue Leu 308 is required for activation of human PXR by rifampin (Tirona et al., 2004), a response not seen with mouse PXR or rat PXR, whereas residues Phe 305 and Asp 318 are required for activation of rat PXR by pregnenolone 16α-carbonitrile (Song et al., 2005), which is not an activating ligand of human PXR. Studies such as these have increased our understanding of the factors that determine species differences in PXR-mediated induction of drug-metabolizing enzymes and help guide the choice of animal model for drug interaction, pharmacokinetic, and toxicological studies during preclinical drug development.

An interesting example of species-specific effects of PXR ligands is provided by the finding that mouse PXR, but not human PXR, is highly responsive to a variety of highly chlorinated PCBs. PCBs that activate mouse PXR *do* bind to human PXR, however, with PCB binding leading to inhibition rather than activation of human PXR transcriptional activity (Tabb et al., 2004). PCBs that bind human PXR may therefore interfere with PXR-regulated metabolism of endogenous steroids and may impair PXR-dependent detoxification of xenobiotics and hepatotoxic bile acids.

Determination of PXR Transcriptional Activity

The characterization of foreign chemicals that bind to and activate PXR has provided a molecular understanding of the actions of many chemicals that modulate drug metabolism. This knowledge may be applied at an early stage of drug development, where lead compounds may be screened to eliminate compounds with a high potential to activate PXR and elicit PXR-based drug–drug interactions or other adverse responses.

Various assays have been developed for the identification and characterization of PXR ligands. These include receptor-binding assays and *trans*-activation assays using PXR-regulated reporter genes (Jones et al., 2002). Reporter gene assays can be modified for high-throughput analysis (Raucy et al., 2002; Zhu et al., 2004) and can be used to predict the rank and potency of PXR activators (El-Sankary *et al.*, 2001). *trans*-Activation assays can be performed by transient transfection of suitable recipient cells, such as the liver-derived cell line HepG2 (Hurst and Waxman, 2004), or may use cells stably transfected with PXR-regulated reporter genes (Lemaire *et al.*, 2004). Luciferase reporter genes offer several important advantages, including high sensitivity and linearity of response over several orders of magnitude, and can be carried out inexpensively and without the use of radiochemicals. The next section describes a cell-based luciferase reporter gene assay for PXR transcriptional activity that is suitable for medium- and high-throughput analysis using a multiwell tissue culture plate format.

Luciferase Reporter Gene Assay for Assaying PXR Transcriptional Activity

Plasmids

Three specific plasmids are required. The first is a PXR expression plasmid, for example, the human PXR expression plasmid pSG5-hPXR (Jones *et al.*, 2002). The second plasmid is a reporter plasmid containing a firefly luciferase cDNA under the control of a PXR-activated regulatory element. The regulatory element may be composed of an isolated, multimerized PXRE sequence, such as (3A4)$_3$-tk-Luc [three copies of an everted repeat with six nucleotide spacing (ER6) derived from the *CYP3A4* gene] or (3A23)$_3$-tk-Luc [three copies of a direct repeat with three nucleotide spacing (DR3) derived from the *CYP3A23* gene] (Hurst and Waxman, 2004; Hustert *et al.*, 2001). Alternatively, it may consist of an intact promoter fragment derived from a PXR-responsive gene, such as *CYP3A4*, which contains multiple PXREs in the context of other natural promoter sequences (Hustert *et al.*, 2001). The third plasmid is a *Renilla* luciferase reporter plasmid, such as pRL-CMV (Promega Corp., Madison, WI), which is used as an internal standard to normalize transfection efficiency between samples.

Cell Culture

1. Plate HepG2 human hepatoma cells (American Type Culture Collection, Manassas, VA) in tissue culture plates containing Dulbecco's modified Eagle medium (DMEM) supplemented with 10% fetal bovine serum (FBS, Sigma-Aldrich Co., St. Louis, MO), 50 U/ml penicillin, and 50 μg/ml streptomycin sulfate (Invitrogen, Carlsbad, CA).
2. Culture cells overnight in a tissue culture incubator at 37° and in a humidified atmosphere of 5% CO_2/95% air.

3. Trypsinize cells, resuspend in DMEM supplemented with 10% FBS, 50 U/ml penicillin, and 50 μg/ml streptomycin sulfate, and pass the suspension through a 20-μl pipette tip three times to obtain single cells.
4. Replate cells at a density of 75,000 cells per 500-μl culture medium per well of a 48-well plate (Corning Inc., Corning, NY).
5. Culture cells at 37° for 24 h in a humidified atmosphere of 5% CO_2/ 95% air.

Transient Transfection

1. Dilute PXR expression plasmid DNA and PXR reporter plasmid DNA in 1× TE buffer (10 mM Tris-EDTA, pH 8) to give working concentrations such that 5 ng of PXR expression plasmid and 90 ng of PXR reporter plasmid are added to each well of a 48-well tissue culture plate. [These plasmid amounts are suitable for HepG2 cells and for the specific PXR expression and reporter plasmids used in this laboratory (Hurst and Waxman, 2004). Plasmid DNA amounts may need to be adjusted when using other plasmid constructs or other recipient cell lines.] Dilute the *Renilla* luciferase reporter plasmid pRL-CMV (final amount of 0.5 ng per well) for use as an internal standard for transfection efficiency. Add salmon sperm DNA (carrier DNA) (Stratagene Inc., La Jolla, CA) to each sample to give 250 ng total DNA per well.
2. For each 10 wells to be transfected, add 3 μl of FuGene 6 transfection reagent (Boehringer-Mannheim, Ingelheim, Germany) to 100 μl of DMEM (not supplemented with FBS), mix gently, and incubate at room temperature for 5 min.
3. Add 10.3 μl of the FuGene 6-DMEM mixture to 1 μl of the plasmid mix containing a total of 250 ng DNA (see step 1) for each well to be transfected.
4. Incubate the mixture of FuGene 6-DMEM and DNA at room temperature for 15 min.
5. Add 11.3 μl of the FuGene 6-DMEM-DNA mixture directly to each tissue culture well containing cells in 250 μl DMEM supplemented with 10% FBS.
6. Culture cells in the presence of the FuGene 6-DMEM-DNA mixture at 37° for 24 h in a humidified atmosphere of 5% CO_2/95% air.

Treatment of Transfected Cells with PXR Ligand

1. Dissolve each compound to be tested for PXR activity in a suitable solvent, for example, dimethyl sulfoxide (DMSO), to obtain a 1000× working stock solution.

2. Dilute the freshly prepared stock solution of each test compound, or vehicle control, 1000-fold into serum-free treatment medium to obtain the final desired concentration [e.g., 10 μM PXR test ligand or 0.1% (v/v) DMSO vehicle control]. Rifampin (10 μM, final concentration) may be used as a positive control for activation of human PXR and pregnenolone (25 μM) may be used as a positive control for mouse PXR. Charcoal-stripped, delipidated FBS can be used in place of regular FBS to decrease basal PXRE-regulated luciferase reporter activity.

3. Aspirate the culture medium containing transfection mix from each well and replace with 250 μl treatment medium containing $1\times$ test compound from step 2.

4. Incubate cells at 37° for 24 h in a humidified atmosphere of 5% CO_2/ 95% air.

Determination of Reporter Activity

1. Aspirate culture medium from each tissue culture well after 24 hr incubation with the test compound.
2. Wash cells once with ice-cold phosphate-buffered saline (pH 7.4).
3. Add 250 μl of cell lysis buffer (passive lysis buffer, Promega Corp.).
4. Incubate cells, shaking gently at 4° for 20 min.
5. Transfer 5 μl of cell extract to a luminometer tube containing 30 μl of luciferase assay reagent (Luciferase Assay Reagent II, Promega Corp.) and measure firefly luciferase activity.
6. Remove tube and add reagent to quench firefly luciferase activity and measure *Renilla* luciferase activity in the same cell lysate using 30 μl Stop and Glo reagent (Promega Dual-Luciferase Reporter Assay System) according to the manufacturer's instructions.

Data Analysis

1. Calculate the normalized reporter activity of each sample by dividing the firefly luciferase activity (reporter activity) by the *Renilla* luciferase activity (internal standard).

2. Calculate the PXR activity of each sample from the ratio of normalized reporter activity (calculated as in step 1) for PXR test ligand-treated cells relative to that of vehicle-treated (control) cells. Data are usually presented as mean activities ± SD for triplicate analyses.

3. Dose–response data may be analyzed using GraphPad Prism software (GraphPad, San Diego, CA) or other suitable software, and EC_{50} values (concentration of test chemical required to produce 50% of the maximal luciferase reporter response) may be calculated.

Concluding Remarks

The discovery of PXR and other nuclear receptors that interact with foreign chemicals has greatly improved our understanding of the signaling pathways and transcriptional events that regulate a large set of genes with diverse biological functions, particularly foreign compound metabolism. In addition to the luciferase-based reporter gene assay for the determination of PXR activity described here, other methods have been developed, including real-time analysis of luciferase reporter gene activity following the introduction of reporters into live animals *in vivo* (Schuetz *et al.*, 2002). The identification of novel endogenous regulators of PXR activity is likely to increase our understanding of the multiple biological roles of this versatile steroid and foreign compound-responsive transcription factor.

Acknowledgments

This work was supported by the Canadian Institutes of Health Research (Grant MOP-42385 to T.K.H.C.) and the Superfund Basic Research Center at Boston University, NIH Grant 5 P42 ES07381 (to D.J.W.).

References

Balasubramaniyan, N., Shahid, M., Suchy, F. J., and Ananthanarayanan, M. (2005). Multiple mechanisms of ontogenic regulation of nuclear receptors during rat liver development. *Am. J. Physiol. Gastrointest. Liver Physiol.* **288**, G251–G260.

Beigneux, A. P., Moser, A. H., Shigenaga, J. K., Grunfeld, C., and Feingold, K. R. (2002). Reduction in cytochrome P-450 enzyme expression is associated with repression of CAR (constitutive androstane receptor) and PXR (pregnane X receptor) in mouse liver during the acute phase response. *Biochem. Biophys. Res. Commun.* **293**, 145–149.

Bertilsson, G., Heidrich, J., Svensson, K., Asman, M., Jendeberg, L., Sydow-Backman, M., Ohlsson, R., Postlind, H., Blomquist, P., and Berkenstam, A. (1998). Identification of a human nuclear receptor defines a new signaling pathway for *CYP3A* induction. *Proc. Natl. Acad. Sci. USA* **95**, 12208–12213.

Blumberg, B., Sabbagh, W., Jr., Juguilon, H., Bolado, J., Jr., van Meter, C. M., Ong, E. S., and Evans, R. M. (1998). SXR, a novel steroid and xenobiotic-sensing nuclear receptor. *Genes Dev.* **12**, 3195–3205.

Chang, T. K. H., Bandiera, S. M., and Chen, J. (2003). Constitutive androstane receptor and pregnane X receptor gene expression in human liver: Interindividual variability and correlation with CYP2B6 mRNA levels. *Drug Metab. Dispos.* **31**, 7–10.

Chen, Y., Ferguson, S. S., Negishi, M., and Goldstein, J. A. (2004). Induction of human *CYP2C9* by rifampicin, hyperforin, and phenobarbital is mediated by the pregnane X receptor. *J. Pharmacol. Exp. Ther.* **308**, 495–501.

Coumoul, X., Diry, M., and Barouki, R. (2002). PXR-dependent induction of human *CYP3A4* gene expression by organochlorine pesticides. *Biochem. Pharmacol.* **64**, 1513–1519.

Echchgadda, I., Song, C. S., Oh, T. S., Cho, S. H., Rivera, O. J., and Chatterjee, B. (2004). Gene regulation for the senescence marker protein DHEA-sulfotransferase by the xenobiotic-activated nuclear pregnane X receptor (PXR). *Mech. Aging Dev.* **125,** 733–745.

El-Sankary, W., Gibson, G. G., Ayrton, A., and Plant, N. (2001). Use of a reporter gene assay to predict and rank the potency and efficacy of CYP3A4 inducers. *Drug Metab. Dispos.* **29,** 1499–1504.

Falkner, K. C., Pinaire, J. A., Xiao, G. H., Geoghegan, T. E., and Prough, R. A. (2001). Regulation of the rat glutathione *S*-transferase A2 gene by glucocorticoids: Involvement of both the glucocorticoid and pregnane X receptors. *Mol. Pharmacol.* **60,** 611–619.

Gardner-Stephen, D., Heydel, J. M., Goyal, A., Lu, Y., Xie, W., Lindblom, T., Mackenzie, P., and Radominska-Pandya, A. (2004). Human PXR variants and their differential effects on the regulation of human UDP-glucuronosyltransferase gene expression. *Drug Metab. Dispos.* **32,** 340–347.

Giguere, V. (1999). Orphan nuclear receptors: From gene to function. *Endocr. Rev.* **20,** 689–725.

Goodwin, B., Moore, L. B., Stoltz, C. M., McKee, D. D., and Kliewer, S. A. (2001). Regulation of the human *CYP2B6* gene by the nuclear pregnane X receptor. *Mol. Pharmacol.* **60,** 427–431.

Goodwin, B., Redinbo, M. R., and Kliewer, S. A. (2002). Regulation of *CYP3A* gene transcription by the pregnane X receptor. *Annu. Rev. Pharmacol. Toxicol.* **42,** 1–23.

Guo, G. L., Moffit, J. S., Nicol, C. J., Ward, J. M., Aleksunes, L. A., Slitt, A. L., Kliewer, S. A., Manautou, J. E., and Gonzalez, F. J. (2004). Enhanced acetaminophen toxicity by activation of the pregnane X receptor. *Toxicol. Sci.* **82,** 374–380.

Handschin, C., Blattler, S., Roth, A., Looser, R., Oscarson, M., Kaufmann, M. R., Podvinec, M., Gnerre, C., and Meyer, C. A. (2004). The evolution of drug-activated nuclear receptors: One ancestral gene diverged into two xenosensor genes in mammals. *Nuclear Receptor* **2,** 7.

Handschin, C., and Meyer, C. A. (2003). Induction of drug metabolism: The role of nuclear receptors. *Pharmacol. Rev.* **55,** 649–673.

Handschin, C., and Meyer, U. A. (2005). Regulatory network of lipid-sensing nuclear receptors: Roles for CAR, PXR, LXR, and FXR. *Arch. Biochem. Biophys.* **433,** 387–396.

Hurst, C. H., and Waxman, D. J. (2004). Environmental phthalate monoesters activate pregnane X receptor-mediated transcription. *Toxicol. Appl. Pharmacol.* **199,** 266–274.

Hurst, C. H., and Waxman, D. J. (2005). Interactions of endocrine-active environmental chemicals with the nuclear receptor PXR. *Toxicol. Environ. Chem.* In press.

Hustert, E., Zibat, A., Presecan-Siedel, E., Eiselt, R., Mueller, R., Fub, C., Brehm, I., Brinkmann, U., Eichelbaum, M., Wojnowski, L., and Burk, O. (2001). Natural protein variants of pregnane X receptor with altered transactivation activity toward *CYP3A4*. *Drug Metab. Dispos.* **29,** 1454–1459.

Jones, S. A., Moore, L. B., Shenk, J. L., Wisely, G. B., Hamilton, G. A., McKee, D. D., Tomkinson, N. C. O., LeCluyse, E. L., Lambert, M. H., Willson, T. M., Kliewer, S. A., and Moore, J. T. (2000). The pregnane X receptor: A promiscuous xenobiotic receptor that has diverged during evolution. *Mol. Endocrinol.* **14,** 27–39.

Jones, S. A., Moore, L. B., Wisely, G. B., and Kliewer, S. A. (2002). Use of *in vitro* pregnane X receptor assays to assess CYP3A4 induction potential of drug candidates. *Methods Enzymol.* **357,** 161–170.

Langmann, T., Moehle, C., Mauerer, R., Scharl, M., Liebisch, G., Zahn, A., Stremmel, W., and Schmitz, G. (2004). Loss of detoxification in inflammatory bowel disease: Dysregulation of pregnane X receptor target genes. *Gastroenterology* **127,** 26–40.

Lehmann, J. M., McKee, D. D., Watson, M. A., Willson, T. M., Moore, J. T., and Kliewer, S. A. (1998). The human orphan nuclear receptor PXR is activated by compounds that regulate *CYP3A4* gene expression and cause drug interactions. *J. Clin. Invest.* **102**, 1016–1023.

Lemaire, G., de Sousa, G., and Rahmani, R. (2004). A PXR reporter gene assay in a stable cell culture system: CYP3A4 and CYP2B6 induction by pesticides. *Biochem. Pharmacol.* **68**, 2347–2358.

Maglich, J. M., Stoltz, C. M., Goodwin, B., Hawkins-Brown, D., Moore, J. T., and Kliewer, S. A. (2002). Nuclear pregnane X receptor and constitutive androstane receptor regulate overlapping but distinct sets of genes involved in xenobiotic detoxification. *Mol. Pharmacol.* **62**, 638–646.

Moore, L. B., Goodwin, B., Jones, S. A., Wisely, G. B., Serabjit-Singh, C. J., Willson, T. M., Collins, J. L., and Kliewer, S. A. (2000a). St. John's wort induces hepatic drug metabolism through activation of the pregnane X receptor. *Proc. Natl. Acad. Sci. USA* **97**, 7500–7502.

Moore, L. B., Parks, D. J., Jones, S. A., Bledsoe, R. K., Consler, T. G., Stimmel, J. B., Goodwin, B., Liddle, C., Blanchard, S. G., Willson, T. M., Collins, J. L., and Kliewer, S. A. (2000b). Orphan nuclear receptors constitutive androstane receptor and pregnane X receptor share xenobiotic and steroid ligands. *J. Biol. Chem.* **275**, 15122–15127.

Pascussi, J. M., Drocourt, L., Fabre, J. M., Maurel, P., and Vilarem, M. J. (2000). Dexamethasone induces pregnane X receptor and retinoid X receptor-α expression in human hepatocytes: Synergistic increase of CYP3A4 induction by pregnane X receptor activators. *Mol. Pharmacol.* **58**, 361–372.

Raucy, J., Warfe, L., Yueh, M. F., and Allen, S. W. (2002). A cell-based reporter gene assay for determining induction of CYP3A4 in a high-volume system. *J. Pharmacol. Exp. Ther.* **303**, 412–423.

Rosenfeld, J. M., Vargas, R., Jr., Xie, W., and Evans, R. M. (2003). Genetic profiling defines the xenobiotic gene network controlled by the nuclear receptor pregnane X receptor. *Mol. Endocrinol.* **17**, 1268–1282.

Schuetz, E., Lan, L., Yasuda, K., Kim, R., Kocarek, T. A., Schuetz, J., and Strom, S. (2002). Development of a real-time *in vivo* transcription assay: Application reveals pregnane X receptor-mediated induction of CYP3A4 by cancer chemotherapeutic agents. *Mol. Pharmacol.* **62**, 439–445.

Song, X., Li, Y., Liu, J., Mukundan, M., and Yan, B. (2005). Simultaneous substitution of phenylanlanine-305 and aspartate-318 of rat pregnane X receptor with the corresponding human residue abolishes the ability to transactivate the *CYP3A23* promoter. *J. Pharmacol. Exp. Ther.* **312**, 571–582.

Staudinger, J. L., Goodwin, B., Jones, S. A., Hawkins-Brown, D., MacKenzie, K. I., LaTour, A., Liu, Y., Klassen, C. D., Brown, K. K., Reinhard, J., Willson, T. M., Koller, B. H., and Kliewer, S. A. (2001). The nuclear receptor PXR is a lithocholic acid sensor that protects against liver toxicity. *Proc. Natl. Acad. Sci. USA* **98**, 3369–3374.

Tabb, M. M., Kholodovych, V., Grun, F., Zhou, C., Welsh, W. J., and Blumberg, B. (2004). Highly chlorinated PCBs inhibits the human xenobiotic response mediated by the steroid and xenobiotic receptor (SXR). *Environ. Health Perspect.* **112**, 163–169.

Tirona, R. G., Leake, B. F., Podust, L. M., and Kim, R. B. (2004). Identification of amino acids in rat pregnane X receptor that determine species-specific activation. *Mol. Pharmacol.* **65**, 36–44.

Traber, M. G. (2004). Vitamin E, nuclear receptors and xenobiotic metabolism. *Arch. Biochem. Biophys.* **423**, 6–11.

Watkins, R. E., Davis-Searles, P. R., Lambert, M. H., and Redinbo, M. R. (2003). Coactivator binding promotes the specific interaction between ligand and the pregnane X receptor. *J. Mol. Biol.* **331**, 815–828.

Waxman, D. J. (1999). P450 gene induction by structurally diverse xenochemicals: Central role of nuclear receptors CAR, PXR, and PPAR. *Arch. Biochem. Biophys.* **369**, 11–23.

Xie, W., Barwick, J. L., Downes, M., Blumberg, B., Simon, C. M., Nelson, M. C., Neuschwander-Tetri, B. A., Brunt, E. M., Guzelian, P. S., and Evans, R. M. (2000). Humanized xenobiotic response in mice expressing nuclear receptor SXR. *Nature* **406**, 435–439.

Xie, W., Radominska-Pandya, A., Shi, Y., Simon, C. M., Nelson, M. C., Ong, E. S., Waxman, D. J., and Evans, R. M. (2001). An essential role for nuclear receptor SXR/PXR in detoxification of cholestatic bild acids. *Proc. Natl. Acad. Sci. USA* **98**, 3375–3380.

Xie, W., Yeuh, M. F., Radominska-Pandya, A., Saini, S. P. S., Negishi, Y., Bottroff, B. S., Cabrera, G. Y., Tukey, R. H., and Evans, R. M. (2003). Control of steroid, heme, and carcinogen metabolism by nuclear pregnane X receptor and constitutive androstane receptor. *Proc. Natl. Acad. Sci. USA* **100**, 4150–4155.

Zhang, H., LeCluyse, E. L., Liu, L., Hu, M., Matoney, L., Zhu, W., and Yan, B. (1999). Rat pregnane X receptor: Molecular cloning, tissue distribution, and xenobiotic regulation. *Arch. Biochem. Biophys.* **368**, 14–22.

Zhu, Z., Kim, S., Chen, T., Lin, J. H., Bell, A., Bryson, J., Dubaquie, Y., Yan, N., Yanchunas, J., Xie, D., Stoffel, R., Sinz, M., and Dickinson, K. (2004). Correlation of high-throughput pregnane X receptor (PXR) transactivation and binding assays. *J. Biomol. Screen.* **9**, 533–540.

[34] Animal Models of Xenobiotic Receptors in Drug Metabolism and Diseases

By Haibiao Gong, Michael W. Sinz, Yan Feng, Taosheng Chen, Raman Venkataramanan, and Wen Xie

Abstract

Drug-metabolizing enzymes, including phase II conjugating enzymes, play an important role in both drug metabolism and human diseases. The genes that encode these enzymes and transporters are inducible by numerous xenobiotics and endobiotics and the inducibility shows clear species specificity. In the past several years, orphan nuclear receptors, such as PXR and CAR, have been established as species-specific "xenobiotic receptors" that regulate the expression of phase I and phase II enzymes and drug transporters. The creation of xenobiotic receptor transgenic and knockout mice has not only provided an opportunity to dissect the transcriptional control of drug metabolizing enzymes, but also offered a unique opportunity to study the xenobiotic receptor-mediated enzyme regulation in both drug metabolism and diseases. "Humanized" hPXR transgenic mice represent a major step forward in the creation and utilization of humanized rodent models for toxicological assessment that may aid in the development of safer drugs.

METHODS IN ENZYMOLOGY, VOL. 400   0076-6879/05 $35.00
Copyright 2005, Elsevier Inc. All rights reserved.   DOI: 10.1016/S0076-6879(05)00034-0

Introduction

Drug Metabolizing Enzymes and Transporters

Metabolism is a major pathway by which drugs are eliminated from the body. The major organ involved in the metabolic elimination of drugs is the liver, followed by the small intestine, lung, and kidney (Sinz and Podell, 2002). Enzymes involved predominantly in the metabolic elimination of drugs can be categorized into two groups: phase I and phase II enzymes. Phase I enzymes are functionalization enzymes that incorporate or uncover a polar functional group by oxidative, hydrolytic, or reductive reactions. The most prevalent oxidative reactions are those mediated by cytochromes P450 (CYP). Phase II enzymes perform conjugative reactions, where a polar endogenous cofactor is bound covalently to the parent drug or metabolite. Phase II reactions include glucuronidation, sulfation, methylation, acetylation, and glutathione and amino acid conjugations. A common metabolic pathway of many drugs is oxidation by CYPs followed by conjugation. Additional information and examples of phase I and II enzymes can be found in numerous excellent reviews (Ioannides, 2002; Lee et al., 2003; Parkinson, 2001; Sinz and Podell, 2002).

Drug uptake and efflux transporters constitute another pathway by which drugs can be absorbed, distributed, or eliminated (Mizuno et al., 2003). Efflux transporters, such as P-glycoprotein (Pgp), can limit the absorption of drugs from the small intestine or across the blood–brain barrier. The organic anion transporting polypeptide (OATP) can facilitate the hepatocyte uptake of drugs from blood, and multidrug resistance-associated protein 2 (MRP2) is involved in the elimination of anionic drugs or metabolites derived from glucuronidation or glutathione conjugation through biliary canalicular membranes (Kim, 2003; Kullak-Ublick, 2003).

Consistent with the significance of drug metabolizing enzymes and transporters, changes in the expression and/or activity of these enzymes or transporters can affect the degree of absorption or elimination of drugs, thereby altering the therapeutic or toxicological response to a drug. These changes are manifested through inhibition, induction, or genetic alterations of the enzymes or transporters. Examples include inhibition of midazolam elimination by the CYP3A4 inhibitor ketoconazole (Olkkola et al., 1994), rifampicin mediated induction of CYP3A4 and loss of efficacy of multiple CYP3A4 substrates, such as oral contraceptives and immunosuppressive agents (Niemi et al., 2003), and reduced elimination of certain drugs or bilirubin through common genetic polymorphisms such as CYP2D6 and UGT1A1 (Gilbert's syndrome), respectively (Bock, 2002; Haining and Yu, 2003). The focus of this chapter is the development of newer animal models to assess metabolism, enzyme induction, and toxicity of drugs and drug candidates.

Challenges for Rodent Models of Drug Metabolism

Primary hepatocytes are a common *in vitro* system to evaluate metabolism, toxicity, and enzyme induction (Tucker *et al.*, 2001; Weaver, 2001). Human hepatocytes are considered the most relevant system to evaluate or predict human metabolism or effects of a new drug. A significant disadvantage of this system is the lack of routine availability of good quality human liver tissue or cells. Other model systems that may provide more consistent access include immortalized hepatocytes or humanized animal models (Mills *et al.*, 2004; Xie and Evans, 2001; Xie *et al.*, 2000a, 2004). Over the years it has been perceived that rodent models have limited utility in predicting drug-related human effects due to significant species differences in drug metabolizing enzymes, transporters, and nuclear hormone receptors. For example, pregnenolone 16α-carbonitrile (PCN) is an effective CYP3A inducer in rat but not humans, and rifampicin induces CYP3A in humans but not in rats (Kocerek *et al.*, 1995). These findings have been attributed to the species differences in the effect of several drugs on CYP3A expression mediated by the nuclear hormone receptor PXR (Jones *et al.*, 2000; Kocerek *et al.*, 1995). These differences across species exemplify the need to develop humanized animal models to evaluate the potential effect of a xenobiotic in humans using animal models.

Nuclear Receptor (NR) and Xenobiotic NRs

The nuclear receptor superfamily includes receptors for hormonal ligands such as steroids, retinoids, thyroid hormone, and vitamin D (Mangelsdorf *et al.*, 1995). A general concept for NR signaling is that in the absence of a ligand, the NR is often associated with a NR corepressor complex, conferring a basal transcription activity of the receptor. Ligand binding to the ligand-binding domain (LBD) of NR induces conformational changes that lead to the release of the corepressor complex and recruitment of the coactivator complex (Glass and Rosenfeld, 2000; Gong and Xie, 2004; Mangelsdorf and Evans, 1998). Coactivator recruitment contributes to chromatin remodeling and subsequent transcriptional activation via specific hormone response elements (HREs) that consist of a minimal core hexameric consensus sequence, 5' AG(G/T)TCA 3', that can be configured into a variety of structured motifs (HRE) (for a review, see Xie and Evans, 2001).

The concept of xenobiotic receptors has developed rapidly since the cloning and characterization of PXR and CAR (Baes *et al.*, 1994; Blumberg *et al.*, 1998; Honkakoski *et al.*, 1998; Kliewer *et al.*, 1998; Swales and Negishi, 2004). It has since been appreciated that PXR and CAR can function as master regulators of phase I and phase II enzymes and drug transporters (Xie *et al.*, 2004). This regulation is achieved by the binding of xenobiotic receptors to their response elements present in the promoter regions of the drug metabolizing enzymes and transporters. The unique complexity of xenobiotic receptor-mediated gene regulation is manifested by the facts that individual xenobiotic NR can regulate multiple genes and multiple xenobiotic NRs can share the same target genes by receptor "cross talk" (Xie *et al.*, 2000b). The phase II enzymes reported to be regulated by PXR and CAR include UGT1A1 (Huang *et al.*, 2003; Sugatani *et al.*, 2001; Xie *et al.*, 2003), SULT (Saini *et al.*, 2004; Sonoda *et al.*, 2003), and GSTs (Falkner *et al.*, 2001). A more complete list of PXR and CAR target genes, including phase I enzymes and drug transporters, can be found elsewhere (Handschin and Meyer, 2003; Wang and LeCluyse, 2003).

Creation of Xenobiotic Receptor Mouse Models

Mouse models are useful tools to study NR-mediated gene regulation and associated pathophysiological events. Homologous recombination (knockout) and transgenic techniques have been used to study the function of xenobiotic receptors. Using PXR as an example, Fig. 1 summarizes strategies to create the loss-of-function knockout model, gain-of-function transgenic models, and the combined "humanized" function models.

Creation of PXR Knockout Mice

Construction of PXR Targeting Vector. Mouse PXR genomic DNA is isolated by screening a 129/Sv library (Stratagene, La Jolla, CA) using a PXR cDNA probe. A targeting vector is generated by replacing the second and third exons of PXR with a PGK-Neo selection marker, in conjunction with a negative selection marker (PGK-TK) (Fig. 2A). The resulting mutant allele has a deletion of two exons spanning nucleotides 339–660 that include amino acids 63–170 of the PXR DBD.

Embryonic Stem (ES) Cell Transfection and Selection. The targeting vector is linearized by *Not* I digestion and transfected into J1 ES cells by electroporation. G418 (200 mg/ml)- and ganciclovir (0.2 mM)-resistant ES clones are screened for designated homologous recombination by Southern blot analysis.

Generation of Chimeric Mice, Germ Line Transmission, and Confirmation of Loss of PXR Expression. PXR$^{+/-}$ ES cells are microinjected into C57BL6/J blastocysts, which are subsequently transplanted into the uteri

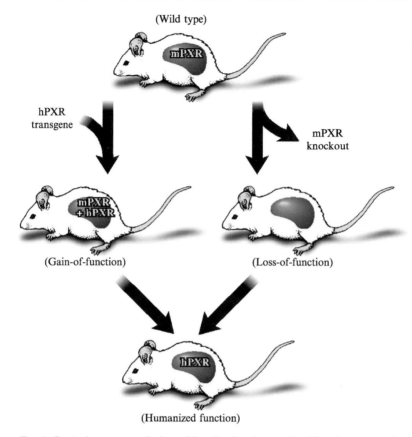

Fig. 1. Strategies to create the loss-of-function knockout, gain-of-function transgenic, and the combined "humanized" function models. Modified from Xie and Evans (2001), with permission of the publisher. (See color insert.)

of pseudopregnant ICR mice. Chimeric male progeny are crossed with C57BL6/J females. Germline transmission of the mutant allele is detected in agouti progeny by Southern blot analysis (Fig. 2B). The absence of PXR mRNA expression in the liver and small intestine of PXR null mice is confirmed by Northern blot analysis (Fig. 2C).

Creation of hPXR and CAR Transgenic Mice

Construction of Transgene. To generate transgenic mice with expression of wild-type hPXR, the hPXR cDNA is cloned downstream of the mouse albumin promoter/enhancer (Pinkert *et al.*, 1987). A SV40 intron/poly(A) sequence is placed downstream of hPXR cDNA (Fig. 3A).

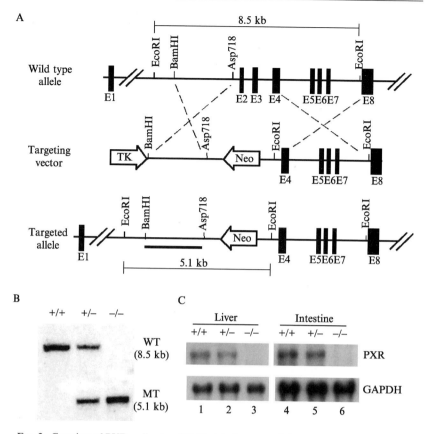

FIG. 2. Creation of PXR null mice. (A) Restriction map of the PXR gene and strategy to generate PXR mutant allele. PXR probes used for Southern blot and expected fragment sizes after *Eco*RI digestion are indicated. E, exon; Neo, neomycin resistance gene; TK, thymidine kinase promoter. (B) Southern blot of *Eco*RI-digested genomic DNA. WT, wild type; MT, mutant. (C) Loss of PXR expression in the liver and small intestine of PXR$^{-/-}$ mice. Adopted from Xie *et al.* (2000a), with permission of the publisher.

Generation of Transgenic Mice and Confirmation of Transgene Expression. The transgene is excised and purified from the plasmid vector and microinjected into the pronuclei of fertilized mouse eggs. The injected eggs are then implanted into the oviduct of pseudopregnant female mice. Transgene positive mice are screened by polymerase chain reaction (PCR) on tail DNA (Xie *et al.*, 2000a). Transgene expression is confirmed by Northern blot analysis (Fig. 3B).

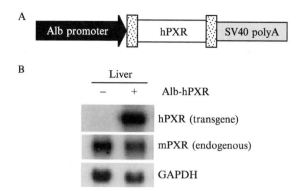

Fig. 3. Creation of PXR transgenic mice. (A) Schematic representation of the Alb-hPXR transgene construct. (B) Northern blot analysis of mouse and human PXR gene expression in the liver of wild-type (−) and transgenic (+) mice. Modified from Xie *et al.* (2000a), with permission of the publisher.

Creation of Tetracycline-Inducible CAR Transgenic Mice. To create a transgenic mouse system that allowed conditional expression of a constitutively activated CAR (VP-CAR) in the liver two lineages of transgenic mice were used as diagrammed in Fig. 4A. First, we created the TetRE-VP-CAR transgene that encodes VP-CAR under the control of a minimal cytomegalovirus promoter and the TetRE (Xie *et al.*, 1999). The TetRE-VPCAR mice were then bred with the Lap-tTA activator line (Jackson Laboratory) to generate bitransgenic animals. Driven by the liver-specific Lap (CCAAT/enhancer-binding protein-β) promoter, the Lap-tTA transgene directs the expression of the tetracycline-responsive transcriptional activator (tTA) exclusively in the hepatocytes (Kistner *et al.*, 1996). We anticipated that tTA bound to TetRE and consequently induced the expression of VP-CAR only in the absence of doxycycline (Dox). The addition of Dox (supplied in drinking at a concentration of 2 mg/ml) results in the displacement of tTA from TetRE and will silence VP-CAR expression (Tet-Off). Northern blot analysis demonstrating the expected transgene expression and Dox regulation is shown in Fig. 4B.

Creation of hPXR "Humanized" Mice

Generation of hPXR Humanized Mice. As diagrammed in Fig. 1, the Alb-hPXR transgene was bred into a PXR-null background, and the resulting PXR-null/hPXR-transgenic mice lacked mPXR but had hPXR transgene expressed in the liver.

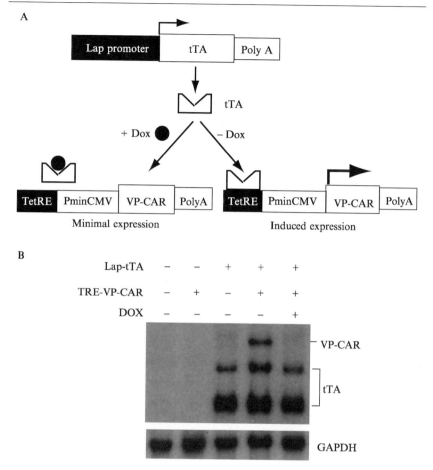

FIG. 4. Creation of transgenic mice that harbor conditional expression of the activated CAR in the liver. (A) Schematic outline of the Lap-tTA/TetRE-VP-CAR two-component Tet-Off transgenic system. The Lap-tTA transgene directs the expression of the tTA activator to the liver. Binding of tTA to the TetRE and induction of the transgene VP-CAR should only occur in the absence of Dox. (B) Liver-specific conditional expression of VP-CAR. Liver RNAs of mice with indicated genotypes were subjected to Northern blot analysis. The mouse in the right-most lane was subjected to 5 days of Dox treatment. Modified from Saini *et al.* (2004), with permission of the publisher.

Humanized Drug Response Profile in "Humanized" Mice. The drug induction of *Cyp3a11* was compared between wild-type and humanized mice by Northern blot analysis (Fig. 5). In humanized mice, *Cyp3a11* was no longer induced by PCN but was efficiently induced by the human-specific

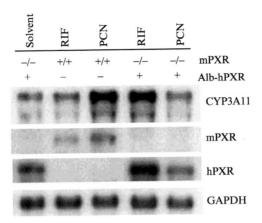

FIG. 5. Drug response profile in humanized mice. Mice with indicated genotypes were treated with a single dose of RIF (5 mg kg^{-1}) or PCN (40 mg kg^{-1}) 24 h prior to liver harvesting. Liver RNA was isolated and subjected to Northern blot analysis with indicated probes. Adopted from Xie *et al.* (2000a), with permission of the publisher.

PXR inducer rifampicin (RIF). In contrast, no induction occurred in wild-type mice by RIF. Expressing hPXR and responding to human-specific PXR inducer exclusively, the "humanized" mice represent a major step toward generating humanized toxicological models to predict xenobiotic enzyme inducibility and drug–drug interactions.

Measurements of Expression and Activity of Drug Metabolizing Enzymes in Animal Models

Detection of Enzyme Expression at mRNA Level

Northern Blot Analysis. Total RNA is extracted using the Trizol reagent (Invitrogen, Calsbad, CA). An equal amount of total RNA (20 μg) is separated on 1.25% agarose–6% formaldehyde gel and transferred to the nylon membrane (Amersham) by capillary action in 10× SSC (0.75 *M* NaCl, 0.075 *M* Na citrate). After UV crosslinking, the membrane is prehybridized at 42° for 2 h in prehybridization buffer containing 50% formamide, 5× SSC, 10 m*M* Na$_3$PO$_4$, 1 m*M* EDTA, 0.5% SDS, 10× Denhardts, and 20 μg/ml of denatured salmon sperm DNA (Sigma). The membrane is then hybridized with a specific probe overnight at 42° in hybridization solution [prehybridization buffer plus 200 μg/ml of denatured salmon sperm DNA, 5% dextran sulfate, and 1 × 10^6 cpm/ml radiolabeled cDNA probe labeled by the random-primed labeling kit (Roche)]. The membrane is washed twice (15 min each) in $2\times$ SSC and 0.1% SDS at 42°, twice (15 min each) in $1\times$ SSC and 0.1% SDS at 65°, and twice (15 min each) in $0.1\times$ SSC and 0.1% SDS at 65°. The hybridization signal is detected by exposing the membrane to X-ray film at −80°. The membranes are stripped and rehybridized with a GAPDH probe for control.

Semiquantitative RT-PCR. Total RNA is treated with RNase-free DNase I to avoid DNA contamination. One microgram of total RNA is reverse transcribed into cDNA by AMV reverse transcriptase (Roche) using oligo(dT) as primer. PCR amplification of cDNA is performed in a $50\text{-}\mu l$ reaction mixture containing 10 mM Tris-HCl buffer (pH 8), 20 mM KCl, 0.1% Triton X-100; 1.5 mM MgCl$_2$; 0.2 mM of each dNTP; 50 pmol of each primer; and 2 U of *Taq* DNA polymerase (Promega). A typical PCR cycle consists of 94° for 30 s, 55° for 40 s, and 72° for 60 s. The amplified products are separated by electrophoresis in 2% agarose gel and visualized with ethidium bromide under UV light. Amplification of the ubiquitously expressed β-actin or GAPDH is performed under the same conditions as a loading control.

Real-Time PCR. Total RNA isolation, DNase I treatment, and first-strand cDNA synthesis are performed as described earlier. Taqman or SYBR Green real-time PCR is performed using the Applied Biosystems 7300 real-time PCR system (Applied Biosystems, Inc., Foster City, CA).

Detection of Enzyme Expression at Protein Level by Western Blot Analysis

Equal amounts of protein are denatured by boiling for 5 min and are loaded onto a 10% Tris-glycine SDS-polyacrylamide gel. Proteins are transferred to nitrocellulose membranes by electroblotting using Tris-glycine buffer with 20% (v/v) methanol. Membranes are blocked for 2 h with 5% nonfat dried milk in phosphate-buffered saline (PBS) containing 0.1% Tween 20 (PBS-T) and are then incubated for 1 h with specific antibody diluted in PBS-T plus 5% nonfat dried milk. After washing with PBS-T three times, the membranes are incubated for 1 h with horseradish peroxidase-labeled secondary antibody in PBS-T plus 5% nonfat dried milk. The blots are then washed three times, and proteins are visualized by the enhanced chemiluminescence (ECL) method (Amersham).

In Vitro Measurement of Phase I and Phase II Enzyme Activity

Microsome Preparation and Measurement of CYP and UGT Activity

MOUSE LIVER MICROSOME PREPARATION. Microsomes are used for the measurement of microsomal enzyme activities, such as those of phase I CYPs and phase II UGTs. Microsomes are prepared from fresh or frozen livers by the differential centrifugation procedure as described previously (Court and Greenblatt, 1997). In brief, liver samples are homogenized by a Teflon grinder in 4 volumes of cold 0.154 M KCl/0.25 M potassium phosphate, pH 7.4, with the addition of butylated hydroxytoluene (10 μl/g tissue) as an antioxidant. The homogenates are spun, and the resultant pellets are reconstituted in 0.25 M sucrose/0.02 M Tris buffer, pH 7.4, and stored at $-80°$. Protein concentration is determined by Lowry's method (Lowry $et\ al.$, 1951).

TOTAL CYPs AND CYP3A ACTIVITY MEASUREMENT. The concentrations of total CYP in the microsomes are determined spectrophotometrically by the method described by Omura and Sato (1964). CYP3A activity is measured by the formation of 6β-hydroxy testosterone from testosterone. Testosterone (250 μM) is incubated in 0.4 mg/ml mice liver microsomes, with MgCl$_2$ (10 mM) and NADPH (1 mM) in 0.1 mM phosphate buffer (pH 7.4). The reaction is allowed to equilibrate in a shaking water bath for 10 min at 37° and is then stopped by cooling the samples quickly by adding 0.5 ml of cold methanol. Samples are mixed and centrifuged at 13,000 rpm for 10 min. Concentrations of 6β-hydroxy testosterone are measured by the reversed-phase HPLC method as described previously (Kostrubsky $et\ al.$, 1999). One hundred microliters of the supernatant is injected onto a LiChrospher 100 RP18 column (4.6 × 250 mm, 5 μm) maintained at 30°, and various components are eluted with a mobile phase of MeOH: H$_2$O (60:40) at a flow rate of 1.2 ml/min. The products are monitored by UV absorption at 242 nm and are quantitated by comparing the absorbance to a standard curve of 6β-hydroxy testosterone. The retention time of 6β-hydroxy testosterone is 5 min.

MEASUREMENT OF UGT ENZYME ACTIVITY

$Glucuronidation\ assay\ using\ HPLC\ fluorescence\ detection.$ The active site of UGTs is localized inside the endoplasmic reticulum. Measurement of the activity of UGTs requires disruption of the membrane barrier using a surfactant such as BRIJ or a pore-forming agent such as alamethicin. Specific substrates are used to measure the isozyme activity of UGTs, such as the formation of estradiol-3-glucuronidation (E-3-G) from estradiol by UGT1A1, acetaminophen-O-glucuronidation (APAPG) by UGT1A6, and morpine-3-glucuronidation (M3G) by UGT2B7. E-3-G is widely used to measure UGT1A1 activity (Fisher $et\ al.$, 2000; Williams $et\ al.$, 2002). In this assay, estradiol (3 μM) is mixed with MgCl$_2$ (0.02 mM), alamethicin (15 μg/ml), 0.1 mM phosphate buffer (pH 7.1), and 0.5 mg/ml microsomal protein. After a 3-min preincubation in a shaking water bath (37°), the metabolic reaction is started by the addition of 10 mM UDP-glucuronic acid (UDPGA). The final incubation volume (250 μl) is allowed to equilibrate in a shaking water bath for 1 h, and the reaction is stopped by adding

25 μl of 6% percholic acid. Dextomethorphan (5 ng/μl) is added as the internal standard. Samples are vortexed and centrifuged at 14,000 rpm for 5 min, and E-3-G concentrations are measured by HPLC as described previously (Alkharfy and Frye, 2002). In brief, 75 μl of supernatant is injected into an Alltima Phenyl column and a mobile phase consisting of 35% acetonitrile and 65% 50 mM ammonium phosphate buffer, pH 3, at a rate of 1 ml/min. Various components are detected using a fluorescence detector set at λ_{ex} 210 nm and λ_{em} 300 nm.

GLUCURONIDATION ASSAY USING THIN-LAYER CHROMATOGRAPHY (TLC). UGT activity is determined using a radiolabeled substrate and UDPGA or an unlabeled substrate and [^{14}C] UDPGA. All substrates are introduced into the reaction mixture in ethanol, and UGTs are activated with alamethiein (60 μg/ml). The UGT1A1 substrate estradiol (100 μM) is preincubated with 20 μg of microsomal protein in 100 mM HEPES-NaOH, pH 7.4, 5 mM MgCl$_2$, and 5 mM saccharolactone at room temperature. The metabolic reactions are initiated by adding UDPGA (4 mM) at 37° for 15 min. The reaction is stopped by the addition of 20 μl ethanol. Samples are applied to TLC plates (Baker Si250, VWR, Irving, TX). Glucuronidated products and untreated substrate are separated by development in chloroform–methanol–glacial acetic acid–water (65:25:2:4, by volume). Radioactive compounds are localized on TLC plates by autoradiography for 3–4 days at −80°. Silica gel in zones corresponding to the glucuronide bands are scraped into scintillation vials, and radioactivity is measured by liquid scintillation counting. An example of the TLC method used to measure estrone glucuronidation can be seen in Xie *et al.* (2003).

Cytosol Preparation and Measurement of SULT and GST Activity

CYTOSOL PREPARATION. Cytosolic preparations are used for *in vitro* measurement of the activity of cytosolic phase II enzymes, such as SULTs and GSTs. The cytosol extract is made by homogenizing 0.1-g tissue sample in 1.5 ml of an appropriate buffer using a 2-ml glass homogenizer. Five millimolar KPO$_4$ buffer (pH 6.5) containing 0.25 M sucrose is used as the homogenization buffer for the SULT assay, and PBS buffer (pH 7.4) is used as the homogenization buffer for the GST assay. After centrifugation at maximum speed at 4° for 10 min, the supernatant is collected and stored at −80° until use.

SULFOTRANSFERASE ASSAY. Specific substrates are used to assess the activity of SULT isozymes, such as dehydroepiandrosterone (DHEA) for SULT2A1 (Mesia-Vela and Kauffman, 2003). In the DHEA sulfation assay (Saini *et al.*, 2004; Sonoda *et al.*, 2002), 5 μM DHEA is mixed with 5–10 μg/ml of cytosolic protein diluted in 5 mM potassium phosphate (pH 6.5) containing 1.5 mg/ml bovine serum albumin and 10 mM dithiothreitol (DTT). Incubation is started by adding 50 μl cocktail buffer, including 50 mM potassium phosphate (pH 6.5), 7.4 mg/ml DTT, and 1.28 μM [^{35}S]3'-phosphoadenosine 5''-phosphosulfate (PAPS), the sulfate donor. After incubation in a shaking water bath for 20 min at 37°, free [^{35}S]PAPS is removed by sequential additions of 50 μl each of 0.1 M Ba(OH)$_2$ and 0.1 M ZnSO$_4$, which precipitates [^{35}S]PAPS. After vortexing and centrifuging, a 300-μl aliquot of supernatant is analyzed for radioactivity using a scintillation counter (e.g., Beckman LS6500).

GLUTATHIONE S-TRANSFERASE (GST) ASSAY. Total GST activity is measured using 1-chloro-2,4-dinitrobenzene (CNDB) as a substrate (Habig and Jakoby, 1981; Labrou et al., 2001). One millimolar GSH is added to GST (0.5–5 μg enzyme; approximately 10 μM active site) in 1 ml of 100 mM potassium phosphate buffer, pH 6.5. After preincubation for 3 min, CDNB (0.1 mM) is added, and activity is measured using an UV/visible spectrophotometer at 25°. The increase of absorbance resulting from the conjugation of dinitrophenyl with glutathione is recorded at 340 nm every 15 s for 3 min. The appropriate controls for any nonenzymatic reactions are performed and subtracted from the catalyzed reaction.

Use of Xenobiotic Receptor Mouse Models for In Vivo Drug Metabolism Studies

In addition to their use in dissecting the transcriptional control of drug metabolizing enzymes and transporters, the use of humanized or knockout mice offers a unique opportunity to study the in vivo relationship between drug concentrations and the corresponding physiological effects. For example, the effects of a drug discovery candidate could be evaluated in such a model by administering several doses of the drug candidate and measuring the effects on the RNA expression or activity of various enzymes. Using this approach, Xie et al. (2000a) demonstrated that increasing concentrations of rifampicin (1–10 mg/kg) lead to corresponding increases in CYP3A11 mRNA levels in hPXR mice. During the drug development process, this type of study could be employed with new drug candidates to assess the upregulation of drug metabolizing enzymes or transporters. By administering the drug candidate to animals for several days at doses yielding plasma exposures similar to those predicted to be efficacious in humans, the effect of the candidate on hPXR-mediated enzyme and transporter expression can be assessed in this in vivo setting.

The study of drug dose (or plasma concentration) and their corresponding pharmacological effects by mathematical modeling is often referred to as pharmacokinetic–pharmacodynamic analysis. This type of modeling has shown utility in comparing the drug concentration–effect relationship in vivo with in vitro assays or in quantitating the time course of drug effects in relation to plasma/organ concentrations (Meibohm and Derendorf, 2002). Humanized PXR or CAR mice offer a vehicle to study the relationship among dose, plasma, and liver concentrations of new and existing drugs to the induction effects on CYP3A and 2B genes, respectively, in a more complex *in vivo* system as compared to current *in vitro* models. In a similar manner, knockout models can be used to assess the involvement of certain nuclear hormone receptors in the expression of enzymes or transporters, changes in physiology, and the toxicity of certain drugs (Qatanani *et al.*, 2004; Wei *et al.*, 2000; Xie *et al.*, 2000a; Zhang *et al.*, 2002).

These newer *in vivo* models offer a dynamic system incorporating drug absorption, distribution, metabolism, and elimination, albeit in mouse, in contrast to the static system of cell culture where cells are exposed continually to drug and perhaps metabolites. In addition, the concentration–effect relationship can be expanded to compare *in vivo* liver/plasma concentrations with concentrations used in the *in vitro* human hepatocyte system and their corresponding *in vivo* and *in vitro* effects on gene transcription, enzyme induction, or toxicity. This type of comparison may actually improve the extrapolation of *in vitro* human hepatocyte concentration–induction effects to the prediction of induction and toxicological responses in patients.

Use of Xenobiotic Receptor Mouse Models to Study Drug Metabolizing Enzyme-Related Liver Diseases and Hormonal Homeostasis

Bile Acid-Induced Cholestasis

Bile acids are cholesterol catabolic byproducts. Excessive bile acids are potentially toxic and may cause cholestatic liver damage (Leuschner *et al.*, 1977). The detoxification of bile acid is mediated by phase I and phase II enzymes as well as drug transporters whose expression is controlled by xenobiotic receptors such as PXR and CAR.

Prevention of Cholestasis by PXR Activation. To examine whether activation of PXR has effects on LCA-induced liver damage, we created Alb-VP-hPXR transgenic mice in which the activated hPXR (VP-hPXR) was expressed in the liver (Xie *et al.*, 2000a). Transgenic mice or their wild-type littermates were dosed with vehicle solvent or LCA (8 mg/day) before liver histological evaluation by H&E staining. The histological liver damage was scored by the appearance of areas of saponifucation/coagulative necrosis. After 4 days of treatment, 58% of wild-type mice exhibited areas of liver damage. In contrast, VP-hPXR transgenic mice were completely resistant to LCA-induced liver damage (Xie *et al.*, 2001), demonstrating that sustained activation of PXR is sufficient to prevent LCA-mediated histological liver damage. The induction of CYP3A11 expression was initially reasoned to be responsible for LCA resistance; however, SUL-T2A9 was later identified as a PXR target gene and was found to be induced in VP-hPXR mice (Sonoda *et al.*, 2002). Because both CYP3A and SULT2A play a role in bile acid detoxification, it is likely that LCA resistance was a combined effect of CYP3A and SULT2A9 induction.

Prevention of Cholestasis by CAR Activation. To examine whether activation of CAR has any effect on LCA sensitivity, TRE-VP-CAR/ Lap-tTA bitransgenic mice or control littermates were dosed with vehicle solvent or LCA for 4 days before liver histological evaluation. Similar to VP-hPXR mice, VP-CAR mice were completely resistant to LCA-induced liver damage (Saini *et al.*, 2004), demonstrating that sustained activation of CAR is sufficient to prevent LCA-mediated cholestasis. This effect is CAR activation dependent, as treatment of Dox blocked protection, presumably due to the silence of VP-CAR transgene expression (Saini *et al.*, 2004). The LCA resistance in VP-CAR mice was associated with an induction of SUL2A9 but not CYP3A11 (Saini *et al.*, 2004).

Differential Effect of Individual and Combined Loss of PXR and CAR on Cholestasis. As both PXR and CAR have been implicated in bile acid detoxification, the effect of individual or combined loss of these two receptors was examined. Following LCA ip (250 mg/kg), wild-type, PXR-null, and CAR-null mice showed no or little hepatotoxicity, suggesting the single knockout mice had sufficient LCA detoxification to prevent hepato-toxicity. In contrast, the livers of LCA-treated double knockout males showed massive liver damage (Uppal *et al.*, 2005). These results suggested that the combined, but not individual, loss of PXR and CAR resulted in a robust sensitivity to LCA hepatotoxicity. Further analysis revealed that the increased sensitivity was male specific and was associated with a profound decrease in the expression of bile acid transporters and defects in bile acid clearance (Uppal *et al.*, 2005). Of note, in an independent study, Zhang *et al.* (2004) reported that double knockout mice were equally sensitive to LCA toxicity as CAR-null mice. Factors such as the dose and route of drug administration and sex of the animal may account for the discrepancies between these two studies.

Hyperbilirubinemia

Bilirubin is the catabolic by-product of heme proteins, such as β-globin and CYP enzymes. Accumulation of bilirubin in the blood is potentially hepato- and neurotoxic. The expression of phase II enzymes and drug transporters responsible for bilirubin clearance is also under the transcriptional control of PXR and CAR.

Prevention of Hyperbilirubinemia by PXR Activation. To evaluate the effect of PXR activation on bilirubin clearance, Alb-VP-hPXR transgenic mice and their wild-type controls were given a single dose of bilirubin (10 mg kg^{-1} body weight) via tail vein injection. Blood samples were collected 1 h after injection and serum was separated by centrifugation at 4000 rpm for 10 min. Total and conjugated bilirubin levels were measured by Antech Diagnostic (Lake Success, NY). One hour after injection, the serum levels of both total and conjugated bilirubin in VP-hPXR transgenic mice were less than half of the wild-type controls (Xie et al., 2003). As UGT1A1 is the principal enzyme that facilitates bilirubin glucuronidation and subsequent clearance, this result is consistent with the finding that PXR regulates UGT1A expression.

Prevention of Hyperbilirubinemia by CAR Activation. VP-CAR mice were used to evaluate the effect of CAR activation on bilirubin clearance. No differences in serum bilirubin levels were seen in untreated VP-CAR or wild-type mice. However, upon bilirubin injection, the levels of total bilirubin in the transgenic mice were less than half of those in wild-type mice (Saini et al., 2005). The protection was VP-CAR dependent, as no protection was seen in mice treated with Dox for 7 days prior to bilirubin injection (Saini et al., 2005). Consistent with their resistance to hyperbilirubinemia, VP-CAR mice had an increased expression of genes encoding bilirubin-detoxifying enzymes and transporters such as UGT1A1, OATP4, GSTA2, OATP2, and MRP2 (Saini et al., 2005).

Differential Effect of Individual and Combined Loss of PXR and CAR on Hyperbilirubinemia. The effect of individual and combined loss of PXR and CAR was examined using PXR and CAR single or double knockout mice. The loss of PXR and/or CAR did not significantly alter the basal level of serum bilirubin. Upon bilirubin infusion, whereas CAR-null and double knockout mice showed similar sensitivity as wild-type mice, PXR-null mice exhibited a surprisingly complete resistance to hyperbilirubinemia (Saini et al., 2005). The increased bilirubin resistance in PXR-null mice was associated with the induction of bilirubin-detoxifying genes, such as UGT1A1, OATP4/SLC21A6, GSTA2, and MRP2. The expression of these genes remained largely unchanged in CAR-null or double knockout mice, consistent with the sensitivity to hyperbilirubinemia found in both genotypes (Saini et al., 2005). To understand the increased bilirubin clearance in PXR-null mice, we proposed that the ligand-free PXR functions as a suppressor to inhibit the constitutive activity of CAR. Deletion of

PXR results in derepression, which leads to the same bilirubin resistance phenotype as seen in VP-CAR mice (Saini *et al.*, 2005).

Hormonal Homeostasis

Steroid hormones such as corticosteroids are substrates for phase II enzymes such as UGT. It is conceivable that genetic or pharmacological activation of xenobiotic receptors and the resultant enzyme regulation will impact the hormonal homeostasis. To measure the effect of PXR activation on corticosterone levels, blood samples were collected from Alb-VP-hPXR transgenic mice and their wild-type controls. Plasma was obtained by centrifuging blood samples at 7000 rpm for 4 min. The 24-h urine was collected using mouse metabolic cages (Nalge, Medina, OH). Corticosterone levels in plasma and urine were measured with a corticosterone ^{125}I-labeled RIA kit from ICN (Biomedical, Irvine, CA). The plasma and urinary corticosterone levels were significantly higher in VP-hPXR mice as compared to their wild-type littermates, indicating that the increased glucuronidation is associated with an increased output of glucocorticoid (Xie *et al.*, 2003).

Closing Remarks

We have observed a rapid growth in our understanding of xenobiotic receptor functions, consistent with the notion that xenobiotic receptor target gene products control the homeostasis of both xenobiotics, such as drugs and environmental toxicants, and endobiotics, such as bile acids and bilirubin. The creation and utilization of xenobiotic receptor mouse models offer a unique and valuable vehicle to dissect the complexity of xenobiotic receptor-mediated gene regulation and the implication of such regulation in drug metabolism and many pathophysiological events.

Acknowledgments

The original research described in this chapter was supported in part by NIH Grants ES012479 and CA107011 and by the University of Pittsburgh Central Research Development Fund.

References

Alkharfy, K. M., and Frye, R. F. (2002). Sensitive liquid chromatographic method using fluorescence detection for the determination of estradiol 3- and 17-glucuronides in rat and human liver microsomal incubations: Formation kinetics. *J. Chromatogr. B Analyt. Technol. Biomed. Life Sci.* **774,** 33–38.

Baes, M., Gulick, T., Choi, H.-S., Martinoli, M. G., Simha, D., and Moore, D. D. (1994). A new orphan member of the nuclear receptor superfamily that interacts with a subset of retinoic acid response elements. *Mol. Cell. Biol.* **14,** 1544–1552.

Blumberg, B., Sabbagh, W., Jr., Juguilon, H., Bolado, J., Jr., van Meter, C. M., Ong, E. S., and Evans, R. M. (1998). SXR, a novel steroid and xenobiotic-sensing nuclear receptor. *Genes Dev.* **12,** 3195–3205.

Bock, K. W. (2002). UDP-Glucuronosyltransferases. *In* "Enzyme Systems That Metabolize Drugs and Other Xenobiotics" (C. Ioannides, ed.), pp. 281–318. Wiley, New York.

Court, M. H., and Greenblatt, D. J. (1997). Molecular basis for deficient acetaminophen glucuronidation in cats: An interspecies comparison of enzyme kinetics in liver microsomes. *Biochem. Pharmacol.* **53,** 1041–1047.

Falkner, K. C., Pinaire, J. A., Xiao, G. H., Geoghegan, T. E., and Prough, R. A. (2001). Regulation of the rat glutathione S-transferase A2 gene by glucocorticoids: Involvement of both the glucocorticoid and pregnane X receptors. *Mol. Pharmacol.* **60,** 611–619.

Fisher, M. B., Campanale, K., Ackermann, B. L., VandenBranden, M., and Wrighton, S. A. (2000). *In vitro* glucuronidation using human liver microsomes and the pore-forming peptide alamethicin. *Drug Metab. Dispos.* **28,** 560–566.

Glass, C. K., and Rosenfeld, M. G. (2000). The coregulator exchange in transcriptional functions of nuclear receptors. *Genes Dev.* **14,** 121–141.

Gong, H., and Xie, W. (2004). Orphan nuclear receptors, PXR and LXR: New ligands and therapeutic potential. *Expert Opin. Ther. Targets* **8,** 49–54.

Habig, W. H., and Jakoby, W. B. (1981). Assays for differentiation of glutathione S-transferases. *Methods Enzymol.* **77,** 398–405.

Haining, R. L., and Yu, A. (2003). Cytochrome P450 pharmacogenetics. *In* "Drug Metabolizing Enzymes" (J. Lee, R. S. Obach, and M. B. Fisher, eds.), pp. 375–419. FontisMedia, Switzerland.

Handschin, C., and Meyer, U. (2003). Induction of drug metabolism: The role of nuclear receptors. *Pharmacol. Rev.* **55,** 649–673.

Honkakoski, P., Zelko, I., Sueyoshi, T., and Negishi, M. (1998). The nuclear orphan receptor CAR-retinoid X receptor heterodimer activates the phenobarbital-responsive enhancer module of the CYP2B gene. *Mol. Cell. Biol.* **18,** 5652–5658.

Huang, W., Zhang, J., Chua, S. S., Qatanani, M., Han, Y., Granata, R., and Moore, D. D. (2003). Induction of bilirubin clearance by the constitutive androstane receptor (CAR). *Proc. Natl. Acad. Sci. USA* **100,** 4156–4161.

Ioannides, C. (2002). "Enzyme Systems That Metabolize Drugs and Other Xenobiotics." Wiley, New York.

Jones, S. A., Moore, L. B., Shenk, J. L., Wisely, G. B., Hamilton, G. A., and McKee, D. D. (2000). The pregnane X receptor, a promiscuous xenobiotic receptor that has diverged during evolution. *Mol. Endocrinol.* **14,** 27–39.

Kim, R. B. (2003). Organic anion-transporting polypeptide (OATP) transporter family and drug disposition. *Eur. J. Clin. Invest.* **33,** 1–5.

Kistner, A., Gossen, M., Zimmermann, F., Jerecic, J., Ullmer, C., Lubbert, H., and Bujard, H. (1996). Doxycycline-mediated quantitative and tissue-specific control of gene expression in transgenic mice. *Proc. Natl. Acad. Sci. USA* **93,** 10933–10938.

Kliewer, S. A., Moore, J. T., Wade, L., Staudinger, J. L., Watson, M. A., Jones, S. A., McKee, D. D., Oliver, B. B., Willson, T. M., Zetterstrom, R. H., Perlmann, T., and Lehmann, J. M. (1998). An orphan nuclear receptor activated by pregnanes defines a novel steroid signaling pathway. *Cell* **92,** 73–82.

Kocerek, T. A., Schuetz, E. G., Strom, S. C., Fisher, R. A., and Guzelian, P. S. (1995). Comparative analysis of cytochrome P4503A induction in primary cultures of rat, rabbit, and human hepatocytes. *Drug Metab. Dispos.* **23**, 415–421.

Kostrubsky, V. E., Ramachandran, V., Venkataramanan, R., Dorko, K., Esplen, J. E., Zhang, S., Sinclair, J. F., Wrighton, S. A., and Strom, S. C. (1999). The use of human hepatocyte cultures to study the induction of cytochrome P-450. *Drug Metab. Dispos.* **27**, 887–894.

Kullak-Ublick, G. A. (2003). ABC transporter regulation by bile acids: Where PXR meets FXR. *J. Hepatol.* **39**, 628–630.

Labrou, N. E., Mello, L. V., and Clonis, Y. D. (2001). Functional and structural roles of the glutathione-binding residues in maize (*Zea mays*) glutathione S-transferase I. *Biochem. J.* **358**, 101–110.

Lee, J., Obach, R. S., and Fisher, M. B. (2003). "Drug Metabolizing Enzymes." FontisMedia, Switzerland.

Leuschner, U., Czygan, P., Liersch, M., Frohling, W., and Stiehl, A. (1977). Morphologic studies on the toxicity of sulfated and nonsulfated lithocholic acid in the isolation-perfused rat liver. *Z. Gastroenterol.* **15**, 246–253.

Lowry, O. H., Rosebrough, N. J., Farr, A. L., and Randall, R. J. (1951). Protein measurement with the Folin phenol reagent. *J. Biol. Chem.* **193**, 265–275.

Mangelsdorf, D. J., Thummel, C., Beato, M., Herrlich, P., Schutz, G., Umesono, K., Blumberg, B., Kastner, P., Mark, M., Chambon, P., and Evans, R. M. (1995). The nuclear receptor superfamily: The second decade. *Cell* **83**, 835–839.

Mangelsdorf, D. J., and Evans, R. M. (1998). The RXR heterodimers and orphan receptors. *Cell* **83**, 841–850.

Meibohm, B., and Derendorf, H. (2002). Pharmacokinetic/pharmacokinetic studies in drug product development. *J. Pharm. Sci.* **91**, 18–31.

Mesia-Vela, S., and Kauffman, F. C. (2003). Inhibition of rat liver sulfotransferases SULT1A1 and SULT2A1 and glucuronosyltransferase by dietary flavonoids. *Xenobiotica* **33**, 1211–1220.

Mills, J. B., Rose, K. A., Sadagopan, N., Sahi, J., and de Morais, S. M. F. (2004). Induction of drug metabolism enzymes and MDR1 using a novel human hepatocyte cell line. *J. Pharm. Expt. Ther.* **30**, 303–309.

Mizuno, N., Niwa, T., Yotsumoto, Y., and Sugiyama, Y. (2003). Impact of drug transporter studies on drug discovery and development. *Pharmacol. Rev.* **55**, 425–461.

Niemi, M., Backman, J. T., Fromm, M. F., Neuvonen, P. J., and Kivisto, K. T. (2003). Pharmacokinetic interactions with rifampicin. *Clin. Pharmacokinet.* **42**, 819–850.

Olkkola, K. T., Backman, J. T., and Neuvonen, P. J. (1994). Midazolam should be avoided in patients receiving the systemic antimycotics ketoconazole or itraconazole. *Clin. Pharmacol. Ther.* **55**, 481–485.

Omura, T., and Sato, R. (1964). The carbon monoxide-binding pigment of liver microsomes. II. Solubilization, purification, and properties. *J. Biol. Chem.* **239**, 2379–2385.

Parkinson, A. (2001). Biotransformation of xenobiotics. *In* "Toxicology, the Basic of Science Poisons" (C. Klaassen, ed.), 6th Ed., pp. 133–224. McGraw-Hill, New York.

Pinkert, C. A., Ornitz, D. M., Brinster, R. L., and Palmiter, R. D. (1987). An albumin enhancer located 10 kb upstream functions along with its promoter to direct efficient, liver-specific expression in transgenic mice. *Genes Dev.* **1**, 268–276.

Qatanani, M., Wei, P., and Moore, D. D. (2004). Alterations in the distribution and orexigenic effects of dexamethasone in CAR-null mice. *Pharm. Biochem. Behav.* **78**, 285–291.

Saini, S. P., Sonoda, J., Xu, L., Toma, D., Uppal, H., Mu, Y., Ren, S., Moore, D. D., Evans, R. M., and Xie, W. (2004). A novel constitutive androstane receptor-mediated and CYP3A-independent pathway of bile acid detoxification. *Mol. Pharmacol.* **65,** 292–300.

Saini, S. P., Mu, Y., Gong, H., Toma, D., Uppal, H., Ren, S., Li, S., Poloyac, S. M., and Xie, W. (2005). A dual role of orphan nuclear receptor PXR in bilirubin detoxification. *Hepatology* **41,** 497–505.

Sinz, M., and Podell, T. (2002). The mass spectrometer in drug discovery. *In* "Mass Spectrometry in Drug Discovery" (M. Sinz and D. Rossi, eds.), pp. 271–335. Dekker, New York.

Sonoda, J., Xie, W., Rosenfeld, J. M., Barwick, J. L., Guzelian, P. S., and Evans, R. M. (2002). Regulation of a xenobiotic sulfonation cascade by nuclear pregnane X receptor (PXR). *Proc. Natl. Acad. Sci. USA* **99,** 13801–13806.

Sugatani, J., Kojima, H., Ueda, A., Kakizaki, S., Yoshinari, K., Gong, Q. H., Owens, I. S., Negishi, M., and Sueyoshi, T. (2001). The phenobarbital response enhancer module in the human bilirubin UDP-glucuronosyltransferase UGT1A1 gene and regulation by the nuclear receptor CAR. *Hepatology* **33,** 1232–1238.

Swales, K., and Negishi, M. (2004). CAR, driving into the future. *Mol. Endocrinol.* **18,** 1589–1598.

Tucker, G. T., Houston, J. B., and Huang, S.-M. (2001). Optimizing drug development: Strategies to assess drug metabolism/transporter interaction potential—toward a consensus. *Pharm. Res.* **18,** 1071–1080.

Wang, H., and LeCluyse, E. (2003). Role of orphan nuclear receptors in the regulation of drug-metabolising enzymes. *Clin. Pharmacokinet.* **42,** 1331–1357.

Weaver, R. J. (2001). Assessment of drug-drug interactions: Concepts and approaches. *Xenobiotica* **31,** 499–538.

Wei, P., Zhang, J., Egan-Hafley, M., Liang, S., and Moore, D. D. (2000). The nuclear receptor CAR mediates specific xenobiotic induction of drug metabolism. *Nature* **407,** 920–923.

Williams, J. A., Ring, B. J., Cantrell, V. E., Campanale, K., Jones, D. R., Hall, S. D., and Wrighton, S. A. (2002). Differential modulation of UDP-glucuronosyltransferase 1A1 (UGT1A1)-catalyzed estradiol-3-glucuronidation by the addition of UGT1A1 substrates and other compounds to human liver microsomes. *Drug Metab. Dispos.* **30,** 1266–1273.

Xie, W., Barwick, J. L., Downes, M., Blumberg, B., Simon, C. M., Nelson, M. C., Neuschwander-Tetri, B. A., Brunt, E. M., Guzelian, P. S., and Evans, R. M. (2000a). Humanized xenobiotic response in mice expressing nuclear receptor SXR. *Nature* **406,** 435–439.

Xie, W., Barwick, J. L., Simon, C. M., Pierce, A. M., Safe, S., Blumberg, B., Guzelian, P. S., and Evans, R. M. (2000b). Reciprocal activation of xenobiotic response genes by nuclear receptors SXR/PXR and CAR. *Genes Dev.* **14,** 3014–3023.

Xie, W., Chow, L. T., Paterson, A. J., Chin, E., and Kudlow, J. E. (1999). Conditional expression of the ErbB2 oncogene elicits reversible hyperplasia in stratified epithelia and up-regulation of TGFalpha expression in transgenic mice. *Oncogene* **18,** 3593–3607.

Xie, W., and Evans, R. M. (2001). Orphan nuclear receptors: The exotics of xenobiotics. *J. Biol. Chem.* **276,** 37739–37742.

Xie, W., Radominska-Pandya, A., Shi, Y., Simon, C. M., Nelson, M. C., Ong, E. S., Waxman, D. J., and Evans, R. M. (2001). An essential role for nuclear receptors SXR/PXR in detoxification of cholestatic bile acids. *Proc. Natl. Acad. Sci. USA* **98,** 3375–3380.

Xie, W., Uppal, H., Saini, S. P., Mu, Y., Little, J. M., Radominska-Pandya, A., and Zemaitis, M. A. (2004). Orphan nuclear receptor-mediated xenobiotic regulation in drug metabolism. *Drug Discov. Today* **9**, 442–449.

Xie, W., Yeuh, M. F., Radominska-Pandya, A., Saini, S. P., Negishi, Y., Bottroff, B. S., Cabrera, G. Y., Tukey, R. H., and Evans, R. M. (2003). Control of steroid, heme, and carcinogen metabolism by nuclear pregnane X receptor and constitutive androstane receptor. *Proc. Natl. Acad. Sci. USA* **100**, 4150–4155.

Zhang, J., Huang, W., Chua, S. S., Wei, P., and Moore, D. D. (2002). Modulation of acetaminophen-induced hepatotoxicity by the xenobiotic receptor CAR. *Science* **298**, 422–424.

Zhang, J., Huang, W., Qatanani, M., Evans, R. M., and Moore, D. D. (2004). The constitutive androstane receptor and pregnane X receptor function coordinately to prevent bile acid-induced hepatotoxicity. *J. Biol. Chem.* **279**, 49517–49522.

[35] Cancer and Molecular Biomarkers of Phase 2

By KIM DALHOFF, KASPAR BUUS JENSEN, and
HENRIK ENGHUSEN POULSEN

Abstract

Associations between genotypes of phase 2 enzymes and cancer risk are extracted from epidemiological studies, namely case–control studies. Variant alleles in glutathione *S*-transferase (GST), UDP-glucuronosyltransferase (UGT), sulfotransferase (SULT), and *N*-acetyltransferase (NAT) have been used as molecular genetic biomarkers of risk. GSTM (my)1 has been associated with an increased risk of colorectal cancer, lung cancer, and bladder cancer and GSTP(pi)1 with prostate cancer. UGT1A1*28 and *37 are both associated with an increased risk of breast cancer as is SULT1A1*2. The presence of UGT1A1*28 results in an increased risk of ovarian cancer and NAT2 of colorectal and lung cancer. A high frequency of SULT1A1*1 has been identified in patients with breast cancer; the role in colorectal cancer is more controversial. This chapter discusses the balance between carcinogen activation and detoxification in relation to phase 2 enzymes.

Introduction

Genetic polymorphisms exist in several phase 2 enzymes that catalyze carcinogen activation or detoxification. Although polymorphisms in glutathione *S*-transferase (GST), UDP-glucuronosyltransferase (UGT),

METHODS IN ENZYMOLOGY, VOL. 400   0076-6879/05 $35.00
DOI: 10.1016/S0076-6879(05)00035-2

sulfotransferase (SULT), and N-acetyltransferase (NAT) are well established, only few data link a particular genotype to an increased or a decreased risk of cancer. An association between a specific genotype and cancer risk is often found by epidemiological studies, that is, case–control studies. We have identified clinical studies using molecular genetic biomarkers (variant alleles) of phase 2 enzymes as markers of susceptibility to the development of cancer (Table I).

Glutathione S-transferase

Glutathione S-transferases are a family of inducible enzymes that are important in carcinogen detoxification. They catalyze the conjugation of a variety of different compounds with the endogenous tripeptide glutathione (GSH). Cytosolic GSTs can be divided into four human families (isozymes) with different but sometime overlapping substrate specificities. They are termed A (alpha), M (my), P (pi), and T (theta).

Both the expression and the protein levels of GST isozymes vary between individuals, making them predisposed to the toxic effects of environmental carcinogens. Also, high levels of GSTs (especially GSTP) have been found in human cancer tumors compared with normal tissue (Chang and Yang, 2000; Kelekar and Thompson, 1998; Wolter *et al.*, 1997). Variation in levels of tumor GST may be associated with resistance or susceptibility to chemotherapeutic agents(Lewis *et al.*, 1989).

One of the most studied GSTs is GSTM1, which is only expressed in 50% of the population (Seidegard *et al.*, 1988). Lack of GSTM1 (GST M1 null genotype) is associated with lung cancer. In a meta-analysis of epidemiological studies (Vineis *et al.*, 1999), it was estimated that both Caucasians and Asians had an increased risk of developing lung cancer (OR = 1.21, 95% CI = 1.06–1.39 and OR = 1.45, 95% CI = 1.23–1.70, respectively). Also, bladder cancer was associated with the GSTM1-null genotype. Caucasians had an increased risk of 1.54 (OR) (95% CI = 1.32–1.80) and Asians 1.71 (OR) (95% CI = 1.09–2.91). Other genetic polymorphic GSTs are associated with cancer. In a Japanese case–control study (Nakazato *et al.*, 2003) of patients with familial prostate cancer, it was shown that the GSTP1-valine variant (A-to-C transition at nucleotide 313) was associated with a significant increased risk of prostate cancer in homozygous individuals (OR = 9.31, 95% CI = 0.47–184).

Not only single polymorphisms within each GST class may be used as biomarkers of cancer risk. Combinations of polymorphisms within different GST classes also influence the risk. Individuals with two high-risk genotypes in M1, P1, or T1 have a significantly higher risk of developing prostate cancer (Nakazato *et al.*, 2003), and the risk of CLL is increased

TABLE I

MOLECULAR BIOMARKERS OF PHASE 2 ENZYMES AND RISK OF CANCER IN CASE–CONTROL
STUDIES [ODDS RATIO (OR) AND 95% CONFIDENCE INTERVALS (CI)]

Disease	Molecular biomarker	OD (95% CI)	Comments
Colorectal cancer	SULT1A1*1	4.4 (1.6–11.8)	Smoking-associated disease
	SULT1A1*1	0.47 (0.27–0.83)	
	NAT2	1.19	Fast metabolizers
	GSTM1	1.78 (1.39–2.17)	
Breast cancer	SULT1A1*2 (both homo- and heterozygous)	2.11 (1.00–4.46)	Premenopausal women smoking >5 cigarettes daily[b]
	SULT1A1*2 (both homo- and heterozygous)	2.83 (1.23–6.54)	Premenopausal women smoking >20 years[b]
	SULT1A1*2	HR[a] 2.9 (1.1–7.6)	Death among tamoxifen-treated women
	UGT1A1*28	1.8 (1.0–3.1)	Invasive disease in premenopausal African-American women
	UGT1A1*37	1.8 (1.0–3.1)	Invasive disease in premenopausal African-American women
Lung cancer	SULT1A1*2 (both homo- and heterozygous)	1.41 (1.04–1.91)	Women and current and heavy smokers
	NAT2	2.0 (1.1–3.7)	Slow-metabolizing Chinese women
	GSTM1	1.21 (1.06–1.39)	Caucasians
	GSTM1	1.45 (1.23–1.70)	Asians
Bladder cancer	NAT2	1.41 (1.23–1.80)	Slow-metabolizing Caucasians
	GSTM1	1.54 (1.32–1.80)	Caucasians
	GSTM1	1.77 (1.09–2.91)	Asians
Familial prostate cancer	GSTP1	9.31 (0.47–1.84)	
Ovarian cancer	UGT1A1*28	7.20 (2.06–25.19)	Mucinous tumors[b]

[a] Hazard ratio.
[b] Case-only study.

2.8-fold if M1, T1, and P1 are "high-risk" genotypes (Yuille *et al.*, 2002). It is even more complicated when high-risk genotypes of other phase II enzymes are combined with GSTs. A polymorphic NAT2 combined with a GSTM1 and T1 null genotype changes the risk of development of breast cancer in women (Lee *et al.*, 2003).

UDP-Glucuronosyltransferase

Conjugation with UDP (uridindiphosphate) by UGT (glucuronidation) is one of the major routes of elimination and detoxification of drugs and endogenous compounds. Carcinogens and their reactive metabolites may also be glucuronidated to harmless intermediates that can be eliminated from the body. However, in some cases, glucuronidation may result in the formation of highly reactive species (Li *et al.*, 1999; Marini *et al.*, 1998) with a potential to create neoplastic development. UGTs consist of two multigene families—UGT1 and UGT2 (Burchell *et al.*, 1991)—on the basis of sequence homology, with distinct but overlapping substrate specificity. The UGT2 family is encoded by separate genes clustered on chromosome 4q13 and consists of the UGT2A and UGT2B subfamilies. UGTs are involved in the inactivation of estradiol and its oxidized and methoxylated metabolites. Variation in activity may therefore modify estrogen exposure and consequently estrogen-related cancer risk. In premenopausal African-American women, genetic epidemiological studies have shown that the UGT1A1*28 (A(TA)7TAA) and the UGT1A1*34 (A(TA)8TAA) promoter alleles were associated with an increased risk of developing invasive breast cancer (OR = 2.1, 95% CI = 1.0–4.2) (Guillemette *et al.*, 2000). However, in a major study of white women within the Nurses' Health Study cohort it was not possible to detect a significant risk in the presence of UGT1A1*28 alleles (Guillemette *et al.*, 2001). Estradiol levels were nevertheless increased among women carrying at least one UGT1A1*28 allele, suggesting a contribution of the glucuronidation pathway in the maintenance of hormone homeostasis. Breast density is a predictor of breast cancer and, in addition, is related to the UGT1A1*28 genotype. However, this association depends on the menopausal status. Premenopausal women homozygote for UGT1A1*28 had significantly lower breast density compared to those with the wild-type genotype (minus 43%). In contrast, postmenopausal women homozygote for the UGT1A1*28 genotype had significantly greater breast density (plus 32%) (Haiman *et al.*, 2003).

Estrogens are somehow also involved in ovarian cancer; however, the exact role is not clear. Cecchin *et al.* (2004) could not detect any difference in prevalence of the UGT1A1*28 genotype between patients with ovarian cancer and controls. However, using a case–case approach it was observed that individuals with a subtype of ovarian cancer (mucinous tumors) had a higher prevalence of the specific genotype compared to patients with the nonmucinous subtype (55% versus 15%) (OR = 7.20, 95% CI = 2.06–25.19), suggesting that UGT1A1*28 could be used as a valuable biomarker when planning ovarian cancer chemotherapy.

Sulfotransferase

Cytosolic sulfotransferase enzymes catalyze the sulfation of a large variety of drugs and endogenous substances. The thermostable phenol SULT (encoded by SULT1A1) is a key enzyme in the metabolism of aryl amines and polycyclic hydrocarbons (PAHs), constituents of tobacco smoke. The gene contains several polymorphic sites with a high prevalence of a G-to-A transition at nucleotide 638 in exon 7, leading to an arginine-to-histidine substitution. It occurs at a frequency of 0.3 (Carlini et al., 2001). The variant SULT1A1*2 allele is associated with decreased activity compared with the wild-type SULT1A1*1 allele.

In a Dutch case–control study (Tiemersma et al., 2004), the association between smoking and colorectal adenomas and the potential of a specific SULT1A1 genotype to modify this association was investigated. In the study, 431 adenoma cases and 432 polyp-free controls were examined and a multivariate analysis, including age, sex, and endoscope indication, as well as alcohol and smoking habits, was performed. The study showed that smoking for more than 25 years more than doubled the adenoma risk (OR = 2.4, 95% CI = 1.4–4.1) compared with never smoking. Presence of the SULT1A1*1 fast sulfation allele modestly increased the risk of smoking-associated colorectal adenomas (from an OR = 3.5, 95% CI = 0.9–12.4 to an OR = 4.38, 95% CI = 1.6–11.8). However, the difference was not significant. Findings of an additive effect of smoking and the SULT1A1 fast sulfation genotype are consistent with results from biochemical studies indicating that SULT1A1 may activate pro-carcinogens in cigarette smokers (Chou et al., 1995; Gilissen et al., 1994; Glatt, 2000). The definite role of SULT in the risk of developing colorectal cancer is, however, not yet clarified. Leukocyte DNA from 226 unrelated cases of histological-confirmed colorectal cancers was genotyped for the SULT1A1 allele and compared with 293 controls (Bamber et al., 2001). When the population was considered overall, there was no significant difference in the occurrence of the SULT1A1*1 and SULT1A1*2 alleles, and there was no association between SULT1A1 genotypes and various clinical parameters, including tumor stage and site. However, when data were analyzed matching for age, it was shown that the most common

SULT1A1*1 allele was associated with a significant reduced risk of colorectal cancer (OR = 0.47, 95% CI = 0.27–0.83), suggesting that the high-activity SULT1A1 enzyme protects against dietary and/or environmental chemicals involved in the pathogenesis of colorectal cancer. In this study, SULT1A1*1 was not a risk-carrying gene as in the aforementioned study. However, patients were not matched for smoking habits, and they were generally older than in the Bamber study. A possible explanation may be related to different functions of phase 2 enzymes under different circumstances. SULT is able to act as a bioactivation enzyme (cigarette smoke) or detoxification enzyme (dietary carcinogens), suggesting that the balance of SULT1A1 function is in favor of detoxification and chemical defense in the latter study and in favor of bioactivation of smoke constituents in the first study.

Interaction between smoking habits and gene polymorphism of SULT1A1 and the risk of breast cancer has also been studied. In a case-only study of 288 women with breast cancer (Saintot et al., 2003), it was shown that a SULT1A1*1/*2 or *2/*2 genotype induces an individual susceptibility to breast cancer among premenopausal currently smoking women. A daily cigarette consumption of more than five, as well as duration of smoking of more than 20 years, increased the risk of breast cancer in women carrying the His SULT1A1 allele (OR = 2.11, 95% CI = 1.00–4.46 and OR = 2.83, 95% CI = 1.23–6.54, respectively). In contrast to colorectal cancers, the variant SULT1A1 genotype is the breast cancer risk-carrying gene in young female smokers, indicating that variation in estrogen metabolism due to differences in sulfation capacity may be an important future biomarker.

A very interesting study from Nowell et al. (2002) found an association between risk of death among breast cancer patients receiving tamoxifen therapy and low SULT1A1 activity. Patients homozygous for the low-activity SULT1A1*2 allele had approximately three times the risk of death [hazard ratio (HR) = 2.9, 95% CI 1.1–7.6] as those who were homozygous or heterozygous for the SULT1A1*1 allele. Among patients who did not receive tamoxifen, there was no association between survival and SULT1A1 genotype (HR = 0.7, 95% CI = 0.3–1.5). SULT catalyzes sulfation of the active metabolite of tamoxifen, 4-OH tamoxifen, which has a much greater affinity for the estrogen receptor than the parent compound. It was quite unexpected that fast sulfation of 4-OH tamoxifen was associated with a clinical benefit. The authors explain their findings by alterations in the bioavailability of the active metabolite or to an undefined estrogen receptor-mediated event. About 13% of the Caucasian population carries the low activity SULT1A1 genotype, consequently possessing a risk of decreased efficacy of tamoxifen. Although this study needs to be confirmed by other studies, changes in strategy in breast cancer therapy may have to be considered.

Very few studies have described the relationship between the SULT1A1*2 polymorphism and the risk of lung cancer. In a case–control study from Wang et al. (2002), 948 Caucasian subjects (463 cases and 485 controls) were available for analysis, suggesting that individuals with the heterozygous variant allele (SULT1A1*1/*2) or the homozygous variant allele (SULT1A1*2/*2) had a modestly increased risk of developing lung cancer of 41% (OR = 1.41, 95% CI = 1.04–1.91). Subgroup analysis showed that women (OR = 1.64, 95% CI = 1.06–2.56), current smokers (OR = 1.74, 95% CI = 1.08–2.79), and heavy smokers (OR = 1.45, 95% CI = 1.05–2.00) were at elevated risk.

N-Acetyltransferase

These enzymes catalyze the transfer of an acetyl group to the xenobiotic or carcinogen that has to be detoxified. The substrates are often strong carcinogens used in industrial processes or are formed during cooking or cigarette smoking. The human NAT genes consist of two different forms–NAT1 and NAT2–of which NAT1 is monomorphically and NAT2 polymorphically distributed. The acetylation polymorphism in humans is regulated at the NAT2 gene locus on the short arm of chromosome 8 (Blum et al., 1990). The NAT2 polymorphism may have a significant effect on individual susceptibility to the development of cancer diseases. Much of the evidence derives from epidemiological studies in which bladder cancer patients and controls with high exposure to aromatic amines were analyzed for their NAT phenotype. Patients with the slow NAT2 phenotype are associated with an increased risk of bladder cancer (Cartwright et al., 1982), suggesting that acetylation is important in the protection of bladder carcinogenesis by an ability to inactivate carcinogens. In nonsmoking Chinese women, the proportion of a slow acetylator genotype in a group of lung cancer patients was higher (38%) compared with controls (24%) (OR = 2.0, 95% CI = 1.1–3.7) (Seow et al., 1999). No effect of NAT2 genotype was seen among smokers. However, the exact role of NAT in human carcinogenesis is not entirely clear. In contrast to bladder cancer, the development of colon cancer appears to be associated with individuals possessing the fast NAT phenotype (Smith et al., 1995). This is in accordance with in vitro studies in Salmonella typhimurium strains where it looks as if NAT activates carcinogens (Grant et al., 1992), presumably by O-acetylation, which appears to be an activating step, instead of the normal inactivating N-acetylation (Hein, 1988). The role of NATs in chemical carcinogenesis is therefore many sided, depending on the type of exposure (type of reaction catalyzed), the type of cancer, and a possible contribution of other metabolic pathways competing with the same substrate as NATs. It has been shown that that hydroxylation of heterocyclic amines is catalyzed by CYP1A2 in addition to NAT (Kadlubar *et al.*, 1992).

Conclusion

Most of the cited studies are case–control studies trying to find associations between metabolic gene polymorphisms and cancer of various sites. Data often derive from populations with very common cancer diseases, such as colorectal or breast cancer. Individual susceptibility to cancer may result from differences in metabolism, DNA repair, altered protooncogene of tumor suppressor gene expression, and nutritional status. Most carcinogens require metabolic activation before binding to DNA, and individuals with elevated or decreased metabolic capacity may have an increased risk. Phase 2 enzymes conjugate metabolites (i.e., detoxification of carcinogens or activated carcinogens) with glucuronide, glutathione, or sulfate to produce hydrophilic nontoxic products excreted easily in urine or bile. A decreased activity of phase 2 enzymes (a poor metabolizer due to a variant allele genotype) may therefore lead to the opposite effect observed by phase 1 metabolism, namely a decreased capacity to detoxify carcinogens and an increased risk of cancer development. Consequently, a very subtle threshold value between activating and detoxifying enzymes of specific carcinogens and a risk of initiation of cancer development may exist. Existence of defect enzymes due to generic polymorphisms may disrupt this balance in one direction or the other depending on the site of the defect.

References

Bamber, D. E., Fryer, A. A., Strange, R. C., Elder, J. B., Deakin, M., Rajagopal, R., Fawole, A., Gilissen, R. A. H. J., Campbell, F. C., and Coughtrie, M. W. H. (2001). Phenol sulphotransferase SULT1A1*1 genotype is associated with reduced risk of colorectal cancer. *Pharmacogenetics* **11**, 679–685.

Blum, M., Grant, D. M., Mcbride, W., Heim, M., and Meyer, U. A. (1990). Human arylamine n-acetyltransferase genes: Isolation, chromosomal localization, and functional expression. *DNA Cell Biol.* **9**, 193–203.

Burchell, B., Nebert, D. W., Nelson, D. R., Bock, K. W., Iyanagi, T., Jansen, P. L. M., Lancet, D., Mulder, G. J., Chowdhury, J. R., Siest, G., Tephly, T. R., and Mackenzie, P. I. (1991). The Udp glucuronosyltransferase gene superfamily: Suggested nomenclature based on evolutionary divergence. *DNA Cell Biol.* **10**, 487–494.

Carlini, E. J., Raftogianis, R. B., Wood, T. C., Jin, F., Zheng, W., Rebbeck, T. R., and Weinshilboum, R. M. (2001). Sulfation pharmacogenetics: SULT1A1 and SULT1A2 allele frequencies in Caucasian, Chinese and African-American subjects. *Pharmacogenetics* **11**, 57–68.

Cartwright, R. A., Rogers, H. J., Barhamhall, D., Glashan, R. W., Ahmad, R. A., Higgins, E., and Kahn, M. A. (1982). Role of N-acetyltransferase phenotypes in bladder carcinogenesis: A pharmacogenetic epidemiological approach to bladder cancer. *Lancet* **2**, 842–846.

Cecchin, E., Russo, A., Corona, G., Campagnutta, E., Martella, L., Boiocchi, M., and Toffoli, G. (2004). UGT1A1*28 polymorphism in ovarian cancer patients. *Oncol. Rep.* **12**, 457–462.

Chang, H. Y., and Yang, X. (2000). Proteases for cell suicide: Functions and regulation of caspases. *Microbiol. Mol. Biol. Rev.* **64**, 821–846.

Chou, H. C., Lang, N. P., and Kadlubar, F. F. (1995). Metabolic activation of N-hydroxy arylamines and N-hydroxy heterocyclic amines by human sulfotransferase(S). *Cancer Res.* **55**, 525–529.

Gilissen, R. A. H. J., Bamforth, K. J., Stavenuiter, J. F. C., Coughtrie, M. W. H., and Meerman, J. H. N. (1994). Sulfation of aromatic hydroxamic acids and hydroxylamines by multiple forms of human liver sulfotransferases. *Carcinogenesis* **15**, 39–45.

Glatt, H. (2000). Sulfotransferases in the bioactivation of xenobiotics. *Chem.-Biol. Interact.* **129**, 141–170.

Grant, D. M., Josephy, P. D., Lord, H. L., and Morrison, L. D. (1992). *Salmonella typhimurium* strains expressing human arylamine N-acetyltransferases: Metabolism and mutagenic activation of aromatic-amines. *Cancer Res.* **52**, 3961–3964.

Guillemette, C., De Vivo, I., Hankinson, S. E., Haiman, C. A., Spiegelman, D., Housman, D. E., and Hunter, D. J. (2001). Association of genetic polymorphisms in UGT1A1 with breast cancer and plasma hormone levels. *Cancer Epidemiol. Biomark. Prevent.* **10**, 711–714.

Guillemette, C., Millikan, R. C., Newman, B., and Housman, D. E. (2000). Genetic polymorphisms in uridine diphospho-glucuronosyltransferase 1A1 and association with breast cancer among African Americans. *Cancer Res.* **60**, 950–956.

Haiman, C. A., Hankinson, S. E., De Vivo, I., Guillemette, C., Ishibe, N., Hunter, D. J., and Byrne, C. (2003). Polymorphisms in steroid hormone pathway genes and mammographic density. *Breast Cancer Res. Treatm.* **77**, 27–36.

Hein, D. W. (1988). Acetylator genotype and arylamine-induced carcinogenesis. *Biochim. Biophys. Acta* **948**, 37–66.

Kadlubar, F. F., Butler, M. A., Kaderlik, K. R., Chou, H. C., and Lang, N. P. (1992). Polymorphisms for aromatic amine metabolism in humans: Relevance for human carcinogenesis. *Environ. Health Perspect.* **98**, 69–74.

Kelekar, A., and Thompson, C. B. (1998). Bcl-2-family proteins: The role of the BH3 domain in apoptosis. *Trends Cell Biol.* **8**, 324–330.

Lee, K. M., Park, S. K., Kim, S. U., Doll, M. A., Yoo, K. Y., Ahn, S. H., Noh, D. Y., Hirvonen, A., Hein, D. W., and Kang, D. (2003). N-Acetyltransferase (NAT1, NAT2) and glutathione S-transferase (GSTM1, GSTT1) polymorphisms in breast cancer. *Cancer Lett.* **196**, 179–186.

Lewis, A. D., Forrester, L. M., Hayes, J. D., Wareing, C. J., Carmichael, J., Harris, A. L., Mooghen, M., and Wolf, C. R. (1989). Glutathione S-transferase isoenzymes in human-tumors and tumor derived cell-lines. *Br. J. Cancer* **60**, 327–331.

Li, Y. Z., Li, C. J., Pinto, A. V., and Pardee, A. B. (1999). Release of mitochondrial cytochrome C in both apoptosis and necrosis induced by beta-lapachone in human carcinoma cells. *Mol. Med.* **5**, 232–239.

Marini, S., Longo, V., Mazzaccaro, A., and Gervasi, P. G. (1998). Xenobiotic-metabolizing enzymes in pig nasal and hepatic tissues. *Xenobiotica* **28,** 923–935.

Nakazato, H., Suzuki, K., Matsui, H., Koike, H., Okugi, H., Ohtake, N., Takei, T., Nakata, S., Hasumi, M., Ito, K., Kurokawa, K., and Yamanaka, H. (2003). Association of genetic polymorphisms of glutathione-S-transferase genes (GSTM1, GSTT1 and GSTP1) with familial prostate cancer risk in a Japanese population. *Anticancer Res.* **23,** 2897–2902.

Nowell, S., Sweeney, C., Winters, M., Stone, A., Lang, N. P., Hutchins, L. F., Kadlubar, F. F., and Ambrosone, C. B. (2002). Association between sulfotransferase 1A1 genotype and survival of breast cancer patients receiving tamoxifen therapy. *J. Natl. Cancer Inst.* **94,** 1635–1640.

Saintot, M., Malaveille, C., Hautefeuille, A., and Gerber, M. (2003). Interactions between genetic polymorphism of cytochrome P450–1B1, sulfotransferase 1A1, catecholo-methyltransferase and tobacco exposure in breast cancer risk. *Int. J. Cancer* **107,** 652–657.

Seidegard, J., Vorachek, W. R., Pero, R. W., and Pearson, W. R. (1988). Hereditary differences in the expression of the human glutathione transferase active on trans-stilbene oxide are due to a gene deletion. *Proc. Natl. Acad. Sci. USA* **85,** 7293–7297.

Seow, A., Zhao, B., Poh, W. T., Teh, M., Eng, P., Wang, Y. T., Tan, W. C., Lee, E. J. D., and Lee, H. P. (1999). NAT2 slow acetylator genotype is associated with increased risk of lung cancer among non-smoking Chinese women in Singapore. *Carcinogenesis* **20,** 1877–1881.

Smith, G., Stanley, L. A., Sim, E., Strange, R. C., and Wolf, C. R. (1995). Metabolic polymorphisms and cancer susceptibility. *Cancer Surv.* **25,** 27–65.

Tiemersma, E. W., Bunschoten, A., Kok, F. J., Glatt, H., De Boer, S. Y., and Kampman, E. (2004). Effect of SULT1A1 and NAT2 genetic polymorphism on the association between cigarette smoking and colorectal adenomas. *Int. J. Cancer* **108,** 97–103.

Vineis, P., d'Errico, A., Malats, N., and Boffetta, P. (1999). Overall evaluation and research perspectives. *IARC Sci. Publ.* **148,** 403–408.

Wang, Y. F., Spitz, M. R., Tsou, A. M. H., Zhang, K. R., Makan, N., and Wu, X. F. (2002). Sulfotransferase (SULT) 1A1 polymorphism as a predisposition factor for lung cancer: A case-control analysis. *Lung Cancer* **35,** 137–142.

Wolter, K. G., Hsu, Y. T., Smith, C. L., Nechushtan, A., Xi, X. G., and Youle, R. J. (1997). Movement of Bax from the cytosol to mitochondria during apoptosis. *J. Cell Biol.* **139,** 1281–1292.

Yuille, M., Condie, A., Hudson, C., Kote-Jarai, Z., Stone, E., Eeles, R., Matutes, E., Catovsky, D., and Houlston, R. (2002). Relationship between glutathione S-transferase M1, T1, and P1 polymorphisms and chronic lymphocytic leukemia. *Blood* **99,** 4216–4218.

Author Index

A

Abdel-Rahman, S. Z., 582
Abdulmessih, A., 465, 470
Abe, I., 93, 96, 97, 101
Abe, T., 532, 534, 535, 537, 538
Abele, R., 419, 434, 436, 437, 444, 446, 461
Abenhaim, L., 162
Abramov, S., 210
Abruzzo, L. V., 487
Ackerley, C. A., 418, 548
Ackermann, B. L., 608
Adachi, M., 174, 521
Adachi, R., 166, 174, 175, 188
Adachi, Y., 15, 123, 418
Adam, P. J., 219
Adam, U., 271
Adams, A., 188, 514
Adesnik, M., 238
Adjaye, J., 192
Adman, E. T., 575
Admon, A., 24
Adsuara, J. E., 30
Aghajari, N., 444
Agostini, R. M., Jr., 560
Agrawal, M. L., 396
Aguilar-Bryan, L., 420
Ahmad, R. A., 624
Ahn, S. C., 171, 172, 174, 180, 188, 521
Ahn, S. H., 621
Aiello, R. J., 416
Aikawa, S., 14, 15
Aishima, T., 318
Akaba, K., 14, 15
Akane, K., 215, 225
Akarsubasi, A., 414
Akashi, K., 452
Akaza, H., 69
Akerboom, T. P., 395, 534
Akinaga, S., 304, 307, 313
Akita, H., 534, 535
Akiyama, F., 307
Akiyama, T. E., 25, 317, 335

Aksoy, I. A., 150, 151, 161, 168, 250
Aksoy, S., 161, 167
Alba, M. M., 30
Albers, S. V., 431
Alberti-Segui, C., 476
Albin, N., 333, 334
Aldinger, C., 416
Aleksunes, L. A., 589
Alkema, W., 30
Alkharfy, K. M., 110, 609
Allan, R. N., 361, 377
Allen, S. W., 84, 591
Allikmets, R., 410, 415, 416, 417, 419, 485, 487, 562
Almgren, P., 420
Almquist, K. C., 396
Alperin, E. S., 274, 289
Alquawi, O., 439
Al-Rohaimi, A., 106, 110
Al-Shawi, M. K., 498
Alvarez, L., 560
Alvarez, M. N., 223
Amachi, T., 565
Amann, E., 238
Amano, I., 264, 266, 270
Amar, M. J., 417
Amar-Costesec, A., 119
Ambrosone, C. B., 623
Ambudkar, S. V., 445, 464, 468, 474, 496, 498
Ames, B. N., 236, 243, 543
Ames, G. F., 435, 437, 445, 469
Amherd, R., 520
Ammenheuser, M. M., 582
Amos, J. A., 414
Amoura, Z., 416
Amsterdam, A., 421
Anan, K. F., 58, 66, 117
Anand, R. P., 210
Ananthanarayanan, M., 27, 30, 515, 517, 546, 547, 560, 590
Anders, M., 194
Anderson, A., 11, 13
Anderson, G. D., 50

Anderson, K., 467
Andersson, P., 168
Andersson, T. B., 250
Ando, J., 298, 299, 303, 304, 314
Ando, M., 14, 234
Ando, Y., 500
Andre, B., 467
Andres, H. H., 235
Andrikovics, H., 494
Angeletti, C., 466, 468
Angelicheva, D., 414
Angelova, M. I., 452, 453
Angenieux, C., 419
Anguiano, A., 414
Anisfeld, A. M., 520
Annilo, T., 410, 417, 418
Antoniak, S. K., 271
Antonio, L., 129
Anwar, A., 67
Anwer, M. S., 546, 547
Anzai, N., 494, 508
Anzulovich, A. C., 522
Aoki, J., 535
Aono, S., 15, 123
Apiwattanakul, N., 544
Appelkvist, E. L., 352
Arand, M., 571, 572, 573, 574, 575, 577, 578, 579, 581
Archelas, A., 577, 578
Archer, I. V., 578
Archer, T. K., 25
Argiriadi, M. A., 573
Arias, I. M., 465, 470, 547, 548, 559
Arlt, V. M., 235, 236
Armstrong, R. N., 141, 571, 576
Arnell, H., 418, 560
Arnould, I., 416, 418
Aronica, E., 396
Arrese, M., 517
Artelt, P., 239
Arthur, L. M., 469
Arts, I. C. W., 326
Artur, Y., 16, 24, 25, 27, 29, 30, 33, 36, 39
Arulpragasam, A., 216
Ashby, S. J., 275, 280, 285, 288
Asman, M., 518, 589
Assem, M., 174, 515, 521
Assmann, G., 416
Assouline, J. A., 61, 63
Astromoff, A., 467

Aten, J., 560
Atmane, N., 216, 217, 219, 221, 223, 224, 225
Attie, A. D., 417
Auborg, P., 414, 415
Audebert, S., 417
Ausmeier, M., 239
Auwerx, J., 523
Auyeung, D. J., 27, 29, 30, 33, 36, 38, 39, 64, 65, 80, 89, 345, 347
Aydogdu, S., 414
Ayrton, A., 592
Ayuso, C., 415
Azarian, S. M., 415

B

Baas, F., 420, 534, 544
Baayen, J. C., 396
Babinet, C., 23, 24, 27
Bacchawat, A. K., 475
Bacchelli, B., 420
Bach, I., 24
Bach, J. P., 23, 24, 27
Bachhawat, A. K., 465, 467, 470
Bacia, K., 452
Backman, J. T., 264, 599, 600
Badary, O. A., 60, 67, 68, 351
Badawi, A. F., 215
Badcock, K., 578
Badger, T. M., 342
Bae, Y., 521
Baek, B. S., 171, 172, 174, 180, 188, 521
Baes, M., 600
Bagla, A., 416
Baharloo, S., 560
Bailey-Dell, K. J., 487
Baird, L., 415
Baird, S., 62
Baird, S. J., 120, 125
Bairoch, A., 2, 3, 6, 14, 58, 92
Baker, A., 560
Baker, S. M., 350
Bakker, C. T., 14, 15, 47, 419, 532
Bakos, E., 471
Balaram, H., 210
Balasubramanian, N., 515, 560, 590
Balcells, S., 415
Balderes, D. A., 464, 466
Baliah, T., 92
Ball, S. E., 47, 48, 49

Ballabio, A., 273, 274
Ballard, P. L., 418
Ballatori, N., 396, 402, 474
Baller, J. F., 517
Ballew, J. D., 420
Balzi, E., 463, 466, 468, 470, 474, 475
Bamber, D. E., 622
Bamforth, K. J., 622
Ban, N., 418
Bancells, C., 414
Bancroft, J. D., 559
Bandiera, S. M., 590
Bandmann, O., 166
Bandoh, K., 300
Bandsma, R. H., 416
Banerjee, R., 2, 18, 47
Bangham, R., 467
Bannai, S., 396, 401
Bansel, S. K., 78
Banting, G., 414
Bao, Y., 272
Baratti, J., 577, 578
Barbacci, E., 24, 33
Barbier, O., 515, 516
Barbirato, F., 575
Barbouti, A., 221, 222
Barhamhall, D., 624
Bark, S., 417
Barker, D. F., 216
Barlage, S., 416
Barnes, N. C., 325
Barnes, R., 417, 562
Barnes, S., 317, 318, 320, 324, 325, 327, 332, 333, 334, 335, 336, 337, 338, 339, 361, 362, 363, 364, 365, 366, 367, 368, 370, 371, 372, 374, 375, 377, 378, 379, 380, 381, 382, 383, 384, 386, 387, 388, 389, 390
Barnett, A. C., 149, 349, 350
Barnouin, K., 534
Barouki, R., 590
Barr, G. C., 47
Barra, J., 23, 24, 27
Barre, L., 58, 59, 118, 127
Barrell, B. G., 578
Barron, D., 264, 468, 473
Barry, C. E. III, 578
Barth, S., 573, 574
Bartsch, H., 532, 534, 563
Bartsch, I., 171, 238, 239, 242, 245, 269
Bartsch, U., 415

Bartuluchi, M., 396
Barut, A., 2, 63
Barwick, J. L., 174, 180, 181, 188, 519, 521, 589, 600, 601, 603, 604, 606, 609, 610, 611, 612
Basham, D., 578
Bassereau, P., 452
Basu, N. B., 47
Basu, N. K., 2, 3, 4, 5, 18, 81, 343, 519
Bates, S. E., 421, 460, 461, 487, 507
Batta, A. K., 514, 515
Battaglia, E., 58, 59, 118, 128, 129, 132, 134
Bauer, B. E., 460, 461, 462, 463, 466, 467, 469, 474
Bauer, J., 223
Baumann, M., 118, 122, 132, 133
Bear, C. E., 402, 440, 445
Beato, M., 600
Beaton, D., 14
Beatty, B. G., 151
Beaudet, M.-J., 11, 13
Beaufay, H., 119
Beaumont, K., 47, 50, 343
Beaune, P., 577
Becker, P. B., 24
Becker, R., 578
Beckman, J., 218, 223, 224
Beckman, K., 415
Bednarek, P., 325
Beer, M., 192
Beetham, J. K., 571, 578, 581
Beier, D., 546
Beigneux, A. P., 590
Beijinen, J. H., 487
Beijnen, J. H., 533
Beil, F. U., 417
Bekaii-Saab, T., 106, 110
Bekri, S., 419
Belanger, A., 2, 3, 6, 14, 58, 92, 106, 108, 264
Belanger, P., 92, 106, 107, 343
Belew, R. K., 200
Belinsky, M. G., 396, 534
Bell, A., 591
Bell, D. A., 215
Bell, D. W., 348
Bell, G. I., 28
Bell, G. W., 28
Bell, P. A., 577
Bellanne-Chantelot, C., 24, 33
Bellingham, D. L., 95

Bellomo, G., 534
Benabdelhak, H., 461, 469
Bend, J. R., 572
Bendaly, J., 14
Benelli, B., 352
Benetazzo, M., 414
Benezra, C., 351
Benfatto, A. M., 364, 365, 378
Benito, R., 467
Ben-Shushan, E., 420
Bentinger, M., 352, 354, 355
Bentley, W. E., 578
Benton, M. R., 336, 338, 339
Berdy, S., 523
Berg, C., 132
Berge, K. E., 417, 562
Berge-Henegouwen, G. P., 559
Bergen, A. A., 415, 420, 534
Berger, J., 415
Berger, R., 559, 560
Berger, S. L., 366, 382, 512
Berger, Y., 67
Bergfors, T., 572, 573, 574, 575, 579
Berkenstam, A., 518, 589
Berman, H. M., 292
Bernard, C. C., 148, 150, 151
Bernard, O., 14, 15, 63, 106, 107, 418, 560, 561, 565
Bernard, P., 16, 24, 25, 27, 29, 30, 33, 36, 39
Bernassola, F., 218
Bernheimer, H., 415
Bernier, F., 150, 151
Bernstein, C., 187
Bernstein, H., 187
Bernstein, P. S., 138, 139, 415
Berry, J., 219
Bertilsson, G., 518, 589
Bertrand, S., 173
Beuers, U., 418, 547
Beutler, E., 14, 15
Beval, B., 396
Bevan, D. R., 62
Beyer, R. E., 353
Bhakta, S., 216
Bhalla, S., 523
Bhardwaj, G., 396
Bhasker, C. R., 149, 162
Bhat, T. N., 292
Bhattacharya, R., 25
Bhattacharyya, A. K., 417

Bhuyan, B., 236
Bidwell, L. M., 149
Bieler, C. A., 235, 236
Biemans-Oldehinkel, E., 431, 432, 434, 436
Bienengraeber, M., 420
Bijleveld, C. M., 374, 560, 561
Billich, A., 274, 299, 303
Billin, A. N., 523
Bilton, R., 224
Bilzer, M., 395, 547
Bingaman, W., 396
Bingham, S. A., 342
Binsack, R., 337
Birch, D. G., 415, 416
Bircher, J., 59
Birchmore, J. L., 466
Birkett, D. J., 149, 162
Bischoff, E. D., 516
Bishop, D. F., 419
Bishop, W. P., 92
Bisi, J., 523
Bissada, N., 417
Bissinger, P. H., 464, 468, 472
Biswal, S., 67, 68
Bito, H., 576
Bizet, C., 215
Bjorkhem, I., 417
Black, H., 25
Blair, J. N. R., 47
Blake, N., 419
Blanch, C. J., 547
Blanchard, R. L., 63, 123, 148, 167, 170, 345, 350
Blanchard, S. G., 514, 560, 589
Blanchet-Bardon, C., 417
Blanckaert, N., 120
Blankenagel, A., 415
Blanquart, C., 515
Blattler, S., 173, 520, 589
Bleau, G., 274
Bledsoe, R. K., 514, 560, 589
Blée, E., 581
Blevins, R. A., 515
Blickrstaff, R. T., 295
Blight, M. A., 461, 469
Bloks, V. W., 416
Blomquist, P., 518, 589
Bluck, L. J. C., 342
Blum, M., 192, 624

Blumberg, B., 514, 518, 519, 589, 590, 591, 600, 601, 603, 604, 606, 610, 611
Blume, R., 59
Blumenfeld, M., 23, 24, 28, 30
Blumrich, M., 264, 271
Blundell, T. L., 199
Bluzat, A., 442, 443
Bocchetta, M., 7
Boche, G., 193
Bock, K. W., 2, 3, 6, 11, 14, 58, 59, 60, 61, 62, 63, 65, 66, 67, 68, 69, 76, 92, 121, 345, 599, 621
Bock, M., 419
Bock-Hennig, B. S., 60, 61, 62, 67, 68, 76
Bocola, M., 577
Bode, J. G., 543
Bodzioch, M., 416
Boehm, C. D., 414
Boeing, H., 251, 348
Boeke, J. D., 467
Boelle, P. Y., 561
Boelsterli, U., 219
Boer, J. M., 420
Boersma, B. J., 324, 336, 337, 338
Boffetta, P., 348, 619
Boger, D. L., 387
Bohen, S. P., 468
Boiocchi, M., 621
Boj, S. F., 25, 28
Bolado, J., Jr., 518, 589, 600
Bolder, U., 361
Boles, J. W., 234
Bolhaar, S. T., 352
Bollileni, J. S., 27, 30
Bollineni, J., 514, 515
Boltes, I., 275, 279, 280, 284, 288
Bolton-Grob, R., 349, 350
Bombieri, C., 414
Bond, C. S., 275, 280, 285, 288
Bonner, R. F., 310, 312, 313
Bonnesen, C., 67
Bookjans, G., 63, 69
Boos, W., 436
Borchardt, R. T., 60
Borlak, J., 24, 25, 26, 28
Boroumandi, S., 193
Borst, P., 174, 396, 402, 418, 419, 421, 440, 460, 461, 463, 471, 485, 487, 521, 532, 534, 544
Bort, R., 38

Borutaite, V., 221
Bosma, P. J., 14, 15, 47, 419, 532, 560, 561, 565
Bosold, F., 193
Bosslet, K., 264, 271
Boteva, K., 414
Bothorel, P., 452, 453
Bottcher, A., 416
Botting, N. P., 324, 336, 337, 338, 339
Bottroff, B. S., 65, 66, 97, 102, 519, 589, 601, 609, 613, 614
Bouadjar, B., 417
Boucher, R. C., 498
Boudreau, F., 23
Boughdene-Stambouli, O., 417
Boukouvala, S., 192
Boulenc, X., 67
Bourne, P. E., 292
Bourrie, M., 67
Boute, O., 561
Boverhof, R., 517
Boyd, C. D., 534
Boyer, J. L., 511, 517, 543, 565
Boyle, J. T., 560
Brachat, A., 476
Bradfield, C. A., 66
Bradford, M. M., 199
Bradley, D. J., 515
Bradley, M. O., 236
Brahimi, N., 215
Brands, A., 61, 63
Brangi, M., 487
Braun, S., 64
Breau, A. P., 106, 108
Brehm, I., 590, 592
Bremer, E., 440
Bremer, J., 364
Brems, J. J., 350
Brendel, M., 464
Brennan, P., 348
Brennan, S. O., 337
Breslow, J. L., 27, 30, 517
Breuning, M., 420
Brewer, B., 417
Brewer, C., 416
Brewer, H. B., 416
Brewer, H. B., Jr., 520
Breymann, C., 418
Brierley, C. H., 47
Briggs, M. R., 521
Brimer, C., 174, 515

Brinkmann, U., 500, 590, 592
Brinster, R. L., 602
Brix, L. A., 149, 349, 350
Broccardo, C., 416
Brockmeier, D., 63, 69
Brodsky, J. L., 472
Brom, M., 397, 532, 561, 565
Broman, K. W., 419
Brooke, E. W., 193, 206, 209, 215, 225
Brooks-Wilson, A., 416, 417
Brosch, R., 578
Brosius, J., 238
Brown, D., 578
Brown, G., 221
Brown, K. K., 520, 523, 589
Brown, P., 463, 475
Broxterman, Q. B., 577
Brück, M., 62, 63, 66, 69, 345
Bruins, A. P., 571
Brune, B., 218
Brunner, B., 577
Brunt, E. M., 589, 600, 603, 604, 606, 610, 611
Bryan, J., 420
Bryson, J., 591
Buchanan, S. G., 469
Buchholz, J. K., 534, 535, 538
Buchholz, U., 396, 397, 401, 532, 534, 538
Buchler, C., 416
Büchler, M., 397, 532
Buck, J., 148, 167, 170, 345, 350
Buckler, F., 2, 11, 66
Buitelaar, M., 487, 533
Bujard, H., 604
Bull, L. N., 418, 515, 559, 560
Bullas, L. R., 238
Bundman, D., 517, 523
Bunschoten, A., 622
Bunting, K. D., 487
Burchell, A., 47
Burchell, B., 2, 3, 6, 14, 47, 48, 50, 51, 52, 53, 54, 55, 58, 61, 62, 92, 110, 117, 118, 120, 125, 126, 129, 343, 344, 345, 347, 544, 621
Burchiel, S. W., 62
Burdelski, M., 560, 561, 563, 565
Burdick, A. D., 62
Burgess, W., 148
Burgstahler, A. D., 546
Burk, O., 519, 563, 590, 592
Burka, L. T., 515
Burley, S. K., 469

Burns, G. R., 274, 298, 303
Burtt, N., 420
Bush, K., 218, 224
Busse, W. W., 264
Bussen, M., 27, 30
Bussey, H., 467
Butcher, N. J., 192, 216
Butler, M. A., 625
Butler-Browne, G., 216
Byrne, C., 621

C

Cabrera, G. Y., 65, 66, 97, 102, 519, 589, 601, 609, 613, 614
Cabrera, M., 194, 196
Cai, H., 174, 515
Calafat, J., 534
Caldwell, J., 375
Callen, D. F., 150, 345
Callizot, N., 415
Camarena, C., 560
Camarero, S., 396
Camici, G., 218, 221
Campagnolo, C., 224
Campagnutta, E., 621
Campanale, K., 608
Campbell, D. A., 47
Campbell, F. C., 622
Campbell, N. R., 235
Cañada, F. J., 328, 329, 330
Caniglia, S., 396
Cannon, R. D., 473
Cano, V., 118, 129
Cantrell, V. E., 608
Cantz, T., 397
Cao, J., 547
Capo, S., 156
Carbone, D. P., 419
Carelli, V., 352
Carlberg, C., 182, 521, 523
Carlini, E. J., 622
Carlton, V. E., 560
Carmichael, J., 619
Carney, J. P., 469
Caro, I., 67
Carpenter, D. S., 192
Carr, B. R., 168
Carraway, K. III, 221
Carrier, J. S., 106, 108

Carta, A., 348
Carter, C. A., 132, 140
Carter, C. W., 252, 256, 260
Carter, S. F., 578
Carter, S. G., 498
Cartwright, R. A., 624
Caruso, D., 522, 523
Carvalho, S., 3, 4, 5, 81, 343, 519
Casamitjana, R., 25, 28
Caselli, A., 218, 221
Casey, R., 414
Cassidy, A., 47
Cassidy, A. J., 344
Castell, J. V., 38
Castellani, L. W., 524
Castro, G., 417
Castro, M. I., 295
Catovsky, D., 621
Catty, P., 470, 474, 475
Cayota, A., 223
Cazenave, J. P., 419
Cecchin, E., 621
Center, M., 396, 397, 534
Cereghini, S., 24, 33, 39, 521
Cerelli, G., 95
Cerundolo, V., 419
Cervenak, J., 494
Cha, S. H., 522, 533, 544
Chalkiadaki, A., 24, 33
Chamagne, A., 216
Chambenoit, O., 416
Chambers, J., 418, 565
Chambon, P., 600
Chan, T. S., 353, 354
Chander, S. K., 299
Chandra, G., 514, 560
Chang, G., 410, 431, 470, 544
Chang, H. C., 332, 334, 336
Chang, H. Y., 619
Chang, T. K. H., 590
Chapdelaine, A., 274
Chapman, H. D., 345
Charlebois, R. L., 431
Chatterjee, B., 167, 168, 170, 171, 172, 173, 174, 180, 183, 188, 521, 589, 590
Cheek, J. M., 572
Chen, C., 344
Chen, D., 445
Chen, F., 2, 3, 4, 15, 515, 517, 560
Chen, G., 119, 132, 138, 140, 350

Chen, H. L., 419
Chen, J., 95, 218, 331, 432, 438, 445, 514, 515, 590
Chen, J.-G., 2
Chen, K., 67, 68
Chen, M., 461
Chen, Q. K., 30
Chen, S., 11, 83, 93, 97, 173
Chen, T., 591
Chen, W., 522
Chen, X., 218
Chen, Y., 589
Chen, Y.-H., 87
Chen, Z. Q., 418
Cheskis, B., 514
Chetrite, G. S., 274, 303
Chi, Y. I., 521
Chiang, J. Y., 522, 523
Chidambaram, A., 415
Childs, S., 409, 418, 560
Chillingworth, T., 578
Chimini, G., 416
Chimizu, C., 168
Chin, E., 604
Ching, M. S., 350
Chion, C., 578
Chittattu, A., 547, 548
Cho, J. W., 2, 3, 4, 5, 81, 343, 519
Cho, S. H., 167, 173, 174, 180, 589, 590
Choi, H.-S., 96, 516, 519, 600
Choi, S. Y., 138, 139
Chojnacki, T., 352, 354, 355
Chou, H. C., 235, 622, 625
Chouard, T., 24, 39
Chow, L. T., 604
Chowdhury, J. R., 2, 3, 6, 14, 15, 92, 621
Chowdhury, N. R., 2
Choy, J. C., 417
Christiani, D. C., 67
Christiano, A. M., 420
Christiansen, A., 149, 162
Christianson, D. W., 573
Christie, J., 514, 515
Christoph, S., 238, 245
Chruszcz, M., 282
Chu, A. M., 467
Chu, Q., 7
Chua, S. S., 514, 517, 523, 601, 611
Chuaqui, R. F., 310, 312, 313
Chung, M., 96, 519

Churcher, C., 578
Churchwell, M. I., 332, 334
Chute, C. G., 350
Cidlowski, J. A., 95
Ciliberto, G., 28
Ciotti, M., 2, 5, 14
Clark, J. B., 396
Clarke, A. R., 210
Clarke, D. B., 324
Clarke, D. J., 47, 48, 344
Claudel, T., 515, 516
Claveau, D., 224
Clayton, R. J., 558
Cleasby, A., 149
Clee, S. M., 416, 417
Clements, P. R., 275, 280, 285, 288
Cline, C. B., 523
Cloarec, S., 24, 33
Clonis, Y. D., 610
Code, E. L., 106, 107
Coffey, M. J., 48
Coffman, B. L., 92, 118
Cohen, B. I., 361
Cohen, J. C., 417, 544, 562
Cohen, N., 210
Cohn, J. A., 414
Colapietro, A. M., 487
Cole, S. P., 396, 397, 401, 402, 430, 432, 439, 485, 534
Cole, S. T., 578
Coleman, R., 547
Colledge, W. H., 414
Collingwood, T. N., 512
Collins, J. A., 416
Collins, J. L., 518, 519, 589, 590
Collison, M. W., 325
Collyer, C. A., 275, 280, 285, 288
Comer, K. A., 148, 151, 171
Comte, G., 468, 473
Conary, J., 298
Condie, A., 621
Congiu, M., 29, 41
Conley, P. B., 24, 31, 35
Connelly, C., 467
Conners, K., 469
Conney, A. H., 2, 61
Connolly, G. C., 396, 474
Connor, R., 578
Connor, W. E., 417
Conseil, G., 468, 473, 534

Consler, T. G., 514, 560, 589
Cook, C. S., 106, 108
Cooke, N. E., 24, 28, 33
Cooke, R. M., 149
Coombes, R. C., 303, 304
Cooper, A. J. L., 396
Cooper, R., 216
Coote, P., 464, 466
Coreneos, E., 220
Cornell, B. A., 354
Corona, G., 621
Corser, R. B., 47
Corsi, A. K., 124
Cortese, R., 24
Corzo, D., 414
Cosgrove, K. E., 420
Costa, R. H., 28
Cote, G. J., 420
Cottrell, J. S., 142
Coughtrie, M. W. H., 47, 48, 51, 52, 53, 55, 148, 149, 151, 167, 170, 238, 240, 245, 251, 345, 348, 349, 350, 622
Coumoul, X., 590
Court, M. C., 60, 63
Court, M. H., 14, 15, 54, 60, 63, 106, 107, 108, 110, 344, 347, 608
Coutinho, P., 129
Coward, L., 318, 320, 324, 325, 332, 333, 334, 335, 336
Coward, W. A., 342
Cox, G. F., 414
Cox, M., 180
Crabtree, G. R., 24, 31, 35
Craggs, G., 219
Craig, L., 469
Crane, A., 420
Cravatt, B. F., 387
Crawford, A. R., 546
Crawford, J. M., 3, 546
Creasy, D. M., 142
Creeke, P. I., 330
Cregar, L., 396
Cremers, F. P., 415
Crespi, C. L., 106, 107
Crestani, M., 522, 523
Cresteil, D., 418, 560, 561, 565
Cribb, A. E., 192
Cristensen, B., 487
Crocenzi, F. A., 547
Cronin, A., 572, 574, 575, 577, 578, 579

Cross, C., 218
Crossman, M. W., 517
Crow, J., 218, 223, 224
Crowe, J. H., 452, 453
Cruse, W. B., 351
Cruz, C., 194, 196
Csiszar, K., 420
Cucullo, L., 396
Cuebas, D. A., 361, 370
Cui, J., 188, 514, 515
Cui, Y., 396, 520, 532, 534, 535, 536, 537, 538, 562, 563
Cui, Z., 467
Cupples, C. G., 194, 196
Currie, R. A., 26
Curtis, C. G., 295
Custer, L. J., 325
Cutting, G. R., 414
Cybularz-Kolaczkowska, A., 464, 468, 472, 473
Czapinska, H., 275, 279, 280, 284, 288
Czech, J., 264, 271
Czernik, P. J., 58, 59, 119, 132, 138, 140
Czich, A., 171, 234, 238, 239, 240, 242, 245, 269
Czuba, B., 378, 379
Czygan, P., 611

D

Dacremont, G., 361, 370
Da Cruz e Silva, O., 48
Dahan, K., 418, 560
Dahl, N., 418, 560
Dahlquist, F. W., 239
Dahms, B. B., 560
Dairou, J., 216, 217, 219, 221, 223, 224, 225
Dajani, R., 149
D'Alessandro, T., 336, 338, 339
Dallas, P. B., 26
Dallner, G., 352, 354, 355
Dallongeville, J., 516
Dalmas, O., 441, 444
Dalton, T. P., 76
Daly, J. W., 571, 579
Daly, M. J., 420
Danesi, R., 500
Danoff, T. M., 47
Dansette, P. M., 2
Darley-Usmar, V. M., 324, 336, 337, 338, 339

Darnel, A. D., 303, 304, 313
Dasaradhi, L., 168
Dassesse, D., 30
Dauwerse, H., 420
Davidson, A. L., 432, 438, 445
Davies, G. J., 129
Davies, R., 578
Davies, S. G., 206, 209
Davis, J. W., 62
Davis, K., 467
Davis, R. G., 521
Davis, R. W., 467
Davis, W., 242
Davis-Searles, P. R., 96, 589
Dawson, P., 517, 533
Day, A. J., 264, 272, 328, 329, 330
Deakin, M., 622
Dean, D. A., 437
Dean, M., 410, 414, 415, 416, 417, 418, 419, 485, 487, 507, 562
de Bakker, P. I., 420
DeBaun, J. R., 231, 236
De Belin, J., 219
De Boeij, W. P., 452
de Boer, A., 14, 15, 47
De Boer, S. Y., 622
de Bont, J. A. M., 572, 573, 575, 578, 579
Debray, D., 561
de Bruin, J., 565
De Bruin, L. S., 196, 198
Decleves, X., 396
Decottignies, A., 460, 465, 468, 470, 472, 473, 474, 475
Deeley, R. G., 396, 397, 401, 534
De Fabiani, E., 522, 523
de Galan-Roosen, T. E. M., 575
Degand, H., 470, 474, 475
Degawa, M., 93, 96, 97, 101
Degli, E. M., 352
de Haas, M., 544
Dehal, S. S., 106, 107
Dehmel, U., 578
Delahodde, A., 464, 470, 475
de la Salle, H., 419
de la Vega, 560
Delbruck, M., 454
del Castillo-Olivares, A., 523
Deleury, E., 129
Deleuze, J. F., 416, 418, 560
Delgoda, R., 193, 216

Deloménie, C., 193, 206, 215, 216, 217
de Longchamp, R. R., 235
del Rio, T., 415
DeLuca, A. J. M., 450
De Magistris, L., 24
Demina, A., 14, 15
de Morais, S. M., 344, 599
Demyan, W. F., 170, 172
Denamur, E., 215
Denefle, P., 416, 418
Deng, Z., 150, 345
den Hollander, A. I., 415
Denicola, A., 223
Denizot, F., 441, 444
Denk, H., 517
Denny, W. A., 193
Denson, L. A., 517, 546
Dent, P., 517
de Pedro, N., 514, 515
Der Doelen, M. A., 560
Derendorf, H., 611
DeRisi, J., 463, 475
Derr, J. E., 450
d'Errico, A., 619
De Simone, V., 24
Desmond, P. V., 29, 41
De Sole, P., 353
de Sousa, G., 590, 592
Desrochers, M., 11, 13, 560
Destro-Bisol, G., 14
Desvergne, B., 16, 24, 25, 27, 29, 30, 33, 36, 39
de Swiet, M., 418, 565
Deters, A., 565
Detmers, F. J., 436
Devanaboyina, U. S., 192, 216
Devaud, C., 418
Devaux, F., 468
Devaux, P. F., 465, 468
De Vivo, I., 621
Devlin, K., 578
De Vree, J. M., 560, 561, 565
de Vries, A. H., 452
de Vries, E. J., 577
de Wet, J., 416, 460, 465, 468, 472, 474, 475
Dey, S., 445, 496
DeYoung, J. A., 559, 560
Dhe-Paganon, S., 521
Diasio, R. B., 362, 363, 364, 375, 377, 378, 379, 380, 382, 383
Díaz, J. C., 328, 329, 330

Diaz, M. C., 560
Dibbelt, L., 274, 298, 303
DiCarlo, F. J., 560
Dickey, M., 469
Dickinson, K., 591
Diederich, W., 416
Dierkesmann, R., 264
Dierks, T., 274, 275, 279, 280, 284, 285, 288
Dieter, M. Z., 76
Dietrich, C. G., 41, 166
Dietrich, F., 467
Dijkstra, B. W., 573
di Pietro, A., 441, 444, 468, 473
Di Rienzo, A., 14
Dirr, A., 195
Diry, M., 590
Divi, R. L., 336
Dixit, B. L., 466
Dixon, P. H., 418, 565
Do-Cao, M. A., 444
Dodgson, K. S., 295
Doehmer, J., 238, 245
Doerge, D. R., 235, 332, 334, 336
Doeven, M. K., 434, 437, 444, 446, 453, 454
Doggett, N. A., 150, 345
Dogra, S., 171, 238, 239, 242, 245, 269
Dohda, T., 26
Dolan, P. M., 67, 68
Dolder, M., 223
Doll, M. A., 192, 202, 216, 621
Dolly, J. O., 295
Dombrowski, F., 547, 548
Domoto, H., 166, 174, 175, 188
Donahee, M., 523
Donaldson, O., 418, 565
Donato, F., 348
Dong, J., 421
Dooley, T. P., 148, 150, 151, 345
Dorin, R. I., 99
Doring, B., 419
Dorko, K., 608
Dormoy, A., 419
Dorwart, M., 469
Douar, A. M., 414, 415
Dow, S. W., 467
Dowdy, D. L., 388
Downes, M., 589, 600, 603, 604, 606, 610, 611
Doyen, A., 23, 24, 25, 27
Doyle, L. A., 487

Drake, R. R., 120, 131, 132, 134
Dranoff, J. A., 546
Dransfeld, O., 547, 548
Dresser, M. J., 544
Drewelow, B., 271
Drexler, I., 419
Driessen, A. J., 431
Dringen, R., 395, 396, 397, 398, 399, 400, 401, 402, 403, 404, 405
Drings, P., 532, 534, 563
Drobnik, W., 416
Drocourt, L., 166, 174, 181, 184, 590
Dror, Y., 210
Drucker, W. D., 297
Druley, T. E., 460, 461
Duan, S. X., 54, 60, 63, 106, 107, 108
Duan, X. J., 206
Duane, W. C., 361, 377
Duanmu, Z., 350
Dubaquie, Y., 591
Dubin, I. N., 561
Dubin, N., 300
Dubuisson, C., 418, 560
Duda, K., 521
Dudou, S., 318
Duez, H., 516
Duffel, M. W., 168, 235, 250, 251, 253, 254, 255, 257
Duggleby, R. G., 149, 349
Duine, J. A., 351
Dumas, N. A., 168, 176, 234
Dumont, M., 418, 474, 560, 561, 565
Dumoulin, F. L., 310
Dunbar, R., 419
Duncan, A. M., 396
Duncan, S. A., 65, 523
Duncliffe, K. N., 36
Dunlop, T. W., 521, 523
Dunn, R. T., 345
DuPont, M. S., 328, 329, 332
Dupret, J. M., 193, 206, 209, 215, 216, 217, 219, 221, 223, 224, 225
Duran-Sandoval, D., 515, 516
Dürk, H., 574, 577, 578
D'Urso, D., 543, 547, 548
Dutta, S., 547, 550
Dutton, G. J., 1, 46, 125
Duurkens, H. H., 444
Duverger, N., 416
Dvorakova, K., 187

E

Ebel, C., 444
Ebert, U., 105
Ebihara, T., 565
Ebner, T., 48
Echchgadda, I., 167, 171, 172, 173, 174, 180, 183, 188, 521, 589, 590
Echevarria, W., 517
Ecker, G., 470
Eckstein, J. A., 65, 66
Edwards, P. A., 515, 520, 524
Eeles, R., 621
Egan-Hafley, M., 611
Egashira, T., 416
Eggertsen, G., 522
Egginton, E., 418, 565
Eggleston, I. M., 67
Eggmann, T., 465, 467
Egner, R., 464, 466, 469, 470, 471, 472, 474
Ehmer, U., 14
Eiberg, H., 560
Eichelbaum, M., 63, 69, 519, 562, 563, 590, 592
Eichler, E. E., 167
Eiglmeier, K., 578
Einhorn, J., 325
Eipper, A., 577
Eiselt, R., 590, 592
Eisenberg, D., 206
Eiserich, J., 218
Elahi, A., 14
El Bakkoury, M., 467
Elbein, A. D., 120, 131, 132, 134
Elder, J. B., 622
Elferink, M. G., 431, 487
Elferink, R. O., 418, 460, 461, 463
Elferink, R. P., 166, 533, 561, 565
Elger, W., 294
Elgin, E., 417
el Hawrani, A. S., 210
Elias, E., 565
Eling, T. E., 198
Elkahloun, A. G., 313
Elling, G., 565
Eloranta, J. J., 516, 517, 518, 522
Elovaara, E., 346
El-Sankary, W., 592
Elzinga, B. M., 517

Emi, M., 416, 575
Emi, Y., 2, 4, 5, 6, 7, 11, 64, 66, 76,
    92, 122
Emmert-Buck, M. R., 310, 312, 313
Emoto, K., 468
Emre, S., 515, 560
Emsley, P., 410
Emtage, S., 469
The ENCODE Project Consortium, 30
Enders, N., 236
Endou, H., 494, 508, 522, 533, 544
Eng, P., 624
Engberts, J. B., 452
Engelke, C. E., 251, 348
Engelking, L. R., 546
Engert, J. C., 417
Engler, M. M., 572
Englert, H., 53
Engstrom, P., 30
Epping, E. A., 469, 470, 471, 472,
    474, 475
Erasmus, E., 376
Eraso, P., 469
Erfle, J. D., 234
Erickson, S. K., 514, 515
Eriksson, L. C., 69
Erion, M. D., 533, 544
Erlander, M. G., 313
Erlinger, S., 418, 560
Erman, M., 284
Ernst, R., 461
Erspamer, V., 60
Escriva, H., 173
Eskinazi, R., 30
Esplen, J. E., 608
Esposti, M. D., 352
Estivill, X., 414
Ethell, B. T., 47, 50, 53, 343
Etzenhouser, B., 351
Eugster, H. P., 577
Evans, D. H., 237
Evans, M. J., 414
Evans, R. H., 600, 601, 603, 604, 606, 609, 610,
    611, 612
Evans, R. K., 132
Evans, R. M., 65, 66, 95, 97, 102, 166, 173,
    174, 175, 180, 181, 188, 514, 515, 518, 519,
    520, 521, 589, 600, 601, 602, 605, 609, 612,
    613, 614
Evans, T. R., 303, 304

Evaristus, E. N., 271
Evers, R., 396, 402, 421, 471, 534, 544
Exley, A., 419
Ezan, E., 498
Ezura, Y., 575

F

Faber, K., 515, 565, 577
Fabre, G., 67
Fabre, J. M., 521, 590
Failing, K., 295
Fakis, G., 192, 216
Falany, C. N., 148, 149, 151, 168, 171, 174,
    176, 234, 238, 239, 240, 245, 250, 294, 298,
    299, 320, 335, 336, 345, 349, 361, 366, 367,
    368, 370, 371, 372, 374, 375, 378, 380, 381,
    382, 383, 384, 386, 387, 389
Falany, J. L., 171, 234, 240, 245, 294, 298,
    299, 374
Falcon-Perez, J. M., 469
Falkner, K. C., 589, 601
Falson, P., 452
Falter, K., 295
Fanburg, B., 219
Faneyte, I. F., 487
Fang, H. L., 174, 350
Fang, J. L., 61
Fang, N., 342
Fang, S., 523
Fang, X., 573
Fantin, A. C., 518
Fantino, J. R., 441, 444
Farr, A. L., 49, 400, 608
Farre, D., 30
Farrington, S., 421
Fasco, M. J., 295, 298, 300
Fattinger, K., 533, 534, 536, 538
Faubion, W. A., 561
Faucon, J. F., 452, 453
Faught, P. R., 560
Faulds, C. B., 328, 329, 330
Faulstich, H., 535
Faust, D. M., 25
Fawole, A., 622
Fehrenbach, T., 535
Fei, P., 167
Feingold, K. R., 590
Feinstein, S. I., 418
Fekete, C. A., 421

Feldman, D., 176, 187
Feltwell, T., 578
Fend, F., 313
Feng, Y., 192, 202, 216
Feng, Z., 292
Fennema, M., 578
Ferguson, R. J., 202
Ferguson, S. S., 589
Fernandes, J., 466
Fernandez, S. B., 471
Fernandez-Checa, J. C., 395, 396
Fernandez-Salguero, P., 66, 572
Fernetti, C., 396
Ferreira, P., 4, 15
Ferreira-Pereira, A., 470, 475
Ferrer, J., 25, 28, 245
Ferrone, S., 419
Feser, W., 194
Feskens, E. J., 420
Fewell, S. E., 472
Fickert, P., 517
Fievet, C., 417
Figura, K., 274, 280
Fillgrove, K. L., 576
Findlay, K. A. B., 48, 110
Finel, M., 118, 122, 132, 133
Fiorentini, D., 353
Fischer, G., 69
Fischer, M., 573, 574
Fischer, P. A., 471
Fischer, R., 561
Fisher, A. B., 418
Fisher, M. B., 105, 110, 599, 608
Fisher, R. A., 80, 89, 600
Fisslthaler, B., 574
Fitzek, M., 239
Fitzpatrick, F., 576
Fitzsimons, E., 419
Fleming, I., 574
Fleury, Y., 318
Flohe, L., 395
Florez, J. C., 420
Flynn, H. C., 348
Foglio, M., 417
Fojo, S., 417
Fojo, T., 421, 487
Folgering, J. H., 452, 453, 454
Forget, S., 560
Forman, B. M., 514, 515
Forrest, H. S., 351

Forrester, L. M., 619
Forsman, T., 346
Forss-Petter, S., 415
Forster, A., 63, 69, 345
Fortier, L.-C., 14, 15, 63
Fortinberry, H., 375, 378, 381, 389
Fortune, J., 546
Foster, B. A., 285
Fouix, S., 215
Fournel-Gigleux, S., 2, 3, 6, 14, 47, 48, 58, 59,
    92, 117, 118, 125, 126, 127, 128, 129, 132,
    134, 345
Fournier, R. E., 25
Foury, F., 467
Fowler, R., 469
Fraenkel, E., 28
Fragner, P., 396
Frain, M., 28
Frame, L. T., 235
Francis, M. C., 236
Francke, U., 31
Frandsen, H., 233, 237, 238, 241, 242, 245
Franey, T., 578
Frank, C., 521, 523
Frank, H., 581
Frank, J., 351
Franke, A. A., 325
Franski, R., 325
Frantz, S. W., 571
Fraser, J. D., 521
Frauca, E., 560
Frayling, T. M., 30
Frech, K., 30
Frede, U., 565
Frederickson, D. S., 417
Frederiksen, H., 233, 238, 242, 245
Fredriksson, C., 468
Freeman, A., 210
Freeman, B., 218, 224
Freeman, L., 417
Frei, E., 235, 236
Freimer, N. B., 559, 560
Freimuth, R. R., 148, 167, 170, 345, 350
Fretland, A. J., 192, 202, 216
Fricke, E., 30
Fricker, D., 419
Fried, M., 518, 522
Friedberg, T., 238, 571, 578
Friedel, G., 264
Friedman, K. J., 414

Friend, S. H., 467
Friesen, R. H., 444
Fritz, P., 264, 271
Fritzgerald, J. F., 560
Frohling, W., 611
Fromm, M. F., 563, 600
Frost, A. R., 168, 176, 234
Fruchart, J. C., 417, 515, 516
Frye, R. F., 110, 609
Fryer, A. A., 622
Fub, C., 590, 592
Fuchsbichler, A., 517
Fuda, H., 167, 168, 172
Fujii, R., 543
Fujii, T., 28, 300
Fujii-Kuriyama, Y., 10, 58, 66, 117
Fujino, M., 543
Fujisawa, Y., 543
Fujisawa-Sehara, A., 10
Fujita, K., 194, 237
Fujita, M., 234
Fujita, Y., 575
Fukami, Y., 317, 335
Fukaya, T., 304, 306
Fukuoka, M., 299, 346, 347, 2987
Fukusumi, S., 543
Fuller, A. P., 313
Fuller, R. S., 466, 468
Funahashi, H., 418
Funke, H., 416
Furstoss, R., 577, 578
Furugori, M., 263, 268, 270
Fuse, S., 300

G

Gabel, C., 416
Gabrilovich, D., 419
Gabrion, J., 396
Gadola, S., 419
Gaedigk, A., 149, 151
Gagel, R. F., 420
Gagne, J. F., 14, 92, 106, 107
Gal, A., 414
Galardi, C., 523
Galaris, D., 221, 222
Galati, G., 353
Galetin, A., 54, 59, 346
Galijatovic, A., 14, 89
Gall, W. E., 132, 138

Gallagher, M. P., 410
Galley, K., 445
Galli, G., 522, 523
Galli, M. C., 353
Gallwitz, W., 170, 172
Galvagni, F., 156
Gamage, N. U., 149
Gamarro, F., 468, 473
Gambaryan, S., 544
Gange, P. V., 106, 107
Gant, N. F., 294
Gantla, S., 14, 15, 47
Gao, F., 420
Gao, X., 469
Gao, Y., 487
Garami, E., 445
Garber, J. E., 348
García-Horsman, J. A., 118, 122, 132, 133
Garcia-Ruiz, C., 395
Gardiner, R. M., 150, 345, 560
Gardner-Stephen, D. A., 36, 59, 346, 589, 590
Garewal, H., 187
Garner, W. H., 221
Garner-Stephen, D. A., 62, 63, 64, 65
Garnier, T., 578
Garrigues, A., 498
Garrison, W. D., 65
Gartung, C., 41, 517, 546
Garza, A., 2, 18
Gas, S., 578
Gaucher, G., 92, 106, 107
Gaudet, D., 420
Gaudet, R., 469
Gautier, J. C., 577
Gavrilkina, M. A., 361
Gay, N. J., 461
Ge, X., 331
Geacintov, N. E., 571
Gebel, T., 578
Gee, A., 437
Gee, J. M., 328, 329, 332
Gee, P., 221
Geertsma, E. R., 453, 454
Geese, W. J., 168
Geffers, R., 30
Gegg, M. E., 396
Gegner, J. A., 239
Geick, A., 519
Geier, A., 41, 166
Gelatti, U., 348

Gelbart, T., 14, 15
Gelboin, H. L., 331
Gelboin, H. V., 572
Gennuso, F., 396
Gentalen, E., 467
Gentles, S., 578
Genzel, Y., 577
Geoghegan, T. E., 589, 601
Georges, E., 439
Gerard, R., 544, 562
Gerber, M., 623
Gerken, M., 264, 271
Gerlach, J. H., 396
Gerloff, T., 41, 560
Germana, M., 273, 274
Germann, U. A., 498
Gerok, W., 121, 547
Gerrard, B., 414, 415
Gerstenecker, C., 395
Gerstner, J., 24
Gervasi, P. G., 621
Geske, D., 14
Gessner, T., 78
Gesson, J.-P., 264, 271
Geuken, M., 515, 565
Geuze, H., 294
Geyer, J., 419
Ghabrial, H., 350
Ghelli, A., 352
Ghislain, M., 472
Gholami, N., 3, 4, 5, 81, 343, 519
Ghosh, D., 23, 24, 30, 274, 275, 279, 280, 282, 284, 286, 287, 288, 290, 298
Ghosh, M., 465
Ghosh, S. S., 121, 122, 123
Giacomini, K. M., 544
Giaever, G., 467
Gibson, G. G., 592
Gibson, R., 415
Gibson, T. J., 199, 429
Gibson, W., 414
Gieselmann, V., 275, 280, 282, 288
Gifford, D. K., 28
Giguere, V., 589
Gil, G., 523
Gilardi, F., 522, 523
Gilbert, L. I., 571, 572
Gileadi, U., 410
Gilissen, R. A. H. J., 622
Gill, D. R., 410

Gill, S. S., 571, 572
Gillessen-Kaesbach, G., 414
Gilliland, G., 292
Gilula, N. B., 387
Girard, H., 14, 15, 63, 110
Girard, P., 452
Girgis, K. R., 419
Glantz, A., 561
Glashan, R. W., 624
Glass, C. K., 600
Glatt, H. R., 151, 171, 231, 232, 233, 234, 235, 236, 237, 238, 239, 240, 241, 242, 243, 244, 245, 250, 251, 269, 348, 349, 352, 578, 622
Glazer, A. N., 543
Gleason, K. J., 193, 209, 217
Glickman, B. W., 194
Glorieux, F. H., 187
Glufke, U., 388, 389, 390
Glutek, S. M., 295, 298
Gnann, A., 469, 472
Gnerre, C., 173, 520, 589
Goda, T., 263, 268, 270
Godeaine, D., 119
Godoy, J. R., 419
Goffeau, A., 460, 463, 464, 465, 466, 468, 470, 472, 473, 474, 475
Going, J. J., 310
Goldhoorn, B. G., 15
Goldkorn, T., 221
Goldman, P., 362
Goldsmith, S., 118
Goldstein, J. A., 589
Goldstein, S. R., 310, 312, 313
Golin, J., 464, 468, 474, 475
Golka, K., 2
Gollan, J. L., 59, 132, 546
Gomez-Lechon, M. J., 38
Gomis, R., 25, 28
Gong, H., 600, 613, 614
Gong, Q.-H., 3, 4, 5, 11, 13, 81, 92, 93, 94, 95, 97, 100, 101, 343, 519
Gong, W. H., 601
Gonzales, L. W., 418
Gonzalez, F. J., 24, 25, 26, 27, 30, 36, 39, 65, 66, 171, 188, 194, 237, 331, 515, 516, 517, 520, 521, 523, 524, 572, 589
Gonzalez-Duarte, R., 415
Gonzalez-Sarmiento, R., 274, 289
Goode, E., 515

Goodfellow, G. H., 192, 193, 202, 206, 209, 216, 217
Goodrow, M. H., 388
Goodsell, D. S., 200
Goodwin, B., 95, 519, 520, 523, 589, 590
Goosen, T. C., 47, 48, 49
Gorboulev, V., 533, 544
Gordon, D. B., 28
Gordon, J. W., 220
Gordon, R., 515, 560
Gordon, S. V., 578
Gores, G. J., 543, 560, 561
Gorokhov, A., 251, 252, 261
Gorter, J. A., 396
Gosh, S. S., 2, 63
Gossen, M., 604
Gossling, E., 30
Goto, T., 378
Gottesman, M. M., 421, 445, 464, 468, 474, 496, 498, 543
Gotz, A., 416
Goudeau, B., 216
Goudonnet, H., 16, 24, 25, 27, 29, 30, 33, 36, 39
Gourley, E., 25, 41
Gouw, A., 560, 561
Goyal, A., 589, 590
Graf, D., 547, 548, 561
Graf, G. A., 417, 544, 562
Graham, J., 202
Graminski, G. F., 141
Grams, B., 64
Granata, R., 601
Grant, A. M., 460, 465, 466, 468, 472, 474, 475
Grant, C. E., 396
Grant, D. F., 571, 572
Grant, D. M., 151, 192, 193, 194, 202, 206, 209, 216, 219, 236, 624
Grant, M. H., 343
Grant, S., 350
Grattan-Smith, P., 415
Graves, M. K., 24, 31, 35
Gray, A. T., 544
Gray, H., 348
Gray, J. M., 418
Gray, K., 202
Green, J., 14
Green, M., 2, 3, 6, 14, 58, 92, 106, 108, 361
Green, R. M., 546
Greenbelt, D. J., 14, 15

Greenberger, L., 487
Greenblatt, D. J., 54, 60, 63, 106, 107, 108, 110, 347, 608
Greene, J. F., 572
Greene, J. W., Jr., 294
Gregori, C., 23, 26, 27, 40
Gregory, P. A., 2, 26, 27, 29, 36, 38, 62, 63, 64, 65
Grider, J. R., 60
Griese, E. U., 63, 69
Griffith, A. P., 325
Griffith, O. W., 399
Griffiths, W. J., 62
Grishin, N. V., 417, 562
Grizzle, W. D., 378
Grodin, J. M., 294
Groen, A. K., 416, 418, 534, 543, 546, 560, 562
Grogan, T. M., 307
Groothuis, G. M., 532
Gros, P., 466
Gross, W. L., 419
Grosveld, G. C., 487
Grout, M., 414
Grove, A. D., 80, 89
Grover, P. L., 572
Groves, J., 218, 219, 223, 224
Grubben, M. J., 187
Grun, F., 590, 591
Grundemann, D., 544
Grundy, S. M., 417
Grüne, S., 546
Grunfeld, C., 590
Gschaidmeier, H., 59, 62, 76, 121, 345
Gu, Y. Z., 66
Guan, J., 421
Gudehithlu, K., 7
Guengerich, F. P., 80, 193, 195, 237, 571
Guenthner, T. M., 572
Guerra, R., 417
Guex, N., 217
Gueze, H., 274
Guggino, W. B., 469
Guhlmann, A., 532, 534
Guillemette, C., 2, 14, 15, 16, 63, 92, 106, 107, 108, 110, 264, 343, 621
Gulick, T., 600
Guo, G., 171, 188
Guo, G. L., 520, 521, 589
Guo, W.-C., 2
Guo, X. H., 59, 346

Guo, Z., 217
Gupta, C. M., 466, 475
Gupta, R. K., 522
Gupta, S., 517
Guss, J. M., 275, 280, 285, 288
Gustafsson, J. A., 62
Gutmann, H. R., 236
Gutterer, J. M., 402, 403
Gutteridge, J. M. C., 217, 218, 219, 221
Guyot, D., 416
Guzelian, P. S., 174, 180, 181, 188, 519, 521,
    589, 600, 601, 603, 604, 606, 609, 610,
    611, 612

H

Ha, S., 131
Haas, J. E., 559
Haase, E., 464
Habata, Y., 543
Haber, D. A., 348
Habes, D., 561
Habib, A. D., 464, 468, 474
Habig, W. H., 610
Hachey, D. L., 377
Hadchouel, M., 23, 24, 27, 418, 560, 561, 565
Hadd, H. E., 295
Haddad, B., 420
Haeggstrom, J. Z., 576, 577
Hagemeier, C., 565
Hagen, M., 151
Hagenbuch, B., 532, 533, 534, 536, 538, 543,
    544, 560, 562
Hagenfeldt, L., 414
Hager, G. L., 25
Hagey, L. R., 360, 361, 374
Hagg, E., 414
Hagiwara, A., 351
Hagiwara, Y., 493
Haiman, C. A., 621
Haimeur, A., 534
Haining, R. L., 599
Haley, B. E., 132, 134, 136
Hall, D., 14
Hall, S. D., 110, 608
Hallberg, B. M., 572, 573, 574, 575, 579
Hallbrucker, C., 547
Hallene, K. L., 396
Halliday, J. A., 194
Halliday, R. S., 200

Hallinan, T., 120, 125
Halliwell, B., 217, 218, 219, 221
Hallström, T. C., 464, 466, 468
Hamada, F. M., 351
Hamilton, G. A., 591, 592, 600
Hamlin, N., 578
Hammans, S., 419
Hammer, R. E., 544
Hammock, B. D., 388, 571, 572, 573, 574,
    578, 581
Hamprecht, B., 396, 397, 398, 399, 400, 401,
    402, 403
Han, Y., 517, 523, 601
Hanai, N., 304
Hanau, D., 419
Handschin, C., 173, 517, 519, 520, 589,
    590, 601
Handschmacher, M. D., 285
Hänel, K., 581
Hanioka, N., 14
Haniu, M., 58, 66, 117
Hankinson, S. E., 621
Hanna, P. E., 193, 202, 209, 217
Hansch, C., 351
Hansen, A. J., 2, 26, 27, 29, 36, 38, 39, 40
Hansen, J. M., 68
Hansen, L. P., 24, 31, 35
Hanson, P. K., 466
Hao, Q., 14, 15, 63, 106, 107, 108, 110
Hara, Y., 14, 15, 263, 268, 270
Harada, M., 543
Harada, N., 298, 299, 303, 304, 314
Harding, B. N., 396
Harding, D., 3, 48, 58, 344
Hardon, E., 28
Harms, A., 64
Harris, A., 23, 26, 33
Harris, A. L., 219, 619
Harris, D., 578
Harris, T. R., 574
Harrison, D. J., 575
Harrop, S. J., 275, 280, 285, 288
Hart, L. K., 417
Hart, R. F., 148
Hart, W. E., 200
Hartshorn, M. J., 410
Hartung, T., 58
Harvey, H. A., 303
Harvey, R. G., 240, 571
Hasada, K., 7

Hasegawa, T., 307
Haser, R., 444
Hashimoto, K., 565
Hasilik, A., 298
Haslewood, G. A. D., 360
Hassan, T., 173
Hassel, B., 487
Hassett, C., 575
Hastings-Smith, D. A., 582
Hasumi, M., 619
Hattori, H., 416
Hattori, K., 69
Haubrock, M., 30
Haumaitre, C., 24, 33
Haung, Y.-H., 93, 97
Hauser, H., 239
Hauser, S. C., 59
Häussinger, D., 543, 546, 547, 548, 550, 561, 563, 565
Haussler, M. R., 166, 174, 175, 188
Hautefeuille, A., 348, 623
Havinga, R., 374
Hawkins-Brown, D., 95, 520, 523, 589
Haws, P., 4, 6, 7, 343
Hayasaka, K., 14, 15
Hayasaka, S., 487
Hayes, J. D., 67, 619
Hayhurst, G. P., 521
Hazard, S., 417, 562
Hazelton, G. A., 132
He, D., 168, 176, 234, 361, 366, 367, 368, 370, 371, 372, 374, 375, 381, 383, 389
He, X.-G., 318
Heales, S. J., 396
Hedge, P. V., 293
Heegsma, J., 515, 565
Heel, H., 67, 76
Hegemann, J. H., 467
Heggie, G. D., 377, 378
Hehl, R., 30
Hehonah, N., 349, 350
Heidrich, J., 518, 589
Heijn, M., 396
Heilig, R., 417
Heim, M., 624
Heimer, S., 417
Hein, D. W., 192, 202, 215, 216, 232, 621, 624
Heinrikson, R. L., 148, 151
Heinzer, A. K., 361, 370
Heizmann, C. W., 30

Hejaz, H. A., 293, 300
Helgason, C. D., 418
Hellier, K., 419
Hemmer, H., 577, 578
Hempel, N., 149, 156, 157, 160, 162
Henderson, C. J., 235, 236
Henderson, P. J., 435, 447
Hendrickson, W. A., 469
Heneghan, J. B., 264
Hengstler, J. G., 240, 242, 578
Hennuyer, N., 417
Henriksen, S. J., 387
Henrissat, B., 129
Henstra, S. A., 23, 26, 33
Hentze, M. W., 565
Henz, M. E., 388, 389, 390
Heppel, L. A., 431
Her, C., 150, 151, 167, 168, 250
Her, S., 170, 172
Herath, A., 219
Herbert, D. C., 173
Hermans, D., 561
Hermelin, B., 561
Hermersdörfer, H., 242
Hermes, M., 577
Hernandez-Guzman, F. G., 274, 275, 279, 280, 282, 286, 287, 288, 298
Hernandez-Martin, A., 274, 289
Herold-Mende, C., 396, 397
Herrero, M. E., 578
Herrlich, P., 600
Herrmann, A., 465
Herrmann, G., 30
Hersey, A., 149
Hertz, G. Z., 30
Hess, H. H., 450
Heuberger, E. H. M. L., 444
Heubi, J. E., 517
Hewer, A., 572
Heydel, J. M., 589, 590
Heyer, A. G., 453
Heyman, M., 420
Heyman, R. A., 517, 523
Hicke, L., 472
Hickman, D., 192, 216
Hidaka, H., 417, 562
Hidaka, K., 303, 304, 313
Hidalgo, I. J., 60
Hierro, L., 560
Hiesberger, T., 25, 41

Higashiyama, T., 274, 275, 279, 280, 282, 286, 287, 288, 298
Higgins, C. F., 410, 430, 432, 439, 461, 485, 565
Higgins, D. G., 199, 429
Higgins, E., 624
Higney, P., 206
Higuchi, H., 543, 560, 561
Hilbert, D. M., 66
Hildebrandt, M. A., 348
Hille, A., 274, 294
Hiller, A., 11, 83
Hills, S., 419
Hincha, D. K., 453
Hindelang, C., 415
Hinnebusch, A. G., 421
Hinnen, A., 577
Hinson, J. A., 236
Hinuma, S., 543
Hinz, W., 577, 578, 581
Hirabayashi, T., 421
Hirakawa, H., 304, 307, 313
Hirano, H., 498, 500
Hirata, D., 467, 468
Hirato, K., 274
Hiratsuka, A., 106, 108, 167
Hirohashi, T., 544
Hiroi, M., 14, 15
Hirose, M., 351
Hirouchi, M., 561
Hirrlinger, J., 395, 396, 397, 400, 401, 402, 403, 404, 405
Hirvonen, A., 215, 621
Hirvonen, J., 118, 122, 132, 133
Hitzl, M., 563
Ho, C., 168
Ho, J., 196
Ho, R. H., 562
Ho, S. N., 178, 179
Ho, Y. C., 138, 139
Hobbs, H. H., 417, 544, 562
Hobkirk, R., 294, 295, 298
Hodgson, D. M., 420
Hoechli, M., 543
Hoekstra, D., 452
Hoffmann, B., 295
Hoffmeyer, S., 500
Hofler, M. W., 311
Hofmann, A. F., 188, 361, 377, 534, 535, 560
Hofmann, J., 468, 473

Hogenesch, J. B., 66
Hogg, N., 218, 224
Holbrook, J. J., 210
Holder, C. L., 332, 334
Holland, I. B., 430, 432, 439, 461, 469, 485
Hollenberg, C. P., 471
Hollenberg, P., 218
Hollenberg, S. M., 95
Hollister, K., 514
Hollman, P. C., 326, 351
Hollo, Z., 471
Holmans, P., 166
Holmberg, B. H., 414
Holmes, A. R., 473
Holmkvist, J., 420
Holroyd, S., 578
Holt, J. A., 523
Holterman, A. X., 28
Holthuis, J. C., 465, 468
Holton, S. J., 216
Holyoak, C., 464, 466
Homburg, S., 574
Homolya, L., 396
Honig, D. M., 331
Honjo, Y., 507
Honkakoski, P., 11, 13, 92, 95, 100, 519, 601
Honma, W., 171, 188, 521
Honma, Y., 166, 174, 175, 188
Hood, A. M., 149, 251
Hooiveld, G. J., 560, 561
Hook, V. Y., 151
Hopfner, K. P., 469
Hopkins, N., 421
Hopkins, P. N., 575
Hoppe, K. L., 416
Hopwood, J. J., 275, 280, 285, 288
Hor, L. I., 437
Horie, T., 396, 397, 532, 534
Horie, Y., 418
Horiuchi, T., 300
Horn, C., 440, 461
Hornhardt, S., 171, 238, 239, 240, 242, 245, 269
Hornischer, K., 30
Hornsby, T., 578
Horton, R. M., 178, 179
Horvath, A., 299
Horvitz, H. R., 416
Horwitz, A. L., 298
Hosagrahara, V. P., 575
Hoshino, M., 274

Hoshita, T., 361, 378
Hosker, B., 224
Hosono, T., 312, 313
Hosoya, M., 543
Hossain, M., 396
Hou, Y., 218
Houlston, R., 621
Housman, D. E., 621
Houston, J. B., 54, 59, 346, 600
Houwen, R. H., 559, 560
Hoyt, R. F., Jr., 417
Hrafnsdottir, S., 466
Hrycyna, C. A., 445, 496
Hrywna, Y., 515
Hsiang, B., 532
Hsieh, C. L., 31
Hsu, Y. T., 619
Hu, C., 26, 33, 435, 445
Hu, M., 95, 331, 589
Hu, X., 420, 534
Huan, L. J., 440
Huang, J.-K., 388
Huang, L., 188, 514, 515
Huang, S.-M., 600
Huang, T., 3, 4, 5, 81, 343, 519
Huang, W., 601, 611, 612
Huang, Y.-H., 11, 14, 83
Huang, Z., 295, 298, 300
Hubacek, J. A., 417
Hubbard, R. E., 410
Huber, C., 419
Huber, M., 532, 534
Huber, R., 533, 534, 536, 538
Huberman, E., 236
Hudson, C., 621
Hudson, J. R., Jr., 313
Hudson, L. G., 62
Hudson, T. J., 417, 420
Huerta, M., 30
Huey, R., 200
Huff, L. M., 507
Huh, N., 312, 313
Hull, M. V., 514, 560
Hum, D. W., 2, 3, 6, 14, 58, 92, 106, 108, 516
Hume, R., 47, 149
Hummel-Eisenbeiss, J., 402, 532, 534, 535, 536, 537, 538, 544, 561, 565
Humphreys, W. G., 193
Hung, J., 507
Hung, L. W., 469

Hung, R. J., 348
Hunt, H. D., 178, 179
Hunt, J. F., 469
Hunt, S. C., 575
Hunter, D. J., 621
Hurst, C. H., 590, 592, 593
Hurst, R. D., 396
Hurst, S., 47, 48, 49
Husain, A., 216
Huse, L. M., 572
Husgafvel-Pursiainen, K., 344
Hussain, K., 420
Hustert, E., 590, 592
Hutchins, L. F., 623
Hutchinson, A., 415, 419, 487
Hutchinson, H. T., 294
Huyer, G., 465
Hwang, M. S., 2
Hwang, S. R., 151
Hyde, S. C., 410
Hyland, R., 47, 48, 49
Hylemon, P. B., 517

I

Iber, F. L., 558
Idle, J. R., 375
Igarashi, P., 25, 41
Igarashi, Y., 467, 468
Igari, I., 120, 132
Iglehart, D. J., 348
Iida, A., 162, 163
Iida, K., 69
Iijima, S., 26
Iijima, T., 33
Ikegami, Y., 487, 493
Ikonen, T. S., 344
Ikushiro, S., 2, 4, 5, 6, 7, 11, 64, 66, 76, 92, 106, 107, 122
Ilett, K. F., 192, 216
Ilias, A., 534
Imada, I., 354
Imai, H., 348
Imai, Y., 506
Inada, J., 378, 379
Inagaki, N., 418, 421
Inatomi, H., 348
Inayoshi, Y., 26
Ingelman-Sundberg, M., 61
Ingendoh, A., 274, 280

Ingram, D. M., 192
Ings, R. M., 106, 108
Innocenti, F., 106, 108, 343
Inokuchi, A., 418
Inoue, Y., 65, 516, 523
Inouye, K., 106, 107
Inui, K., 494, 508
Ioannides, C., 599
Iocco, P., 148, 151
Irie, T., 312, 313
Iscan, M., 171
Ishibe, N., 621
Ishida, J., 317, 335
Ishida, T., 291, 304, 307, 313
Ishidate, M. J., 194, 236
Ishigami, T., 575
Ishihara, M., 416
Ishii, J., 416
Ishii, Y., 2, 26, 27, 29, 36, 38, 40, 59, 124
Ishikawa, E., 506
Ishikawa, T., 395, 464, 486, 487, 493, 494, 498,
    500, 508, 534, 598
Ishiwata, I., 294, 295, 297, 298, 299
Ishiyama, T., 33
Ishizuka, H., 174
Ismair, M. G., 533, 534, 536, 538
Ito, A., 487, 493
Ito, K., 532, 534, 535, 537, 538, 619
Ito, M., 14
Ito, N., 351
Itoh, H., 452
Itoh, K., 67, 68, 69, 388
Itoh, N., 317, 335
Itoh, T., 543
Iwahashi, M., 546
Iwai, M., 14
Iwamori, M., 294, 295, 297, 298, 299
Iwamori, Y., 294, 295, 297, 298, 299
Iwamoto, M., 521
Iwano, H., 118, 347
Iwata, H., 194, 237
Iyanagi, T., 2, 3, 4, 5, 6, 7, 11, 14, 58, 64, 66, 76,
    92, 106, 107, 117, 122, 621
Iyer, L., 106, 108, 343
Izumi, T., 576

J

Jack, R. W., 436
Jackson, M. R., 3, 47, 48, 58, 127

Jackson, P. L., 324, 336, 337, 338
Jacobs, M. N., 317
Jacq, C., 463, 475
Jacquemin, E., 418, 534, 560, 561, 565
Jaehrling, J., 2, 11, 66
Jagels, K., 578
Jager, B., 532, 534, 563
Jager-Krikken, A., 560, 561
Jakobsen, I. B., 429
Jakoby, W. B., 235, 242, 250, 251, 350, 610
Jalal, S., 348
James, E. R., 417
Jan, C.-R., 388
Janigro, D., 396
Jansen, H., 534
Jansen, P. L., 14, 15, 47, 59, 121, 515, 532, 534,
    544, 560, 561, 565, 621
Janssen, D. B., 571, 573, 577, 578
Jantz, H., 53
Jara, P., 560
Jareborg, N., 30
Jault, J. M., 441, 444, 468, 473
Javitt, N. B., 168
Javq, C., 468
Jeanmougin, F., 429
Jedlitschky, G., 129, 396, 397, 401, 402, 493,
    534, 544
Jencks, W. P., 216, 225
Jendeberg, L., 518, 589
Jeong, W., 218, 221
Jerecic, J., 604
Jeremiah, S. J., 344
Jerina, D. M., 2, 571, 579
Jeronimusstratingh, C. M., 571
Jestin, J. L., 210
Jhoti, H., 149
Ji, G.-P., 335, 336
Jiang, Y., 321
Jidenko, M., 444
Jin, F., 622
Jingami, H., 516
Jinno, H., 14
Jobard, F., 417
Jobsis, A. C., 274
Johansson, E., 193
Johl, U., 193
Johnson, A. E., 124
Johnson, F. B., 561
Johnson, I. T., 328, 329, 332
Johnson, K., 414

Johnson, M. R., 361, 362, 363, 374, 375, 378, 379, 380, 382, 383
Johnson, N., 192
Johnson, R., 521
Johnston, D. G., 565
Jones, A. L., 151
Jones, B. C., 47, 48, 49
Jones, D. P., 68
Jones, D. R., 608
Jones, H., 559
Jones, K., 318
Jones, S. A., 518, 519, 520, 523, 589, 590, 591, 592, 600, 601
Jones, T., 467
Jones, T. A., 572, 573, 574, 575, 579
Jones, T. J., 612
Jonker, J. W., 485, 487, 533
Jordan, J. T., 378, 379
Jordanova, A., 414
Jorgensen, B. R., 62, 63, 64, 65
Joseph, J., 224
Josephy, P. D., 193, 194, 195, 196, 197, 198, 200, 202, 236, 237, 624
Journault, K., 92, 106, 107, 108, 343
Jover, R., 38
Jowell, P. S., 414
Joyce, C. W., 417
Joyeux, H., 67
Jude, A. R., 119
Juguilon, H., 518, 589, 600
Juijn, J. A., 559, 560
Jukema, J. W., 417
Jumel, K., 461
Jung, D., 418, 517, 518
Jung, M. H., 173

K

Kaderlik, K. R., 625
Kadlubar, F. F., 192, 193, 202, 215, 235, 622, 623, 625
Kaelin, C. M., 348
Kaestner, K. H., 522
Kagalwalla, A. F., 560
Kagamiyama, H., 210
Kagan, V. E., 353
Kage, K., 506
Kageyama, G., 419
Kahl, R., 67
Kählberer, K., 67

Kahn, A., 23, 26, 27, 40
Kahn, E., 560
Kahn, M. A., 624
Kahnert, A., 275, 279, 280, 284, 288
Kahya, N., 452
Kaji, K., 264
Kajiyama, Y., 25
Kakizaki, S., 11, 13, 92, 93, 94, 95, 97, 100, 101, 601
Kaku, T., 54, 106, 108
Kakuta, Y., 149, 251, 252, 253, 256, 260, 261
Kakyo, M., 532
Kalin, N., 466
Kalinichenko, V. V., 28
Kalk, K. H., 573
Kalow, W., 507
Kalpana, G. V., 2, 63, 121, 122, 123
Kalyanaraman, B., 218, 224
Kamakura, S., 14
Kamataki, T., 28, 192, 194, 237
Kamihira, M., 26
Kaminski, D. L., 350
Kaminski, D. R., 350
Kaminski, P. A., 210
Kaminski, W. E., 416
Kaminsky, L. S., 295, 298, 300
Kamisako, T., 123, 532, 538
Kamiya, A., 65
Kampman, E., 622
Kan, Y. W., 67, 68
Kanagawa, S., 419
Kanai, Y., 544
Kanaya, E., 516
Kandaswami, C., 317
Kane, R. E., 350
Kaneko, C., 303, 304, 307, 313
Kaneoka, H., 26
Kang, D., 621
Kaniwa, N., 14
Kannan, R., 395
Kano, J., 33
Kanou, M., 29
Kansler, T. W., 67, 68
Kapeghian, J. C., 351
Kaplowitz, N., 395, 396
Kaptein, E., 149, 349
Kapur, S., 351
Karami, K. J., 350
Karanth, S., 350
Karas, D., 30

Karas, H., 30
Karcher, A., 469
Karger, A. B., 420
Karin, M., 517, 523
Karl, D. W., 498
Karpen, S. J., 514, 517, 546
Karpowich, N., 469
Karrer, F., 559
Kartenbeck, J., 397, 532, 547, 561, 565
Kasai, N., 106, 107
Kasamatsu, S., 487, 493
Kasper, C. B., 577
Kasper, S. C., 65, 66
Kast, H. R., 515, 520
Kastelein, J. J., 420
Kastner, P., 600
Katan, M. B., 351
Kathmann, E. C., 420
Kato, R., 148, 192, 234, 345, 487, 493
Kato, T., 234
Kato, U., 468
Katoh, M., 14, 15
Katoh, T., 348
Katrakili, N., 25, 26
Katzera, H. G., 414
Katzmann, D. J., 464, 469, 470, 471, 472
Kauffman, F. C., 609
Kaufmann, M. R., 173, 520, 589
Kaur, R., 465
Kaushal, G. P., 132
Kawabata, S., 487
Kawai, K., 69
Kawakami, M., 307
Kawamata, Y., 543
Kawamoto, T., 92, 100, 561
Kawamura, A., 193, 202, 216
Kawana, K., 166, 174, 175, 188
Kawanami, O., 418
Kawano, S., 419
Kazlauskas, A., 76
Kean, L. S., 466, 468
Kearins, M. C., 469
Kearney, W. R., 118
Kebede, B., 578
Keifer, F., 233
Keil, A., 563
Keino, H., 15
Keitel, V., 546, 561, 563, 565
Kel, A. E., 30
Kelekar, A., 619

Kellner, R., 577, 578
Kelly, S. L., 220
Kel-Margoulis, O. V., 30
Kelsey, K. T., 67
Kemp, S., 414
Kemper, B., 521
Kemper, J. K., 521, 523
Kempner, E. S., 364, 365
Keniry, M. A., 354
Kennard, O., 351
Kenngott, S., 418
Kensler, T. W., 67, 68
Kent, U., 218
Kepa, J. K., 67
Kepes, A., 436
Keppler, D., 395, 396, 397, 400, 401, 402, 403,
    404, 405, 493, 520, 532, 534, 535, 536, 537,
    538, 543, 544, 547, 548, 561, 562, 563, 565
Kerb, R., 500, 563
Kerhoas, L., 325
Kern, A., 471
Kerr, T. A., 523
Kertesz, M. A., 275, 279, 280, 284, 288
Kessler, F. K., 2, 27, 29, 30, 33, 36, 38, 39, 64,
    65, 76, 80, 89, 345, 347
Kessler, M. R., 345, 347
Kester, M. H., 349
Keung, M. W., 300
Khan, S., 353
Kharasch, E., 174
Khavari, P. A., 24
Khelifi-Touhami, F., 351
Kholodovych, V., 590, 591
Khorasanizadeh, S., 512
Kibitz, R., 547, 548
Kidd, J. F., 402, 440
Kido, R., 378, 379
Kight, K., 396
Kiguchi, K., 294, 295, 297, 298, 299
Kihara, A., 467, 468
Kihira, K., 378
Kii, I., 487, 493
Kikuyama, S., 421
Killenberg, P. G., 364, 378, 379
Kim, C. S., 517, 523
Kim, D. S., 170, 172, 195
Kim, J. M., 171
Kim, M. H., 571
Kim, P. M., 2
Kim, R., 595

Kim, R. B., 515, 523, 562, 591, 599
Kim, S., 591
Kim, S. C., 173
Kim, S. H., 469
Kim, S. K., 41
Kim, S. M., 206
Kim, S. U., 621
Kim do, G., 361, 370
Kimmel, A. R., 366, 382
Kimura, A., 418
Kimura, M., 304, 307, 313, 378, 379
Kimura, S., 2, 3, 4, 15, 66, 167, 572
Kimura, T., 14, 15
Kinae, N., 263, 264, 266, 268, 270
King, C. D., 61, 63, 92
King, C. M., 231, 236
King, J., 298
King, L. III, 388, 389, 390
King, R. S., 251
Kingma, J., 571, 577
Kinoshita, E., 318
Kinoshita, S., 300
Kinosita, K., Jr., 452
Kiontke, S., 461
Kipp, H., 547, 548
Kira, J., 415
Kirby, L. C., 517
Kirch, W., 105
Kirchgessner, T. G., 532
Kirima, K., 272
Kirk, M., 318, 320, 324, 325, 332, 333, 334,
    335, 336, 337, 338, 388, 389, 390
Kirkpatrick, R. B., 361
Kirshna Murthy, H. M., 285
Kishi, T., 353
Kishimoto, J., 307
Kispal, G., 419
Kistner, A., 604
Kitada, H., 171, 188, 521
Kitakaze, M., 14
Kitamoto, T., 162, 163
Kitamura, K., 318
Kitamura, Y., 162, 163
Kitanova, S. A., 353
Kitto, J. L., 516
Kiuchi, Y., 544
Kivistö, K. T., 264, 600
Klaassen, C. D., 233, 234, 344, 345, 520,
    523, 589
Klein, I., 485, 486

Klein, K., 562, 563
Klein, M., 465, 467
Klein, P. D., 377
Klem, A. J., 235
Klempnauer, J., 24, 25, 26, 28
Kliewer, S. A., 95, 96, 100, 514, 518, 519, 520,
    523, 560, 589, 590, 591, 592, 601
Klomp, L. W., 559, 560
Kloos, D. U., 30
Klucken, J., 416
Klünemann, C., 493, 534
Knapper, C., 416
Kneepkens, C. M., 374
Knehr, M., 578
Kneip, S., 2, 63
Knisely, A. S., 418, 515, 559, 560
Knodo, K., 388
Knoers, N. V., 415
Knol, J., 435, 442, 443, 444, 447
Knosp, B. M., 118
Knowles, M. R., 414
Kobatake, E., 487, 493
Kobayashi, K., 534
Kobayashi, T., 106, 107, 415, 468
Kocarek, T. A., 174, 350, 595
Kocerek, T. A., 600
Koch, C., 416
Koch, M., 264, 271
Kocoshis, S. A., 560
Kodofuku, T., 274
Koeller, D. M., 419
Koepsell, H., 533, 544
Kogan, I., 402
Kohane, I. S., 30
Köhle, C., 2, 11, 60, 65, 66, 67, 68
Kohn, A. B., 151
Kohno, K., 418
Kohno, S., 487
Koike, H., 619
Koiwai, O., 7, 15, 123
Koizumi, N., 300
Kojima, H., 11, 13, 92, 93, 94, 95, 97, 100, 101,
    417, 562, 601
Kok, F. J., 622
Kolaczkowska, A., 464, 468
Kolaczkowski, M., 464, 468, 472, 473, 474, 475
Kolble, K., 310
Kole, L., 2, 18
Koller, B. H., 520, 523, 589
Kölling, R., 471, 472

Komamura, K., 14
Komatsu, M., 534, 535, 536, 537, 538
Komoroski, B. J., 515
Kondo, A., 415
Kondo, S., 419
Kong, A. N., 7, 167
Kong, A.-N. T., 7, 67
König, J., 396, 397, 400, 401, 403, 404, 405,
    520, 532, 534, 535, 536, 537, 538, 544,
    562, 563
Koning, H., 544
Koning, J. H., 560, 561
Konings, W. N., 431, 436, 439, 440, 451, 464,
    468, 472, 473
Konno, T., 565
Konorev, E., 218, 224
Kool, M., 402, 420, 421, 534, 544
Koprubasi, F., 414
Koreeda, M., 571
Kosakarn, P., 194
Koshland, D. E., Jr., 469, 474
Kostiainen, R., 118, 122, 132, 133
Kostrubsky, V. E., 608
Kosykh, V., 515, 516
Kote-Jarai, Z., 621
Kotov, A., 294
Koup, J. R., 47, 48, 49
Koval, M., 418
Kovarova, M., 2, 18
Koynova, G. M., 353
Kozarsky, K. F., 523
Koziarz, P., 193
Kralli, A., 464, 468, 469, 472, 474, 475
Kranich, O., 396, 399, 400, 401
Krapf, G., 466
Krasnikov, V., 453, 454
Krauss, N., 275, 280, 282, 288
Krausz, K., 331
Kremer, M., 313
Kress, C., 23, 24, 27
Krishnamurthy, S., 466
Krishnan, A. V., 176
Krishnaswamy, S., 54, 60, 106, 107, 108, 110
Kroemer, H. K., 264, 271
Kroetz, D. L., 572
Krogmann, T., 487
Krone, B., 14
Kroon, P. A., 328, 329, 330
Kruh, G. D., 396, 534
Ktistaki, E., 26, 33

Kubitz, R., 543, 546, 547, 548, 550, 563, 565
Kubota, S., 2, 3, 4, 5, 18, 81, 343, 519
Kubushiro, K., 294, 295, 297, 298, 299
Kuchel, O., 162
Kuchinski, N., 420
Kuchler, K., 430, 432, 439, 460, 461, 462, 463,
    464, 465, 466, 467, 468, 469, 470, 471, 472,
    474, 475, 485
Kudlow, J. E., 604
Kudo, A., 487, 493
Kudo, G., 572
Kufardjieva, A., 414
Kühlkamp, T., 547, 550, 563, 565
Kuhlow, A., 251
Kuhn, U. D., 350
Kuil, A., 534
Kuiper, J. M., 452
Kuipers, F., 374, 416, 487, 515, 516, 517, 533,
    544, 560, 561
Kujiraoka, T., 416
Kullak-Ublick, G. A., 41, 511, 516, 517, 518,
    522, 533, 534, 536, 538, 543, 547, 608
Kulling, S. E., 331
Kulozik, A. E., 565
Kumagai, S., 419
Kumagai, Y., 69
Kumar, R., 512
Kumar, S., 429
Kume, T., 54, 106, 107
Kunst, M., 534
Kuo, C. J., 31
Kuramoto, T., 378
Kurbegov, A. V., 559
Kurie, J. M., 514, 517
Kurkela, M., 118, 122, 132, 133
Kuroda, Y., 348
Kurokawa, K., 619
Kuroki, S., 361
Kuroki, T., 312, 313
Kurosumi, M., 307
Kurz, A. K., 547, 548, 561
Kurz, E. U., 396
Kurz, G., 534
Kusaka, H., 303, 304, 313
Kushida, H., 194, 237
Kuss, E., 274, 298, 303
Kussmaul, L., 399, 400, 402, 403
Kutsche, K., 414
Kuwano, M., 418, 565
Kwak, M. K., 67, 68

Kwakye, J. B., 364, 375, 378, 379, 380
Kwiterovich, P., 417, 562
Kwon, J., 218, 221
Kwon, K., 218, 221
Kyaw, M., 272

L

Labrie, F., 150, 151
Labrou, N. E., 610
Lachaud, A. A., 11, 13
Lacourciere, G. M., 571
Laemmli, U. K., 137, 139
Laffitte, B. A., 515
Lagutina, I., 487
Lahr, G., 310, 311
Lai, L., 402, 514, 515
Lai, R., 421
Laidler, P., 282
Lakhdar, H., 417
Lam, P., 418
Lamb, J. G., 7
Lamb, R. F., 310
Lamba, J., 515
Lamba, V., 523
Lambert, G., 417, 515, 516, 520, 521
Lambert, I. B., 193
Lambert, L., 466, 470, 474, 475
Lambert, M. H., 96, 518, 589, 591, 592
Lambros, N., 221, 222
Lamers, W. H., 532, 534
Lammert, F., 517, 561
Lamph, W. W., 515
Lamprecht, A., 470, 475
Lan, L. B., 523, 595
Lancet, D., 2, 3, 6, 14, 58, 92, 621
Land, S., 30
Landi, L., 353
Landmann, L., 533, 534, 536, 538, 543, 560
Lane, D., 192
Lanfermeijer, F. C., 436
Lang, F., 547
Lang, N. P., 192, 215, 235, 622, 623, 625
Lang, T., 418, 563
Lange, H., 419
Langenbach, R., 236
Langmann, T., 416, 417, 590
Lapetina, E. G., 218
Laplanche, J. L., 396
Lapunzina, P., 560

Larrauri, J., 560
Larrinaga-Shum, S., 575
Lasaroms, J. J. P., 326
Laskowski, R. A., 200
Lasserre, F., 521
Lassmann, H., 415
Latham, C. F., 149
LaTour, A., 520, 523, 589
Laub, M., 467
Laudet, V., 173
Launay, J. M., 396
Lavesque, E., 106, 108
Lavie, J., 420
Law, M., 303, 304
Lawrence, G., 419
Lazarowski, A., 396
Lazarus, P., 14, 61
Lazzaro, D., 24
Leafasia, J., 419
Leak, D. J., 578
Leake, B. F., 523, 562, 591
Leakey, J., 60
Learned, R. M., 514, 560
Lebioda, L., 282
Leblanc, G., 150, 151, 435, 447
Leblond, B., 300
Lebo, R., 95
Lebwohl, M. G., 420
LeCluyse, E. L., 149, 156, 157, 160, 162, 233, 518, 519, 589, 591, 592, 601
Lecureur, V., 515
Ledvina, P. S., 435
Lee, A. T., 410, 470
Lee, B. K., 67
Lee, C., 218, 221
Lee, E. J. D., 624
Lee, E. W., 300
Lee, H. A., 330
Lee, H. P., 624
Lee, H. Y., 514, 517
Lee, J., 599
Lee, K. M., 621
Lee, M. H., 417, 562
Lee, S., 218, 221
Lee, S. S. T., 66
Lee, S. W., 2, 63, 121, 122, 123, 218, 221
Lee, T. K., 402
Lee, W., 523, 562
Lee, Y. C., 168
Lee, Y. H., 27, 36, 39, 521, 572

Lee, Y.-K., 96, 517, 519, 523
Lee, Y. M., 532, 534, 563
Leeder, J. S., 66
Leese, M. P., 299, 300
Lefebvre, P., 25
Lefevre, C., 417
Leff, M. A., 192, 202, 216
Legare, D., 460
Leggas, M., 174, 521
Leggett, B., 349
Lehmann, J. M., 96, 100, 514, 518, 560, 589, 601
Lehmköster, T., 63, 66, 69, 76
Lehr, P., 274
Lei, W., 7
Leidy, C., 452, 453
Leier, I., 396, 397, 401, 493, 520, 532, 534, 535, 538
Leifeld, L., 310
Leiter, E. H., 345, 375, 378, 381, 389
Lemaire, G., 590, 592
le Maire, M., 444
Lemoine, C., 418
Lenaz, G., 352
Lenhard, B., 30
Lenoir, G., 452
Leomoine, Y., 464, 465, 468, 474
LePage, G., 414
Lépine, J., 264
L'Episcopo, F., 396
Lepoittevin, J. P., 351
Lerner, R. A., 387
Le Sage, G., 543
Le Saux, O., 420, 534
Leslie, E. M., 402
Lester, R., 120, 132, 134
Leterme, S., 468, 475
Letschert, K., 534, 535, 536, 537, 538, 562
Leuschner, U., 561, 611
Le Vangie, R., 313
Levin, W., 2
Levy, D., 442
Levy, G., 92
Lew, J. L., 514, 515
Lewicki-Potapov, B., 30
Lewinski, K., 282
Lewinsky, R. H., 36, 62, 63, 64, 65
Lewis, A. D., 619
Lewis, D. F., 317
Lewis, H. A., 469

Lewis, K. F., 47
Lewis, M. C., 523
Lewis, R. A., 415
Lezzar, A., 351
Li, A. P., 350
Li, C., 224, 402, 445
Li, C. J., 621
Li, D., 514, 5175
Li, H., 514, 515
Li, J., 331
Li, L., 402
Li, P. K., 293
Li, S., 559, 613, 614
Li, T., 523
Li, W. P., 544, 562
Li, X., 517
Li, X. Z., 460
Li, Y., 415, 591
Li, Y. Z., 621
Li, Z. S., 465, 470, 474
Lian, L.-Z., 318
Liang, H., 467
Liang, S., 611
Liang, W. J., 435, 447
Liao, H., 467
Liao, M., 559, 560
Liberato, D. J., 132
Liddle, C., 589
Lidholt, K., 119
Liebhaber, S. A., 24, 28, 33
Liebisch, G., 416, 590
Liebl, R., 299
Liehr, J. G., 271
Liersch, M., 611
Liggins, J., 342
Li-Hawkins, J., 544
Lila, M. A., 321
Lilienblum, W., 59, 61, 67
Lill, R., 419
Lima, M., 25
Lin, G.-F., 2
Lin, H., 95, 218
Lin, J. H., 591
Lin, L.-Z., 318
Lin, R., 174
Lin, S. X., 168, 171
Lin, W. R., 396
Lin, Y., 174
Lin, Z., 350
Lin, Z. P., 396

Lindahl, U., 119
Lindblom, T., 589, 590
Lindenau, J., 396, 397, 400, 401, 403, 404, 405
Linder, A., 264
Lindhout, D., 14, 15, 47
Lindros, K. O., 344
Lindt, T., 119
Ling, V., 409, 418, 496, 560
Lingappa, V. R., 125
Linnane, A. W., 352
Linton, K. J., 418, 432, 565
Liotta, L. A., 310, 312, 313
Lipp, H. P., 62
Lippa, S., 353
Liska, D. J., 166
Liskamp, R. M., 466
Litman, T., 460, 461, 487
Littarru, G. P., 353
Little, J. M., 58, 59, 119, 132, 138, 140, 600, 601
Liu, G., 465, 467, 470
Liu, J., 233, 591
Liu, K. J., 62
Liu, L., 209, 219, 233, 589
Liu, M. C., 168, 172, 300
Liu, P. Q., 435, 445, 469
Liu, S. Y., 24, 25, 26
Liu, X., 416
Liu, Y., 419, 520, 523, 589
Liu, Z., 218
Liyou, N. E., 149, 349, 350
Lizun, S., 543, 547, 548
Llanos, D. E., 350
Lloyd, A. S., 324
Lloyd, R. S., 582
Lo, J. L., 188, 514
Loader, J. A., 219
Loayza, D., 471, 472
Locher, K. P., 410, 470
Locker, J., 23, 24, 25, 30
Lockwood, C. R., 30
Löffler, F., 398
Lohmer, R., 577
Loirat, C., 24, 33
Lokmane, L., 24, 33
Lombard, M., 578
Lombardi, R., 465, 468
Lomri, N., 559, 560
Longo, M., 452, 453
Longo, V., 621

Longuemaux, S., 215
Looser, R., 173, 520, 589
Lopez, C., 218
Lopez, J., 414, 415
Lopez, S., 26
Lopez-Collazo, E., 560
Lord, H. L., 194, 236, 624
Lorentzen, L. J., 62
Lorenzen, J., 41
Lorimer, D., 469
Losko, S., 472
Lou, Y., 350
Loubser, O., 416, 417
Louisot, P., 2, 3, 6, 14, 58, 92
Lovenberg, W., 60
Lowe, G., 224
Lowery, R. G., 118
Loweth, A., 224
Lowry, O. H., 49, 400, 608
Lu, D.-R., 2
Lu, K., 417, 562
Lu, T. T., 166, 174, 175, 188, 523
Lu, X., 543
Lu, Y., 589, 590
Lu, Y. P., 465, 470, 474
Lubahn, D. B., 95
Lubbert, H., 604
Lubieniecki, F., 396
Luc, P. V., 23, 24, 30
Luciani, M. F., 416
Luckow, V. A., 488
Luczynski, J., 464, 468
Luisetti, M., 414
Luk, A., 514, 560
Lukatela, G., 275, 280, 282, 288
Lunetta, K. L., 348
Luo, G., 523
Luscher, T. F., 519
Lustig, K. D., 514, 560
Lutjohann, D., 417, 544
Luukkanen, L., 118, 122, 132, 133, 346
Luu-The, V., 150, 151

M

Ma, L., 251, 517
Ma, M., 7
Ma, Q.-W., 2
Mac Arthur, M. W., 200
Mac Donald, P. C., 294

Machida, T., 95, 100
MacKenzie, K. I., 520, 523, 589
Mackenzie, P. I., 2, 3, 6, 14, 15, 26, 27, 29, 36,
    38, 39, 40, 54, 58, 59, 60, 62, 63, 64, 65, 92,
    117, 120, 123, 125, 127, 132, 346, 589,
    590, 621
Mackie, J. E., 396
MacMillan-Crow, L., 218, 224
Macrina, N., 298, 299
Madon, J., 560
Maeda, H., 419
Maeda, K., 298, 299, 303, 304, 314
Maestro, M. A., 25, 28
Magdalou, J., 2, 3, 6, 14, 58, 59, 92, 117, 118,
    125, 126, 127, 128, 129, 132, 134, 138
Magid, M. S., 515, 560
Maglich, J. M., 95, 589
Magnolato, D., 318
Maguire, A., 419
Mahé, Y., 464, 465, 466, 468, 470, 471,
    474, 475
Maher, T. A., 414
Mai, K. H., 67, 68
Maichele, A., 577
Maier, Y., 418
Maintoux-Larois, C., 418
Maiti, S., 350
Maji, D., 59, 124
Majumdar, D., 168, 173
Makan, N., 624
Maki, K., 14, 15
Makino, N., 396, 401
Makino, Y., 418
Makishima, M., 166, 174, 175, 188, 514, 515,
    517, 523, 560
Makkonen, H., 521, 523
Makundan, M., 591
Malats, N., 619
Malaveille, C., 348, 623
Malecaze, F., 216, 217, 223, 224
Maliepaard, M., 487
Malik, N., 11, 76
Malinovskii, V. A., 350
Malkoski, S. P., 99
Mallampalli, M. P., 465
Malone, L., 466
Maloney, P. C., 469
Maloney, P. R., 523
Mamnun, Y. M., 463, 464, 465, 466, 467, 475
Manao, G., 218, 221

Manautou, J. E., 589
Manchee, G. R., 149
Mancini, M. A., 173
Mandel, J. L., 414, 415
Mangat, S., 187
Mangelsdorf, D. J., 166, 173, 174, 175, 180,
    188, 514, 515, 517, 523, 560, 600
Maniwa, F., 565
Manns, M. P., 2, 14, 63, 64, 69
Manouvrier, S., 561
Mansfield, T. A., 523
Marc, P., 463, 475
Marchetti, B., 396
Marcil, M., 416
Marco, S., 470, 475
Mardon, H. J., 192
Marigo, C., 414
Marinescu, V. D., 30
Marini, S., 621
Mark, M., 600
Markowitz, J., 560
Marks, J. R., 348
Maron, D. M., 236, 243
Marrone, A., 14
Marroni, M., 396
Marschall, H. U., 561
Marsden, C. D., 166
Marshall, A. D., 235, 250, 251, 350
Marshall, T., 264
Martella, L., 621
Martin, J. L., 149
Martin, M. V., 80
Martin, Y. N., 348
Martinasevic, M. K., 61, 63
Martineau, I., 14
Martinek, V., 235, 236
Martinez, V., 521
Martinez-Burgos, M., 469
Martinez-Mir, A., 415
Martinoia, E., 465, 467
Martinoli, M. G., 600
Maruo, Y., 14
Maruta, M., 298, 299, 303, 304, 314
Maruya, E., 419
Marwood, T. M., 196
Marzocchini, R., 218, 221
Masaki, T., 388
Masdeu, C., 24, 33
Mashford, M. L., 29, 41
Masimirembwa, C. M., 250

Mason, D. L., 465, 467, 470
Mason, J. I., 309
Masseguin, C., 396
Masuno, H., 166, 174, 175, 188
Mata, N. L., 415, 416
Matas, N., 215
Matern, H., 121
Matern, S., 41, 121, 561
Mathew, P. M., 420
Matoba, N., 361
Matoney, L., 589
Matsui, H., 619
Matsui, M., 122
Matsumoto, N., 354
Matsumoto, Y., 299, 346, 347, 2987
Matsuno, S., 532
Matsuzaki, S., 307
Matsuzaki, Y., 561
Matsuzawa, T., 92
Mattei, M. G., 24
Matthias, P., 34
Matutes, E., 621
Matwyshyn, G. A., 7
Matys, V., 30
Mauerer, R., 417, 590
Maurel, P., 166, 174, 181, 184, 521, 590
Maw, G., 310
May, A., 419
May, L., 464, 468, 474
Mayer, R., 561
Mazon, M. J., 469
Mazzaccaro, A., 621
Mcbride, W., 624
McCarthy, L. C., 47
McCarthy, L. R., 47
McCarthy, M. I., 565
McCarthy, V. A., 23, 26, 33
McClellan, M., 264
McClure, M., 546
McCord, J., 223
McCoy, E. C., 194
McDermott, G., 469, 474
McDonagh, A. F., 543
Mcelroy, R. C., 294
McElroy, W. D., 450
McGee, J., 576
McGill, A., 218
McGruk, K., 47
McIntosh, D. B., 460, 465, 468, 472, 474, 475

McKee, D. D., 96, 100, 518, 523, 589, 591, 592, 600, 601
McKusick, V. A., 558
Mclaughlan, R., 328, 329, 330
McLennan, H., 352
McLeod, H. L., 14, 15
McLoughlin, L., 560
McManus, B. M., 417
McManus, M. E., 148, 149, 150, 151, 156, 157, 160, 162, 349, 350
McMorris, T., 515
McMullen, G. L., 352
McNally, B. A., 419
McNeill, Y. Y., 523
McNeish, J., 416
McPhail, T., 66
McPhie, P., 235, 250, 251, 350
McQueen, C. A., 192, 202
Mealey, K. L., 419
Mecky, A., 577, 578
Meech, R., 2, 59, 117, 123, 125, 127
Meerman, J. H., 238, 240, 243, 244, 245, 622
Mehner, G., 63, 69
Meibohm, B., 611
Meier, P. J., 402, 511, 515, 516, 517, 519, 522, 532, 533, 534, 535, 536, 538, 543, 544, 546, 547, 560, 562
Meijer, D. F., 532
Meijer, J., 575
Meinke, P. T., 514
Meinke, T., 188
Meinl, W., 233, 238, 239, 240, 242, 243, 244, 245, 250, 251, 348, 349
Méléard, 452, 453
Melino, G., 218
Mellar, F., 264
Mello, L. V., 610
Mellon, F. A., 328, 332
Meloche, C. A., 168, 176, 234
Meluleni, G., 414
Mendel, D. B., 24, 31, 35
Mendoza, L., 30
Menzel, H., 395
Mercier, C., 396
Merikhi, A., 543
Merino, G., 562
Merkler, D., 361, 388, 389, 390
Merkler, K. A., 388, 389, 390
Merkus, J., 575
Mermelstein, R., 194, 233

Mesia-Vela, S., 609
Messeguer, X., 30
Metz, R. P., 27, 30, 36, 38, 39, 63
Metzger, E., 188, 514
Metzler, M., 331
Meunier, V., 67
Meyer, C. A., 589, 590
Meyer, D., 196
Meyer, U. A., 173, 192, 507, 517, 519, 520, 577, 601, 624
Meyers, S., 475
Michael, H., 30
Michaelis, S., 462, 463, 465, 467, 470, 471, 472
Michalski, C., 562
Michel, J., 514, 515
Michel, S., 194
Mickley, L., 487
Middleton, E., Jr., 317
Midtvedt, T., 62
Mieli-Vergani, G., 560
Mierau, G. W., 559
Mietinnen, T., 417
Mihalik, S. J., 361, 370
Mihm, M., 223
Miki, Y., 303, 304, 307, 313, 506
Mikkola, J., 346
Milder, I. E. J., 326
Milewski, S., 466
Millen, L., 469
Miller, A. B., 521
Miller, D. M., 436, 437
Miller, E. C., 61, 193, 231, 236
Miller, J. A., 61, 193, 231, 236, 251
Miller, J. H., 194, 196
Miller, L. K., 488
Miller, M. A., 192
Millikan, R. C., 621
Mills, J. B., 600
Milne, A. M., 55
Milson, D. W., 295
Milunsky, A., 414
Mimmack, M. M., 410
Minami, H., 14
Minami, M., 576
Minami, Y., 33
Minchin, R. F., 192, 216
Miners, J. O., 54, 59, 60, 63, 346
Mioshi, A., 59
Mirabelli, F., 534
Misawa, E., 578

Mishima, C., 162, 163
Misra, S., 559
Mita, S., 534, 535
Mitchell, C., 23, 27, 40
Mitchell, G., 414
Mitchelmore, C., 41
Mitchison, H. M., 150, 345
Mitomo, H., 487, 493
Mitra, P. S., 2, 18
Mitro, N., 522, 523
Miwa, M., 93, 95, 96, 97, 98, 100, 101, 543
Miyahara, K., 467, 468
Miyakawa, T., 467, 468
Miyake, K., 26, 487
Miyata, H., 452
Miyata, M., 171, 188, 515, 516, 521, 572
Miyoshi, A., 124
Mizeracka, M., 132
Mizobuchi, S., 353
Mizuno, N., 600
Mizunuma, M., 467
Mizutani, T., 29
Modi, S., 149
Moehle, C., 417, 590
Moffit, J. S., 589
Moghaddam, M. F., 572
Mohr, S., 218
Mohrenweiser, H. W., 167
Mohri, K., 53
Möhrle, B., 67
Mol, C. A., 534
Mol, O., 515, 565
Molano, J., 469
Molday, R. S., 415
Molhuizen, H. O., 416
Moller, J. V., 444
Molzer, B., 415
Monaci, P., 24
Monaghan, G., 47
Moneti, G., 218, 221
Monier, S., 238
Monk, B. C., 473
Monkiewicz, M., 282
Monneret, C., 264, 271
Monnerjahn, S., 236
Monske, M. L., 62
Monterde, M. I., 578
Montgomery, J., 419
Montminy, V., 92, 106, 107
Moody, J. E., 469

Mooghen, M., 619
Moon, J.-H., 272
Moore, D. D., 96, 174, 514, 515, 516, 517, 519,
    521, 523, 560, 600, 601, 605, 609, 611, 612
Moore, J. T., 95, 96, 100, 518, 520, 521, 589,
    591, 592, 601
Moore, L. B., 518, 519, 523, 589, 590, 591,
    592, 600
Moore, M., 534
Moore, R., 11, 13, 95, 336, 338, 339, 519
Morale, M. C., 396
Moran, E., 26
Mor-Cohen, R., 565
Morelle, C., 239
Moreton, K. M., 210
Morgan, D. J., 350
Morgan, M. R. A., 264, 272, 328, 329, 330
Morgenstern, R., 239, 349
Mori, K., 353
Mori, M., 95, 100
Moridani, M. Y., 351
Mörike, K., 63, 69
Morimoto, H., 354
Morinobu, A., 419
Morisaki, K., 507
Morishita, Y., 33
Morisseau, C., 573, 574
Moritz, A., 419
Moriya, T., 291, 303, 304, 307, 313
Morlot, S., 414
Mornhinweg, E., 563
Morohoshi, T., 418
Morris, G. M., 200
Morris, J. J., 487
Morriseau, C., 388
Morrison, L. D., 194, 236, 624
Mörsky, S., 118, 122, 132, 133
Mosbach, E. H., 361
Moseley, J., 559
Moser, A. H., 590
Moser, H. W., 414
Moshage, H., 544
Moss, D. S., 200
Mosser, J., 414, 415
Motohashi, H., 67, 68
Mottino, A. D., 547
Mouchel, N., 23, 26, 33
Mowbray, S. L., 572, 573, 574, 575, 579
Moye-Rowley, W. S., 463, 464, 465, 466, 467,
    468, 469, 470, 471, 472, 474, 475

Moynault, A., 416
Mu, Y., 521, 600, 601, 605, 609, 612,
    613, 614
Muallem, S., 565
Muccio, D., 324, 336, 337, 338
Muckel, E., 233, 237, 238, 241, 242, 245
Mueller, L., 84
Mueller, R., 590, 592
Mueller, S., 221
Mueschenborn, S. S., 414
Mühlbauer, M., 464, 466
Mühlfeld, A., 547, 548
Muir, M., 294
Mukhopadhyay, S., 546, 547
Mulder, G. J., 47, 236, 242, 250, 621
Mulder, M., 352
Mulholland, D. J., 180
Muller, D., 350
Müller, F., 577, 578, 581
Muller, M., 493, 534, 544, 560, 561, 565
Muller, M. M., 34, 515
Muller, U., 311
Mulugeta, S., 418
Mulvaney, A. W., 206, 209
Munch, R., 30
Munroe, P. B., 150, 345
Münzel, P. A., 2, 11, 61, 62, 63, 66, 67, 69,
    76, 345
Murakami, H., 313
Murakami, S., 469
Murakami-Murofushi, K., 468
Murawec, T., 294
Mürdter, T. E., 264, 271
Muro, K, 106, 108
Murphy, N. H., 92
Murphy, P. A., 318, 323
Murray, H. L., 28
Murty, C. V., 168, 173
Muscat, J. E., 14
Muscillo, M., 274
Mushtaq, A., 193, 202, 206, 207, 209, 215, 216,
    225, 229
Mutoh, H., 576
Myara, A., 561

N

Nadaf, S., 419
Nader, J., 470, 475
Nadon, N., 310

Nagai, F., 122
Nagakura, M., 498, 500
Nagano, M., 416
Nagar, S., 63, 123
Nagata, K., 148, 171, 188, 252, 345, 521, 572
Nagengast, F. M., 187
Nagura, H., 304, 306, 309
Naitoh, T., 532
Najib-Fruchart, J., 417
Nakagawa, S., 317, 335
Nakagomi, R., 532
Nakahara, K., 272
Nakai, D., 532
Nakajima, M., 14, 15, 54, 106, 107
Nakajima, N., 544
Nakajo, S., 421
Nakamura, H., 468
Nakamura, K., 419, 473
Nakamura, T., 171, 188, 418, 521, 565
Nakamura, Y., 162, 163
Nakane, M., 506
Nakanishi, T., 419
Nakao, H., 348
Nakashima, R., 469
Nakata, K., 498
Nakata, S., 619
Nakata, T., 303, 304, 307, 313
Nakatomi, K., 487
Nakauchi, H., 487
Nakayama, T., 264, 266, 270, 351
Nakazato, H., 619
Nakshatri, H., 26
Nakshatri, P., 26
Namura, S., 396
Nangju, N. A., 192, 216
Nanno, T., 15
Napier, I., 420
Narayanaswami, V., 534
Nardini, M., 573
Narkewicz, M., 559
Nasser, S. M., 348
Nathans, J., 415
Nathanson, M. H., 546
Naudin, L., 416
Nauerth, A., 298
Nauta, H., 121
Nauta, N., 59
Nave, K. A., 415
Naviliat, M., 223

Neal, M., 150, 345
Nebert, D. W., 2, 3, 6, 14, 58, 66, 76, 92, 518, 621
Nechushtan, A., 619
Needs, P. W., 324, 328, 332
Neff, M. W., 419
Negishi, M., 10, 11, 13, 92, 93, 94, 95, 96, 97, 98, 100, 101, 149, 156, 157, 160, 162, 168, 251, 252, 253, 255, 256, 260, 261, 519, 520, 589, 601
Negishi, Y., 65, 66, 97, 102, 519, 589, 601, 609, 613, 614
Nei, M., 429
Neidle, S., 193
Neimark, E., 515, 517, 560
Nelson, C., 180
Nelson, D. R., 621
Nelson, E. B., 148
Nelson, M. C., 520, 589, 600, 603, 604, 606, 610, 611, 612
Nelson, W. L., 236
Nelson-Piercy, C., 418, 565
Német, K., 494
Nemeth, A., 560
Nesher, Y., 420
Ness, G. C., 562
Netter, P., 58, 59, 118, 127, 129
Neu, H. C., 431
Neu, M., 149
Neufeld, E. D., 417
Neuhaus, P., 562
Neuschwander-Tetri, B. A., 589, 600, 603, 604, 606, 610, 611
Neuvonen, P. J., 599, 600
Neu-Yilik, G., 565
Newcomer, M. E., 206, 576
Newman, B., 621
Newman, H., 295
Newman, J. W., 388, 574
Newman, N. J., 448
Newman, S. P., 299, 303
Ng, P. Y., 361, 377
Ngo, A., 352
Nguyen, B. L., 298
Nguyen, C. M., 515
Nguyen, L., 562
Nguyen, N., 11, 14, 83, 93, 97
Nichols, J. W., 460, 465, 466, 468, 472, 474, 475
Nicol, C. J., 589

Nicosia, A., 24
Nicotera, P., 534
Nielsen, I. M., 560
Nielsen, K., 421
Niemi, M., 600
Nies, A. T., 396, 397, 402, 520, 532, 534, 538, 544, 561, 562, 563, 565
Niimi, M., 473
Niino, Y., 312, 313
Nikaido, H., 437, 460, 469, 474
Nikaido, K., 435, 445, 469
Nill, K., 60, 67, 68
Nilsson, T., 127
Nischalke, H. D., 310
Nishio, T., 532
Nishitani, S., 93, 96, 97, 98, 100, 101
Nishiyama, T., 62, 63, 64, 65, 106, 108
Nishizawa, M., 123
Nissen, R. M., 421
Nissim, I., 396
Nitta, H., 307
Niwa, T., 600
Noble, M. E., 193, 216
Noe, J., 516, 534, 543
Noel, H., 274
Nogee, L. M., 418
Noguchi, H., 93, 96, 97, 101
Noguchi, M., 33
Noh, D. Y., 621
Nohmi, T., 194, 236, 237
Nokhbeh, M. R., 193
Noll, G., 519
Nomura, H., 532
Noonan, D. J., 515
Noone, P. G., 414
Nordlund, P., 576
Noren, O., 41
Northrup, D. B., 252
Notarfrancesco, K. L., 418
Nourani, A., 470, 475
Novikoff, A. B., 119
Novikoff, P. M., 119
Nowak, G., 132, 140
Nowell, S., 134, 235, 623
Nozawa, R., 264, 266, 270
Nozawa, S., 294, 295, 297, 298, 299
Nuessler, A. K., 562
Nugier, J., 498
Nundy, S., 517
Nussbaumer, P., 274, 299, 303

O

Oates, R. D., 414
Obach, R. S., 599
Obermayer-Straub, P., 2, 63, 64
Obermoeller, R. D., 345
O'Brien, P. J., 351, 353, 354
O'Cochlain, F., 420
O'Connell, N. C., 560
Oda, K., 561
Odom, D. T., 28
Oekonomopulos, R., 53
Oelkers, P., 517
Oertel, R., 105
Oesch, F., 238, 570, 571, 572, 573, 574, 575, 577, 578, 579, 581
Ogawa, J., 418
Ogawa, K., 351
Ogawara, H., 317, 335
Ogura, K., 106, 108, 167
Oguri, K., 59, 124
Oh, T., 171, 172, 174, 180, 188, 521
Oh, T. S., 167, 173, 174, 180, 589, 590
O'Hare, M. J., 219
Ohashi, N., 54, 106, 107
Ohgiya, S., 59, 118, 124, 347
Ohkubo, I., 29
Ohlsson, R., 518, 589
Ohlsson-Wilhelm, B., 303, 304
Ohno, Y., 299, 2987
Ohnuma, T., 106, 108
Ohta, A., 468
Ohta, K., 299, 2987
Ohta, M., 106, 107
Ohtake, N., 619
Ohtsu, A., 14
Ohuchi, N., 291
Ohyama, H., 378, 379
Oinonen, T., 344
Oka, M., 487
Okada, H., 263, 268, 270
Okada, M., 209
Okai, T., 274
Okairi, Y., 294, 295, 297, 298, 299
Okamoto, A. Y., 210, 514, 560
Okamoto, T., 353
Oku, A., 533
Okubo, K., 318
Okubo, M., 419
Okuda, H., 167

Okugi, H., 619
Okuno, E., 378, 379
Oldfield, M. F., 324
Oliveira, E. J., 343
Oliver, B. B., 518, 601
Oliviero, S., 156
Olkkola, K. T., 599
Olson, A. J., 200
Olson, J. S., 436, 437
Olson, T. M., 420
Olthof, M. R., 351
Omiecinski, C. J., 575
Omura, T., 608
Ong, E. S., 95, 518, 520, 589, 600, 612
Onishi, Y., 498, 500
Ookhtens, M., 395, 396
Oomen, L. C., 534
Oostra, B. A., 14, 15, 47
Oosumi, K., 498, 500
Oradei, A., 353
O'Reilly, D. R., 488
Orelle, C., 441, 444
Orlowski, S., 498
Ornitz, D. M., 602
Oro, A. E., 95, 515
Orrenius, S., 534
Orso, E., 416
Orth, U., 414
Ortiz, D. F., 465, 470, 559
Ortlund, E., 282
Orzechowsky, A., 61
Osada, H., 467
Osada, Y., 348
Osamura, R. Y., 307
Osawa, M., 487
Osawa, S., 162, 163
Osawa, Y., 274, 275, 279, 280, 282, 286, 287, 288, 298
Osborne, M. R., 235, 236
Oscarson, M., 173, 589
Ose, L., 417
Oshihi, N., 576
Osterloh-Quiroz, M., 240, 242
Ostrow, J. D., 92, 396
O'Sullivan, A., 53
Ota, K., 571
Otake, Y., 89
Ott, C. M., 125
Ottenhoff, R., 396, 416, 534
Otter, M., 419, 532

Otterness, D. M., 167, 168, 250
Oude Elferink, R. P., 14, 15, 47, 396, 419, 485, 487, 532, 534, 543, 560, 562
Oue, S., 210
Ouellet, M., 224
Ouellette, M., 460
Ourlin, J. C., 166, 174, 181, 184, 520, 521
Ouzzine, M., 58, 59, 117, 118, 125, 126, 127, 128, 129
Overkamp, K. M., 573, 575
Overton, G. C., 30
Owens, I. S., 2, 3, 4, 5, 6, 11, 13, 14, 15, 18, 47, 58, 76, 81, 92, 93, 94, 95, 97, 100, 101, 117, 343, 519, 601
Owsianik, G., 472
Oyasu, R., 69
Ozalp, C., 523
Ozawa, S., 14, 148, 234, 235, 299, 345, 2987
Ozawa, Y., 318
Ozbas-Gercerer, F., 396
Ozbilen, O., 168, 173
Özvegy-Lacka, C., 494

P

Pabel, U., 233, 238, 240, 242, 245, 251
Packer, L., 353
Pagano, G., 219
Pai, T. G., 168, 172
Paigen, B., 417
Paigen, K., 264, 271
Paine, M. F., 105
Pakhomova, S., 576
Pal, K., 572
Palcidi, D., 348
Palmer-Toy, D., 313
Palmiter, R. D., 602
Paloma, E., 415
Pan, L. X., 514, 515
Panayiotidus, M., 222
Pandeya, N., 419
Pandjaitan, R., 464, 466, 470, 471, 475
Pangborn, W., 274, 275, 279, 280, 286, 287, 288
Paniagua, A., 25, 28
Panwar, S. L., 468
Papadopoulou, B., 460
Papafotiou, G., 25, 26
Pappin, D. J., 142
Parc, R., 561

Pardee, A. B., 621
Parenti, G., 273, 274
Parikh, A., 195, 237
Park, J., 361, 370
Park, S. K., 621
Parker, M. G., 274, 298
Parkhill, J., 578
Parkinson, A., 28, 599
Parkinson, G. N., 193
Parks, D. A., 337
Parks, D. J., 514, 560, 589
Parle-McDermott, A., 464, 470, 475
Parrish, D. D., 351
Parrizas, M., 25, 28
Parviz, F., 523
Paschalis-Thomas, D., 221, 222
Pascolo, L., 396
Pascussi, J. M., 166, 174, 181, 184, 521, 590
Pasqualini, J. R., 274, 298, 303
Pasquali-Romchetti, I., 420
Pastan, I., 445, 496, 498, 543
Pastuszak, I., 132
Patel, R. P., 324, 336, 337, 338, 339
Patel, S. B., 562
Paternostre, M. T., 442, 443
Paterson, A. J., 604
Paterson, E. S., 193
Patil, S. D., 192, 216
Patten, C. J., 60, 63, 106, 107, 108
Patzlaff, J. S., 447, 449, 450
Paul, A., 216, 217, 221, 223
Paul, M., 561, 565
Paul, P., 132
Paul-Abrahamse, M., 15
Pauli-Magnus, C., 418
Paulusma, C. C., 396, 419, 532, 534
Pauly, K., 234, 238, 240, 245
Paumgartner, G., 547
Pawlikowska, L., 559, 560
Pawlowska, J., 560
Pawlowski, P. G., 403
Payne, C. M., 187
Payton, M., 192, 193, 206, 207, 209, 215, 216, 229
Pearce, S. R., 410
Pearson, W. R., 619
Pease, L. R., 178, 179
Pecheur, E. I., 452
Pecreaux, J., 452
Pedersen, L., 149

Pedersen, L. C., 168, 251, 252, 253, 255, 256, 260, 261
Pedersen, L. G., 251, 252, 253, 256, 260, 261
Pedersen, N. C., 419
Peehl, D. M., 176
Pegoraro, R., 417, 562
Pei, Z., 361, 370
Peiffer, A., 415
Peitsch, M. C., 217
Pelaez, F., 514, 515
Peluffo, G., 223
Pennington, M. W., 3, 4, 5, 81, 343, 519
Pentenga, M., 571
Perchermeiser, M. M., 233
Percival, M., 224
Perez-Victoria, J. M., 468, 473
Perkins, D. N., 142
Perlmann, T., 515, 518, 601
Perlmutter, D. H., 26, 33
Pero, R. W., 619
Perussed, L., 14, 15, 63, 110
Peterkin, V., 47, 48, 49
Peters, W. H., 14, 59, 121, 532, 534, 575
Petersen, K. U., 351
Peterson, A., 236
Peterson, G., 325
Peterson, K. M., 416
Peterson, P. A., 127
Peterson, T. G., 320, 335, 336
Petrotchenko, E. V., 168, 251, 252, 253, 255, 256, 260, 261
Petrowski, M. L., 516
Petullo, D. M., 65, 66
Petzinger, E., 419
Petzl-Erler, M. L., 14
Pezzella, F., 219
Pfeiffer, B., 396, 398, 399
Pflugrath, J. W., 436, 437
Philippsen, P., 476
Phillips, B., 231, 236
Phillips, D. H., 235, 236, 238, 242, 245
Phillips, J., 547
Phillips, M. J., 418, 548
Phylactides, M., 23, 26, 33
Picard, B., 215
Piccoli, D. A., 560
Pichard, A. L., 23, 26, 27, 40
Piée-Staffa, A., 240
Pier, G. B., 414
Pieraccini, G., 218, 221

Pierce, A. M., 519, 601
Pierpoint, W. S., 351
Pietrzak, E., 294
Pietta, P. G., 317
Piette, J. C., 416
Pignatti, P. F., 414
Pijnenborg, A. C., 487, 534
Pike, J. W., 187
Pinaire, J. A., 589, 601
Pineau, T., 66, 521
Pinkert, C. A., 602
Pinto, A. V., 621
Piper, P., 464, 466
Pircher, P. C., 516
Pizzagalli, F., 533, 534, 536, 538
Pizzey, J., 168
Plant, N., 592
Plante, L., 274
Plass, J. R., 515, 565
Pleasure, D., 396
Plebani, A., 419
Pleiss, J., 573, 574
Plemper, R. K., 469, 470, 472
Pless, D., 118, 129
Plewniak, F., 429
Plomp, A. S., 420
Plosch, T., 487, 533
Plumb, G. W., 328, 329, 330
Podoll, T., 599
Podust, L. M., 591
Podvinec, M., 173, 183, 520, 589
Poelarends, G., 439
Poellinger, L., 76
Pogge von Strandmann, E., 24
Poh, W. T., 624
Pohl, A., 465
Pohlmann, R., 476
Pojer, C., 517
Pokorny, N., 193
Pokrovskaya, I., 119
Polly, P., 182
Poloyac, S. M., 613, 614
Polyak, K., 348
Pomata, H., 396
Pomorski, T., 465, 468
Pompeo, F., 193, 206, 209, 215, 225
Pomplun, D., 251
Pompon, D., 577, 578
Pongratz, I., 76
Ponka, P., 420

Pontoglio, M., 23, 24, 25, 27, 28, 30, 41
Poolman, B., 431, 432, 433, 434, 435, 436, 437,
    440, 442, 443, 444, 445, 446, 447, 448, 449,
    450, 452, 453, 454
Pope, C. N., 350
Pope, F. M., 420
Pope, J., 192, 216
Pope, M., 571
Popescu, N. C., 3, 4, 5, 81, 343, 519
Popowski, K., 522
Porru, S., 348
Porsch-Ozcurumez, M., 416
Porteu, A., 23, 26, 27, 40
Portillo, F., 469
Post, A., 354
Post, K. W., 469
Postlethwait, E. M., 582
Postlind, H., 77, 83, 518, 589
Potter, B. V., 274, 293, 298, 299, 300, 303
Potter, C., 3, 4, 5, 14, 81, 343, 519
Poulos, A., 415
Poupon, R., 561
Povey, S., 344
Powell, G. M., 295
Powell-Oliver, F. E., 95
Pozzi, E. J., 547
Prades, C., 416, 418
Prasad, R., 466
Prasain, J. K., 317, 318, 327, 336, 337,
    338, 339
Pratt, C., 132
Presecan-Siedel, E., 590, 592
Pretorius, P. J., 376
Price, E. M., 498
Price, K. R., 330
Price, R. R., 523
Pries, F., 571
Pritchard, M., 128
Probst, P., 150, 345
Prospero-Garcia, O., 387
Prossnitz, E., 437
Prough, R. A., 589, 601
Pujol, A., 414, 415
Pullen, J. K., 178, 179
Puranen, T., 284
Purdie, D., 419
Purinton, S. C., 295
Purohit, A., 274, 293, 298, 299, 300, 303
Purvis, I. J., 47
Putman, M., 451

Q

Qatanani, M., 601, 611, 612
Qin, X. F., 264
Qin, Y.-Q., 2
Qiu, H., 421
Qu, Q., 440, 441
Quaglino, D., 420
Quandt, K., 30
Quattrochi, L. C., 77, 83, 84
Quinn, P. J., 353
Quintanilla-Martinez, L., 313
Quiocho, F. A., 432, 435, 436, 437, 438, 445

R

Rabl, W., 420
Rachmel, A., 132
Radi, R., 218, 223, 224
Radominska, A., 118, 120, 128, 129, 131, 132,
    134, 138, 171
Radominska-Pandya, A., 58, 59, 62, 63, 64,
    65, 66, 97, 102, 118, 119, 132, 138, 140,
    519, 520, 589, 590, 600, 601, 609, 612,
    613, 614
Radu, R. A., 416
Rafter, J., 62
Raftogianis, R. B., 150, 151, 168, 250, 622
Rahman, M. A., 517
Rahmani, R., 590, 592
Raijmakers, M. T. M., 575
Rainey, W. E., 168
Rajagopal, R., 622
Rajashankar, K. R., 469
Ramachandra, M., 445, 496
Ramachandran, V., 515, 608
Ramagli, L. S., 167
Raman, J., 210
Ramirez, J., 106, 108, 343
Ramjeesingh, M., 402, 440, 445
Ramkema, M., 396
Ramos, C., 414
Ramos-Gomez, M., 67, 68
Ramponi, G., 218, 221
Rance, D. J., 47, 50, 343
Randall, R. J., 49, 400, 608
Rando, R. R., 138, 139
Rao, G., 464, 466
Rao, S. I., 251
Rao, T. R., 170, 172

Raschko, F. T., 2, 11, 66
Raskind, W. H., 419
Rastinejad, F., 512
Ratain, M. J., 106, 108, 343
Ratcliff, R., 414
The Rat Genome Sequencing Project
    Consortium, 30
Ratta, M., 352
Raub, T. J., 60
Raucy, J., 84, 591
Ravaud, S., 444
Ravid, T., 221
Ray, K., 24, 28, 33
Raymond, G. V., 414
Rea, P. A., 465, 467, 470, 474
Read, J., 180
Real, J., 416
Reardon, I. M., 148, 151
Rebbeck, T. R., 622
Rebbeor, J. F., 396, 402, 474
Rebischung, C., 476
Recksiek, M., 274, 285
Redder, T., 523
Reddy, B. S., 187
Reddy, S., 218
Redeker, S., 396
Redfield, B. G., 60
Redinbo, M. R., 96, 518, 589
Reed, J. E., 300
Reed, M. J., 274, 293, 298, 299, 300, 303
Rees, D. C., 410, 470
Reetz, M. T., 577
Regina, A., 396
Rehse, P. H., 168, 171
Reichel, C., 418
Reid, G., 174, 521, 534
Reimann, A., 25, 41
Reinehr, R., 543, 561
Reinhard, J., 520, 523, 589
Remaley, A. T., 416, 417
Ren, S., 521, 601, 605, 609, 612, 613, 614
Rennie, P., 180
Renola, L., 274
Renskers, K. J., 148
Repa, J. J., 514, 523, 560
Repasch, M. H., 47
Resibois, M., 30
Ressler, B., 414
Rettie, A. E., 575
Reuter, I., 30

Rey-Campos, J., 24, 39
Reyes, J., 560
Reymond, J.-L., 583
Rhee, K. H., 24
Rhee, S., 218, 221
Rhei, E., 348
Rhodes, H. E., 417
Rhodes, M. J. C., 329, 330
Ricci, M. J., 271
Richard, K., 149
Richards, A., 420
Richards, R. I., 15
Richardson, D. R., 420
Richarme, G., 436
Richi, A. K., 487
Richie, J. P., Jr., 14
Richter, C., 223
Richter, T., 311
Ricker, J. V., 452, 453
Ridder, I. S., 573
Ridderstrom, M., 250
Riddle, B., 216, 225
Riedel, H., 221
Riedel, J., 53
Riezman, H., 465, 468
Rigaud, J. L., 441, 442, 443, 444, 452, 470, 475
Rijnbout, S., 274, 294
Riley, E., 349, 350
Riley, R. J., 48, 54, 347
Rinaldi, N. J., 28
Ring, B. J., 608
Ring, J. A., 350
Ringeisen, F., 23, 24, 28, 30
Ringer, D. P., 310
Ringpfeil, F., 420
Rink, R., 573, 578
Riordan, J. R., 469, 472
Rios, G. R., 61, 63, 92
Risch, A., 192, 215, 532, 534, 563
Ritter, J. K., 2, 3, 4, 6, 13, 14, 15, 27, 29, 30, 33, 36, 38, 39, 58, 63, 64, 65, 66, 76, 80, 89, 92, 345, 347
Ritz, U., 419
Rius, M., 402
Riva, A., 30
Rivera, F., 25, 28
Rivera, J. I. S., 2, 18
Rivera, M., 300
Rivera, O. J., 167, 173, 174, 180, 589, 590
Roach, M. L., 416

Robberecht, P., 30
Roberts, K. D., 274
Roberts, R. C., 151
Roberts, R. L., 562
Robertson, K. R., 419
Robertson, R. N., 354
Robey, R. W., 487, 507
Robinson, G., 313
Robinson, J. J., 300
Rockstroh, J. K., 310
Rodbourn, L., 2, 15
Rodrigues-Lima, F., 193, 206, 215, 216, 217, 219, 221, 223, 224, 225
Rodriguez, D., 223
Roe, A. L., 76
Roelofsen, H., 544
Roest, T. J., 349
Rogan, P. K., 66
Rogers, H. J., 624
Röhrdanz, E., 69
Rohrschneider, K., 415
Rolfe, P. A., 28
Rollini, P., 25
Roma, M. G., 547
Romain, Y., 162
Romalo, G., 294
Romano-Spica, V., 487
Roomp, K., 416, 417
Roques, C., 67
Rose, F. A., 295
Rose, K. A., 600
Rosebrough, N. J., 49, 400, 608
Rosello, L., 30
Rosen, B. P., 465
Rosen, J., 469
Rosenberg, N., 565
Rosenfeld, J. M., 174, 180, 181, 188, 521, 589, 601, 609, 612
Rosenfeld, M. G., 95, 600
Rosenkranz, H. S., 194, 233
Rosenthal, F. E., 464, 468, 469, 472, 474, 475
Rosenthal, H. E., 294
Roses, A. D., 47
Roset, R., 30
Rosier, M., 416, 418
Rosing, H., 487, 533
Rosmorduc, O., 561
Ross, D., 67
Ross, D. D., 487
Rossi, A., 218

Rossi, S. S., 361
Rost, D., 547, 562
Rost, M., 350
Rotert, S., 30
Roth, A., 173, 589
Roth, C. B., 410, 431, 470
Roth, J. A., 148, 560
Roth, M. E., 523
Roundtree, M., 543
Rousste, M., 67
Roux, F., 396
Rowlands, M. G., 303, 304
Rowley, J. Y., 231, 236
Roy, A. B., 295, 297
Roy, A. K., 168, 170, 171, 172, 173, 174, 180,
    183, 188, 521
Roy Chowdhury, J., 47, 58, 63, 119, 121,
    122, 123
Roy Chowdhury, N., 47, 63, 121, 122, 123
Royer, L. J., 416
Royo, I., 515
Rozeboom, H. J., 573
Rubbo, H., 223
Rudikoff, S., 66
Ruebner, B. H., 558
Ruedig, C., 400
Ruetz, S., 466
Ruis, M., 534, 544
Runge-Morris, M., 174, 350
Runswick, M. J., 461
Ruschitzka, F., 519
Russell, D. W., 518, 523
Russell, P. L., 440, 441
Russell, R. R., 431
Russo, A., 621
Rust, S., 416
Rustan, T. D., 202
Rutter, M. E., 469
Ryan, M., 47
Ryffel, G. U., 24
Ryu, J. I., 238
Rzhetsky, A., 410, 485

S

Saarikoski, S. T., 344
Sabbagh, W., Jr., 518, 589, 600
Saborowski, M., 522
Sacchiero, R. J., 517
Sachsenmeir, K. F., 219

Sa-Cunha, A., 521
Sadagopan, N., 600
Sadeghpour, A., 522
Sadler, V. M., 62
Saeki, M., 14
Saeki, S., 523
Saeki, T., 304
Saenger, W., 275, 280, 282, 288
Safe, S., 519, 601
Saffman, P. G., 454
Safra, N., 419
Sagara, J., 396, 401
Saha, N., 547, 548, 550
Saha, T., 2, 18
Sahi, J., 600
Sai, K., 14
Sai, Y., 494, 508, 533
Saijo, N., 14
Saini, S. P., 65, 66, 97, 102, 519, 521, 589, 600,
    601, 605, 609, 612, 613, 614
Saintot, M., 623
Saito, H., 274
Saito, K., 192
Saito, S., 162, 163
Saito, T., 300
Saito, Y., 14
Saji, H., 419
Saka, N., 264, 266, 270
Sakaki, T., 106, 107
Sakakibara, Y., 168, 172, 300
Sakalian, M., 384, 386, 387, 389
Sakamoto, G., 307
Sakisaka, S., 418, 565
Sakuma, N., 417, 562
Sala-Trepat, J. M., 28
Salavaggione, O. E., 348
Salazar, J. J., 582
Salem, M., 418
Salen, G., 514, 515, 562
Sali, A., 199
Salisbury, S. A., 351
Salomon, D. S., 304
Salvioli, R., 274
Salwinski, L., 206
Samokyszyn, V. M., 132, 138, 140
Sampath, J., 487
Samuelsson, B., 576, 577
Sanchez, E., 47
Sanchez-Sabate, E., 560
Sandaltzopoulos, R., 24

Sandberg, A. A., 294
Sandberg, M., 575
Sandelin, A., 30
Sandy, J., 193, 216
Saner, K. J., 168
Sanglard, D., 464, 468, 469, 472, 474, 475
Sano, K., 487
Sansom, L. N., 148, 150, 151
Santen, R. J., 303, 304
Santner, S. J., 303, 304
Sappal, B. S., 2, 63, 121, 122, 123
Sar, M., 95
Sarabia, S. F., 271
Saraste, M., 461
Sarde, C. O., 414, 415
Sarkadi, B., 396, 485, 486, 494, 498, 534
Saruta, M., 307
Sasaki, A., 14, 15
Sasaki, H., 304, 306
Sasaki, M., 532, 534, 535, 537, 538
Sasano, H., 168, 291, 303, 304, 306, 307,
    309, 313
Sasseville, V., 532
Sato, H., 7, 14, 15, 29, 123, 534
Sato, K., 575
Sato, M., 264, 266, 270
Sato, R., 120, 608
Sato, T., 274
Satsukawa, M., 167
Sauder, J. M., 469
Sauer, B., 27
Sauer, F., 234
Sauerbruch, T., 310
Sauna, Z. E., 498
Saunders, A. M., 47
Savazin, G., 264
Savov, A., 414
Sawada, J.-I., 14
Sawada, S., 487
Sawai, T., 309
Sawicki, M., 284
Saxel, H., 30
Saxena, A., 254
Scalbert, A., 264
Scarborough, G. A., 498
Schachner, M., 415
Schachter, H., 2, 3, 6, 14, 58, 92
Schaefer, F. V., 310
Schaefer, K., 523
Schaffner, S. F., 420

Schaffner, W., 34
Schantz, S. P., 14
Schareck, W., 345
Scharl, M., 590
Scharschmidt, B. F., 559, 560
Schatz, B., 298
Schaub, T., 493, 534
Schauer, W., 475
Schedin, S., 352
Scheer, M., 30
Scheffer, G. L., 396, 419, 487, 532, 533, 534,
    561, 565
Schekman, R., 124
Schellens, J. H., 487
Scheper, R. J., 396, 419, 487, 532, 533, 534
Scherfeld, D., 452
Scherrmann, J. M., 396
Scheurer, U., 518
Schilders, G. W., 575
Schimitt, B. M., 533
Schinkel, A. H., 485, 487, 533, 534
Schippers, I. J., 544
Schlaeger, C., 196
Schliess, F., 543
Schlosser, M. J., 351
Schlotterer, C., 15, 16
Schmeiser, H. H., 235, 236
Schmid, R. D., 573, 574
Schmidt, B., 274, 275, 279, 280, 284, 285,
    288, 294
Schmidt, W. K., 472
Schmitt, H. P., 396, 397
Schmitt, L., 440, 461, 469
Schmitt, M., 543, 547, 548
Schmitt, W., 419
Schmitz, G., 417, 590
Schmohl, S., 2, 11, 66, 67
Schneider, B. L., 27, 30
Schneider, M., 523
Schoonjans, K., 523
Schorling, S., 466
Schreiber, E., 34
Schreiber, J., 28
Schreiner, E. P., 299
Schrem, H., 24, 25, 26, 28
Schrenk, D., 61, 63, 69
Schreyer, S. A., 417
Schriml, L. M., 417, 487
Schröder, A., 543
Schteingart, C. D., 361, 377

Schueck, N. D., 419
Schuele, S., 517
Schuetz, E. G., 174, 515, 521, 523, 595, 600
Schuetz, J. D., 174, 487, 515, 521, 534, 595
Schuetzer-Muehlbauer, M., 470
Schug, J., 30
Schuler, G., 295
Schüller, C., 460, 461, 462, 463, 464, 465, 466, 467
Schulman, I. G., 516
Schultz, J., 417, 562
Schulz, J. B., 396, 397, 400, 401, 402, 403, 404, 405
Schulze, B., 577
Schuster, M., 466
Schütte, J. K., 14
Schutz, G., 600
Schutze, K., 310, 311
Schuurman, E. J., 420
Schuurman-Wolters, G. K., 445
Schwab, M., 67
Schwab, M. H., 415
Schwartz, M., 436
Schwarz, L. R., 61
Schwarz, M., 517, 523
Schwatz, B. S., 67
Schweikert, H. U., 294
Schwierzok, A., 238, 245
Schwille, P., 452
Schwinger, E., 414
Scobie, H., 351
Scotto di Carlo, A., 348
Sczepanski, J., 464, 468, 474
Seddon, J. M., 415
Seddon, R., 47
See, C. G., 344
Sefler, A., 521
Segura-Aguilar, J., 353
Sehayek, E., 27, 30, 517
Seidegard, J., 619
Seidel, A., 62, 236, 238, 240, 245, 581
Seidel, J., 274, 294
Seidl, T. L., 534
Sekine, A., 162, 163
Sekine, T., 522, 533, 544
Selassie, C. D., 351
Selcer, K. W., 293
Seliger, B., 419
Seligsohn, U., 565
Selivanov, V. A., 420

Selmer, T., 274, 275, 280, 282, 285, 288
Senay, C., 118, 129, 132, 134
Sengstag, C., 577
Senior, A. E., 498
Senkel, S., 24
Senn, T., 174
Sensen, C. W., 431
Seol, W., 96, 516, 519
Seow, A., 624
Serabjit-Singh, C. J., 519, 590
Serbinova, E. A., 353
Servos, J., 464
Sessions, R. B., 210
Setchell, K. D., 325, 559, 560
Seth, P., 348
Sevlever, G., 396
Seyama, Y., 576
Sfakianos, J., 332, 333
Sfakianos, M. K., 361, 384, 386, 387, 389
Sgroi, D. C., 313
Shah, P. P., 362
Shahid, M., 590
Shan, B., 417, 514, 523, 560, 562
Shani, M., 565
Shao, X., 25, 41
Shapiro, L., 274, 289
Sharma, K. G., 465, 467, 470
Sharma, S., 432, 438, 445
Sharma, V., 168, 235, 250, 251, 253, 254, 255, 257
Sharom, F. J., 440, 441
Sharp, P., 415
Sharpe, M. A., 396
Shatz, O., 420
Shaw, D. R., 361, 365, 366, 367, 371
Shaw, P. M., 168
Shearer, B., 521
Sheen, Y. Y., 2, 3, 4, 15
Shefer, S., 27, 30, 514, 515, 562
Sheinfil, A., 187
Shen, J., 465
Shen, J.-H., 2
Sheng, J., 254
Shenk, J. L., 591, 592, 600
Shenoy, S., 350
Shentu, S., 543
Shepherd, S. R. P., 120, 125
Sheridan, J., 419
Shevtsov, S., 251, 252, 261
Shi, H., 62

Shi, Y., 131, 520, 589, 612
Shibutani, S., 168
Shibuya, M., 317, 335
Shigenaga, J. K., 590
Shih, D. Q., 27, 30
Shih, S. C., 472
Shimada, K., 514, 515
Shimada, M., 148, 171, 188, 521
Shimane, M., 533
Shimazui, T., 69
Shimizu, A., 419
Shimizu, C., 167, 168, 172
Shimizu, M., 299, 346, 347, 2987
Shimizu, T., 576
Shimizu, Y., 274
Shimoi, K., 263, 264, 266, 268, 270
Shimomura, I., 166, 174, 175, 188
Shimozawa, N., 415
Shin, D. S., 469
Shindo, Y., 234
Shindyalov, I. N., 292
Shinkyo, R., 106, 107
Shinnoh, N., 415
Shinohara, A., 192
Shintani, Y., 543
Shioda, S., 418, 421
Shiotsu, Y., 303, 304, 313
Shiozawa, K., 487
Shirai, T., 351
Shiraki, T., 516
Shirao, K., 14
Shirmoi, K., 264, 266, 270
Shitara, Y., 534
Shively, J. E., 58, 66, 117
Shneider, B. L., 515, 517, 560
Shoda, J., 561
Shoelson, S. E., 521
Shoemaker, D. D., 467
Shonsey, E. M., 361, 374
Shore, V., 562
Shriver, S., 469
Shroyer, N. F., 415
Shulenin, S., 417, 418, 562
Shulkes, A., 350
Shult, P., 264
Shuman, H., 418, 437
Siciliano, M. J., 150, 167, 345
Siegel, D., 67
Sies, H., 395, 534
Siest, G., 47, 48, 128, 621

Signorelli, K., 396
Siiteri, P. K., 294
Silhavy, T. J., 436
Silverman, L. M., 414
Sim, E., 192, 193, 202, 206, 207, 209, 215, 216, 219, 220, 225, 229, 624
Simat, T. J., 331
Simha, D., 96, 519, 600
Simon, C. M., 519, 520, 589, 600, 601, 603, 604, 606, 610, 611, 612
Simon, F. R., 546
Simons, K., 465
Sims, P., 572
Sinal, C. J., 171, 188, 515, 516, 517, 520, 521, 524, 572
Sinclair, J. C., 193, 207, 215, 216, 220
Sinclair, J. F., 608
Sindelar, P., 53
Singaraja, R. R., 417
Singh, N., 415
Singh, R. J., 224
Sinko, E., 534
Sinsheimer, J. E., 571
Sinz, M., 591, 599
Sipilä, J., 251
Sirvent, A., 515, 516
Sisodiya, S. M., 396
Siuzdak, G., 387
Sjollema, K., 442, 443
Sjostrom, H., 41
Sjövall, J., 62
Slatkin, M., 419
Slaunwhite, W. R., Jr., 294
Slavin, J. L., 29, 41
Slitt, A. L., 589
Slomczynska, M., 170, 172
Sloper-Mould, K. E., 472
Smallwood, R. A., 350
Smelt, V., 192, 216
Smids, M., 578
Smit, J. J., 534
Smith, B. A., 236
Smith, B. J., 198
Smith, C. A. D., 575
Smith, C. L., 619
Smith, D., 47
Smith, D. A., 47, 48, 49
Smith, D. J., 347
Smith, G., 624
Smith, G. E., 488

Smith, K. D., 361, 370
Smith, M., 325, 361, 366, 367, 368, 370, 371, 372
Smith, P., 219
Smith, P. C., 469
Smith, P. D., 310, 312, 313
Smith, R. L., 375
Smith, S., 386
Smith-Johnson, M., 318
Smriti, 466
Smudlers, Y. M., 420
Snyder, G. D., 573
Snyder, R., 61
Soars, M. G., 47, 48, 54, 65, 66, 347
Socoshis, S. A., 418
Soda, H., 487
Sogawa, K., 58, 66, 117
Sokal, E., 418, 560, 561, 565
Sokol, M. J., 5
Sokol, R. J., 559
Solache, I. L., 151
Soléau, S., 452, 453
Solis, W. A., 76
Song, C. S., 167, 170, 171, 172, 173, 174, 180, 183, 188, 521, 589, 590
Song, W. C., 300
Song, X., 591
Song, Y. H., 24, 28, 33
Sonke, T., 577
Sonoda, J., 174, 180, 181, 188, 521, 601, 605, 609, 612
Sonodo, J., 521
Sopko, B., 235, 236
Soroka, C. J., 517, 565
Sorrentino, B. P., 487
Soucek, P., 80
Soucy, P., 150, 151
Soumillion, J. P., 464, 468, 472, 473
Soutoglou, E., 25, 26
Sowadski, J. M., 285
Sowers, L. C., 514
Spaliviero, M., 348
Sparla, F., 352
Sparreboom, A., 500
Sparrow, A., 572
Specht, K., 311
Spector, A., 221
Spector, A. A., 573
Speigelman, D., 621
Speilberg, S. P., 192

Spelberg, J. H. L., 577
Spengler, U., 310
Sperker, B., 264, 271
Spitz, M. R., 67, 624
Spitzer, J., 449
Spitzer, T., 521
Spodsberg, N., 41
Spring, H., 397, 532, 534, 535, 538, 561, 565
Springfield, J. R., 236
Spurr, N. K., 47
Squier, M. W., 396
Squires, J. E., 350
Srisuma, S., 67, 68
Srivastava, A. K., 562
Stacey, M., 215
Stachowiak, O., 223
Staels, B., 515, 516
Staines, A. G., 51, 52, 53
Stalenhoef, A. F., 417
Stallcup, M. R., 512
Stamler, J. S., 219
Stamoudis, V., 571, 572
Stamps, A. C., 219
Stanley, E. L., 149
Stanley, L. A., 192, 216, 624
Staple, E., 362
Star, R. A., 313
States, J. C., 216
Staudinger, J. L., 344, 518, 520, 523, 589, 601
Stauffer, D., 415
Stavenuiter, J. F. C., 622
Steegers, E. A. P., 575
Steers, G., 219
Stein, C., 274, 294
Stein, W. D., 460, 461
Steinbach, J. H., 361
Steinberg, S. J., 361, 370
Steiner, H. H., 396, 397
Steinfels, E., 441, 444
Steinreiber, A., 577
Stellaard, F., 560, 561
Stephenson, S. H., 575
Steppan, C. M., 25
Stern, J., 396
Sterner, D. E., 512
Stewart, A. J., 396
Stewart, C., 414
Stewart, V. C., 396
Stiborova, M., 235, 236

Stieger, B., 511, 515, 516, 517, 533, 534, 535, 536, 538, 543, 546, 547, 560
Stiehl, A., 611
Stier, G., 24
Stimmel, J. B., 514, 560, 589
Stobiecki, M., 325
Stocker, R., 543
Stoffel, M., 27, 30, 591
Stoltz, C. M., 95, 520, 589
Stolz, A., 517
Stone, A. N., 54, 623
Stone, E., 621
Stone, R., 396
Stormo, G. D., 30
Storms, D., 572
Stoytchev, T. S., 353
St Pierre, M. V., 465, 470
Straney, R., 521
Strange, R. C., 622, 624
Strassburg, C. P., 2, 14, 26, 47, 58, 63, 64, 69, 78, 92
Stratford, F., 440
Straub, P., 7
Straub, R., 264, 271
Strauss, J. F. III, 168
Strautnieks, S., 418, 515, 560, 561, 565
Stravitz, R. T., 517
Strelevitz, T. J., 105
Stremmel, W., 221, 590
Stricker, N. L., 560
Strom, S. C., 174, 515, 521, 595, 600, 608
Strott, C. A., 167, 168, 172, 521
Stroup, D., 522
Stuart, M. C., 448, 449
Stubbs, M. T., 469
Stumptner, C., 517
Sturley, S. L., 464, 466
Sturm, E., 516, 517, 560, 561, 565
Subba Rao, G., 475
Suchy, F. J., 27, 30, 515, 517, 546, 547, 560, 590
Suck, D., 24
Sudo, K., 388
Sueyoshi, T., 10, 11, 13, 92, 93, 94, 95, 96, 97, 98, 100, 101, 519, 601
Sugahara, T., 168, 172
Sugai, S., 419
Sugatani, J., 11, 13, 92, 93, 94, 95, 96, 97, 98, 100, 101, 601
Sugimoto, Y., 506, 533, 534

Sugita, M., 419
Sugiyama, Y., 532, 533, 534, 535, 537, 538, 544, 561, 600
Suiko, M., 168, 172, 300
Sumathy, K., 210
Sumino, Y., 487, 493, 494, 508
Summers, M. D., 488
Summerscales, J. E., 194, 195, 196, 197, 200, 202
Sun, D., 174, 521
Sun, H., 415
Sun, W., 533, 544
Sun, Z., 421
Surh, Y.-J., 251
Sutcliffe, I. C., 431
Sutherland, E., 546
Sutherland, G. R., 15
Sutherland, L., 48
Sutter, G., 419
Suzuki, A., 194, 237
Suzuki, H., 532, 533, 534, 535, 537, 538, 544, 561
Suzuki, K., 619
Suzuki, M., 263, 268, 270, 532, 533, 534
Suzuki, N., 168
Suzuki, T., 168, 274, 291, 303, 304, 306, 307, 309, 313
Svensson, K., 518, 589
Svoboda, M., 30
Swaisland, A., 572
Swales, K., 601
Swart, J., 420
Swarts, H. J., 573
Swedmark, S., 168, 239, 349
Sweeney, C., 221, 623
Sweeny, D. J., 377, 378
Swiezewska, E., 352, 354, 355
Swindell, E. C., 421
Sydow-Backman, M., 518, 589
Szabo, S. M., 235
Szczypka, M., 465, 467, 469, 470, 474
Szmelcman, S., 436

T

Taavitsainen, P., 346
Tabb, M. M., 590, 591
Tablin, F., 452, 453
Taboury, J., 561
Tachibana, O., 14, 15

Tachikawa, T., 312, 313
Taglicht, D., 462, 463
Taha, R. A., 351
Tainer, J. A., 469
Takada, D., 416, 575
Takada, M., 419
Takagi, H., 95, 100
Takagi, T., 30
Takahashi, K., 307
Takahashi, S., 351
Takahashi, T., 353
Takahashi, Y., 28
Takaishi, M., 312, 313
Takakura, K., 218, 224
Takase, S., 263, 268, 270
Takashima, S., 304
Takeda, S., 124
Takei, T., 619
Takenaka, O., 534
Takeuchi, S., 298, 299, 303, 304, 314
Takeuchi, Y., 14
Takikawa, H., 396
Talalay, P., 67, 68
Talianidis, I., 24, 25, 26, 33
Tam, A., 472
Tamai, I., 533
Tamaki, T., 272
Tampe, R., 419, 432, 434, 436, 437, 444, 446, 461
Tamura, K., 429
Tan, T. H., 67
Tan, T. M., 402
Tan, W. C., 624
Tanabe, S., 14, 15, 487
Tanaka, E., 106, 107
Tanaka, K., 294, 295, 297, 298, 299
Tanaka, M., 59, 124
Tanaka, N., 561
Tanaka-Kagawa, T., 14
Tangirala, R. K., 516
Taniguchi, K., 418
Tanner, M. S., 560
Tannickal, V., 219
Tao, D., 7
Taratuto, A. L., 396
Tardy, M., 396
Tarr, P. T., 520
Tarui, S., 498, 500
Taskinen, J., 118, 122, 132, 133, 149, 251, 346
Tatsuno, M., 346, 347

Tautz, D., 15, 16
Tchernof, A., 14
Tector, J., 350
Teh, M., 624
Teisserenc, H., 419
Teitel, C. H., 192
Tekaia, F., 578
Tekkaya, C., 271
Telatar, M., 414
ten Brink, J. B., 420
Teng, S., 313, 353
Tenopoulou, M., 221, 222
Tephly, T. R., 2, 3, 6, 14, 47, 58, 61, 63, 92, 118, 361, 621
Terao, J., 272
Terashima, I., 168
Terashita, S., 544
Terauchi, Y., 28
Terrett, J. A., 219
Terrier, N., 129, 132, 138
Terry, S., 420
Testa, N., 396
Teubner, W., 238, 239, 242, 245, 251
Teusner, J. T., 149, 162
Theis, K., 275, 280, 282, 288
Thelin, A., 352
Theoharides, T. C., 317
Thermann, R., 565
Thevananther, S., 514, 517, 523
Thibodeau, P. H., 469
Thiele, D. J., 465, 467, 469, 470, 474
Thiele, S., 30
Thijssen, J. H. H., 291
Thimmupalla, R. K., 67, 68
Thines-Sempoux, D., 119
Thom, M., 396
Thomas, H., 578
Thomas, N. L., 348
Thomas, P. J., 469
Thomas, P. M., 420
Thompson, C. B., 619
Thompson, E. B., 95, 512
Thompson, E. T., 80, 89
Thompson, J. D., 199, 429
Thompson, J. R., 515
Thompson, R., 560, 565
Thompson, R. J., 515, 560, 561
Thompson, S., 464, 466
Thong, N. Q., 105
Thony, B., 30

Thorgeirsson, S. N., 236
Thornquest, A. D., Jr., 521
Thornton, J. M., 200
Thottassery, J., 174
Thummel, C., 600
Thummel, K. E., 110, 174
Thunnissen, M., 576
Thygesen, P., 215
Tian, H., 417, 562
Tian, T. G., 578
Tibboel, D., 349
Tiermersma, E. W., 622
Tijmes, N., 415
Timms, C. W., 572
Tinkelenberg, A. H., 464, 466
Tint, G. S., 514, 515, 562
Tipton, F. K., 58
Tipton, K. F., 2, 3, 6, 14, 92
Tiribelli, C., 396
Tirolo, C., 396
Tirona, R. G., 523, 562, 591
Toell, A., 182
Toffoli, G., 621
Toh, H., 576
Toh, S., 418
Tohkin, M., 515, 516, 572
Toide, K., 28
Tokoi, T., 14, 15
Tokui, T., 532
Tolmie, J., 419
Tolun, A., 414
Toma, D., 521, 601, 605, 609, 612, 613, 614
Toma, Y., 274
Tomita, Y., 351
Tomkinson, N. C. O., 591, 592
Tonda, E., 93, 96, 97, 101
Tongio, M. M., 419
Toniolo, P. G., 300
Ton-Nu, H. T., 361, 377
Tontonoz, P., 520
Toomes, H., 264
Topp, J., 2, 63
Tordai, A., 494
Torii, A., 307
Tornovsky, S., 420
Torra, I. P., 515, 516
Torre, C., 577
Touchstone, J. C., 294
Towbin, H., 132
Tozawa, A., 171, 188, 521

Traber, M. G., 590
Traber, P. G., 23
Traina, V. M., 351
Tran, H. M., 2, 3, 4, 15
Trauner, M., 511, 517, 543
Traver, R. D., 67
Travers, K. J., 472
Travis, G. H., 415, 416
Treat, S., 132
Tresillian, M., 149, 349, 350
Trip, M. D., 420
Troelsen, J. T., 41
Tronche, F., 23, 24, 28, 30
Troost, D., 396
Tsai, R., 514, 515
Tschuch, C., 420
Tserng, K., 377
Tsiftsoglou, S. A., 202
Ts'o, P. O. P., 62
Tsoi, C., 239, 349
Tsolas, O., 222
Tsou, A. M. H., 624
Tsuchida, T., 272
Tsuchiya, E., 467
Tsuchiya, K., 272
Tsuda, H., 307
Tsuda, M., 59, 124, 544
Tsuji, A., 494, 508, 533
Tsuji, M., 416
Tsukada, N., 548
Tsukahara, S., 506
Tsukazaki, K., 294, 295, 297, 298, 299
Tsukino, H., 348
Tsunoda, T., 30
Tsuruda, K., 59, 124
Tsuruo, T., 506
Tsuzuki, T., 148
Tsvetanov, S., 149
Tsvetkova, N. M., 452, 453
Tu, H., 514, 560
Tuchweber, B., 418
Tucker, G. T., 600
Tukey, R. H., 2, 7, 11, 14, 26, 47, 58, 63, 64, 65,
    66, 77, 78, 83, 87, 92, 93, 97, 102, 519, 589,
    601, 609, 613, 614
Tukey, R. T., 69
Tuma, P. L., 544, 562
Tunge-Morris, M., 174
Tuomi, T., 420
Turgeon, D., 106, 108

Turi, T., 416
Turina, M., 519
Turk, B., 193
Turk, D., 193
Turk, V., 193
Turunen, M., 352
Tyndale, R. F., 507
Tyson, C. A., 544
Tyson, H., 414
Tyson, K. L., 219
Tytgat, G. N., 14, 15, 47, 560
Tyurin, V. A., 353
Tzameli, I., 96, 514, 519
Tzekov, R. T., 415, 416

U

Uchaipichat, V., 59, 346
Uchida, T., 318
Uchida, Y., 418
Uchino, H., 533
Uchiumi, T., 418, 565
Uchiyama, A., 415
Udenfriend, S., 60
Ueda, A., 11, 13, 92, 93, 94, 95, 97, 100,
    101, 601
Ueda, K., 565
Ueda, M., 148
Ueng, Y., 80
Ueno, K., 14
Uesawa, Y., 53
Ueyama, H., 29
Uitto, J., 420
Ujhazy, P., 559
Ullmer, C., 604
Ullrich, D., 59, 69
Ulmanen, I., 344
Umeda, M., 468
Umemura, S., 575
Umesono, K., 600
Unadkat, J. D., 192, 216
Unamuno, P., 274, 289
Une, M., 361, 378
Unno, M., 532
Uppal, H., 521, 600, 601, 605, 609, 612,
    613, 614
Upton, A., 192
Urban, P., 577, 578
Urban, Z., 420, 534
Urbatsch, I. L., 498

Urnov, F. D., 512
Usón, I., 275, 279, 280, 284, 288
Usui, T., 29
Utsumi, T., 298, 299, 303, 304, 314
Utsunomiya, H., 561
Uyama, E., 15

V

Vaccaro, A. M., 274
Vagelous, P. R., 362
Vaisanen, M. M. S., 521, 523
Vaisman, B. L., 417
Vallee, B. L., 576, 577
Van, D. B. I., 560
Van Bergmann, K., 417
Van Breda, A. J., 274
van Dam, M., 416
van Deemter, L., 534
Van den Berg, F. M., 274
van den Bogaart, G., 453, 454
VandenBranden, M., 110, 608
van den Eeckhout, E., 571
van den Hazel, B., 463, 475
van de Pol, D. J., 415
Van Der, B. G., 560
van der, L. M., 544
van der, S. W., 192
van der Does, C., 432
van der Heide, T., 431, 433, 444, 447, 448,
    449, 450
van der Heijden, I., 534
Van der Loos, C. M., 274
van der Lugt, N. M., 534
van der Meer, R., 546
van der Rest, M., 464, 468, 472, 473
van der Valk, M. A., 487, 533
van der Valk, P., 396
van der Vliet, A., 218
van der Werf, M. J., 572, 573, 575, 579
van der Westhuizen, F. H., 376
Van Der Woerd, W. L., 560
van De Vijver, M. J., 487
van Dijk, C. H., 349
van Driel, M., 415
Van Dyck, L., 468, 475
van Eijk, M. J., 559, 560
van Es, H. H., 15
van Gastelen, M. A., 487
van Geel, B. M., 414

van Geer, M. A., 396
van Ginkel, C. J., 352
Van Goor, H., 560, 561
van Haren, F. J., 415
van Helvoort, A., 418
Van Laethem, J. L., 30
van Loo, B., 577
Van Loon, J. A., 235
van Meer, G., 465, 466, 468
van Meter, C. M., 518, 589, 600
van Mil, S. W., 559, 560
van Munster, I. P., 187
van Ooteghem, N. A., 559
van Poelje, P. D., 533, 544
Vanpouderoyen, G., 571
van Roon, M. A., 534
van Soest, S., 420
Vanstapel, F., 120
van Strien, A. E., 293
van Veen, H. W., 440, 451
van Vliet, E. A., 396
van Waes, M. A., 124
Varadi, A., 396, 485, 486, 534
Vargas, R., Jr., 589
Varin, C., 298
Vásadi, A., 494
Vath, G. M., 193, 202, 209, 217
Vaughan, J., 166
Vaxillaire, M., 25
Vaz, W. L., 465
Vederas, J. C., 388, 389, 390
Veenhoff, L. M., 435, 444, 447
Veggi, L. M., 547
Venema, D. P., 326
Venitt, S., 242
Venkatachlam, K. V., 350
Venkataramanan, R., 515, 608
Verde, A., 297
Verdoes, J. C., 575, 578
Verkade, H. J., 517
Veronese, M. E., 148, 150, 151
Verwiel, P. E., 351
Vessey, D. A., 364, 365, 378, 379
Vestweber, C., 378
Veyhl, M., 544
Vicart, P., 216
Vickers, R. J., 206, 209
Vielemeier, A., 295
Vihko, P., 274, 284
Vilageliu, L., 415

Vilarem, M. J., 166, 174, 181, 184, 521, 590
Villeneuve, L., 14, 15, 63
Vincent, T. S., 351
Vineis, P., 619
Viollet, B., 25
Visser, H., 578
Visser, T. J., 149, 349
Voet, M., 464
Vogel, A., 14
Voigt, S., 41
Volckaert, G., 464
Volkert, T. L., 28
vom, D. S., 547, 550
von Angerer, E., 299
von Bergmann, K., 417, 544
von Bülow, R., 275, 279, 280, 284, 288
von Dahl, S., 543, 547, 548
von dem Bussche, A., 310
von Figura, K., 274, 275, 279, 280, 282, 284, 285, 288, 294, 298
Vonk, R. J., 374
von Moltke, L. L., 14, 15, 54, 60, 63, 106, 107, 108, 110
Vorachek, W. R., 25, 619
Vore, M., 547
Vos, T. A., 544
Vreugdenhil, S., 578
Vu, T. P., 77, 83
Vyhlidal, C. A., 66

W

Wach, A., 476
Wacke, R., 271
Wada, H., 576
Wada, M., 418, 494, 508, 565
Wade, L., 518, 601
Wagenaar, E., 487, 533, 534
Wagner, C. R., 193, 202, 209, 217
Wagner, M., 475
Wahala, K., 317, 326
Waheed, A., 274, 294
Wahl, G. M., 366, 382
Wahler, D., 583
Wahli, W., 16, 24, 25, 27, 29, 30, 33, 36, 39
Wakabayashi, N., 67, 68
Wakabayashi, T., 14, 15
Wakamatsu, K., 388
Walboomers, J. M. M., 274
Walch, A., 311

Walker, C. H., 572
Walker, D., 131
Walker, J. E., 353, 461
Walker, S., 131
Wallace, A. C., 200
Walle, T., 89, 351
Walle, U. K., 89, 351
Wallimann, T., 223
Walter, C. A., 173
Walter, G., 299
Walts, A. D., 417
Wanders, R. J., 361, 370, 414
Wang, C.-C., 317, 327
Wang, D. Y., 523
Wang, H., 149, 156, 157, 160, 162, 193, 209, 217, 514, 518, 519, 601
Wang, H.-J., 318, 323
Wang, I. X., 435, 445, 469
Wang, J., 245, 294, 361, 366, 367, 368, 370, 371, 372
Wang, K., 218
Wang, L., 517, 523, 565
Wang, M., 468, 475
Wang, M. Y., 578
Wang, P., 218, 220
Wang, R., 418
Wang, T. H., 514, 515
Wang, X., 28
Wang, Y., 402, 445
Wang, Y. F., 624
Wang, Y. T., 624
Wang, Z., 532
Ward, A., 192
Ward, J. M., 66, 515, 516, 520, 521, 589
Wareing, C. J., 619
Warfe, L., 591
Warnes, G., 418, 565
Warshawsky, L., 298
Warskulat, U., 543, 547, 548, 563, 565
Washburn, K. A., 11, 13, 95, 519
Waskell, L., 218
Wasserman, W. W., 30
Watabe, T., 106, 108, 167, 572
Watanabe, M., 194, 236, 237, 354
Watanabe, R., 59, 124
Watanabe, S., 58, 66, 117, 317, 335
Watanabe, T., 7
Watanabe, Y., 54, 106, 107
Wate, C., 419
Wate, J., 419
Waterham, H. R., 414

Watkins, P. A., 361, 370
Watkins, R. E., 96, 518, 589
Watson, D. G., 343
Watson, M. A., 96, 100, 518, 523, 589, 601
Watson, W. H., 68
Watt, A. J., 65
Waxman, D. J., 76, 520, 589, 590, 592, 593, 612
Weaker, F. J., 173
Weatherill, P. J., 47
Weaver, C. M., 318, 321
Weaver, J., 418, 565
Weaver, R. J., 600
Webb, S. J., 192, 216
Weber, W. W., 192, 202, 235
Webster, C. R., 546, 547
Weerasekera, N., 418, 565
Wegman, J. J., 420
Wehr, K., 565
Wei, P., 517, 523, 611
Weiergräber, O., 543
Weinberger, C., 95, 515
Weinshilboum, M., 151, 167, 168, 170, 345, 350
Weinshilboum, R. M., 148, 150, 151, 161, 165, 167, 235, 250, 348, 622
Weintraub, N. L., 573
Weir, L. E., 354
Weisburger, J. H., 187
Weiss, M. C., 25
Weiss, R. A., 310, 312, 313
Weissbach, H., 60
Weissenbach, J., 417
Weissig, H., 292
Weissman, J. S., 472
Wells, P. G., 2, 344
Welsh, W. J., 590, 591
Welti, D., 318
Wemmie, J. A., 465, 467, 469, 472
Wen, P., 25
Weng, J., 415, 416
Werner, H., 415
Werner, T., 30
Werner, U., 271
Wernet, D., 67
Wert, S. E., 418
Westbrook, J., 292
Westerhoff, H. V., 546
Westerman, P. W., 354
Westin, S. K., 516
Westwood, I. M., 209, 216
Wetterholm, A., 576, 577

Wettstein, M., 543, 547, 548, 550
Wharton, B., 464, 466
Wheeler, J. B., 361, 365, 366, 367, 368, 370, 371, 372, 374
Whetsell, L., 310
White, C. R., 337
White, J. H., 174
White, M. B., 414
Whitfield, G. K., 166, 174, 175, 188
Whitsett, J. A., 418
Wibo, M., 119
Wickliffe, J. K., 582
Widdowson, D. A., 578
Wiebel, F. J., 233
Wieben, E. D., 167, 348, 350
Wielinga, P., 534
Wiencke, J. K., 67
Wiepert, M., 350
Wiersma, D. A., 452
Wijnholds, J., 402, 421, 534, 544
Wilborn, T. W., 148, 151
Wilcox, B. J., 388, 389, 390
Wilcox, L. J., 464, 466
Wild, D., 194, 195
Wildfang, J., 59
Wiley, D. C., 469
Wilkinson, A. P., 317, 328, 330, 332
Wilkinson, G. N., 342
Williams, C. N., 375
Williams, J. A., 47, 48, 49, 608
Williams, S. H., 23, 26, 33
Williams, S. N., 84
Williams, S. P., 518, 521
Williamson, C., 565
Williamson, G., 264, 272, 317, 328, 329, 330, 332
Williamson, K. C., 572
Willinger, B., 470
Willson, T. M., 96, 100, 514, 518, 519, 520, 521, 523, 560, 589, 590, 591, 592, 601
Wilson, J. G., 523
Wilson, J. X., 353
Wilson, L. S., 318, 336, 337, 338, 384, 386, 387, 389
Wilson, T. H., 448
Windmill, K. F., 149, 162, 349, 350
Wingender, E., 30
Winkler, L., 581
Winski, S. L., 67
Winterbourn, C. C., 337
Winters, M., 623

Winzeler, E. A., 467
Wisely, G. B., 518, 519, 521, 590, 591, 592, 600
Wiseman, H., 324
Witek, S., 464, 468
Witkowska, H. E., 386
Witkowski, A., 386
Witzleben, C. L., 560
Wohlk, N., 420
Wojnowski, L., 590, 592
Wojtaszek, P., 325
Wolf, C. R., 235, 236, 572, 619, 624
Wolf, D. H., 469, 470, 472
Wolf, L., 294
Wolf, M. J., 132
Wolff, B., 299
Wolff, M. S., 300
Wolffe, A. P., 512
Wolfger, H., 460, 463, 464, 465, 467
Wolford, R. G., 25
Wolkers, W. F., 452, 453
Wolkoff, A. W., 534, 546
Wolter, K. G., 619
Wolters, H., 517
Wonacott, A. J., 149
Wong, A. K., 419
Wong, C. K., 300, 349, 350
Wong, P., 218
Woo, L. W., 293, 299, 300, 303
Wood, A. W., 2
Wood, C. E., 295
Wood, J. M., 449
Wood, N. W., 166
Wood, T. C., 161, 167, 168, 250, 622
Woodland, M. P., 578
Wooster, R., 48
Word, J. B., 582
Wright, S. D., 188, 514, 515
Wrighton, S. A., 65, 66, 105, 110, 608
Wu, L. J., 207, 215
Wu, L. L., 575
Wu, Q., 4, 6, 7, 343
Wu, R. R., 533, 544
Wu, W., 174
Wu, X. F., 624
Wu, Y., 532
Wu, Y. C., 416
Wubbolts, M. G., 577
Würgler, F. E., 577
Wyckoff, H. W., 285
Wynder, E. L., 187
Wysock, W., 464

X

Xenarios, I., 206
Xi, X. G., 619
Xia, X., 543
Xian, M., 218
Xiang, C.-Q., 2
Xiang, L., 523
Xiao, G. H., 192, 216, 589, 601
Xie, D., 591
Xie, W., 62, 63, 64, 65, 66, 97, 102, 166, 174,
    175, 180, 181, 188, 519, 520, 521, 589, 590,
    600, 601, 602, 603, 604, 605, 606, 609, 610,
    611, 612, 613, 614
Xie, X., 245, 361, 366, 367, 368, 370, 371, 372
Xin, Y. P., 575
Xu, C. F., 47
Xu, F., 168, 172
Xu, G., 514, 515
Xu, J., 119
Xu, L., 25, 521, 601, 605, 609, 612
Xu, X., 67

Y

Yaciuk, P., 26
Yaffe, S. J., 92
Yagi, H., 2
Yamada, H., 124
Yamada, S., 166, 174, 175, 188
Yamada, T., 415
Yamada, Y., 15
Yamaguchi, A., 469
Yamagushi-Nagamatsu, Y., 59, 124
Yamakawa, K., 93, 95, 96, 97, 98, 100, 101
Yamamato, M., 67, 68
Yamamoto, H., 263, 268, 270
Yamamoto, K., 14
Yamamoto, K. R., 468
Yamamoto, M., 67, 68, 69
Yamamoto, Y., 543
Yamanaka, H., 14, 15, 619
Yamane, M., 10
Yamano, G., 418
Yamashita, E., 469
Yamashita, J., 14, 15
Yamauchi, Y., 300
Yamazaki, H., 28
Yamazoe, Y., 148, 171, 188, 234, 238, 245,
    252, 345, 521

Yan, B., 589, 591
Yan, N., 591
Yanada, J., 351
Yanai, H., 167, 168, 172
Yanaihara, A., 274
Yanaihara, N., 274
Yanaihara, T., 274
Yanchunas, J., 591
Yang, K., 218, 221
Yang, L., 7
Yang, S., 514, 515
Yang, T., 419
Yang, T. J., 572
Yang, W., 487
Yang, W. P., 532
Yang, X., 619
Yang, Y., 62, 76, 522
Yaniv, M., 23, 24, 25, 27, 28, 30, 39
Yano, K., 468
Yano, T., 210
Yasuda, K., 174, 515, 521, 595
Yasui, Y., 7, 123
Yasumura, S., 14, 15
Yates, J. E., 473
Yawo, H., 532
Ybazeta, G., 14
Yeatman, M. T., 2, 3, 4, 5, 15
Yeh, H. Z., 361
Yergey, A. L., 132
Yeuh, M. F., 65, 66, 83, 97, 102, 519, 589, 601,
    609, 613, 614
Yi, J. R., 395
Yogalingam, G., 125
Yokoi, T., 54, 106, 107
Yokota, H., 118, 120, 347
Yoo, K. Y., 621
Yoshida, H., 487, 543
Yoshida, I., 418
Yoshida, T., 14
Yoshihara, M., 571
Yoshikawa, M., 487
Yoshimura, N., 298, 299, 303, 304, 314
Yoshinari, K., 11, 13, 92, 93, 94, 95, 96, 97, 98,
    100, 101, 601
Yoshisue, K., 59, 124
Yoshizumi, M., 272
Yost, S. C., 544
Yotsumoto, Y., 600
Yotumoto, N., 118
Youle, R. J., 619

Young, R. A., 28
Yousef, G. G., 321
Yousef, I. M., 418
Yu, A., 599
Yu, E. W., 469, 474
Yu, H., 417, 562
Yu, J., 188, 514, 515
Yu, L., 416, 417, 544, 562
Yu, R., 67
Yu, S., 342
Yu, T. W., 207, 215
Yu, Z., 572
Yuasa, A., 118, 120, 347
Yudkoff, M., 396
Yueh, M.-F., 11, 93, 97, 591
Yuille, M., 621

Z

Zabriskie, N. A., 415
Zahn, A., 590
Zaidi, T., 414
Zalatoris, J. J., 63, 123
Zaman, G. J., 419, 532
Zamora, I., 250
Zanger, U. M., 563
Zatloukal, K., 517
Zavacki, A. M., 514, 515, 560
Zawada, G., 132, 138
Zech, B., 218
Zeiss, C., 414
Zelcer, N., 174, 418, 521, 534
Zeldin, D. C., 572
Zelko, I., 92, 96, 100, 519, 601
Zeller, H. D., 578
Zemaitis, M. A., 600, 601
Zenz, R., 517
Zerweck, E., 378
Zetterstrom, R. H., 518, 601
Zgurskaya, H. I., 469, 474
Zhan, J., 106, 108
Zhan, Z., 487
Zhang, H., 218, 589
Zhang, J., 601, 611, 612
Zhang, J. Y., 106, 108, 611
Zhang, K. R., 624
Zhang, L. H., 416, 544
Zhang, M., 469, 522
Zhang, P., 141
Zhang, R., 364, 378

Zhang, S., 608
Zhang, T., 4, 6, 7, 343, 515
Zhang, X., 467, 468
Zhang, Y., 352, 524
Zhang, Z., 218
Zhanger, U. M., 562
Zhao, A., 188, 514, 515
Zhao, B., 624
Zhao, L., 421
Zhao, L. X., 418
Zhao, X., 469
Zhen, R. G., 474
Zheng, J., 23, 24, 30
Zheng, P., 419
Zheng, W., 622
Zheng, Z., 14, 61
Zhou, B., 218
Zhou, C., 421, 590, 591
Zhou, F., 337
Zhou, H., 195
Zhou, M., 168, 171
Zhou, N. Y., 578
Zhou, S., 487
Zhu, B. T., 271
Zhu, W., 589
Zhu, X., 148, 149, 150, 151, 162,
    349, 350
Zhu, Y., 23, 532
Zhu, Z., 591
Zhuang, Z., 310, 312, 313
Zibat, A., 590, 592
Ziccardi, W., 464, 468, 474
Zimmer, J., 419
Zimmerman, T. L., 514, 517
Zimmermann, F., 415, 604
Zimmermann, R., 418
Zimniak, P., 132, 134, 171
Zingman, L. V., 420
Ziurys, J. C., 59
Zivelin, A., 565
Zizlsperger, N., 28
Zodan-Marin, T., 418
Zollner, G., 517
Zolman, K. L., 67
Zou, J. Y., 572, 573, 574, 575, 579
Zuo, Z., 67
Zuther, E., 453
Zwarts, K. Y., 417
Zweibaum, A., 67
Zwinderman, A. H., 417, 420

Subject Index

A

ABC transporters, *see* ATP-binding cassette transporters
Acetyl CoA: arylamine *N*-acetyltransferases, *see* *N*-Acetyltransferases
N-Acetyltransferases
 cancer-associated alleles, 624–625
 catalytic mechanism, 217
 functional overview, 192
 human NAT2 variant screening with detection of active variant enzymes by reversion assay/mutagen activation activity assay, 198–199
  bacteria strains, 196
  initial screening, 197, 200
  materials, 196–197
  overview, 196
  sequencing, 198
  structure modeling of mutation effects, 199–200, 202, 204, 206–210
 isoenzymes, 192
 kinetic mechanism, 216
 lacZ mutagenicity assay, 194, 197–198
 mutagen bioactivation, 193–194, 231–232
 mutagenicity testing
  cofactors and activity in subcellular activating systems, 234–236
  natural expression of enzymes in target cells of test systems, 233–234
  recombinant systems
   cofactor supply, 242–243
   expression level optimization and characterization, 239–242
   historical perspective, 236–238
   positive controls, 243–245
   prospects, 246
   vectors, 237–239
 oxidative inhibition of NAT1
  activity assays, 219–221
  endogenous expression in cells, 219
  hydrogen peroxide inhibition of recombinant enzyme, 221–223

 overview, 217–219
 peroxynitrite inhibition of cellular enzyme, 223–225
 recombinant protein expression and purification, 219
 structures, 193, 199–200, 202, 216–217
 tissue distribution, 192
Adrenaline test, epoxide hydrolase assay, 583–584
Adrenoleukodystrophy, *ABCD1* mutations and pathophysiology, 414–415
Ah receptor, *see* Aryl hydrocarbon receptor
Androgen receptor, sulfotransferase 2A1 regulation, 173–174
Apical sodium-dependent bile acid transporter, farnesoid X receptor regulation, 517–518
Aryl hydrocarbon receptor
 cytochrome P450 gene induction, 76
 cytosolic complex, 76
 UDP-glucuronosyltransferase 1A1 gene regulation
  activity assay, 78–79
  cell culture and treatment, 77–78
  cell lines, 83–84
  cloning of regulatory regions, 81–82
  electrophoretic mobility shift assay, 87–90
  materials, 77
  Northern blot analysis of transcripts, 79–80
  response element deletion analysis and site-directed mutagenesis, 84, 86, 89–90
  transient transfection assay, 82–83
  Western blot, 80–81
 UDP-glucuronosyltransferase 1A6 gene regulation, 66
Arylamine *N*-acetyltransferases, *see* *N*-Acetyltransferases
ASBT, *see* Apical sodium-dependent bile acid transporter

ATP-binding cassette transporters
anionic conjugate transport in liver,
533–534
hepatic transporters, *see* Multidrug
resistance-associated protein 2
ATP-binding cassette transporters, *see also*
specific transporters
ABCG2 human transporter studies
activity high-speed screening, 493–494
baculovirus–insect cell expression system
expression vector, 487–488
insect cell culture, 490
recombinant virus amplification and
titration, 489–490
transfection, 489
virus purification by plaque assay, 489
discovery, 487
functions, 487
gene structure, 487
plasma membrane vesicle preparation,
490–491
single nucleotide polymorphism
screening, 494, 496
Western blot, 491, 493
cancer stem cell expression, 422
classification, 485–486
detergent-solubilized enzyme studies
nucleotide-binding domain ligand
studies, 440–441
purification
chromatography, 433–435
detergent solubilization, 435
substrate-binding protein
substrate binding analysis, 435–439
translocator complex, 439–440
essential genes, 421
functional overview, 430
genes
human gene types and features,
410–412, 485
organization, 410
membrane reconstitution
approaches, 442
ATP-regenerating system for studies,
446–447
detergent removal, 445
equilibration with detergent, 443–444
examples, 445–446
giant unilamellar vesicle transporter
studies

vesicle preparation, 452–453
fluorescence correlation spectroscopy,
453–455
large unilamellar vesicle transporter
studies
fluorescent substrates, 451–452
stoichiometry, 449–450
substrate import versus export,
448–449
lipids, 447–448
lipid-to-protein ratio, 455
optimization, 442–443
mutation and diseases
adrenoleukodystrophy, 414–415
cholestasis, 418–419
complex diseases, 421
cystic fibrosis, 412–414
dilated cardiomyopathy, 420
familial persistent hyperinsulinemic
hypoglycemia of infancy, 420
immune deficiency, 419
ivermectin sensitivity, 419
lamellar ichthyosis type 2, 417–418
macular degeneration, 415–416
pseudoxanthoma elasticum, 420
sideroblastic anemia, 419–420
sitosterolemia, 417, 562
surfactant deficiency, 418
Tangier disease, 416–417
phase III detoxification, 486–487
single nucleotide polymorphism array
detection of polymorphisms, 505–508
structure, 410, 430–432, 461
yeast transporters
functions, 460–461
genes, 460, 462
genetic mutants and phenotypic
characterization, 466–468
overexpression and purification,
475–477
pleiotropic drug resistance
functional assays, 472–475
pump types, 462–464
vacuolar sequestration, 465–466
prospects for study, 477–478
substrate specificity and
structure–function analysis,
468–470
trafficking studies, 470–472
7-Azido-4-methylcoumarin

photoaffinity labeling of UDP-
glucuronosyltransferases, 137–138
synthesis, 136
5-Azido-UDP-glucuronic acid
photoaffinity labeling of UDP-
glucuronosyltransferases, 136–137
synthesis, 134–136

B

BAL, *see* Bile acid CoA ligase
BAT, *see* Bile acid CoA: amino acid
N-acyltransferase
Bile acid CoA: amino acid N-acyltransferase
activity assays, 378–379
function, 374
glycine-conjugated bile acid metabolism,
see Peptidylglycine α-amidating
monooxygenase
mass spectrometry analysis of products
peptidylglycine α-amidating
monooxygenase products, 389–390
rat and mouse bile samples, 388–389
simultaneous detection of
acyltransferase and thioesterase
activities, 389
purification
biotinylated enzyme, 384–385
human liver enzyme, 380–381
recombinant enzyme cloning and
expression, 380–384
relationship to other amino acid
conjugating enzymes, 375–376
sequence homology between species, 382,
386–387
species distribution, 378
standards for product analysis
chemical synthesis, 377–378
commercial availability, 376–377
enzymatic synthesis, 378
substrate specificity, 375
Bile acid CoA ligase
activity assay, 365
history of study, 360–361
liquid chromatography-mass
spectrometry analysis of products,
371–372
microsome preparations, 364–365
polyclonal antibody generation, 370–371
purification

rat liver microsome preparations, 365–366
recombinant enzyme cloning and
expression, 366–368
rat liver enzyme sequencing and
relationship to other enzymes,
368–370
standards for analysis of bile acid CoA
thioesters
chemical synthesis, 362–363
commercial sources, 361
enzymatic synthesis, 364
substrate specificity, 361
Bile salt export pump
farnesoid X receptor transactivation, 515
hepatobiliary transport, 544
progressive familial intrahepatic
cholestasis
defects, 560
diagnostics, 565
BSEP, *see* Bile salt export pump

C

Cancer, phase II enzyme polymorphisms
as markers
N-acetyltransferase, 624–625
glutathione S-transferase, 619, 621
overview, 618–620, 625
sulfotransferase, 622–624
UDP-glucuronosyltransferase, 621–622
CAR, *see* Constitutive androstane receptor
Carbamazepine, UDP-
glucuronosyltransferase substrate
specificity, 52–53
Cholestasis, ATP-binding cassette
transporter mutations and
pathophysiology, 418–419
Cholestatic syndromes, *see* Dubin–Johnson
syndrome; Progressive familial
intrahepatic cholestasis; Sitosterolemia
Coenzyme Q
antioxidant activity of analogs, 352–354
degradation pathways, 352–355
functions, 352
glucuronidation, 354
isoprenyl chain conjugation, 354
phosphorylation of metabolites, 354
sulfation, 354
CoMFA, *see* Comparative molecular
field analysis

Comparative molecular field analysis
  principles, 250
  sulfotransferases
    applications, 259–261
    overview, 251–253
    rat SULT1A1 substrates, 253–254
    rat SULT2A3
      alignment rule application and model
        construction, 257, 259
      homology modeling of catalytically
        competent complex, 255–256
      substrates and kinetic constants,
        258–259
Confocal laser scanning microscopy,
  subcellular localization of canalicular
  transporters, 547–548
Constitutive androstane receptor
  cytochrome P450 regulation, 519
  knockout mouse
    cholestasis studies, 612
    hyperbilubirubinemia studies, 613–614
  nuclear receptor interactions in bile acid
    and drug metabolism transcriptional
    regulation, 520–521
  target genes, 601
  transcriptional activity, 512–514,
    600–601
  transgenic mice expressing human gene
    cholestasis prevention by
      activation, 612
    drug response profile studies, 61
    hyperbilubirubinemia prevention, 613
    microinjection and transgene
      verification, 603
    tetracycline-inducible system, 604
    transgene construction, 602
UDP-glucuronosyltransferase 1A1
  gene regulation
    deletion assays, 97–99
    electrophoretic mobility shift assay,
      99–100
    overview, 92–93, 100–102
    plasmids, 95
    site-directed mutagenesis studies, 97
    transcript analysis with reverse
      transcriptase–polymerase chain
      reaction, 94
    transient transfection and luciferase
      assays, 96
    Western blot, 94–95

UDP-glucuronosyltransferase 1A6 gene
  regulation, 65–66
CoQ, *see* Coenzyme Q
CYP, *see* Cytochrome P450
Cystic fibrosis, transmembrane conductance
  regulator mutations and
  pathophysiology, 412–441
Cytochrome P450
  aryl hydrocarbon receptor and gene
    induction, 76
  assay of mouse liver microsomes
    microsome preparation, 607–608
    testosterone as substrate, 608
  constitutive androstane receptor
    regulation, 519
  phase II metabolism overview, 486–487,
    511–512, 599
  pregnane X receptor regulation, 518–519

D

DAVERMA, *see* Detection of active
  variant enzymes by reversion
  assay/mutagen activation
Deglucuronidation, *see* β-Glucuronidase
Detection of active variant enzymes by
  reversion assay/mutagen activation,
  human *N*-acetyltransferase 2 variant
  screening with detection of
  active variant enzymes by reversion
  assay/mutagen activation
  activity assay, 198–199
  bacteria strains, 196
  initial screening, 197, 200
  lacZ mutagenicity assay, 194, 197–198
  materials, 196–197
  overview, 196
  sequencing, 198
  structure modeling of mutation effects,
    199–200, 202, 204, 206–210
Dilated cardiomyopathy, *ABCC9* mutations
  and pathophysiology, 420
DNase I footprinting,
  sulfotransferase 2A1 regulatory
  regions, 184–186
Dubin–Johnson syndrome, multidrug
  resistance-associated protein 2
  mutations, 561–562
Dulcin, UDP-glucuronosyltransferase
  substrate specificity, 52–53

E

Electrophoretic mobility shift assay
  hepatocyte nuclear factor 1 binding to
      UDP-glucuronosyltransferase genes,
      35, 39–41
  sulfotransferase 2A1 regulatory regions,
      186–187
  UDP-glucuronosyltransferase 1A1 gene
  aryl hydrocarbon receptor interactions,
      87–90
  nuclear receptor interactions, 99–100
EMSA, *see* Electrophoretic mobility shift
  assay
Epoxide hydrolases
  assays
      adrenaline test, 583–584
      9,10-*cis*-epoxystearic acid, 581–583
      styrene 7,8-oxide as substrate, 579–581
  biocatalytic synthesis applications, 577
  functions
      bacteria metabolism, 573
      detoxification, 570–572
      signal molecule processing, 572–573
  recombinant protein expression
      expression systems, 577–578
      refolding from inclusion bodies, 578–579
  soluble versus membrane forms, 32
  structure and catalytic mechanisms
      FosX, 576
      α/β hydrolase fold enzymes, 573–575
      leukotriene A_4 hydrolase, 576–577
      limonene epoxide hydrolase, 575–576
Estrogen sulfatase, *see* Estrone/
  dehydroepiandrosterone sulfatase
Estrone/dehydroepiandrosterone sulfatase
  catalytic mechanism, 284–285
  deficiency and X-linked ichthyosis,
      289–290
  estrogen sulfatase activity
      assay, 295–297
      estrogen activity effects and modulation,
          298–300
      estrogen sulfoconjugates, 293–294
      tissue distribution, 297–298
  function and distribution, 274
  membrane anchoring, 286–288
  oligomerization, 288
  substrate-binding modes, 282–284
  substrate specificity, 290–291, 303

tissue distribution
  immunohistochemistry, 304–307
  laser capture microdissection and
      reverse transcriptase–polymerase
      chain reaction, 310–314
  rationale for study, 303–304
  in situ hybridization, 307–310

F

Familial persistent hyperinsulinemic
  hypoglycemia of infancy, *ABCC8*
  mutations and pathophysiology, 420
Farnesoid X receptor
  bile acid homeostasis
      bile acid uptake system negative
          regulation, 517–518
      nuclear receptor role, 514–515
  nuclear receptor interactions in bile acid
      and drug metabolism transcriptional
      regulation, 520–521
  progressive familial intrahepatic
      cholestasis defects, 560
  small heterodimer partner as mediator of
      transcriptional regulation, 516
  target genes, 515–516
  transcriptional activity, 512–514
Farnesol, UDP-glucuronosyltransferase
  substrate specificity, 52–53
Flow cytometry, sodium taurocholate-
  cotransporting polypeptide subcellular
  localization, 552, 554
Fluorescence correlation spectroscopy,
  ATP-binding cassette transporters in
  giant unilamellar vesicles, 453–455
FosX, structure and catalytic mechanism, 576
FXR, *see* Farnesoid X receptor

G

Giant unilamellar vesicle, *see* ATP-binding
  cassette transporters
Glucocorticoid receptor, UDP-
  glucuronosyltransferase 1A1 gene
  regulation
  deletion assays, 97–99
  electrophoretic mobility shift assay, 99–100
  overview, 92–93, 100–102
  plasmids, 95
  site-directed mutagenesis studies, 97

Glucocorticoid receptor, UDP-glucuronosyltransferase 1A1 gene regulation *(cont.)*
  transcript analysis with reverse transcriptase–polymerase chain reaction, 94
  transient transfection and luciferase assays, 96
  Western blot, 94–95
β-Glucuronidase
  flavonoid deglucuronidation, 264
  function, 264
  inflammation studies
    Caco-2 cell activity, 267
    luteolin glucuronidation status assay with high-performance liquid chromatography, 269–270
    neutrophil activity
      assay of release, 268, 271
      overview, 266–267
    plasma activity in mice, 265–266, 271
Glucuronidation, *see* UDP-glucuronosyltransferase
Glutathione, metabolism, 395
Glutathione *S*-transferase
  assay of cytosolic preparations, 609–610
  cancer-associated alleles, 619, 621
GR, *see* Glucocorticoid receptor
GST, *see* Glutathione *S*-transferase

H

Hepatocyte nuclear factor-1
  coactivators, 26
  dimerization, 24
  expression profiles of isoforms, 24
  gene screening for binding elements, 30–31
  homology between isoforms, 23–24
  mechanisms of HNF1α action, 24–26
  mutation and disease, 41
  tissue distribution, 23
  UDP-glucuronosyltransferase gene regulation
    cell lines for study, 33–34
    deletion construct studies, 36
    electrophoretic mobility shift assay, 35, 39–41
    luciferase reporter assays, 31–32, 36
    network of interactions, 28–30
    nuclear extract preparation, 34–35

    null cell line studies, 36
    promoter binding sites, 26–28
    prospects for study, 41
    reporter systems, 38–39
    site-directed mutagenesis studies of promoters, 32–33, 38
    UGT1A6, 63–65
    Western blot of HNF1α expression, 34, 38
Hepatocyte nuclear factor-4α
  bile acid and drug metabolism regulation, 522–524
  DNA binding, 521
  glucose homeostasis role, 521–522
High-performance liquid chromatography
  bile acid CoA ligase product analysis, 371–372
  luteolin glucuronidation status assay in inflammation, 269–270
  phytoestrogens in food samples, 327–328
  UDP-glucuronosyltransferase assays, 49–50, 112–113
HMR1098, UDP-glucuronosyltransferase substrate specificity, 52–53
HNF1, *see* Hepatocyte nuclear factor-1
HNF-4α, *see* Hepatocyte nuclear factor-4α
HPLC, *see* High-performance liquid chromatography
Hydrogen peroxide
  extracellular generation and quantification, 399, 402–405
  recombinant *N*-acetyltransferase 1 inhibition, 221–223
Hyperbilirubinemia
  nuclear receptor knockout mice studies, 613–614
  nuclear receptor transgenic mice and prevention
    constitutive androstane receptor, 613
    pregnane X receptor, 613

I

Immune deficiency, TAP protein mutations and pathophysiology, 419
Immunohistochemistry, estrone/dehydroepiandrosterone sulfatase localization, 304–307
Inflammation
  β-glucuronidase studies
    Caco-2 cell activity, 267

luteolin glucuronidation status assay with high-performance liquid chromatography, 269–270
neutrophil activity
assay of release, 268, 271
overview, 266–267
plasma activity in mice, 265–266, 271
phytoestrogen metabolism in inflammatory cells
chlorinated phytoestrogen synthesis, 336–338
nitrated phytoestrogen synthesis, 336–338
product formation and analysis, 338–339
ISH, *see In Situ* hybridization
Isoflavones, *see* Phytoestrogens
Ivermectin sensitivity, *ABCB1* mutations and pathophysiology, 419

L

Lactase phlorizin hydrolase, phytoestrogen conjugate metabolism, 328–329
lacZ mutagenicity assay, *see N*-Acetyltransferases
Lamellar ichthyosis type 2, *ABCA12* mutations and pathophysiology, 417–418
Large unilamellar vesicle, *see* ATP-binding cassette transporters
Laser capture microdissection, estrone/ dehydroepiandrosterone sulfatase localization, 310–314
LCA, *see* Lithocholic acid
LCM, *see* Laser capture microdissection
Leukotriene A4 hydrolase, structure and catalytic mechanism, 576–577
Limonene epoxide hydrolase, structure and catalytic mechanism, 575–576
Lithocholic acid, sulfotransferase 2A1 induction, 187–188
Luteolin, glucuronidation status assay with high-performance liquid chromatography in inflammation, 269–270

M

Macular degeneration, *ABCA4* mutations and pathophysiology, 415–416
Mass spectrometry

bile acid amino acid conjugate analysis
peptidylglycine α-amidating monooxygenase products, 389–390
rat and mouse bile samples, 388–389
simultaneous detection of acyltransferase and thioesterase activities, 389
bile acid CoA ligase product analysis, 371–372
photoaffinity-labeled peptides, 141–142
phytoestrogen metabolite analysis, 321–323, 327–328, 333, 335, 337–339
UDP-glucuronosyltransferase substrate specificity studies, 50–51
MDR1, *see* Multidrug resistance protein 1
MDR3, *see* Multidrug resistance protein 3
Mrp1, *see* Multidrug resistance-associated protein 1
Mrp2, *see* Multidrug resistance-associated protein 2
Mrp3, *see* Multidrug resistance-associated protein 3
Multidrug resistance-associated protein 1, glutathione export
astrocyte studies
cell viability assay, 400
glutathione quantification, 399
hydrogen peroxide extracellular generation and quantification, 399, 402–405
incubation conditions, 400–402
inhibitors, 397, 401–402
materials, 397–398
protein localization, 396
overview, 396
Multidrug resistance-associated protein 2
anionic conjugate transport in liver, 534
Dubin–Johnson syndrome mutations, 561–562
functional analysis in MDCKII cells
cell culture and transfection, 535
double transfection with organic anion transporters, 536–538
inhibitors, 538–539
substrate specificity, 538
transport assays, 535–536
subcellular localization
confocal laser scanning microscopy, 547–548
differential centrifugation, 548, 550

Multidrug resistance-associated protein 2
    (*cont.*)
    sucrose gradient centrifugation, 550, 552
Multidrug resistance-associated protein 3,
    single nucleotide polymorphisms,
    562–563
Multidrug resistance protein 1
    ATPase activity high-speed screening,
        498–500
    cancer expression, 496
    polymorphisms versus ATPase
        activity, 500
    quantitative structure–activity relationship
        analysis of substrate specificity, 500,
        502–505
    structure, 498
    tissue distribution, 496
Multidrug resistance protein 3
    mutation in cholestasis and cholelithiasis,
        561
    progressive familial intrahepatic
        cholestasis defects, 560–561
Mutagenicity, *see* N-Acetyltransferases;
    Sulfotransferases

N

NATs, *see* N-Acetyltransferases
Neutrophil, β-glucuronidase activity
    assay of release, 268, 271
    overview, 266–267
NMR, *see* Nuclear magnetic resonance
Northern blot
    drug-metabolizing enzyme transcript
        analysis in animal models, 606–607
    sulfotransferase 2A1, 177
    UDP-glucuronosyltransferase 1A1
        transcripts, 79–80
Nrf2, UDP-glucuronosyltransferase 1A6
    gene regulation, 67–69
NTCP, *see* Sodium taurocholate-
    cotransporting polypeptide
Nuclear magnetic resonance,
    phytoestrogen metabolite analysis,
    321, 324, 336

O

OAT, *see* Organic anion transporters
Oct-1, *see* Octamer transcription factor-1

Octamer transcription factor-1, hepatocyte
    nuclear factor 1 coactivation, 26
Organic anion transporters
    conjugate transport in liver, 531–532
    functional analysis in MDCKII cells
        cell culture and transfection, 535
        double transfection with multidrug
            resistance-associated protein 2,
            536–538
        inhibitors, 538–539
        substrate specificity, 538
        transport assays, 535–536
    hepatobiliary transport, 544
    hepatocyte nuclear factor-4α regulation of
        OAT2, 522
    human liver types, 532–533
    single nucleotide polymorphisms, 562

P

PAM, *see* Peptidylglycine α-amidating
    monooxygenase
Paracetamol, UDP-glucuronosyltransferase
    1A6 substrate, 59–60
Pdr5p
    overexpression and purification, 475–477
    substrate specificity and structure–function
        analysis, 469
    trafficking studies, 472–473
Peptidylglycine α-amidating monooxygenase
    activity assay, 388
    glycine-conjugated bile acid metabolism,
        387
    mass spectrometry analysis of products,
        389–390
    substrate specificity, 387–388
Peroxynitrite, inhibition of cellular
    N-acetyltransferase 1, 223–225
PFIC, *see* Progressive familial intrahepatic
    cholestasis
P-glycoprotein, *see* Multidrug resistance
    protein 1
Phase I metabolism, overview, 486–487,
    511–512
Phase II metabolism, overview, 486–487,
    511–512, 599
Phase III metabolism, ATP-binding cassette
    transporters, 486–487
Photoaffinity labeling
    catalytic mechanism and residues, 127–131

UDP-glucuronosyltransferases
  direct labeling, 131–132, 138–139
  labeling conditions, 136–140
  modified residue identification
    liquid chromatography–tandem mass
      spectrometry of peptides,
      141–142
    proteolytic digestion, 141
    site-directed mutagenesis verification,
      142
  principles, 131
  probes
    7-azido-4-methylcoumarin synthesis,
      136
    5-azido-UDP-glucuronic acid
      synthesis, 134–136
    overview, 132
    retinoic acid probe synthesis, 136
  recombinant human protein preparation
    cloning, 133
    expression in baculovirus–insect cell
      system, 133–134
    membrane isolation, 134
    nickel affinity chromatography, 134
Phytoestrogens
  estrogenic activity, 317
  extraction from foods
    acidified solvent extraction, 323, 325
    enzymatic hydrolysis of extracts,
      325–326
    lignan extraction, 326–327
    methanol extraction, 325
  glycoside conjugates
    foods, 318
    gastric tract hydrolysis studies
      cannulated everted sac model,
        329–330
      lactase phlorizin hydrolase, 328–329
    structures, 319
  inflammatory cell metabolism
    chlorinated phytoestrogen synthesis,
      336–338
    nitrated phytoestrogen synthesis,
      336–338
    product formation and analysis,
      338–339
  isoflavone metabolism in mammary cell
    lines, 335–336
  liver metabolism studies
    glucuronidation

conjugate analysis and isolation,
  332–333
  in vitro, 332
  in vivo, 332
microsomal hydroxylation, 331
reversed-phase high-performance liquid
  chromatography of food samples,
  327–328
standards for analysis, 320–321
sulfation studies, 334–335
types, 317–318, 320
variability in foods, 318
Polymerase chain reaction, see Reverse
  transcriptase–polymerase chain
  reaction
Pregnane X receptor
  cytochrome P450 regulation, 518–519
  interindividual variability, 590
  knockout mouse generation
    breeding, 602
    cholestasis studies, 612
    embryonic stem cell transfection and
      selection, 601
    hyperbilirubinemia studies, 613–614
    targeting vector construction, 601
  ligands, 590
  nuclear receptor interactions in bile acid
    and drug metabolism transcriptional
    regulation, 520–521
  species specificity of ligand activation,
    590–591
  target genes, 589, 601
  tissue distribution, 589–590
  transcriptional activity
    assay approaches, 591–592
    luciferase reporter assay
      cell culture, 592–593
      data analysis, 594
      detection, 594
      ligand treatment, 593–594
      plasmids, 592
      transient transfection, 593
    overview, 512–514, 600–601
  transgenic mice expressing human gene
    breeding, 604
    cholestasis prevention by activation,
      611–612
    corticosterone levels, 614
    drug response profile studies, 605–606,
      610–611

Pregnane X receptor *(cont.)*
hyperbilubirubinemia prevention, 613
microinjection and transgene
verification, 603
transgene construction, 602
UDP-glucuronosyltransferase 1A1 gene
regulation
deletion assays, 97–99
electrophoretic mobility shift assay,
99–100
overview, 92–93, 100–102
plasmids, 95
site-directed mutagenesis studies, 97
transcript analysis with reverse
transcriptase–polymerase chain
reaction, 94
transient transfection and luciferase
assays, 96
Western blot, 94–95
UDP-glucuronosyltransferase 1A6 gene
regulation, 65–66
xenosensor activity, 589
Progressive familial intrahepatic cholestasis
diagnostics, 563–565
types and gene mutations, 558–562
Pseudoxanthoma elasticum, *ABCC6*
mutations and pathophysiology, 420
PXR, *see* Pregnane X receptor

Q

Quantitative structure–activity
relationships, *see* Comparative
molecular field analysis; Multidrug
resistance protein 1
Quinines, redox activity, 351
Quinones
biological functions, 350–351
detoxification, 351–352

R

Reverse transcriptase–polymerase chain
reaction
drug-metabolizing enzyme transcript
analysis in animal models, 607
estrone/dehydroepiandrosterone sulfatase
localization, 310–314
UDP-glucuronosyltransferase 1A1, 94

S

Serotonin, UDP-glucuronosyltransferase
1A6 substrate, 60–61
SHP, *see* Small heterodimer partner
Sideroblastic anemia, *ABCB7* mutations and
pathophysiology, 419–420
Single nucleotide polymorphism
ATP-binding cassette transporters
ABCG2 human transporter screening,
494, 496
array detection of polymorphisms,
505–508
multidrug resistance protein 1
polymorphisms and ATPase
activity, 500
cancer-associated alleles
N-acetyltransferase, 624–625
glutathione *S*-transferase, 619, 621
sulfotransferase, 622–624
UDP-glucuronosyl transferase, 621–622
sinusoidal transporter proteins, 562–563,
565
SULT1A genes, 162, 163, 348
UDP-glucuronosyltransferases
UDP-glucuronosyltransferase 1A6
promoter, 63
UGT1 genes, 2, 15–16, 47
Site-directed mutagenesis
sulfotransferase 2A1 promoters, 178–180
UDP-glucuronosyltransferase genes
hepatocyte nuclear factor 1 binding
sites, 32–33, 38
photoaffinity label verification, 142
UDP-glucuronosyltransferase 1A1 gene
regulation
aryl hydrocarbon receptor, 84, 86,
89–90
nuclear receptors, 97
Sitosterolemia, ATP-binding cassette
transporter mutations and
pathophysiology, 417, 562
In Situ hybridization, estrone/
dehydroepiandrosterone sulfatase
localization, 307–310
Small heterodimer partner, nuclear receptor
interactions, 516, 521
SNP, *see* Single nucleotide polymorphism
Snq2p, overexpression and purification,
475–477

Sodium taurocholate-cotransporting
polypeptide
farnesoid X receptor regulation, 517
function, 543–544
short-term regulation of canalicular
transport, 546
single nucleotide polymorphisms, 562
subcellular localization
FLAG-tagged protein flow cytometry,
552, 554
sucrose gradient centrifugation,
550, 552
Steroid sulfatase, *see* Estrone/
dehydroepiandrosterone sulfatase
Sulfatases, *see also* Estrone/
dehydroepiandrosterone sulfatase
catalytic mechanism, 284–285
classification, 274
oligomerization, 288
sequence conservation
catalytic machinery, 280–282
comparison of enzyme types, 273–274,
278–279
structures, 275–276, 279–280
Sulfotransferases
assay of cytosolic preparations, 609–610
cancer-associated alleles, 622–624
catalytic mechanism, 250
classification, 167, 251, 345, 348
comparative molecular field analysis
overview, 251–253
applications, 259–261
rat SULT1A1 substrates, 253–254
rat SULT2A3
alignment rule application and model
construction, 257, 259
homology modeling of
catalytically competent
complex, 255–256
substrates and kinetic constants,
258–259
functional overview, 166
mutagen bioactivation, 231–232
mutagenicity testing
cofactors and activity in subcellular
activating systems, 234–236
natural expression of enzymes in target
cells of test systems, 233–234
recombinant systems
cofactor supply, 242–243

expression level optimization and
characterization, 239–242
historical perspective, 236–238
positive controls, 243–245
prospects, 246
vectors, 237–239
ortholog identification, 350
phenol sulfation, 345, 348–350
phytoestrogens as substrates, 334–335
SULT1A genes
evolution, 150
isoenzymes, 148–149
loci, 149
promoter studies
cloning, 150–153
deletion constructs, 160, 162
flanking sequences, 157–158, 160
luciferase reporter assays, 157
transfection of constructs, 153,
156–157
single nucleotide polymorphisms, 162,
163, 348
structure, 150
transcription factors, 162
SULT2A1
assay, 172–173
lithocholic acid induction, 187–188
nuclear receptor regulation
androgen receptor regulation,
173–174
cell lines, 176
DNase I footprinting, 184–186
electrophoretic mobility shift assay,
186–187
Northern blot, 177
overview, 174–175
promoter constructs and site-directed
mutagenesis, 178–180
regulatory element analysis, 180–184
vitamin D receptor regulation,
174–176, 180–188
Western blot, 177–178
recombinant enzyme expression,
171–172
substrate specificity, 168, 170
tissue distribution, 168
SULT2B1 features, 167–168
SULT1A genes, *see* Sulfotransferases
Surfactant deficiency, *ABCA3* mutations and
pathophysiology, 418

T

Tangier disease, *ABCA1* mutations and
    pathophysiology, 416–417

U

UDP-glucuronosyltransferase
  assay of mouse liver microsomes
    high-performance liquid
        chromatography with fluorescence
        detection, 608–609
    microsome preparation, 607–608
    thin-layer chromatography, 609
  cancer-associated alleles, 621–622
  catalytic mechanism and residues, 127–131
  deglucuronidation, *see* β-Glucuronidase
  functional overview, 1–2
  gene families, *see UGT1* genes;
        UGT2 genes
  homology with other
        glucuronosyltransferases, 129–130
  membrane-bound enzymes
    endoplasmic reticulum association,
        119–121
    interactions with other drug-
        metabolizing enzymes, 124
    oligomerization analysis, 121–124
    signal sequence, 117
    structure
      active site architecture and
          organization, 127–131
      determinants for membrane
          association, residency, and
          latency, 124–127
      topology prediction, 125–126
    subcellular localization, 118–119
  phenol glucuronidation, 343–347
  phosphorylative regulation, 2
  photoaffinity labeling
    direct labeling, 131–132, 138–139
    labeling conditions, 136–140
    modified residue identification
      liquid chromatography–tandem mass
          spectrometry of peptides,
          141–142
      proteolytic digestion, 141
      site-directed mutagenesis verification,
          142
    principles, 131

  probes
    7-azido-4-methylcoumarin synthesis,
        136
    5-azido-UDP-glucuronic acid
        synthesis, 134–136
    overview, 132
    retinoic acid probe synthesis, 136
  recombinant human protein preparation
    cloning, 133
    expression in baculovirus–insect cell
        system, 133–134
    membrane isolation, 134
    nickel affinity chromatography, 134
  phytoestrogen glucuronidation studies
    conjugate analysis and isolation,
        332–333
    in vitro, 332
    in vivo, 332
  probe substrates for isoforms
    applications, 105–106
    gene polymorphism considerations,
        106, 110
    glucuronidation assay
      data analysis, 112, 114
      high-performance liquid
          chromatography, 112–113
      incubation conditions, 110–112
    overview of types, 54–55, 105,
        107–108, 110
    specificity criteria, 106
  substrate specificity
    assays
      high-performance liquid
          chromatography with
          radiochemical detection, 49–50
      kinetic parameter determination, 52
      liquid chromatography–tandem mass
          spectrometry, 50–51
      quantitative analysis, 51–52
    carbamazepine, 52–53
    drug glucuronidation assays, 49
    dulcin, 52–53
    enzyme preparation. 48–49
    farnesol, 52–53
    HMR1098, 52–53
    isoenzymes in drug metabolism, 47–48
UDP-glucuronosyltransferase 1A1
  aryl hydrocarbon receptor regulation
      of gene
    activity assay, 78–79

cell culture and treatment, 77–78
cell lines, 83–84
cloning of regulatory regions, 81–82
electrophoretic mobility shift assay, 87–90
materials, 77
Northern blot analysis of transcripts, 79–80
response element deletion analysis and site-directed mutagenesis, 84, 86, 89–90
transient transfection assay, 82–83
Western blot, 80–81
nuclear receptor regulation of gene
deletion assays, 97–99
electrophoretic mobility shift assay, 99–100
overview, 92–93, 100–102
plasmids, 95
site-directed mutagenesis studies, 97
transcript analysis with reverse transcriptase–polymerase chain reaction, 94
transient transfection and luciferase assays, 96
Western blot, 94–95
transmembrane domain, 126–127
UDP-glucuronosyltransferase 1A6
gene expression regulation
aryl hydrocarbon receptor, 66
carcinogenesis, 69
hepatocyte nuclear factor 1, 63–65
modes of expression, 62–63
Nrf2, 67–69
nuclear receptors, 65–66
promoter polymorphisms, 63
oligomerization studies, 62
structure, 58–59
substrates
arylamines and aryl hydrocarbons, 61–62
paracetamol, 59–60
serotonin, 60–61
tissue distribution, 58
UGT, see UDP-glucuronosyltransferase
UGT1 genes
classification, 3–9, 343–344
comparative genomics between human, mouse, and rat loci, 3–9
hepatocyte nuclear factor 1 binding, see Hepatocyte nuclear factor 1
inducers, 9–12, 344
isoenzyme origins, 18

mutation and disease, 2, 13, 15–16
polymorphisms, 2, 15–16, 47
UGT1A1, see UDP-glucuronosyltransferase 1A1
UGT1A6, see UDP-glucuronosyltransferase 1A6
UGT2 genes
classification, 16–17
comparative genomics between human and rat loci, 16–17
hepatocyte nuclear factor 1 binding, see Hepatocyte nuclear factor 1

V

VDR, see Vitamin D receptor
Vitamin D receptor
sulfotransferase 2A1 regulation
overview, 174–176
regulatory element analysis, 180–184
DNase I footprinting, 184–186
electrophoretic mobility shift assay, 186–187
Northern blot, 177
Western blot, 177–178
promoter constructs and site-directed mutagenesis, 178–180
cell lines, 176
xenobiotic-metabolizing enzyme regulation, 166–167

W

Western blot
ABCG2 human transporter, 491, 493
drug-metabolizing enzyme analysis in animal models, 607
hepatocyte nuclear factor 1α expression, 34, 38
sulfotransferase 2A1, 177–178
UDP-glucuronosyltransferase 1A1, 80–81, 94–95

Y

Ycf1p
overexpression and purification, 475–477
substrate specificity and structure–function analysis, 469
trafficking studies, 471, 473
trafficking studies, 471

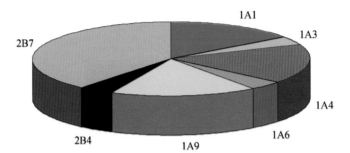

BURCHELL *ET AL.*, CHAPTER 3, FIG. 2. Relative contributions to the glucuronidation of drugs by human liver UGTs. Based on data reported by Williams *et al.* (2004), but specifically modified to focus on hepatic glucuronidation.

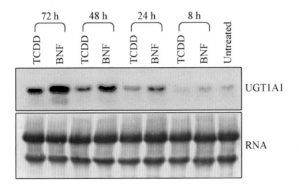

YUEH *ET AL.*, CHAPTER 5, FIG. 2. UGT1A1 transcript levels after TCDD and BNF treatment. Northern blot of UGT1A1 RNA in HepG2 cells after treatment with DMSO (untreated), 10 n*M* TCDD, and 20 μ*M* BNF for 8, 24, 48, and 72 h. RNA (15 μg) was separated on a formaldehyde agarose gel, transferred to a nitrocellulose membrane, and incubated with a 423-bp probe from *UGT1A1* cDNA. RNA was visualized by a phosphoimager.

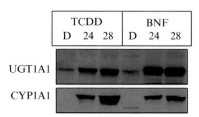

YUEH *ET AL.*, CHAPTER 5, FIG. 3. Protein analysis of UGT1A1 and CYP1A1. Microsomal fractions were collected from HepG2 cells treated with DMSO (D), 10 n*M* TCDD, or 20 μ*M* BNF for 48 and 72 h. Western blots were performed using 10 μg protein and blotted using anti-UGT1A1 or anti-CYP1A1 antibodies.

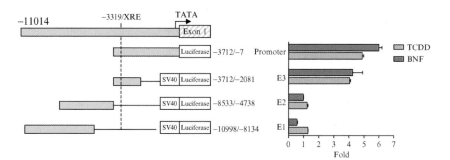

YUEH *ET AL.*, CHAPTER 5, FIG. 4. Promoter activity of *UGT1A1* in response to TCDD and BNF by luciferase reporter assay. Four DNA fragments that cover the entire *UGT1A1* promoter and enhancer (11 kb) were cloned from a BAC library and subcloned into the pGL3 reporter plasmid. HepG2 cells were transiently transfected with the reporter plasmids, and firefly luciferase activity was measured in the cytosolic fraction after 48 h of treatment. Values were normalized to renilla luciferase activity and shown as fold over DMSO treatment.

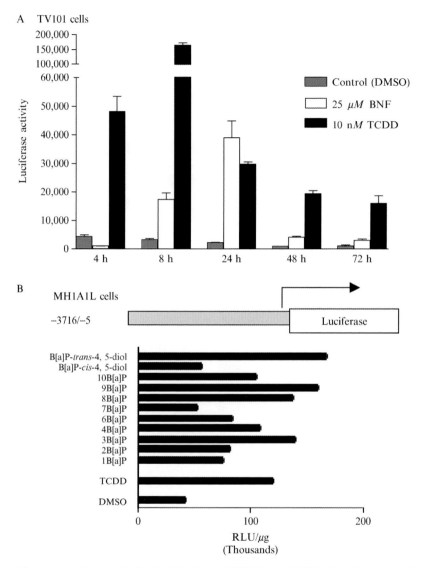

YUEH *ET AL.*, CHAPTER 5, FIG. 5. Induction of CYP1A1-and UGT1A1-luciferase in stably transfected HepG2 cells. (A) The *CYP1A1* promoter luciferase plasmid that contains multiple XREs was used to establish TV101 cells. Luciferase activity was measured at various times after treatments. Activity is expressed as relative light units (RLU)/μg of protein. (B) HepG2 cells were stably transfected with a 3.7-kb *UGT1A1* promoter luciferase plasmid to generate the MH1A1L cell line. Treatment of cells was carried out for 48 h with 5 μM of each B[a]P metabolite. Luciferase activity is expressed as RLU/μg of protein.

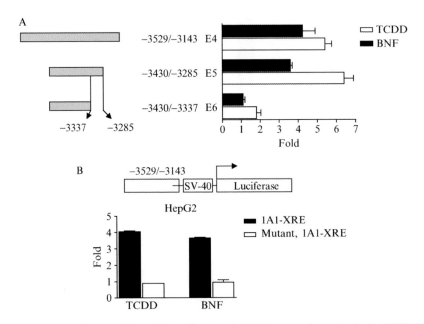

YUEH *ET AL.*, CHAPTER 5, FIG. 6. Identification of a TCDD-responsive region in the *UGT1A1* promoter. (A) Using plasmid E3 as a DNA template, E4, E5, and E6 were generated by progressive truncation of the 5′ and 3′ ends. Transient transfections were conducted with these truncated plasmids, and promoter responsiveness to 48 h treatment with TCDD was determined by luciferase expression. (B) The core sequence of the XRE, CACGCA, was mutated to ACCGCA by site-directional mutagenesis PCR. The luciferase reporter plasmid containing either wild-type or mutated *UGT1A1*-XRE was transiently transfected into HepG2 cells. Cells were treated for 48 h with 10 nM TCDD and 20 μM BNF. Results shown are fold activity over DMSO.

A

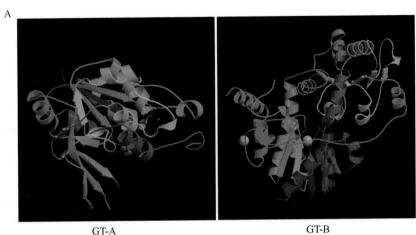

GT-A   GT-B

B

GT representative member	• SpsA (*Bacillus substilis*), production of mature spore coat • GlcAT-I (man), glycosaminoglycan synthesis	• GtfB (bacteria), antibiotics synthesis • MurG (bacteria), peptidoglycan synthesis
Fold	2 Rossmann-like a/b/a domains tightly associated (N- and C-terminal domains, which bind the donor and the acceptor substrate, respectively)	2 Rossmann-like a/b/a domains less tightly associated (N- and C-terminal independent domains, which bind the acceptor and the donor substrate, respectively)
Metal activation	Strict, with a DXD motif (Mn^{2+}, Mg^{2+})	Less strict (Mn^{2+}, Mg^{2+})
Catalytic base	Asp or Glu	Asp or Glu
PDB code	SpsA, 1QG8	GtfB, 1IIR

RADOMINSKA-PANDYA *ET AL.*, CHAPTER 8, FIG. 1. Structural and functional comparison of crystallized glycosyltransferases of GT-A and GT-B families.

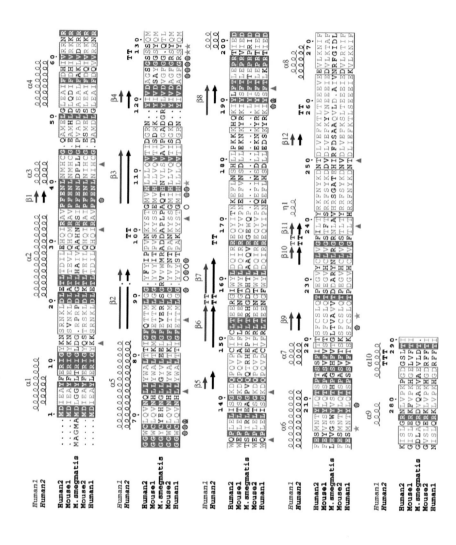

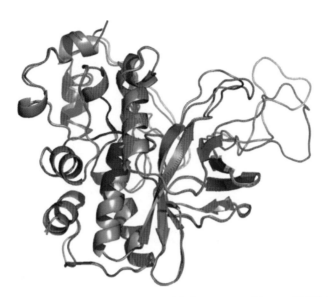

SAVULESCU *ET AL.*, CHAPTER 11, FIG. 6. Modeled structures of human NAT1 (gray) and NAT2 (magenta) are overlaid. The structures of the human NATs were modeled using the methods described in the text.

SAVULESCU *ET AL.*, CHAPTER 11, FIG. 2. Amino acid sequence alignment of NATs, indicating the secondary structural elements from the models. Numbering follows the human NAT2 sequence. The sequences of NAT from *M. smegmatis* and mouse (NAT1 and NAT2) are shown for comparison. Sites of the mutations in the 14 variants investigated in this study are indicated with purple triangles. Residues calculated to be interacting (within 6Å) of the docked ligands are shown with green stars (2-AF) and black circles (IQ).

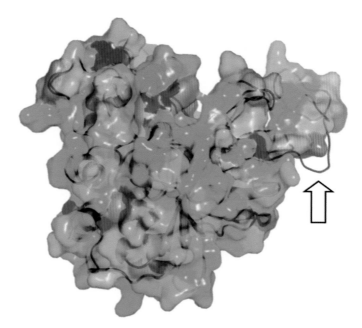

Savulescu *et al.*, Chapter 11, Fig. 7. The same structures shown in Fig. 6, with addition of a transparent surface colored according to Modeller violation values (RGB, blue ~0, green ~0.5, and red >1). The structures are similar, except in the vicinity of the interdomain region (residues 190–210), seen as a patch of red (arrow).

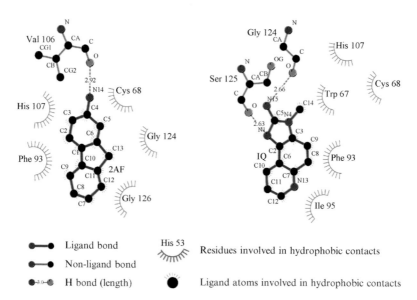

Savulescu *et al.*, Chapter 11, Fig. 8. Predicted nonbonded interactions of AF (left) and IQ (right) with NAT2, determined using AutoDock. The aromatic rings from both ligands form hydrophobic interactions with the phe37, phe93, and tyr190 residues, with each of these residues playing a differing role when binding IQ or AF. The amino groups of both substrates face toward the active site cysteine (indicated by orange lettering).

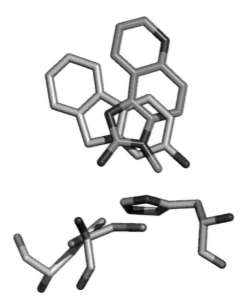

SAVULESCU *ET AL.*, CHAPTER 11, FIG. 9. The docked positions of AF (gray) and IQ (green) are overlaid. Catalytic triad residues are shown.

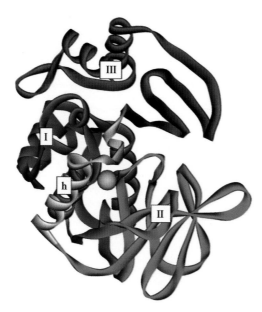

SAVULESCU *ET AL.*, CHAPTER 11, FIG. 10. The structure of *S. typhimurium* NAT (1E2T.pdb) is shown. The catalytic cysteine sulfur atom is represented as a ball. The three domains are indicated by Roman numerals, and the helix connecting domains I and II is represented by the letter "h".

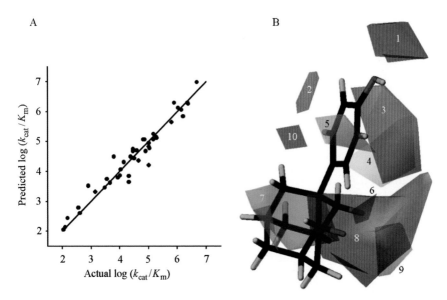

SHARMA AND DUFFEL, CHAPTER 14, FIG. 2. CoMFA results for catalytic specificity of rSULT1A1. (A) Quantitative results for a total of 44 substrates are shown (data from Sharma and Duffel, 2002). The line indicates a perfect agreement between actual and predicted values and can be used to estimate the error in the predicted values; it is not a regression line. (B) The contour coefficient map from CoMFA of rSULT1A1 is overlaid on the structure of 4-(1-adamantyl)-phenol. Sterically favored regions for substrates (numbered 7 and 9) are shown in green, whereas sterically unfavorable regions (numbered 4 and 5) are shown in yellow. Regions favoring a positive charge on the substrate are shown in red (numbered 2, 3, and 6), and regions favoring a negative charge on the substrate are in blue (numbered 1, 8, and 10).

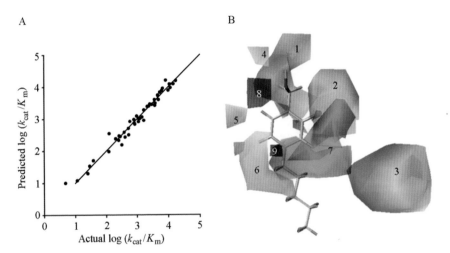

Sharma and Duffel, Chapter 14, Fig. 3. CoMFA results for catalytic specificity of rSULT2A3. (A) Quantitative results from CoMFA of substrates in Tables I and II. The line indicates a perfect agreement between actual and predicted values and can be used to estimate the error in the predicted values; it is not a regression line. (B) Contour coefficient map from CoMFA of rSULT2A3 substrates from Tables I and II is shown overlaid on the structure of (4-pentylphenyl)methanol. A sterically favored region for substrates (numbered 7) is shown in green, whereas sterically unfavorable regions (numbered 1–6) are shown in yellow. Regions shown as favoring a negative charge on the substrate (numbered 8 and 9) are in blue, and a single small region favoring a positive charge is seen in red.

A   B

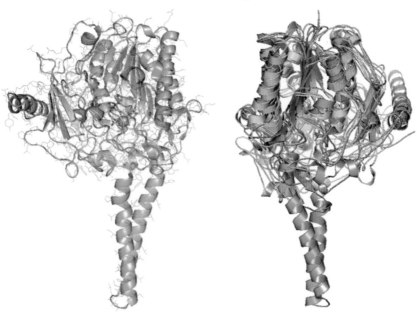

GHOSH, CHAPTER 16, FIG. 1. (A) The crystal structure of ES showing the course of the polypeptide backbone through α helices, β sheets, and loop regions. Side chains are shown as thin bonds. Atoms are color coded: carbon, green; nitrogen, blue; oxygen, red; and sulfur, yellow/orange. The bivalent cation Ca^{2+} at the active site is shown as a grayish purple sphere. Side chain ligands to the metal atom are shown as thick bonds. (B) A least-square superposition of four crystal structures of sulfatases: ES, green; ARSA, cyan; ARSB, pink; and PAS, yellow. The view is roughly 180° rotated about a vertical axis with respect to the view in A.

A   B

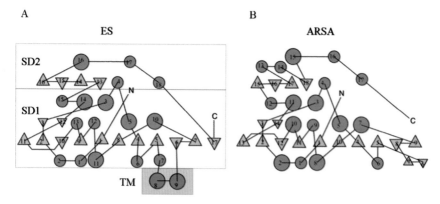

GHOSH, CHAPTER 16, FIG. 2. Folding topologies of (A) ES and (B) ARSA representing the soluble sulfatases created with the topology cartoon server TOPS software (University of Glasgow, UK; http://www.tops.leeds.ac.uk). Helices are shown as circles and strands as triangles. Subdomains SD1 and SD2 and the putative transmembrane domain TM in ES are highlighted.

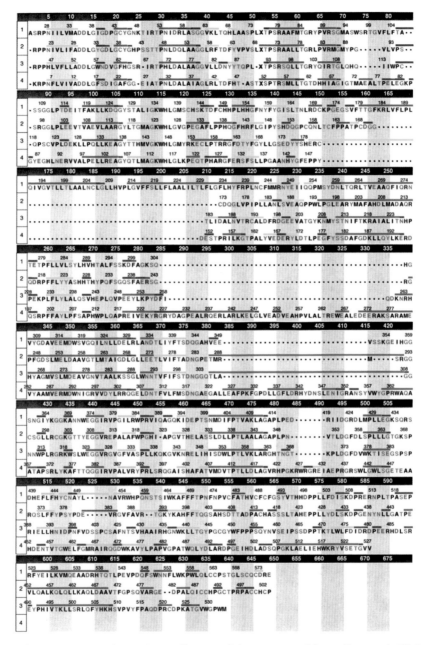

Ghosh, Chapter 16, Fig. 3. An alignment of the sequences of four sulfatases—1, ES; 2, ARSA; 3, ARSB; and 4, PAS—and their known secondary structures. Helical regions are marked in red lines, sheets in yellow, and loops in blue. Residues are color coded: acidic (DE), red; basic (RKH), blue; hydrophobic (AVILMFPW), green; and hydrophilic (GSTNQCY), cyan; X, Formylglycine.

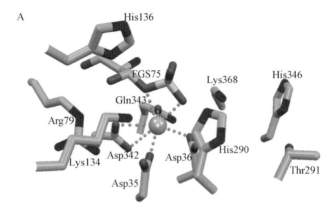

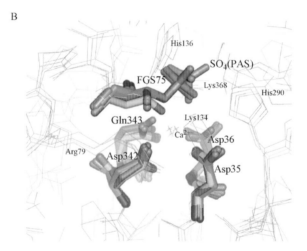

GHOSH, CHAPTER 16, FIG. 4. (A) Active site catalytic residues in ES and the coordination of Ca^{2+} as described in the text. (B) A least-square superposition of four ligand side chains in crystal structures of sulfatases. The ligand side chains and formylglycine residues, along with their sulfate groups, are shown as thick bonds. The sulfate group in PAS not linked covalently to the side chain is labeled. Carbon atoms have the same colors as the colors used for the main chains (ES, green; ARSA, cyan; ARSB, pink; and PAS, yellow). Four bivalent cations shown as crosses for clarity have the same color codes as the main chains. Other side chains are shown as thin bonds.

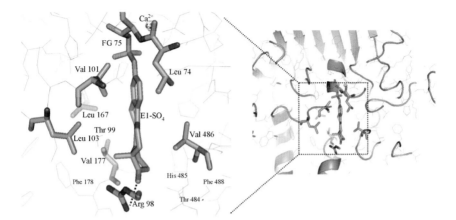

GHOSH, CHAPTER 16, FIG. 5. A view of the active site of ES with a modeled E1 sulfate molecule linked covalently to FG 75 at an intermediate step, as described in the text. (Right) The boxed region showing a broader view of the active site cleft, including a turn of a transmembrane helix, is magnified (left). Important leucine and valine residues, as well as the Arg 98 side chain, are shown as thick bonds. Other side chains of interest are shown as thin bonds and are labeled.

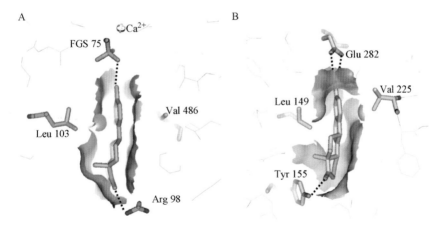

GHOSH, CHAPTER 16, FIG. 6. Similarity of recognition of the estrogenic A ring by ES and 17HSD1: (A) E1 complex of ES as observed in a crystal structure and (B) crystal structure of the 17BHSD1–equilin complex. Residues of the Leu-Val sandwich are shown as thick bonds and are labeled. Molecular surfaces of the proteins are drawn in the same color code as the ligands.

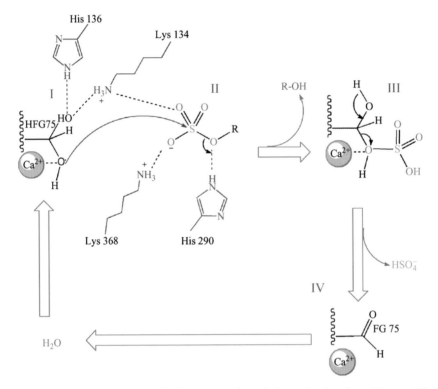

GHOSH, CHAPTER 16, FIG. 7. The proposed catalytic mechanism for sulfatases. The catalytic side chain labeled as HFG75 is hydroxylformylglycine 75. See text for an explanation of the four steps of catalysis.

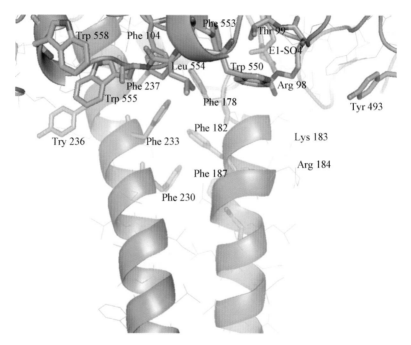

Gʜᴏsʜ, Cʜᴀᴘᴛᴇʀ 16, Fɪɢ. 8. Large hydrophobic residues at the putative protein–lipid interface in the crystal structure of ES. Some of these side chains are responsible for the hydrophobic "tunnel" described in the text. A bound E1 sulfate molecule at the active site and two transmembrane helices are partially visible.

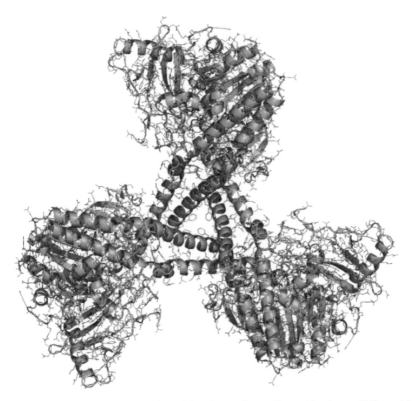

GHOSH, CHAPTER 16, FIG. 9. A view of the observed crystallographic trimer of ES roughly along the threefold rotation axis.

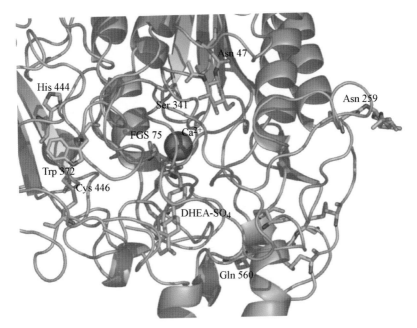

GHOSH, CHAPTER 16, FIG. 10. Point mutation sites in X-linked ichthyosis displaying amino acids of the wild-type enzyme (Ser 341, Trp 372, His 444, Cys 446, and Gln 560). The catalytic residue FG 75, the bivalent cation Ca^{2+}, and a covalently linked modeled DHEA-sulfate are shown at the active site. Two observed glycosylation sites (Asn 47 and Asn 259), each with one molecule of N-acetyl glucosamine, are also shown.

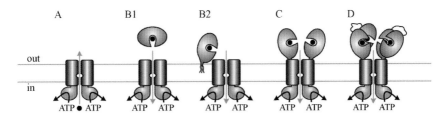

POOLMAN *ET AL.*, CHAPTER 25, FIG. 1. Schematic representation of the domain organization of ABC transporters. (A) Efflux system. (B) Conventional SBP-dependent uptake system with periplasmic (B1) or lipid-anchored SBP (B2). The chimeric substrate-binding/translocator systems with two and four substrate-binding sites per functional complex are shown in C and D, respectively. "'Out" and "in" indicate the extra- and intracellular side of the membrane, respectively; the translocator and NBD subunits are in gray and orange, respectively.

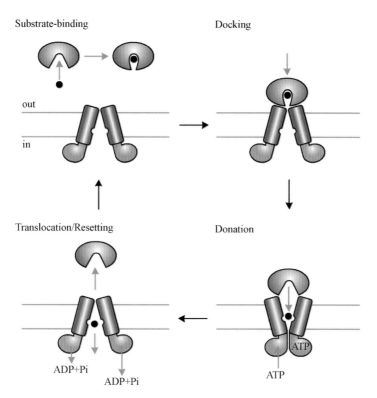

Substrate-binding

Docking

out

in

Translocation/Resetting

Donation

ADP+Pi

ADP+Pi

ATP

ATP

POOLMAN *ET AL.*, CHAPTER 25, FIG. 2. Model for the transport cycle of a SBP-dependent ABC transporter. For explanation, see text.

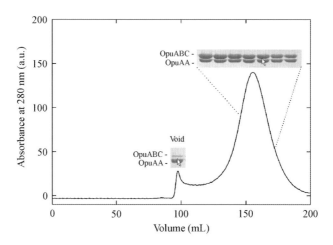

POOLMAN *ET AL.*, CHAPTER 25, FIG. 3. Size-exclusion chromatography of OpuA from *L. lactis* (unpublished experiment). OpuA was purified by Ni-NTA chromatography, essentially as described (van der Heide *et al.*, 2000), concentrated to 20 mg/ml on Amicon Ultra (100,000 MWCO,) and 40 mg of protein was loaded onto Sephadex S300 in 10 m*M* Na-HEPES, pH 8.0, 100 m*M* NaCl, 15% (v/v) glycerol, and 0.15 % (w/v) CyMal5. (Inset) A Coomassie brilliant blue-stained SDS–polyacrylamide gel of the individual protein fractions.

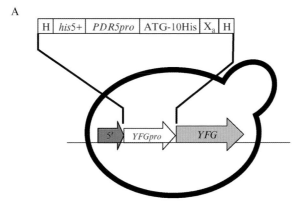

A

| H | *his5+* | *PDR5pro* | ATG-10His | X_a | H |

Strain: Haploid *S.c.* chromosomal *PDR1-3*

B

| 5′ | *his5+* | *PDR5pro* | ATG-10His | X_a | YFG |

ERNST *ET AL.*, CHAPTER 26, FIG. 3. Tagging cassette for genomic tagging and over-expression of yeast genes. (A) A specialized, haploid *S. cerevisiae* strain with a *PDR1-3* background is transformed with a PCR product carrying the cassette for overexpression and affinity tagging of your favorite gene (YFG) using a short homology flanking region of about 40 bp (H) on both sides of the cassette. The cassette contains the *Schizosaccharomyces pombe his5+* (*his5+*) module as the selection marker, the sequence of the *PDR5* promotor (*PDR5pro*), a start codon (ATG), and a deca-histidine affinity tag (10His), as well as a factor X_a cleavage site (X_a). (B) The correctly targeted cassette substitutes the promotor of the gene of interest *(YFG)*, driving overexpression from the *PDR5* promoter. The expression protein of interest is fused to a removable N-terminal deca-histidine tag, which allows for rapid and efficient purification by chelating metal affinity chromatography.

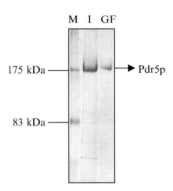

ERNST *ET AL.*, CHAPTER 26, FIG. 4. Affinity purification of Pdr5p via a deca-histidine affinity tag. Cells overexpressing Pdr5p with an N-terminal deca-histidine tag were harvested from a 10-liter culture and disrupted using a cell rupture machine. Pdr5p was extracted and solubilized in detergent from a plasma membrane-enriched fraction. Pdr5p was purified with immobilized metal affinity chromatography (I) followed by gel filtration chromatography (GF). A Coomassie-stained 7.5% SDS–PAGE, containing highly purified Pdr5p after immobilized metal affinity chromatography (I) and gel filtration (GF), is shown. M indicates protein molecular mass marker in kilodaltons. Further details of the purification scheme will be reported elsewhere (Ernst *et al.*, in preparation).

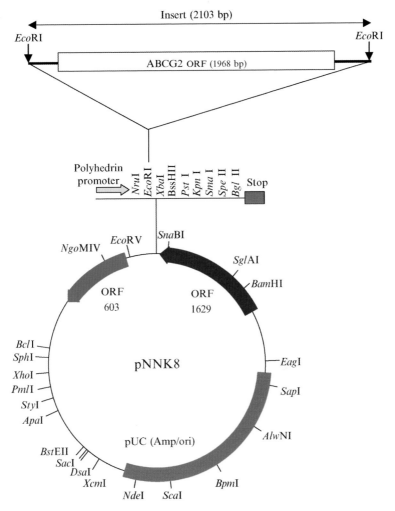

ISHIKAWA *ET AL.*, CHAPTER 27, FIG. 2. Schematic diagram of the pNNK8 transfer vector for the expression of human ABCG2 in insect cells.

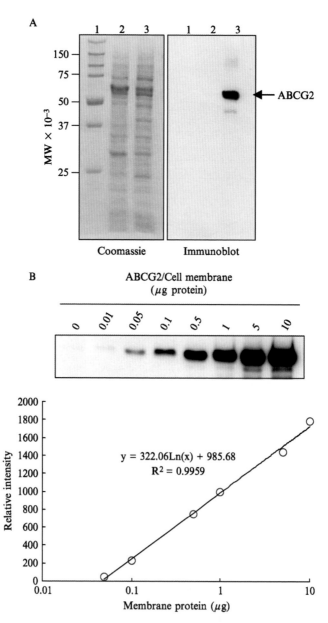

ISHIKAWA *ET AL.*, CHAPTER 27, FIG. 4. Detection and quantitative analysis of human ABCG2 expressed in the plasma membrane of insect cells. (A) Coomassie staining and immunoblot analyses of membrane proteins. Plasma membrane proteins (2 μg) were separated by SDS–PAGE (10% acrylamide gel), and ABCG2 was detected by immunoblotting. (B) Quantitative analysis of ABCG2 protein in the plasma membrane preparation. Different amounts of plasma membrane proteins (0, 0.01, 0.05, 0.1, 0.5, 1, 5, and 10 μg) were subjected to SDS–PAGE, and ABCG2 protein was detected by immunoblotting. The signal intensity of immunoblots is expressed as a function of the logarithmic value of the amount of membrane proteins applied to SDS–PAGE.

A

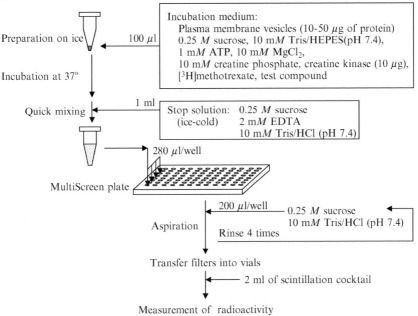

Preparation on ice → 100 µl → Incubation medium:
 Plasma membrane vesicles (10-50 µg of protein)
 0.25 M sucrose, 10 mM Tris/HEPES(pH 7.4),
 1 mM ATP, 10 mM MgCl$_2$,
 10 mM creatine phosphate, creatine kinase (10 µg),
 [^{3}H]methotrexate, test compound

Incubation at 37°

Quick mixing ← 1 ml ← Stop solution: 0.25 M sucrose
 (ice-cold) 2 mM EDTA
 10 mM Tris/HCl (pH 7.4)

280 µl/well

MultiScreen plate

200 µl/well → 0.25 M sucrose
Aspiration   10 mM Tris/HCl (pH 7.4)
   Rinse 4 times

Transfer filters into vials

2 ml of scintillation cocktail

Measurement of radioactivity

B

Ishikawa *et al.*, Chapter 27, Fig. 5. High-speed assay of ATP-dependent MTX transport mediated by ABCG2 in plasma membrane vesicles. (A) Procedure for the vesicle transport assay using 96-well MultiScreen plates. (B) High-speed screening system (BioTec EDR384S) used for the vesicle transport assay.

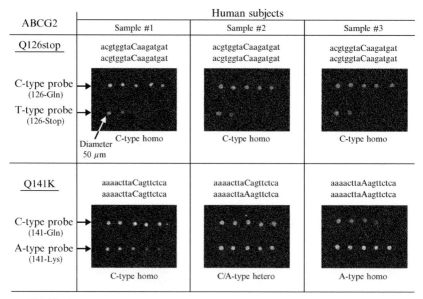

ABCG2	Human subjects		
	Sample #1	Sample #2	Sample #3
<u>Q126stop</u>	acgtggtaCaagatgat acgtggtaCaagatgat	acgtggtaCaagatgat acgtggtaCaagatgat	acgtggtaCaagatgat acgtggtaCaagatgat
C-type probe (126-Gln) → T-type probe (126-Stop) →	Diameter 50 μm C-type homo	C-type homo	C-type homo
<u>Q141K</u>	aaaacttaCagttctca aaaacttaCagttctca	aaaacttaCagttctca aaaacttaAagttctca	aaaacttaAagttctca aaaacttaAagttctca
C-type probe (141-Gln) → A-type probe (141-Lys) →	C-type homo	C/A-type hetero	A-type homo

Hybridization

Weak ⟵⟶ Strong

ISHIKAWA *ET AL.*, CHAPTER 27, FIG. 13. Detection of nonsynonymous polymorphisms (Q126stop and Q141K) of ABCG2 with the SNP array.

A

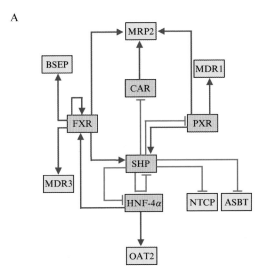

B

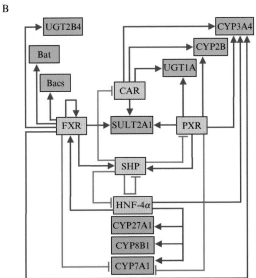

ELORANTA *ET AL.*, CHAPTER 28, FIG. 1. Nuclear receptors FXR, SHP, PXR, CAR, and HNF-4α regulate a complex network of genes involved in (A) transport and (B) metabolism of drugs and bile acids. Nuclear receptors are shown within blue boxes, bile acid transporter genes in yellow boxes, and phase I and phase II metabolizing genes in light green boxes. Green and red lines indicate transcriptional activation and suppression, respectively.

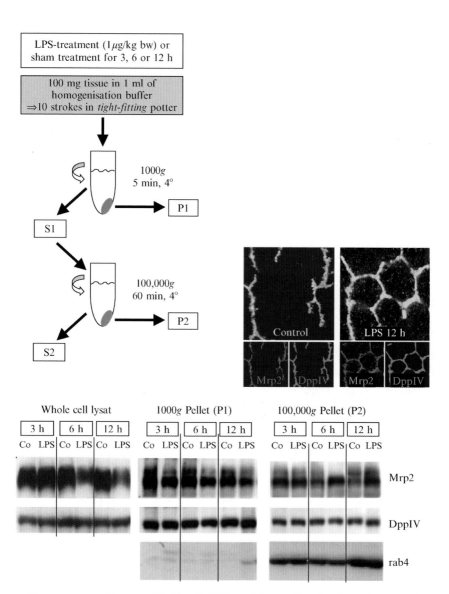

KUBITZ *ET AL.*, CHAPTER 30, FIG. 2. Differential centrifugation in order to enrich canalicular proteins. Liver tissue was homogenized by tight-fitting pottering, and fractions enriched in canalicular proteins (P1) or endomembranes (P2) were collected as illustrated. LPS treatment induced a retrieval (see confocal pictures) and a reduction in Mrp2. Neither the localization nor the amount of the canalicular marker protein dipeptidylpeptidase IV (DppIV) was altered by LPS treatment. Despite the decrease in total Mrp2 by LPS treatment (whole cell lysate), Mrp2 in the endomembrane fraction increased slightly (100,000g pellet, P2). No changes were observed for DppIV. Note that the 1000g pellet was almost free of the endocytic marker protein rab4. S1/S2, supernatants.

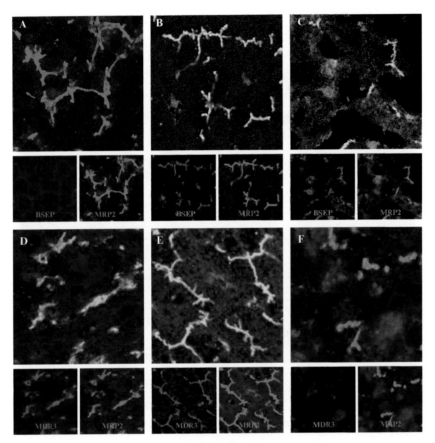

KUBITZ *ET AL.*, CHAPTER 31, FIG. 2. Immunofluorescence of livers from patients with PFIC. Livers from one patient with PFIC-2 (A, D) and two patients with PFIC-3 (B and E; C and F) were examined by immunofluorescence and confocal laser-scanning microscopy. In the upper row, the bile salt export pump BSEP (red), which is mutated in PFIC-2, was counterstained by the multidrug resistance protein 2 (MRP2, green). In PFIC-2 patients, no canalicular BSEP staining was detectable. In contrast, some patients have MDR3 mutations (B, E; middle), which display a normal canalicular MDR3-staining (MDR3 in red; MRP2 in green). Other MDR3 mutations (such as patient of C and F) have absent canalicular MDR3 (right).

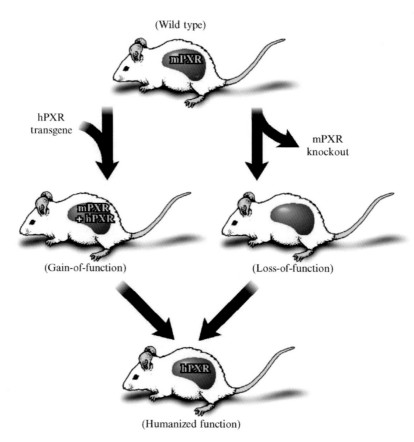

GONG *ET AL.*, CHAPTER 34, FIG. 1. Strategies to create the loss-of-function knockout, gain-of-function transgenic, and the combined "humanized" function models. Modified from Xie and Evans (2001), with permission of the publisher.